Process Safety in Coal Mining

Process Safety in Coal Mining

Editors

Feng Du
Aitao Zhou
Bo Li

MDPI

Basel • Beijing • Wuhan • Barcelona • Belgrade • Novi Sad • Cluj • Manchester

Editors
Feng Du
School of Emergency Management and Safety Engineering
China University of Mining and Technology-Beijing
Beijing, China

Aitao Zhou
School of Emergency Management and Safety Engineering
China University of Mining and Technology-Beijing
Beijing, China

Bo Li
State Key Laboratory Cultivation Base for Gas Geology and Gas Control
Henan Polytechnic University
Jiaozuo, China

Editorial Office
MDPI
St. Alban-Anlage 66
4052 Basel, Switzerland

This is a reprint of articles from the Special Issue published online in the open access journal *Processes* (ISSN 2227-9717) (available at: https://www.mdpi.com/journal/processes/special_issues/5S8KX057H8).

For citation purposes, cite each article independently as indicated on the article page online and as indicated below:

Lastname, A.A.; Lastname, B.B. Article Title. *Journal Name* **Year**, *Volume Number*, Page Range.

ISBN 978-3-0365-8660-1 (Hbk)
ISBN 978-3-0365-8661-8 (PDF)
doi.org/10.3390/books978-3-0365-8661-8

Contents

About the Editors

Feng Du

Feng Du, Associate Professor, is mainly engaged in mine safety engineering, coal rock dynamic disaster prevention and control, safety and emergency management, gas flow theory in coal and other aspects of research. In recent years, more than 40 academic papers of his have been published. Among them, as the first author/correspondent author, 25 academic papers were published in important journals such as Energy, Fuel, Process Safety and Environmental Protection, Science of the Total Environment, International Journal of Coal Science & Technology, and Coal Journal, and 7 invention patents were applied for or authorized. He is currently chairing the sub-topics of the key projects of the National Natural Science Foundation of China, the youth project of the National Natural Science Foundation of China, the general project of the China Postdoctoral Science Foundation, the open fund of the National Key Laboratory (key project) and the open fund of the Key Laboratory of the Ministry of Education (general project). Providing support for research, he participated in five projects including the sub-topics of the National Key Research and Development Plan and the general project of the National Natural Science Foundation of China. He served as the guest editor of two SCI journals and the regular reviewer of more than 10 SCI journals. He was awarded the National Excellent Doctoral Dissertation Award of Safety Science and Engineering Discipline of China Occupational Health Association, and the National Scholarship of Excellent Graduates and Doctoral Students in Beijing.

Aitao Zhou

Aitao Zhou, Professor is a visiting scholar of Pennsylvania State University, is a nationally registered safety engineer, winner of the 'Coal Youth Science and Technology Award', and part of the first batch of 'Yueqi Young Scholars' from China University of Mining and Technology (Beijing). He is involved in the Young Standing Committee of Ventilation Safety and Health Professional Committee of China Occupational Safety and Health Association and is a member of the Young Expert Committee of 'Industrial and Mining Automation' and 'Coal Science and Technology'. He has contributed to the International Journal of Rock Mechanics and Mining Sciences and the work of other domestic and foreign journal reviewers. He presided over 3 projects of the National Natural Science Foundation of China, 2 sub-projects of the National Key R&D Program, 6 open projects of key laboratories, and more than 10 enterprise cooperation projects. In recent years, more than 60 papers of his have been published in domestic and foreign journals, more than 40 papers have been included in SCI, EI, etc., 6 monographs and textbooks have been edited, 7 invention patents have been obtained, and 14 scientific research awards have been awarded at the provincial and ministerial levels.

Bo Li

Bo Li, Professor, is one of the Selected Central Plains Talent Program (Yucai series)—Central Plains young top talents, outstanding post-doctoral researchers in Henan Province, young backbone teachers in Henan Province, Henan Province young talent lift project, Henan Province high-level talents, and Henan Polytechnic University outstanding young people. He has long been engaged in the teaching of and research on coal and rock gas dynamic disasters, gas extraction and other aspects. He presided over more than 10 projects, including the general project of the National Natural Science Foundation of China, the Youth Fund of the National Natural Science Foundation of China, the basic and cutting-edge technology research project of Henan Province, the doctor-oriented fund, the university doctoral fund project and the enterprise-entrusted project. More than 10 national invention patents of his were approved, and more than 40 papers were published, including 28 SCI/EI. His research results have won one second prize for scientific and technological progress in Henan Province, one third prize for scientific and technological progress in Henan Province, and two second prizes and one third prize for te China Coal Industry Association. He was awarded the honorary title of excellent master's thesis instructor of Henan Polytechnic University, excellent educator of Henan Polytechnic University, and advanced individual of spiritual civilization of Henan Polytechnic University.

Preface

Coal is an important part of natural resources and the focus of social energy and economic development. With the increase in the energy demand, the amount of coal mining is also increasing, which leads to many safety problems. Improving the elevels of safety in the coal mining process while ensuring that coal production is carried out at an acceptable level is a problem that we must attach great importance to. In this book, we collect, revise and collate the latest work of the main researchers, covering the theory and technology of coal mine disaster prevention and control, safety management and control in the process of coal mining, mineral processing and intelligent/ green mining engineering, mining environmental protection and sustainable development, and other aspects of safety technology research, focusing on coal mine production safety.

The book contains a total of 27 articles, and discusses and studies the progress of safety research in theory, experiments and simulation. In the research of coal mine production safety related-fields, the contents include mine ventilation, fire and explosion prevention, gas drainage, coal dust prevention, safety equipment and so on. In addition, there are studies on coal mine safety, production safety, management safety and ancient building safety. These articles have solved the outstanding challenges related to production safety in different fields from the perspectives of theory, experiment and application.

The book was collected, collated and revised by Du Feng, Zhou Aitao and Li Bo, who aim to provide some guidance and a reference for production safety in the field of coal mines, and hope to inspire scientific research workers in the field of coal mine and geology. This work was supported by the National Natural Science Foundation of China (Nos. 52130409 and 52004291). This work has been strongly supported and guided by the editorial department. Here together to express my heartfelt thanks.

Due to the limited experience of some editors, there may be omissions and errors in the book. Readers are kindly requested to criticize and correct them.

Feng Du, Aitao Zhou, and Bo Li
Editors

processes

MDPI

Editorial

Special Issue "Process Safety in Coal Mining"

Feng Du [1,*], Aitao Zhou [1] and Bo Li [2]

1 School of Emergency Management and Safety Engineering, China University of Mining and Technology-Beijing, Beijing 100083, China; cumtbzat@126.com
2 State Key Laboratory Cultivation Base for Gas Geology and Gas Control, Henan Polytechnic University, Jiaozuo 454000, China; anquanlibo@163.com
* Correspondence: fdu@cumtb.edu.cn

As an important natural resource, coal plays a critical role in social and economic development. Due to increasing energy demands, the amount of coal mining is also increasing, precipitating many safety problems. Determining a method whereby the safety level of the coal-mining process can be improved while ensuring acceptable levels of coal production is a problem to which we must attach great importance. However, in the current instantiation of the coal-mining process, coal mine disasters occur frequently, and the levels of coal mine safety management and control are still very insufficient. Therefore, there is an urgent need to further understand the process safety control related to and new technologies used in coal mining. Process safety in coal mining refers to the implementation of measures and practices to identify, prevent, and mitigate potential risks and hazards associated with mining operations. It focuses on ensuring the safety of workers, protecting the environment, and preventing accidents that could result in injuries, fatalities, or damage to equipment and infrastructure.

This Special Issue of *Processes*, titled "Process Safety in Coal Mining", collects the latest work of the main researchers in this field, covering the theory and technology behind coal mine disaster prevention and control, safety management and control in the process of coal mining, mineral processing and green-mining engineering, mining-related environmental protection and sustainable development, and other aspects of safety technology research, focusing on coal mine production safety. Despite the interdisciplinary nature of the different studies included, the ultimate theme of security closely links the different studies contained within this issue.

Coal mine production safety concerns the measures taken to prevent accidents in mines and address and control accidents quickly and effectively in order to ensure the safety and health of the miners. The safety of coal mine production involves many aspects, including mine ventilation, fire and explosion prevention, gas drainage, coal dust prevention, the safety equipment employed, etc.

Due to the complex stress of the surrounding rock, the support and arrangement of roadways play a key role in ensuring the safety of coal mining and the mines themselves. There is a close relationship between changes in surrounding rock stress and mine support. Linlin Chen and coworkers [1] used the equivalent circle method to solve a problem regarding a loose circle of a rectangular roadway and used numerical calculations to obtain the deformation and stress distribution laws of the surrounding rock under the excavation conditions of large section whole coal cavern groups (WCCGs). Guosheng Xu and coworkers [2] studied the variation law of surface dynamic motion based on the surveying and mapping line above the working face of Beigou Coal Mine. The prediction function of the subsidence velocity of the main surface sections and overlying strata in Peigou Coal Mine was established. Using Dongliang Coal Mine as a model for engineering, Li Li and coworkers [3] studied three typical forms of basic roof movement. They obtained the critical fracture conditions of three forms using theoretical calculations. Adopting the resulting state of roof strata movement as the different overlying boundary conditions of the

Citation: Du, F.; Zhou, A.; Li, B. Special Issue "Process Safety in Coal Mining". *Processes* **2023**, *11*, 2271. https://doi.org/10.3390/pr11082271

Received: 20 July 2023
Accepted: 26 July 2023
Published: 28 July 2023

top coal, the influence on the recovery rate was simulated, and the interaction mechanism between the roof base and the top coal was explored. Chen Li and coworkers [4] expounded on the mechanism behind strong rock behavior. They proposed a treatment method for the manifestation of strong strata behaviors by using hydraulic fracturing technology to break the key stratum. Zhengzheng Xie and coworkers [5] revealed the preconditions required for such damage and a bidirectional deterioration mechanism for the deformation and stress of the surrounding rocks. They built an anchorage mechanical model for the thick layer of a roadway roof and proposed a cross-boundary anchor-grouting (CBAG) differential support technique. Xiao-He Wang and coworkers [6] conducted a field test to monitor surface displacement. They used the FLAC3D numerical simulation software to analyze the main stress levels, the shapes of the plastic zones, deformation variables, and the connectivity between the plastic zone of the roadway and the plastic zone of the residual coal pillar in different positions. Qiang Fu and coworkers [7] studied the instability mechanism of cross-section coal pillars and roadways under the dual effects of the bottom support pressure of the remaining coal pillars in the overlying coal seam and the mining of the lower coal seam's working face. They proposed a stability control method for long anchor cable reinforcement support.

The content of coalbed methane directly affects the amount of coalbed methane and mine gas emitted, which is critical to the appropriate design of mine ventilation, gas drainage, and outburst risk assessments. Coal is a kind of porous medium with a complex structure. Adsorption pore structure is a critical parameter influencing the recovery rate of coalbed methane. Gas extraction is critical to the prevention and control of mine gas disaster. Hongyan Lei and coworkers [8] developed a mathematical model for the rapid prediction of coal seam gas content. Wendi Wang and coworkers [9] tested the adsorption pore structure parameters of four coal samples with different metamorphic degrees and discussed the main factors affecting the complexity of pore structure and the influence of fractal characteristics on coal adsorption performance. Hongyan Lei and coworkers [10] used COMSOL software to carry out a three-dimensional numerical simulation, explained the mechanism behind gas migration under negative pressure drainage, and studied the borehole spacing when multi-holes were used to extract gas. Xiaoyang Cheng, and coworkers [11] proposed a data-driven fine control method for gas mining. The technical methods they provided allowed for the fine control of pumping hole design, construction, measurement, and repair. Shuanlin Wang and coworkers [12] combined rock mechanics and fracture mechanics to analyze the influence of the shadow effect of the stress field between cracks on the evolution of equivalent stress and the plastic zone. Zhie Wang and coworkers [13] analyzed the deformation characteristics of boreholes under three typical coal–rock conditions and analyzed the stress, strain, and plastic deformation of rocks around boreholes with different diameters. They simulated the effect of casings on borehole stability.

Mine fires are among the principal disasters that threaten the safety of coal production. It is critical to study coal-related spontaneous combustion and mine ventilation systems to facilitate the prevention and control of mine fires. Jueli Yin and coworkers [14] constructed a large-scale acoustic emission test system for the spontaneous combustion of coal and experimentally tested the acoustic emission signals during the spontaneous combustion of coal. Lina Qu and coworkers [15] constructed a macromolecular model of coal gas in the Huainan mining area. Haiyan Wang and coworkers [16] constructed two typical mine ventilation networks: a parallel branch and a diagonal branch.

Coal and gas outburst and coal–rock composite dynamic disasters refer to the phenomenon wherein coal and gas outbursts occur in the working face of a coal mine due to the interaction between the coal seam and surrounding rock during coal mining, which is accompanied by rock mass instability, mine roof cave-ins, and other rock dynamics disasters. Dongling Sun and coworkers [17] theoretically analyzed the microscopic formation process of a burst shock wave. They deduced the relationship expression between outburst gas pressure and outburst shock wave intensity and proposed a concept concerning the equivalent sound velocity of coal gas flow. Feng Li and coworkers [18] revealed the vibration response characteristics of the interface between coal and rock under an impact load. Haitao Sun and coworkers [19] further clarified the occurrence and developmental mechanism of coal–rock gas composite dynamic disasters in an engineering-scale mining environment and provided guidelines for the prevention and control of coal–rock gas dynamic disasters in deep mines. Haibo Sun and coworkers [20] observed the law behind the evolution in the number of cracks in the process of coal specimen instability via DEM-based numerical simulation.

Safety evaluation management plays an important role in coal mine safety. It is a comprehensive and systematic assessment and analysis of all aspects of a coal mine production system that is conducted to identify potential safety risks and hidden dangers, provide a scientific basis and decision support, and promote the implementation of effective safety management measures in coal mines, thereby improving the safety level of coal mines. Xidi Jiang and coworkers [21] used the Python language to analyze the main body of 'people' exhibiting unsafe behavior. Twenty-four attribute factors were selected from five aspects of individuals, namely, their emotions, motivations, abilities, personalities, and stresses, to construct a comprehensive model of human behavior dubbed SMAPP (sentiment, motivation, ability, personality, and pressure). Zhiliu Wang and coworkers [22] developed an interval trapezoidal fuzzy soft set method to realize the dynamic risk assessment of a high slope. Jing Li and coworkers [23] constructed a coal mine noise simulation experiment system and concluded that the noise level should be reduced to 90 dB to reduce miners' propensity to cause accidents.

Large-scale, high-intensity coal mining in the western mining areas of China has led to reduction in groundwater resources. The implementation of a coal mine in an underground reservoir is an effective means of realizing the protection and utilization of water resources in China's western mining areas. One of the important standards for the safety of underground reservoirs in coal mines is to monitor the development of cracks in coal pillar dams. Bao Zhang and coworkers [24] studied the optimum width of a coal pillar dam using a mechanical model, numerical calculations, and a similar simulation method. Aiming to address the safety problems caused by goaf and water accumulation in open-pit mines, Sheng Zhang and coworkers [25] combined a shallow three-dimensional seismic method with a transient electromagnetic method to detect the distribution range and water-forming conditions related to goafs. Combined with a determination of the special terrain conditions of open-pit mines, they proposed a reasonable data interpretation method, which greatly improved predictive capacity in relation to goaf and water accumulation conditions.

In addition to the safety of coal mining, many ancient architectural structures have different degrees of peeling and foundational problems on the inner surface of their retaining structure due to their long history of construction. To solve the problem regarding the accumulation of moisture inside palace buildings, using an ancient wooden palace building in Beijing as a model, Fang Liu and coworkers [26] used the control variable method to simulate and analyze the influence of outdoor relative humidity, soil moisture, wall moisture, and other factors on the indoor heat and moisture transfer of ancient buildings.

The above articles have furthered the discussion and research on theories, experimentation, and simulation with respect to coal-mining safety research. These articles have solved the outstanding challenges related to production safety in different fields from the perspectives of theory, experimentation, and application, and there is still much to be explored in the future. The Special Issue covers a broad range of topics consistent with the mission outlined by *Processes* to become a highly visible outlet for publishing novel studies in systems modeling, process engineering, and associated applications. The journal will continue to solicit high-quality contributions in these domains.

We thank all contributors and editors for their enthusiastic support of this Special Issue as well as the editorial staff of *Processes* for their efforts.

Funding: This work was supported by the Innovative Research Group Project of the National Natural Science Foundation of China (Nos. 52130409 and 52004291).

Data Availability Statement: The data used to support the findings of this study are available from the corresponding author upon request.

Conflicts of Interest: The authors declare no conflict of interest.

References

1. Chen, L.; Wang, Z.; Wang, W.; Zhang, J. Study on the Deformation Mechanisms of the Surrounding Rock and Its Supporting Technology for Large Section Whole Coal Cavern Groups. *Processes* **2023**, *11*, 891. [CrossRef]
2. Xu, G.; Li, D.; Zhang, Y.; Li, H. Overlying Strata Dynamic Movement Law and Prediction Method Caused by Longwall Coal-Mining: A Case Study. *Processes* **2023**, *11*, 428. [CrossRef]
3. Li, L.; Zhang, X.; Luo, J.; Hu, B. Theoretical Analysis of the Movement Law of Top Coal and Overburden in a Fully Mechanized Top-Coal Caving Face with a Large Mining Height. *Processes* **2022**, *10*, 2596. [CrossRef]
4. Li, C.; Xin, D.; Liu, Y.; Chen, T. A Case Study on Strong Strata Behaviors Mechanism of Mining Reserved Roadway and Its Prevention Techniques. *Processes* **2023**, *11*, 1341. [CrossRef]
5. Xie, Z.; Wang, J.; Zhang, N.; Guo, F.; He, Z.; Xiang, Z.; Zhang, C. Study on Time-Dependent Failure Mechanisms and CBAG Differential Support Technology of Roadway in Steeply Inclined Coal Seam. *Processes* **2023**, *11*, 866. [CrossRef]
6. Wang, X.-H.; Zhang, H.-H.; Wu, Z.; Li, X.-L.; Sui, Y.; Gao, R.-Q. Selection and Optimization Mechanism of the Lower Return Roadway Layout in the near Residual Coal Pillar Area. *Processes* **2022**, *10*, 2471. [CrossRef]
7. Fu, Q.; Yang, K.; He, X.; Liu, Q.; Wei, Z.; Wang, Y. Destabilization Mechanism and Stability Control of the Surrounding Rock in Stope Mining Roadways below Remaining Coal Pillars: A Case Study in Buertai Coal Mine. *Processes* **2022**, *10*, 2192. [CrossRef]
8. Lei, H.; Dai, L.; Cao, J.; Li, R.; Wang, B. Experimental Study on Rapid Determination Method of Coal Seam Gas Content by Indirect Method. *Processes* **2023**, *11*, 925. [CrossRef]
9. Wang, W.; Liu, Z.; Zhang, M.; Yang, H. Experimental Study on Fractal Characteristics of Adsorption Pore Structure of Coal. *Processes* **2023**, *11*, 78. [CrossRef]
10. Du, F.; Cui, W.; Wang, K. Study on Gas Migration Mechanism and Multi-Borehole Spacing Optimization in Coal under Negative Pressure Extraction. *Processes* **2023**, *11*, 259. [CrossRef]
11. Cheng, X.; Sun, H. A Data-Driven Fine-Management and Control Method of Gas-Extraction Boreholes. *Processes* **2022**, *10*, 2709. [CrossRef]
12. Wang, S.; Luo, J. Study on the Shadow Effect of the Stress Field around a Deep-Hole Hydraulic-Fracturing Top-Cutting Borehole and Process Optimization. *Processes* **2023**, *11*, 367. [CrossRef]
13. Wang, Z.; Yang, X.; Wang, G.; Gong, H. Study on Instability Characteristics of the Directional Borehole on the Coal-Seam Roof: A Case Study of the Tingnan Coal Mine. *Processes* **2023**, *11*, 1675. [CrossRef]
14. Yin, J.; Shi, L.; Liu, Z.; Lu, W.; Pan, X.; Zhuang, Z.; Jiao, L.; Kong, B. Study on the Variation Laws and Fractal Characteristics of Acoustic Emission during Coal Spontaneous Combustion. *Processes* **2023**, *11*, 786. [CrossRef]
15. Qu, L.; Liu, L.; Chen, J.; Wang, Z. Molecular Model Construction and Optimization Study of Gas Coal in the Huainan Mining Area. *Processes* **2023**, *11*, 73. [CrossRef]
16. Wang, H.; Xu, Z.; Wang, L.; Fan, C.; Zhang, Y. Experimental Study on the Effect of Air-Doors Control Adjacent to the Fire Source on the Characteristics of Smoke Back-Layering. *Processes* **2022**, *10*, 2496. [CrossRef]
17. Sun, D.; Cao, J.; Dai, L.; Li, R.; Liu, Y. Investigation of Formation Process and Intensity of Coal and Gas Outburst Shockwave. *Processes* **2023**, *11*, 659. [CrossRef]
18. Li, F.; Wang, G.; Xiang, G.; Tang, J.; Ren, B.; Chen, Z. Vibration Response of the Interfaces in Multi-Layer Combined Coal and Rock Mass under Impact Load. *Processes* **2023**, *11*, 306. [CrossRef]
19. Sun, H.; Dai, L.; Lu, J.; Cao, J.; Li, M. Analyzing Energy Transfer Mechanism during Coal and Gas Protrusion in Deep Mines. *Processes* **2022**, *10*, 2634. [CrossRef]
20. Sun, H.; Zhang, B.; Song, Z.; Shen, B.; Song, H. Mechanics-Seepage Experimental and Simulation Study of Gas-Bearing Coal under Different Load Paths. *Processes* **2022**, *10*, 2255. [CrossRef]
21. Jiang, X.; Zhong, S.; Liang, J. Simulation of Unsafe Behavior in Mine Operation Based on the SMAPP Model. *Processes* **2023**, *11*, 1732. [CrossRef]
22. Wang, Z.; Hu, M.; Zhang, P.; Li, X.; Yin, S. Dynamic Risk Assessment of High Slope in Open-Pit Coalmines Based on Interval Trapezoidal Fuzzy Soft Set Method: A Case Study. *Processes* **2022**, *10*, 2168. [CrossRef]
23. Li, J.; Cai, Z.; Liu, H.; Xin, Y. Experimental Research on the Influence of Short-Term Noise Exposure on Miners' Physiology. *Processes* **2023**, *11*, 425. [CrossRef]
24. Zhang, B.; Ni, W.; Hao, X.; Li, H.; Shen, Y. A Study on the Development and Evolution of Fractures in the Coal Pillar Dams of Underground Reservoirs in Coal Mines and Their Optimum Size. *Processes* **2023**, *11*, 1677. [CrossRef]

25. Zhang, S.; Guan, W.; Chang, S.; Meng, Q.; Dong, Y.; Chen, Q. Integrated Geophysical Prediction of Goaf and Water Accumulation in Pingshuo Dong Open-Cut Mine with Ultrashallow and High Drops. *Processes* **2023**, *11*, 1653. [CrossRef]
26. Liu, F.; Zhang, X.; Zeng, J.; Li, Y.; Wang, G. The Numerical Study on Indoor Heat and Moisture Transfer Characteristics of an Ancient Palace Building in Beijing. *Processes* **2023**, *11*, 1900. [CrossRef]

processes

MDPI

Article

The Numerical Study on Indoor Heat and Moisture Transfer Characteristics of an Ancient Palace Building in Beijing

Fang Liu, Xiaofen Zhang, Jiarui Zeng, Yafei Li and Gang Wang *

School of Environmental and Energy Engineering, Beijing University of Civil Engineering and Architecture, Beijing 100044, China; liufang@bucea.edu.cn (F.L.)
* Correspondence: wanggang@bucea.edu.cn

Abstract: The heritage of ancient buildings is an important part of the world's history and culture, which has extremely rich historical–cultural value and artistic research value. Beijing has a large number of ancient palace buildings, and because of the age of their construction, many of them have problems with varying degrees of peeling and molding on the inner surfaces of the envelope. To solve the problems of damp interiors of palace buildings, a mathematical model of indoor heat and moisture transfer was established based on an ancient wooden palace building in Beijing. The model was validated by fitting the measured and simulated data. And the effects of outdoor relative humidity, soil moisture, wall moisture, and other factors on indoor heat and moisture transfer of ancient buildings were simulated and analyzed via the control variables method. The results showed that the measured and simulated data are within the error range, which verifies the accuracy of the model. And the simulation of indoor humidity matched the measured humidity. Thus, the simulation results were consistent with the actual situation. The variable trend of the relative humidity of the indoor environment with the outdoor humidity is inconsistent from plane to plane, i.e., it increases or remains constant with the increase in the outdoor humidity. Indoor ambient relative humidity increased with increasing wall moisture. And the indoor average temperature is 24.5 °C, and indoor relative humidity ranged between 87.4% and 92.4%. Soil moisture and wall moisture were the main factors affecting indoor relative humidity.

Keywords: ancient building; heat transfer moisture transfer; simulation

Citation: Liu, F.; Zhang, X.; Zeng, J.; Li, Y.; Wang, G. The Numerical Study on Indoor Heat and Moisture Transfer Characteristics of an Ancient Palace Building in Beijing. *Processes* **2023**, *11*, 1900. https://doi.org/10.3390/pr11071900

Academic Editors: Feng Du, Aitao Zhou, Bo Li and Udo Fritsching

Received: 10 May 2023
Revised: 8 June 2023
Accepted: 20 June 2023
Published: 24 June 2023

1. Introduction

Ancient buildings are an important part of the world's historical and cultural heritage, which is an important carrier for inheriting national traditional culture and promoting cultural exchanges among nations and cannot be restored once they are damaged [1]. Nowadays, the protection of cultural relics and ancient buildings is gaining more and more attention. "Preserve the original appearance" is the basic principle for the protection of ancient buildings, so higher requirements for the technology of protection and restoration of ancient buildings have been proposed. With the advancement of technology, the conservation and repair of ancient buildings have gradually incorporated modern simulation and digital means from a purely manual approach [2,3] to seek the root mechanism of problem solving.

To better preserve its original appearance and historical value, revitalize the social value of ancient buildings, and enhance the comfort of visitors, many researchers have been conducting various explorations in recent years [4–7], such as ancient building wall repair work and ancient architecture of digital protection, etc. The typical problems of ancient building damage are the degradation of the anti-corrosion paint surface and the moldy wood structure. It was found that the deterioration and mold problems of ancient buildings are influenced by the coupling of multiple factors, such as the thermal and humid environment around the building, air flow, underground soil moisture, and the building envelope, etc. [8–10]. Exploring the patterns of influence of different factors on the deterioration of ancient buildings is an important prerequisite for taking effective conservation measures. At present, numerical simulation methods investigating the effects of airflow, the envelope structure, soil, precipitation, and other factors on the heat and moisture migration of ancient buildings are a low-cost and highly efficient way to grasp the influence law. And the degree of ancient building damage can put forward effective protection schemes for the problems such as dampness. Therefore, it is especially important to carry out more in-depth research and analysis based on numerical simulation methods for ancient buildings and propose reasonable conservation plans.

Currently, some scholars have conducted single-factor or multi-factor coupled analyses in numerical simulation studies. For ordinary residential buildings, Hong et al. [11] used the software CFD to study the indoor wet environment and linearly fitted the factors affecting humidity and concluded that the predicted humidity distribution was in good agreement with the experiment. Teodosiu et al. [12] investigated numerical models for assessing thermal comfort and developed CFD numerical models for airflow modeling and humidity modeling of indoor air, then analyzed and demonstrated the good potential for correctly estimating the indoor environment under stable and uniform thermal conditions. In addition to the relevant research on heat and humidity transfer in residential buildings, existing literature has also investigated building types such as libraries, temples, churches, grottoes, museums, etc. Liu [13] focused on and studied the insulation technology of ancient buildings in the Amu River basin, using AIRPAK software to simulate and analyze the indoor temperature of ancient buildings and explore the effect of active heating measures. Bi et al. [14] conducted an experimental and numerical analysis of the moisture and heat transfer in the cave walls of the Mogao Caves using a comparison of simulation results with experimental temperature and relative humidity values, finding that temperature changes may lead to condensation of moisture in the air, and the application of the one-dimensional model to simulate the cave wall has limitations. Balocco et al. [15] adopted CFD software to carry out a three-dimensional transient simulation of the natural ventilation system of a historical building library in Italy and explored the operation mode of the natural ventilation system inside the building, which confirmed that the ancient building could create a good natural ventilation environment. Li et al. [16] analyzed the hygrothermal environment of the Luohan Hall of Baosheng Temple located in the southeast of China and concluded that the air temperature near the sculpture had different fluctuation modes in different directions, and the relative air humidity fluctuated violently. Cao et al. [17] used the CFD software Fluent to numerically simulate the air movement, temperature, and relative humidity of the micro-environment in the museum and concluded that the large gradient of temperature and humidity distribution was not conducive to the preservation of artifacts. In addition, some scholars have studied numerical simulations for the cultural heritage of wooden structures. Huijbregts et al. [18] conducted a two-dimensional multi-area building simulation on a wooden cabinet in a Dutch castle to study the influence of indoor climate changes on the heat and humidity of the cabinet and concluded that indoor climate conditions were controlled by humidity rather than restricted by the average water content of the room. Napp et al. [19] studied the different indoor climate control schemes of a church in Estonia via field measurement and simulation and concluded that dehumidification measures could be used to prevent mold growth and protect wood parts from cracking.

Summarizing previous research results, it is known that the wet and cold interior of buildings only takes into account the influence of outdoor climatic conditions without exploring the influence of soil moisture and wall moisture on the situation of large indoor humidity. In this paper, soil moisture and wall moisture added to the influence of indoor humidity in ancient buildings.

The object of this study is the world's most complete surviving palace-type ancient building with a construction history of 600 years. The three-dimensional heat and moisture transfer characteristics of palace-style wooden ancient buildings in Beijing are studied. A mathematical model of indoor heat and moisture transfer is established, and the accuracy of the model is verified via comparison with simulated data using actual measurements of indoor relative humidity. The influences of outdoor humidity environment, soil moisture, and wall humidity on the indoor heat and moisture transfer characteristics of ancient buildings are analyzed. And the correlation analysis of these three factors and the main protection measures of ancient buildings are proposed. The research results provide effective theoretical guidance for the conservation of palace-style wooden ancient buildings in Beijing.

2. Indoor and Outdoor Heat and Humidity Environment Testing of Ancient Buildings

2.1. Overview of Ancient Buildings

The palace-style buildings in Beijing are one of the largest and best-preserved wooden ancient buildings in the world, which have a variety of architectural structures with more than 70 different building types [20]. In this study, a palace-style wooden ancient building in Beijing is selected as the research object. The building is located on the west side of the palace-style building complex in Beijing, which was built during the Jiajing period of the Ming Dynasty.

2.2. Outdoor Environmental Parameters

According to the data statistics of a weather station in Beijing (https://www.xihe-energy.com/, accessed on 1 January 2023), the outdoor temperature and humidity change curve of the ancient building is shown in Figure 1. Through the investigation and analysis of the environmental monitoring of ancient buildings, the relative humidity of the indoor environment of the ancient building is high in summer. The original appearance of the building is damaged by the humid indoor environment. Therefore, the outdoor ambient temperature and humidity for the July period from 2020 to 2022 were selected and averaged as the outdoor meteorological parameter of choice. The outdoor temperature is 26.3 °C, the outdoor relative humidity is 67.17%, and the atmospheric pressure is 101.35 kPa. The moisture content is 14.49 g/kg $_{\text{dry air}}$, and the water vapor partial pressure is 2.3073 kPa.

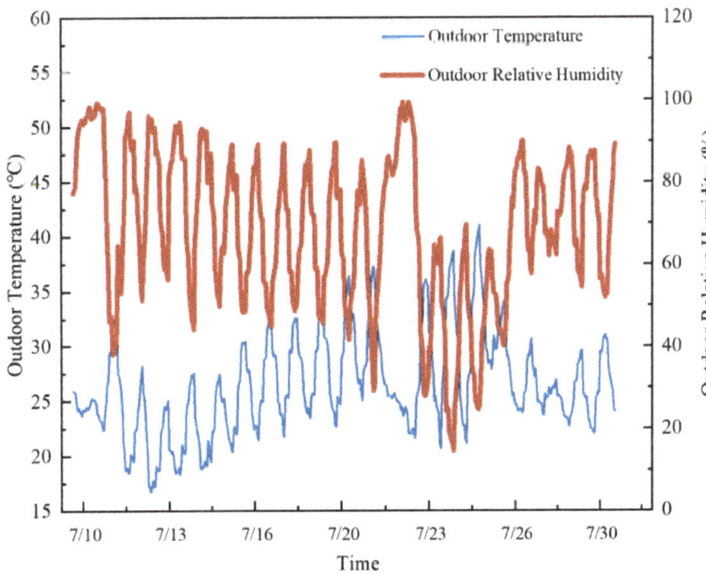

Figure 1. Outdoor temperature and humidity change curve.

2.3. Indoor Environmental Parameters

Based on the relevant research [21], the indoor air temperature is 24.7 °C, the indoor relative humidity is 75%, the moisture content is 14.67 g/kg $_{\text{dry air}}$, and the water vapor partial pressure is 2.3347 kPa. The indoor relative humidity of the interior walls is 92%, the moisture content is 18.09 g/kg $_{\text{dry air}}$, and the water vapor partial pressure is 2.8639 kPa.

2.4. Indoor Flooring Parameters

Soil temperature and humidity can have an impact on the coupled heat and moisture transfer [22]. The ground temperature is set at 23.3 °C, and the water absorption rate of ancient grey bricks generally ranges from 14% to 25% [23], so the water absorption rate of grey bricks is set at 14% in this work. Beijing, located in the North China Plain, is sandy soil. According to relevant studies [21,24], the soil volumetric moisture content of ancient buildings in the area is 0.09 m^3/m^3. The field capacity of sandy soils in the North China Plain is 10–14%. The soil bulk density value of this study is selected from the Second National Soil Survey of China (SNSSC), and the reference value is 0.95 g/cm^3. The conversion relationship between the soil's relative humidity and the soil's volumetric water content is expressed as follows [25]:

$$\theta_V = R \times f_C \times S_C \tag{1}$$

where θ_V is the soil's volumetric water content, m^3/m^3; R is the soil's relative humidity, %; f_C is the soil's field capacity (taken as 10%); and S_C is the soil's bulk density, kg/m^3.

Consequently, it is calculated that the soil's relative humidity of the indoor building ground is 81.5%, the moisture content is 15.97 g/kg $_{\text{dry air}}$, and the water vapor partial pressure is 2.5371 kPa.

3. Numerical Simulation

3.1. Physical Model

The physical graphic of the ancient building is shown in Figure 2. The main area of the ancient building is 26.63 m^2, and the height of the hard roof is 2.73 m. Its building envelope structure mainly includes a roof, walls, doors and windows, and indoor floors composed of a variety of building structures and components of different materials. The specific structure and its adjustment effect on the environment are shown in Table 1. Table 2 shows the dimensions of the ancient building envelope.

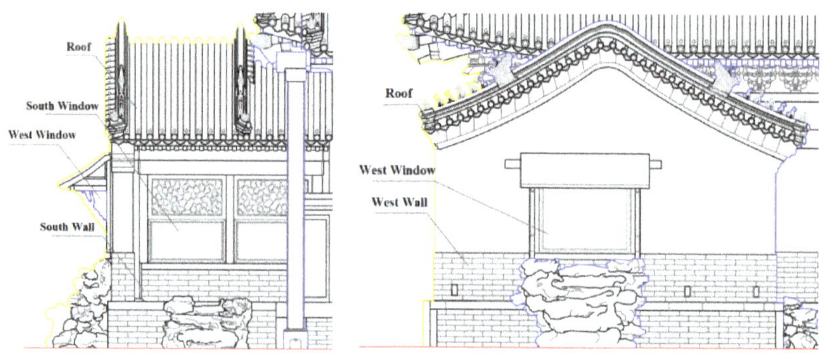

(a) (b)

Figure 2. Elevation of a palace-style wooden ancient building in Beijing: (**a**) south elevation and (**b**) west elevation.

Table 1. Composition and role of the ancient architecture building envelope.

Building Envelope Types	Structural Composition	Environmental Functions
Roof	Tiles, tile mud, bottom tile, clip ridge ash, gray back, watch board, etc.	It can buffer the external temperature, heat insulation, heat preservation, windproof, rainproof, and so on.
Interior wall	Papering, gold cladding earthen wall, oil-painting ground layer, etc.	Built of grey brick, the exterior paint is made of water-resistant properties. It has the functions of being windproof, waterproof, and blocking solar radiation. It can buffer the change in external temperature, has thermal insulation performance, and blocks indoor and outdoor air circulation.
Exterior wall	Wall brick masonry, plaster layer, and bonding material	
Doors and windows	Embrace frame, Geshan door, glass, threshold, metal components, finish coat layer, and internal papering	The doors and windows composed of wood and glass can reduce indoor and outdoor ventilation and buffer temperature changes to a certain extent.
Interior flooring	There is a kang under the indoor brick of the building.	The indoor ground plays an important role in isolating underground moisture; the indoor ground of the kang system has the effect of heat conduction.
Roof	Tiles, tile mud, bottom tile, clip ridge ash, gray back, watch board, etc.	It can buffer the external temperature, heat insulation, heat preservation, windproof, rainproof, and so on.

Table 2. Envelope-sized building table.

Name	L/mm	W/mm	H/mm
Building body	5366	5554	6816
Window 1 (west side)	1982	2088	---
Window 2 (south side)	3812	2261	---
Floor tiles/pc	550	400	53
	517	510	53

Note: The thickness of the east–west wall is 505 mm; the thickness of the north–south wall is 624 mm.

In the room, the release of water vapor comes mainly from the ground and walls, and water vapor can be used as a medium between the two. Due to differences in indoor and outdoor pressure and no mechanical equipment, the air is only penetrated by opening the window, and the density of air in and out is small. The wet environment near the window is close to the wet outdoor environment, and the wet air in the room penetrates the outdoors through the window.

After simplifying the physical graphic, the building structure of the physical model is shown in Figure 3.

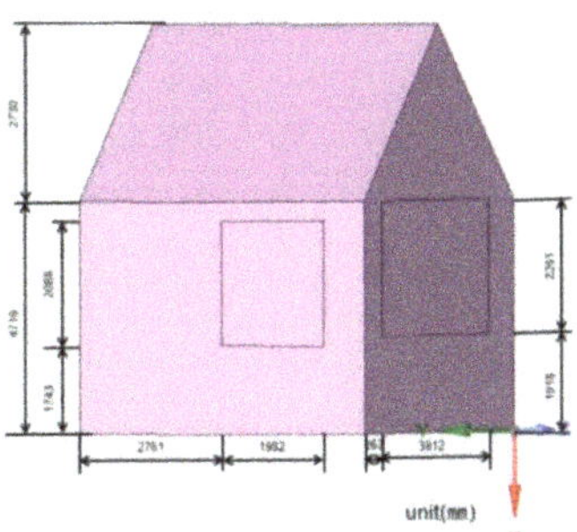

Figure 3. Structure of a palace-style wooden ancient building in Beijing.

3.2. *Mathematical Model*

3.2.1. Basic Assumptions

In this study, the ground and wall are set as the wet source, and the effects of ground soil moisture and wall moisture on the relative humidity inside the room are simulated, respectively. To simplify the simulation, the actual physical process is assumed as follows:

(1) The soil and the wall are wet sources;
(2) The ground and the wall are the water vapor flow inlet, and the two windows are the gas outlets;
(3) The building interior is natural convection, and the indoor airflow is a low-speed flow, which is regarded as an incompressible fluid [26], and satisfies the ideal gas equation of state;
(4) The water vapor mass fraction of the windows is set as the outdoor water vapor mass fraction.

3.2.2. Governing Equations

Based on the physical model and basic assumptions, the governing equations in this paper are outlined as follows.

Continuity equation:

$$div(U) = 0 \tag{2}$$

Momentum equation:

$$div(uU) = -\frac{1}{\rho}\frac{\partial p}{\partial x} + div(\nu \boldsymbol{grad} u) \tag{3}$$

$$div(vU) = -\frac{1}{\rho}\frac{\partial p}{\partial y} + div(\nu \boldsymbol{grad} v) \tag{4}$$

$$div(\omega U) = -\frac{1}{\rho}\frac{\partial p}{\partial z} + div(\nu \boldsymbol{grad} \omega) \tag{5}$$

Energy equation:

$$div(\rho TU) = div\left(\frac{\lambda}{c_p}\boldsymbol{grad} T\right) + S_T \tag{6}$$

The ideal gas equation of state:

$$\mathrm{p} = \mathrm{\rho RT} \tag{7}$$

where ρ is the fluid density, g/cm^3; c_p is the specific heat capacity of fluid, J/(kg·K);

λ is the thermal conductivity; W/(m·K) and **U** are the velocity vectors; T is the thermodynamic temperature; K. ν is the kinematic viscosity of the fluid; m^2/s and S_T are the viscous dissipation terms.

Due to the existence of the moat, the ground dissipates moisture. The windows of ancient buildings are made of wooden materials, which have poor sealing performance, resulting in a pressure difference between indoors and outdoors, causing some moisture to dissipate and diffuse from the outside to the inside. Therefore, in the simulation procedure, it is considered that the airflow in the ancient building blows upwards from the ground. The standard model is used for simulation analysis [27]. In the near wall region, the flow state is laminar and has a low Reynolds number, with a ground average Reynolds number of 1.2109, which is processed using the wall function method.

The floor and walls are made of grey bricks. The thermal conductivity of grey brick is 0.265 W/(m·K). The water vapor diffusion coefficient is 2.92×10^{13} kW·kg/(Pa·m·s). The heat flux of the building envelope and the parameters of the indoor environment are shown in Tables 3 and 4.

Table 3. Heat flux values of the building envelope.

Designation		Heat Flux (W/m^2)
Ground		1.50
Wall	North–South Wall	0.68
	East–West Wall	0.84
	Average	0.76

Table 4. Indoor environmental parameters table.

Environment		Temperature/°C	Relative Humidity/%	Moisture Content/g/kg$_{dry\ air}$	Water Vapor Partial Pressure/kPa	Mass Fraction of Water Vapor/%
Indoor	Center	24.7	75	14.67	2.3347	0.0145
	Ground	23.3	81.5	15.97	2.5371	0.0157
	Interior wall	24.7	92	18.09	2.8639	0.0167

3.3. Grid Partitioning and Irrelevance Verification

3.3.1. Grid Division

Grid generation is very important in CFD (Computational Fluid Dynamics) simulations, which seriously affects the accuracy of CFD simulations by the quality of the generated grid [17]. In this work, a combination of tetrahedral and hexahedral grid types is used to mesh the building. Due to the high accuracy and quality of structural grids, a combination of tetrahedral and hexahedral grids is chosen as the structural grid. The length (X) of the computational region is 6.816 m, the width (Y) is 5.554 m, and the height (Z) is 5.366 m. The number of grids in the X-direction is 97, the number of grids in the Y-direction is 79, and the number of grids in the Z-direction is 77. The average value of this grid quality is 0.77, the grid cell size is 0.07 m, and the number of grids is 1,085,324. At the same time, boundary conditions and local encryption were used. The grids of the ground and the south corner of the palace-style building were encrypted, and the number of grids after encryption is 3,307,497. The results of local encryption are shown in Figure 4a. The results of grid division are shown in Figure 4b.

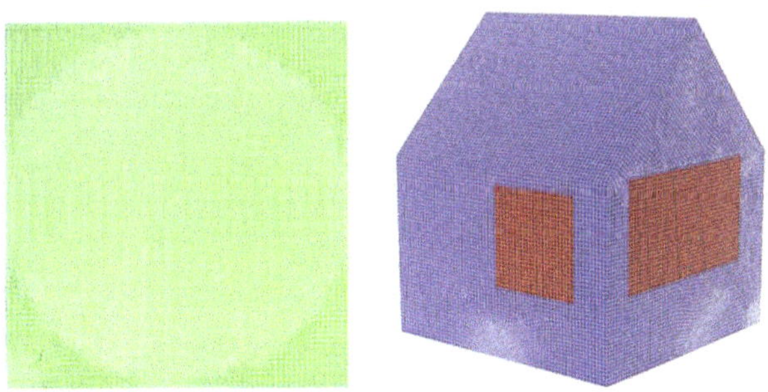

(a) **(b)**

Figure 4. Grid division of a palace-style ancient building model in Beijing: (**a**) local encrypted grid division and (**b**) overall meshing diagram.

3.3.2. Grid Independence Verification

To ensure the accuracy of the calculation results, the generated grids are verified for independence [26]. The following grids all converge monotonically, which are presented in Table 5. Comparing the data in Table 5, the grid independence verification is performed with the value of the average relative humidity on the ground as the reference variable. Grade 3 is closer to grade 1 than the other grades in terms of average relative humidity. In addition, the calculation time of grade 3 is much less than that of grade 1, which can reduce the calculation time to a certain extent. Therefore, grade 3 is selected for the simulation study in this work. At this time, the number of grids is about 1.08 million, the grid cell size is 0.07 m, and the grid quality is good, which can meet the requirements of calculation accuracy.

Table 5. Data statistics of mesh verification.

Grid Classification Levels	Grid Size/m	Number of Meshes/Pieces	Ground Average Temperature/°C	Ground Average Relative Humidity/%
1	0.05	17,315,932	23.3	79.69638
2	0.06	1,594,340	23.3	79.63364
3	0.07	1,085,324	23.3	79.66715
4	0.08	779,490	23.3	79.60327
5	0.09	583,620	23.3	79.59776

3.4. Initial Working Conditions

The outdoor relative humidity range is 45.45% to 87.89%.

The range of relative humidity is simulated by changing the water vapor mass fraction, with a gradient of 0.02%. The effects of ground soil relative humidity, outdoor ambient temperature, and wall relative humidity on indoor relative humidity are investigated. The boundary conditions are set as follows:

(1) The west window and south window are set as the pressure outlets. The outdoor humidity is 67.17%, and the water vapor mass fraction is 0.0142%.
(2) The ground and wall are set as the mass flow inlet, and the mass flow rate is 0.3 m/s.
(3) The wall temperature is set to 26.3 °C, and the ground temperature is 23.3 °C. The water vapor mass fraction of ground and wall are 0.0145% and 0.0167%.

4. Results and Discussion

4.1. Mathematical Model Accuracy Verification

According to the relevant standards [28], the indoor area is less than 30 m^2, the measurement points should be placed on a plane of (1.3 ± 0.1) m height, and the humidity measurement points should be no less than three. In this paper, six relative humidity measurement points are set up, and the measurement points are indicated by the numbers 1–6. The locations of the indoor relative humidity measurement points are shown in Figure 5a, and the coordinates of the measurement points are 1 (−1.2 m, −1.6 m, −1.5 m), 2 (−1.2 m, −3.2 m, −1.5 m), 3 (−1.2 m, −1.6 m, −3 m), 4 (−1.2 m, −3.2 m, −3 m), 5 (−1.2 m, −1.6 m, −4.5 m), and 6 (−1.2 m, −1.6 m, −4.5 m). At this point, multiple measurements are taken at each measurement point on a particular day in summer, and the average is taken as the actual indoor RH data. The model's accuracy is verified by simulating relative humidity at each measurement point. A comparison graph of measured and simulated data is shown in Figure 5b.

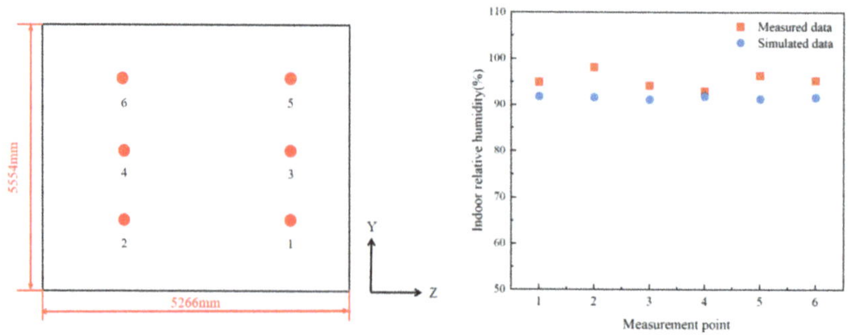

(**a**) (**b**)

Figure 5. Location of measurement points and data map. (**a**) Location of indoor relative humidity measurement points and (**b**) comparison chart of measured and simulated data.

As seen from Figure 5b, the measured and simulated interior average relative humidity at each measurement point ranged from 90% to 100%. Clearly, the relative humidity is large. The maximum relative error between measured and simulated data was calculated to be 6%, and the minimum relative error was 1.2%. The results show that the data from the simulation are consistent with the measured data. The accuracy of the simulation results is verified based on the comparison of the measured and simulated data.

4.2. The Indoor Temperature and Humidity Distribution in Initial Working Conditions

In this work, under the initial working conditions of ancient buildings (outdoor temperature of 26.3 °C and outdoor relative humidity of 67.17%), the indoor temperature and humidity distribution of ancient buildings are simulated. For the purpose of the following analysis, detailed cross-sections in the horizontal and vertical directions are shown in Figure 6 below.

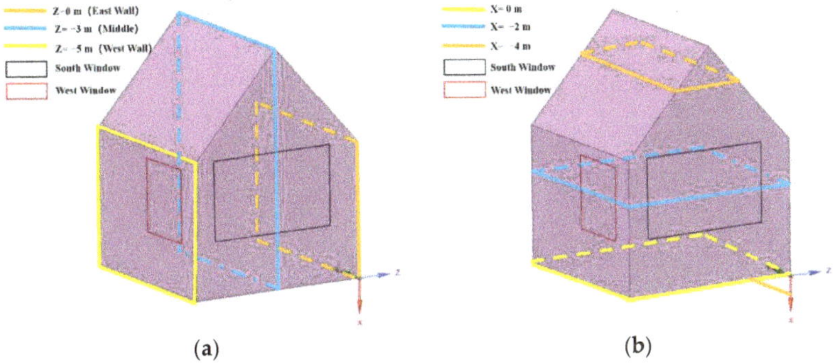

Figure 6. Cross-sectional view in different directions. (**a**) Cross-section in the vertical direction and (**b**) cross-section in the horizontal direction.

The distribution of temperature and relative humidity at Z = 0 m, Z = −3 m, and Z = −5 m in the vertical direction are shown in Figure 7, respectively. As seen in Figure 7a,d, the temperature and relative humidity distribution at Z = 0 m (east wall) remain essentially constant. This is due to the fact that there are no windows in the east wall, and there is no heat dissipation to dissipate humidity. However, there are smaller temperature and humidity fluctuations at ground level. This is due to the fact that the wall root zone is less exposed to sunlight, which causes its humidity to rise slightly. The temperature was around 24.5 °C, and the relative humidity was 90.2%.

As shown in Figure 7b,e, the temperature and relative humidity in the middle part of the room remain more or less constant at Z = −3 m. The temperature is 24.5 °C, and the relative humidity is about 84.3%, mainly due to its more even distribution of moisture content. Fluctuations in temperature and humidity occur on the lower side (floor) and the right side (around the south window). The relative humidity is lower near walls and the ground than inside, mainly because sunlight in summer and poor heat storage capacity of windows can make the part near windows and walls warmer than inside and contain less moisture.

As shown in Figure 7c,f, at Z = −5 m (west wall), fluctuations in temperature and humidity occur mainly in the west window. The temperature and relative humidity at the west window are lower than at the same vertical plane, and the temperature and relative humidity gradually increase from the middle of the window to the surrounding area. This is due to the fact that the windows are wooden and will have gaps, resulting in infiltration and moisture dispersion, resulting in low moisture content. The overall temperature is at 24.4 °C–24.7 °C, and the relative humidity is at 75.9–85.8%.

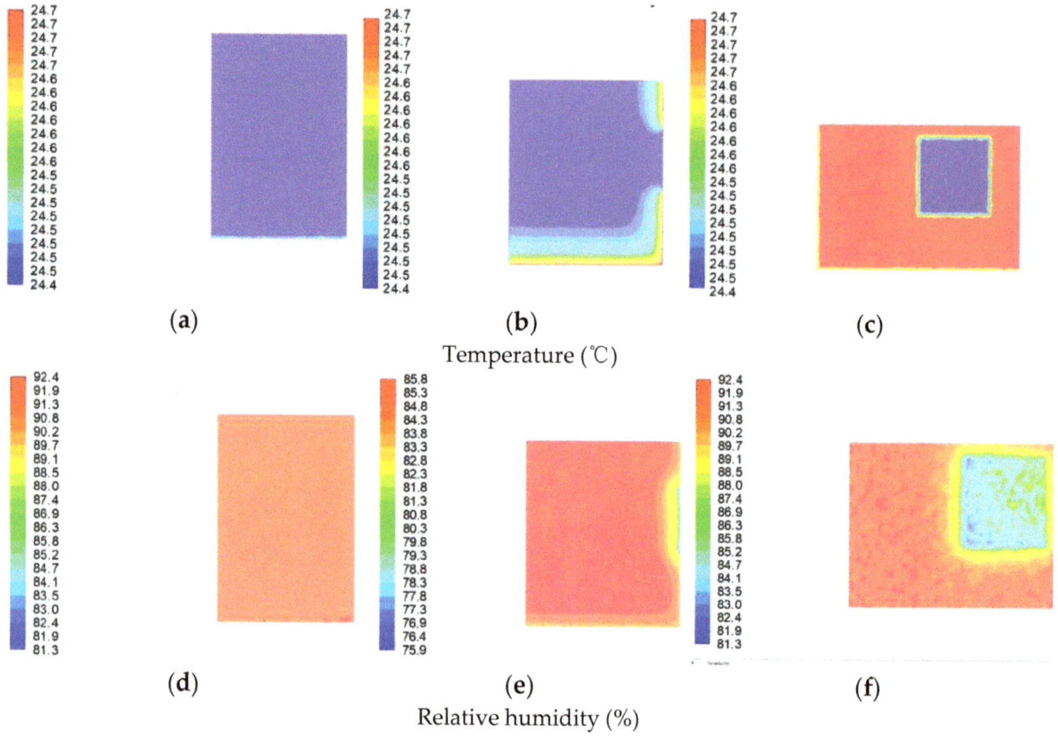

Figure 7. Indoor temperature and relative humidity distribution in the vertical direction. (**a**) Indoor temperature distribution at Z = 0 m, (**b**) indoor temperature distribution at Z = −3 m, (**c**) indoor temperature distribution at Z = −5 m, (**d**) indoor relative humidity distribution at Z = 0 m in the vertical direction, (**e**) indoor relative humidity distribution at Z = −3 m, and (**f**) indoor relative humidity distribution at Z = −5 m.

The distribution of temperature and relative humidity in the horizontal direction at X = 0 m, X = −2 m, and X = −4 m are shown in Figure 8, respectively. Analysis of Figure 8a–c above shows that the room temperature is maintained at 23.3 °C and remains the same. From Figure 8d, it can be seen that the relative humidity of the ground remains unchanged at 91.7%. This is mainly due to the presence of a moat on the ground, which makes its moisture content evenly distributed. From Figure 8e, it can be seen that the relative humidity values of the south and west windows are larger than those of the rest of the plane, which is due to the infiltration effect of the windows that makes their humidity greater. From Figure 8f, it can be seen that the relative humidity on the roof remains constant, and the indoor relative humidity value is higher.

The comprehensive analysis reveals that the indoor average temperature is 24.5 °C, and the indoor relative humidity range is between 87.4% and 92.4%. Overall, the temperature fluctuations in the interior of the ancient building are small, while the relative humidity fluctuates widely. Therefore, the following section focuses on the effect of each factor on indoor humidity. The control variate method is taken to investigate changes in humidity.

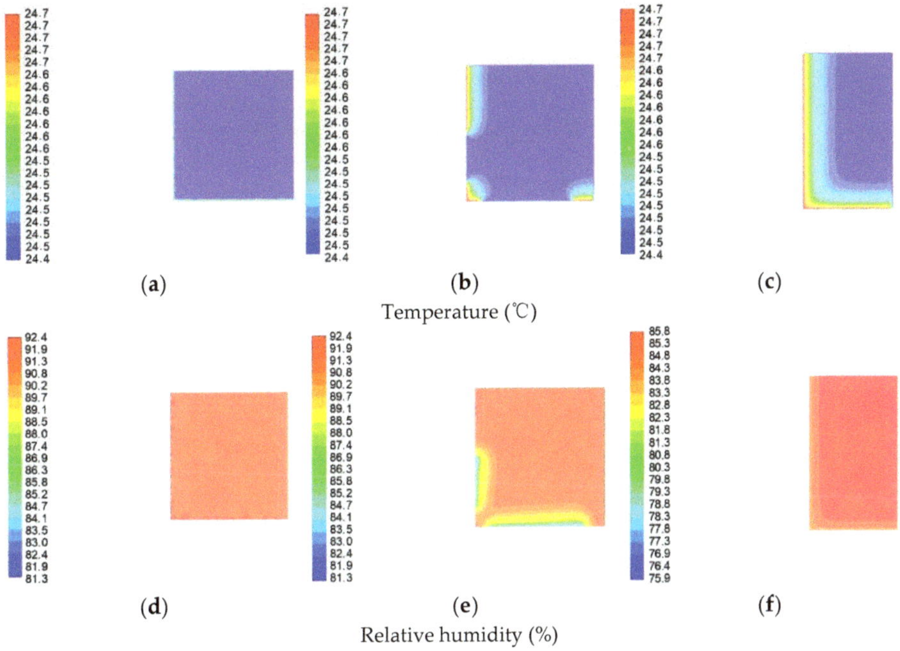

Figure 8. Indoor temperature and humidity distribution in the horizontal direction. (**a**) Indoor temperature distribution at X = 0 m, (**b**) indoor temperature distribution at X = −2 m, (c) indoor temperature distribution at X = −4 m, (**d**) indoor relative humidity distribution at X = 0 m in the vertical direction, (**e**) indoor relative humidity distribution at X = −2 m, and (**f**) indoor relative humidity distribution at X = −4 m.

4.3. Effect of Outdoor Humidity on Indoor Moisture Transfer

The humidity of the soil and walls is set as the initial state. With the difference in outdoor humidity, the indoor humidity is shown in Figure 9a,b. Due to the ancient age of this historic building, outdoor air enters the interior through windows, which are made of a paper material and have a significant degree of damage. Thus, the outdoor humid environment is an important factor affecting the indoor humidity of ancient buildings.

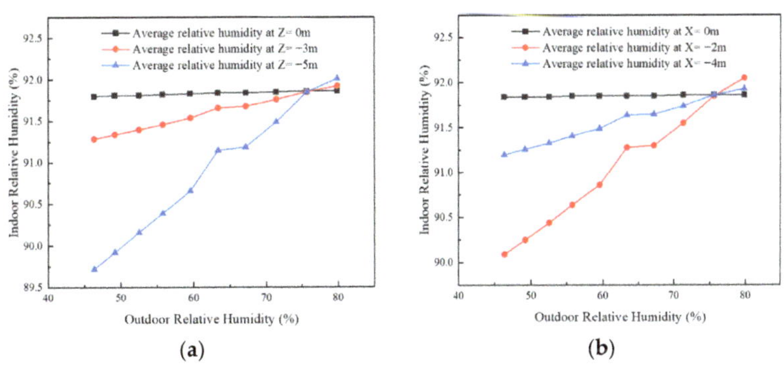

Figure 9. Effect of outdoor humidity on the humidity in the horizontal and vertical direction of the room. (**a**) Trends in the effect of outdoor ambient relative humidity on average relative humidity in the vertical direction inside the room and (**b**) trends in the effect of outdoor ambient relative humidity on average relative humidity in the horizontal direction of the room.

As seen from Figure 9a, under a given outdoor humidity, the indoor relative humidity is observed to change in the vertical direction from east to west (Z = 0 m to Z = −5 m). When the outdoor humidity is less than 75.6%, the average indoor relative humidity shows a trend of gradual decrease. And when the outdoor humidity is greater than 75.6%, the average indoor relative humidity shows a gradually increasing trend. When located in the same vertical plane, with the increase in outdoor humidity, the average indoor relative humidity increases significantly at Z = −5 m. At Z = −3 m, the trend of the average indoor relative humidity increases more slowly than that at Z = −5 m. At Z = 0 m, the average indoor relative humidity remains unchanged.

As seen from Figure 9b, under a given outdoor humidity, the change in indoor relative humidity in the horizontal direction from the ground to the roof (X = 0 m to X = −4 m) is observed. When the outdoor humidity is less than 75.6%, the average indoor relative humidity shows a trend of first decreasing and then gradually increasing, and the change in decreasing is larger. At outdoor humidity greater than 75.6%, the average indoor relative humidity showed a trend of first increasing and then gradually decreasing. Therefore, it was concluded that the average relative humidity at ground level was the highest. When located in the same horizontal plane, with the increase in outdoor humidity, the average indoor relative humidity is the highest and remains the same at X = 0 m. At X = −2 m, the average indoor relative humidity increases significantly. At X = −4 m, the average indoor relative humidity shows a slowly increasing trend.

4.4. *Effect of Soil Moisture on Moisture Transfer*

Xu et al. [22] carried out experiments on the effect of coupled heat and moisture transfer on soil heat storage systems and confirmed that ignoring moisture migration and temperature dependence of soil thermal conductivity would lead to a low predicted value of the numerical model, and the influence of initial soil moisture on coupled heat and moisture transfer should be considered. In this paper, the influence of soil moisture on the humidity in different directions in the room was simulated by changing the mass fraction of groundwater vapor (soil moisture).

As seen from Figure 10a, under a given soil moisture, changes in indoor relative humidity in the vertical direction from east to west (Z = 0 m to Z = −5 m) are observed. The average indoor relative humidity decreases gradually, but the decrease is small. When located in the same vertical plane, the average indoor relative humidity gradually increased with the increase in soil moisture, and the trend of the increase was almost the same on each plane.

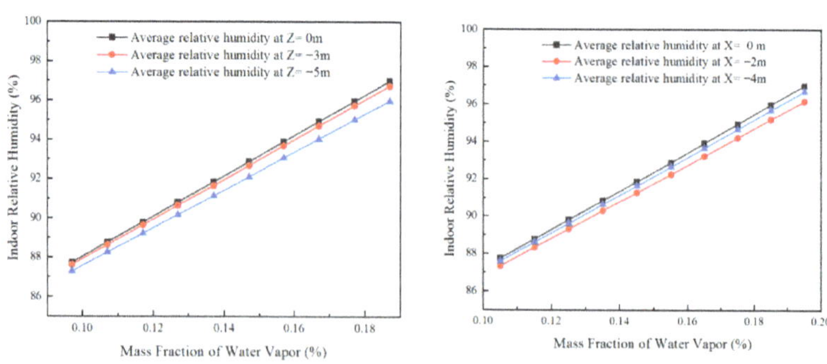

(a) (b)

Figure 10. Effect of soil moisture on the humidity in the horizontal and vertical direction of the room. (**a**) Trends in the effect of soil moisture on average relative humidity in the vertical direction inside the room and (**b**) trends in the effect of soil moisture on average relative humidity in the horizontal direction of the room

As seen from Figure 10b, under a given soil moisture, the average indoor temperature shows a first decreasing and then increasing trend when observing the change in indoor temperature and humidity in the horizontal direction from ground to roof (X = 0 m to X = −4 m), but the change is smaller. When located in the same horizontal plane, the average indoor relative humidity increases gradually with the increase in soil humidity.

4.5. Effect of Wall Moisture on Indoor Humidity Transfer

Over time, the porosity of the ancient building envelope will become larger, and its moisture storage capacity will also be enhanced. In addition, excessive humidity in the building envelope can lead to damage to building materials and mold growth [29]. It is also a factor that produces wall mold, as it migrates into the indoor environment. Therefore, the influence of wall humidity on the humidity in different directions of the room was simulated by changing the water vapor mass fraction (changing wall moisture) of the wall.

As seen from Figure 11a, the indoor relative humidity values on different planes are the same, so the relative humidity of the whole room can be analyzed by the above figure. From Figure 11a,b, it can be seen that as the humidity of the walls increases, the relative humidity in the room also gradually increases. This is due to the moisture transfer from the walls, which makes the humidity in the room also increase gradually.

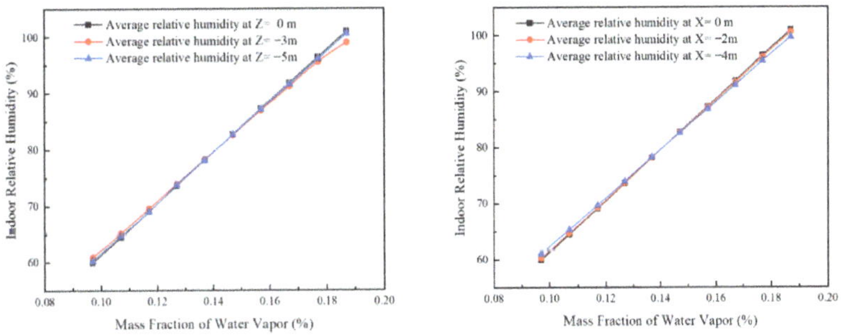

Figure 11. Effect of wall humidity on the average humidity in the horizontal and vertical direction of the room. (**a**) Trends in the effect of wall humidity on average relative humidity in the vertical direction inside the room and (**b**) trends in the effect of wall humidity on average relative humidity in the horizontal direction of the room.

From the analysis of indoor relative humidity law in 4.4 and 4.5, it can be seen that when the ground or wall water vapor mass fraction decreases (ground soil or wall moisture decreases), the average indoor relative humidity decreases when the conditions of building protection and maintaining the original appearance of the building can be met. Therefore, the relative humidity of the indoor environment can be reduced by lowering the ground soil moisture or wall moisture.

4.6. The Sensitivity of Factors

While the effect patterns of outdoor environmental humidity, soil moisture, and wall humidity on indoor relative humidity have been obtained via simulation, the sensitivity of each factor requires correlation analysis that takes statistical methods [30], with indoor relative humidity as the target (significant test level of 0.01). Outdoor ambient relative humidity, soil moisture, and wall moisture are set as the reference series X_0 and the indoor ambient relative humidity is set as the comparison series $X_i(k)$, respectively. Based on the existing correlation empirical equations [31,32], the calculation steps for correlation analysis are as follows.

Relation coefficient:

$$\gamma_i(k) = \frac{min_i min_k \Delta_i(k) + \xi max_i max_k \Delta_i(k)}{\Delta_i(k) + \xi max_i max_k \Delta_i(k)} \tag{8}$$

where $\Delta_i(k) = \left| X'_0(k) - X'_i(k) \right|, \xi \in (0,1)$ is the resolution factor, which usually takes the value of 0.5.

Correlation degree:

$$R_i = \frac{1}{n} \sum_{k=1}^{n} \gamma_i(k), i = 1,2,3,\ldots,m; k = 1,2,3,\ldots,n \tag{9}$$

From Table 6, it can be seen that the correlation between outdoor ambient humidity to indoor relative humidity is weak; while the correlation between soil humidity and wall humidity to indoor relative humidity is strong, and the correlation between wall and indoor relative humidity is lower than that of soil by 0.002. This paper mainly focuses on the problem of wet building floors and high indoor relative humidity. Therefore, it can be concluded that soil moisture and wall moisture are the main factors affecting indoor relative humidity.

Table 6. Correlation analysis result.

Dependent Variable	Indoor Relative Humidity
Correlation coefficient with outdoor environmental humidity	0.486
Correlation coefficient with soil moisture	0.995
Correlation coefficient with wall moisture	0.993

4.7. Indoor Temperature and Humidity Distribution after Moisture-Proof Treatment

According to related studies [33–35], antimicrobial and hydrophobic coatings can be used to reduce room humidity by coating and protecting moldy areas of bricks without affecting the appearance of ancient buildings, such as SiO_2-TiO_2 hybrid fluorinated B-72, lime putty mortar (ASPL and ASPL/PP series), etc. This coating has better resistance to acid, alkali, salt, and UV light as well as the inhibition of damp proofing. Ground and walls are as examples of wet sources, obtained from Figure 12. When the water vapor mass fraction is 0.0105% (ground soil relative humidity of 59.02%) and the wall water vapor mass fraction is 0.0107% (wall relative humidity of 60%), the relative humidity of the ancient building interior is 60.2% to 78.8%, in line with the requirements of indoor relative humidity under summer working conditions. Indoor ground temperature and humidity distribution are shown in Figure 12, and the temperature at the ground is 23.3 °C, with an average relative humidity of 60.32%. As seen in Figure 12a, after the damp-proofing treatment, the relative humidity of the ancient building interior was reduced from 90.9% to 94.0% to 60.2% to 78.8%, and the relative humidity decreased and met the requirements of indoor relative humidity under summer working conditions.

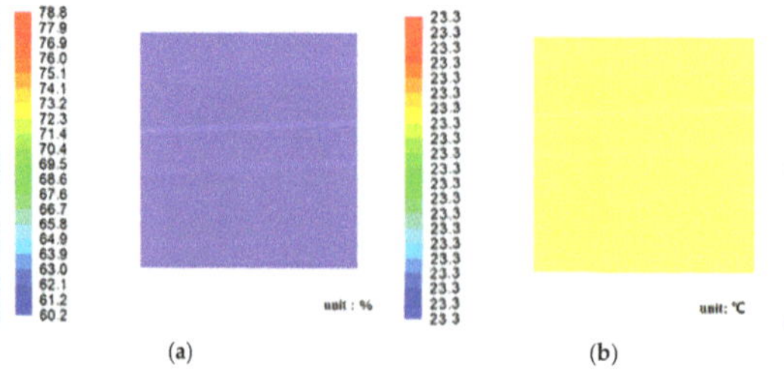

Figure 12. Indoor ground temperature and humidity distribution picture: (**a**) Relative humidity distribution at ground level and (**b**) temperature distribution at ground level.

5. Conclusions

To solve the problem of mold growth causing damage to the architectural style of an ancient palace in Beijing, a simulation study of indoor heat and humidity transfer characteristics of ancient buildings was carried out via an ANSYS numerical simulation research method. The results of the parameters affecting the indoor humidity and the proposed protective measures were obtained as follows.

(1) In the initial state, the overall room temperature varied between 24.4 °C and 24.7 °C, and the relative humidity fluctuated between 87.4% and 92.4%, with high relative humidity.
(2) The indoor relative humidity changes with the outdoor relative humidity and increases with the increase in soil moisture and wall moisture.
(3) Soil moisture and wall moisture are the main factors affecting the change in indoor relative humidity.
(4) By adding hydrophobic coatings to the floor and walls, the room temperature was 23.3 °C, and the relative humidity ranged from 60.2% to 78.8%, which had the effect of reducing humidity.

Author Contributions: All authors contributed to this paper. Conceptualization, F.L. and X.Z.; methodology, writing—original draft preparation, resources, supervision, F.L.; software, J.Z.; validation, X.Z. and J.Z.; formal analysis, Y.L.; writing—original draft preparation, supervision, F.L.; writing—review and editing, G.W. All authors have read and agreed to the published version of the manuscript.

Funding: This research received no external funding.

Data Availability Statement: No new data were created or analyzed in this study.

Acknowledgments: We also would like to thank the anonymous reviewers for their valuable comments and suggestions that lead to a substantially improved manuscript.

Conflicts of Interest: The authors declare no conflict of interest.

References

1. Wei, K.; Yang, W.Y.; Zhou, B.; Wang, K.; Sun, J.Z.; Sun, X.K.; Xu, M.; Chen, Q.Q.; Qiu, B.; Wang, W.; et al. The color change analysis of historic wooden remains after fire-suppression by fluorinated chemical gases. *Herit. Sci.* **2021**, *9*, 93. [CrossRef]
2. Qiu, Y.H. 3D Reconstruction and Intelligent Digital Conservation of Ancient Buildings Based on Laser Point Cloud Data. *J. Electr. Comput. Eng.* **2022**, *2022*, 7182018. [CrossRef]
3. Rubeis, T.D.; Nardi, I.; Muttilloa, M.; Paoletti, D. The restoration of severely damaged churche-Implications and opportunities on cultural heritage conservation, thermal comfort and energy efficiency. *J. Cult. Herit.* **2020**, *43*, 186–203. [CrossRef]
4. Santosuosso, P. A satisfaction-based model for risk indexing in cultural heritage conservation. *J. Cult. Herit.* **2022**, *57*, 173–183. [CrossRef]
5. Fico, D.; Rizzo, D.; Casciaro, R.; Corcione, C.E. Historically Accurate Reconstruction of the Materials and Conservation Technologies Used on the Facades of the Artistic Buildings in Lecce (Apulia, Italy). *Materials* **2022**, *15*, 3658. [CrossRef]
6. Wang, X.Y.; Wang, J.; Wang, J.H.; Sheng, G.H. Experimental and Numerical Simulation Analyses of Flame Spread Behaviour over Wood Treated with Flame Retardant in Ancient Buildings of Fuling Mausoleum, China. *Fire. Technol.* **2022**. [CrossRef]
7. Huai, C.P.; Xie, J.C.; Liu, F.; Du, J.T.; Chow, D.H.C.; Liu, J.P. Experimental and Numerical Analysis of Fire Risk in Historic Chinese Temples: A Case in Beijing. *Int. J. Archit. Herit.* **2022**, *16*, 1844–1858. [CrossRef]
8. Talbot, R.; Picco, M.; Greenfield, D.; Ashton, P.; Arbuthnot, E.; Hashemi, A. Historic Churches and Their Hygrothermal Environment: A Review of Criteria Related to Building Fabric, Artefacts, Artwork and Occupants. *Sustainability* **2022**, *14*, 7822. [CrossRef]
9. Steeman, H.J.; Belleghem, M.V.; Janssens, A.; Paepe, M.D. Coupled simulation of heat and moisture transport in air and porous materials for the assessment of moisture related damage. *Build. Environ.* **2009**, *44*, 2176–2184. [CrossRef]
10. Belleghem, M.V.; Steeman, M.; Willockx, A.; Janssens, A.; Paepe, M.D. Benchmark experiments for moisture transfer modelling in air and porous materials. *Build. Environ.* **2011**, *46*, 884–898. [CrossRef]
11. Huang, H.; Kato, S.; Hu, R.; Ishida, Y. Development of new indices to assess the contribution of moisture sources to indoor humidity and application to optimizati-on design: Proposal of CRI(H) and a transient simulation for the prediction of i-ndoor humidity. *Build. Environ.* **2011**, *46*, 1817–1826. [CrossRef]
12. Teodosiu, C.; Hohota, R.; Rusaouën, G.; Woloszyn, M. Numerical prediction of ind-oor air humidity and its effect on indoor environment. *Build. Environ.* **2003**, *38*, 655–664. [CrossRef]

13. Liu, W.Q. The Evolution of Cold Adaptation Technology within Ancient Buildings in Amur River Basin Viewed from Archaeology. *Int. J. Environ. Res. Public Health* **2022**, *19*, 14470. [CrossRef] [PubMed]
14. Bi, W.B.; Yan, Z.F.; Zhang, Z.M.; Yao, S.S.; Zhang, J.J.; Wang, X.D. Modeling and numerical simulation of heat and mass transfer in the cave wall of the Mogao -Grottoes in China. *Build. Environ.* **2021**, *201*, 108003. [CrossRef]
15. Balocco, C.; Grazzini, G. Numerical simulation of ancient natural ventilation syste-ms of historical buildings. A case study in Palermo. *J. Cult. Herit.* **2009**, *10*, 313–318. [CrossRef]
16. Li, S.; Xie, H.R.; Ma, Y.; Hokoi, S.C.; Li, Y.H. Assessing the deterioration risk of polychrome clay sculptures based on the hygrothermal environment: A case study of Baosheng temple, China. Case Stud. *Constr. Mater.* **2022**, *17*, e01287. [CrossRef]
17. Cao, L.N.Y.; Cao, J.J.; Lee, S.C.; Zhang, Y.W.; Tie, X.X. Numerical Simulation of the Micro Environment in the Han Yang Mausoleum Museum. *Aerosol. Air. Qual. Res.* **2012**, *12*, 544–552. [CrossRef]
18. Huijbregts, Z.; Schellen, H.; Schijndel, J.V.; Ankersmit, B. Modelling of heat and moisture induced strain to assess the impact of present and historical indoor climate conditions on mechanical degradation of a wooden cabinet. *J. Cult. Herit.* **2015**, *16*, 419–427. [CrossRef]
19. Napp, M.; KalameesA, T. Energy use and indoor climate of conservation heating, dehumidification and adaptive ventilation for the climate control of a mediaeval c-hurch in a cold climate. *Energy. Build.* **2015**, *108*, 61–71. [CrossRef]
20. Wei, L.; Ma, Y.; Guo, Z.M.; Ding, J.J.; Jin, G.W.; Gu, A.; Lei, Y. Application of Advanced Analytical Techniques in Organic Cultural Heritage: A Case Study of Ancient Architecture Relics in the Palace Museum (Beijing). *Coatings* **2022**, *12*, 636. [CrossRef]
21. Zhang, X.G.; Zhang, S.; Chen, H.Y. Research and Analysis of the Building Environment in the Conservation of Yang Xin Hall Area in the Palace Museum. *Tradit. Chin. Archit. Gard. Prot. Cult. Herit.* **2018**, *2*, 31–36. (In Chinese)
22. Xu, Y.S.; Zeng, Z.T.; Sun, D.A. Experimental and numerical investigation on the effect of heat and moisture coupling migration of unsaturated lateritic clay for the soil thermal storage system. *Energy Build.* **2022**, *276*, 112499. [CrossRef]
23. Zhao, P.; Zhang, X.Z.; Qin, L.; Zhang, Y.S.; Zhou, L.Z. Conservation of disappearing traditional manufacturing process for Chinese grey brick: Field survey and laboratory study. *Constr. Build Mater.* **2019**, *212*, 531–540. [CrossRef]
24. Chen, X.Y.; Ye, J.C.; Lu, G.H.; Qin, F.X. Study on field capacity distribution about soil of China. *Water Resour. Hydropower Eng.* **2004**, *35*, 113–116, 119. (In Chinese) [CrossRef]
25. *SL 364-2015*; Specifications for soil moisture monitoring. Ministry of Water Resources of the People's Republic of China: Beijing, China, 2015. (In Chinese)
26. Xiong, J.; Li, A.G.; Liu, C.P.; Dong, J.G.; Yang, B.; Cao, J.J.; Ren, T. Probing the historic thermal and humid environment in a 2000-year-old ancient underground tomb and enlightenment for cultural heritage protection and preventive conservation. *Energy Build.* **2021**, *251*, 111388. [CrossRef]
27. Vereecken, E.; Roels, S. Review of mould prediction models and their influence on mould risk evaluation. *Build. Environ.* **2011**, *51*, 296–310. [CrossRef]
28. *GB/T 18883-2022.2022-07-11*; Standards for indoor air quality. National Health Commission of the People's Republic of China: Beijing, China, 2022. (In Chinese)
29. Li, Y.H.; Kong, Z.Y.; Xie, H.R.; Ma, Y.; Mu, B.G.; Hokoi, S.C. Construction type influences features of rising damp of blue-brick masonry walls, Constr. *Build. Mater.* **2021**, *284*, 122791. [CrossRef]
30. Jacob, K.S. Statistical Methods in Psychiatric Research and SPSS. *Indian J. Psychiatry* **2016**, *58*, 356. [CrossRef]
31. Ashraf, J.; Luo, L.Q.; Anser, M.K. Do BRI policy and institutional quality influence economic growth and environmental quality? An empirical analysis from South Asian- countries affiliated with the Belt and Road Initiative. *Environ. Sci. Pollut. Res.* **2022**, *29*, 8438–8451. [CrossRef]
32. Mausam, K.; Pare, A.; Ghosh, S.K.; Tiwari, A.K. Thermal performance analysis of hybrid-nanofluid based flat plate collector using Grey relational analysis (GRA): An approach for sustainable energy harvesting. *Therm. Sci. Eng. Prog.* **2023**, *37*, 101609. [CrossRef]
33. Vacher, S.; Hernandez, C.; Bärtschi, C.; Poussereau, N. Impact of paint and wall-paper on mould growth on plasterboards and aluminum. *Build. Environ.* **2010**, *45*, 916–921. [CrossRef]
34. Wang, K.Y.; Bu, N.J.; Zhen, Q.; Liu, J.B.; Bashir, S. Modified nano-SiO2/TiO2 hybrid fluorinated B-72 as antimicrobial and hydrophobic coatings for the conservation of ancient bricks. *Constr. Build. Mater.* **2023**, *365*, 130090. [CrossRef]
35. Branco, F.G.; Belgas, M.D.L.; Mendes, C.; Pereira, L.; Ortega, J.M. Characterization of Fresh and Durability Properties of Different Lime Mortars for Being Used as Masonry Coatings in the Restoration of Ancient Constructions. *Sustainability* **2021**, *13*, 4909. [CrossRef]

processes

Article

Simulation of Unsafe Behavior in Mine Operation Based on the SMAPP Model

Xidi Jiang *, Shuheng Zhong and Jialu Liang

School of Energy and Mining Engineering, China University of Mining and Technology (Beijing), Beijing 100083, China; zhsh@cumtb.edu.cn (S.Z.); jialu@student.cumtb.edu.cn (J.L.)
* Correspondence: jiangxidicumtb@126.com; Tel.: +86-152-1057-2805

Abstract: Mine accidents are mostly caused by human unsafe behavior. To reduce the unsafe behavior of mine operation and reduce the accident of mine operation, the main body of unsafe behavior 'people' is analyzed, and 24 attribute factors are selected from five aspects of people's emotions, motivation, ability, personality, and pressure to construct the comprehensive model of human behavior SMAPP (sentiment–motivation–ability–personality–pressure). The program tool for recording, saving, and executing the mutual and interactive influence relationship of 24 attribute factors under different state values and the simulation process framework of SMAPP was constructed by using 1071 rule statements written in Python language. The fuzzy rules are used to simulate different scenarios. The simulation results are consistent with the actual research results, which shows the reliability and scientificity of the model. In addition, additional events are added to the simulation process to make the model more realistic. Through the simulation results, the influence of employees' emotions, motivations, abilities, personalities, pressures, and additional events on the unsafe behavior of mine operations is analyzed and predicted, and the measures to reduce the unsafe behavior of mine operations are further proposed.

Keywords: SMAPP model; unsafe behavior; fuzzy rules; simulation framework

Citation: Jiang, X.; Zhong, S.; Liang, J. Simulation of Unsafe Behavior in Mine Operation Based on the SMAPP Model. *Processes* **2023**, *11*, 1732. https://doi.org/10.3390/pr11061732

Academic Editor: Carlos Sierra Fernández

Received: 18 May 2023
Revised: 4 June 2023
Accepted: 5 June 2023
Published: 6 June 2023

1. Introduction

Mine operation is a high-risk industry. Although the number of mine production accidents and deaths has decreased significantly in recent years, it still occasionally occurs. According to statistics and analysis of coal mine production safety accidents in China from 2010 to 2019, statistics of major gas explosion accidents in coal mines from 2011 to 2020, statistical analysis and prevention countermeasures of coal mine accidents from 2008 to 2020, it is found that the number of accidents and deaths is the least in winter, and the number of accidents and deaths is the most in summer. The main reasons are as follows: the production days in winter are less, and the safety management of coal mines during the Spring Festival is more strict; in summer, the weather is hot and humid, and there are many mosquitoes, which affect the rest of the employees. Employees are prone to being impetuous and sleepy, and their safety awareness is relaxed. The unsafe behavior of employees is affected by emotions, the working environment, and the management system, which leads to mine accidents [1–5]. Therefore, predicting people's unsafe behavior and taking corresponding preventive measures is one of the effective means to reduce accidents.

A large number of scholars at home and abroad have studied the unsafe behavior of people. Dana Willmer summarizes the degree of various factors of 338 accidents in the United States. Human factors are the most important cause of accidents, and management is the second most important reason [6]. Donald et al. found that the personality characteristics of miners are closely related to the occurrence of accidents [7]. Glenn Legault et al. believed that the miners' working environment, unsafe psychological state, working pressure, and coal mine shift system would hurt and harm miners' unsafe behavior [8–10].

Hu Zhe et al. established a cognition-based model of workers' safety behavior ability to study the effect of safety training on the improvement of safety behavior ability [11]. Zhang Han et al. analyzed coal mine production accidents and their causes from 2001 to 2015 and proposed a new idea of production safety management based on risk pre-control management [12]. Zhang Denghao et al. pointed out through a questionnaire survey and analysis that emotional burnout played a complete mediating role between job demands and unsafe behaviors, and behavior style moderated the indirect effect of job demands on unsafe behaviors [13]. Yang Dong et al. studied the relationship between safety awareness and unsafe behavior of electric power employees with different personality traits and found that neuroticism was significantly positively correlated with unsafe behavior. Agreeableness, extraversion, openness, and conscientiousness were significantly negatively correlated with unsafe behavior. Openness has a negative predictive effect on unsafe behavior, and neuroticism and agreeableness have a significant regulatory effect on safety awareness of unsafe behavior [14]. Based on the theoretical framework of social cognition, Yang Zhenhong et al. established the regulating effect of accident experience on miners' unsafe behavior under the regulating effect of a safe atmosphere [15]. Zhou Jianliang et al. found that psychological adjustment can effectively control workers' unsafe behaviors [16]. Li Guangli et al. found that tension, infirmity, anxiety, depression, irritability, boredom, and drowsiness are the key unsafe emotions that miners experience more frequently in daily life. Anger and tension have certain differences among different marital statuses and different types of work, and depression has certain differences among different types of work and working years [17]. Zhang Yuliang studied the influence of outport employees' emotional changes on production safety in coal enterprises and proposed corresponding management countermeasures [18]. Liu Jialun et al. found that when workers have more safety knowledge and work experience, the cognitive process of dangerous accidents tends to be objective and rational, and the risk sensitivity increases accordingly [19], and Tian Shuicheng adopted the entropy weight TOPSIS (Technique for Order Preference by Similarity to Ideal Solution) method to comprehensively evaluate the unsafe state of front-line miners and concluded that the importance of family support was the largest, while the importance of self-efficacy was the least [20].

As a complex individual, there are many factors affecting human behavior. Some scholars have clarified the attributes of human beings. There is no unified view on which characteristics and factors of human beings will affect human behavior. Ren Hao believes that physiological factors, such as the human nervous system and endocrine system, psychological factors, such as human motivation, feeling, ability, character, etc., and family, media, public policy, and other social factors, can affect people's behavior [21]. Zhang Jian also pointed out that people's emotions are related to people's performance, and proved this with the famous Yex–Dodson's law [22].

This paper attempts to make a comprehensive analysis of the subject 'person' of unsafe behavior, based on fuzzy rules (fuzzy rules are binary fuzzy relations R defined on $X \times Y$). There are two explanations for $A \rightarrow B$: one is A coupling B, and the other is A leading to B. Based on these two explanations and different operators, fuzzy rules can have a variety of legal calculation formulas). According to the existing research results, this paper constructs the SMAPP (sentiment–motivation–ability–personality–pressure) model from five aspects—emotion, motivation, ability, personality, and pressure—which is referred to as the human behavior model. A simulation framework suitable for multiple scenarios and prediction groups is proposed to predict the behavior of employees, and different scenarios are simulated to analyze the behavior of mine workers. Additional events (optimization of mining enterprise management system, safety culture, humanistic care, skills training, etc.) are added to the scene where employees' unsafe behavior has an increasing trend. The direct impact of additional events on employee attributes is fuzzy-assigned. We analyzed the impact of additional events on other attributes of employees and changes in employee behavior. Measures to reduce employee unsafe behavior are proposed from the aspects of improving the working environment, reducing work intensity, and strengthening skills training. Measures to prevent mine safety accidents are summarized from the aspects of safety culture construction and management reform.

2. Construction of SMAPP Model

The sentiment is people's attitude towards objective things and corresponding reaction behavior. Ortony, Clore, and Collins proposed the OCC model, which divides emotions into 11 pairs of basic emotions [23,24]. In employee emotion and management, Zhang Jian introduced five important employee emotion phenomena, namely, emotional intelligence, emotional work, occupational stress, emotion and creativity, and emotion and decision. This paper mainly refers to Wang Chunxue's research on the impact of emotions on safety sign recognition, and defines emotions as positive (happiness), neutral, and negative (sadness, anger, and fear), assuming that positive emotions promote the performance of safety behaviors [25].

Motivation is a kind of internal power, which affects the direction, intensity, and endurance of individual behavior. It is formed under self-regulation so that the internal needs of individuals and external incentives match. According to American psychologist Herzberg's theory of motivation–health care in 1959, motivation is divided into motivating factors and healthcare factors. Health factors mainly include policy and management, supervision, wages, colleague relations, and working conditions. These are non-work factors that, if satisfied, can eliminate dissatisfaction and maintain productivity, but do not motivate people to behave more positively. Motivating factors include achievement, promotion, development, etc. If these factors are satisfied, they can generate great incentives for people; if they are not satisfied, they will not generate dissatisfaction such as health factors. This paper defines motivation as progress, development, achievement, salary, and policy.

Ability is the degree to which an individual can perform a task or task as required. Ren Hao believes that a person's ability includes IQ, EQ, creativity, and moral quotient; Stephen believes that ability can be divided into intellectual ability and physical ability. Intellectual ability mainly includes language understanding, logical reasoning, understanding and expression, vocabulary use, associative memory, imagination, etc. Physical abilities are mainly strength, vitality, endurance, and coordination. According to the existing research, abilities are divided into physiological, cognitive, and technical levels, as well as learning, creativity, and social abilities.

Personality mainly refers to the relatively stable characteristics that affect people's behaviors. According to John Holland's personal–job fit theory, personality is divided into six parts [26]. According to Hippocrates' humoral dominance theory, personality is divided into four parts: bilious, sanguine, mucous, and melancholic. In this paper, according to the big five personalities of the ocean theory of personality, personality is defined as openness, neuroticism, conscientiousness, extroversion, and agreeableness. See Figure 1 for the manifestations of each type of personality.

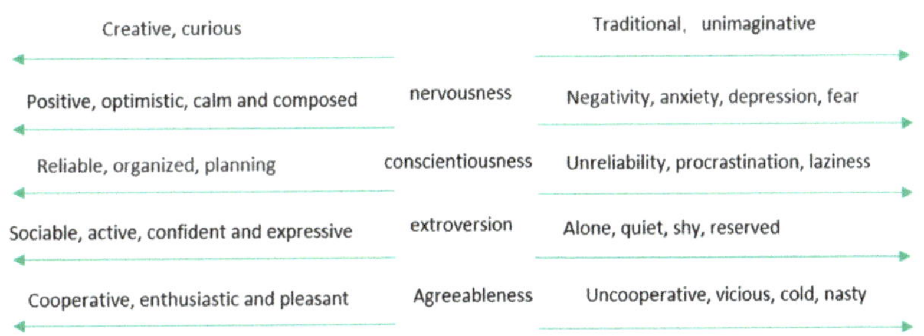

Figure 1. Big Five personality.

With the quickening pace of society, today, most people live under high pressure. While the right amount of stress is good for people's performance, excessive stress can certainly take a toll on people's physical and mental health. There are many reasons for people's stress. Su Yong divided stress into four parts: task-related, role-related, interpersonal, and working conditions [27]; or, simply stated: work, family, society, relationships, and health.

The human behavior model is established according to emotion, motivation, ability, personality, pressure, and the corresponding secondary influencing factors selected, as shown in Figure 2.

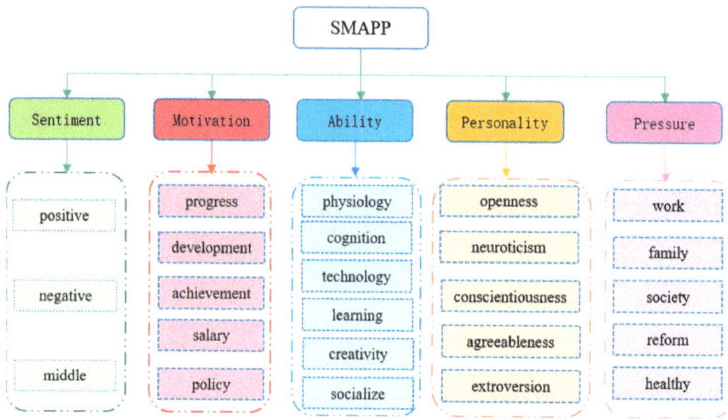

Figure 2. Human behavior model.

3. Build SMAPP Model Simulation Process Framework

In the aspect of the realization of human model simulation, there are three main methods. The first is a mathematical method, using mathematical formulas to predict human behavior, but this traditional simulation method can only get a simplified description of the whole system. It does not fully emphasize the complex hierarchical structure of the system, and sometimes it is difficult to extract formulas [28]. The second model is a rules-based system, which is a mature and widely used approach, as we only need to extract the interaction between different attributes and describe it in the form of rules [29]. Thirdly, agent-based modeling (ABM), seeks the simplest explanation for the observed complex system, modeling individuals such as people, organizations, and enterprises as agents. Agents have their state and rules of behavior. Through the behavior and interaction of micro-agents, sudden phenomena, dynamic equilibrium, and nonlinear results can be reproduced to deduce the macroscopic phenomena and operation results of the system [30]. There is no universally recognized definition of an agent. Currently, the cognitive architecture BDI (belief–desire–intention) model [31] is widely used in agent modeling (generally, it is believed that the thinking state of an agent includes belief, desire, and intention) and is too rational for describing complex and changeable human beings. For example, ACT-R believes that the cognitive process of human beings requires the participation of different modules, which correspond to the physiological structure of the human brain. The modules work independently and are coordinated by the central production system, and it is understandable to take rule-based reasoning as its core part to predict the output [32]. In this paper, the simulation framework of the SMAPP model is built based on rules; see Figure 3. The specific process is as follows:

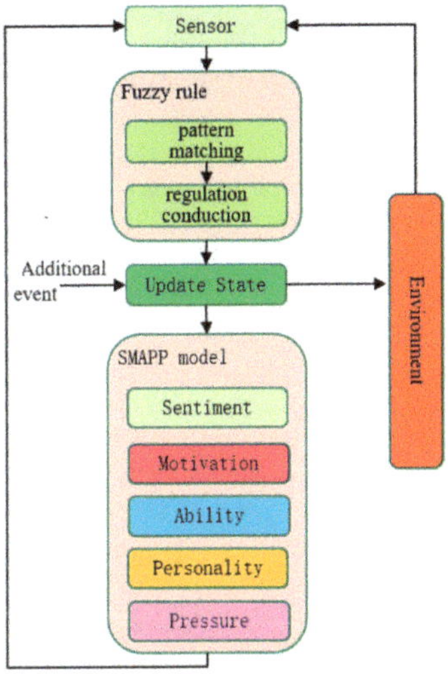

Figure 3. SMAPP model simulation process framework.

(1) The sensor obtains the SMAPP model attribute factor and environment status value at the same time, and fuzzy processing the status value of the attribute factor. In this paper, the fuzzy states of factors are defined into five types: very high, high, medium, low, and very low corresponding score intervals are shown in Table 1;
(2) According to the state of SMAPP model factors and environment, the fuzzy rules are mapped to build rules that conform to the synergistic influence between all factors. According to the existing research results, the mutual and interactive influence relationship between 24 attribute factors in the model is studied, and 1071 rule statements written in Python language are constructed to record, save and execute the mutual and interactive influence relationship between attribute factors under different state values;
(3) Execute these rules to update the state of SMAPP model factors and environment. During the execution of model rules, five update states are defined: sharp increase, increase, maintain, decrease, and sharp decrease; and the corresponding increments are 3%, 1%, 0, −1%, and −3% of the original parameter values, respectively [33]. Due to the complexity and uncertainty of people, random variables of −1 to 1 are added during the execution of model rules;
(4) Accidents often occur in the actual work, so additional events are added during the execution of the model rules. If an emergency occurs, SMAPP model factors and environment state can be updated directly;
(5) Enter the first step or end the simulation. With the implementation of the rule, the state of people and the environment is constantly updated. The number of updates is denoted as variable "T". Then, new rules are created to change the state of people or environment, such as conditions related to "T". When T = 72, it reduces (unsafe behavior); that means that the unsafe behavior decreased after 72 runs of the rule, probably as a result of improved skills and cognition with learning.

Table 1. Corresponding score interval of fuzzy state.

Fuzzy State	Very High	High	Medium	Low	Very Low
Value	>90	70–90	40–70	20–40	<20
Level	I	II	III	IV	V

Different application scenarios have different definitions of environment. To verify the simulation results of the model on unsafe behaviors, the environment is defined as a task, and three attributes are defined for it. The first is Difficulty, which is divided into five levels as before, but with different names: Very Hard, Hard, Medium, Easy, and Very Easy. The second property is the status value; it records the output value of the task, used to reflect the performance of the employee, and will affect the state of the employee. The third is unsafe behavior, which is consistent with the corresponding score of fuzzy state in Table 1 above, and divided into five levels: extremely unsafe (more than 90 points), unsafe (70–90 points), generally safe (40–70 points), safe (20–40 points) and very safe (less than 20 points). In the unsafe behavior of mining operations, "T" represents the unit of time in weeks.

According to the simulation architecture of the SMAPP model, Python language rules are used to describe the relationship between different attributes. For example:

if (isHigh(self.management) and isHigh(self.ability)):
self. progress = ActionUtil.increase (self. progress)
if (isHigh(self.openness) and isHigh(self.sentiment)):
self.creativity = ActionUtil.increase(self.creativity)
if (isVeryLow(self.promotion) and isVeryLow(self.extraversion)):
self.positive = ActionUtil.highDecrease(self.positive)
if (isVeryLow(self.promotion) and isLow(self.extraversion)):
self.positive = ActionUtil.highDecrease(self.positive).

The condition of the above rule is the state of the person and the environment, and the execution part is to change the state value in five update modes. The rule contains all possible states related to unsafe behavior in mine operations. The synergetic influence relationship of the SMAPP model is shown in Figure 4. The output of different factors is represented by different colors: purple for stress, green for sentiment, yellow for personality, red for motivation, blue for ability, light blue for safety tasks instead of the external environment, and black lines for human output. The different colored lines indicate the influence of factors on other factors, such as extraversion promoting sociability, with a yellow line going from extraversion to sociability. Ability will affect the individual's progress but will also affect the sense of achievement; there is a blue line pointing to progress and sense of achievement; and emotion will affect people's creativity and directly affect individual behavior (unsafe behavior). There is a green line pointing to creativity and unsafe behavior; motivation will promote individuals to improve their technical level; and a red line pointing to technology.

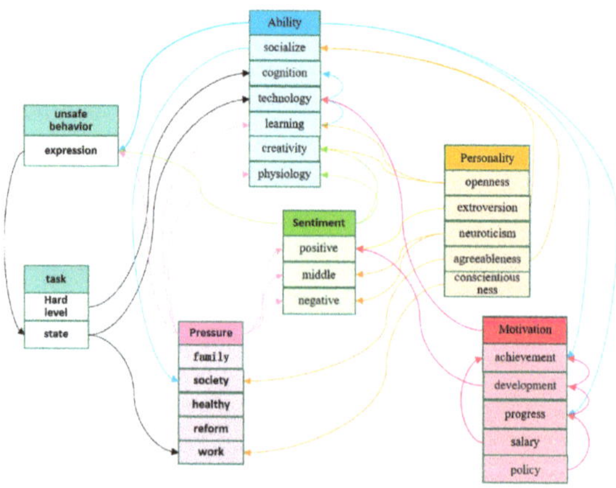

Figure 4. Synergistic influence diagram of SMAPP model attribute factors.

4. Simulation of Miners' Behavior Based on the SMAPP Model

A. Simulation of miners' behaviors under different scenarios

According to the SMAPP model and its simulation framework, we simulate three scenarios for mine production employees. (1) Level III for employees with 60 points in ability, emotion, and personality. (2) Weak ability, do not want to learn and change, and mood, personality, and ability of 30 points for the poor performance of the staff corresponding to Grade V. (3) Poor behavior becomes excellent behavior. First, the behavior was set to 30 points. In the 30th week, additional events were added, and its motivation and emotion were set to 80 points to represent the change of the person. Like a person who wants to learn, eager to make changes in real life. The specific results are as follows:

(1) The personality, emotion, motivation, and ability of the employees were set as 60 points; that is, the corresponding level III employees. Regardless of the environment and organizational changes, the employees were simulated according to the corresponding fuzzy rules of the SMAPP model simulation framework. The horizontal axis represented the times of the implementation and updating of the rules, and the vertical axis represented the ability, achievement, motivation, pressure, emotion, and unsafe behavior trends, respectively. From Figures 5–10, it can be seen that the ability of level III employees increases with the increase in training times and the enhancement of their subjective learning awareness. Among them, the first change is the individual's learning ability, and the last change is the individual's social ability. With the improvement of ability, personal emotions will become more positive, while work motivation will also be greatly improved, and personal pressure will also decrease as emotions become more positive. Therefore, the probability of unsafe behavior of employees at work will be reduced, thus ensuring the safe production of mines;

(2) The employee's ability, motivation, and emotion score are set at 30 points, indicating that the person has weak ability, poor mood, and subjective reluctance to study and work; the corresponding level is V. Trends in competence, motivation, stress, emotion, and unsafe behavior are shown in Figures 10–14. From Figures 10–14, it can be seen that when the individual's comprehensive ability is weak and the subjective learning awareness is not high, and they do not actively ask for them to change, and the organizational environment and the policy family, and other conditions do not change, their job motivation opportunities will decline and, at a lower level, and their negative emotions will increase. With time, when the personal ability is getting worse and worse, work pressure increases greatly, and the score is five times as much as in the beginning. Therefore, the probability of unsafe behavior of employees at work will increase exponentially, which will have a greater impact on mine safety production;

(3) If an employee is a person with low ability, emotion, and motivation at the time of entry, but with the occurrence of additional events such as marriage and family changes, accident experience training, management system, and promotion policy changes, they realize that they need to learn and improve their ability to change the status quo, and their work motivation and positive emotion become high [10,12,34]. In the simulation rules, the ability, personality, and emotion of employees are first set as low (30 points, corresponding to Level V in this paper), and then events are added (for example, training, improvement of working environment, optimization of the control system, an increase in salary, etc.), the corresponding value of learning, emotion, and motivation of employees was set as high at the 30th update (80 points, corresponding to level II in this paper), and the results were shown in Figures 15–19. It can be seen from Figures 15–19 that, due to employees' low ability, emotion, and motivation at the beginning, the overall quality of individuals is relatively low. In the case of certain subjective learning awareness and other external factors, their personal stress and unsafe behavior do not immediately decrease but will increase slightly. However, with continuous learning and the occurrence of additional events, the technical level and cognition of employees are also increasing, their emotions are in a positive state

for a long time, and the pressure is also reduced. Therefore, the unsafe behavior of employees gradually decreases with the increase in ability, motivation, and emotion and the decrease in pressure, to ensure the safe production of mines.

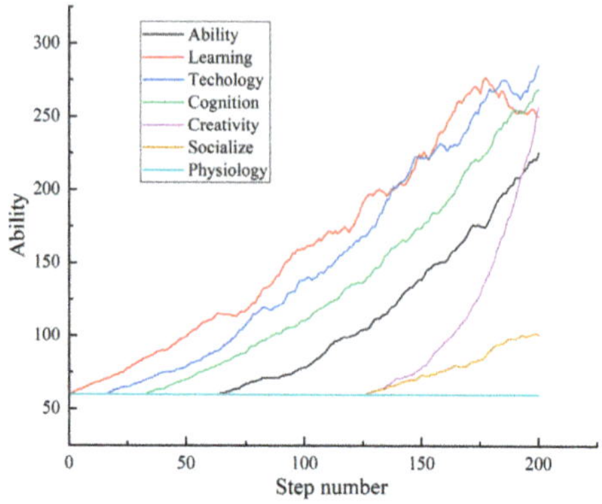

Figure 5. Ability trend curve of employees level III.

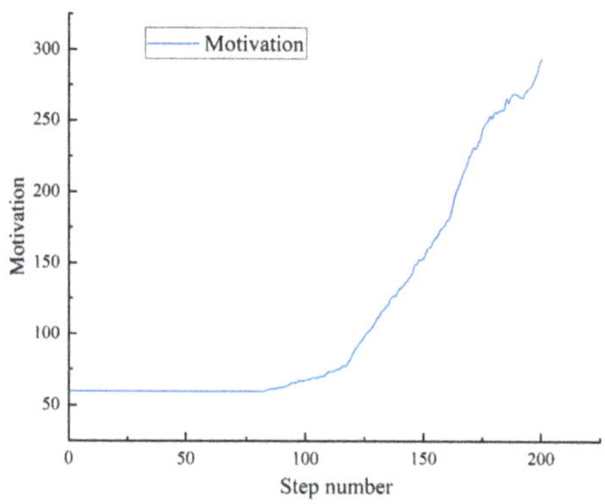

Figure 6. Motivation trend of employees level III.

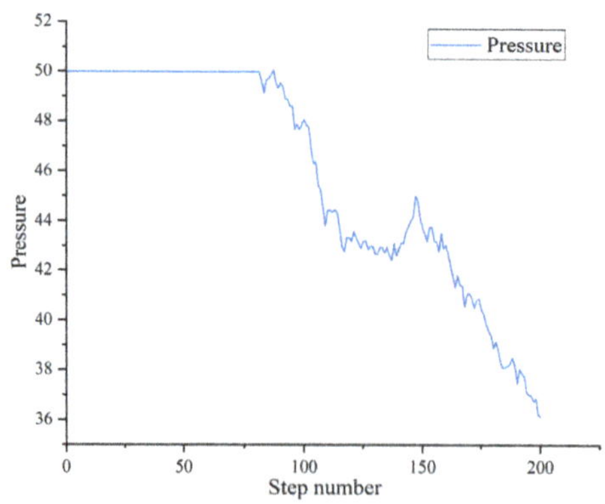

Figure 7. Stress trend of employees level III.

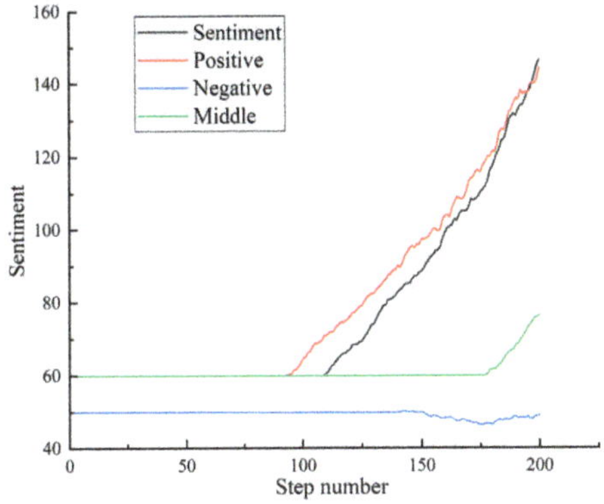

Figure 8. Emotional trend curve of employees level III.

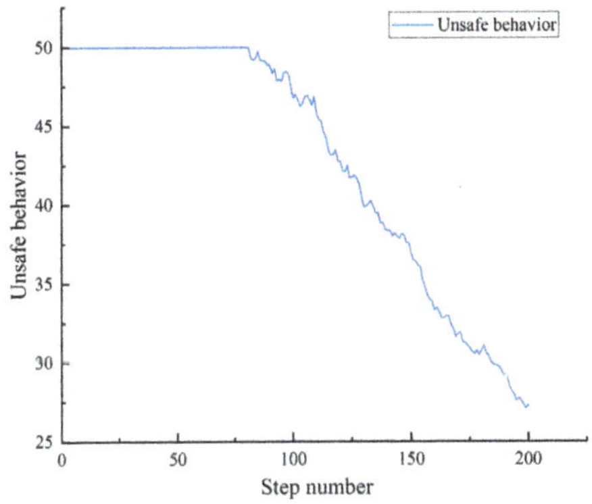

Figure 9. Unsafe behavior trend of employees level III.

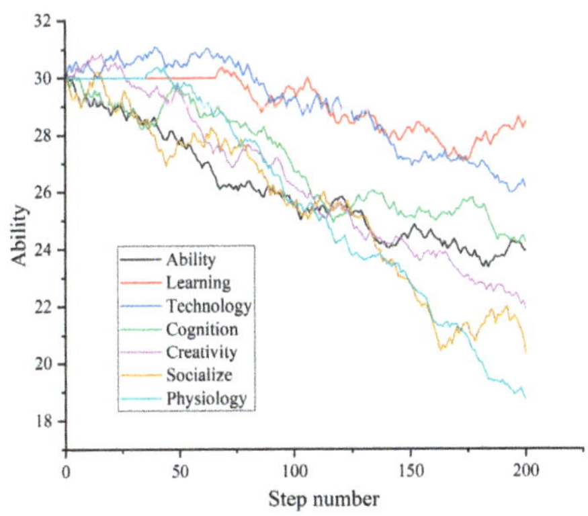

Figure 10. Ability trend curve of employees level V.

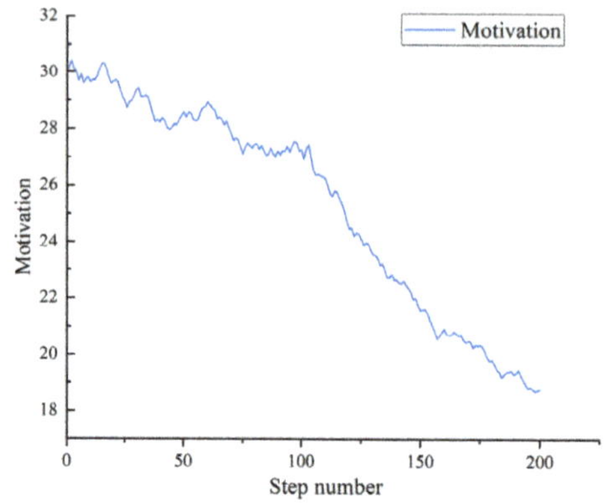

Figure 11. Motivation trend of employees level V.

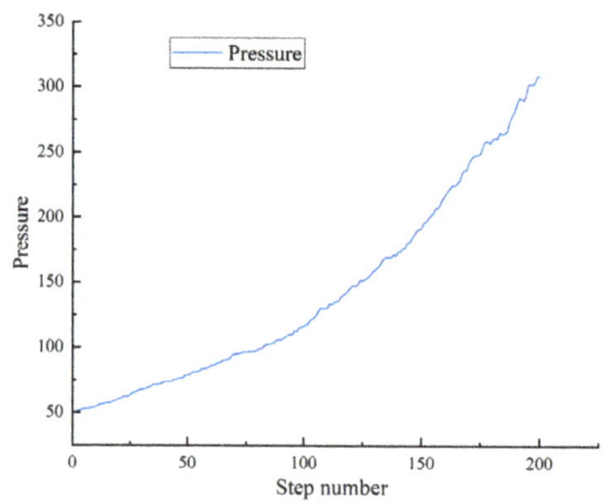

Figure 12. Stress trend of employees level V.

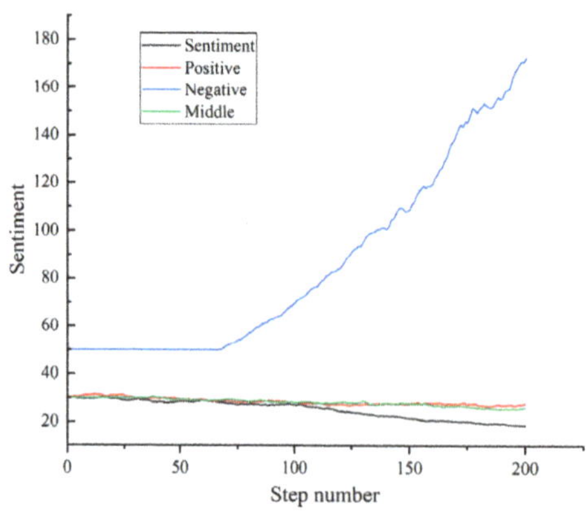

Figure 13. Emotional trends of employees level V.

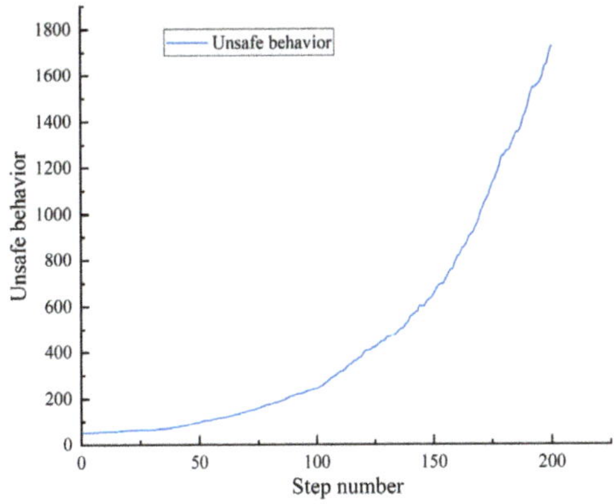

Figure 14. Unsafe behavior trend of employees level V.

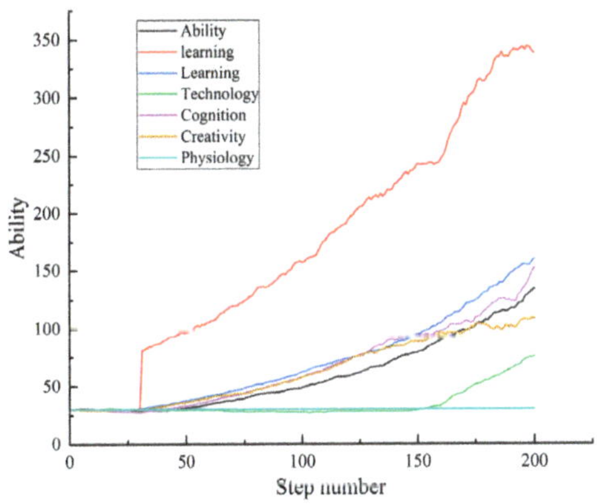

Figure 15. Ability trend of employees from level V to level II.

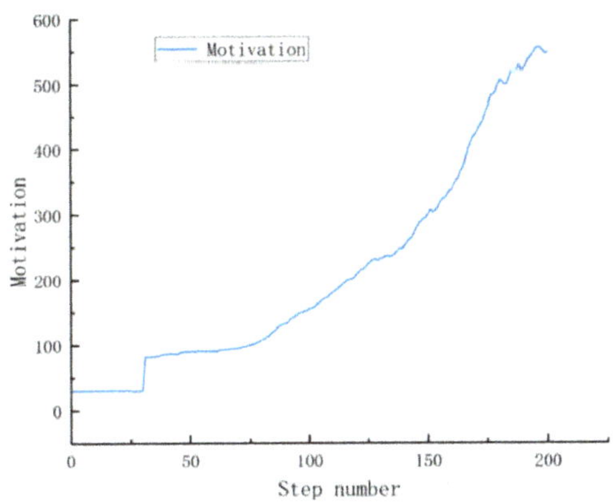

Figure 16. Motivation trend of employees from level V to level II.

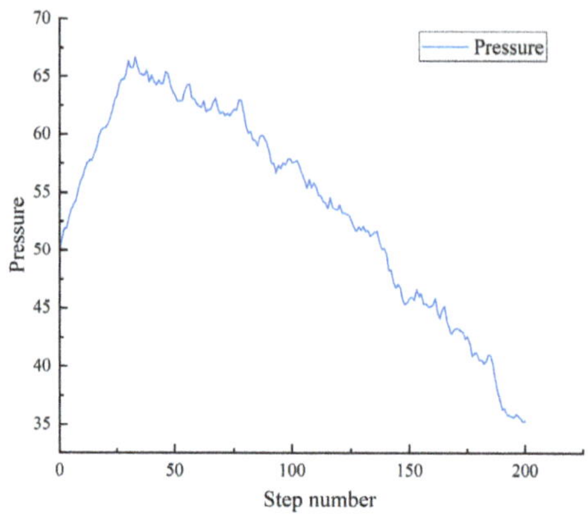

Figure 17. Stress trend of employees from level V to level II.

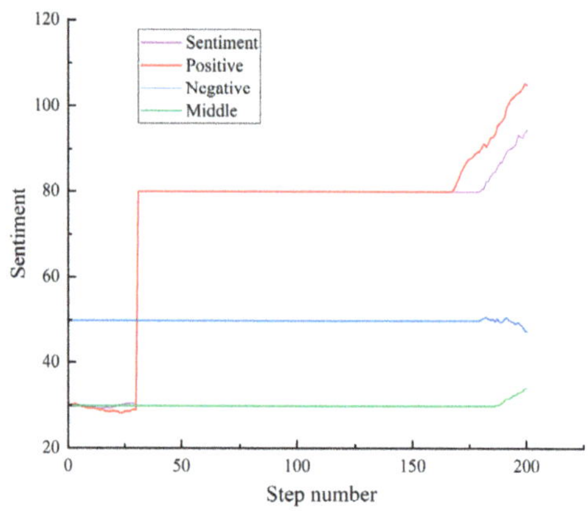

Figure 18. Emotional trend of employees from level V to level II.

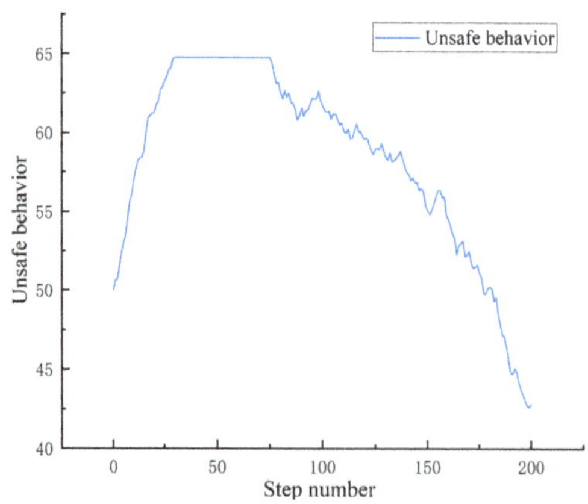

Figure 19. Unsafe behavior trend of employees from level V to level II.

B. Analysis of unsafe behaviors of miners and preventive measures

Through the simulation results, it can be seen that employees' emotions, motivations, abilities, personalities, and pressures directly or indirectly affect employees' behavior (unsafe behavior). The higher the scores of emotion, motivation, ability, and personality, the less unsafe behavior; the greater the pressure, the more unsafe behavior. Additional events such as employee family happiness, accident experience training, management system optimization, and promotion system improvement can change employee motivation, learning, and emotion, thus indirectly reducing employee unsafe behavior. The accident causation theory points out that human unsafe behavior is the direct cause of accidents. Therefore, reducing the occurrence of human unsafe behavior can avoid accidents to a large extent. Humans, machines, the environment, and management are indispensable. These four elements influence each other and promote each other, thus ensuring the smooth progress of mine safety production.

As a mining production enterprise where accidents occasionally occur, it must conduct questionnaire tests on employees before entry, after induction training, weekly, and after regular training, to timely understand the situation and changes in employees' sentiment, pressures, motivations, and abilities, and formulate corresponding improvement countermeasures for employees whose emotions and motivations are significantly reduced and pressures are significantly increased, thereby preventing a reduction of capacity and an increase in unsafe behavior. In terms of the preparation of the test questionnaire, the personality scale applicable to various enterprises was compiled by referring to the Big Five personality questionnaire; the mood and stress scale was compiled by referring to Li Guangli's miners' unsafe emotions questionnaire [17], Fan Wei's employees' mental resilience scale and emotion scale questionnaire [34], and Di Hongxi's employees' work stress questionnaire, and the ability scale was compiled according to the actual work needs and job types of enterprises [35]. The motivation scale was compiled by referring to the "1327" unsafe behavior control system of Bulianta Coal Mine [31] and Zhang Junjun's work motivation structure questionnaire of knowledge workers in the transition period [36].

According to the questionnaire test results (scores) of employees, the SMAPP model simulation framework was substituted for simulation to predict the changing trend of employees' unsafe behaviors, and corresponding measures were taken from the following aspects for employees with an increasing tendency of unsafe behaviors: (1) optimize the management system, develop a sound management system, promotion system, reward and punishment system, working environment, and other aspects to improve employees' job satisfaction or sense of achievement, so as to improve work motivation [37,38]; (2) strengthen the construction of corporate emotional safety culture, including emotional safety communication, emotional safety atmosphere, emotional safety measures, corporate identity, reasonable degree of safety regulations, etc., to improve the emotional stability of employees and reduce insecure emotions [39,40] (negative emotions); (3) pay attention to the training and assessment of employees' professional skills and safety skills, strengthen the training of employees' professional skills and safety skills (according to previous studies, they can be trained once a week), conduct assessment, and strictly carry out the work with certificates. Conduct regular tests on people with unsafe tendencies, and record the impact of different types of measures implemented on unsafe behaviors of employees. Because of the complexity and variability of employees, the trend curve model of various attribute factors of the SMAPP model will be studied and recorded in the future to reduce the unsafe behaviors of mine employees in a targeted way. Start from the main body of the accident to reduce the occurrence of mine accidents.

5. Conclusions

(1) The emotion, motivation, ability, personality, and stress were analyzed, and 24 attribute factors were selected to establish the SMAPP model. Analyze the mutual and interactive influence relationship of 24 attribute factors in the SMAPP model under different state values, and construct 1071 rule statements written in Python

language to record, save, and execute the mutual and interactive influence relationship of attribute factors under different state values. Build a simulation framework for multiple scenarios;

(2) According to the SMAPP model and its simulation framework, three scenarios of mine production employees are simulated in the programming tool written in Python language, and the employee behavior (unsafe behavior trend) under different scenarios is predicted. The simulation results are consistent with the existing research results, indicating that the SMAPP model and simulation framework are reliable under the set rules;

(3) Through the analysis of the simulation results, the higher the employee's emotion, motivation, ability, and personality score, the less unsafe behavior, the greater the pressure, and the more unsafe behavior. Through the model, the trend of unsafe behavior is predicted. From the aspects of improving employees' work motivation and job satisfaction, improving employees' emotional stability, reducing employees' stress and negative emotions, enhancing employees' professional skills, safety skills training, and implementing the certificate system, the preventive measures for unsafe behavior of mine production employees are put forward to prevent mine production safety accidents.

Author Contributions: Conceptualization, X.J. and S.Z.; methodology, X.J. and S.Z.; software, X.J.; validation, X.J. and S.Z.; formal analysis, X.J. and S.Z.; investigation, X.J. and S.Z.; data curation, X.J. and J.L.; writing—original draft preparation, X.J. and S.Z.; writing—review and editing, X.J., S.Z. and J.L; supervision, S.Z.; funding acquisition, S.Z. All authors have read and agreed to the published version of the manuscript.

Funding: This research received no external funding.

Data Availability Statement: Not applicable.

Acknowledgments: The authors are thankful to the anonymous reviewers for their kind suggestions.

Conflicts of Interest: The authors declare no conflict of interest.

References

1. Wu, J.S.; Zhang, X.Y.; Yin, S.X.; Xin, M.; Lian, H.Q. Statistics and analysis of water hazards in Coal mines in Recent 20 Years. *Coal Technol.* **2020**, *41*, 86–89. (In Chinese)
2. Meng, Y.; Xie, D.H.; Su, B.; Gu, A.X. Statistics and Analysis of coal mine production safety accidents in China from 2010 to 2019. *Min. Eng. Res.* **2020**, *35*, 27–33.
3. Shen, Z.Y.; Wang, S.E. Statistical analysis of heavy and large gas explosion accidents in coal mines from 2010 to 2020. *Inn. Mong. Coal Econ.* **2022**, *2*, 93–96.
4. Zhang, P.S.; Li, F.X.; Zhu, H.C.; Niu, H.; Li, X.H. Analysis and Countermeasures of Coal Mine Accidents from 2008 to 2020. *Saf. Environ. Prot. Min. Ind.* **2020**, *49*, 128–134.
5. Fang, Y.Z.; Liu, M.J.; Wei, J.P.; Liu, Q.J. Study on Characteristics and Countermeasures of Coal Mine Accident in Our Country. *Coal Technol.* **2007**, *16*, 3–5.
6. Willmer, D.; Haas, E.; Steine, L. *Human Factors Considerations in Addressing Mining Occupational Illnesses, Injuries, and Fatalities*; National Institute for Occupational Safety and Health, Mining Program: Rockville, MD, USA, 2020.
7. Lan, D.; Stephen, Y. Managing safety: An attitudinal based approach to improving safety in organizations. *Leadersh. Organ. Dev. J.* **1996**, *17*, 13–20.
8. Legault, G.; Clement, A.; Kenny, G.P.; Hardcastle, S.; Keller, N. Cognitive consequences of sleep deprivation, shiftwork, and heat exposure for underground miners. *Appl. Ergon.* **2017**, *58*, 144–150. [CrossRef]
9. Li, Y.; Yue, X.J.; Chen, X.J. Analysis of association rules on characteristics of miners' unsafe behavior. *Saf. Coal Mine* **2022**, *53*, 247–252.
10. Iraj, M.; Fakhradin, G.; Kalatpour, O.; Moghimbeigi, A. Constructing a Bayesian network model for improving the safety behavior of employees at workplaces. *Appl. Ergon.* **2017**, *58*, 35–47.
11. Hu, Z.; Hu, H.; Xu, F.; Wang, W. Simulation of Construction workers' safety behavior Ability based on cognitive style. *J. Saf. Environ.* **2022**, *22*, 194–200.
12. Zhang, H.; Wang, F. Analysis and Countermeasures of Coal Mine Production Accidents Based on Miners' Unsafe Behavior. *Coal Eng.* **2019**, *51*, 177–180. (In Chinese)

13. Zhang, D.H.; Wang, Y.C.; Zheng, Q.Y.; Dong, Y. The Effect of job requirements on the unsafe behavior of coal miners: The role of emotional exhaustion and behavior style. *Chin. J. Clin. Psychol.* **2020**, *30*, 573–577+553.
14. Yang, D.; Yang, Y.X.; Xu, R. Research on the Relationship between Safety awareness, Big Five Personality, and Unsafe Behavior of Electric Power Employees. *J. Southwest Univ.* **2021**, *43*, 129–136.
15. Yang, Z.H.; Li, P.; Wang, W.J. Study on the Influence of Accident Experience on Miners' unsafe behavior based on Social Cognition Theory. *J. Saf. Environ.* **2021**, *21*, 1607–1613.
16. Zhou, J.L.; Hu, F.X.; Xing, Y.D.; Gao, J.R. The Influence of Job Stress and Job Burnout on Unsafe Behavior of construction workers. *China Saf. Sci. J.* **2022**, *32*, 14–22.
17. Li, G.L.; Tian, S.C.; Yan, Y.Z.; Chen, Y.G.; Wu, Z.Y. Research on Identification of Critical Unsafe Emotions in Miners. *J. Xi'an Univ. Sci. Technol.* **2021**, *41*, 793–799.
18. Zhang, Y.L. The Influence of town employees' Emotional Change on production safety in coal enterprises. *Coal Mine Saf.* **2018**, *49*, 251–253.
19. Liu, J.L.; Han, Y.; Zhang, J.J.; Feng, Z.D. The difference in risk cognition results of construction workers with different working years. *J. Civ. Eng. Manag.* **2018**, *35*, 175–180.
20. Tian, S.C.; Feng, S.; Gao, X.Y. A gleam of miners' unsafe state of entropy TOPSIS evaluation method research. *J. Saf. Environ.* **2023**, 1–10. [CrossRef]
21. Ren, H.; Zhen, J.; Zhang, J.G. *Organizational Behavior—A Modern Perspective*, 2nd ed.; Tsinghua University Press: Beijing, China, 2020; pp. 56–68, 77–89, 130–131.
22. Zhang, J. *Employee Emotion and Management*; Tsinghua University Press: Beijing, China, 2009; pp. 36–121.
23. Ortony, A.; Clore, G.L.; Collins, A. The Cognitive structure of emotions: Factors affecting the intensity of emotions. *Contemp. Sociol.* **1988**, *18*, 2147–2153.
24. Wang, Z.L.; Qiao, X.I.; Wang, C. Research on Emotion Modeling Based on Custom Space and OCC Model Computer Engineering. *Comput. Eng.* **2007**, *33*, 189–192.
25. Wang, C.X.; Lv, S.R. Research on the Influence of Emotion on Safety Sign Recognition. *China Saf. Sci. J.* **2016**, *26*, 15–20.
26. Stephen, P. *Robbins Essentials of Organizational Behavior*; China Renmin University Press: Beijing, China, 2018; pp. 36–42, 70–78.
27. Su, Y.; He, Z.M. *Modern Organizational Behavior*, 3rd ed.; Tsinghua University Press: Beijing, China, 2021; pp. 49–58.
28. Andrew, I.I. *Artificial War: Multi-Agent Based Simulation of Combat*; Publishing House of Electronics Industry: Beijing, China, 2010; p. 55.
29. Cai, Z.X.; Durkin, J.; Gong, T. *Advanced Expert Systems: Principles, Design and Applications*; Science Press: Beijing, China, 2005; pp. 159–181.
30. Liu, Y.L.; Yao, J.S. Simulation of ABM based on Mesa. *Comput. Telecommun.* **2021**, *295*, 5–7+36.
31. Li, T.X.; Wang, W.P.; Wang, T.; Li, X.B. BDI model of strategic agent based on Knowledge graph. *Syst. Eng. Electron.* **2023**, *45*, 119–126.
32. Zhu, T.; Hu, Y.Q.; Zhu, S.H.; Liu, W.; Liu, H.X. Cognitive Simulation of Driver Distraction and Traffic Flow Characteristics. *J. Chang'an Univ. (Nat. Sci. Ed.)* **2018**, *38*, 87–93.
33. Ni, L.; Xiang, L.; Kang, Y. Human Performance Model Based on Fuzzy Rules. In Proceedings of the 2014 IEEE Chinese Guidance, Navigation and Control Conference, Yantai, China, 8–10 August 2014; pp. 2225–2230.
34. Fan, W. *The Effect of Employee Resilience on Task Adaptation Performance—The Interaction between Emotional Dimension and Emotional Motivation*; East China Normal University: Shanghai, China, 2021; pp. 150–153.
35. Di, H.X. *Study on the Structure, Propagation Law of Working Pressure and Its Influence on Unsafe Behavior of Coal Mine Workers*; Xi'an University of Science and Technology: Xi'an, China, 2017; pp. 154–159.
36. Chen, Z. Research on the control system of unsafe behavior in 10 million tons mine. *Coal Sci. Technol.* **2021**, *49* (Suppl. S2), 216–223.
37. Zhang, J.J. *Research on the Work Motivation Structure and Incentive System Adjustment Strategy of Knowledge Workers in the Transition Period*; Yanshan University: Yanshan, China, 2019; pp. 88–95.
38. Ni, G.D.; Li, H.K.; Cao, M.X.; Wang, K.D. Study on inducing factors and intervention strategies of unsafe behavior of New Generation construction workers. *Saf. Environ. Eng.* **2022**, *29*, 8–16.
39. Ni, G.D.; Gao, L.; Wang, X.R.; Gu, T.T.; Wang, T.X. Study on the correction mechanism of Job Satisfaction on Unsafe Behavior of Construction Workers. *China Work Saf. Sci. Technol.* **2021**, *17*, 156–162.
40. Liu, B. *Study on the Influence of Safety Emotion Culture on Miners' Insecurity*; Xi'an University of Science and Technology: Xi'an, China, 2018; pp. 32–43.

processes

MDPI

Article

A Study on the Development and Evolution of Fractures in the Coal Pillar Dams of Underground Reservoirs in Coal Mines and Their Optimum Size

Bao Zhang [1,2,*], Wei Ni [2], Xiuqiang Hao [2], Huiqiang Li [2] and Yupeng Shen [1,*]

1 School of Energy and Mining Engineering, China University of Mining and Technology-Beijing, Beijing 100083, China
2 CHN Energy Technology & Economics Research Institute, Beijing 102211, China
* Correspondence: 13716180380@163.com (B.Z.); cumtbsyp@163.com (Y.S.)

Abstract: The western mining areas of China, which are rich in coal resources, lack water resources. Large-scale and high-intensity coal mining in China's western mining areas has led to the loss of groundwater resources. Underground reservoirs in coal mines are an effective means of achieving the protection and utilization of water resources in these western mining areas. One of the important standards for the safety of an underground reservoir in a coal mine involves checking whether the development of cracks in the coal pillar dam body, under the dual stress conditions of overlying strata and lateral water pressure, passes through the coal pillar dam body or its top and bottom plates, forming a seepage channel for mine water. This article focuses on the safety issues associated with coal pillar dams in the underground reservoirs of coal mines. From the perspectives of overlying rock pressure and lateral water pressure on coal pillar dams, mechanical models, numerical calculations, and similar simulation methods were used to analyze macroscopic deformation, displacement, and crack development in coal pillar dams of different sizes under vertical and horizontal stress and to study the optimum width of coal pillar dams. Our research results indicated that the optimal width of the coal pillar dam body can be determined via numerical simulation based on the deformation and stress state in a given dam. When the horizontal stress increases, the smaller the coal pillar width is, the greater the increment of syy and sxx becomes, and the more likely the coal pillar is to be damaged. Similar simulations showed that the smaller the size of the coal pillar is, the easier it is to generate stress concentration, and the more likely this stress is to "eat away" the coal pillar dam body. There is a certain relationship between the size of the coal pillar dam and the range of crack development The larger the coal pillar size is, the less obvious the stress concentration effect becomes, and the less likely the crack is to penetrate the internal and external parts of the reservoir. Taking the Shangwan mine as an example, it was determined that the maximum water head height that could be carried by the 15-m coal pillar dam body was 50 m. A comprehensive study of the development and evolution of cracks in the coal pillar dam of an underground reservoir in a coal mine, and the characteristics of sliding instability, was conducted to determine the optimal size and maximum water storage height of a coal pillar that does not penetrate the inner and outer parts of the reservoir. The development and evolution of cracks are important factors affecting the stability of coal pillar dams. This study can expand and improve the basic theory of underground reservoirs in coal mines, provide a scientific basis for determining the optimum size of a coal pillar dam, guarantee the long-term safe and stable operation of the coal pillar dams of underground reservoirs in coal mines, and continuously save mine water resources and increase the economic benefits of coal mines. These implications are of great significance for the long-term stable operation of underground reservoirs in coal mines under similar geological conditions.

Keywords: coal mine underground reservoir; coal pillar dam; fissure development and evolution law; optimum size of coal pillar; maximum water level

Citation: Zhang, B.; Ni, W.; Hao, X.; Li, H.; Shen, Y. A Study on the Development and Evolution of Fractures in the Coal Pillar Dams of Underground Reservoirs in Coal Mines and Their Optimum Size. *Processes* **2023**, *11*, 1677. https://doi.org/10.3390/pr11061677

Academic Editor: Ján Piteľ

Received: 5 April 2023
Revised: 21 May 2023
Accepted: 26 May 2023
Published: 31 May 2023

1. Introduction

Coal currently has the most stable energy supply and the highest independent guaranteed supply rate, and is the most economic basic energy, in China, and has always guaranteed the energy security of China [1–3]. The overall distribution pattern of coal resources in China is "more in the north and less in the south, more in the west and less in the east, deep in the east and dry in the west" [4,5]. Following years of high-intensity mining, the coal resources in the east and central regions have tended to be exhausted, and coal production has been transferred to the western provinces, which contributed 81% of China's coal production in 2022. However, the water resources in the western mining area are extremely short, accounting for only 3.9% of the country's water resources, and the ecological environment in the western area is fragile. According to the relevant statistics, every 1 ton of coal mined in China produces approximately 2 tons of mine water; however, to date, the rate of mine water utilization has only been approximately 25% [6]. Most of the produced mine water is discharged to the ground, not only causing the waste of water resources due to evaporation but also polluting the environment [7,8]. The Shendong mining area is the world's only 200-million-ton mining area, with seven mines producing an annual output of more than 10 million tons. The Shendong mining area is located on the edge of the arid and water-scarce Maowusu Desert, where the ecological environment is extremely fragile and water resources are extremely scarce [9]. Coal mining causes the destruction of water resources, accelerates the loss of surface water, and leads to more and more prominent problems of water use in mining areas. After years of exploration and practice, more than 30 underground reservoirs of coal mines have been successfully built in the Shendong Mining area, and this has largely solved the problem of water resource protection and utilization in the mining area and protected the area's ecological environment. With the continuous popularization and application of underground reservoir technology in coal mines, more and more attention has been paid to the long-term operational safety of underground reservoirs in coal mines—especially the safety of coal pillar dams—and this has become key to ensuring the long-term safe operation of underground reservoirs in coal mines.

Based on the 20 years of technical exploration and engineering practice in the Shendong mining area, academician Gu Dazhao [10] put forward the theoretical framework and technical system of underground reservoirs for coal mines with "transport, storage and use" as the core themes. The idea of using the goaf as an underground water storage space was thus conceived, creating a new way to protect and utilize mine water resources in the process of coal mining. A large number of engineering practices have since been carried out and good results have been achieved [11,12]. Based on theoretical analysis and numerical simulation, Pang Yihui [13] comprehensively analyzed the feasibility of building an underground reservoir in a coal mine under the condition of large mining height. In their study of coal mining stress fields, B. Unver and Zhang Guojun [14,15] established a three-dimensional calculation model to analyze the movement law of overlying strata and its mining response mechanism in thick coal seam mining. Rajendra Singh [16] studied the variation law of mining stress and the movement law of overlying strata in thick coal seam mining through field testing. Zhang Nong [17] analyzed the failure characteristics and the stability of roof roadways in relief mining. Kong Xiangsong and Shan Renliang [18] established the theoretical expression of stress waves and the dynamic constitutive model of coal, and studied the classification of surrounding rock and the theory of dynamic pressure resistance. In a study on coal pillar dams, Ashok Jaiswal and B.K. Shrivastva [19] established a numerical calculation model and used the strain softening constitutive model to study the strengths of coal pillars. Robert Bertuzzi et al. [20] collated coal pillar detection data from Australia, South Africa, the United States, and other countries and analyzed the strengths of coal pillars under different conditions by using various methods. Petr Waclawik et al. [21] studied the stability of coal pillars by monitoring and analyzing the stress states of coal pillars. Academician Gu Dazhao [22] analyzed the stability of coal pillars in underground reservoirs of coal mines. Song Yimin and Zhang Jie et al. [23,24] analyzed the deformation field and energy evolution characteristics of coal pillar instability failure and studied the instability characteristics of coal pillars and their overlying rocks. Zhang Cun [25] analyzed the stability of residual coal pillars in underground reservoirs of mines under mining-induced water flooding. Chen Dongdong and Zhao Hongbao et al. [26–28] studied the distribution characteristics of the lateral stress field in coal pillar goaf under different conditions. Tu Min and Wu Baoyang [29,30] used a variety of research methods to study the rational arrangement of coal pillar dams in distributed underground reservoirs of coal mines. In a study on fracture detection, Cao Zhiguo [31] analyzed the distribution model and water flow characteristics of the main channels of water-conducting fractures in mining-induced overlying rocks. Xie Hongxin [32] studied the damage characteristics of water-bearing coal samples under cyclic loading–unloading. Wu Baoyang and Wang Fangtian et al. [33,34] studied the damage and failure law in coal pillar dams under various conditions. Wang Zhenwei [35] numerically analyzed the stress wave propagation law and its influencing factors on coal rock mass. Reed Guy et al. [36] studied the stability of the coal pillar system based on an evaluation method that accounted for the interaction between the coal pillar and the overlying strata. Based on a large number of experimental studies, Yilmaz found that decrease in the friction angle under water infiltration was the main reason for the strength weakening of water-soaked coal rock mass, and that compared with what was found in dry water-soaked coal rock mass, Poisson's ratio increased while Young's modulus decreased [37]. Through component analysis and mechanical tests, Verstrynge et al. found that the transition from metastable creep failure to accelerated failure in water-soaked sandstone was related to the reaction between water and mineral components [38]. E. Ozdemir et al. found that increase in water content significantly reduced the tensile and compressive strengths of rocks [39].

In the past, the domestic methods of water conservation coal mining were limited height mining, strip mining, and filling mining, all of which affected mining efficiency. For Shendong, with its shallow buried working face with a large mining height, limited mining and strip mining would lead to the waste of coal resources. Large-mining-height filling mining has the disadvantages of high cost and poor effectiveness, while water-preserved coal mining, or the large discharge of mine water, will cause a large amount of evaporation and high loss of water resources, and so underground reservoir technology is the best choice to solve the problem of mine water protection and utilization. With the popularization and application of underground reservoir technology, increasing attention has been paid to the safety of coal pillar dams. Increasing numbers of scholars are studying the coal pillar dams in underground reservoirs of coal mines, but at present, there are few studies on the development and the evolution law of cracks in coal pillar dams and the critical conditions of instability and slip. These factors are key to the safety of coal pillar dams and can determine the optimum size of a coal pillar and the height of water storage while being of great significance to the long-term stable operation of underground reservoirs in coal mines.

Underground reservoirs in coal mines are extremely valuable for the mining industry. They offer an effective way to protect mine water and can reduce the pumping of mine water, save electricity and energy, purify mine water, and realize the green mining of coal. The construction of underground reservoirs in coal mines could reduce global carbon dioxide emissions and the pollution caused by mine water in the environment, effectively contributing to addressing the global climate problem.

The main purpose of this paper is to determine the optimum width of coal pillar dams in the underground reservoir of the Shangwan coal mine so as to reduce and prevent the seepage from the coal pillar dam and improve the safety of underground reservoirs in coal mines.

The research plan of this paper is to analyze the mechanical conditions of crack propagation inside the coal pillar dam under different stress conditions from the perspective of overlying rock pressure and lateral water pressure and study the crack development law and safety of coal pillar dams with different widths, using numerical simulation and similar simulation, so as to obtain the optimum width of the coal pillar dam in the Shangwan coal mine.

2. The Mechanical Mechanism of Fracture Development and Evolution in the Coal Pillar Dam

The coal rock mass is a complex porous solid, and the primary fractures are formed in the coal rock body due to geological tectonic action; the overlying rock layer on the working face is in equilibrium when mining is not carried out, and the primary fractures exist in isolation. With the remining of the working face, the original stress balance is broken, and a stress increase zone and stress reduction zones are formed in the coal pillar dam and in the top and bottom plate, respectively, as shown in Figure 1. Stress changes cause primary fractures to propagate and create secondary fractures. As the overlying rock layer on the working face continues to separate, fracture, and balance, the stress is transmitted to the coal pillar dam and to the top and bottom plate so that the stress field also changes. The repeated stress changes make the secondary fractures continue to increase and form a network through each other and then form macroscopic mining cracks [40–42]. If a macroscopic mining fracture has a high degree of development, resulting in the formation of a seepage fracture channel between the underground reservoir of the coal mine and the adjacent roadway outside, the phenomenon of water seepage will result. Therefore, the degree of fracture development in the coal pillar dam is a key factor affecting the safety of underground reservoirs in coal mines, and the fracture development characteristics of the coal pillar dam are inseparable from the stress state in which it is located.

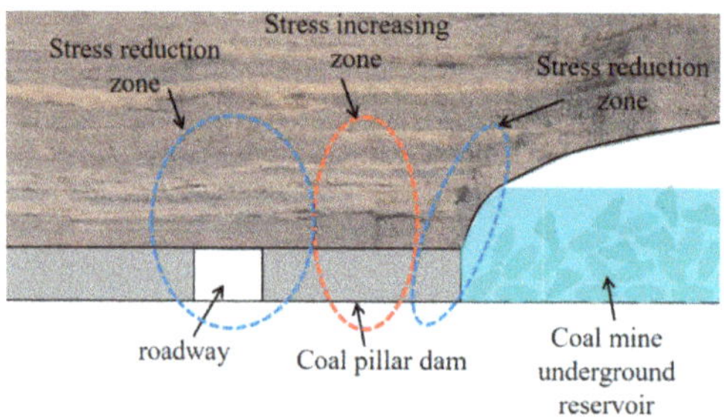

Figure 1. Stress distribution of the coal pillar dam.

In the coal pillar dam's stress increase zone, as the stress increases from the original rock stress to the peak of lateral support pressure, the coal rock mass is subjected to compressive stress. According to the Griffith strength theory, the cracks in the coal rock body propagate due to tensile stress, which is measured by uniaxial tensile strength, Rt, and the strength criterion of fracture propagation is as follows:

$$\begin{cases} \frac{(\sigma_1-\sigma_3)^2}{8(\sigma_1+\sigma_3)} \geq \mathrm{R_t} & if \frac{\sigma_3}{\sigma_1} \geq -\frac{1}{3} \\ \sigma_3 = -R_t & if \frac{\sigma_3}{\sigma_1} < -\frac{1}{3} \end{cases} \tag{1}$$

The angle between the propagation direction of the secondary crack and the long axis of the crack is 2β when $\frac{\sigma_3}{\sigma_1} \geq -\frac{1}{3}$, $\beta = \frac{1}{2} \arccos \frac{\sigma_1-\sigma_3}{2(\sigma_1+\sigma_3)}$, and the propagation direction of the secondary crack is parallel to the direction of the long axis of the crack when $\frac{\sigma_3}{\sigma_1} \geq -\frac{1}{3}$ (Figure 2).

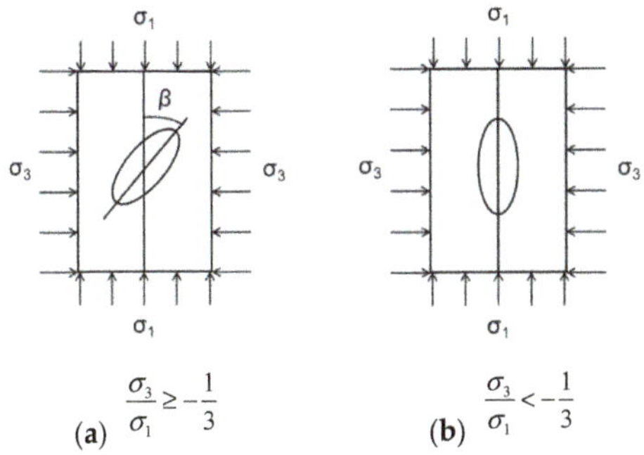

Figure 2. Expansion direction diagram of secondary crack tensile fracture.

In the coal pillar dam's stress increase zone, when the stress is reduced from the peak of the lateral support pressure to the original rock stress, the lateral pressure change is not obvious, and the vertical stress decreases significantly (this is the pressure relief process; Figure 3. The main form of crack expansion is reverse slip [43], and its critical stress is derived thus:

$$\sigma_1 = \frac{\frac{\sigma_2 \sin 2\theta}{2} - \mu(\sigma_3 - \frac{8G_0\alpha}{k+1})\cos^2\theta - 2(\tau_c + \mu\sigma_{1m}\sin^2\theta)}{\cos\theta\sin\theta + \mu\cos^2\theta} \tag{2}$$

where μ is the friction coefficient of coal rock, τ_c is cohesion, $\sigma_{1\mathrm{m}}$ is the axial lateral stress at the starting point of unloading, θ is the angle between crack and principal stress direction, and G_0 is the shear modulus of coal rock.

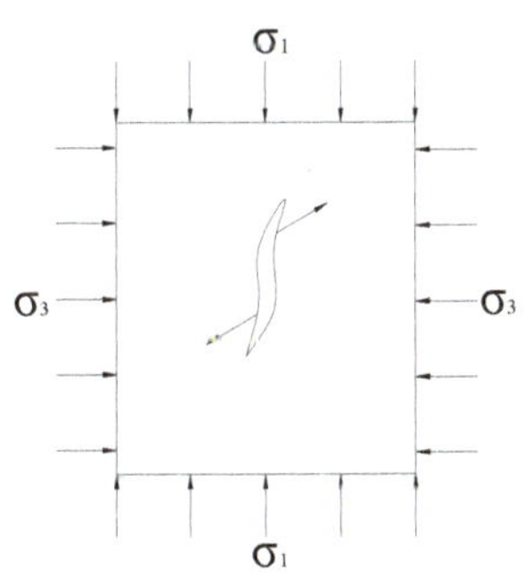

Figure 3. Expansion direction diagram of secondary crack reverse slip.

In the stress reduction zone of the coal pillar dam, the stress is significantly reduced, a large number of secondary fractures are generated by the development of primary fractures, and the secondary fractures of the coal rock mass penetrate due to the tensile-shear composite failure. As shown in Figure 4, the propagations of any two fractures AB and CD produce secondary fractures AF and CE, and the secondary fractures penetrate with tension crack EF to form a fracture network. The penetration strength between fractures is depicted below [44,45]:

$$\sigma_1 = \frac{h\sigma_t(\sin\alpha + f_r\cos\alpha) - 4lc_r + B\sigma_3}{A} \tag{3}$$

where h is the vertical distance between fractures, σ_t is the uniaxial tensile strength of the rock, α is the inclination angle of the tensile crack, fr is the friction coefficient of the rock, c_r is the rock cohesion, l is the length of the secondary fracture, and A and B are the penetration functions.

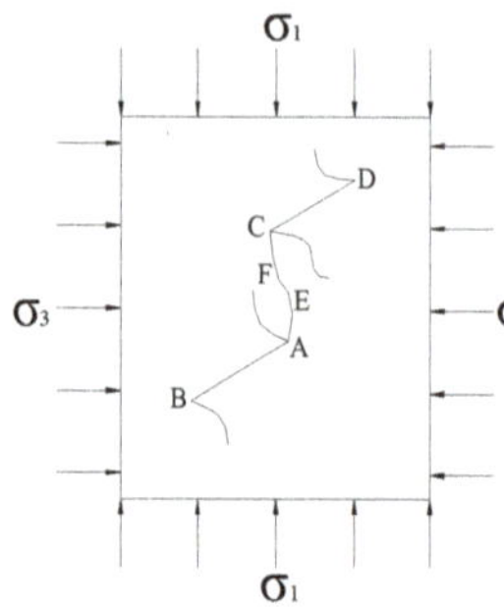

Figure 4. Diagram of transfixion between secondary cracks.

3. Numerical Simulation of Coal Pillar Dam Width in the Shangwan Mine

The special ecological environment of the Shendong area has created the diversity and complexity of the ecological system of the region, and the problem that needs to be solved urgently at this stage is the lack of water resources. However, the coal seam in the Shendong mining area has good storage conditions, large reserves, and a high output, and there is a high loss of surface water and groundwater resources due to the transportation of surrounding rock as a consequence of coal mining, exacerbating the water shortage in the region. Therefore, the utilization of the underground reservoir in the coal mine is an important measure to achieve a win–win situation of energy development and water resource protection and utilization.

The typical underground reservoir of a Shendong coal mine is exemplified in the Shangwan coal mine. The Shangwan coal mine is located in Ejin Horo Banner, Ordos City, Inner Mongolia Autonomous Region, China, and was completed and put into operation in 2000. It covers an area of 61.8 square kilometers and has nine coal seams that can be mined, geological reserves of 1.23 billion tons, recoverable reserves of 830 million tons, an approved production capacity of 16 million tons/year, and a service life of 53 years. The coal is long-flame coal and non-bonded coal, and is a source of high-quality power, coal-to-oil, and chemical and metallurgical coal. The geological structure is simple, and the hydrogeological type is classified as medium. The rock mass of the top and floor of the coal seam in the mining area is of good quality and is relatively complete, but its mechanical strength is low, as it contains mainly weak rocks, and the stability of the top and floor rocks of the coal seam is poor.

3.1. Numerical Simulation Construction and Research Scheme Design of the Underground Reservoir Dam in the Coal Mine

In a study of underground reservoirs, understanding the size and stability of coal pillar dams is the top priority. The general stratigraphic column of the Shangwan coal mine engineering is shown in Figure 5, and the width of the coal pillar dam in the mining process of the UDEC discrete element and the internal stress field, displacement field, and fracture development and change characteristics of the coal pillar dam under the action of lateral water pressure in the goaf area have been constructed by using UDEC discrete elements. This construction has obtained the optimum size of the coal column dam body, providing a theoretical basis for the retention of the coal pillar dam in the Shangwan coal mine.

Based on the general stratigraphic column of the Shangwan coal mine, the numerical simulation is as shown in Figure 6. The model size is defined as length (working face layout direction) × height (vertical direction) = 236 m × 280 m, with a 2-2 coal roof plate of 74 m, 2-2 coal seam and 3-1 coal seam spacing 100 m, and 3-1 coal floor of 50 m. The boundary impact area on both sides of the model is of 40 m, the excavation area studied in the middle is of 200 m, and the model contains two layers of coal in the form of 2-2 coal and 3-1 coal. The upper coal seam of 2-2 coal simulates the excavation face in the central area and its two side roadways and the lower coal seam, 3-1 coal, simulates the excavation of two roadways and one working face on the left and right sides of the central area. The two layers of coal are arranged with 30-m and 15-m coal column dam bodies, respectively, and the spatial relationship between the excavation areas of the upper and lower coal seams is shown in Figure 6.

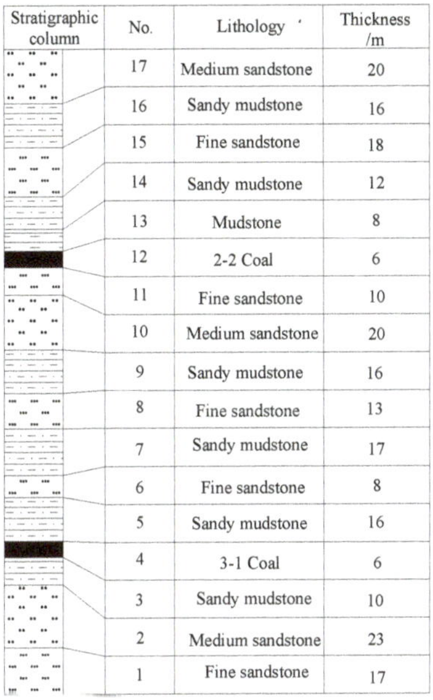

Figure 5. General stratigraphic column of the Shangwan coal mine working face.

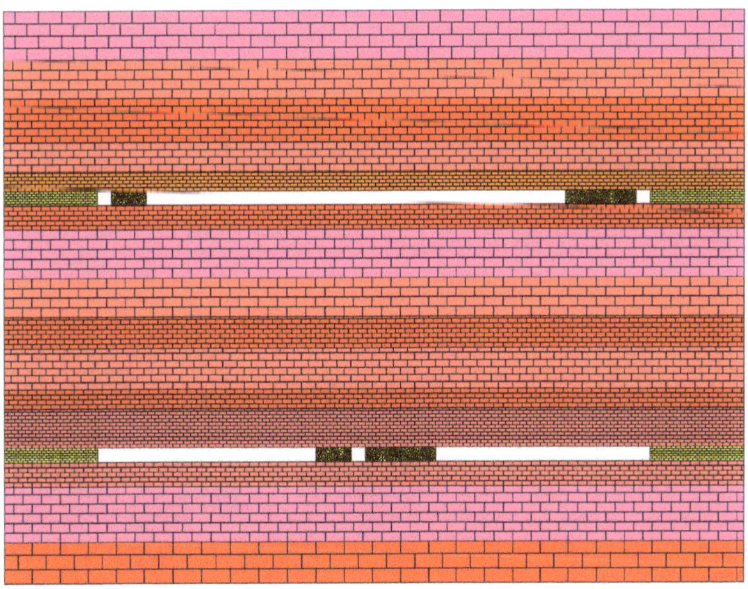

Figure 6. Spatial distribution characteristics of the simulated excavation area.

In order to study the internal variation characteristics of the coal pillar dam, a measuring line with an interval of 1 m is arranged in the middle of the studied coal pillar dam to monitor its deformation and stress. The objective of the simulation is to excavate 2-2 coal first and then excavate 3-1 coal. Horizontal stress is applied to the 3-1 coal pillar dam to effectively replace the influence of goaf water pressure. During this process, the width and roadway height of the coal pillar dam on the same side of 2-2 coal and 3-1 coal are consistent, and the difference is only limited to the coal pillar dam. There is no lateral horizontal stress. Due to limited space, it is assumed that the optimum coal pillar width for 2-2 coal excavation is equivalent to that for 3-1 coal excavation without lateral horizontal stress. Therefore, the main focus of the simulation study is on the changes of the coal pillar dam body under the conditions of 3-1 coal excavation with horizontal stresses of 0 MPa and 10 MPa.

3.2. Analysis of Simulation Results

(1) Variation characteristics of coal pillar dam body without lateral horizontal stress

From a longitudinal comparison of the distribution of internal stress and deformation under the condition of changing the width of the coal pillar, the vertical stress distribution characteristics and horizontal deformation amount of the coal pillar dam body under only the roof load are obtained using the FISH language as shown in Figures 7 and 8. Figure 7 shows that the widths of the left coal pillar are 15 m, 20 m, and 25 m, respectively, from top to bottom, and the widths of the right coal pillar are 30 m, 35 m, and 40 m, respectively. From the horizontal deformation amount of the coal pillar, it can be concluded that as the size of the coal pillar decreases, the deformation amount on both sides of the coal pillar shows an increasing trend, with a large increment on the free side of the coal pillar and a small increment on the roadway side of the coal pillar. The internal deformation amount of the 15–25-m coal pillar increases approximately linearly, while the internal deformation amount of the 30–40-m coal pillar is basically consistent and small. By dividing the deformation amount by using a color scale and analyzing the range of coal pillar deformation using color scale area of displacement of less than 0.1 m as a reference, it can be concluded that the corresponding areas of coal pillars that are of 15 m, 20 m, 25 m, 30 m, 35 m, and 40 m are 0 m, 5 m, 10 m, 17 m, 22 m, and 29 m, respectively, accounting for 0%, 25%, 40%, 57%, 62%, and 72% of coal pillar width, respectively. The proportion of this area increases with the increase in coal pillar width, with increments of 25%, 15%, 17%, 5%, and 10%, respectively. From these results, it can be concluded that the stable area inside the coal pillar under the condition of 30 m has a larger increment compared to under the conditions of 25 m and 35 m. From the perspective of the distribution characteristics of vertical stress inside the coal pillar, it can be seen that the vertical stress on the left and right sides of the coal pillar increases as the size of the coal pillar decreases. Between them, the vertical stress peak value of the 30-m coal pillar is the highest and the bearing stress is the largest. Considering that the internal deformation of the coal pillar is small, it can be concluded that the 30-m coal pillar has the optimal size of the coal pillar dam.

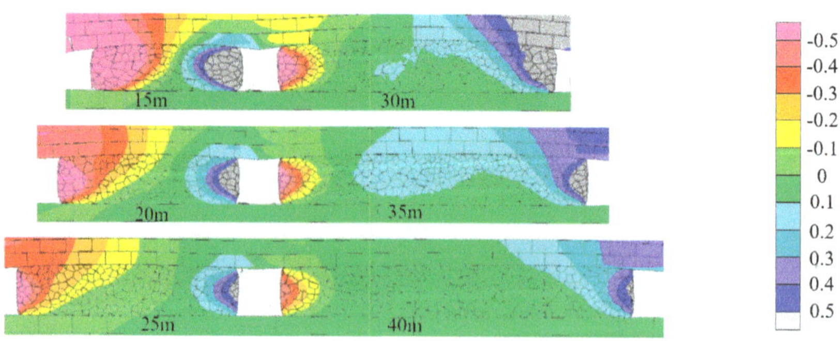

Figure 7. Cloud Chart of Horizontal Displacement of Coal Pillar Dam/m.

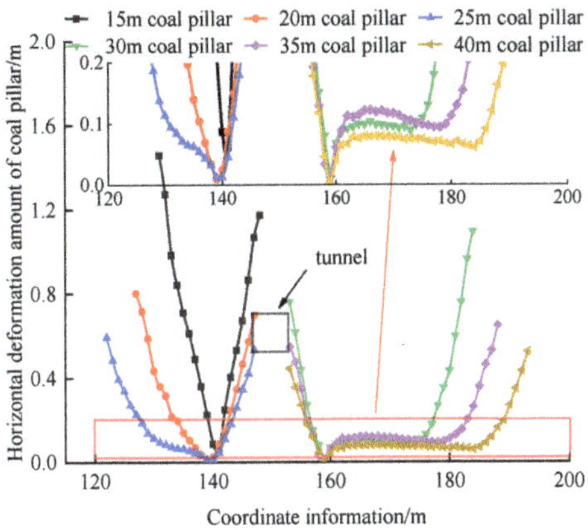

(a) Horizontal displacement of measuring point

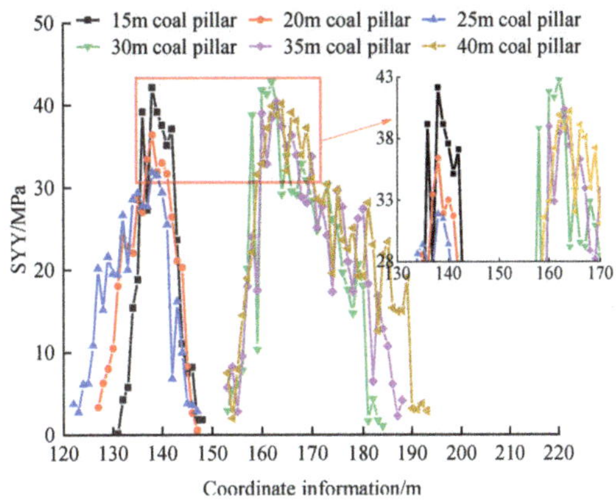

(b) Vertical stress distribution characteristics of measuring point

Figure 8. Change characteristics of a coal pillar dam body without lateral horizontal stress.

(2) Variation characteristics of coal pillar dam considering lateral horizontal stress

Based on the above pressure variation characteristics of coal pillar dams without horizontal lateral pressure, it is necessary to supplement the variation characteristics of coal pillar dam stability under goaf water filling conditions in order to further study the stability of coal pillar dams in underground reservoirs. Due to the large amount of data from multiple groups of comparative analysis, when analyzing only 15 m and 30-m coal pillars, horizontal stress is applied to both sides of the 3-1 coal seam, and the variation rules of sxx and syy stresses under the conditions of 0 MPa, 5 MPa, and 10 MPa are shown in Figure 9.

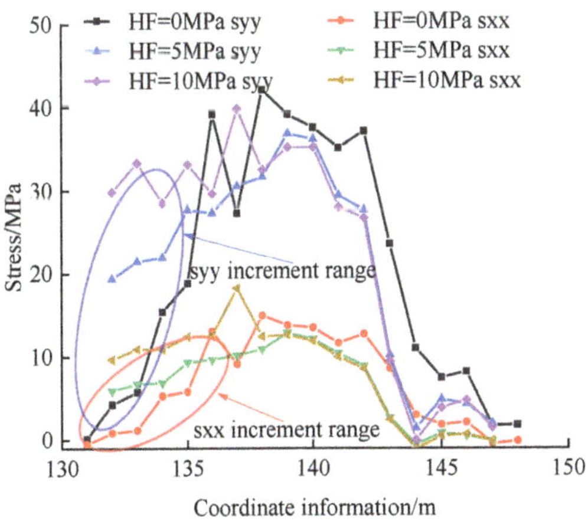

(a) Under the condition of 15-m coal pillar

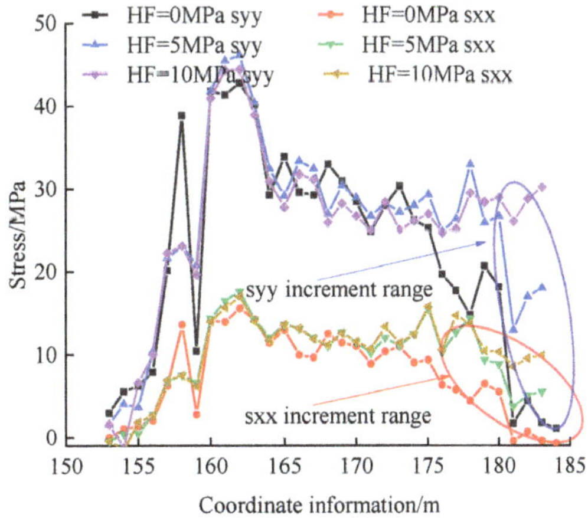

(b) Under the condition of 30-m coal pillar

Figure 9. *Cont.*

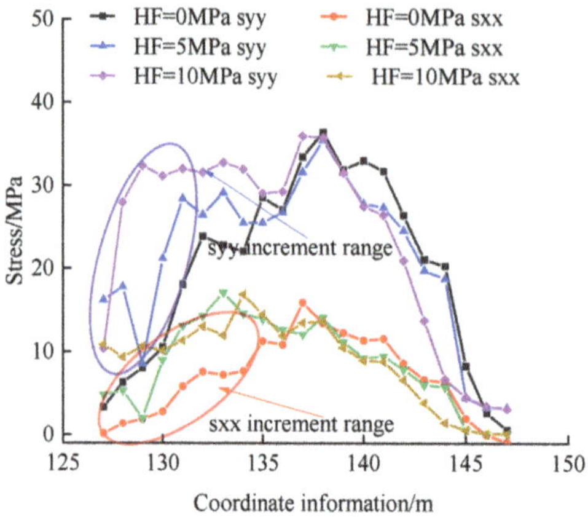

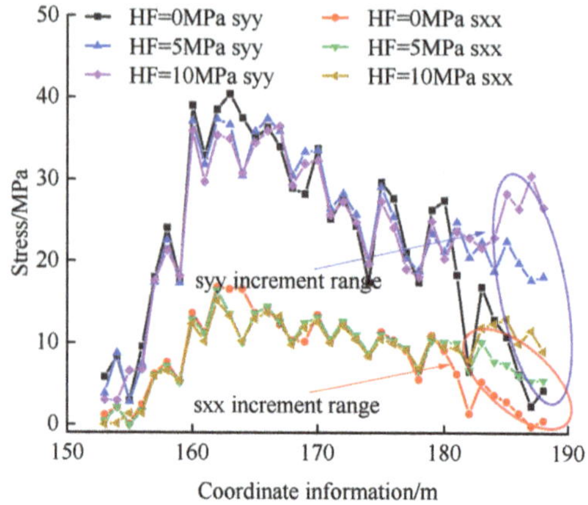

(**c**) Under the condition of 20-m coal pillar (**d**) Under the condition of 35-m coal pillar

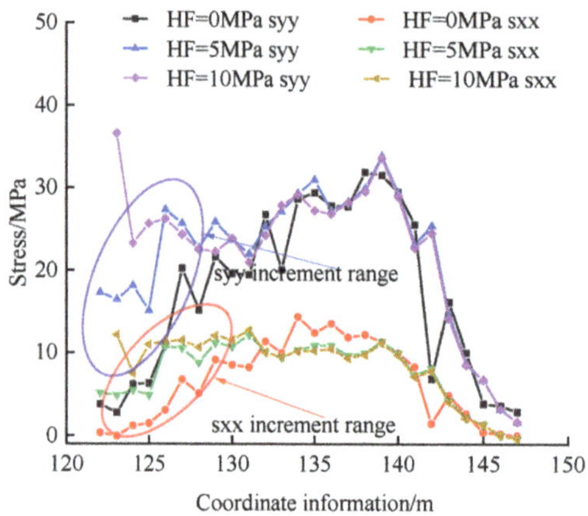

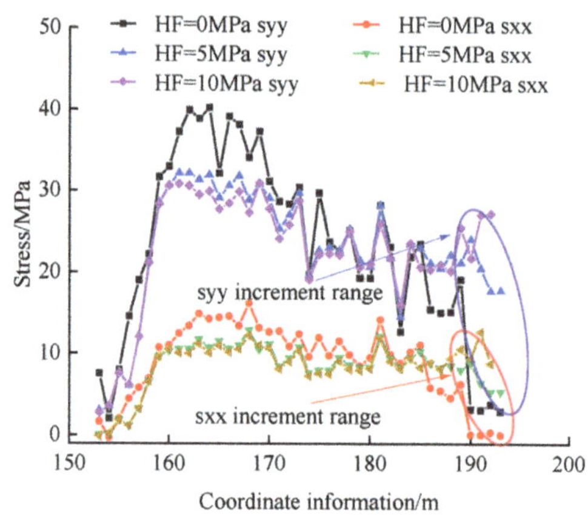

(**e**) Under the condition of 25-m coal pillar (**f**) Under the condition of 40-m coal pillar

Figure 9. The influence of the change in the width of the coal pillar dam on its stress variation characteristics.

As shown in Figure 9, under the action of horizontal stress applied to the goaf side of the coal pillar dam, the stress of syy and sxx on the goaf side of the coal pillar dam, which is positively correlated with the horizontal stress, increases. The smaller the size of the coal pillar is, the greater the increment of stress in syy and sxx is. After comparing and analyzing the differences in the changes of coal pillar syy and sxx under the action of horizontal stress, it can be concluded that horizontal force has an impact on the stability of coal pillars. With the increase in coal pillar width, the influence range of horizontal force on the goaf side of the coal pillar dam becomes smaller and smaller. When the coal pillar width is 15 m or 20 m, the application of horizontal stress causes changes in sxx and syy in all areas inside the coal pillar, thus indicating that horizontal force is transmitted from the goaf side to the roadway side. Horizontal force has a significant impact on the stress state of the coal pillar, indicating that the stability and bearing capacity of the coal pillar decrease under the action of horizontal force. When the width of the coal pillar is 25–40 m, the variation areas of the syy of the coal pillar after the horizontal force is applied are 12 m, 8 m, 8 m, and 7 m, and for those values, the sxx variation areas are 16 m, 10 m, 8 m, and 7 m, respectively. Considering the influence range of the horizontal force on the coal pillar stress in the goaf side when the width of the coal pillar is 30–40 m, it can be found that the influence range of the horizontal force on the coal pillar is basically the same. When the width of the coal pillar is 30 m, the influence of the horizontal force on the coal pillar on the goaf side no longer changes significantly. From the comparison of the bearing capacity and stress state of the coal pillar, it can be concluded that a coal pillar of 30 m is the optimum width.

As shown in Figure 10, by comparing the internal stress variation characteristics of 15–25 m and 30–40-m coal pillars under the same horizontal force conditions, when there is no horizontal stress on the coal pillar dam body, the fluctuation range of syy and sxx stress in the middle area of both coal pillars and the side of the goaf is relatively large. When a 5-MPa horizontal stress is applied to the goaf side of the coal pillar dam, the fluctuation range of syy and sxx stress in the middle area of the 15–25 m and 30–40-m coal pillars and the side of the goaf is significantly reduced. When a 10-MPa horizontal stress is applied to the goaf side of the coal pillar dam, the syy and sxx stresses in the middle areas of the 15–25 m and 30–40-m coal pillars and the side of the goaf are similar. The coal pillar width of 15–25 m is significantly larger than the area affected by the horizontal force of 30–40-m coal pillars. Therefore, when the horizontal stress on the side of the gob of the coal pillar dam increases, the smaller the coal pillar width is, the greater the stress change on syy and sxx on the side of the gob of the coal pillar dam becomes. Therefore, as the horizontal stress increases, the size of the coal pillar decreases, and the stress increment of syy and sxx becomes more significant. The goaf side of the coal pillar is the starting point for the stress increment of syy and sxx. The larger the stress increment is, the more easily the coal pillar is damaged, indicating that small coal pillars are more sensitive to horizontal pressure.

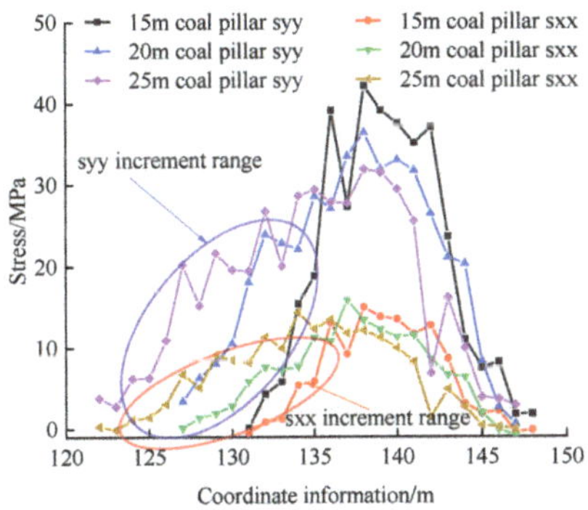

(**a**) 15–25-m coal pillar without horizontal stress

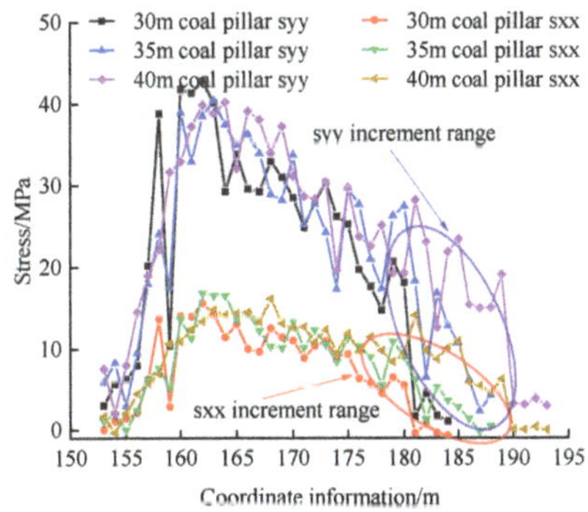

(**b**) 30–40-m coal pillar without horizontal stress

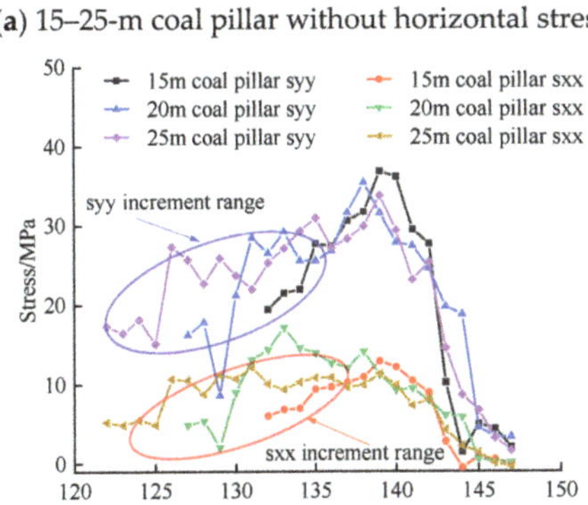

(**c**) Horizontal stress of 15–25-m coal pillar = 5 MPa

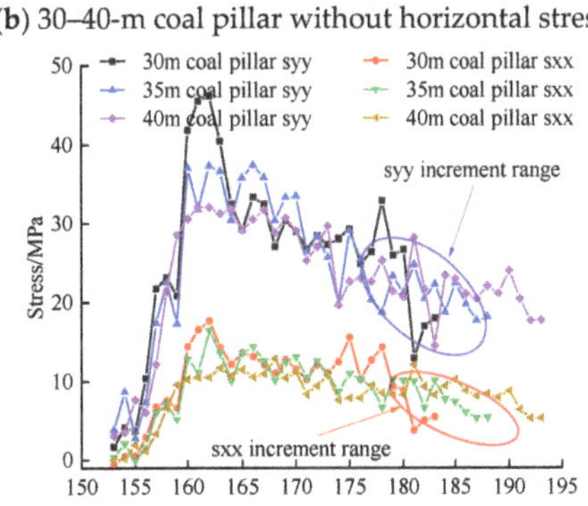

(**d**) Horizontal stress of 30–40-m coal pillar = 5 MPa

Figure 10. *Cont.*

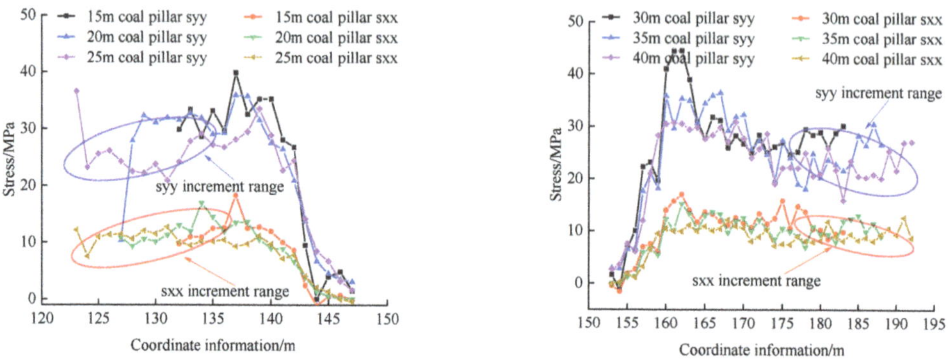

(**e**) Horizontal stress of a 15–25-m coal pillar = 10 MPa (**f**) Horizontal stress of a 15–25-m coal pillar = 10 MPa

Figure 10. Stress variation characteristics of a coal pillar dam body under lateral horizontal stress conditions.

4. Experimental Study on Crack Development in a Coal Pillar Dam

Based on the numerical simulation results for crack development in a coal pillar dam body of different sizes in the Shangwan mine, the coal pillar dam is of the best size when the width of the coal pillar is 30 m. In this section, similar simulation experiments that were used to study coal pillar dams with two widths—30 m and 15 m—are described.

4.1. Similar Simulation Experiment Research Scheme

This simulation experiment, which is similar to the previous one, analyzed the macroscopic fracture development in and slip instability characteristics of a coal pillar dam under different stress conditions and used the "Multi-seam Mining Coal Mine Underground Reservoir Simulation Test Platform" to carry out the relevant experiments. Above this platform are seven servo motors, which can load vertical force, that can simulate the self-weight of the overlying strata and achieve local loading, and two servo motors, which can load horizontal force, that can simulate the lateral water pressure of underground reservoirs of coal mines. The simulation test platform can be laid on a simulation test platform with dimensions of 2100 mm × 1800 mm × 300 mm. The similarity ratio of this experiment was 1:150, and the physical and mechanical parameters of each rock layer were consistent with the numerical simulation parameters. In this simulation, the process of mining a 2-2 coal face and a 3-1 coal face in Shangwan coal mine and forming an underground reservoir was simulated. The two layers of coal were arranged with coal pillar dams of 30 m and 15 m, respectively. As shown in Figure 11, the 2-2 coal layout had a 178-m-wide working face and a 15 m and a 30-m coal pillar dam body were arranged on both sides. The 3-1 coal layout had two 132-m working faces, and the two adjacent coal pillar dam bodies, which were of 15 m and 30 m, were arranged in the middle.

In order to monitor the stress change characteristics inside the coal column dam, a coordinate system was established, with X in the transverse direction and Y in the longitudinal direction, and the stress measurement points were arranged. The scheme, as shown in Figure 12, indicated that one stress measuring point in the Y direction was uniformly arranged inside the two coal pillars of 2-2 coal seam along the X direction, and two stress measuring points in the Y direction were uniformly arranged at the upper boundary along the X direction (Figure 12a,b). One stress measuring point in the X direction was uniformly arranged inside the two coal pillars of 3-1 coal along the Y direction, and one stress measuring point in the Y direction was arranged at the upper boundary (Figure 12c,d).

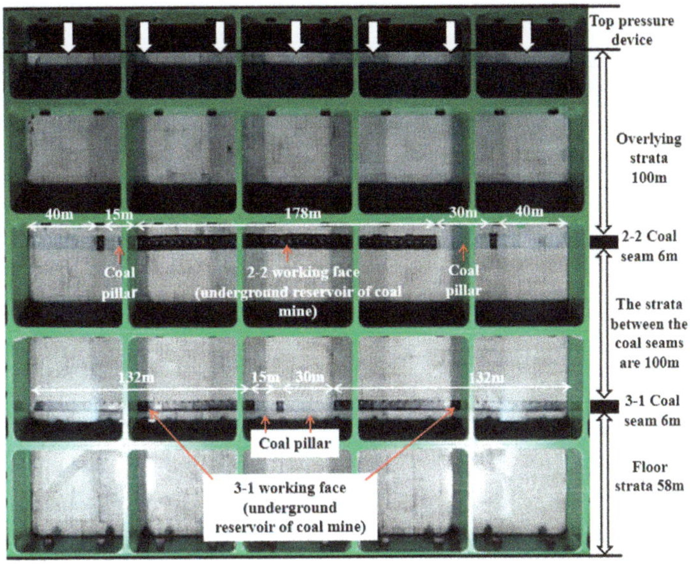

Figure 11. Similar simulation experiment scheme.

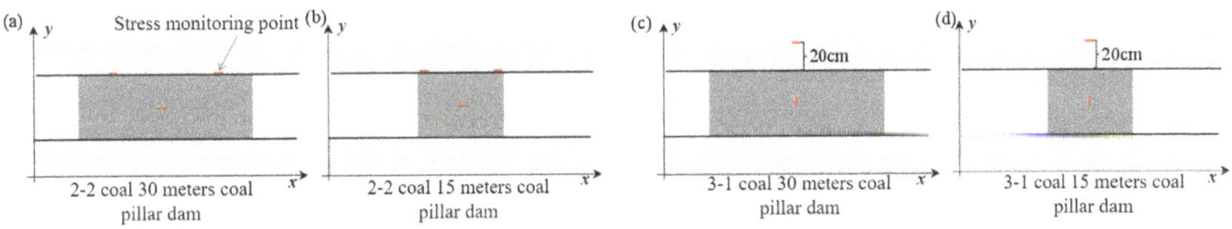

Figure 12. Arrangement of stress monitoring points of a coal pillar dam.

4.2. Analysis of the Fracture Development Characteristics of a Coal Pillar Dam under the Stress of Overlying Strata

In order to analyze the vertical bearing capacities of coal pillar dams of different sizes, after mining the 2-2 coal seams, the load was applied on top of the two coal pillars. The stress monitoring sheet embedded in the coal column was used to monitor the stress of the coal pillar dam and its top and bottom plate. A high-definition camera was used to take regular pictures of the surface of the coal pillar dam to monitor the development characteristics of and displacement of cracks on the surface of the coal pillar dam. According to the observation results of stress, displacement, and fracture development, the fracture development evolution law and the failure instability characteristics of the coal pillar dam were further analyzed.

(1) Analysis of the stress failure characteristics of a 15-m 2-2 coal pillar

The overlying rock layer of the 2-2 coal 15-m coal pillar was locally compressed, as shown in Figure 13; the stress inside the coal pillar and the top monitoring point increased linearly with the increasing upper load. When the stress of the monitoring point adjacent to the reservoir side at the top of the coal pillar reached 0.62 MPa, the stress value dropped sharply, and no obvious damage occurred on the roadway side, while the roof plate of the reservoir side was completely destroyed, resulting in the stress loss of pressure at the monitoring point. The stress at the monitoring points inside the coal pillar and adjacent to the roadway side at the top continued to fluctuate and increase linearly, and subsequently showed the characteristics of slow growth, during which the original fracture of the coal pillar dam expanded and the fracture opening and length increased. When the stress of the monitoring point reached 0.69~0.73 MPa, the cracks between the top and bottom plates of the coal pillar dam (reservoir side) gradually expanded and penetrated each other. When the stress of the monitoring point reached 0.78~0.89 MPa, the coal body of the coal pillar dam (reservoir side) formed a semi-oval spalling body, which was separated from the coal pillar, and the crack could be observed macroscopically at up to about 5 mm. The coal body on the other side of the roadway also produced long cracks that ran through the top and bottom plates and basically formed a semi-oval spalling. When the stress of the monitoring point reached 0.85–0.91 MPa, the coal body of the coal pillar dam (reservoir side) had a wall caving, which was completely scattered in the roadway, and the coal body spalling on the other side was separated from the coal pillar, forming a crack of about 4 mm. When the stress of the monitoring point reached 0.96–1.04 MPa, the coal body on both sides of the roadway had a wall caving, and after cleaning the roadway, it could be seen that the effective support width of the coal pillar dam was getting smaller, and the stress of the monitoring point also tended to be stable, and the coal pillar dam recovered stability. With the continuous increase in the upper load, the stress at the monitoring point suddenly dropped sharply from 0.9 MPa directly to approximately 0.3 MPa.

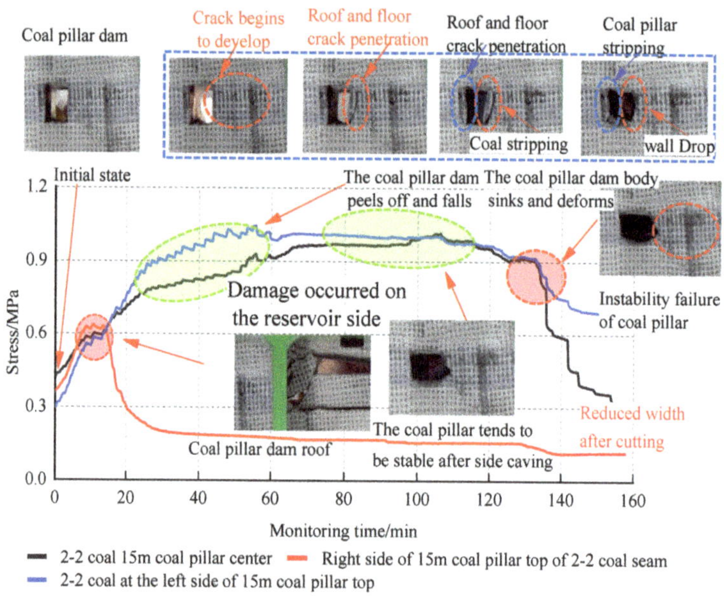

Figure 13. Stress failure characteristics of a 15-m coal pillar of 2-2 coal.

As shown in Figure 14, externally, the coal pillar dam produced obvious deformation, indicating that at this time, the internal cracks of the coal pillar dam had a high degree of development. The internal structure had been completely destroyed, as it could not bear the load of the overlying rock layer, and the load was transmitted to the outer coal pillar and the gravel in the reservoir, so far indicating that the coal pillar was unstable and damaged. The final effective bearing width of the coal pillar dam body was 11.5 m.

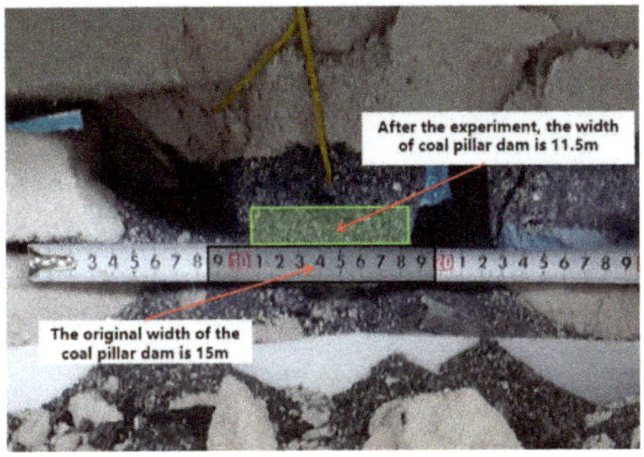

Figure 14. Final width of the coal pillar dam after instability and failure.

(2) Analysis of stress failure characteristics of a 30-m 2-2 coal pillar

The overlying rock layer of the 30-m coal pillar of 2-2 coal was locally compressed, as shown in Figure 15, and the stress in the internal and top monitoring points of the coal pillar dam increased linearly over a period of time. After reaching 0.62–0.69 MPa, with the continuous increase in the upper load, the stress of the monitoring point did not change significantly, and after a long time until the upper unloading, obvious cracks and wall caving did not appear in the coal pillar. Only a certain deformation appeared in the coal pillar. These results showed that the 30-m coal pillar could remain stable for a long time under the state of top pressure, that it did not strip the wall caving, and that its bearing capacity was better than that of the 15-m coal pillar. The increase in the width of this coal pillar weakened the stress concentration benefit, the coal pillar was not easily "eroded" by stress and cracks, the overall bearing capacity was enhanced, and the coal pillar dam was safer.

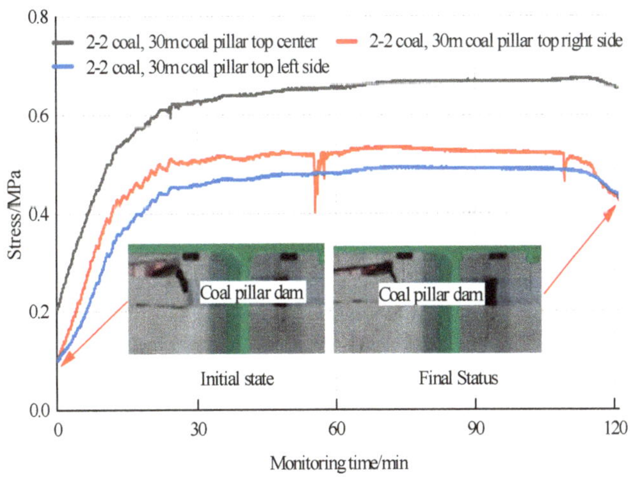

Figure 15. Stress failure characteristics of a 30-m coal pillar of 2-2 coal.

(3) Analysis of the crack development characteristics of a coal pillar dam

According to the experimental results, the fracture development characteristics of the 15-m coal pillar dam of the 2-2 coal seam were more obvious. Therefore, taking the 15-m coal pillar dam body of the 2-2 coal seam as an example, the digital image processing and analysis technology DAVIS was used to further analyze the fracture development characteristics of the coal pillar dam under vertical load conditions. As shown in Figure 16, under the vertical load at the top, the coal pillar dam body went through four stages—fracture generation, propagation, roadway roof failure, and coal pillar dam damage—from the initial stage before compression. In the fracture generation stage, with the increasing pressure at the top, cracks began to appear in the top corner and floor of the roadway of the coal pillar dam. In the fracture propagation stage, visible cracks also appeared directly above the roadway, and cracks were generated in the middle of the coal pillar. In the stage of roof failure of the roadway, cracks appeared continuously in the top rock mass, flakes appeared at the top corner, and cracks in the middle of the coal pillar gradually developed towards the bottom corner. In the damage stage of the coal pillar, the roof plate collapsed sporadically, the coal pillar appeared to have wall caving, the crack expanded to the upper rock formation of the roadway, and the crack in the coal pillar developed and extended to the side of the goaf.

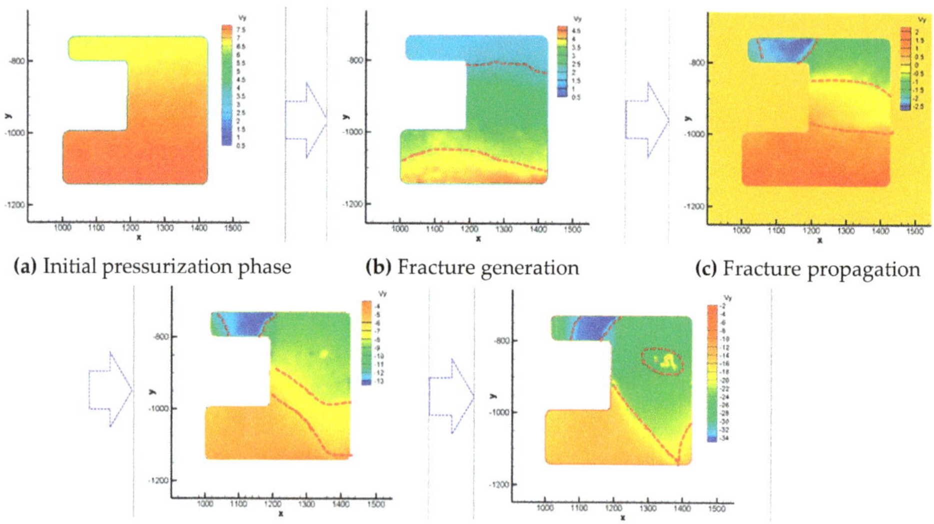

(a) Initial pressurization phase **(b)** Fracture generation **(c)** Fracture propagation

(d) Roadway roof failure **(e)** Coal pillar damage

Figure 16. Characteristics of crack development in a coal pillar dam.

Based on the above analysis, the following points can be noted. (1) The smaller the size of a coal pillar dam is, the smaller the pressure that can carry the overlying rock layer is, the easier it is to form a stress concentration area in the coal pillar, and the more likely the stress concentration benefit is to produce the phenomenon of an "eroded" coal pillar, resulting in a decrease in the coal pillar's effective width. (2) According to the experimental results, after the continuous compression of the coal pillar dam, the roof plate near the reservoir side is the first to be damaged, followed by the coal body on the roadway side. Then, the crack development degree of the coal pillar dam increases and the smaller size of the coal pillar will eventually become damaged. (3) The larger the size of the coal pillar dam is, the less obvious the stress concentration effect is and the stronger its bearing capacity is; moreover, there is no obvious stripping of the wall caving, thus ensuring the long-term safe operation of the reservoir.

4.3. Analysis of the Fracture Development Characteristics of a Coal Pillar Dam under the Dual Action of the Overlying Rock Layer and Lateral Water Pressure

In order to analyze the lateral bearing capacities of different sizes of coal pillar dam, after mining in 3-1 coal seams, the lateral load was applied to two coal pillars to simulate the lateral water pressure. The linear displacement sensor was used to monitor the horizontal displacement of the coal pillar, and the surface of the coal pillar dam was photographed regularly to monitor the fracture development characteristics of the coal pillar dam. According to the observed results of applied load, displacement, and fracture development, the fracture development and evolution law and the failure slip characteristics of the coal pillar dam were further analyzed.

(1) Analysis of the stress-failure slip characteristics of a 15-m 3-1 coal pillar

The lateral load was applied to the 15-m coal pillar of 3-1 coal, and the displacement of the roadway outside the coal pillar dam was recorded. As shown in Figure 17, after the loading started, the load increased linearly, but the amount of displacement was always 0. After loading for 5 min, the load did not increase, and was stable at about 480 kg. At this time, which was the elastic deformation stage, the coal pillar dam had not yet been displaced. When loading was done for approximately 6 min, the crack began to develop and the coal pillar produced obvious deformation. At the same time the coal pillar produced horizontal displacement, which increased linearly. When loading was done for approximately 12 min, the load suddenly decreased, and then the displacement also stopped at 9.15 mm. At this point, the coal pillar crack had been highly developed for the plastic failure stage, and the coal pillar roof plate had obvious separation cracks. When loaded to 23 min, the load dropped abruptly again, the coal pillar displacement suddenly increased linearly, and the coal pillar dam body was unstable and slipped.

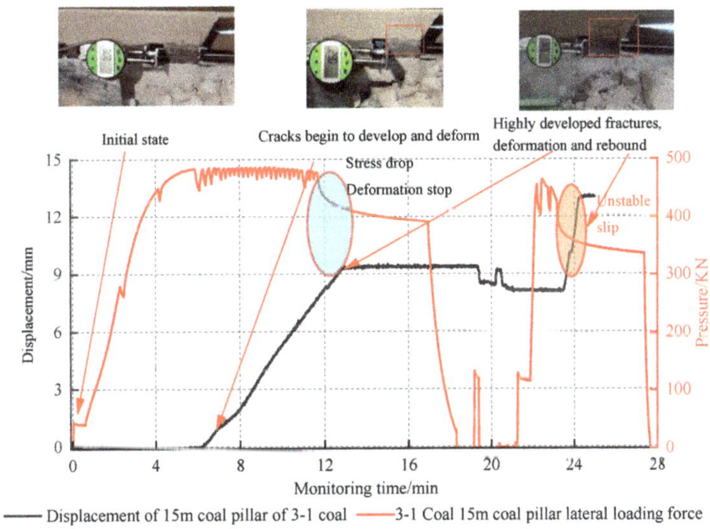

Figure 17. Stress failure characteristics of a 15-m coal pillar of 3-1 coal.

(2) Analysis of stress-failure slip characteristics of a 30-m 3-1 coal pillar

We apply the load laterally to the 30-m coal pillar of 3-1 coal. As shown in Figure 18, after the start of loading, the load increased in volatile linear growth, but the amount of displacement was always 0. After 10 min of loading, the load reached a peak of approximately 1250 kg, after which the coal pillar dam was displaced; this was the elastic deformation stage. When the displacement reached about 7 mm, the crack of the roof of the coal pillar dam had been preliminarily penetrated and the coal pillar had obvious deformation. Within 10–12 min of loading, the load continued to decrease, and the displacement increased linearly with a large increase; this was the plastic failure stage. When loaded to 12 min, the load dropped, the coal pillar dam suddenly slipped and became unstable, and the cracks between the top and bottom plates of the coal pillar had been completely penetrated.

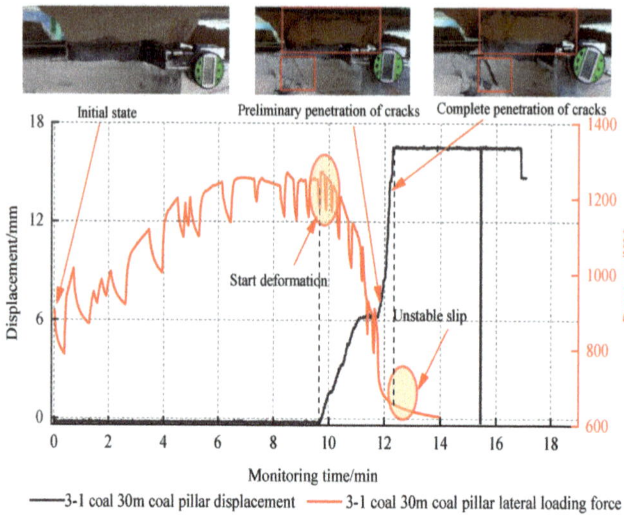

Figure 18. 3-1 Stress failure characteristics of a 30-m coal pillar.

Based on the above analysis, the following can be seen: (1) The size of a coal pillar dam is proportional to the lateral maximum water pressure it can bear. (2) Under the action of lateral water pressure, the coal pillar dam undergoes the process of "elastic deformation–plastic failure–instability slip", and the fracture develops most rapidly in the plastic failure stage—generally until the top and bottom plate cracks are completely penetrated—and instability slip occurs. (3) Experimentally, the maximum water level that the coal pillar dam can withstand can be roughly determined. For example, the maximum water level that can be withstood by a 15-m coal pillar is 50 m, which is the safe water storage height.

5. Conclusions

In this paper, the fracture evolution law and the critical conditions of failure instability of a coal pillar dam under different load conditions were studied, and the optimum size and safe water storage height of coal pillar dams were studied by establishing mechanical models, numerical simulations, and similar simulations. The following conclusions were obtained:

(1) The smaller the size of the coal pillar dam is, the smaller the pressure of the overlying rock layer is, and thus, the easier it is to form a stress concentration area in the coal pillar. Here, the stress concentration benefit is more likely to cause the coal pillar to be "eroded," resulting in a decrease in its effective width. The larger the size of the coal pillar dam is, the less obvious the stress concentration effect is and the stronger its bearing capacity is, and there is no obvious stripping phenomenon, thus ensuring the long-term safe operation of the reservoir.

(2) According to the experiment that used a similar simulation, the process of being "eroded," for the 15-m coal pillar dam in Shangwan coal mine, was as follows. First, with the increase in stress, obvious cracks first appeared in the roof and floor of the roadway, extending to the interior of the coal pillar dam until the cracks in the roof and floor were connected, forming a semi-elliptical coal body in the coal pillar dam body and separating from the coal pillar dam body. Subsequently, the wall spalling and the width of the coal pillar dam body reduced. The "eroded" process would be repeated many times during the operation of the coal pillar dam, and so it could be noted that the safety of small coal pillars is not as effective as that of large coal pillars.

(3) The size of the coal pillar dam is proportional to the lateral maximum water pressure it can bear. The coal pillar dam undergoes the process of "elastic deformation–plastic

damage–instability slip" under the action of lateral water pressure. By using the experimental method, the maximum water level that the coal pillar dam can withstand can be roughly determined. The maximum water level that a 15-m coal pillar can withstand in the Shangwan coal mine is 50 m, which is the safe water storage height.

(4) Through interpreting the results of numerical simulation, it was determined that the optimal size of a coal pillar dam in the Shangwan mine is 30 m, and that the fracture development of the coal pillar at this size does not penetrate inside and outside the reservoir under the action of the overlying rock layer and lateral water pressure, thus ensuring the long-term safe operation of the underground reservoir in the coal mine.

6. Future Prospects and Possible Research Directions

Underground reservoirs in coal mines are an effective means to develop China's coal mining in the direction of environmental protection. They not only reduce the waste of water resources and carbon dioxide emissions, but also save costs and improve the economic benefits of coal mines. In this paper, one of the factors affecting the stability of coal pillar dams in underground reservoirs of coal mines has been discussed: that is, the development and the evolution law of fissures, and some useful suggestions on the size of coal pillars, have been put forward. Future research on underground reservoirs in coal mines may be directed toward the following topics:

(1) Theoretical aspects: investigating (i) whether coal mining causes the coal mine water migration law in overburden rock, (i) the purification mechanisms and effectiveness of underground reservoirs in coal mines, and (iii) whether the underground reservoirs in coal mines affect the stability of sewage systems and water inflow prediction;

(2) Construction technology: in-depth research on the construction technology used for underground reservoirs in coal mines under complex geology, large burial depth, high gas levels, rock burst, and other conditions;

(3) Construction technology: the artificial dam body and its joint may easily produce mine water leakage. The construction of the artificial dam body should be improved, and appropriate grouting or anti-seepage materials to ensure an anti-seepage effect at the joint should be researched and developed;

(4) Safety monitoring system: no systematic safety evaluation system has been established for underground reservoirs of coal mines. Monitoring equipment and integrated systems for underground reservoirs in coal mines need to be developed and a scientifically proven and safe monitoring system needs to be established.

Author Contributions: Conceptualization, B.Z.; Methodology, B.Z. and W.N.; Investigation, B.Z. and X.H.; Writing—original draft, B.Z., W.N. and H.L.; Formal analysis, B.Z., H.L. and Y.S.; Writing—review and editing, B.Z. and H.L. All authors have read and agreed to the published version of the manuscript.

Funding: This research received no external funding.

Institutional Review Board Statement: Not applicable.

Informed Consent Statement: Not applicable.

Data Availability Statement: The data used to support the findings of this study are available from the corresponding author upon request.

Conflicts of Interest: The authors declare no conflict of interest.

References

1. Dong, S.; Xu, B.; Yin, S.; Han, Y.; Zhang, X.; Dai, Z. Technology and engineering development strategy of water protection and utilization of coal mine in China. *J. China Coal Soc.* **2021**, *46*, 3079–3089.
2. Zhang, G.; Li, Q.; Xu, Z.; Zhang, Y. Roof Fractures of Near-Vertical and Extremely Thick Coal Seams in Horizontally Grouped Top-Coal Drawing Method Based on the Theory of a Thin Plate. *Sustainability* **2022**, *14*, 10285. [CrossRef]
3. Yan, J.; Zhang, X.; Wang, K.; Song, X.; Yue, S.; Hou, J. Experimental Study on Creep Characteristics and Long–Term Strength of Anthracite. *Processes* **2023**, *11*, 947. [CrossRef]

4. Wang, X.-H.; Zhang, H.-H.; Wu, Z.; Li, X.-L.; Sui, Y.; Gao, R.-Q. Selection and Optimization Mechanism of the Lower Return Roadway Layout in the near Residual Coal Pillar Area. *Processes* **2022**, *10*, 2471. [CrossRef]
5. Shi, X. Research progress and prospect of underground mines in coal mines. *Coal Sci. Technol.* **2022**, *50*, 216–225.
6. Yu, Y.; Ma, L. Application of roadway backfill mining in water-conservation coal mining: A case study in northern shaanxi, China. *Sustainability* **2019**, *11*, 3709. [CrossRef]
7. Ju, J.; Xu, J.; Li, Q.; Zhu, W.; Wang, X. Progress of water—Preserved coal mining under water in China. *Coal Sci. Technol.* **2018**, *46*, 12–19.
8. Gu, D.; Zhang, J. Modern Coal Mining Affected to Underground Water Deposit Environment in West China Mining Area. *Coal Sci. Technol.* **2012**, *40*, 114–117.
9. Xiong, Z.; Wang, L. The Pollution Status of Ground Water and Treatment Methods in China. *World Sci. Res. J.* **2021**, *7*, 227–232.
10. Gu, D. Theory framework and technological system of coal mine underground reservoir. *J. China Coal Soc.* **2015**, *40*, 239–246.
11. Gu, D. Water Resource Protection and Utilization Engineering Technologyof Coal Mining in "Energy Golden Triangle" Region. *Coal Eng.* **2014**, *46*, 34–37.
12. Gu, D.; Zhang, Y.; Cao, Z. Technical progress of water resource protection and utilization by coal mining in China. *Coal Sci. Technol.* **2016**, *44*, 1–7.
13. Pang, Y.; Li, P.; Zhou, B. Underground reservoir construction technical feasibility analysis in 8.0m large mining height working face. *Coal Eng.* **2018**, *50*, 6–15.
14. Unver, B.; Yasitli, N.E. Modelling of strata movement with a special reference to caving mechanism in thick seam coal mining. *Int. J. Coal Geol.* **2006**, *66*, 227–252. [CrossRef]
15. Zhang, G.; Li, Q.; Zhang, Y.; Du, F. Failure characteristics of roof in working face end based on stress evolution of goaf. *Geomech. Geophys. Geo-Energ. Geo-Resour.* **2021**, *53*, 7. [CrossRef]
16. Singh, R.; Mandal, P.K.; Singh, A.K.; Singh, T.N. Cable-bolting-based semi-mechanised depillaring of a thick coal seam. *Int. J. Rock Mech. Min. Sci.* **2001**, *38*, 245–257. [CrossRef]
17. Zhang, N.; Yuan, L.; Wang, C.; Kan, J.; Xu, X. Deformation characteristics and stability analysis of roof roadway in destressed mining. *J. China Coal Soc.* **2011**, *36*, 1784–1789.
18. Kong, X.; Shan, R.; Ju, T. Model experiment of deformation and failure mechanism of coal roadway surrounding rock and its engineering application. *J. Min. Saf. Eng.* **2017**, *34*, 464–471.
19. Jaiswal, A.; Shrivastva, B.K. Numerical simulation of coal pillar strength. *Int. J. Rock Mech. Min. Sci.* **2009**, *46*, 779–788. [CrossRef]
20. Bertuzzi, R.; Douglas, K.; Mostyn, G. An Approach to model the strength of coal pillars. *Int. J. Rock Mech. Min. Sci.* **2016**, *89*, 165–175. [CrossRef]
21. Waclawik, P.; Ptacek, J.; Konicek, P.; Kukutsch, R.; Nemcik, J. Stress-state monitoring of coal pillars during room and pillar extraction. *J. Sustain. Min.* **2016**, *15*, 49–56. [CrossRef]
22. Gu, D.; Yan, Y.; Zhang, Y.; Wang, E.-Z. Experimental study and numerical simulation for dynamic response of coal pillars in coal mine underground reservoir. *J. China Coal Soc.* **2016**, *41*, 1589–1597.
23. Song, Y.; Yang, X. Evolution characteristics of deformation and energy fields during coal pillar instability. *J. Min. Saf. Eng.* **2013**, *30*, 822–827.
24. Zhang, J.; Wang, B. Stability of isolated coal pillar and overburden instability in shallow-buried interval gob. *J. Min. Saf. Eng.* **2020**, *37*, 936–942.
25. Zhang, C.; Han, P.; Wang, F.; He, X. The stability of residual coal pillar in underground reservoir with the effect of mining and water immersion. *J. China Coal Soc.* **2021**, *50*, 220–227.
26. Chen, D.; Wu, Y.; Xie, S.; He, F.; Sun, Y.; Shi, S.; Jiang, Z. Study on the first fracture of the main roof plate structure with one side goaf and elastic-plastic foundation boundary. *J. China Coal Soc.* **2021**, *46*, 3090–3105.
27. Zhao, H.; Cheng, H.; Wang, L.; Liu, Y.; Ji, D.; Zhang, Y. Distribution characteristics of deviatoric stress field and failure law of roadway surrounding rock under non-hydrostatic pressure. *J. China Coal Soc.* **2021**, *46*, 370–381.
28. Qi, X.; Deng, G.; Huang, K. Stress transfer law of coalpillar floor in room-pillar area of close distance coal seam. *J. Xi'an Univ. Sci. Technol.* **2021**, *41*, 649–656.
29. Tu, M.; Lin, Y.; Zhang, X.; Pu, Q.; Dang, J.; Zhao, G. Evolution of overburden structure and reasonable width of section coal pillar in large space isolated island stope. *J. Min. Saf. Eng.* **2021**, *38*, 857–865.
30. Wu, B. Reasonable Arrangement Mode About Coal Pillar Dam of Distributed Groundwater Reservoir in Coal Mine. *Saf. Coal Mines* **2018**, *49*, 68–72.
31. Cao, Z.; Ju, J.; Xu, J. Distribution model of water-conducted fracture main channel and its flow characteristics. *J. China Coal Soc.* **2019**, *44*, 3719–3728.
32. Xie, H.; Yao, Q.; Yu, L.; Shan, C. Study on damage characteristics of water-bearing coal samples under cyclic loading-unloading. *Sustainability* **2022**, *14*, 8457. [CrossRef]
33. Wu, B.; Wang, W.; Guo, D. Strength damage and AE characteristics of fractured sandstone under the influence of water intrusion times. *J. Min. Saf. Eng.* **2020**, *37*, 1054–1060.
34. Wang, F.; Liang, N.; Li, G.; Zhao, B. Failure evolution mechanism of coal pillar dams in complex stress environment. *J. Min. Saf. Eng.* **2019**, *36*, 1145–1152.

35. Wang, Z.; Ma, K.; Tian, H.; Li, Q. Numerical analysis of stress wave propagation law of coal and rock mass and its influencing factors. *Coal Sci. Technol.* **2019**, *47*, 66–72.
36. Guy, R.; Kent, M.; Russell, F. An assessment of coal pillar system stability criteria based on a mechanistic evaluation of the interaction between coal pillars and the overburden. *Int. J. Min. Sci. Technol.* **2017**, *27*, 11–17.
37. Yilmaz, I. Influence of water content on the strength and deformability of gypsum. *Int. J. Rock Mech. Min. Sci.* **2009**, *47*, 342–347. [CrossRef]
38. Verstrynge, E.; Adriaens, R.; Elsen, J.; Van Balen, K. Multi-scale analysis on the influence of moisture on the mechanical behavior of ferruginous sandstone. *Constr. Build. Mater.* **2014**, *54*, 78–90. [CrossRef]
39. Özdemir, E.; Sarici, D.E. Combined Effect of Loading Rate and Water Content on Mechanical Behavior of Natural Stones. *J. Min. Sci.* **2018**, *54*, 931–937. [CrossRef]
40. Du, J.; Meng, X. *Mining Science*; University of Mining and Technology Press: Beijing, China, 2009; pp. 389–393.
41. Qian, M.; Shi, P. *Mining Pressure and Strata Control*; China University of Mining and Technology Press: Beijing, China, 2003; pp. 108–109.
42. Zhang, Y.; Xu, L.; Liu, K. Formation mechanism and evolution laws of gas flow channel in mining coal and rock. *J. China Coal Soc.* **2012**, *37*, 1444–1446.
43. Zhou, X.; Zhang, Y. *Beijing: Constitutive Theory and Application of Unloading Rock Mass [M]*; Science Press: Beijing, China, 2007; pp. 55–72.
44. Chen, W. *Analysis Theory and Engineering Applications of Fissured Rock Mass Stability in Underground Engineering*; Science Press: Beijing, China, 2007; pp. 233–234.
45. Yang, J.; Hu, J.; Wu, Y.; Zhang, B. Numerical simulation of seepage and stability of tailing dams: A case study in ledong, China. *Sustainability* **2022**, *14*, 12393. [CrossRef]

processes

MDPI

Article

Study on Instability Characteristics of the Directional Borehole on the Coal-Seam Roof: A Case Study of the Tingnan Coal Mine

Zhie Wang [1], Xin Yang [2,3,4,*], Gongda Wang [3,4,*] and Haiwen Gong [5]

1 School of Management Engineering, Capital University of Economics and Business Beijing, Beijing 100083, China; 12021210006@cueb.edu.cn
2 Research Institute of Macro-Safety Science, University of Science and Technology Beijing, Beijing 100083, China
3 School of Civil and Resources Engineering, University of Science and Technology Beijing, Beijing 100083, China
4 State Key Laboratory of the Ministry of Education of China for High-Efficient Mining and Safety of Metal Mines, University of Science and Technology Beijing, Beijing 100083, China
5 School of Emergency Management and Safety Engineering, China University of Mining and Technology (Beijing), Beijing 100083, China; sqt191010116@student.cumtb.edu.cn
* Correspondence: b2081146@ustb.edu.cn (X.Y.); b1936282@ustb.edu.cn (G.W.)

Abstract: Directional long drilling on the roof is an effective gas control measure in the goaf, but there is little research on the stability of the surrounding rock. In this study, the geological conditions of the #4 coal seam in the Tingnan Coal Mine, Shaanxi Province, China taken as the application background, and the deformation characteristics of boreholes under four typical coal and rock conditions were first analyzed based on the Universal Distinct Element Code (UDEC) numerical simulation. Secondly, the stress, strain, and plastic deformation of the rock surrounding the borehole with different diameters were carried out using the Fast Lagrangian Analysis of Continua 3D (FLAC 3D). The effect of the casing on the stability of the borehole was also simulated. The results showed that the borehole stability of coal and mudstone was lower than that of fine-grained sandstone and coarse-grained sandstone. The larger the borehole diameter, the lower the stability. The borehole tended to be unstable, especially when the diameter was 160 mm and 200 mm. Traditional pipes can provide some protection, but for large boreholes, the protection is poor. Based on the above research, uniaxial compression tests were carried out on various internal support tubes, such as 'line-shaped', 'Y-shaped', and 'cross-shaped'. The results showed that the cross-shaped pipe had the highest compressive strength, which was 4–5 times that of the other types of protective pipe and had a good protective effect. The research results can provide reliable technical support for the protection of directional boreholes on roofs through strata and have important implications for the popularization and application of the directional long borehole technique.

Keywords: directional long borehole on the roof; borehole stability; lithology; borehole diameter; internal support borehole protection pipe

Citation: Wang, Z.; Yang, X.; Wang, G.; Gong, H. Study on Instability Characteristics of the Directional Borehole on the Coal-Seam Roof: A Case Study of the Tingnan Coal Mine. *Processes* **2023**, *11*, 1675. https://doi.org/10.3390/pr11061675

Academic Editor: Raymond Cecil Everson

Received: 13 April 2023
Revised: 11 May 2023
Accepted: 14 May 2023
Published: 31 May 2023

1. Introduction

The gas disaster is one of the major disasters that threaten coal mine safety production. With the development of coal resources gradually extending to the depths, the difficulty of gas control is increasing. Borehole drainage is the basic gas control measure. At present, gas pre-drainage mainly includes the borehole drilling in the bedding coal seam and the penetrating borehole in the bottom roadway, which can penetrate the coal seam so that the gas easily flows into the boreholes along the bedding plane, which can effectively reduce the gas content in the coal seam [1–4]. Buried pipe drainage, high-level drilling, and roof directional drilling are mainly used for gas control in the goaf during mining. Because the roof directional borehole is located in the fracture zone of the coal seam, where the horizon is high and less affected by mining disturbance, the borehole and gas drainage life cycle is long, which can effectively solve the problem of gas overflow in the upper corner of the goaf [5–8].

However, the openings of roof directional boreholes are often located in the coal seam, and they have to pass through mudstone and sandstone before entering the fracture zone. Due to the large diameter of the borehole (typically around 160 mm), the boreholes are susceptible to loss of stability resulting in deformation and collapse. Therefore, the stability of the directional borehole on the roof through the strata directly affects its drainage effect. Borehole diameter, lithology, and other parameters are important parameters affecting borehole stability [9,10].

Research on the influencing factors of borehole stability and borehole failure modes in the bedded coal seam has been conducted by scholars at home and abroad. The influencing factors of borehole stability mainly include rock strength, gas pressure, stress and strain, burial depth, lateral pressure coefficient, and anisotropic permeability [11–14]. Yang et al. [15] indicated that rock mechanics is the main theoretical basis for ensuring good stability, sand production, or casing damage. Karatela et al. [12] investigated the stability of boreholes in fractured rock using the discrete element method. The results show that the stability of boreholes depends largely on the strength of the rock. The tensile and shear failure of the borehole increases with increasing fluid velocity and pore pressure. Ding et al. [16] found that the large difference in permeability of rock layers around the borehole would lead to the change in stress state, damage area, collapse pressure, and fracture pressure around the borehole. Zhao et al. [17] studied the instability characteristics of the borehole under steady vertical load using the gas drainage borehole collapse dynamic monitoring devices. The attenuation of the borehole circumferential strain is an important symbol for the prediction and warning of borehole instability and collapse. A borehole may be damaged resulting from the integrated effect of stratigraphic and structural factors. Katanov et al. [18] proposed a model based on neural simulation to analyze the deformation of rock layers with different strength characteristics. Ma et al. [19] solved the problem of severe borehole deviation in coal mine gas drainage by summarizing the borehole deviation law and improving the precision directional drilling tool. Dychkovsky et al. [20] simulated the influence of geological faults on the stress and deformation state of rock mass by FLAC 5.00 and established a three-dimensional network visualization by computer simulation results and data interpolation method. The research shows that in the Lviv-Volyn coal basin, the geological fault of up to 3 m distance has a great influence on the stress and deformation state of the rock mass. Based on the parameters of the geo-mechanical model developed and confirmed, Petlovanyil et al. [21] have determined the reasonable range of inclination angle and key parameters of the radius of curvature when using underground gasification technology to develop thin coal seams. Zhang et al. [22] numerically simulated the deformation characteristics of boreholes with different burial depths, and the results showed that the deeper the burial depth, the more obvious the deformation. The form of instability and failure was the collapse of the upper part, and the fracture of the left and right sides formed the fracture area. Qu et al. [23] found that the stability of coal seam borehole was affected by the time lag effect based on field tests and numerical simulation; they found that the change of pore pressure was the main factor affecting the time lag effect, and the rich cleat was the internal factor. Niu et al. [24,25] studied the monitoring and evaluation of borehole stability through experiments and simulations, proposed an index to calculate the degree of borehole damage based on the residual area, and fitted the functional equation between the relative pressure of the sensor (the difference between the real-time pressure of the sensor and the coupling pressure of the borehole wall) and the degree of borehole damage. Combined with the amount of gas extracted, it was verified that the borehole deformation first increased and then became stable with time, and then increased and then decreased with depth. Zhang et al. [26] established the gas migration channel zoning model and determined the parameters of the optimal directional long borehole in the working face by using UDEC and COMSOL software. Xu et al. [27] found that when the borehole is shallow, the friction resistance between the drill pipe and the borehole wall increases linearly. With the increase of borehole depth, the friction resistance gradually develops into an exponential relationship. In addition, Liu et al. [28] concluded that the effective drainage radius of a directional long drilling hole has an exponential relationship with the distance from the drilling opening. Yuan et al. [29,30] conducted the combined drilling method of curtain grouting in underground deep wells, which can significantly reduce the risk of water inrush in deep mining with complex hydrogeological conditions. A mathematical model to describe the unstable pressure dynamics in stress-sensitive coalbed methane reservoirs was proposed by Wang et al. [31]. Wang et al. [32] proposed the gradient recognition and memory-cutting method for the continuous advancement of non-uniform coal seams, such as coal seams with folded structures on long-arm working-face.

Moreover, scholars have conducted a great number of field tests and research on a series of borehole protection technologies, such as borehole reinforcement, regional solidification, and screen protection. Xue et al. [33] found that increasing the casing strength and thickness can effectively control the borehole instability and greatly improve the gas extraction effect through field tests. To improve the stability of the borehole. Zhai et al. [34] conducted research on the technology of screen pipe protection. By comparing the maximum AE event technology and energy dissipation rate, it was found that the screen pipe can effectively resist external stress disturbance, prevent hole collapse, and improve the drainage effect of the borehole. Di et al. [35] proposed the regional solidification pore formation method for soft coal seams, which can solidify the strength of the rock surrounding the borehole and improve the pore formation rate. Qi et al. [36] tested the full-hole deep screen mesh tube drainage technology to solve the problem of internal collapse and negative pressure loss of deep coal seam drainage boreholes, which can effectively control the collapse and deformation of boreholes and reduce the negative pressure loss. Compared with conventional drainage, after 90 days, the gas drainage concentration increased by 101% and the gas flow increased by 97%, so the gas drainage rate increased significantly. Li et al. [37] proposed to integrate the technology of borehole digging, protection, and sealing in the construction of the borehole site, so as to strengthen the stability of the borehole and solve the problems of difficult borehole formation, poor drainage effect and high danger of coal seam explosion in soft coal seams.

Currently, the research on borehole stability mainly focuses on the surrounding rock stress, burial depth, lateral pressure coefficient, anisotropic permeability, and other influencing factors, and the failure characteristics of the borehole in the bedding coal seam, while the research on the stability and borehole protection technology of roof directional long borehole is still lacking. Compared with the traditional high-position alley gas extraction technology, the roof directional long borehole has the advantages of a short construction period, low investment, and long extraction period. Therefore, the stability of the roof-directional long borehole with different lithology and borehole diameters was analyzed by numerical simulation in this paper. The compression experiments were carried out to study the protective effect of different internal support structures in the directional borehole, which is beneficial to realize the effective hole formation of the directional long borehole on the roof and high-efficiency gas extraction in the goaf treatment of the coal mine face. This paper will provide some guidance for the popularization and application of directional long boreholes on the roof technology, and also gradually realize the "replacing alley with borehole" in goaf gas treatment. The rest of this study is organized as follows. Section 2 analyzes the deformation characteristics of boreholes under different coal and rock conditions, and simulates the stress, strain, and plastic deformation of rocks around boreholes with different diameters. In Section 3, the experiment on the influence of the internal support hole protection tube on the stability of the hole wall was carried out. Section 4 analyzes the mechanism of numerical simulation and laboratory experiment results. Finally, the conclusions are presented in Section 5.

2. Numerical Simulation of the Borehole Stability

Lithology, borehole diameter, and borehole protection tubing have a great influence on the stability of the borehole; for example, the strength of the rock will affect the rate of hole formation and the durability of the borehole. The larger the diameter of the borehole, the closer it is to the uniaxial compressive strength of the rock and the easier it is to destroy the stability of the borehole in the rock formation. The shear strength of the casing is also closely related to the effectiveness of the casing [38–40]. Therefore, this section takes the 208-working face of Tingnan Coal Mine in Xianyang, Shaanxi Province, China as an object to conduct numerical simulation research to analyze the influence of lithology, borehole diameter, borehole protection pipe, and other factors on borehole stability. Tingnan Mine Field is mainly covered by Quaternary loess and Tertiary red soil, and the Lower Cretaceous Luohe Formation is exposed in major valleys along the Heihe River and Jinghe River. The strata in the minefield of Tingnan Coal Mine are, from top to bottom, Holocene (Q_4), Upper Pleistocene Malan Formation (Q_3), Quaternary-Middle Pleistocene Lishi Formation (Q_2), Huachi Formation (K_{1h}), Luohe Formation (K_{11}), Lower Cretaceous Yijun Formation (K_{1y}), Anding Formation (J_{2a}), Zhiluo Formation (J_{2z}) and Middle Cretaceous Yanan Formation. The Jurassic Yan'an Formation is a coal-bearing layer, and the Triassic Hujiahe Formation is a direct or indirect sedimentary basement. The surface and overlying strata are basically the same, and the aquifer of the Luohe Formation is in the upper part. The minefield is located in the middle part of the anticline (Lujia–Xiaolingtai), and the stratum in the middle is close to the level. The terrain in the south wing is gentler than that in the north wing, and the dip angle in the north wing is 4°~6°. The back of the syncline is Mengcun, which is connected with the north wing of the anticline from Lujia to Xiaolingtai. The strike of the layer is N 20° E and the dip angle is about 2°. The north wing crosses the Great Buddha Temple to the south and the occurrence of the southeast corner of the layer changes. The anticline structure affects the change of layer thickness, coal seam thickness, and occurrence in the minefield. Tingnan coal mine is a high-gas mine, and the 4# coal seams belong to the type II spontaneous combustion coal seams, and the coal dust is explosive. The coal seam thickness of the 208-working face varies greatly, with the average coal seam thickness being 11.5 sm. The base roof of the working face is coarse-grained sandstone, which is grey and mainly composed of quartz and feldspar. The immediate roof is mudstone, dark grey and lumpy, containing a large number of plant fossils and locally a small number of calcite veins. The direct floor is aluminum mudstone, light grey, dense, and contains plant root fossils. The base is fine-grained sandstone, light grey, chestnut, and hard, with sandy muddy breccias. The 208-working face is located in the west wing of the second panel, with a total length of 2527 m. The description of formation lithology characteristics is shown in Table 1.

Table 1. The description of formation lithology in 208 working-face.

Stratigraphic Category	Rock Category	Average Thickness	Characteristics of the Lithology
Basic roof	Coarse-grained sandstone	11.63~49.24 30.5	Coarse-grained sandstone, gray, mainly composed of quartz and feldspar, sub-round, argillaceous cementation, wavy bedding, mixed with thin siltstone, locally containing siderite nodules, which is obviously in contact with the underlying.
Direct roof	Mudstone	1.3~5.06 2.07	Mudstone, gray, dark gray, lumpy, containing a large number of plant fossils, locally containing a small number of calcite veins, with a slip surface on the upper part and obvious contact with the underlying.
Direct floor	Aluminous mudstone	0.94~3.2 1.98	Aluminous mudstone, light gray, dense, containing plant root fossils, with a sliding surface at the broken part.
Basic floor	Fine-grained sandstone	2.8~19.8 9.51	Fine-grained sandstone, light gray, reddish brown, relatively hard, containing sandy argillaceous breccia.

2.1. Analysis of the Influence of Lithology on the Borehole Stability

2.1.1. Model Building

Given that the roof-directed long borehole has a large span in the roof, the horizon and rock lithology through which the hole passes is complex. The borehole stability of coal, mudstone, fine-grained sandstone, and coarse-grained sandstone in the working face rock mass was investigated using the Universal Distinct Element Code (UDEC). The model is 300 m long and 180 m high as shown in Figure 1a. According to the histogram distribution of rock layers, joints are set and grids are divided, and the number of grids is 2,976,224. At the upper boundary of the model, the self-weight stress of the overlying strata is imposed, and the simulated mining depth is 430 m. According to the actual situation, it is assumed that the left and right sides are mined together, and the distance between the mining boundary and the left and right boundaries of the model is 60 m, each mining is 10 m, and the mining length is 180 m. To facilitate simulation and calculation, the model is partially simplified: The change in coal seam dip angle and thickness is ignored, and the calculation is based on the average coal seam thickness. Considering the large span of high-level directional drilling in horizontal and vertical directions, the strata and rock lithology through which the drilling passes are also complicated. Therefore, these study models and analyzes the stability of high-level directional drilling under different lithologies based on mining disturbance. Considering the different lithologic horizons that are encountered in drilling practice, four main rock layers are selected for study in the model, namely coal rock, mudstone, fine-grained sandstone, and coarse-grained sandstone. Given that the roof-directed long borehole has a large span in the roof, the horizon and rock lithology through which the hole passes is complex. The borehole stability of coal, mudstone, fine-grained sandstone, and coarse-grained sandstone in the working face rock mass was investigated using the UDEC. A 1 m × 1 m rock mass model with a 130 mm aperture is created as shown in Figure 1b,c. In this model, Voronoi is used to generate joints, and the blocks are distributed in triangular mesh elements. The upper boundary is set as the vertical downward self-weight stress boundary, the lower boundary is fixed as a constraint, and the left and right boundaries are restricted horizontal displacements, as shown in Figure 2. When the calculation model is set, the model should be assigned according to the constitutive model. The Coulomb slip model is selected for numerical simulation, and the rock mechanics parameters of 208 working faces in the Tingnan Coal Mine are shown in Table 2.

Table 2. The rock mechanics parameters.

Lithology	Bulk Modulus/(GPa)	Shear Modulus/(GPa)	Cohesive Strength/(MPa)	Internal Friction/(°)	Density/(kg/m^3)	Tensile strength/MPa	Source
Coal	1.42	0.57	1.2	28	1400	0.64	Lab measurement
Mudstone	4.54	4.31	2.08	32	2560	1.32	Lab measurement
Coarse-grained sandstone	4.58	4.42	2.57	34	2530	1.28	Lab measurement
Fine-grained sandstone	4.64	4.32	4.57	35	2540	1.35	Lab measurement

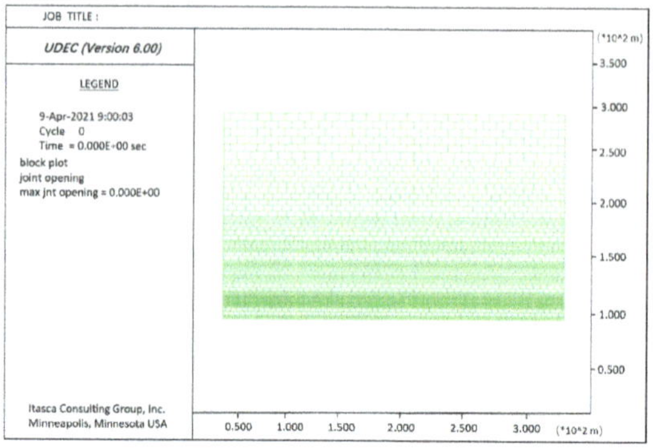

(**a**) Coal seam mining model of directional drilling structure

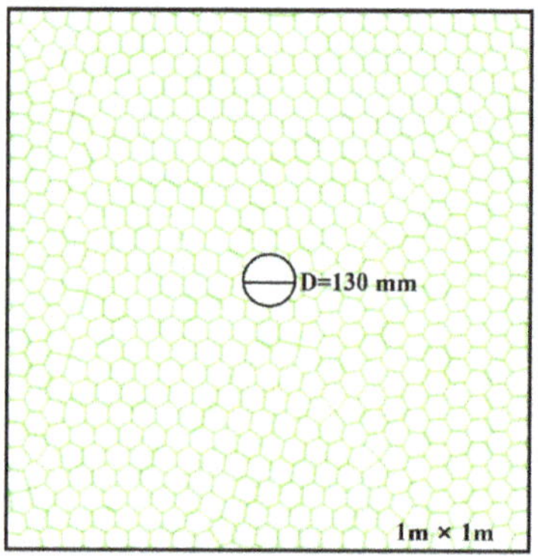

(**b**) Joint distribution

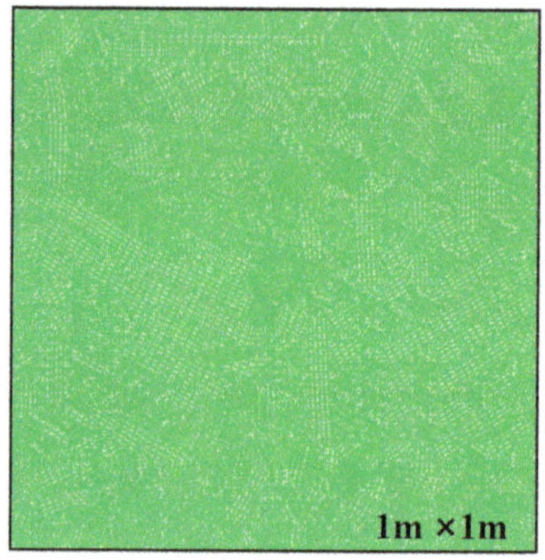

(**c**) Grid division

Figure 1. The Numerical computation models. (**a**) The model of the coal seam mining model of directional drilling structure; (**b**) the joint distribution of the model; (**c**) the grid division of the model.

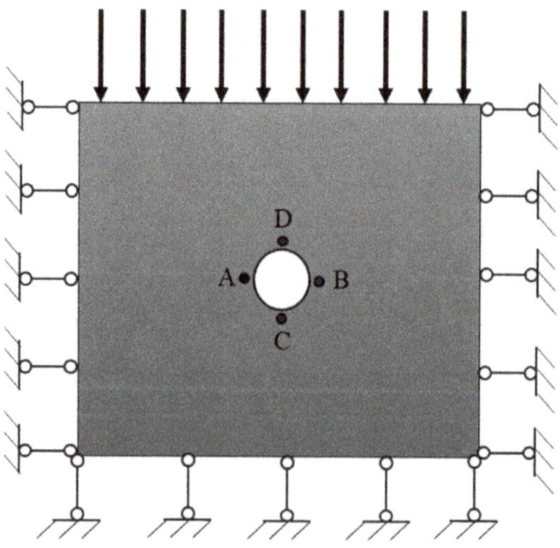

Figure 2. Schematic diagram of boundary constraints.

2.1.2. Analysis of the Simulation Results

(1) Variation of the Borehole Stress

The stress variation range of the surrounding rock is shown in Figure 3. The stress around the borehole in coal and rock ranges from 1.92 MPa to 35.57 MPa, that which around the borehole in mudstone ranges from 1.77 MPa to 29.84 MPa, that which around the borehole in fine-grained sandstone ranges from 1.49 MPa to 27.58 MPa, and that which is around the borehole in coarse-grained sandstone ranges from 1.71 MPa to 27.20 MPa. From the stress range, it can be seen that coal and mudstone have large stress peaks and stress fluctuations, while the stress ranges of fine-grained sandstone and coarse-grained sandstone are relatively small and proximate to each other.

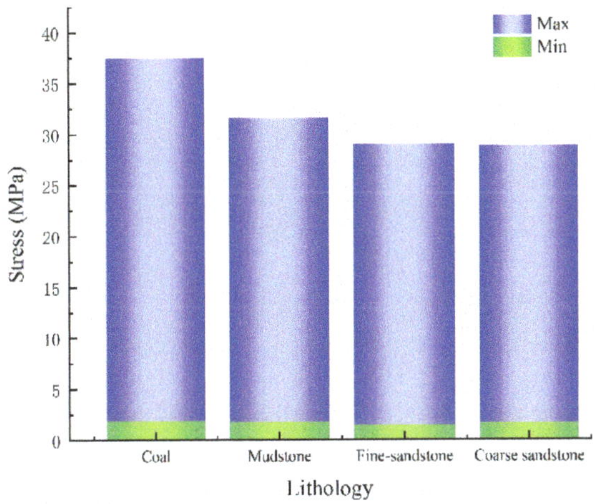

Figure 3. Stress range under different lithologies.

The maximum principal stress contour of the borehole is shown in Figure 4. The peak value of the maximum principal stress is mainly distributed in some borehole walls near the left and right sides of the borehole, while the rest of the rock mass is evenly distributed with lower stress. The maximum principal stress results of different lithologies show that the stress concentration of coal and mudstone is higher, the high-stress area of rock mass around the borehole is larger, and the stress distribution of fine-grained sandstone and coarse-grained sandstone is relatively more uniform. It indicates that the stress state of coal and mudstone is unstable.

(2) Variation of the borehole displacement

The displacement contour of the boreholes with different lithologies is shown in Figure 5. As can be seen from Figure 5, the overall displacement trend of the borehole under the four lithologies is similar, and the displacement is all in the direction of the borehole center. The displacement of the rock mass around the borehole is small, and the displacement of the rock mass on the upper and lower sides of the borehole is larger than that on the left and right sides. However, there are still some differences in the borehole displacement of different lithologies. The borehole displacement of coal and mudstone is obviously larger than that of fine-grained sandstone and coarse-grained sandstone, especially on the upper and lower sides of the borehole. The block with the largest displacement is coal, followed by mudstone. The four lithologies are similar on the left and right sides.

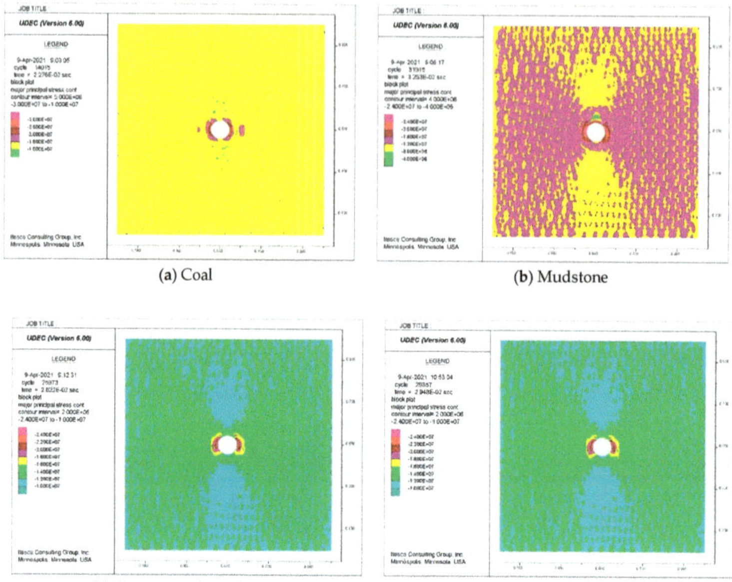

(a) Coal (b) Mudstone (c) Fine-grained sandstone (d) Coarse-grained stone

Figure 4. Maximum principal stress contours. (**a**) The maximum principal stress contour of the coal; (**b**) the maximum principal stress contour of the mudstone; (**c**) the maximum principal stress contour of the fine-grained sandstone; (**d**) the maximum principal stress contour of the coarse-grained sandstone.

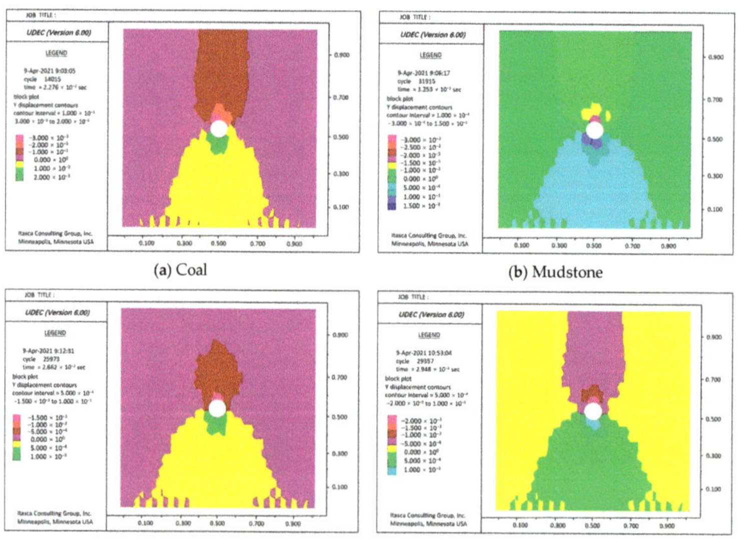

(a) Coal (b) Mudstone (c) Fine-grained sandstone (d) Coarse-grained stone

Figure 5. Displacement cloud map in Y direction. (**a**) The displacement cloud map in Y direction of the coal; (**b**) the displacement cloud map in Y direction of the mudstone; (**c**) the displacement cloud map in Y direction of the fine-grained sandstone; (**d**) the displacement cloud map in Y direction of the coarse-grained sandstone.

During the simulation, four measuring points A, B, C, and D were set at the top, bottom, left, and right sides of the borehole. The displacement change of the measuring point is shown in Figure 6. The displacement of each monitoring point first increases and then tends to be stable within the hole. The maximum displacement of the four points is 3.15 mm of the coal at point D, and the minimum is 0.49 mm of the fine-grained sandstone at point A, which shows that the rock mass around the borehole of each lithology is only slightly displaced and remains unchanged. The results of the four measuring points show that the displacement of the coal rock is significantly larger than that of the other three lithologies, and the displacement of the mudstone is also larger than that of the other two lithologies. The displacement changes of fine-grained sandstone and coarse-grained sandstone are similar, and that of coarse-grained sandstone is slightly larger than that of fine-grained sandstone. The results show that there are differences in rock mass displacement around the borehole under different lithologies, which are caused by different rock mechanical parameters. The density, bulk modulus, shear modulus, and tensile strength of coal and mudstone are all low, which leads to cracking or even fracturing of the surrounding rock mass after the borehole is drilled, and then the displacement changes.

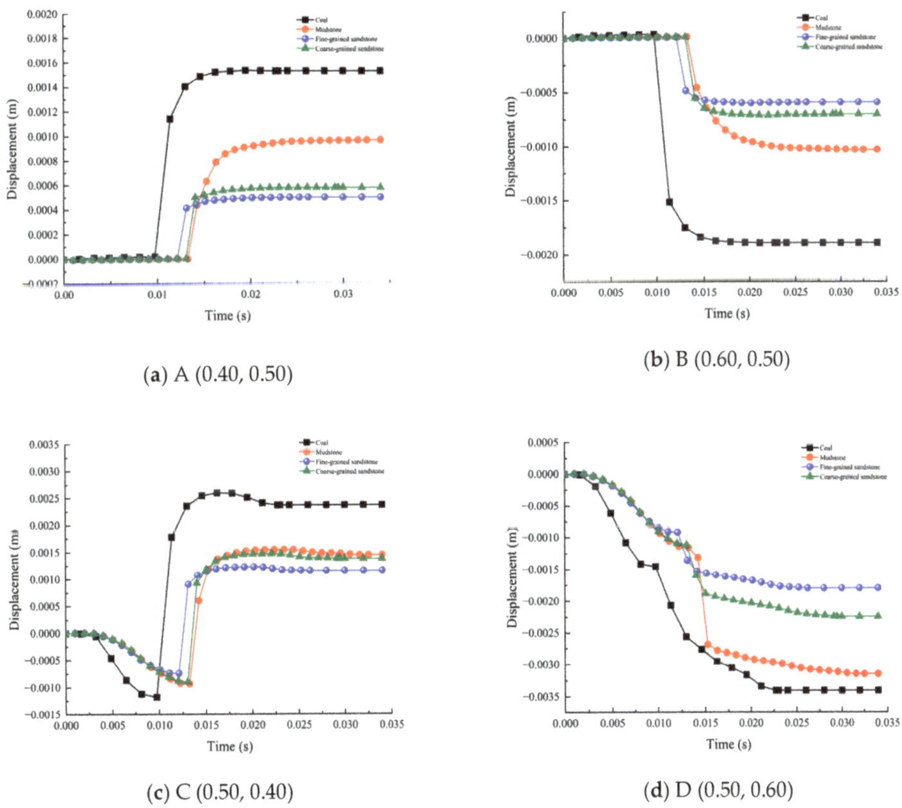

(a) A (0.40, 0.50)

(b) B (0.60, 0.50)

(c) C (0.50, 0.40)

(d) D (0.50, 0.60)

Figure 6. Displacement variation under different lithologies. (**a**) The displacement variation in the A (0.40, 0.50); (**b**) the displacement variation in the A (0.60, 0.50); (**c**) the displacement variation in the A (0.50, 0.40); (**d**) the displacement variation in the D (0.50, 0.60).

In summary, the stress and displacement changes of the rock mass around the borehole are obviously larger in the coal and mudstone than in the fine-grained sandstone and coarse-grained sandstone, and the stress and displacement of the rock mass after the borehole is drilled are strongly influenced by the lithology, which also shows that the borehole stability is high in the fine-grained sandstone and coarse-grained sandstone, while that in the coal and mudstone is poor.

2.2. Analysis of the Influence of Borehole Diameter on Borehole Stability

A three-dimensional borehole model is created based on FLAC 3D and the borehole stability of four diameters, namely 100 mm, 130 mm, 160 mm, and 200 mm, is investigated. The borehole model is shown in Figure 7. The Moore-Coulomb constitutive model is selected for the simulation, which is suitable for solving rock mechanics and excavation problems. The initial stress is $S_{xx} = -1.1985 \times 10^7$ Pa, $S_{yy} = -1.1985 \times 10^7$ Pa, $S_{zz} = -1.1985 \times 10^7$ Pa, and the lateral pressure coefficient is 1.0. The velocity of the left and right boundaries in the X direction is set to 0; the velocity of the bottom boundary in the Y direction is set to 0; and the top boundary is set to the gravity of the overlying strata. Then the excavation simulation was started after reaching the equilibrium level.

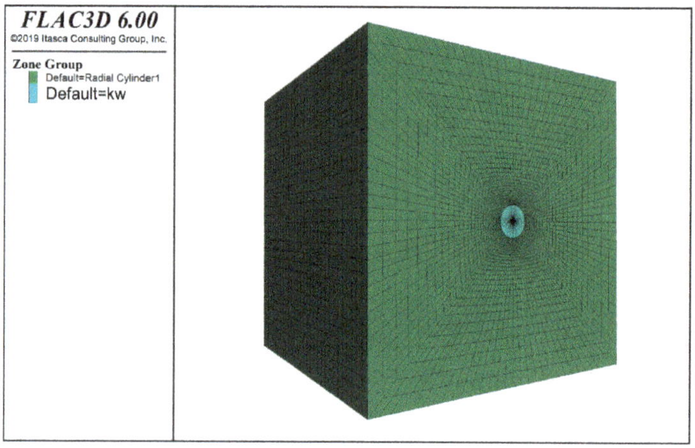

Figure 7. The borehole model.

2.2.1. Analysis of the Stress and Displacement

The maximum principal stress contour of boreholes of different diameters is shown in Figure 8. The maximum principal stress nephogram at 100 mm and 130 mm is distributed in a regular and uniform circle around the borehole, and it begins to deform into an ellipse at 160 mm, while the contour is completely deformed at 200 mm, and the stress expands to the boundary region, resulting in stress equilibrium and instability of the surrounding rock. Thus, the results indicate that when the borehole diameter exceeds 160 mm, the maximum principal stress distribution of the borehole begins to deform, the stress balance tends to be destroyed, and the stress concentration area is proportional to the borehole diameter.

The displacement monitoring of the rock mass around the borehole is shown in Figure 9. According to the curve, the displacement range of the left side of the borehole is 0~3.5 mm, and the displacement ranges of the upper and lower sides are 0~12 mm and 0~25 mm, respectively. The displacement of the left side of the borehole is the smallest, and the displacement of the upper and lower sides of the borehole is much larger than that of the left side, indicating that the borehole damage is mainly caused by the large deformation of the upper and lower sides. The displacement of each monitoring point increases with the hole diameter, and the displacement of the rock mass around the hole changes more. Boreholes with different diameters produce displacements at almost the same time after drilling and then gradually stabilize.

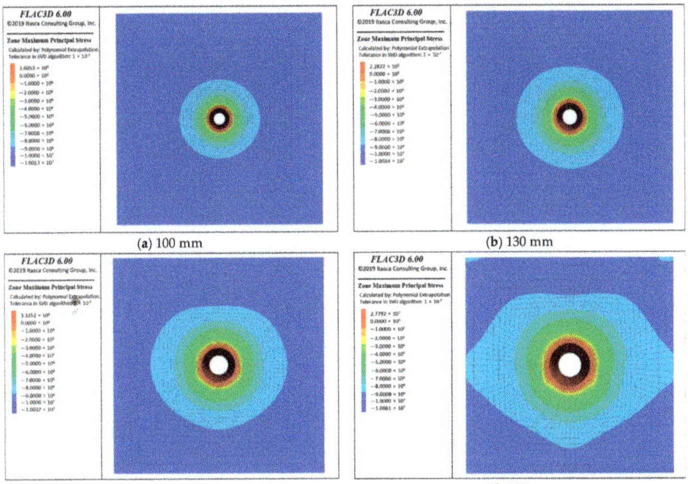

(**a**) 100 mm (**b**) 130 mm (**c**) 160 mm (**d**) 200 mm

Figure 8. Maximum principal stress contours. (**a**) The maximum principal stress contour of the borehole of diameter of 100 mm; (**b**) the maximum principal stress contour of the borehole of diameter of 130 mm; (**c**) the maximum principal stress contour of the borehole of diameter of 160 mm; (**d**) the maximum principal stress contour of the borehole of diameter of 200 mm.

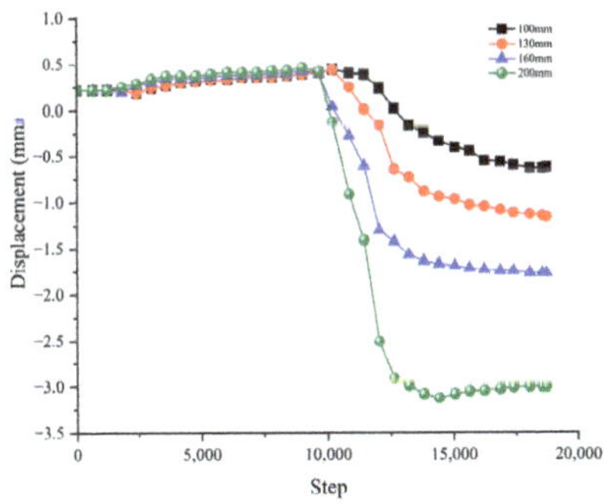

(**a**) Borehole left surrounding rock

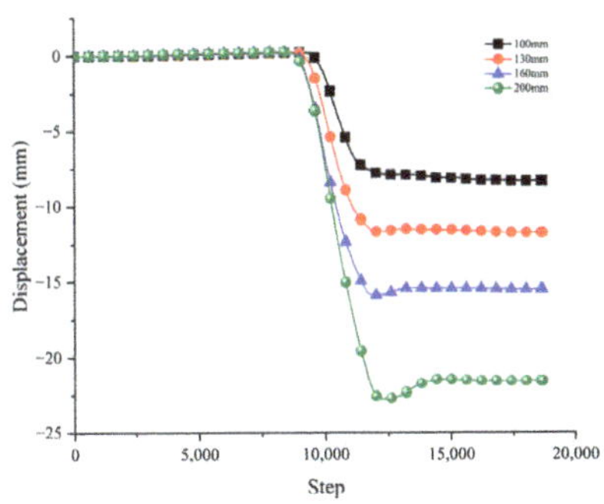

(**b**) Borehole upper surrounding rock

Figure 9. *Cont.*

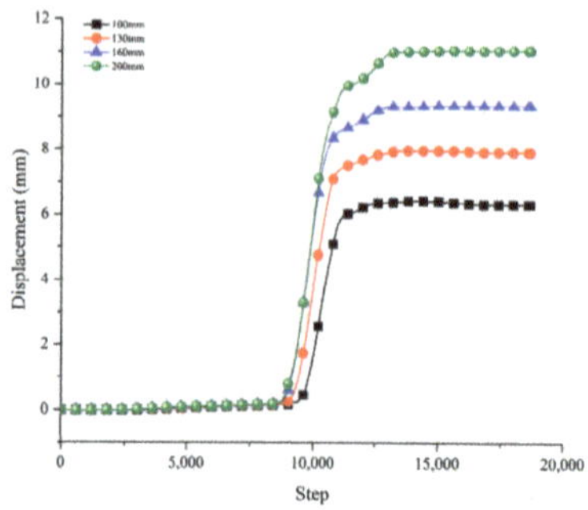

(c) Borehole under a surrounding rock

Figure 9. The variation curve of surrounding rock displacement with different diameters. (**a**) The variation curve of borehole left rock displacement with different diameters; (**b**) the variation curve of borehole upper rock displacement with different diameters; (**c**) the variation curve of borehole under rock displacement with different diameters.

2.2.2. Analysis of the Plastic Failure

The S-distribution and volume of plastic failure of the surrounding rock mass of boreholes are shown in Figures 10 and 11. The distribution of plastic failure zones is similar to that of the vertical stress and is symmetrical, extending around the hole wall. The volumes of plastic failure zones with four diameters of 100 mm, 130 mm, 160 mm, and 200 mm are 0.007 m^3, 0.012 m^3, 0.018 $m^{3,}$ and 0.03 m^3 respectively. As the borehole diameter increases, so does the extent and volume of the plastic failure zone. The volume of the plastic failure zone can be divided into shear failure and tensile failure. The volume of tensile failure changes very little with the increase in pore diameter as can be seen from Figure 11. On the contrary, the volume of shear failure increases linearly with the increase in pore diameter and is much larger than that of tensile failure in all four-borehole diameters. Therefore, it can be concluded that the failure mode of the boreholes is mainly shear failure.

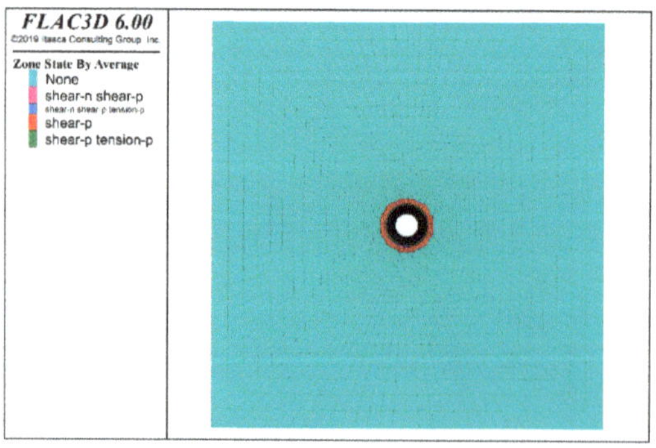

(**a**) 100 mm

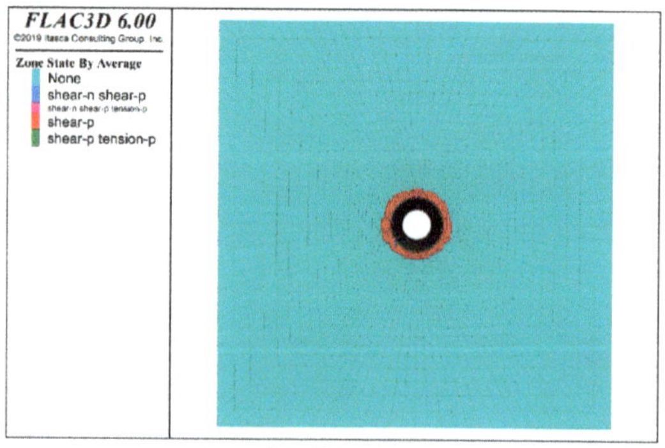

(**b**) 130 mm

Figure 10. *Cont.*

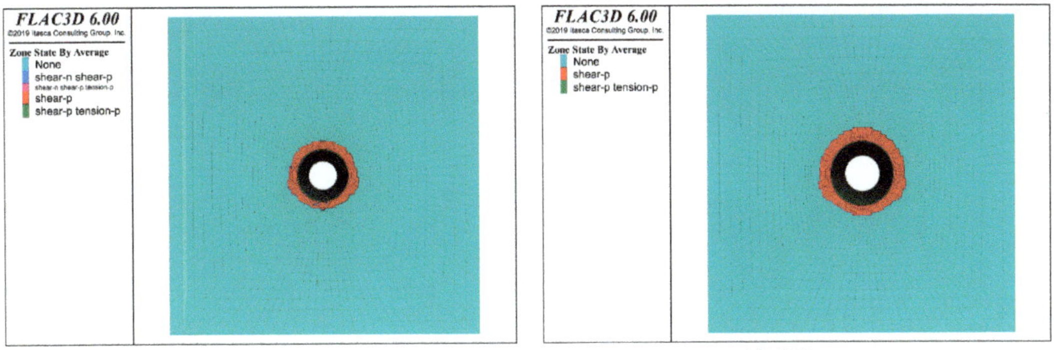

(**c**) 160 mm (**d**) 200 mm

Figure 10. Distribution of plastic failure of boreholes with different diameters. (**a**) The distribution plastic failure of the borehole of diameter of 100 mm; (**b**) the distribution plastic failure of the borehole of diameter of 130 mm; (**c**) the distribution plastic failure of the borehole of diameter of 160 mm; (**d**) the distribution plastic failure of the borehole of diameter of 200 mm.

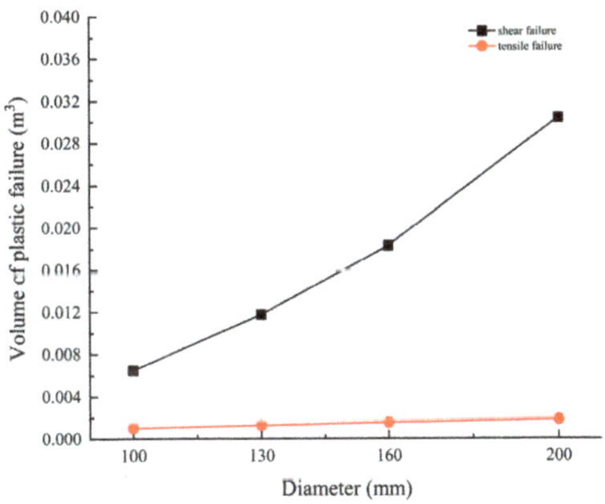

Figure 11. The volume of plastic failure zone with different diameters.

In summary, the analysis of the stress state and displacement of the borehole with four diameters of 100 mm, 130 mm, 160 mm, and 200 mm showed that the larger the diameter, the more unstable they are, and the plastic failure volume is directly proportional to the borehole diameter. Therefore, it can be inferred from the comprehensive stress, displacement, and plastic zone that the stability of the borehole decreases as the borehole diameter increases, and the borehole tends to be unstable when the diameters are 160 mm and 200 mm.

2.3. Analysis of Protection Effect of the Hole Protection Pipe

The stability of boreholes is investigated by simulating the application of a borehole protection tube to analyze the protection effect. The modeling process of boreholes is the same as in Section 2.2. The mechanical parameters of the casing are shown in Table 3; the casing is simulated after the borehole is excavated. The model of the casing is shown in Figure 12.

Table 3. Mechanical parameters of the borehole protection pipe.

	Bulk Modulus/(GPa)	Shear Modulus/(GPa)	Cohesive Strength /(MPa)	Internal Friction/ (°)	Density/(kg/m^3)	Tensile Strength/MPa
borehole protection pipe	42.6	33.8	4.8	34	2500	3.9

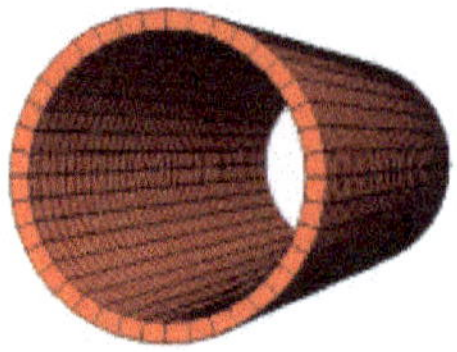

Figure 12. Model of the borehole protection pipe.

2.3.1. Analysis of the Stress and Displacement

The maximum principal stress contour of the borehole before and after protection is shown in Figure 13. It can be seen from the contour that the stress concentration at the 160 mm and 200 mm borehole walls is greatly reduced compared to the borehole without the casing. The distribution range of stress concentration is obviously reduced compared to that before protection, especially the stress range before and after borehole protection of 200 mm boreholes changes greatly. This indicates that the borehole protection tube can provide effective support and change the stress state around the borehole wall.

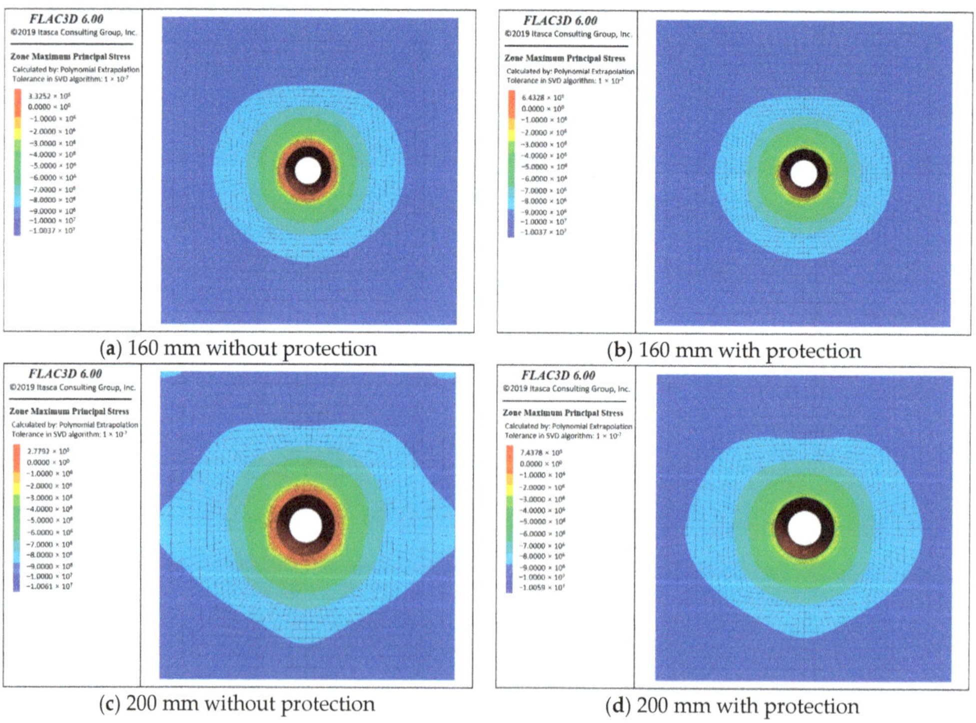

Figure 13. Maximum principal stress contours with and without protection.

The displacement monitoring curve of the rock mass around the borehole before and after protection is shown in Figure 14. The displacement changes in the left, upper, and lower protection holes of the 160 mm borehole are 0.2 mm, 3.1 mm, and 2.2 mm respectively. The displacement changes in the left, top, and bottom protection holes of the 200 mm hole are 0.3 mm, 5.4 mm, and 2.7 mm respectively. The results show that the displacement of each monitoring point has obviously changed before and after hole protection, especially the displacement of the upper side of the borehole has the biggest difference. It also shows that the borehole protection pipe can weaken the displacement and deformation caused by borehole excavation, and improve the support and stability of boreholes.

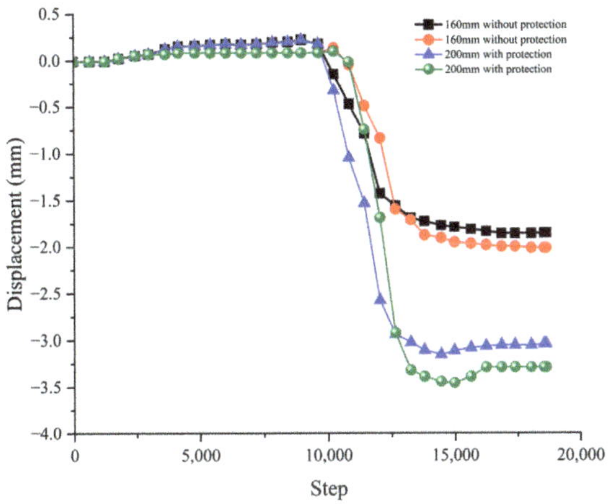

(**a**) Borehole left surrounding rock

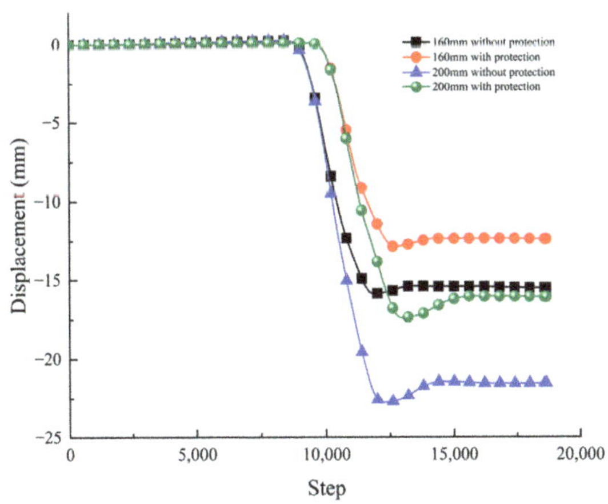

(**b**) Borehole upper surrounding rock

Figure 14. *Cont.*

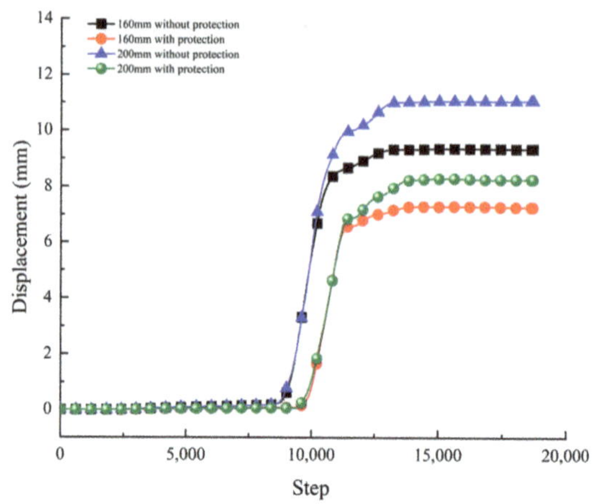

(c) Borehole under a surrounding rock

Figure 14. The variation curve of surrounding rock displacement of the borehole with time.

2.3.2. Analysis of the Plastic Failure

The distribution of plastic failure zones of the boreholes before and after protection is shown in Figure 15. From the plastic distribution diagram, it can be seen that the area of plastic damage of rock mass in 160 mm and 200 mm boreholes are obviously reduced after the application of borehole protection tubes; especially, the area of plastic damage of rock mass in the upper and lower sides of boreholes is the largest. The reduction of the plastic failure zone indicates that the failure instability of the rock mass around the borehole is alleviated, the degree of displacement and deformation is reduced, and the stability of the rock mass around the borehole is strengthened.

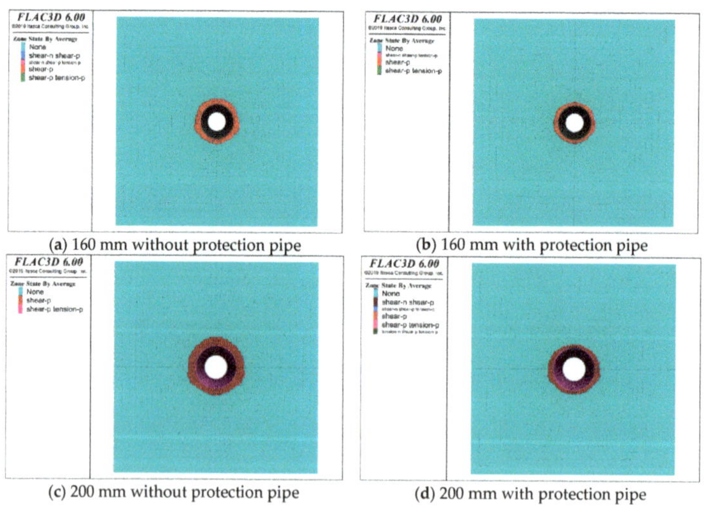

(a) 160 mm without protection pipe (b) 160 mm with protection pipe (c) 200 mm without protection pipe (d) 200 mm with protection pipe

Figure 15. Distribution of plastic failure of boreholes with different diameters. (**a**) The distribution plastic failure of the borehole of diameter of 100 mm without protection pipe; (**b**) the distribution plastic failure of the borehole of diameter of 160 mm with protection pipe; (**c**) the distribution plastic failure of the borehole of diameter of 200 mm without protection pipe; (**d**) the distribution plastic failure of the borehole of diameter of 200 mm with protection pipe.

Therefore, it is found that the stress concentration around the borehole is relieved according to the FLAC 3D simulation, the displacement of rock mass around the borehole is reduced, and the plastic damage is reduced, which indicates that the borehole protection tube can provide some protection support. However, for the large boreholes of 160 mm and 200 mm, the strong stress concentration and plastic damage obviously still exist after the protection, and the stability of the borehole needs to be further strengthened.

3. Experimental Study of the Internal Support Borehole Protection Pipe

According to the above analysis of borehole stability, the stability of the directional long borehole on the roof needs to be further improved. The borehole protection pipe is the main measure to change the stability of the borehole, but after using the conventional borehole protection pipe, there is still serious plastic damage and stress concentration phenomenon in directional long boreholes of large diameters. Therefore, the conventional pipe can only improve the stability of the directional long hole in the roof to a certain extent. Therefore, various internal supporting structures are designed to enhance the effect of hole protection, and their effects are studied through compression experiments.

3.1. Experimental Apparatus and Procedures

Three types of borehole protection pipes with internal support structures: "line-shaped", "Y-shaped", and "cross-shaped", are designed as shown in Figure 16. Two common borehole diameters of 100 mm and 160 mm are selected for experimental research.

(**a**) Line-shaped (**b**) Y-shaped (**c**) Cross-shaped

Figure 16. Three internal support structures of the borehole protection pipe.

This experiment is based on the GCTS Rock Mechanics Test System as shown in Figure 17. The system is a hydraulic servo-mechanical system produced by GCTS Company in the United States, which meets the requirements of the ISRM triaxial rock test of the International Society of Rock Mechanics and the American standard ASTM D2664-04. It is mainly used to test the mechanical properties and seepage characteristics of rock, concrete, and coal under complex loading conditions. The test accuracy is high and the performance is stable.

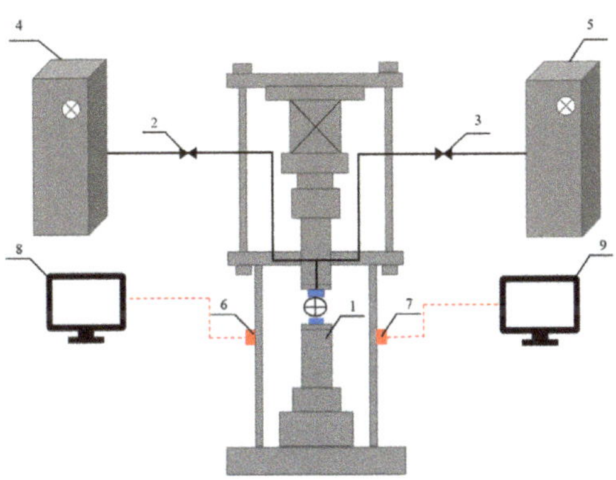

Figure 17. Schematic of the GCTS Rock Mechanics Test System (1 Compression experimental apparatus; 2/3 Controller valve; 4 Control cabinet of confining pressure volume; 5 Control cabinet of pore pressure volume; 6 Radical compressive force sensor; 7 Displacement sensor; 8/9 Storage computer).

The experimental steps of the GCTS rock mechanics test system are as follows:

(1) First, start up the test system and controller, adopt the feedback mode of "displacement control" and debug the system;
(2) Place the sample of the internal support borehole protection tube on the test bench and control the pressure head to move down to fix the sample;
(3) Set the load rate of 10 mm/min uniform speed for the compression test, and then stop after the radial compressive force reaches its peak.

3.2. Test Results

The compression test diagrams of the casing with different internal support structures are shown in Figure 18. All the internal support structures have obvious deformation during the compression process, and the deformation of the internal support structure reduces the deformation of the borehole protection tube so that it can provide good support to the borehole.

(**a**) Line-shaped (**b**) Y-shaped (**c**) Cross-shaped

Figure 18. Three internal support structures of the borehole protection pipe.

The variation of the radial compressive force of the borehole protection tube of different internal support structures of 100 mm and 160 mm is shown in Figures 19 and 20. The peak values of the radial compressive force of 100 mm different support structures are 5.59 kN, 21.03 kN, 11.79 kN, and 29.43 kN respectively. The peak values of the radial compressive force of different 160 mm support structures are 7.29 kN, 18.31 kN, 13.89 kN, and 31.30 kN respectively. The radial compressive force of different internal support structures of the two types of openings are "cross-shaped" > "line-shaped" > "Y-shaped" > Conventional pipe, and the peak compressive force of 160 mm borehole protection pipe is slightly greater than 100 mm.

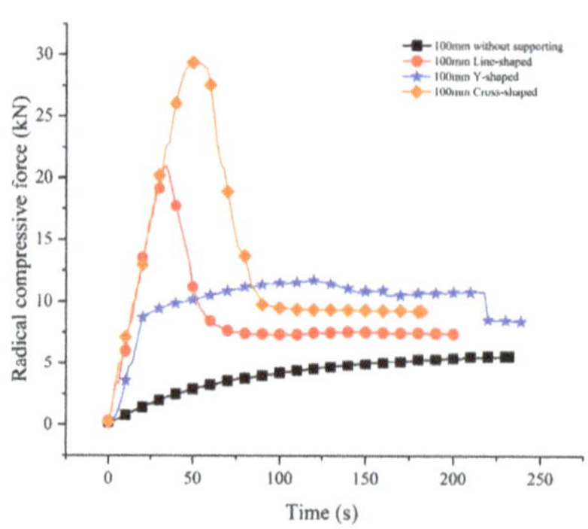

Figure 19. The radial compressive force of 100 mm borehole protection pipe with different internal support structures.

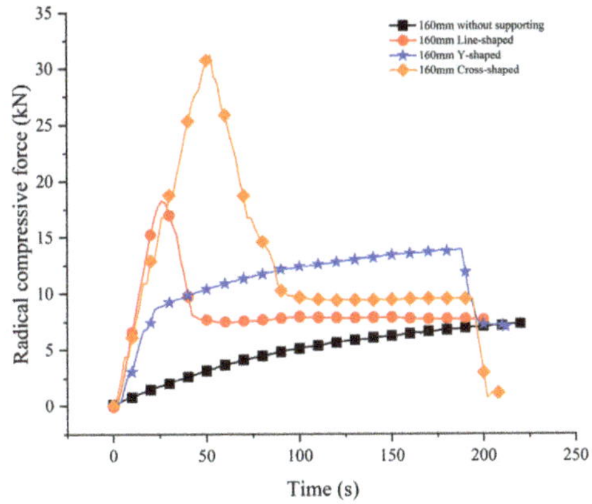

Figure 20. The radial compressive force of 160 mm borehole protection pipe with different internal support structures.

The displacement of different internal support structures in the radial compression test is shown in Figure 21. The displacement trends of 10 mm diameter and 160 mm diameter are basically the same, and the displacement of 160 mm of the same kind of internal support structure is slightly larger than 100 mm; smaller deformation of hole protection tubes are line-shaped and cross-shaped, and the cross-shaped is the smallest.

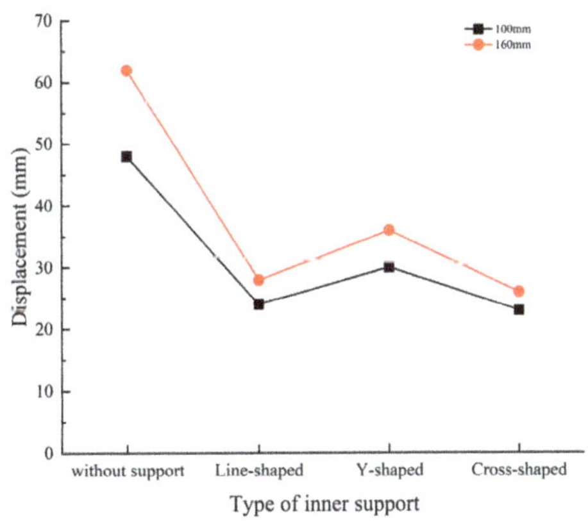

Figure 21. Displacement of borehole protection pipes with different internal support structures at peak compression force.

Therefore, according to the comprehensive analysis of the radial compressive force and deformation of the casing, it can be concluded that all three internal support structures can improve the compression resistance of the casing. Considering that the cross-shaped internal support structure can not only effectively support the vertical compression to resist compression, but also support the pipe wall in the horizontal direction to prevent deformation. Thus, the borehole protection effect of the cross-shaped internal support structure of the borehole protection pipe is the best.

4. Discussion

The directional long borehole on the roof has the characteristics of penetrating more layers and larger borehole diameter. Prior to drilling, the coal seam is in a state of equilibrium under the combined action of overburden pressure, horizontal pressure, and formation pore pressure. However, the stress balance of the original coal seam is disturbed by the formation of the well, the stress of the coal seam is redistributed, and the pore and fracture structure of the surrounding rock is changed. When the stress in a certain part of the surrounding rock of the borehole exceeds the maximum load that the coal rock can bear, the fractures will spread and penetrate rapidly, and the coal body will be squeezed into the borehole, resulting in instability and collapse of the borehole. From the numerical simulation results, it can be seen that both lithology and borehole diameter significantly influence the stability of roof-directional boreholes. On the one hand, borehole stability is significantly lower in coal and mudstone than in coarse and fine sandstone. Because the mechanical strength of coal and mudstone is lower than that of sandstone, the stress on the rock surrounding the borehole is different, and the radial deformation around the borehole is greater. When the radial deformation reaches the limit, the inner ring coal near the borehole wall fractures and the coal in the fractured area collapses and falls into the borehole. On the other hand, borehole diameter is inversely proportional to borehole stability, and when the borehole diameter exceeds 160 mm, the plastic failure and stress concentration of the borehole are severe and tend to be unstable. Therefore, to avoid instability and collapse of the borehole in soft rock such as coal and mudstone, it is essential to fully consider the rock mass distribution and the selected borehole diameter when designing and constructing directional long boreholes on the roof. Effective hole protection measures must be taken to prevent the failure of critical holes due to the oversized hole diameter.

According to the compression test results, the internal support structure can effectively improve the compression resistance of the conventional borehole protection pipe; especially, the cross-shaped internal support pipe has the best compression resistance which can provide strong support from two directions perpendicular to each other, and the internal support can bear part of the pressure of the borehole protection pipe to keep it in a relatively stable state. The internal supporting structure casing is suitable for roof-directional long holes. Improving the stability of the roof-directional long borehole by the internal support structure borehole protection pipe is beneficial to the efficient gas extraction in the goaf and can promote the popularization and application of the roof-directional long borehole technology. However, the compressive effect of the internal supporting casing cannot be matched with the field effect. Therefore, it is necessary to conduct systematic field experiments in various mines to further investigate borehole protection effects.

5. Conclusions

Borehole construction is the prerequisite for gas extraction in drilling. However, during the drilling construction, factors such as lateral pressure factor, the mechanical strength of rock, confining pressure, and so on are influenced, so that the drilling hole is easily deformed and collapsed, so that the drilling gas extraction performance is poor, and brings hidden hazards to the coal mine safety production. Research on the borehole collapsing law and corresponding borehole protection techniques has great practical significance for preventing borehole distortion and collapsing, increasing borehole stability, and improving gas drainage. Numerical simulations and laboratory experiments have been carried out to investigate the stability of directional long boreholes on the roof, and the main conclusions are as follows:

(1) Under different lithologies, the variation in borehole stress and displacement is significantly greater in coal and mudstone than in coarse and fine sandstone, indicating that borehole stability is strong in fine and coarse sandstone, but weak in coal and mudstone.
(2) With the increase of the borehole diameter, the stress, displacement, and plastic failure volume of the four borehole diameters of 100 mm, 130 mm, 160 mm, and 200 mm increase, that is, the stability of the borehole gradually decreases. The borehole tends to be unstable when the hole diameter is 160 mm and 200 mm.
(3) Stress, displacement, and plastic damage to the rock around the borehole are reduced after the conventional casing is installed. The results show that the conventional

tubing does provide some support to the borehole. However, severe plastic damage and stress concentration still exist when applied to large-diameter borehole protection, so the borehole protection effect is weak.

(4) Compared with the conventional wellbore protection pipe, the peak value of the compressive force of the wellbore protection pipe with line-shaped, Y-shaped, and cross-shaped internal support structures is significantly increased and the displacement is reduced. In particular, the peak value of the compressive force of the cross-shaped internal support structures is four times that of the conventional ones, and the displacement change is half that of the conventional ones. It indicates that the internal support structure can improve the borehole protection effect, and the cross-shaped has the best borehole protection effect.

Due to the limited computing power, the numerical simulation is mainly based on two-dimensional simulation, and many simplifications were made. A closer combination of drilling and mining should be considered for analysis in the future. Moreover, the production cost of the internal support hole protection pipe is higher than that of the common hole protection pipe. Further research should improve and reduce the cost, so as to facilitate the subsequent popularization and application.

Author Contributions: Methodology, Z.W.; conceptualization, X.Y.; formal analysis, X.Y., Z.W., and G.W.; investigation, X.Y. and H.G.; writing—original draft preparation, Z.W. and H.G.; writing—review and editing, X.Y., and H.G.; translation, Z.W.; funding acquisition, X.Y. and G.W. All authors have read and agreed to the published version of the manuscript.

Funding: This research was funded by the key control factors and influence mechanism of coal seam nitrogen injection critical breakthrough pressure of the National Natural Science Foundation of China (Grant No. 52204196) and Study on Cyclically Variable Flow Nitrogen Flushing Coal Seam Gas Technology and its Key Controlling Parameters of the National Natural Science Foundation of China (Grant No. 51974161).

Data Availability Statement: The data that support the findings of this study are available from the corresponding author upon reasonable request.

Conflicts of Interest: The authors declare that they have no known competing financial interests or personal relationships that could have appeared to influence the work reported in this paper.

References

1. Liu, Z.; Cheng, Y.; Jiang, J.; Li, W.; Jin, K. Interactions between coal seam gas drainage boreholes and the impact of such on borehole patterns. *J. Nat. Gas Sci. Eng.* **2017**, *38*, 597–607. [CrossRef]
2. Cheng, Z.; Pan, H.; Zou, Q.; Li, Z.; Chen, L.; Cao, J.; Zhang, K.; Cui, Y. Gas flow characteristics and optimization of gas drainage borehole layout in protective coal seam mining: A case study from the Shaqu coal mine, Shanxi province, China. *Nat. Resour. Res.* **2020**, *30*, 1481–1493. [CrossRef]
3. Lin, H.; Huang, M.; Li, S.; Zhang, C.; Cheng, L. Numerical simulation of influence of Langmuir adsorption constant on gas drainage radius of borehole in coal seam. *Int. J. Min. Sci. Technol.* **2016**, *26*, 377–382. [CrossRef]
4. Wei, P.; Huang, C.; Li, X.; Peng, S.; Lu, Y. Numerical simulation of boreholes for gas extraction and effective range of gas extraction in soft coal seams. *Energy Sci. Eng.* **2019**, *7*, 1632–1648. [CrossRef]
5. Zhang, X.; Gao, J.; Jia, G.; Zhang, J. Study on the influence mechanism of air leakage on gas extraction in extraction boreholes. *Energy Explor. Exploit.* **2022**, *40*, 1344–1359. [CrossRef]
6. Wang, G.; Fan, C.; Xu, H.; Liu, X.; Wang, R. Determination of long horizontal borehole height on the roofs and its application to gas drainage. *Energies* **2018**, *11*, 2647. [CrossRef]
7. Li, H.; Liu, Y.; Wang, W.; Liu, M.; Ma, J.; Guo, X.; Guo, H. The integrated drainage technique of directional high-level borehole of super large diameter on the roof replacing roof extraction roadway: A case study of the underground Zhaozhuang Coal Mine. *Energy Rep.* **2020**, *6*, 2651–2666. [CrossRef]
8. Wang, J. Research and application of pressure relief gas drainage technology by long borehole along large diameter from roof fracture zone. *IOP Conf. Ser. Earth Environ. Sci.* **2019**, *358*, 032011. [CrossRef]
9. Duan, H.; Wang, Y.; Xiao, Q.; Wang, J.; Peng, D. Gas extraction technology and application of near horizontal high directional borehole. *Energy Rep.* **2022**, *8*, 1326–1333. [CrossRef]
10. Li, T.; Wu, B.; Lei, B. Study on the optimization of a gas drainage borehole drainage horizon based on the evolution characteristics of mining fracture. *Energies* **2019**, *12*, 4499. [CrossRef]

11. Yao, X.; Cheng, G.; Shi, B. Analysis on gas extraction borehole instability and control method of pore-forming in deep surrounding-rock with weak structure. *J. China Coal Soc.* **2010**, *35*, 2073–2081. [CrossRef]
12. Karatela, E.; Taheri, A. Three-dimensional hydro-mechanical model of borehole in fractured rock mass using discrete element method. *J. Nat. Gas Sci. Eng.* **2018**, *53*, 263–275. [CrossRef]
13. Yurum, Y.; Bozkurt, D.; Yalcin, M.N. Change of the structure of coals from the Kozlu K20 G borehole of Zonguldak basin with burial depth 2. macromolecular structure. *Energy Sources* **2001**, *23*, 521–527. [CrossRef]
14. Zhang, N.; Xue, F.; Zhang, N.; Feng, X. Patterns and security technologies for co-extraction of coal and gas in deep mines without entry pillars. *Int. J. Coal Sci. Technol.* **2015**, *2*, 66–75. [CrossRef]
15. Yang, C.; Liu, J. Petroleum rock mechanics: An area worthy of focus in geo-energy research. *Adv. Geo-Energy Res.* **2021**, *5*, 351–352. [CrossRef]
16. Ding, L.; Wang, Z.; Liu, B.; Lv, J.; Wang, Y. Borehole stability analysis: A new model considering the effects of anisotropic permeability in bedding formation based on poroelastic theory. *J. Nat. Gas Sci. Eng.* **2019**, *69*, 102932. [CrossRef]
17. Zhao, H.; Li, J.; Liu, Y.; Wang, Y.; Wang, T.; Cheng, H. Experimental and measured research on three-dimensional deformation law of gas drainage borehole in coal seam. *Int. J. Min. Sci. Technol.* **2020**, *30*, 397–403. [CrossRef]
18. Katanov, Y.; Vaganov, Y.; Cheymetov, M. Neural simulation-based analysis of the well wall stability while productive seam penetrating. *Min. Miner. Depos.* **2021**, *15*, 91–98. [CrossRef]
19. Ma, Y.; Xu, Y. Research into technology for precision directional drilling of gas-drainage boreholes. *Min. Miner. Depos.* **2022**, *16*, 27–32. [CrossRef]
20. Dychkovskyi, R.O.; Lozynskyi, V.H.; Saik, P.B.; Petlovanyi, M.V.; Malanchuk, Y.Z.; Malanchuk, Z.R. Modeling of the disjunctive geological fault influence on the exploitation wells stability during underground coal gasification. *Arch. Civ. Mech. Eng.* **2018**, *18*, 1183–1197. [CrossRef]
21. Petlovanyi, M.; Lozynskyi, V.; Saik, P.; Sai, K. Predicting the producing well stability in the place of its curving at the underground coal seams gasification. *E3S Web Conf.* **2019**, *123*, 01019. [CrossRef]
22. Zhang, X.; Wang, W.; Yang, M. Study on deformation and destabilization characteristics and modes of drainage borehole. *Energy Sources* **2020**, *42*, 2448–2459. [CrossRef]
23. Qu, P.; Shen, R.; Fu, L.; Wang, Z. Time delay effect due to pore pressure changes and existence of cleats on borehole stability in coal seam. *Int. J. Coal Geol.* **2011**, *85*, 212–218. [CrossRef]
24. Niu, Y.; Song, X.; Li, Z.; Wang, E.; Liu, Q.; Zhang, X.; Cai, G.; Zhang, Q. Experimental study and field verification of stability monitoring of gas drainage borehole in mining coal seam. *J. Pet. Sci. Eng.* **2020**, *189*, 106985. [CrossRef]
25. Niu, Y.; Zhang, X.; Wang, E.; Li, Z.; Cheng, Z.; Duan, X.; Li, H.; Wei, Y.; Qian, J.; Cai, G.; et al. A new method of monitoring the stability of boreholes for methane drainage from coal seams. *Measurement* **2020**, *154*, 107521. [CrossRef]
26. Zhang, Q.; Wang, E.; Li, Z.; Wang, H.; Xue, Z. Control of directional long borehole on gas drainage and optimal design: Case study. *J. Nat. Gas Sci. Eng.* **2022**, *107*, 104766. [CrossRef]
27. Xu, B.L.; Liu, S.Y.; Li, H.S. Drill string's axial force transfer law in slide directional drilling in underground coal mine. *Tunn. Undergr. Space Technol.* **2022**, *130*, 104701. [CrossRef]
28. Liu, J.; Wu, Z.; Lu, P.; Liu, Z.; Su, M. Study on Effective Extraction Radius of Directional Long Borehole and Analysis of the Influence Mechanism. *ACS Omega* **2023**, *8*, 2344–2356. [CrossRef]
29. Yuan, S.; Han, G. Combined Drilling Methods to Install Grout Curtains in a Deep Underground Mine: A Case Study in Southwest China. *Mine Water Environ.* **2020**, *39*, 902–909. [CrossRef]
30. Yuan, S.; Sun, B.; Han, G.; Duan, W.; Wang, Z. Application and Prospect of Curtain Grouting Technology in Mine Water Safety Management in China: A Review. *Water* **2022**, *14*, 4093. [CrossRef]
31. Wang, H.; Kou, Z.; Guo, J.; Chen, Z. A semi-analytical model for the transient pressure behaviors of a multiple fractured well in a coal seam gas reservoir. *J. Pet. Sci. Eng.* **2020**, *198*, 108159. [CrossRef]
32. Wang, S.; Wang, S. Longwall mining automation horizon control: Coal seam gradient identification using piecewise linear fitting. *Int. J. Min. Sci. Technol.* **2022**, *32*, 821–829. [CrossRef]
33. Xue, F.; Zhang, N.; Feng, X.; Zheng, X.; Kan, J. Strengthening borehole configuration from the retaining roadway for greenhouse gas reduction: A case study. *PLoS ONE* **2015**, *10*, e0115874. [CrossRef] [PubMed]
34. Zhai, C.; Xu, Y.; Xiang, X.; Yu, X.; Zou, Q.; Zhong, C. A novel active prevention technology for borehole instability under the influence of mining activities. *J. Nat. Gas Sci. Eng.* **2015**, *27*, 1585–1596. [CrossRef]
35. Di, C.; Li, Q.; Sun, C.; Ni, G.H.; Yang, W. Analysis on borehole instability and control method of poreforming of hydraulic fracturing in soft coal seam. *J. China Coal Soc.* **2012**, *37*, 1431–1436. [CrossRef]
36. Qi, Q.; Jia, X.; Zhou, X.; Zhao, Y. Instability-negative pressure loss model of gas drainage borehole and prevention technique: A case study. *PLoS ONE* **2020**, *15*, e0242719. [CrossRef]
37. Li, H.; Wang, W.; Liu, Y.; Ma, J.; Gao, H. An integrated borehole, protection and sealing technology for improving the gas drainage effect in soft coal seams. *Energy Rep.* **2020**, *6*, 2030–2043. [CrossRef]
38. Xu, H.; Fang, Z.; Sang, S.; Liu, H.; Ding, H.; Dou, X.; Liu, Q. Analysis of the shear failure of surface methane capture boreholes for improving the drainage period efficiency: A lithological perspective. *Energy Explor. Exploit.* **2019**, *38*, 92–100. [CrossRef]

39. Meier, T.; Rybacki, E.; Reinicke, A.; Dresen, G. Influence of borehole diameter on the formation of borehole breakouts in black shale. *Int. J. Rock Mech. Min. Sci.* **2013**, *62*, 74–85. [CrossRef]
40. Ji, Q. Research and application of auger-air borehole and sieve pipe borehole protection in soft outburst-prone coal seams. *Procedia Eng.* **2014**, *73*, 283–288. [CrossRef]

processes

Article

Integrated Geophysical Prediction of Goaf and Water Accumulation in Pingshuo Dong Open-Cut Mine with Ultrashallow and High Drops

Sheng Zhang [1,2,*], Wenyuan Guan [1], Suoliang Chang [1], Qinggang Meng [3], Yinping Dong [4] and Qiang Chen [1]

1 Department of Earth Science and Engineering, Taiyuan University of Technology, Taiyuan 030024, China
2 Shanxi Institute of Geological Survey, Taiyuan 030006, China
3 No.2 Oil Production Plant, Xinjiang Oilfield Company, PetroChina, Karamay 834008, China
4 Coal Geological Geophysical Exploration Surveying & Mapping Institute of Shanxi Province, Jinzhong 030600, China
* Correspondence: zhangsheng@tyut.edu.cn

Abstract: To address safety problems caused by goaf and water accumulation in open-pit mines, the shallow three-dimensional seismic method and transient electromagnetic method were integrated and applied to detect the mining goaf distribution scope and the water accumulation conditions. In view of the special topographic conditions of an ultrashallow layer and high drop in an open-pit mine, we proposed utilizing bin homogenization and multidomain joint denoising to improve the reliability of seismic data. By using seismic-sensitive attributes to predict the goaf, the transient electromagnetic method was employed to further predict the water accumulation in the goaf. The results show that the shallow seismic method could clearly reflect the reflected-wave variation features of the goaf. The features of a junction between a normal seam and goaf vary obviously, and the prediction effect of the goaf boundary with high resolution is in line with the actual situation. Furthermore, taking the goaf scope prediction with the shallow three-dimensional seismic method as a base, targeted transient electromagnetic detection was deployed, with a detailed analysis of the survey-line repeated areas of the shallow seismic and transient electromagnetic methods. Making full use of the advantages of the shallow seismic and transient electromagnetic methods, we propose a reasonable data interpretation method in combination with the special topographic conditions of open-pit mines, which greatly improves the prediction effect of goaf and water accumulation conditions.

Keywords: integrated geophysical prospecting; goaf prediction; water accumulation; open-cut mine

Citation: Zhang, S.; Guan, W.; Chang, S.; Meng, Q.; Dong, Y.; Chen, Q. Integrated Geophysical Prediction of Goaf and Water Accumulation in Pingshuo Dong Open-Cut Mine with Ultrashallow and High Drops. *Processes* **2023**, *11*, 1653. https://doi.org/10.3390/pr11061653

Academic Editors: Feng Du, Aitao Zhou and Bo Li

Received: 16 March 2023
Revised: 15 May 2023
Accepted: 19 May 2023
Published: 29 May 2023

1. Introduction

Since the founding of New China in 1949, nationwide construction has spurred demand for coal and the exploitation of coal resources. Currently, coal production in China accounts for almost half of the world's total production [1]. Since 2005, to improve the efficient utilization and safe mining of coal resources, China has carried out the effective integration of coal resources [2]. However, most coal mines with integrated resources exhibit problems, such as a serious lack of data on the old goaf, unclear distribution of the boundary area, unclear water area and quantity in the goaf, and difficulty carrying out geophysical prospecting work underground in small closed-pit coal mines, which have exposed serious hidden dangers to coal mine safety. The Pingshuo Dong open-cast coal mine is a typical example. Due to long-term disorderly mining, no-filling mining, rich mining, and poor abandonment of surrounding small coal mines, many underground goafs exist in the Pingshuo Dong open-cut coal mine, and most of these goafs have no recorded specific information, such as location and size. If the roof thickness of the goaf is less than the minimum safe thickness, a roof collapse can occur, which can cause surface subsidence or cracking, seriously threatening the safety of personnel and equipment working above the roof [3,4].

Geophysical prospecting methods commonly used in goaf and water accumulation include the shallow seismic method [5,6], transient electromagnetic method [7–9], high-density electrical method [10,11], magnetotelluric method [12,13], and ground-penetrating radar [14,15]. Various geophysical methods have achieved certain effects in detecting goaf and water accumulation, but each geophysical method has its own applicable conditions and advantages. The shallow seismic exploration method allows continuous measurement, and the range of goaf detected is more accurate and reliable, but it can hardly reflect the accumulated water of the goaf area [16]. The transient electromagnetic method is especially sensitive to low resistance and has a good effect in detecting goafs with accumulated water, and the detection depth is relatively large, but it is greatly affected by electromagnetic interference. The high-density resistivity method has a better detection effect on goaf water accumulation, but it is affected by topography and grounding resistance [17]. The magnetotelluric method is effective at detecting water accumulation in goafs, and the detection depth is large, but it can be affected by static effects and near-field effects. Because a single geophysical method is limited, the geophysical interpretation often exhibits multiple solutions and low reliability [18,19]. In recent years, many studies and applications have been carried out for integrated geophysical methods [20–23]. Integrated geophysical prospecting can obtain more physical information, verify and supplement other methods, reduce the multiple solutions of a single geophysical prospecting method, and improve the resolution of exploration and the reliability of interpretation results. However, for open-pit mining areas, the terrain conditions are more complicated, the coal seams are very shallow, and there are high-drop steps (transport lines). For goaf distribution and water accumulation prediction, the high-density electric method cannot meet the condition of the line layout, the ground-penetrating radar detection depth is insufficient, and the shallow seismic acquisition coverage is uneven. How to achieve higher resolution and better exploration results has not been systematically discussed in depth. This paper takes the prediction of the goaf distribution and water accumulation of the Pingshuo Dong open-cast coal mine as an example. Aiming at this type of problem, this paper proposes combining small-array high-density acquisition, multidomain joint denoising, and bin homogenization, weakening the influence of high-drop steps on shallow seismic exploration, and then using the multiple-attribute method to predict the goaf. Based on the electromagnetic environment analysis and topographical features of the study area, the transient electromagnetic method was used to identify goafs. The practical application shows that the high-density 3D seismic and transient electromagnetic comprehensive interpretation can effectively identify the goaf and water accumulation characteristics of an ultrashallow high-altitude open-pit mine. This proves the feasibility and application of the proposed method and can provide a meaningful reference for research in similar areas.

2. Geological Background and Engineering Geological Conditions

The study area is located in the northeastern margin of the Ningwu coalfield in Shanxi, China. Most of the surface is covered by Cenozoic strata, which are typical loess hilly landforms. There are sporadic strata bedrock outcrops near the coal-seam outcrops in the northeastern part of this area and at the bottoms of major gullies in this area. The study area is located on the southwest side of the eastern open-cast coal mine and its surrounding area, and the exploration area is approximately 0.35 km^2, the scope is delineated by a red circled. as shown in Figure 1.

The Pingshuo Dong open-cast coal mine is located in the low hills of the Pingshuo Platform in the Shanxi Loess Plateau. The whole area is mostly covered with loess. The loess plateau in this area has been strongly eroded and cut. The valley is relatively developed in a "V" shape with a cutting depth of 40–70 m. The highest point in the area is located in the northern part of Shifeng, with an altitude of 1537.99 m. The lowest point is located at the Maguan River in western Xuegaodeng village, with an altitude of 1213.8 m, and the maximum relative height difference is 324.19 m. The ground surface in the study area has been stripped, showing a step-like shape, with an elevation of 1190–1380 m, a minimum of 1190 m in the northeast corner, a height of 1380 m in the south, and a relative height difference of 190 m. Some areas are the main roads in the southern part of the mine, and the traffic is more intensive, as shown in Figure 2.

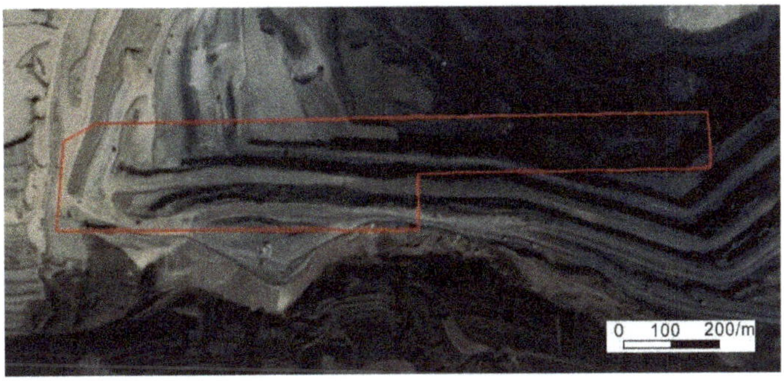

Figure 1. Topographic map of the Pingshuo Dong open-cast coal mine.

Figure 2. Topographical photos of the study area, (**a**) high drop, (**b**) narrow platform, (**c**) high voltage line, (**d**) complex noise environment.

3. Methods

3.1. Attribute-Based Predicted Multiparameter Analysis for Goaf Prediction

The geophysical characteristics of the goaf and its upper strata have changed compared with those before mining, including differences in density, velocity and other elastic parameters, which provide the physical basis for detecting goafs by the seismic method [24]. A large amount of data proves that the existence of goafs changes the characteristics of coal-seam reflection waves, which is manifested in the distortion, dislocation, and even disappearance of coal-seam reflection-wave groups, the reduction in seismic amplitude, frequency, phase reversal, and waveform changes. These characteristics are the main signs for identifying goafs. Currently, seismic attributes are among the most effective ways to identify goafs.

The extraction of different seismic attributes is essentially the selection of different attribute parameters. Different geological characteristics correspond to different physical parameters. Therefore, selecting reasonable seismic attributes in a wide range of seismic attributes depends on the characteristics of geological bodies or geological phenomena. To improve the accuracy of identifying goafs, we analyzed the geological characteristics of goaf development and its seismic response and selected several layer attributes that are extremely sensitive to goafs and small faults from more than 20 attributes, including average amplitude energy, instantaneous frequency, dominant frequency, instantaneous phase, and waveform-clustering attributes.

Vakman compared various methods for extracting instantaneous parameters of signals and concluded that the analytical signal method is the most reasonable [25]. The analytical signal method uses the Hilbert transform to obtain the corresponding analytical signal to form a complex signal and then extracts the instantaneous parameters of the signal.

Assuming that the real signal is $x(t)$, the corresponding imaginary part can be obtained by the Hilbert transform [26]:

$$y(t) = HT[x(t)] = \frac{1}{\pi}\int_{-\infty}^{\infty}\frac{x(\tau)}{t-\tau}d\tau \tag{1}$$

$HT[\bullet]$ represents the Hilbert transform. Then, we can construct an analytic signal $s(t)$ by $x(t)$ and $y(t)$:

$$s(t) = x(t) + iy(t) = A(t)e^{i\theta(t)} \tag{2}$$

According to the analytic signal, the instantaneous parameters of the original signal $x(t)$ can be extracted, including

$$A(t) = \sqrt{x^2(t) + y^2(t)} \tag{3}$$

$$phi(t) = \tan^{-1}[y(t)/x(t)] \tag{4}$$

$$\omega(t) = \frac{1}{2\pi}\frac{dphi(t)}{dt} = \frac{1}{2\pi}\frac{x(t)y'(t) - y(t)x'(t)}{x^2(t) + y^2(t)} \tag{5}$$

where $A(t)$, $\phi(t)$, and $\omega(t)$ represent the instantaneous amplitude, instantaneous phase, and instantaneous frequency, respectively.

The instantaneous amplitude is a measure of the reflective intensity, which is the square root of the total energy of the real and imaginary parts of the analytical signal. It mainly reflects the change in energy and can highlight the change in special rock strata. The instantaneous phase is a measure of the continuity of seismic events in seismic profiles, and its phase can be displayed regardless of the intensity of energy, even if the weak-amplitude effective wave can be well displayed on the instantaneous phase diagram. When the wave propagates in an anisotropic homogeneous medium, its phase is continuous when the wave is in an abnormal medium. Its phase will change significantly in the abnormal position, and it is obviously discontinuous in the profile. The instantaneous frequency is the rate of time variation of the phase, which can reflect the lithological changes in the stratum and help to identify the stratum.

For the same detection object, the obvious change in three instantaneous pieces of information in the same location may reflect a change in the physical properties of the detection object. Among the three parameters, the instantaneous phase spectrum has the highest resolution. The change in the instantaneous frequency and amplitude spectrum is also more intuitive and can be used to determine the general location of underground anomalies and the instantaneous phase to determine the stratified boundary.

The C3 algorithm is a coherence algorithm based on feature structure analysis [27]. The calculation of coherence values in the algorithm can use multichannel seismic records.

When calculating the coherence value at the sampling point (x, y, t), first define a data body containing the point (controlled by a two-dimensional spatial window and a one-dimensional time window), assuming that the time window contains J channels. For the seismic record, the number of sampling points in the time window is 2 N + 1; let $d_n^T(x, y, t) = (d_{n1}, d_{n2}, \cdots, d_{nJ}), n = -N, -N+1, \cdots, N-1, N$. The seismic trace in the data body can be represented as matrix.

$$\mathbf{D}(x, y, t) = [d_{-N}, d_{-N+1}, \cdots, d_N]^T \tag{6}$$

The correlation matrix is calculated as follows:

$$\mathbf{C}(x, y, t) = \mathbf{D}^T\mathbf{D} = \sum_{n=-N}^{n=N} d_n d_n^T = \begin{pmatrix} \sum_{n=-N}^{n=N} d_{n1}^2 & \sum_{n=-N}^{n=N} d_{n1}d_{n2} & \cdots & \sum_{n=-N}^{n=N} d_{n1}d_{nJ} \\ \sum_{n=-N}^{n=N} d_{n2}d_{n1} & \sum_{n=-N}^{n=N} d_{n2}^2 & \cdots & \sum_{n=-N}^{n=N} d_{n2}d_{nJ} \\ \vdots & \vdots & \ddots & \vdots \\ \sum_{n=-N}^{n=N} d_{nJ}d_{n1} & \sum_{n=-N}^{n=N} d_{nJ}d_{n2} & \cdots & \sum_{n=-N}^{n=N} d_{nJ}^2 \end{pmatrix} \tag{7}$$

The coherence value E_c (x, y, t) at point (x, y, t) is defined as

$$E_c(x, y, t) = \frac{\lambda_{\max}}{Tr(C)} = \frac{\lambda_{\max}}{\sum_{j=1}^{J} C_{jj}} = \frac{\lambda_{\max}}{\sum_{j=1}^{J} \lambda_j} \tag{8}$$

where λ_{max} is the maximum value in λ_j.

It is easy to know from Equation (7) that the correlation matrix is positive definite or semipositive definite, so the value of E_c falls in the range of 1/J–J. When the waveforms of different traces are the same, E_c is equal to 1; when the waveforms are different, E_c is less than 1, and the greater the difference between waveforms, the smaller the value of Ec. The difference between waveforms can reflect the changes in geological bodies, so E_c can depict the discontinuities of geological bodies.

3.2. TEM-Based Apparent Resistivity Prediction for Water Accumulation Identification

The transient electromagnetic method (TEM) is a geophysical method that uses the principle of electromagnetic induction [28,29]. It emits pulsed electromagnetic waves as an excitation field source (called 'primary field') through an ungrounded return line or ground-line-source electric dipole. It also observes and studies the spatial distribution and time characteristics of the electromagnetic field (called 'secondary field') of the stratum or geological target body (underground goaf) generated by the excitation field (called 'primary field'), which can be used to infer and explain the spatial structure and physical characteristics of the stratum or goaf water accumulation.

Transient electromagnetic measurements are carried out by using the secondary eddy-current field excited by a step-pulse current in a large fixed source loop [30,31]. A rectangular emitter loop is placed on the surface of a homogeneous isotropic ground with magnetic conductivity and conductivity. The area of the loop is S, and the inner loop provides a step-pulse current. Before disconnecting the current, the emission current produces a stable magnetic field in the space around the loop and in the ground, as shown in Figure 3a.

When the current is turned off, the magnetic field generated by the emission current also disappears immediately. This change is quickly transmitted to the ground around the return line in the primary field with the aid of air and underground conductive medium and induces a current in the earth to maintain the magnetic field before the emission current is disconnected, so that the magnetic field of the space does not immediately disappear. However, because of the ohmic loss of the medium, the magnetic field generated by it will decay rapidly with the decay of the induced current. The rapid decay of the magnetic field will induce a new weaker eddy current in the surrounding medium. This process continues until the earth's ohmic losses have exhausted the magnetic field energy (see Figure 3b), forming a geodetic electromagnetic field as the target of detection.

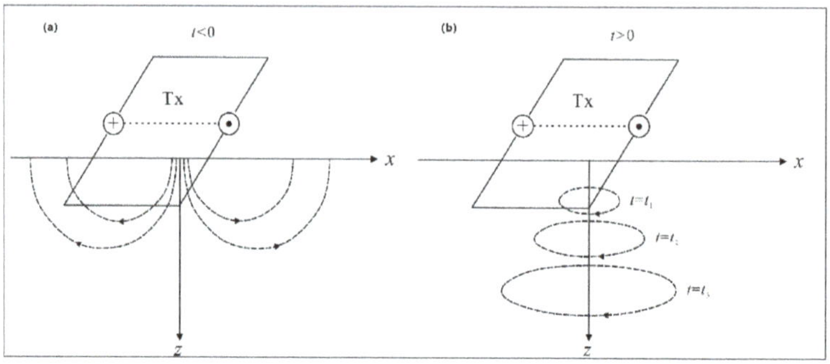

Figure 3. Rectangular loop lines and equivalent eddy-current ring, (**a**) establishment of stable magnetic field; (**b**) attenuation of magnetic field (Tx: transmission line frame. t: decay time).

4. Application and Discussion

The high-density three-dimensional seismic and transient electromagnetic combined detection method is employed to identify the goafs in the detection area. Three-dimensional seismic design has 5843 survey points in the whole area, including 2399 shot points and 3444 detection points. The exploration area is 0.25 km^2, and the coverage area of 24 times is 0.2 km^2. The normal distance of the transient electromagnetic method is 26~35 m. The engineering survey line is arranged along the design roadway and its surroundings. The point distance of key areas above the roadway is 10 m, some areas are densified to 5 m, and the point distance of other areas is 20 m. The survey lines are arranged along the east–west axis, and a total of six survey lines are arranged. Two survey lines are arranged above the No. 4 coal roadway, numbered N0 and N35 from south to north. Four survey lines are arranged along the No. 9 coal roadway, numbered l0, L33, l59, and L85 from south to north, and the survey points are numbered 0, 20, and 40 from west to east according to the point distance, as shown in Figure 4.

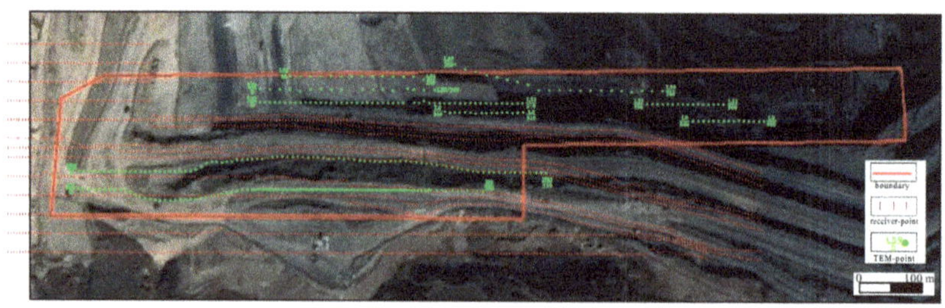

Figure 4. Layout of 3D seismic and transient electrical detection.

The buried depth of Coal Seam 4 is 0–150 m, with an average of 70 m. The 9th coal seam has a buried depth of 40–210 m, with an average of about 110 m. The extremely shallow burial depth of the target layer and the small survey area led to the limited length of survey lines. Therefore, seismic exploration adopts the method of small array and high-density acquisition, with high shot-point density and high coverage times. After the field collection test, the shot interval has changed from 24 m to 6 m, and the coverage times have changed from 24 to 64. Meanwhile, due to the restriction of the field acquisition and construction conditions (vehicle walking platform), the distribution of shot and detection points is extremely uneven (Figure 5a), resulting in extremely uneven coverage of the study area (Figure 5b).

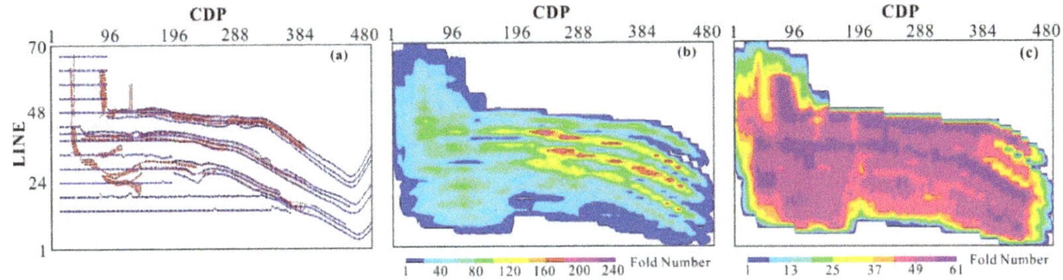

Figure 5. The distribution of shot and receiver points (**a**), the actual coverage times (**b**), and coverage times after bin homogenization (**c**) in the study area.

Due to the limitation of terrain conditions, such as large drop and steep slopes between the platforms, the original coverage times are extremely uneven. Therefore, the surface homogenization method is used to eliminate the adverse effects of uneven coverage, including static bin homogenization and dynamic bin homogenization. Static bin homogenization is used to make only one track in each offset group in a bin. If there are N tracks in an offset group in this bin, the N tracks in the offset are weighted and superimposed to represent the offset group; if there are N tracks in an offset group in the bin, then the N tracks are weighted and superimposed, and the superimposed track represents the offset group. Dynamic surface equalization is mainly used to find the offset group in the adjacent element. Define a large rectangular bin centered on the current bin. Any missing offset set in the current bin searches for the corresponding offset set in the adjacent bin within a circle, where the radius of the circle is a linear function of the offset set and its range is limited to the large bin.

During the processing, the seismic trace is edited on the single-shot record to eliminate the strong interference channel and the invalid redundant channel. On this basis, dynamic panel homogenization is used to supplement the missing offset set, and static panel homogenization is used to make the coverage times and energy more balanced (Figure 5c).

Figure 6 shows the CDP gathers before and after bin homogenization, where Figure 6a,b show the processing effects of bin-homogenizing the number of insufficient superpositions traced to the average number of superpositions. Figure 6c, d show the processing effects of bin-homogenizing the number of excessive superpositions traced to the average number of superpositions. From the processing effects, we find that the basic shape of the seismic profile has not changed. The overall energy of the seismic section is more balanced.

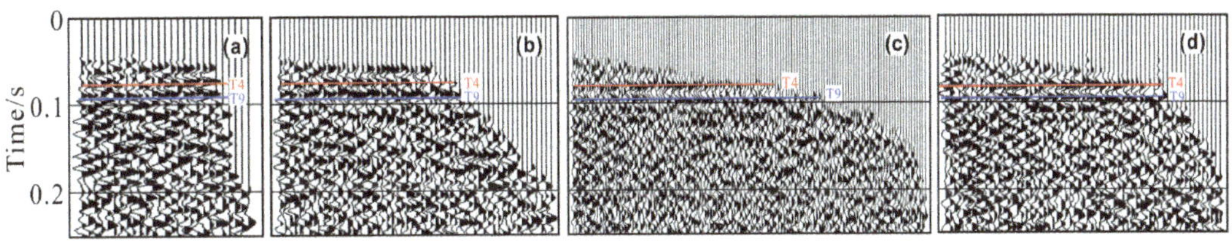

Figure 6. Comparison of stacked gathers before and after bin homogenization, (**a**) before bin homogenization, (**b**) after bin homogenization, (**c**) before bin homogenization, (**d**) after bin homogenization.

Figure 7 shows the comparison of stacked gathers before and after bin homogenization. We can determine that the energy of the missing trace is recovered (at the red elliptical mark), and the amplitude of the homogenized bin can reflect the real situation of the underground media.

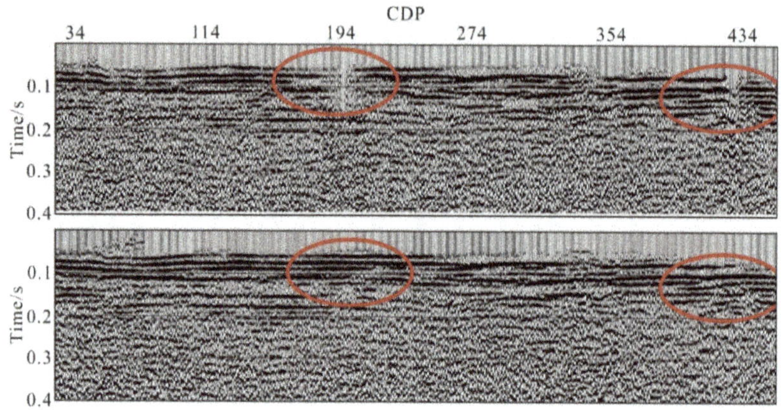

Figure 7. Comparison of stacked gathers before and after bin homogenization.

Due to the influence of engineering construction, noise has a great influence on shallow seismic exploration. Suppressing noise is very important for the accurate identification of goafs. Data analysis shows that the main noise sources include linear interference, surface waves, and abnormally strong amplitudes. In this study, a multidomain joint denoising method is employed through experiments to classify and eliminate different types of noise. This follows the law of progressive, step-by-step denoising and gradually improves the signal-to-noise ratio. For linear interference, the apparent speed and frequency range of linear interference is analyzed, and the noise attenuation method with less waveform modification is adopted to eliminate the noise. For the outliers and abnormal amplitudes, the regional anomalous amplitude attenuation method and the frequency-division abnormal amplitude attenuation method are used for suppression. The strategy of multiple-trace statistics, single-trace denoising, and frequency-division suppression is adopted to suppress random abnormal amplitude noise and surface waves. Strong energy interference in seismic records is identified automatically in different frequency bands, and the spatial position of noise is determined. According to the defined threshold value and attenuation coefficient, time-varying and space-varying methods are adopted to suppress random abnormal amplitude noise. Figure 8 shows the comparison of single-shot records before and after suppressing noise. Within the scope of the red rectangular box, we can determine that the surface waves and other noises are effectively suppressed, and the energy of the primary reflection wave is obviously enhanced. Figure 9 is the comparison of stacked gathers before and after suppressing noise. Within the scope of the red elliptical frame, it can be seen that the weak waveform after noise suppression is displayed, which can more effectively reflect the subtle changes in underground media.

This study identifies goaf and water accumulation characteristics by three steps: seismic forward modeling assistant interpretation, goaf seismic-wave-anomaly characteristics interpretation, and seismic- and electrical-method comprehensive-anomaly interpretation. First, a goaf geological model suitable for the exploration area is established, and wave forward modeling is carried out to study the seismic response characteristics of various goafs (Figure 10) and assist in anomaly interpretation.

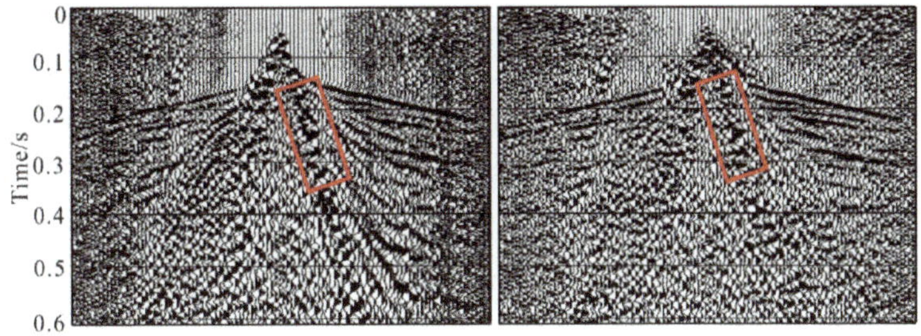

Figure 8. Comparison of single-shot records before (**left**) and after (**right**) noise suppression.

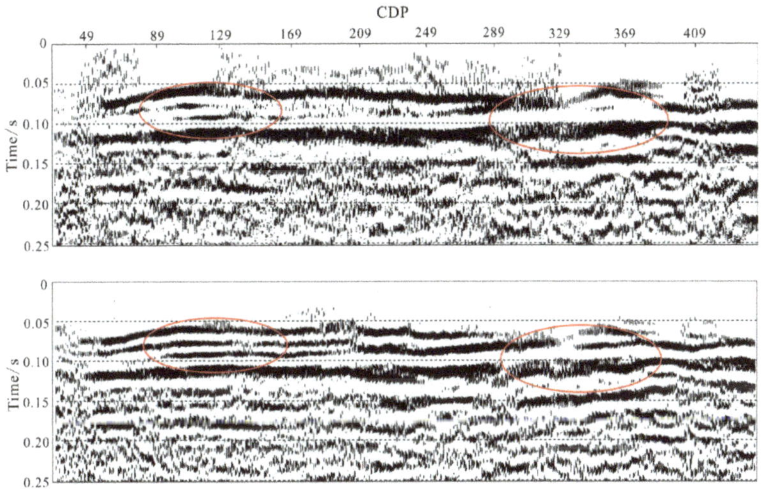

Figure 9. Comparison of stacked gathers before (**upper**) and after (**lower**) suppressing noise.

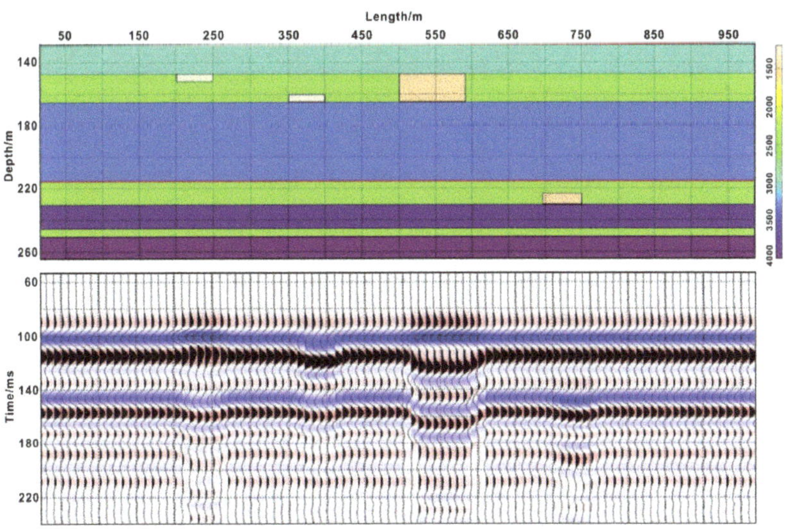

Figure 10. Forward modeling and seismic profile of goaf of different types and in different positions.

Due to the air or water content in the goaf, the absorption coefficient of the goaf is significantly higher than that of the normal coal seam, resulting in no reflection or weak reflection at the bottom interface of the goaf and substratum [32]. There is a fault in the wave group; the reflected wave is characterized by energy enhancement, frequency reduction, and an obvious concave phase (Figure 11a), or the energy becomes weaker and the continuity becomes worse (Figure 11b).

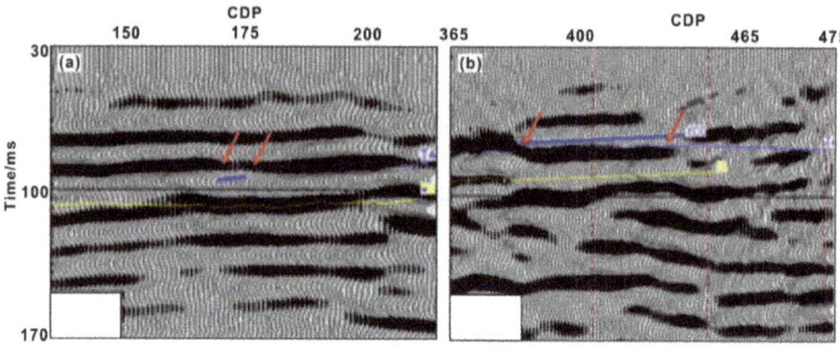

Figure 11. Seismic response of goaf in actual seismic profiles. (**a**) Increased energy and low frequency, and (**b**) weak energy and worse continuity.

Comparing the reflected waves, seismic attributes, and drilling verification results of coal seams in the study area, a total of four coal-seam goafs are explained. The goaf labeled ① is located at the northern boundary of the exploration area, and the goaf anomaly is irregular. Combined with the interpretation results of the northern transient electromagnetic method, the goaf is produced by mining from north to south in the Yuling coal mine, and the abnormal area in the seismic work area is approximately 1359 m^2.

The goaf marked with ② is located outside the southeast exploration boundary, in front of the Xiaoxiyao mine roadway, and the boundary of the abnormal area is irregular. The goaf is caused by mining from east to west in the Xiaoxiyao mine, and the abnormal area in the area is approximately 7352 m^2.

The goaf marked with ③ and ④ is located southeast of the exploration boundary, and the abnormal boundary is irregular. The goaf is formed by the westward mining of the Xiaoxiyao mine. However, the northern part of the abnormal area is greatly disturbed by the raw coal transportation channel, and the southern part may be affected by the deterioration of data quality due to its close proximity to the boundary of the detection area, so the reliability of the results needs to be verified by more drilling results.

There is a seismic anomaly zone of coal-seam structure change in coal seam 9, which is located in the west of the exploration area. The plane shape is an irregular polygon, and the east side is connected with the seismic anomaly of the collapse column. The scope is delineated by a dotted line in Figure 12. The drilling data confirmed that the reason was that coal seams 9–1 and 9–2 merged into one layer.

A collapse column was identified via the drilling and log data from coal seam 4 (Figure 13). The lithology below K2 sandstone, as the marker bed of the No. 4 coal roof, is breccia, accompanied by a small amount of coal chips and blocks, with poor cementation. The shape of the collapse column on the plane is an approximate ellipse, especially in the instantaneous frequency attribute, and the characteristics of the collapse column are the most obvious. The collapsed column is a good water channel and water-accumulating area, so it is very important to identify the development of the coal-seam collapse column.

Based on the seismic interpretation of the development characteristics of special geological bodies such as goafs, faults, and collapse columns, the transient electromagnetic method is employed to predict the water accumulation in the key areas of the study area. The transient electromagnetic survey lines covering this area are N0 and N35. Due to terrain limitations, we adopted small coils and multiple circles of central loop to detect water accumulation in the goaf using the sensitivity of electrical resistivity to water.

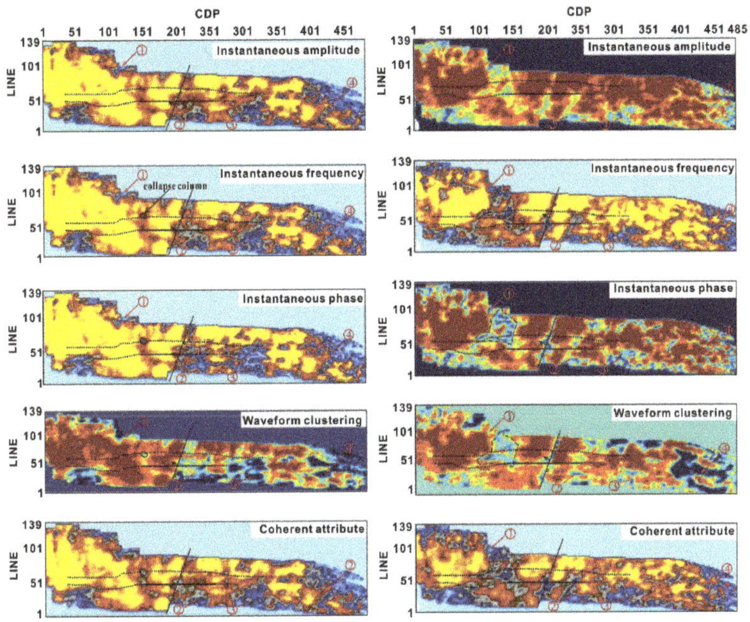

Figure 12. Distribution plan of seismic-interpretation abnormal areas of coal seams 4 (**left**) and 9 (**right**), ①–④ are the number of the goafs.

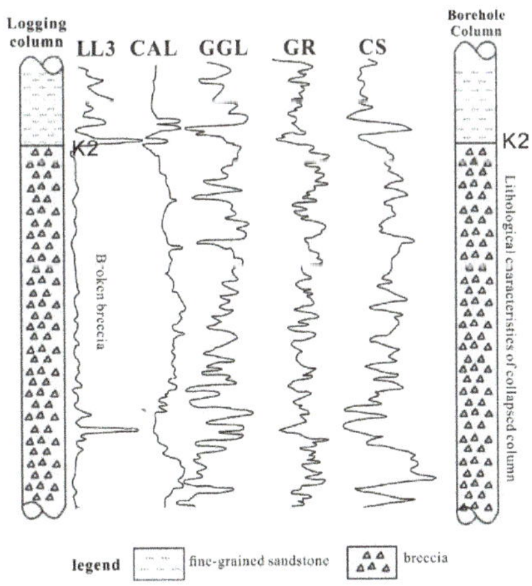

Figure 13. Comparison of logging and drilling properties of collapsed columns.

The length of survey line N35 is 910 m, and it is located on the 1320 platform. There are two small-scale low-resistance anomalies with relatively small ranges and amplitudes, which are located near point 310 and point 570 (at the red circle of the Figure 14). Among them, the low-resistivity anomaly of measuring point 570 is also shown near the corresponding position of line N0, and the anomaly is caused by the enhanced water yield of the formation near the fault. The abnormal position of point 310 is consistent with the position of the collapse column interpreted by the seismic attribute, which is speculated to be caused by the increased water abundances of the collapse column.

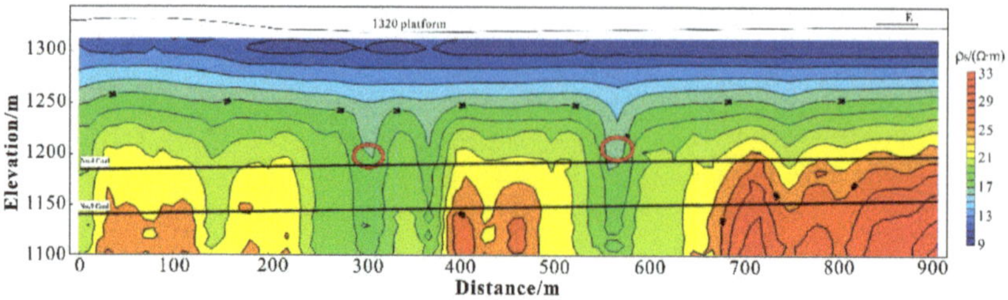

Figure 14. One-dimensional inversion profile of the apparent resistivity of the N35 line.

The length of survey line N0 is 790 m, which is located on the 1350 platform. The apparent resistivity profile east of survey point 530 (in the direction of the large point) shows a large low-resistivity anomaly (at the red circle of Figure 15). Combined with the comprehensive interpretation of adjacent survey line N35 and three-dimensional seismic data, it is speculated that the anomaly here is caused by the fragmentation of the formation near the fault and the enhancement in water abundances.

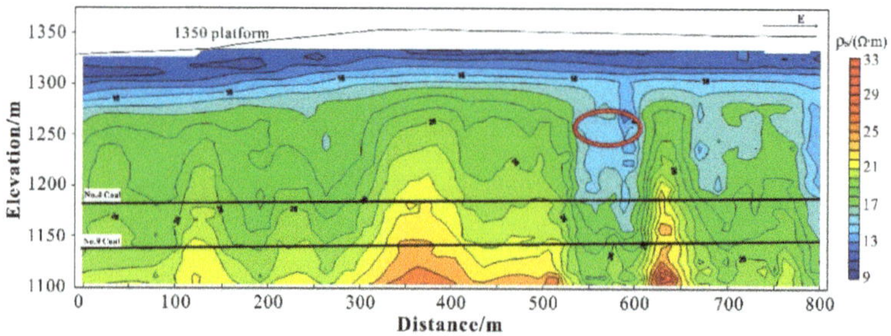

Figure 15. One-dimensional inversion profile of the apparent resistivity of the N35 line.

Integrating the results of the seismic interpretation of the goaf and the transient electromagnetic interpretation of the low-resistance abnormal area of each survey line, the scope of the goaf and water accumulation, collapse column, and fault water-rich anomaly displayed in the detection area is delineated on the plane. The subsequent drilling and testing results are basically consistent with the prediction results; in particular, the accurate prediction of the scope of the goaf plays a vital role in safe mining and personnel safety. Figure 16 shows the verification hole DLT-18-6 passing through goaf No. 4. The verification hole DLT-18-6 in the goaf with number 4, compared with the information of the adjacent well, revealed that the section where no coal core was found happened to be the depth section of coal 4 and 9. Due to the presence of goaf, the diameter value of the coal-seam section shows significant fluctuations, and the gamma–gamma value is significantly lower than that of the No.3 coal seam. Combined with the prediction of water accumulation, relevant coal-mining measures can be effectively formulated to avoid production risks (Figure 17).

The local part of the study area is particularly affected by the on-site construction conditions, such as the mining party's heavy vehicles, gas stations, drilling areas, and blasting areas, which are close to the geophysical exploration work area or even overlap and have a large interference with the seismic signal. In addition, the local coverage times are low, and the signal-to-noise ratio of the data is poor, which directly leads to the poor reliability of the interpretation results. For example, the southwest corner of the study area is a typical disturbed area, with a large area of abnormal areas. However, we predicted the possible results in advance, and the results were confirmed by a very small amount of drilling. Therefore, the judgment of the reliability of data is a very important link.

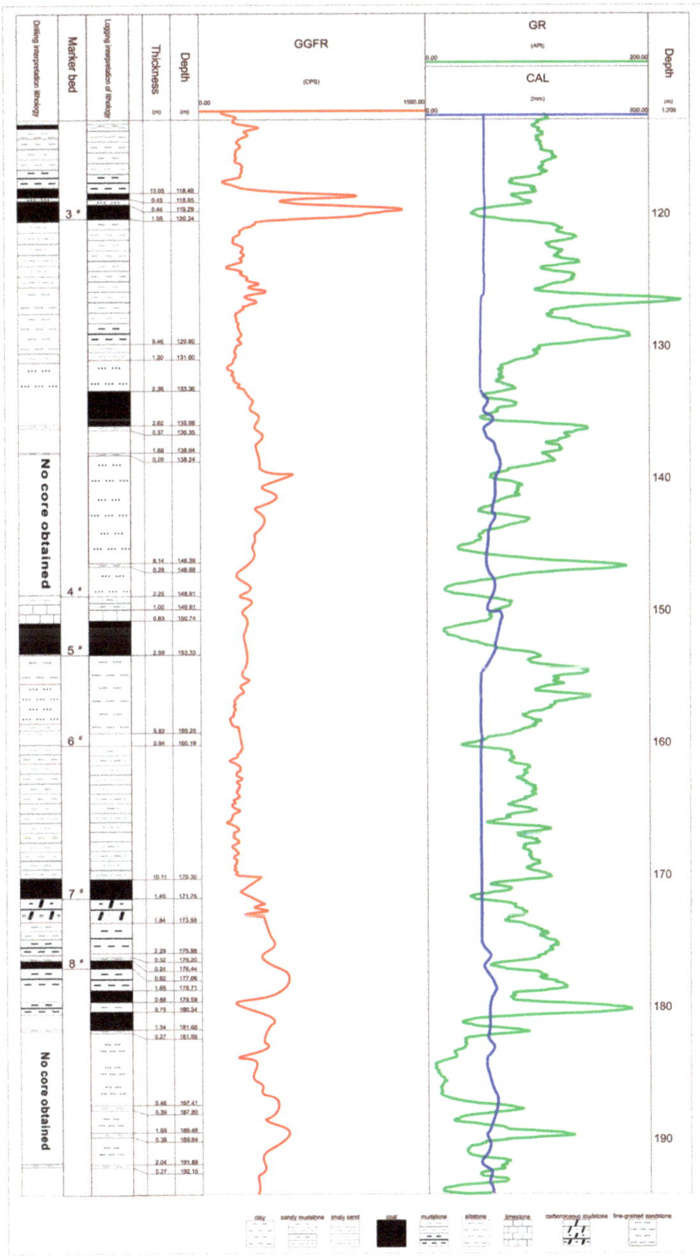

Figure 16. Borehole Histogram of the DLT-18-6 (The red line is GGFR, the blue line is CAL, and the green line is CAL).

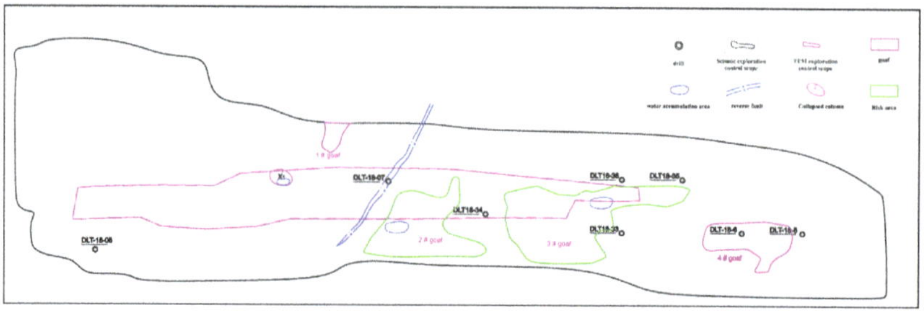

Figure 17. Plan distribution map for predicting risk areas of goaf and water accumulation.

To ensure safe and efficient mining, it is recommended to deploy ground drilling or advance drilling and geophysical exploration in the abnormal area and poorly controlled area in advance during the mining process to further determine the nature and location of the abnormality and ensure safety [33]. Combined with relevant geological data such as tunneling and drilling in different stages, comprehensive geological research should be strengthened to further improve the reliability and accuracy of interpretation results.

5. Conclusions

For the shallow-buried, deep, and large-gap open-pit mine goaf and water accumulation prediction problem, we adopt an integrated geophysical prospecting strategy. To solve the problems of uneven coverage and noise caused by ultrashallow layers and high drops, we propose a direct bin homogenization and multidomain joint denoising method. The results show that the shallow three-dimensional seismic method can clearly reflect the reflected-wave variation features of the seam. The features of a junction between a normal seam and goaf vary obviously, the effect of the set goaf boundary and scope is good, and the resolution is high. With a detailed analysis of the survey-line repeated area of the shallow three-dimensional seismic and transient electromagnetic methods, along with the apparent resistivity threshold scope of the transient electromagnetic method in the normal seam, goaf and goaf water-accumulation area are determined and can be applied to guide the explanation work. The transient electromagnetic method could especially reflect sensitivity to the low resistance, judge the water accumulation in the goaf, and effectively avoid the limit of the shallow three-dimensional seismic method to judge the water accumulation in the goaf. These results provided useful references for predicting goaf and water accumulation of open-pit mines with ultrashallow and high-drop conditions.

Author Contributions: Conceptualization, S.Z.; methodology, S.Z. and W.G.; validation, S.C. and Q.C.; investigation, Q.C. and Q.M.; writing—original draft preparation, S.Z. and Q.M.; writing—review and editing, S.Z. and S.C.; visualization, W.G. and Y.D.; supervision, S.C.; project administration, S.C. and Q.M.; funding acquisition, S.Z. and S.C. All authors have read and agreed to the published version of the manuscript.

Funding: This work was financially supported by the National Natural Science Foundation of China (42102212), the unveiling projects of the Department of Science and Technology of Shanxi Province, grant number (20191101018); and the Research and Development Project of Yangquan Coal Industry (Group) Co., Ltd. (GY18027), the National Natural Science Foundation of China (U21A20107).

Institutional Review Board Statement: Not applicable.

Informed Consent Statement: Not applicable.

Data Availability Statement: Not applicable.

Conflicts of Interest: The authors declare no conflict of interest.

References

1. Gui, H.; Qiu, H.; Qiu, W.; Tong, S.; Zhang, H. Overview of goaf water hazards control in China coalmines. *Arab. J. Geosci.* **2018**, *11*, 49. [CrossRef]
2. Zhang, J.; Fu, M.; Geng, Y.; Tao, J. Energy saving and emission reduction: A project of coal-resource integration in Shanxi Province, China. *Energy Policy* **2011**, *39*, 3029–3032. [CrossRef]
3. He, H.; Yu, W.; Wang, J.; Wen, K. Application research of high density resistivity method to karst geological exploration in mining area. In Proceedings of the 3rd International Conference on Environmental & Engineering Geophysics, Wuhan, China, 15–18 June 2008; pp. 385–388.
4. Hu, G.; Su, J.; Zhao, N. Application of three dimensional laser scanning technology in Pingshuo open-pit mine goaf detection. *Opencast Min. Technol.* **2015**, *7*, 61–63, (In Chinese with English abstract).
5. Yang, D.; Guo, W.; Tan, Y. Application of magnetotelluric method to the detection of overburden failure height in shallow seam mining. *Arab. J. Geosci.* **2018**, *11*, 350. [CrossRef]
6. Lu, T.; Liu, S.D.; Wang, B.; Wu, R.; Hu, X. A review of geophysical exploration technology for mine water disaster in China: Applications and trends. *Mine Water Environ.* **2017**, *36*, 331–340. [CrossRef]
7. Chang, J.; Yu, J.; Liu, Z. Three-dimensional numerical modeling of full-space transient electromagnetic responses of water in goaf. *Appl. Geophys.* **2016**, *13*, 539–552. [CrossRef]
8. Su, B.; Yu, J.; Sheng, C. On the borehole electromagnetic method for exploration of coal mining goaf. *Elektron. Elektrotech.* **2016**, *22*, 37–40. [CrossRef]
9. Yan, S.; Xue, G.; Qiu, W.; Li, H.; Zhong, H. Feasibility of central loop TEM method for prospecting multilayer water-filled goaf. *Appl. Geophys.* **2016**, *13*, 587–597. [CrossRef]
10. Dam, J.; Meulenkamp, J. Some results of the geo-electrical resistivity method in ground water investigations in The Netherlands. *Geophys. Prospect.* **2010**, *15*, 92–115. [CrossRef]
11. Liang, K.; Zhu, Y.; Shen, L. Application of high density resistivity method in groundwater exploration in Weifang. *Ground Water* **2013**, *35*, 112–113.
12. Ueda, T.; Mitsuhata, Y.; Uchida, T.; Marui, A.; Ohsawa, K. A new marine magnetotelluric measurement system in a shallow-water environment for hydrogeological study. *J. Appl. Geophys.* **2014**, *100*, 23–31. [CrossRef]
13. Wang, N.; Wang, Z.; Sun, Q.; Hui, J. Coal Mine Goaf Interpretation: Survey, Passive Electromagnetic Methods and Case Study. *Minerals* **2023**, *13*, 422. [CrossRef]
14. Francke, J.; Utsi, V.; Singh, K.K.K. The design of an intrinsically safe ground penetrating radar for detecting abandoned workings in underground coal mines. In Proceedings of the IEEE 14th International Conference on Ground Penetrating Radar (GPR), Shanghai, China, 4–8 June 2012; pp. 125–130.
15. Kotyrba, A.; Stańczyk, K. Sensing underground coal gasification by ground penetrating radar. *Acta Geophys.* **2017**, *65*, 1185–1196. [CrossRef]
16. Brocher, T.; Ewing, J. A comparison of high-resolution seismic methods for determining seabed velocities in shallow water. *J. Acoust. Soc. Am.* **1998**, *79*, 286–298. [CrossRef]
17. Wilkinson, P.; Chambers, J.; Meldrum, P.; Ogilvy, R.; Mellor, C.; Caunt, S. A comparison of self-potential tomography with electrical resistivity tomography for the detection of abandoned mineshafts. *J. Environ. Eng. Geophys.* **2005**, *10*, 381–389. [CrossRef]
18. Li, S.; Zhou, Z.; Ye, Z.; Li, L.; Zhang, Q.; Xu, Z. Comprehensive geophysical prediction and treatment measures of karst caves in deep buried tunnel. *J. Appl. Geophys.* **2015**, *116*, 247–257. [CrossRef]
19. Zhang, S.; Huang, H.; Dong, Y.; Yang, X.; Wang, C.; Luo, Y. Direct estimation of the fluid properties and brittleness via elastic impedance inversion for predicting sweet spots and the fracturing area in the unconventional reservoir. *J. Nat. Gas Sci. Eng.* **2017**, *45*, 415–427. [CrossRef]
20. Lelièvre, P.; Oldenburg, D. A comprehensive study of including structural orientation information in geophysical inversions. *Geophys. J. Int.* **2010**, *178*, 623–637. [CrossRef]
21. Liu, B.; Liu, Z.; Li, S.; Nie, L.; Su, M.; Sun, H. Comprehensive surface geophysical investigation of karst caves ahead of the tunnel face: A case study in the xiaoheyan section of the water supply project from songhua river, Jilin, China. *J. Appl. Geophys.* **2017**, *144*, 37–49.
22. Chang, S.; Qu, Z.; Zhang, S.; Fu, Y.; Chen, Q. Suppression of multiple reflection–refraction waves in the shallow low-velocity zone via vertical source combination. *Acta Geophys.* **2021**, *69*, 1799–1811. [CrossRef]
23. Zhang, S.; Huang, H.; Zhu, B.; Li, H.; Zhang, L. Seismic facies-controlled prestack simultaneous inversion of elastic and petrophysical parameters for favourable reservoir prediction. *Explor. Geophys.* **2018**, *49*, 655–668. [CrossRef]
24. Zhang, X.; Ma, J.; Li, L. Monitoring of coal-mine goaf based on 4D seismic technology. *Appl. Geophys.* **2020**, *17*, 54–66. [CrossRef]
25. Vakman, D. On the analytic signal, the teager-kaiser energy algorithm, and other methods for defining amplitude and frequency. *IEEE Trans. Signal Process.* **2002**, *44*, 791–797. [CrossRef]
26. Taner, M.; Koehler, F.; Sheriff, R. Complex seismic trace analysis. *Geophysics* **1979**, *44*, 1041–1063. [CrossRef]
27. Gersztenkorn, A.; Marfurt, K.J. Eigenstructure-based coherence computations as an aid to 3-d structural and stratigraphic mapping. *Geophysics* **2012**, *64*, 1468–1479. [CrossRef]
28. Fitterman, D. Transient electromagnetic sounding for groundwater. *Geophysics* **1986**, *51*, 995–1005. [CrossRef]

29. Shi, R.; Darcherif, A.; Sabonnadiere, J. Computation of transient electromagnetic fields radiated by a transmission line: An exact model. *IEEE Trans. Magn.* **1995**, *31*, 2423–2431. [CrossRef]
30. Chen, J.; Pi, S.; Zhang, Y.; Lin, T. Weak coupling technology with noncoplanar bucking coil in a small-loop transient electromagnetic system. *IEEE Trans. Ind. Electron.* **2021**, *69*, 3151–3160. [CrossRef]
31. Qi, T.; Zhang, F.; Pei, X.; Feng, G.; Wei, H. Simulation research and application on response characteristics of detecting waterflled goaf by transient electromagnetic method. *Int. J. Coal Sci. Technol.* **2022**, *9*, 17–26. [CrossRef]
32. Xue, G.; Pan, D.; Yu, J. Review the applications of geophysical methods for mapping coal-mine voids. *Prog. Geophys.* **2018**, *33*, 2187–2192. (In Chinese)
33. Wang, B.; Li, W.; Ye, Q.; Ma, K. A nonparametric spectrum estimation method for dispersion and attenuation analysis of borehole acoustic measurements. *Pet. Sci.* **2022**, *20*, 241–248. [CrossRef]

processes

MDPI

Article

A Case Study on Strong Strata Behaviors Mechanism of Mining Reserved Roadway and Its Prevention Techniques

Chen Li *, Delin Xin, Yan Liu and Tuantuan Chen

CCTEG Wuhan Engineering Company, Wuhan 430064, China
* Correspondence: lichen@zmwhy.com.cn

Abstract: The underground roadway of the Buertai Coal Mine adopts the double-roadway layout. Double-roadway layout mode has a roadway that is affected by repeated mining, called reserved roadways. The reserved roadway is strongly affected by mining, and the strata behaviors appear violently. This paper studies the strata behaviors that occur in auxiliary haulage roadway (AHR) during the mining of panel 42106. By analyzing geological conditions, mining influencing factors, and roadway layout, the mechanism of strong rock behavior has been clarified. Then, based on the theoretical analysis, we put forward the treatment method for the manifestation of strong strata behaviors by using hydraulic fracturing technology to break the key stratum. In this way, the high stress of the surrounding rock can be reduced by forcing the hard roof to be broken. After the application of hydraulic fracturing technology, we monitored the deformation of roadways and the periodic weighting law of the working face. The strength of strata behaviors has significantly weakened, and the application of this technology ensures the safe production of coal mines.

Keywords: roadway failure; strong strata behaviors; stress transfer; hydraulic fracturing; field application

1. Introduction

The degree of roof collapse after coal mining is related to the strength and structure of the overlying rock [1–3]. Overlying rocks with stronger rock strength and more complete structure are not easy to collapse. Therefore, the cantilever beam structure is formed at the edge of the goaf, leading to the phenomenon of strong strata behaviors in the roadway and working face [4,5]. The key stratum breakage block size and form determine the drastic degree of manifestation of strata behaviors [6,7]. Strong strata behaviors' manifestation is often accompanied by severe deformation of the roadway [8,9], wall caving [10], rock bursts [11], and other disasters [12,13].

In the traditional practice of prevention and control of the manifestation of strong strata behaviors, disasters such as roof collapse and coal wall caving are often prevented passively by optimizing support parameters [14,15]. However, passive support cannot effectively prevent hazards, such as large deformation of the roadway and comprehensive cutting of the roof of the mining goaf [16]. These strong strata behaviors' manifestation is caused by the rapid release of enormous potential energy stored in the rock layers during the crushing of high-strength and large overhanging overlying rocks [17,18]. Therefore, it is necessary to adopt the method of "active elimination" to reduce the potential energy stored in rock strata [19]. Hydraulic fracturing has been promoted for its effectiveness in dealing with various dynamic hazards [20,21]. Nguyen et al. [22] provide methods for hydraulic fracturing modeling, such as grid generation, execution time, convergence, and numerical integration problems, and indicate that Newton–Cotes orthogonality must be used for cohesive interface elements of secondary flow, at least for the proposed problems. Wu et al. [23] investigated the effect law of primary laminar orientation and horizontal stress difference magnitude on hydraulic fracture initiation pressure, extension pattern,

Citation: Li, C.; Xin, D.; Liu, Y.; Chen, T. A Case Study on Strong Strata Behaviors Mechanism of Mining Reserved Roadway and Its Prevention Techniques. *Processes* **2023**, *11*, 1341. https://doi.org/10.3390/pr11051341

Academic Editors: Feng Du, Aitao Zhou and Bo Li

Received: 21 February 2023
Revised: 23 April 2023
Accepted: 25 April 2023
Published: 26 April 2023

hydraulic pressure–time curve, and fracture volume by large true triaxial directional hydraulic fracture test with a high-energy industrial CT scan. Cheng et al. [24] studied the propagation process of multi-well fractures through physical model experiments, simulated the dynamic fracture initiation and propagation process of cross-hole fractures during directional hydraulic fracturing, and analyzed the evolution law of pore elastic stress and pore pressure between multiple wells.

At present, the technology to solve the strata behaviors after coal mining in the Shendong Mining Area is more inclined to carry out strong support in the working face and roadway, and there is no precedent to use hydraulic fracturing technology to weaken the key layer of overlying rock in strong mining face [25,26]. Therefore, it is crucial to study the prevention and control principles of hydraulic fracture based on the analysis of the mechanism and dominance of strong strata behaviors' manifestation. In this paper, we take panel 42107 of the Buertai Coal Mine as the engineering background. Based on the theoretical analysis, the prevention and control principles and technical solutions of hydraulic fracturing technology were studied. The practical application is remarkable.

2. Project Profile

2.1. Panel Profile

Buertai Coal Mine is the largest underground mining coal mine in China with an annual output of 20 Mt. The mine is located in the central part of Dongsheng Coalfield in Ordos City, southern Inner Mongolia Autonomous Region. The mining roadway in Buertai Coal Mine adopts the double-roadway layout mode, i.e., two mining roadways are excavated at the same time, one of which is affected by repeated mining and needs to serve two working faces (called reserved roadways). The geographic location, related photos, and mining layout of the Buertai Coal Mine are shown in Figure 1.

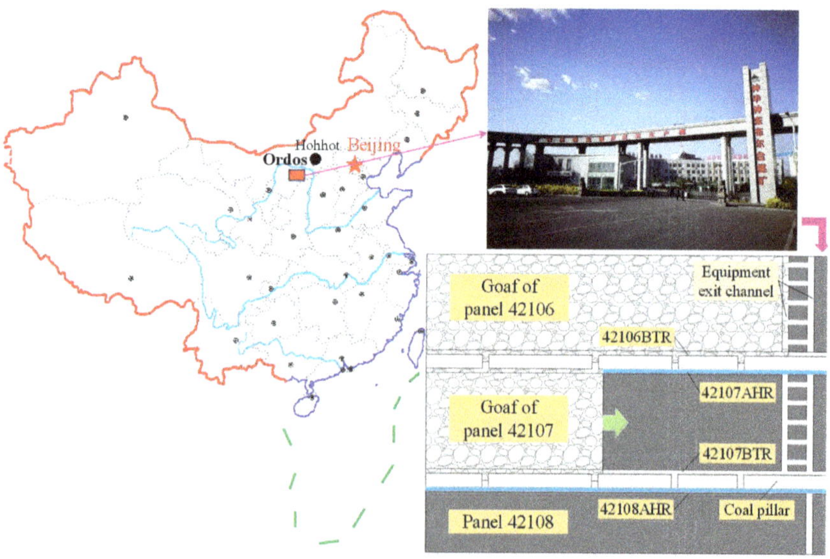

Figure 1. Basic overview thumbnail of Buertai Coal Mine.

Comprehensive mechanized top-coal caving is adopted in panel 42107 of coal seam 4-2 in Buertai Coal Mine. The length of the working face is 300 m, the average depth of the coal seam is 457 m, the thickness of the coal seam is 7.6 m, the dip angle of the coal seam is 1~3°, and the length of the panel is 4808 m. The average distance between coal seam 4-2 and overlying coal seam 2-2 is 75 m. The upper part of the coal seam 4-2 roadway is coal seam 2-2 goaf. There are two key strata in the overlying strata between coal seam 4-2 and coal seam 2-2, which are fine sandstone subcritical stratum 1 with a thickness of 8.6 m and siltstone subcritical stratum 2 with a thickness of 22.6 m. Rock parameters of the roof and floor of 4-2 coal seam are given in Table 1.

Table 1. Geological parameters of coal and rock.

Lithology	Burial Depth/m	Thickness/m	Cohesion/MPa	Compressive Strength/MPa	Notes
Sandy sandstone	390.9	19.2	6.6	59	
Clip coal	391.6	0.7	1.9	12	
Sandy sandstone	406.8	15.2	5.4	43	
Siltstone	428.4	22.6	6.6	54	Subcritical stratum 2
Sandy sandstone	436.1	5.7	4.2	39	
Fine Sandstone	444.7	8.6	6.2	52	Subcritical stratum 1
Sandy sandstone	451.0	6.4	3.6	34	
4-2 Coal seam	457.4	7.6	2.5	17	
Sandy sandstone	465.0	6.1	3.5	33	

2.2. Strong Strata Behaviors Survey

Panel 42106 of coal seam 4-2 is adjacent to panel 42107, as shown in Figure 1. During the mining of panel 42106, the leading area of 42106AHR (reserved roadway of panel 42105) appeared to manifest strong strata behaviors, and the influence range could reach 100 m ahead. Strong strata behaviors' manifestation seriously affected the normal mining of the working face, as shown in Figure 2.

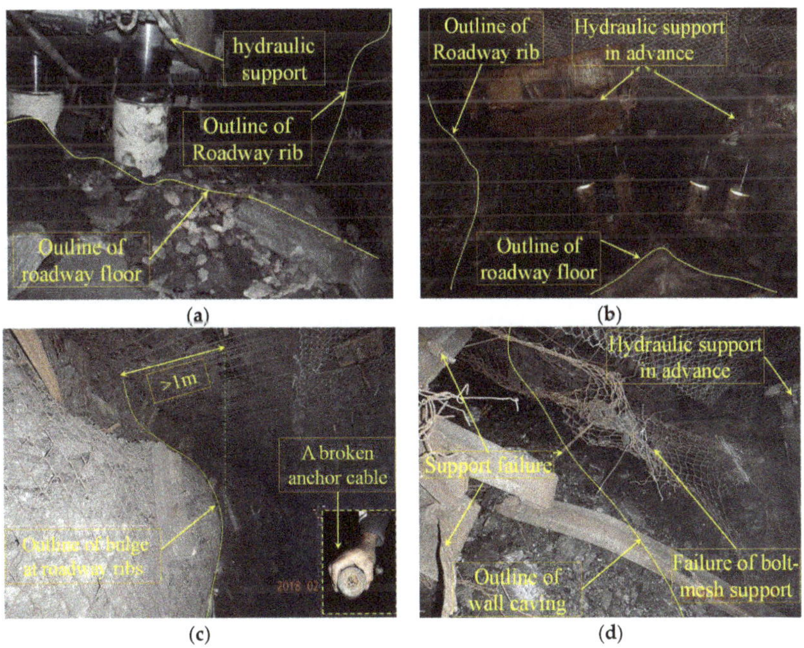

Figure 2. Strong strata behaviors' manifestation: (**a**) the hydraulic bracket was buried, (**b**) heaving floor and hydraulic bracket tilt, (**c**) ribs bulge and anchor broken, and (**d**) nets bag and leakage of gangue.

The working resistance of the hydraulic bracket was greater than 23,000 kN, and the coal wall flakes were serious during periodic weighting of the working face. The floor vibrated and caused the coal cutter to pop up and led to the overhead hydraulic bracket being buried, as shown in Figure 2a. The roadway in front of the working face shows the heaving floor and hydraulic bracket tilt, as shown in Figure 2b. The roadway adjacent to the goaf (coal pillar 25 m) was severely deformed within 40 m in front of the working face, with ribs bulging greater than 2 m and the heaving floor greater than 1.5 m, as shown in Figure 2c. The anchor net broke, resulting in the nets bag and leakage of gangue, as shown in Figure 2d.

3. Mechanism and Treatment Principle of Strong Strata Behaviors

3.1. Main Controlling Factors of Strong Strata Behaviors

(1) Geological conditions

The roof of coal seams 4-2 in Burtai Coal Mine is thicker and stronger, with strong self-sustaining ability. The upper part of the goaf is prone to forming a cantilever roof, which increases the fracturing step of the basic roof. This leads to an increase in the elastic energy accumulated before the roof breaks. In the process of a hard roof breaking or sliding, the elastic energy stored in the roof is prone to suddenly release, inducing strong strata behaviors disaster [27]. When the buried depth of the coal seam is greater than 350 m, the frequency and intensity of strong strata behaviors will gradually increase. The buried depth of coal seam 4-2 is 457 m, which is the internal factor of strong strata behaviors' manifestation.

(2) Influence of mining stress

Under the influence of repeated mining, the reserved roadway is located in an extremely complex superimposed stress field composed of original rock stress, abutment pressure, mining dynamic load, and expansion pressure in the plastic zone [28]. When the roof periodically intensifies, the position of the fracture line will affect the manifestation of strata behaviors. If the fracture line is located in the goaf, a cantilever beam is formed at the edge of the pillar. The pressure from the insufficient span of the overlying strata in the goaf is transmitted to the coal pillar through a cantilever beam. If the fracture line is located inside the coal pillar, the rotational deformation of the basic roof on the coal pillar will transmit several times the initial stress to the coal pillar. All these are not conducive to the stability and maintenance of the roadway.

When 42106AHR (reserved roadway) is subjected to secondary mining, it is 3~5 times of original rock stress under the influence of advanced abutment pressure, abutment pressure of side mined-out area, and concentrated stress of residual coal pillars in coal seam 2-2. Under the disturbance of multiple stress superposition, it will cause rapid deformation and failure of the mining roadway, which can easily lead to the occurrence of strong strata behaviors' manifestation [29,30].

(3) The impact of mining spatial relationship (coal pillar in overlying goaf)

The goaf of coal seam 2-2 is located 75~80 m above panel 42106 of coal seam 4-2. The siltstone between the two coal seams is thick and hard. The horizontal projection distance between 42106AHR and the overlying coal pillar (the coal pillar between panel 22104 and panel 22105) is 43.5 m. The fault protection pillar of coal seam 2-2 also has an important influence on the stress of the lower strata [5]. Stress concentration occurs in a certain range below the coal pillar, and the difference within stress distribution is greater as it gets closer to the edge of the pillar. The spatial location relationship of mining is shown in Figure 3.

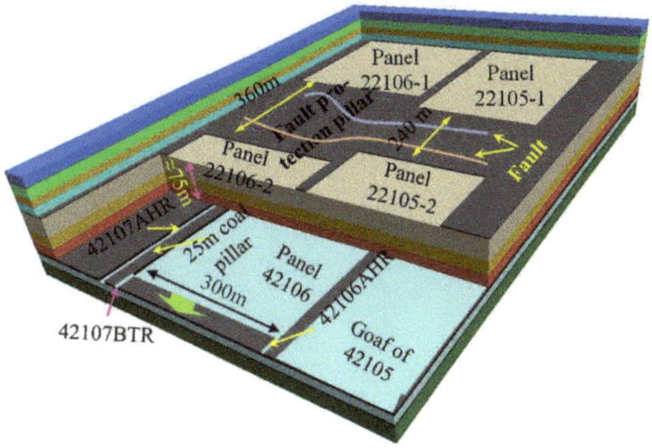

Figure 3. The spatial position relationship between coal seam 4-2 and coal seam 2-2.

3.2. Technical Principle of Disaster Prevention and Control of Strong Strata Behaviors

The integrated hydraulic fracturing of the coal seam roof is similar to the artificial creation of a "relief layer" in advance. After fracturing above the working face, the overlying strata produce fractures within the fracturing range. These fractures destroy the previously intact overburden structure within the fracture range. When the overburden load acts on the fractured area, the fractured rock mass is compressed to fill the fracture, which reduces the elastic deformation of the fractured rock mass. Therefore, after fracturing, the support of the upper roof is reduced due to the deformation of rock mass within the fracturing range, as shown in Figure 4. Pan et al. have carried out a detailed theoretical derivation according to the ground prefracturing coal seam roof, and the research results can be a certain reference value for the paper [31].

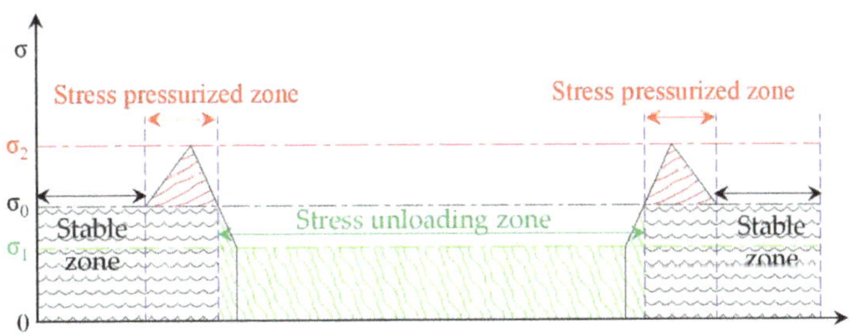

Figure 4. Stress boundary conditions of coal seam in working face after fracturing, where σ_0 is the stress of primary rock, σ_1 is the stress in the unloading zone of the fracturing range, and σ_2 is the maximum stress in the pressurized zone at the fracturing boundary.

In order to obtain the stress control effect of hydraulic fracturing, a numerical mining model (FLAC 3D) was established, and the calculation results are shown in Figure 5. When hydraulic fracturing technology is adopted, the goaf fully collapses, and the caving line overlaps with the pressure relief line, which is distributed along the hydraulic fracturing hole. The maximum vertical stress in the area affected by mining is 56.15 MPa. When hydraulic fracturing is not implemented, we set a 20 m overburden overhanging area at the goaf boundary. The maximum vertical stress in the area affected by mining is 64.09 MPa. After hydraulic fracturing, the peak vertical stress was reduced by 7.94 Pa. The vertical stress in the coal pillar decreases obviously.

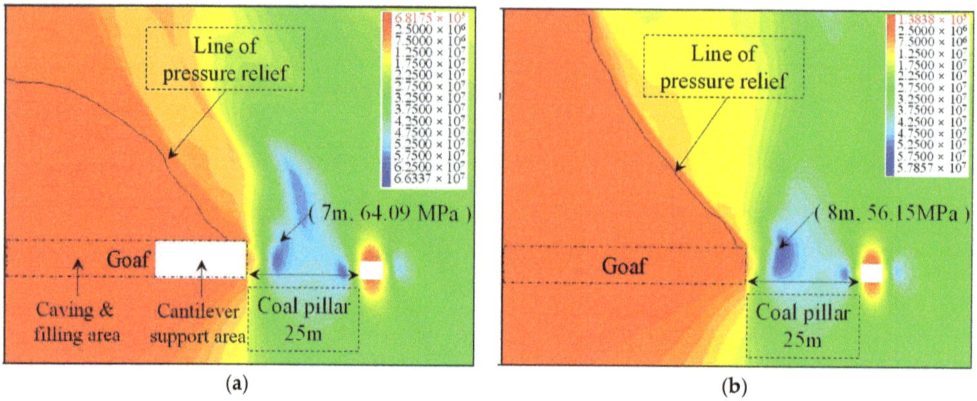

Figure 5. Distribution characteristics of regional stress field before and after hydraulic fracturing: (**a**) before hydraulic fracturing and (**b**) after hydraulic fracturing.

Compared with before fracturing, artificial cracking caused by a thick hard roof on the coal seam of the working face roof can make long beams become short beams and large pieces become smaller. Thus, the hard roof no longer has the cantilever function. The impact risk of the dynamic load caused by roof fracture with an increased overhanging area is reduced. At the same time, the hard roof can collapse and fill the goaf in time after fracturing, which reduces the high stress concentration in the coal pillar and also reduces the concentrated dynamic load caused by the roof pressure. It can be seen that hydraulic fracturing can significantly weaken the risk of a dynamic disaster of strong strata behaviors.

3.3. Technical Basis for Controlling Strong Strata Behaviors

To avoid the occurrence of strong strata behaviors' manifestation in 42107AHR and 42107 working face during the mining of panel 42107, we decided to adopt hydraulic fracturing technology to relieve concentrated stress according to the above analysis. According to the relevant literature and our rich engineering practice experience, hydraulic fracturing roof strata to reduce the strong mining pressure in stope should be based on the key stratum theory of rock strata control. The working condition of the project site is very complicated (the overlying coal seam 2-2 is mined out, but there is 300~360 m fault protection coal pillar locally, the coal pillar is 20 m in the section of the coal working face of 2-2 coal, and the overlying strata are also affected by the fault, all of which lead to a complex overlying rock stress environment. In addition, it is difficult to establish a realistic theoretical model due to the large number, thickness variation, and strength difference of rock strata), as shown in Figure 3. Therefore, we only establish a simplified theoretical model under ideal conditions for schematic analysis. There are some differences between the calculation results of this model and the actual working conditions, but it can be used for reference.

The key strata control the deformation movement of local and whole strata, and the mechanical properties of the key stratum are important factors affecting the periodic breaking step. The stress state of the key stratum before breaking and turning was simplified into a pure bending fracture model of the beam structure, as shown in Figure 6. According to Figure 4, the relationship between the tensile strength of the key stratum and its periodic breaking step can be qualitatively analyzed.

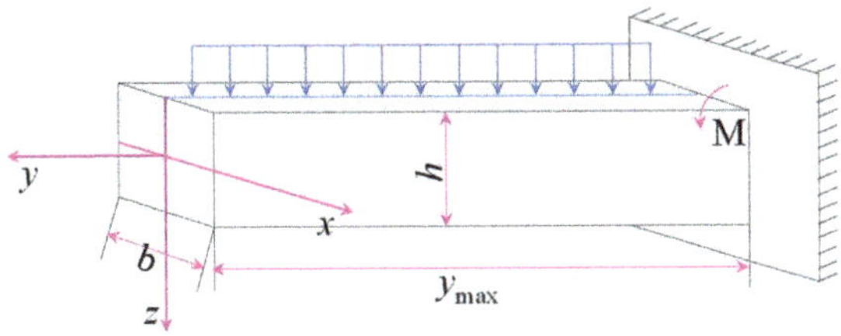

Figure 6. Mechanical model of the elastic beam structure.

According to the mechanics of materials, the maximum normal stress of a pure bending beam occurs at the furthest distance from the neutral axis. The relationship between the maximum tensile stress and the maximum bending moment can be obtained as shown in Equation (1).

$$\sigma_{\max} = M_{\max} h / (2 I_z) \tag{1}$$

where $\sigma_{\max}$ is the maximum tensile stress (MPa), $M_{\max}$ is the bending moment (kN·m), Iz is the moment of inertia for the Z-axis in the rectangular section (m4), and h is the thickness of the key stratum (m).

$$M_{\max} = -q y_{\max}^2 / 2 \tag{2}$$

where q is the interface compressive stress on the key stratum (MPa), and $y_{\max}$ is the maximum roof spacing (m).

$$I_z = bh^3 / 12 \tag{3}$$

where b is the width of the rectangle interface (m). Take b = 1 m, substitute Equations (2) and (3) into Equation (1), and the relationship between the periodic breaking step distance of the key stratum and its tensile strength can be obtained. If $\sigma_{\max}$ = RT, then:

$$y_{\max} = h\sqrt{\frac{RT}{3q}} \tag{4}$$

where RT is the tensile strength of the key stratum (MPa).

According to Equation (4), the larger the tensile strength of the key stratum is, the larger the maximum suspended roof distance is. In the process of panel propulsion, too large overhead distance will make the hydraulic bracket and coal body bear more load. This will lead to working face pressure frame, wall caving, and roadway deformation. If the strength of the key stratum can be weakened, the strong strata behaviors can be controlled to a certain extent. Therefore, the selection of a hydraulic fracturing horizon is the key to controlling strong strata behaviors.

In the mining process of panel 42107, subcritical stratum 1 and 2 will form the fracture structure of "cantilever beam + masonry beam", which will affect the stope strata behaviors. According to the geological parameters of the overlying strata of coal seam 4-2 combined with Equation (4), it can be calculated that the periodic weighting distance of subcritical stratum 1 is 36.75 m, and that of subcritical stratum 2 is 53.25 m. Subcritical stratum 2 is the key to influencing the stope strata behaviors.

4. Hydraulic Fracturing Technology and Scheme

4.1. Design Basis of Hydraulic Fracturing Program

Many scholars have studied the law of strata behaviors in the process of advancing the working face [32–35]. They obtained the "square position" characteristic of the working face [36]. That is, when the advancing length of the working face is an integer multiple of the width of the working face, the strata behaviors will be more severe [37,38].

From the perspective of "square position", we should design a long fracturing borehole when the panel advances to integral multiples of the width of the working face. The design can promote the full collapse of the hard roof and weaken the "square position". The vertical height of the long fracturing borehole should be close to the upper boundary of the siltstone (subcritical stratum 2) to ensure its collapse. The horizontal width of the long fracturing borehole should ensure that the fracture range covers the width of the working face. The short fracturing borehole is designed to prevent cantilever beams from forming in the hard rock above the lateral coal pillar. The short fracturing borehole shall be at least 10 m within the siltstone bed. The horizontal length of the borehole is determined according to the depth of the borehole and the construction angle of the drilling equipment.

The spacing of the fracturing borehole is set according to the effective fracturing radius of the borehole [39,40]. The fracturing radius of the borehole is 3.2~4.3 M when the water injection pressure is 12~16 MPa under the condition of crustal stress and surrounding rock strength. Therefore, 1/3 (7 m) of the average periodic weighting was selected as the fracturing borehole spacing. In this way, sufficient fracturing can ensure that the edge of the goaf caved in time after mining, so as to reduce the stress concentration.

4.2. Parameter Design of Implementation Scheme

Based on the above analysis, hydraulic fracturing drilling holes were arranged to the roof through 42107AHR, 42107BTR, and 42108 AHR. The vertical depths of the long and short holes are 38 m and 30 m, respectively. The drilling hole opening is about 2.2 m away from the floor.

Single-hole multiple fracturing technology is adopted, and the part of drilling in the coal seam is not fractured. The fracturing boreholes are arranged before mining, and the fracturing operation is carried out ahead of the working face in the process of mining. The sealing pressure of hydraulic fracturing is 12~16 MPa. The hydraulic fracturing roof fracturing process is divided into three steps, that is, drilling, sealing, and fracturing, respectively.

The hydraulic fracture drill holes were drilled by ZDY1200S fully hydraulic drilling rig along the right rib of 42107AHR (Fracturing drilling A), the left rib of 42107BTR (Fracturing drilling C), and 42108AHR (Fracturing drilling B) to the roof of panel 42107, respectively, with a diameter of 56 mm, as shown in Figure 7.

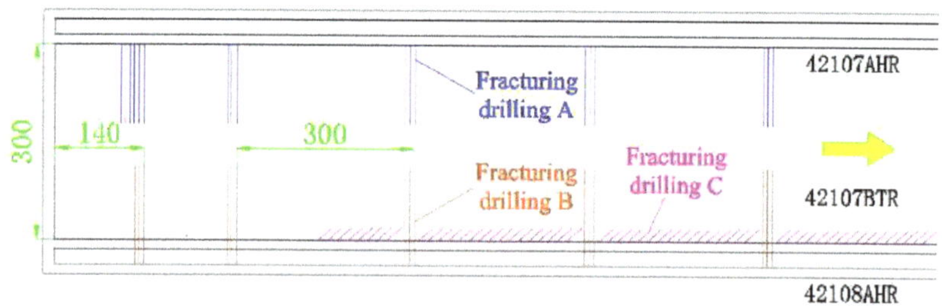

Figure 7. Hydraulic fracturing drill layout plan of hydraulic fracture drill holes.

In 42107AHR, a total of 18 hydraulic fracturing drill holes with a length of 150 m are arranged. The six drill holes in the first group are arranged within the range of 80~140 m from the open-off cut of panel 42107, as shown in Figure 6a. In 42108AHR, a total of 15 hydraulic fracturing drill holes with a length of 150 m are arranged. The three drill holes in the first group are arranged within the range of 120~140 m from the open-off cut of 42107, as shown in Figure 6b.

The second group of three drill holes is 300 m away from the open-off cut, and then each group is 300 m away from the other. The spacing of the drill holes in the group is 7 m, the drilling angle is 17°, and the vertical height of the drill holes is 38 m. In 42107BTR, a total of 65 hydraulic fracturing drill holes with a length of 43 m are arranged in the range of 560~1080 m away from the open-off cut, the drilling angle is 51°, and the vertical height of the drill holes is 30 m, as shown in Figure 8c. The area where the borehole is located in the coal seam will not be fractured (red section). To prevent the loose gangue from leaking in front of the hydraulic bracket, the yellow section of the borehole will not be fractured. The green part is fractured every 3 m, and the fracturing duration is 10–15 min.

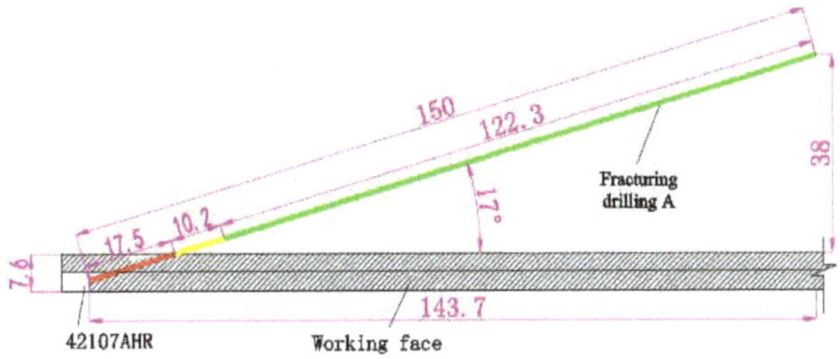

(a)

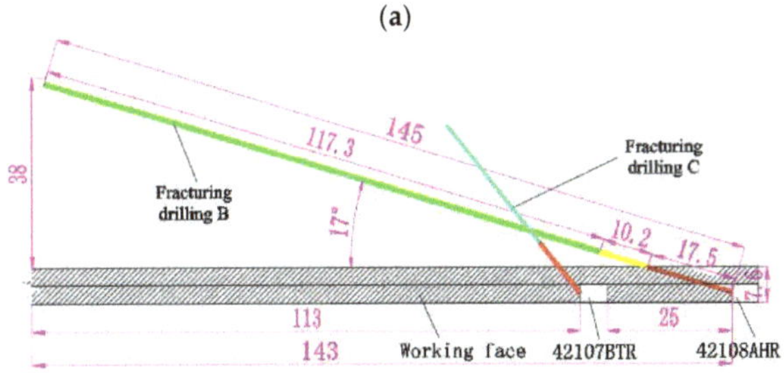

(b)

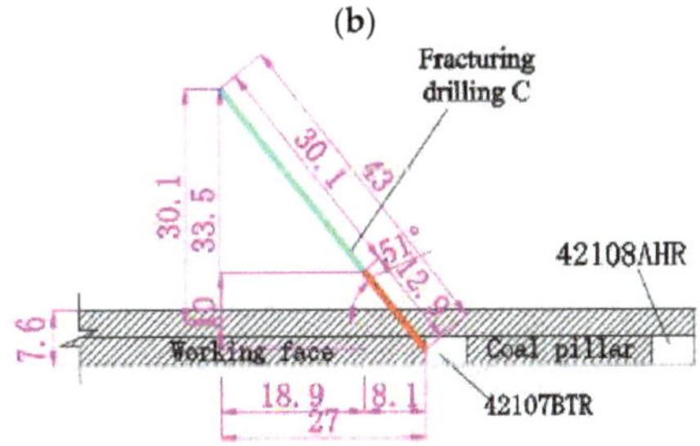

(c)

Figure 8. Schematic diagram of hydraulic fracturing technical scheme parameters: (**a**) parameters of hydraulic fracture drill holes in 42107AHR, (**b**) parameters of hydraulic fracture drill holes in 42108AHR, and (**c**) parameters of hydraulic fracture drill holes in 42107BTR.

5. On-Site Industrial Test Monitoring

5.1. Monitoring of Roadway Deformation

After hydraulic fracturing, the integrity of 42107AHR at the front of the working face has been significantly improved compared to 42106AHR, as shown in Figure 9. To monitor the stability control effect of the roadway after hydraulic fracturing, two deep base point monitoring stations are set up at 42107AHR (100 m, 125 m, and 150 m in front of the working face). Deep base point separators are installed in the roof, left rib (on the side of the coal pillar), and right rib (on the side of the working face) of each station, with base point installation depths of 3 m and 8 m. The measured value at No. 3 is the largest, and its monitoring data are plotted as a curve, as shown in Figure 10.

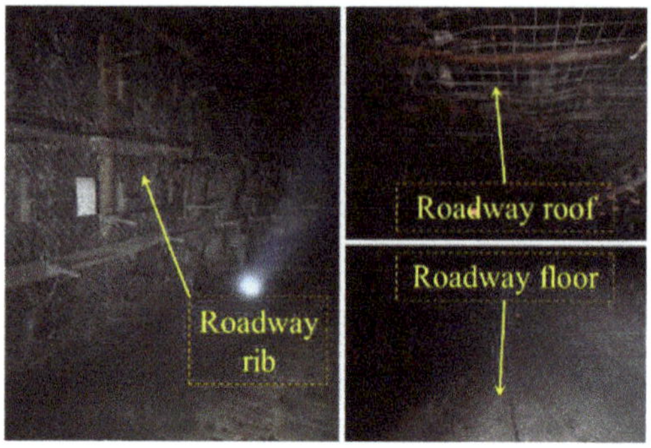

Figure 9. Picture of roadway control effect.

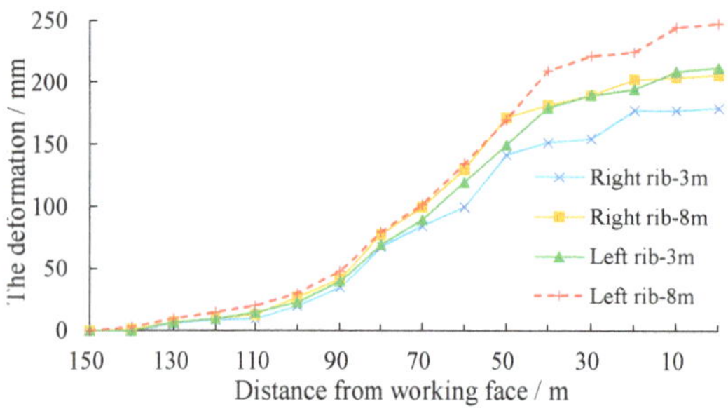

(a)

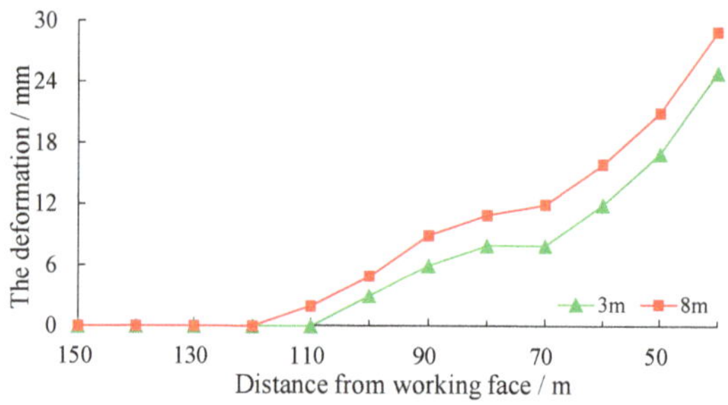

(b)

Figure 10. Deep base point displacement monitoring: (**a**) deformation monitoring curve of rib wall rock and (**b**) deformation monitoring curve of the roof.

From Figure 10, t it can be seen that the measured deformation of the left rib is greater than that of the right rib. During repeated mining of 42107AHR, significant deformation began to occur approximately 100 m in front of the working face. The closer the working face is to the monitoring point, the more deformation occurs, and the deformation rate is faster and faster. When the measuring point is 40 m away from the working face, the deformation of the left rib at the 8 m base point of the deep surrounding rock is about 200 mm, and the deformation of the right rib is about 180 mm. The roadway 40 m in front of the working face is reinforced with an advanced hydraulic bracket, which damages the roof monitoring points. The rib monitoring points are supported by hydraulic support ribs to prevent continuous deformation of the ribs. In general, compared to the large deformation of surrounding rock before hydraulic fracturing, the variation of roadway has been effectively controlled at the present stage. The failure depth of surrounding rock is mainly concentrated within 3 m of the surrounding rock, accounting for more than 80% of the total deformation.

5.2. Monitoring of Periodic Weighting Parameters

To monitor the control effect of hydraulic fracturing on the stope strata behaviors, we analyzed the hydraulic bracket data in the mining process of panel 42106 and panel 42107. The monitoring data hydraulic bracket is shown in Figures 11 and 12.

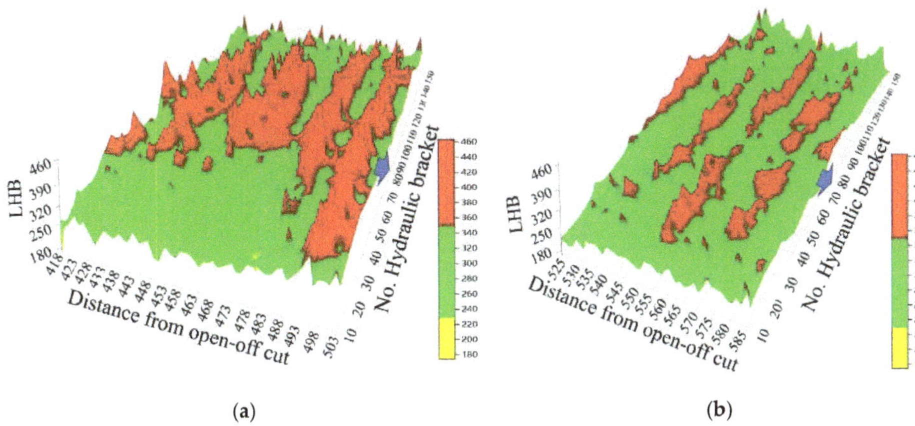

Figure 11. Monitoring data of hydraulic bracket on working face: (**a**) hydraulic bracket data of panel 42106 and (**b**) hydraulic bracket data of panel 42107.

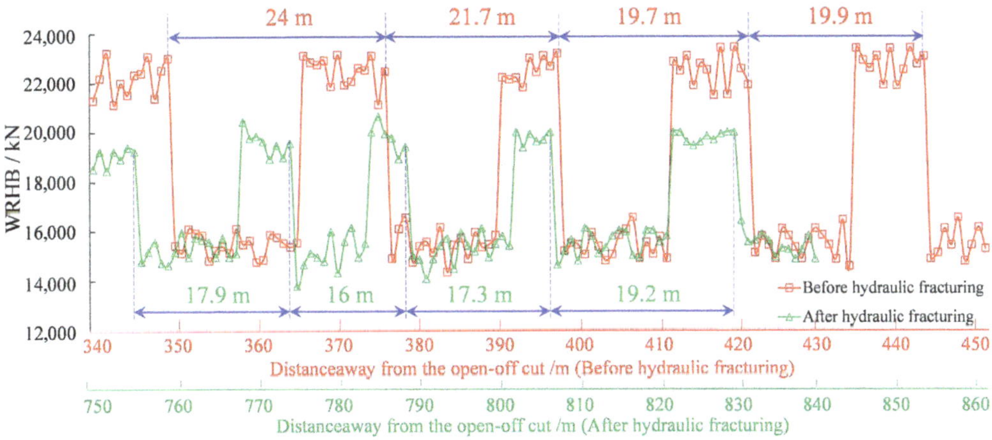

Figure 12. Periodic weighting comparison of working face before and after hydraulic fracturing.

From Figures 11 and 12, the following conclusions can be drawn. The average step distance of periodic weighting of panel 42106 is approximately 21.6 m, with an average duration of 7.5 m. The normal load of the hydraulic bracket (LHS) is about 289 bar when the working face is advancing normally, and the average final working resistance of the hydraulic bracket (WRHB) is 15,207 kN. During the periodic weighting process, the average LHS is approximately 395 bar, and the average final WRHB is 22,535 kN. The average step distance of periodic weighting of panel 42107 is about 16.9 m, with an average duration of 4.9 m. The normal LHS is about 265 bar when the working face is advancing normally, and the average final WRHB is 15,146 kN. The average LHS is about 354 bar during the periodic weighting, and the average final WRHB is 19,599 kN, which is less than the rated resistance of the hydraulic bracket of 21,000 kN.

Based on the above analysis, it can be inferred that after implementing hydraulic fracturing measures in panel 42107, the roof fracture line is close to the coal pillar. Hydraulic fracturing can achieve stress unloading on both sides of the working face, promoting the transfer of high stress in the surrounding rock of the roadway. The use of hydraulic fracturing technology promotes timely collapse of the roof, avoids high concentration of stress, and effectively controls the behavior of strong rock layers in the working face.

6. Conclusions

(1) The characteristics of strong strata behaviors of AHR were obtained by field observation. The specific manifestations of the strong strata behaviors include a large mining impact range, large roadway deformation, severe anchor cable damage, large inclination angle of roadway hydraulic support, and early warning of hydraulic support resistance in the working face.

(2) Based on various factors, the mechanism of strong strata behaviors in the roadway of the Burtai 4-2 coal seam has been obtained. The thicker and harder the roof strata, the greater the elastic energy accumulated before the roof fractures. The reserved roadway was greatly affected by the repeated mining of the working face when the double roadway was arranged, and the stress superposition was obvious. The coal pillar in the goaf of the overlying 2-2 coal seam promotes stress concentration in the 4-2 coal seam, resulting in more complex regional stress field conditions.

(3) Based on the theoretical analysis, the implementation scope of hydraulic fracturing was determined. Combined with the site situation, the pressure relief control technical scheme of hydraulic fracturing in the 42107 working face is proposed. After adopting a reasonable hydraulic fracturing scheme, the average step distance and load of periodic weighting were significantly reduced, the deformation and failure of the roadway were weakened, and high stress transmission in the working face area was achieved.

Author Contributions: All authors contributed to this paper. Conceptualization, C.L. and D.X.; methodology, C.L. and Y.L.; validation, C.L. and T.C.; formal analysis, C.L.; data curation, Y.L.; writing—original draft preparation, T.C.; writing—review and editing, C.L. and D.X.; validation, Y.L. and C.L.; funding acquisition, C.L. All authors have read and agreed to the published version of the manuscript.

Funding: This work is supported by the China Postdoctoral Science Foundation (2021M700427) and the Innovation Project of CCTEG Wuhan Engineering Company (ZZYF202319 and ZZYF202321).

Institutional Review Board Statement: Not applicable.

Data Availability Statement: All data and/or models used in the study appear in the submitted article.

Conflicts of Interest: The authors declared no potential conflict of interest with respect to the research, authorship, and publication of this article.

References

1. Wang, S.; Luo, J. Study on the shadow effect of the stress field around a deep-hole hydraulic-fracturing top-cutting borehole and process optimization. *Processes* **2023**, *11*, 367. [CrossRef]
2. Wang, S.; Zhu, G.; Zhang, K.; Wang, Y. Research on short wall continuous mining technology with full roof falling and roof control measures. *Geotech. Geol. Eng.* **2021**, *39*, 4877–4886. [CrossRef]
3. Cai, M. Key theories and technonogies for surrounding rock stability and ground control in deep mining. *J. Min. Strat. Control Eng.* **2020**, *2*, 5–13. [CrossRef]
4. Yu, X.; Bian, J.; Liu, C. Determination of energy release parameters of hydraulic fracturing roof neargoaf based on surrounding rock control of dynamic pressure roadway. *J. Min. Strat. Control Eng.* **2022**, *4*, 25–34.
5. Liao, Z.; Feng, T. Mechanism and application of layered grouting reinforcement for fractured coal and rock roadway. *Appl. Sci.* **2023**, *13*, 724. [CrossRef]
6. Wang, Y.G.; Guo, W.B.; Bai, E.H.; Zhang, Z.Y.; Kang, Y.P.; Chai, H.B.; Chen, J.J. Characteristics and mechanism of overlying strata movement due to high-intensity mining. *J. China Coal Soc.* **2018**, *43*, 28–35. [CrossRef]
7. Hu, Q.; Cui, X.; Liu, W.; Ma, T.; Geng, H. Law of overburden and surface movement and deformation due to mining super thick coal seam. *J. Min. Strat. Control Eng.* **2020**, *2*, 31–39. [CrossRef]
8. Li, C.; Wu, Z.; Zhang, W.; Sun, Y.; Zhu, C.; Zhang, X. A case study on asymmetric deformation mechanism of the reserved roadway under mining influences and its control techniques. *Geomech. Eng.* **2020**, *22*, 449–460. [CrossRef]
9. Li, J.; Ren, J.; Li, C.; Zhang, W.; Tong, F. Failure mechanism and stability control of soft roof in advance support section of mining face. *Minerals* **2023**, *13*, 178. [CrossRef]

10. Kang, H.; Xu, G.; Wang, B.; Wu, Y.; Jiang, P.; Pan, J.; Ren, H.; Zhang, Y.; Pang, Y. Forty years development and prospects of underground coal mining and strata control technologies in China. *J. Min. Strat. Control Eng.* **2019**, *1*, 7–39. [CrossRef]
11. Liu, W.; Qian, D.; Yang, X.; Wang, S.; Deng, J.; Cui, Q.; Li, Z. Stress relief and support for stability of deep mining roadway with thick top coal in Hujiahe Coal Mine with the risk of rock burst. *Shock Vib.* **2021**, *2021*, 3822336. [CrossRef]
12. Zhang, N.; Han, C.; Xie, Z. Theory of continuous beam control and high efficiency supporting technology in coal roadway. *J. Min. Strat. Control Eng.* **2019**, *1*, 48–55. [CrossRef]
13. Du, F.; Wang, K.; Zhang, X. Experimental study of coal-gas outburst: Insights from coal-rock structure, gas pressure and adsorptivity. *Nat. Resour. Res.* **2020**, *29*, 2481–2493. [CrossRef]
14. Kang, H. Spatial scale analysis on coalmining and strata control technologies. *J. Min. Strat. Control Eng.* **2020**, *2*, 5–30. [CrossRef]
15. Du, F.; Wang, K.; Dong, X.; Wei, J. Numerical simulation of damage and failure of coal-rock combination based on CT three-dimensional reconstruction. *J. China Coal Soc.* **2021**, *46*, 253–262. [CrossRef]
16. Yuan, C.; Fan, L.; Cui, J.; Wang, W. Numerical simulation of the supporting effect of anchor rods on layered and nonlayered roof rocks. *Adv. Civ. Eng.* **2020**, *2020*, 4841658. [CrossRef]
17. Cai, W.; Dou, L.; Si, G.; Hu, Y. Fault-induced coal burst mechanism under mining-induced static and dynamic stresses. *Engineering* **2021**, *7*, 687–700. [CrossRef]
18. Duan, C.; Zheng, Q.; Xue, J.; Yu, G.; Luo, Y. Experimental study on zonal failure discontinuous deformation of deep surrounding rock under different working conditions. *J. Min. Strat. Control Eng.* **2021**, *3*, 76–84. [CrossRef]
19. Moradi, A.; Tokhmechi, B.; Rasouli, V.; Fatehi Marji, M. A Comprehensive numerical study of hydraulic fracturing process and its affecting parameters. *Geotech. Geol. Eng.* **2017**, *35*, 1035–1050. [CrossRef]
20. Gao, F. Influence of hydraulic fracturing of strong roof on mining-induced stress-insight from numerical simulation. *J. Min. Strat. Control Eng.* **2021**, *3*, 5–13. [CrossRef]
21. Jendryś, M.; Hadam, A.; Ćwiękała, M. Directional hydraulic fracturing (dhf) of the roof, as an element of rock burst prevention in the light of underground observations and numerical modelling. *Energies* **2021**, *14*, 562. [CrossRef]
22. Nguyen, V.; Lian, H.; Rabczuk, T.; Bordas, S. Modelling hydraulic fractures in porous media using flow cohesive interface elements. *Eng. Geol.* **2017**, *225*, 68–82. [CrossRef]
23. Wu, Y.; Yang, J. True tri-axial directional hydraulic fracturing test on sandstone with transverse grooves in coal mine. *J. China Coal Soc.* **2020**, *45*, 927–935. [CrossRef]
24. Cheng, Q.; Huang, B.; Zhao, X. Numerical investigation on the mechanism of rock directional fracturing method controlled by hydraulic fracturing in dense linear mul-tiholes. *Shock Vib.* **2020**, *2020*, 6624047. [CrossRef]
25. Whittles, D.; Lowndes, I.; Kingman, S. Influence of geotechnical factors on gas flow experienced in a UK longwall coal mine panel. *Int. J. Rock Mech. Min. Sci.* **2005**, *43*, 369–387. [CrossRef]
26. Yang, D.; Zhang, Y.; Wang, S.; Niu, C.; Chai, J. Analysis of the influence of hidden fault dip angle on ground pressure behavior in shallow seam roof. *J. Min. Strat. Control Eng.* **2020**, *2*, 72–82. [CrossRef]
27. Dong, Y.; Yao, W.; Liu, Y. Strong mining pressure treatment in double lane layout working face with large mining depth and high ground stress. *Coal Sci. Technol.* **2018**, *46*, 103–107.
28. Li, C.; Li, P.; Lu, S.Y.; Luo, M.; Liu, Y.L.; Zuo, M. Failure analysis and control of retained roadway at working face under protection coal pillar of the faults. *J. Min. Sci. Technol.* **2020**, *5*, 519–527. [CrossRef]
29. Andrews, P.; Butcher, R.; Ekkerd, J. The geotechnical evolution of deep-level mechanized destress mining at South Deep mine. *J. S. Afr. Inst. Min. Metall.* **2020**, *120*, 33–40. [CrossRef]
30. Du, F.; Wang, K.; Guo, Y.; Wang, G.; Wang, L.; Wang, Y. The mechanism of rock burst-outburst coupling disaster considering the coal-rock combination: An experiment study. *Geomech. Eng.* **2020**, *22*, 255–264. [CrossRef]
31. Pan, J.; Kang, H.; Yan, Y.; Ma, S.; Ma, W.; Lu, C.; Lv, D.; Xu, G.; Feng, M.; Xia, Y.; et al. The Method, Mechanism and Application of Preventing Rock Burst by Artificial Liberation Layer of Roof. *J. China Coal Soc.* **2023**, *48*, 636–648. [CrossRef]
32. Fu, Q.; Yang, K.; He, X.; Wei, Z.; Yang, Q. Characteristics of Strata Behavior and Differentiated Control of Fully Mechanized Mining Working Face with Abnormal Roof. *Sustainability* **2022**, *14*, 13354. [CrossRef]
33. Yin, W.; Bin, Z.; Deng, Z.; Kun, L.; Wang, H. Research on behavior of underground pressure in shallow coal seam with three-face goaf working face. Front. *Energy Res.* **2022**. [CrossRef]
34. Rong, H.; Cheng, T. Research and Control of Underground Pressure Law of Coal Mining Face in Soft Rock Strata Based on Neural Network under the Background of Wireless Network Communication. *Mob. Inf. Syst.* **2021**, *2021*, 6560713. [CrossRef]
35. Zhang, Q.; Zhang, J.X.; Kang, T.; Sun, Q.; Li, W. Mining pressure monitoring and analysis in fully mechanized backfilling coal mining face-A case study in Zhai Zhen Coal Mine. *J. Cent. South Univ.* **2015**, *22*, 1965–1972. [CrossRef]
36. Li, D.; Jiang, F.; Wang, C.; Tian, Z.; Wang, Y. Study on the mechanism and prevention technology of "square position" and "stress breakdown effect" inducing rock burst. *J. Min. Safety Eng.* **2018**, *35*, 1014–1021.
37. Rong, H.; Yu, S.; Wang, Y.; Pan, L.; Guo, K.; Li, N. Analysis of structural instability movement of hard overburden and its influence mechanism on rockburst. *J. Min. Strat. Control Eng.* **2022**, *4*, 063026. [CrossRef]
38. Hao, X.; Han, G.; Xie, J.; Zhang, Y.; Han, Z. Rock burst mechanism of roadway excavation along goaf with small coal pillar in Ordos mining area. *J. Min. Strat. Control Eng.* **2023**, *5*, 023017. [CrossRef]

39. Jia, J.; Wang, D.; Li, B. Study on influencing factors of effective fracturing radius of hydraulic fracturing. *J. Saf. Sci. Technol.* **2022**, *18*, 58–64. [CrossRef]
40. Li, Q.; Deng, Y.; Hu, Q.; Zhang, Y.; Song, M.; Liu, J.; Shi, J. Mesoscopic law of stress and fracture evolution of coal seams hydraulic fracturing. *Coal Geol. Explor.* **2022**, *50*, 32–40. [CrossRef]

processes MDPI

Article
Experimental Study on Rapid Determination Method of Coal Seam Gas Content by Indirect Method

Hongyan Lei [1,2], Linchao Dai [1,2,*], Jie Cao [1,2], Rifu Li [1,2] and Bo Wang [1,2]

1 State Key Laboratory of the Gas Disaster Detecting, Preventing and Emergency Controlling, Chongqing 400037, China
2 China Coal Technology and Engineering Group, Chongqing Research Institute, Chongqing 400037, China
* Correspondence: 20212001047g@cqu.edu.cn; Tel.: +86-023-6523-9611

Abstract: In view of the problems of heavy work, long cycle, high cost and low efficiency in the process of indirect determination of coal seam gas content, the basic gas parameters and coal quality indexes of 24 coal samples from 5 coal mines in the Hancheng area of Shanxi Province are measured by the laboratory measurement method. The raw coal gas content–gas desorption index of drilling cuttings (W–K_1) relationship model is characterized by logarithmic function. Using SPSS data analysis software, a stepwise multiple linear regression method is used for statistical analysis. The results show that the factors that have a significant impact on the regression slope C in the W–K_1 relationship model are gas adsorption constant (a), apparent density (ARD), initial velocity of gas diffusion (Δp) and consistent coefficient (f). The factors that have a significant impact on the regression constant D are Δp and atmospheric adsorption (Q). Then, the mathematical model of rapid prediction of coal seam gas content is determined. Compared with the measured values, the average absolute error rate is 12.84%, which meets the prediction requirements and provides a simple and easy method for rapid determination of coal seam gas content in coal mines in the Hancheng area.

Keywords: coal seam gas content; W–K_1 relation model; multiple linear regression; gas pressure

1. Introduction

Coal seam gas content directly affects the amount of coal seam gas and the amount of mine gas emission, which are of great significance for the correct design of mine ventilation, gas drainage and outburst risk assessment [1–4]. Coal seam gas content is an important parameter to characterize the occurrence of coal seam gas. Accurate prediction of coal seam gas content plays a vital role in the prevention and control of coal mine gas disasters. The determination method of coal seam gas content is divided into the direct method and indirect method. The direct method [5–7] is to measure the amount of desorbable gas and normal pressure gas in the underground field and laboratory and then calculate the gas loss in the sampling process. The sum of the three parts is the gas content of the coal seam. The indirect method [8–10] is to measure the basic gas parameters such as gas adsorption constant, industrial analysis, density and porosity in the laboratory. Combined with the measured coal seam gas pressure in the underground field, the adsorbed gas amount and free gas amount of coal are calculated by the Langmuir equation. The sum of the two parts is the coal seam gas content. Because there are many factors affecting the gas content of coal seams [1,11–15], and the gas occurrence has the characteristics of complexity, nonlinearity, dynamics and random uncertainty, it is difficult to accurately determine the gas content of a coal seam. In recent years, for the prediction of coal seam gas content, Zhang et al. [16] used the method of multiple linear regression analysis. Chen et al. [17–19] adopted the grey theory method. Zhang et al. [20,21] used the neural network analysis method. Wang et al. [22,23] used a machine learning algorithm. They studied different mathematical models for predicting coal seam gas content. The prediction model mainly focuses on the

Citation: Lei, H.; Dai, L.; Cao, J.; Li, R.; Wang, B. Experimental Study on Rapid Determination Method of Coal Seam Gas Content by Indirect Method. *Processes* **2023**, *11*, 925. https://doi.org/10.3390/pr11030925

Academic Editor: Qingbang Meng

Received: 25 February 2023
Revised: 15 March 2023
Accepted: 16 March 2023
Published: 17 March 2023

relationship between the influencing factors such as coal seam depth, coal seam thickness, floor elevation, fault distance, coal rock dip angle and coal seam gas content. These scholars are using direct methods to predict coal seam gas content. However, there are few studies on predicting coal seam gas content by indirect methods in a laboratory. Li [24] used the drilling cuttings gas desorption index method to predict coal seam gas content. Through the study, the exponential mathematical model for predicting coal seam gas content was obtained, which avoided the determination of coal seam gas pressure. However, the regression coefficient of the exponential mathematical model is not accurately determined. Compared with the direct method, the gas basic parameters measured by the indirect method are all measured values, and the gas pressure of the coal seam is also measured. There are fewer influencing factors in the measurement process, the measurement error is small, and the measurement data is relatively reliable. However, the disadvantage of using an indirect method to measure coal seam gas content is that it is necessary to measure coal seam gas pressure underground. This is relatively heavy work, and the measurement period is long (about one month). Especially in the gently inclined coal seam or the coal seam with poor surrounding rock density, it is difficult to measure the gas pressure of the coal seam. In short, it is common that the results of a direct method are low, while the results of indirect methods in many mines are close to reality [25]. To solve the problem of accurate determination of coal seam gas pressure in the indirect determination of coal seam gas content, 24 coal samples from 5 coal mines in the Hancheng area of Shanxi Province are selected. The relationship model between gas basic parameters, coal quality index, drilling cuttings gas desorption index and gas pressure of these coal samples was measured in the laboratory. The SPSS data analysis software was used, and the stepwise multiple linear regression analysis method was used. A mathematical model for predicting the gas content of coal seams in the Hancheng area unrelated to gas pressure was established. This method can quickly and accurately determine the gas content of the coal seam in this area, and provide technical guidance and reference for mine gas disaster prevention, coal and tile outburst prediction, gas emission prediction and so on.

2. Indirect Method to Determine the Coal Seam Gas Content

2.1. Coal Seam Gas Content Calculation

The most commonly used indirect determination method of coal seam gas content at home and abroad is to calculate the coal seam gas content according to the known coal seam gas pressure and the gas adsorption constant value of coal measured in the laboratory [26]. The calculation formula is:

$$W = \frac{abp(100 - A_{ad} - M_{ad})}{100(1 + bp)(1 + 0.31M_{ad})} + \frac{10kp}{ARD} \tag{1}$$

where W is the raw coal gas content, m^3/t. a is gas adsorption constant, cm^3/g. b is the gas adsorption constant, MPa^{-1}. p is coal seam gas pressure, MPa. A_{ad} is the ash content of coal, %. M_{ad} is the moisture of coal, %. k is the porosity of coal, %. ARD is the apparent density of coal, t/m^3.

2.2. Coal Sample Taking and Test

Four and five coal samples are taken from different sampling sites of 3# coal seam and 5# coal seam, respectively, in the Xiangshan Coal Mine in the Hancheng area of Shanxi Province. Two and three coal samples are taken from 2# coal seam and 3# coal seam, respectively, in the Xiayukou Coal Mine. Three and two coal samples are taken from 3# coal seam and 5# coal seam, respectively, in the Xinglong Coal Mine. Three coal samples are taken from 3# coal seam in the Sangbei Coal Mine. Two coal samples are taken from 3# coal seam in the Sangshuping Coal Mine. A total of 24 coal samples are taken. According to the GB/T482–2008 'coal seam coal sample taking method', GB/T474–2008 'coal sample preparation method', GB/T477–2008 'coal sample screening test method' and other national standards, about 5 kg mixed coal samples are taken from freshly exposed coal seams, indicating the sampling location, and the packaging is strictly sent to the laboratory for drying, crushing and screening, and coal samples of different particle sizes are prepared for inspection, according to the MT/T752–1997 'Determination method of methane adsorption in coal', GB/T212–2008 'Industrial analysis method of coal', GB/T 217–2008 'Truth relative density determination method of coal', GB/T6949–2008 'Apparent relative density determination method of coal', AQ1080–2009 'Determination method of initial velocity index (Δp) of gas emission in coal', GB/T23561.12–2010 'Determination method of firmness coefficient of coal' and other coal industry standards and national standards. The HCA-type high-pressure capacity method gas adsorption device, industrial analysis tester, density tester, gas emission initial velocity tester, WTC gas outburst parameter tester and other instruments and equipment are used. In the laboratory, the moisture M_{ad}, ash A_{ad}, volatile V_{daf}, true density TRD, apparent density ARD, porosity k, atmospheric adsorption Q and gas adsorption constant a and b of 24 coal samples from 5 coal mines in the Hancheng area are measured. The results of gas basic parameters of 24 coal samples are summarized.

The variation ranges of each parameter of 24 coal samples measured in Table 1 are as follows: M_{ad} is 0.54~1.59%, A_{ad} is 4.30~26.87%, V_{daf} is 13.10~18.61%, TRD is 1.32~1.57 g·cm^{-3}, ARD is 1.29~1.51 g·cm^{-3}, k is 2.05~5.84%, Q is 2.1624~3.8969 cm^3·g^{-1}, a is 17.0599~26.6249 cm^3·g^{-1}, and b is 0.8217~1.7279 MPa^{-1}. From the distribution characteristics of the measured data, the selected coal samples are universal and extensive.

Table 1. Summary of basic gas parameters of 24 coal samples measured.

Coal Sample Number	Sampling Site	Moisture M_{ad}/%	Ash A_{ad}/%	Volatile V_{daf}/%	True Density TRD/(g·cm^{-3})	Apparent Density ARD/(g·cm^{-3})	Porosity k/%	Atmospheric Adsorption Capacity Q/(cm^3·g^{-1})	Gas Adsorption Constant a/(cm^3·g^{-1})	Gas Adsorption Constant b/MPa^{-1}
M01	Xiangshan mine 3# coal seam 22301 return airway	0.79	26.87	17.59	1.57	1.51	3.82	3.0595	24.5904	0.9459
M02	Xiangshan Mine 5# coal seam track downhill	0.93	16.67	14.26	1.48	1.43	3.38	3.4328	25.8505	1.1127
M03	21220 mining face of 2# coal seam in Xiayukou Mine	0.65	9.36	16.20	1.39	1.34	3.60	3.0190	22.8960	1.1560
M04	Xiayukou mine 3# coal seam 23306 Yunshun	0.81	4.30	14.44	1.32	1.29	2.27	3.3520	26.6249	0.9581
M05	21301 open-off cut of 3# coal seam in Xinglong Coal Mine	0.70	11.73	18.61	1.46	1.43	2.05	2.4276	17.0599	1.2944
M06	Xinglong Mine 5# coal seam 4501 Yunshun	0.65	15.64	17.90	1.50	1.45	3.33	3.1282	22.3545	1.251 5
M07	Sangbei Mine 3# coal seam 11308 transport trough	0.68	12.08	14.22	1.41	1.37	2.84	3.0605	19.7706	1.2854
M08	Sangbei Mine 3# coal seam 11308 return air trough	0.82	10.06	14.95	1.39	1.35	2.88	3.2195	22.7722	1.2786
⋮	⋮	⋮	⋮	⋮	⋮	⋮	⋮	⋮	⋮	⋮
M23	4301 control roadway of 3# coal seam in Sangshuping Mine	0.73	9.61	14.89	1.41	1.37	2.84	3.0570	22.4181	1.0547
M24	South belt roadway of 3# coal seam in Sangshuping Mine	0.84	15.53	15.20	1.48	1.44	2.70	3.7367	22.1534	1.5578

3. K_1–p Relation Model

The drilling cuttings gas desorption index K_1 characterizes the characteristic parameters of the coal gas desorption speed. It reflects the coal seam gas content and the size of the initial pressure relief gas desorption rate. Scholars at home and abroad have carried out extensive and in-depth research on the law of gas desorption from drilling cuttings in the laboratory [27,28]. They mainly studied the influence of mathematical fitting of gas desorption law and gas pressure on the gas desorption index of drilling cuttings. The research results show that the gas desorption law of drilling cuttings is the law of gas desorption amount of drilling cuttings changing with time. It can not only reflect the gas adsorption pressure and desorption gas content in coal but also reflect the damage degree of coal; that is, it can be used for outburst prediction. It is also used to determine the gas pressure and tile content of a coal seam. According to Zhao's research results [29], there is a power function relationship between the gas desorption index of drilling cuttings and the gas pressure of adsorption equilibrium:

$$K_1 = mp^n \tag{2}$$

where K_1 is the gas desorption index of drilling cuttings, mL·g^{-1}·min$^{-0.5}$. p is the adsorption equilibrium gas pressure, MPa. m and n are undetermined constants, $0 < n < 1$.

The instrument used to determine K_1 in the underground field and laboratory is a WTC gas outburst parameter instrument. The underground site is measured according to the AQ/T 1065–2008 drilling cuttings gas desorption index determination method. The laboratory determination of the K_1–p relationship model test process is based on the test method adopted by Lei [30]. According to Equation (2), the key to determine the gas desorption index of drilling cuttings is to determine the undetermined constants m and n. The essence of laboratory determination of the K_1–p relationship model is to determine the undetermined constants m and n. The measured K_1–p relationship model results of 24 coal samples, gas basic parameters such as gas initial velocity Δp, coal firmness coefficient f and coal quality index measurement results are summarized in Table 2 below.

Table 2. Measured 24 coal samples relational model and other parameters summary table.

Coal Sample Number	Sampling Site	Initial Velocity of Gas Diffusion Δp/mmHg	Consistent Coefficient of Coal f	K_1–p Relation Model	$(K_1)_{p=1}$	W–K_1 Relation Model
M01	Xiangshan mine 3# coal seam 22301 return airway	21	0.49	$K_1 = 0.5086\ p^{0.5761}$	0.5086	$W = 5.408\ \ln K_1 + 11.39$
M02	Xiangshan Mine 5# coal seam track downhill	25	0.30	$K_1 = 0.6598\ p^{0.5652}$	0.6598	$W = 6.310\ \ln K_1 + 11.94$
M03	21220 mining face of 2# coal seam in Xiayukou Mine	20	0.32	$K_1 = 0.4093\ p^{0.6163}$	0.4093	$W = 6.053\ \ln K_1 + 15.23$
M04	Xiayukou mine 3# coal seam 23306 Yunshun	26	0.10	$K_1 = 0.9097\ p^{0.5427}$	0.9097	$W = 7.722\ \ln K_1 + 11.35$
M05	21301 open-off cut of 3# coal seam in Xinglong Coal Mine	14	0.22	$K_1 = 0.4400\ p^{0.6535}$	0.4400	$W = 4.000\ \ln K_1 + 10.46$
M06	Xinglong Mine 5# coal seam 4501 Yunshun	12	0.43	$K_1 = 0.3438\ p^{0.6871}$	0.3438	$W = 4.911\ \ln K_1 + 14.35$
M07	Sangbei mine 3# coal seam 11308 transport trough	14	0.39	$K_1 = 0.2696\ p^{0.6147}$	0.2696	$W = 5.015\ \ln K_1 + 14.96$
M08	Sangbei mine 3# coal seam 11308 return air trough	13	0.39	$K_1 = 0.2704\ p^{0.6040}$	0.2704	$W = 5.755\ \ln K_1 + 17.00$
⋮	⋮	⋮	⋮	⋮	⋮	⋮
M23	4301 control roadway of 3# coal seam in Sangshuping Mine	10	0.44	$K_1 = 0.2646\ p^{0.6634}$	0.2646	$W = 5.271\ nK_1 + 12.04$
M24	South belt roadway of 3# coal seam in Sangshuping Mine	16	0.31	$K_1 = 0.2678\ p^{0.7174}$	0.2678	$W = 4.368\ \ln K_1 + 14.84$

The variation ranges of each parameter of 24 coal samples measured in Table 2 are as follows: Δp is 7~37 mmHg, f is 0.10~0.55, K_1–p relationship model coefficient m is 0.217~0.9371, and index n is 0.4431~0.7740.

4. Test Result Analysis

4.1. Relationship between K and Coal Quality Index

Coal quality indexes mainly include gas emission initial velocity Δp, firmness coefficient f, volatile matter V_{daf}, ash A_{ad}, etc. They macroscopically reflect some essential characteristics related to coal and gas desorption. Among them, Δp and f are commonly used in regional outburst risk prediction. Because Δp, f, V_{daf}, A_{ad}, etc. are all coal quality indicators unrelated to gas pressure, the relationship between them and the values under certain gas pressure conditions can only be considered. In Table 2, the K_1 value is calculated when the gas pressure p = 1 MPa. Through nonlinear fitting of the data in Tables 1 and 2, the relationship between $(K_1)_{p=1}$ and Δp, f, V_{daf}, A_{ad} is obtained, as shown in Figure 1.

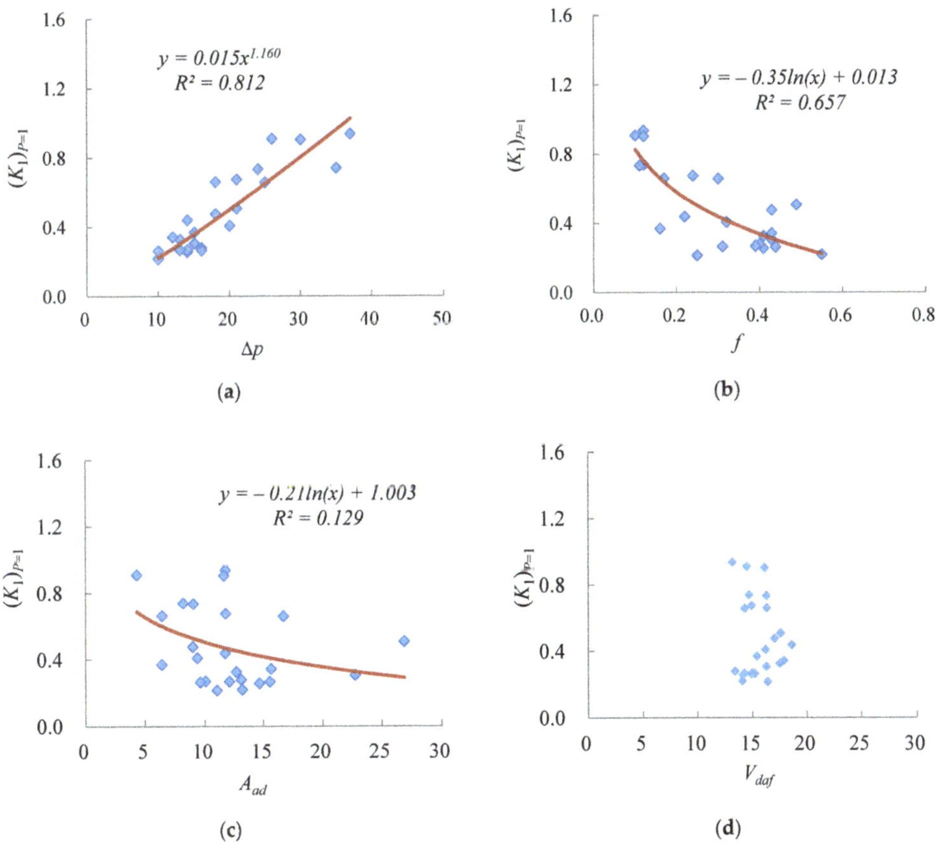

Figure 1. Relationship between $(K_1)_{p=1}$ and coal quality index. (**a**) Relationship between $(K_1)_{p=1}$ and Δp. (**b**) Relationship between $(K_1)_{p=1}$ and f. (**c**) Relationship between $(K_1)_{p=1}$ and A_{ad}. (**d**) Relationship between $(K_1)_{p=1}$ and V_{daf}.

It can be seen from Figure 1a that the relationship between $(K_1)_{p=1}$ and Δp conforms to the power function increasing relationship. With the increase in Δp, the value of $(K_1)_{p=1}$ also increases correspondingly. The correlation coefficient R^2 is 0.812, indicating that there is a significant correlation between K_1 and Δp, and Δp significantly affects the positive

$(K_1)_{p=1}$. It can be seen from Figure 1b that $(K_1)_{p=1}$ is in accordance with the logarithmic function attenuation relationship. As f increases, the value of $(K_1)_{p=1}$ decreases accordingly. The correlation coefficient R^2 is 0.657, indicating that there is also a certain correlation between $(K_1)_{p=1}$ and f, and f affects negative $(K_1)_{p=1}$, but the significance is lower than Δp. It can be seen from Figure 1c that the distribution of 24 coal samples represented by 24 points in the figure is relatively disordered. The relationship between $(K_1)_{p=1}$ and A_{ad} is an approximately logarithmic function, and the correlation coefficient R^2 is only 0.129, indicating that the correlation between $(K_1)_{p=1}$ and A_{ad} is not close. It can be seen from Figure 1d that 24 points are concentrated in the range of 13.10~18.61 of V_{daf}, and $(K_1)_{p=1}$ has no correlation with V_{daf}.

From the above analysis, it can be seen that under the same gas pressure condition, the value of K_1 mainly depends on Δp. In theory, K_1 is the same as Δp, which is an index to reflect the risk of outburst by the amount of initial gas desorption. Their fundamental difference is only the difference of adsorption gas pressure; that is, K_1 index reflects the change of adsorption gas pressure more than Δp. The firmness coefficient f mainly reflects the ability of coal to resist damage. In many cases, the lower the strength of a coal seam, the greater the initial gas desorption. The regression analysis shows that under the condition of certain gas pressure, K_1 decreases with the increase in f in a negative exponential law. The volatile matter V_{daf} of coal reflects the metamorphic degree of coal, and the ash A_{ad} reflects the yield of effective carbon. Both of them are not closely related to the initial gas desorption amount. In short, K_1 is most affected by Δp and less affected by V_{daf} and A_{ad}. Therefore, Δp is the main influencing factor of K_1, and f, V_{daf} and A_{ad} are secondary influencing factors.

4.2. The Relationship between Coal Seam Gas Content W and K_1

The coal seam gas content W in Equation (1) and the drilling cuttings gas desorption index K_1 in Equation (2) are directly related to the gas pressure p, and as the gas pressure increases, K_1 and W increase. To find the relationship between K_1 and W, gas pressure p is set to 0.1, 0.5, 1.0, 2.0, 3.0, 4.0, 5.0 MPa. Equations (1) and (2) are used to calculate 7 K_1 and 7 W corresponding to 7 gas pressures, and then, the nonlinear fitting is carried out. The W–K_1 relationship model is obtained, and the correlation coefficient R^2 is greater than 0.9900. The results are summarized in Table 2. One coal sample each (limited to the length of the article) was selected from the Xiangshan Coal Mine, Xiayukou Coal Mine, Xinglong Coal Mine, Sangbei Coal Mine and Sangshuping Coal Mine for nonlinear fitting. The fitting relationship curve is shown in Figure 2.

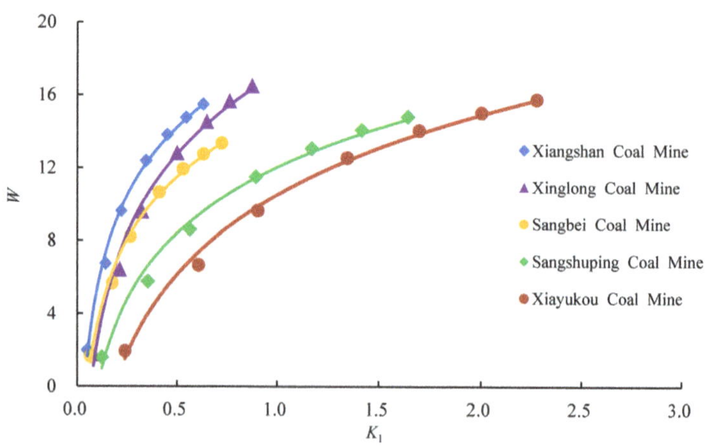

Figure 2. The relationship between W and K_1.

It can be seen from Figure 2 that there is a significant correlation between coal seam gas content W and drilling cuttings gas desorption index K_1, which is characterized by logarithmic function. The W–K_1 relationship model can be expressed by Equation (3):

$$W = C \ln K_1 + D \quad (3)$$

where W is the gas content of raw coal, m^3/t. K_1 is the gas desorption index of drilling cuttings, $mL \cdot g^{-1} \cdot min^{-0.5}$. C is regression slope. D is regression constant.

In the process of adsorption and desorption of coal, the change of coal seam gas content W and drilling cuttings gas desorption index K_1 is caused by the change of coal seam gas pressure. This change eventually leads to the continuous change of diffusion coefficient in the process of coal desorption, which leads to the unique nonlinear relationship between W and K_1. Figure 2 shows that when the gas desorption index K_1 of the five coal mines is the same, the coal seam gas content W of the Xiangshan Coal Mine is the largest, and the coal seam gas content W of the Xiayukou Coal Mine is the smallest. From Table 2, the W–K_1 relationship model coefficient C is 4.000~7.722, and D is 9.829~17.98.

4.3. Determination of Regression Coefficient of W–K_1 Relation Model

It can be seen from Equation (3) that the key is to determine the regression coefficients C and D, whether it is to determine the coal seam gas content W through the drilling cuttings gas desorption index K_1 or to determine K_1 through W. Although the regression coefficients C and D of the W–K_1 relationship model of a specific sampling site in a coal mine can be obtained through the laboratory, combined with the measured data of the underground site, this is the result based on a large amount of experimental data such as 11 gas basic parameters and coal quality indexes measured in the laboratory and the measured coal seam gas pressure and drill cuttings gas desorption index K_1 in the coal mine. The determination of these parameters is heavy, long cycle, high cost and low efficiency. For the coal mines in the Hancheng area, only by finding out the universal law of the regression coefficients C and D in the W–K_1 relationship model and establishing a regression coefficient relationship model suitable for the area can the coal seam gas content be predicted quickly and accurately, and a simple and easy method for predicting the coal seam gas content in the Hancheng area is provided.

Taking 24 coal samples from 5 coal mines in the Hancheng area as samples, 24 regression coefficients C and 24 regression coefficients D (as shown in Table 2) obtained by the W–K_1 relationship model are taken as dependent variables, respectively. M_{ad}, A_{ad}, V_{daf}, TRD, ARD, k, Q, a, b, Δp, f and another 11 gas basic parameters and coal quality indexes corresponding to coal samples are used as independent variables. The SPSS data analysis software is used. The stepwise multiple linear regression method is used for statistical analysis. According to the order of input parameters, 11 independent variables are introduced into the regression formula one by one. The regression results show that for the dependent variable C, after eliminating the seven parameters that cannot have a significant impact, the final model introduces four parameters a, ARD, Δp and f. Similarly, for the dependent variable D, after eliminating the nine parameters that cannot have a significant impact, the final model introduces two parameters such as Δp and Q. Through SPSS data analysis, the multiple linear regression model of regression coefficients C and D in the W–K_1 relationship model is:

$$C = 11.031 + 0.148a + 0.084\Delta p - 7.919ARD + 2.263f \quad (4)$$

$$D = 4.724 - 0.339\Delta p + 4.725Q \quad (5)$$

4.4. Multiple Linear Regression Model Test

The prerequisites for the establishment of multiple linear regression models (4) and (5) are that there is no serious autocorrelation between series, there is no strong multicollinearity between independent variables and the residual distribution basically obeys the normal distribution. Only when these three assumptions are satisfied at the same time can it be proved that (4) and (5) have certain validity and reliability. SPSS data analysis software is used for statistical analysis of the output results (as shown in Table 3). The coefficient of determination of the regression model (4) in Table 3 is 0.833, indicating that the four independent variables a, ARD, Δp and f can explain 83.3% of the change of the dependent variable C, R^2 is close to 1, and the goodness of regression fitting is better. That is to say, there is a very close linear correlation between C and a, ARD, Δp and f. The coefficient adjustment R^2 of the regression model (5) is only 0.631, indicating that the two independent variables such as Δp and Q can explain 63.1% of the change in the dependent variable D, and there is a certain linear correlation between D and Δp and Q.

Table 3. Model summary.

Regression Model	Adjusted R^2	Sig	Durbin-Wastson
(4)	0.833	0.038	2.082
(5)	0.631	0.001	1.377

Regression model (4) significance level (t statistics corresponding probability value) $Sig = 0.038 < 0.05$. The probability that 11 independent variables cannot have a significant impact on the dependent variable C is 0. Rejecting the null hypothesis, there are at least four independent variables, a, ARD, Δp and f, that have a significant impact on C in the 11 independent variables. Among them, a, Δp and f significantly affect positive C, and ARD significantly affects negative C. The significance level of regression model (5) (the probability value corresponding to t statistics) $Sig = 0.001 < 0.05$, indicating that there are at least two independent variables, Δp and Q, that have a significant impact on D, where Δp significantly affects negative D and Q significantly affects positive D.

Regression models (4) and (5) Durbin-Wastson in Table 3. are 2.082 and 1.377, respectively, which are close to 2. In statistics, when the Durbin-Wastson value is significantly close to 0 or 4, the sequences are not independent of each other, and there is a serious autocorrelation. Therefore, it can be considered that the series in regression models (4) and (5) are independent of each other and there is no serious autocorrelation.

In statistics, it is generally believed that there is no strong multicollinearity between independent variables when the variance inflation factor VIF is less than 5. The output result of statistical analysis using SPSS data analysis software is that the variance expansion factor VIF of the four independent variables of a, ARD, Δp and f in the regression model (4) is 1.761, 1.771, 3.262, and 3.133, respectively, all less than 5. The variance expansion factor VIF of the two independent variables, such as Δp and Q, are 1.488 and 1.488, respectively, which are also less than 5. It shows that there is no strong multicollinearity between a, ARD, Δp and f, between Δp and Q; that is, there is no strong correlation.

From the standard P–P diagram and scatter diagram of the regression standardized residuals of the dependent variables C and D in Figures 3 and 4, it can be seen that the standardized residuals are basically distributed around the asymptote, the scatters are basically linear, and the data and models are basically matched. The distribution of sample points is scattered and irregular, the residuals are random, and there is no heteroscedasticity. This shows that the residual basically obeys the normal distribution.

The main factors that significantly affect the regression coefficient C are a, ARD, Δp and f, and the main factors that significantly affect the regression coefficient D are Δp and Q. There is no serious autocorrelation between the tested series, and there is no strong multicollinearity between the independent variables. The residual distribution is basically normal distribution. Therefore, the C and D regression models (4) and (5) established by SPSS data analysis software have certain validity and reliability.

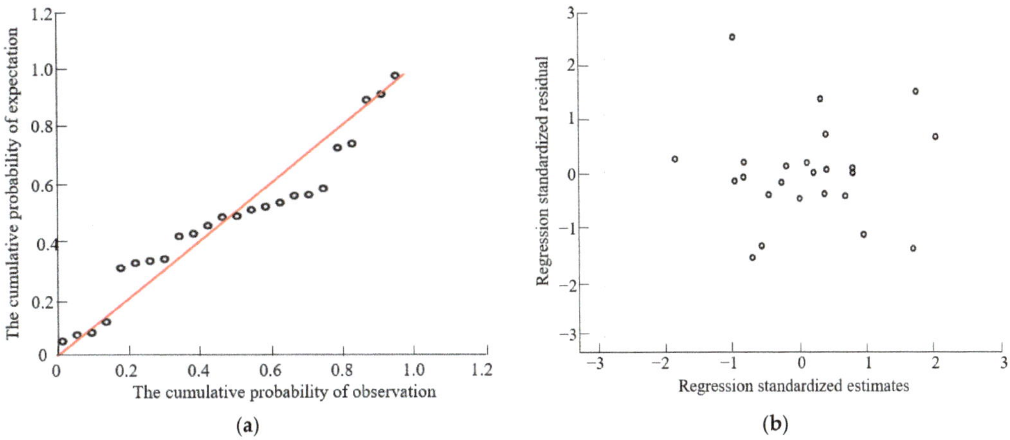

Figure 3. Standard *P–P* diagram and scatter plot diagram of regression normalized residual of dependent variable *C*. (**a**) Cumulative probability of observation. (**b**) Regression standardized predicted value.

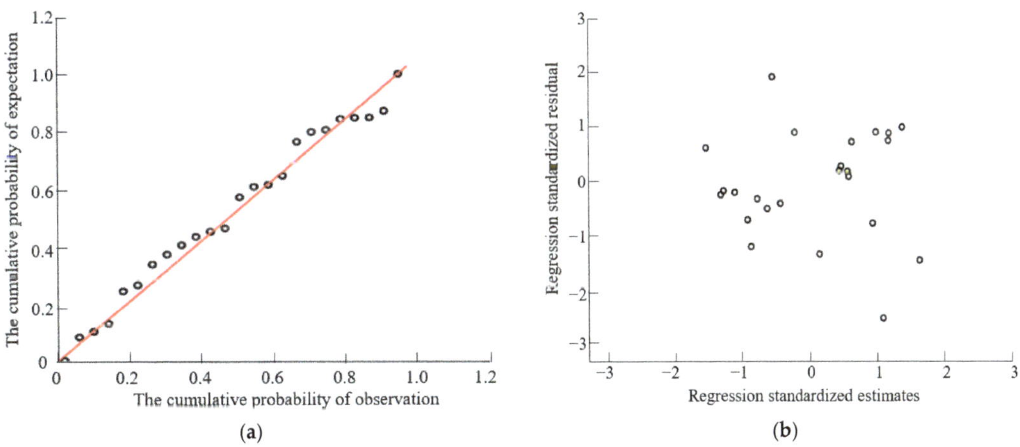

Figure 4. Standard *P–P* diagram and scatter plot diagram of regression normalized residual of dependent variable *D*. (**a**) Cumulative probability of observation. (**b**) Regression standardized predicted value.

5. Mathematical Model Prediction Correction of Coal Seam Gas Content

Through the above research and analysis, the regression models (4) and (5) of regression coefficients *C* and *D* are established. Substituting (4) and (5) into Equation (3), the mathematical model (6) for predicting coal seam gas content in coal mines in the Hancheng area is finally obtained:

$$\begin{aligned} W' = & (11.031 + 0.148a + 0.084\Delta p - 7.919ARD + 2.263f)\ln K_1 \\ & +(4.724 - 0.339\Delta p + 4.725Q) \end{aligned} \tag{6}$$

where W' is the predicted value of coal gas content, m^3/t.

Three coal samples are taken from the Xiangshan Coal Mine, Xiayukou Coal Mine and Sangbei Coal Mine in the Hancheng area. Eleven gas basic parameters and coal quality indexes of M_{ad}, A_{ad}, V_{daf}, TRD, ARD, F, Q, a, b, Δp, f, etc. of three coal samples are measured in the laboratory. At the same time, the K_1–p relationship model of three coal samples is measured by a WTC gas outburst parameter instrument [31]. The gas content of the coal seam is predicted by mathematical model (6). The gas pressure p is set as 0.5, 1.0, 2.0, 3.0, 4.0 and 5.0 MPa, respectively. The predicted value W' of the mathematical model of coal seam gas content is compared with the measured value W (as shown in Table 4).

Table 4. Comparison of predicted and measured values of the mathematical model of coal bed methane content.

Gas Pressure p/MPa	Xiayukou Coal Mine 23208 Working Face of 2# Coal Seam			Sangbei Coal Mine 3# Coal Seam 11308 Transport Trough			Xiangshan Coal Mine 5# Coal Seam South Wing Belt Transport Roadway		
	Measured Value (W)/($m^3 \cdot t^{-1}$)	Predicted Value (W′)/($m^3 \cdot t^{-1}$)	Error Rate/%	Measured Value (W)/($m^3 \cdot t^{-1}$)	Predicted Value (W′)/($m^3 \cdot t^{-1}$)	Error Rate/%	Measured Value (W)/($m^3 \cdot t^{-1}$)	Predicted Value (W′)/($m^3 \cdot t^{-1}$)	Error Rate/%
0.5	4.9836	3.7021	−25.71	4.8044	4.1224	−14.20	5.4155	5.4867	1.31
1.0	7.4337	5.7095	−23.19	7.0595	6.1035	−13.54	7.8719	7.5497	−4.09
2.0	9.9104	7.7169	−22.13	9.3171	8.0845	−13.23	10.2810	9.6128	−6.50
3.0	11.2080	8.8912	−20.67	10.5292	9.2434	−12.21	11.5566	10.8196	−6.38
4.0	12.0432	9.7243	−19.25	11.3446	10.0656	−11.27	12.4092	11.6758	−5.91
5.0	12.6496	10.3760	−17.97	11.9672	10.7033	−10.56	13.0581	12.3400	−5.50

Remarks: error rate = (W′–W)/W × 100%.

It can be seen from Table 4 that when the gas pressure is 0.5 MPa, 1.0 MPa, 2.0 MPa, 3.0 MPa, 4.0 MPa and 5.0 MPa, the variation range of the predicted value W' of the mathematical model of coal seam gas content and the measured value W is 3.7021~13.0581 m^3/t. The predicted value W' and the measured value W increase with the increase in gas pressure. When the gas pressure is in the range of 0.5~5.0 MPa, the maximum deviation between the predicted value W' and the measured value W is −2.3189 m^3/t when the gas pressure is 4.0 MPa in the Xiayukou Coal Mine. The minimum deviation is 0.0711 m^3/t corresponding to the gas pressure of 0.5 MPa in the Xiangshan Coal Mine. The average deviation is −1.2178 m^3/t. The maximum error rate is −25.71%. The minimum error rate is 1.31%. The average error rate is −12.84%.

In Figure 5, the trend of the predicted value W' of the mathematical model of coal seam gas content in the three coal mines is generally consistent with the measured value W, which conforms to the characteristics of logarithmic function. The predicted value W' of the Xiangshan Coal Mine is very close to the measured value W, and the change trend of the Xiayukou Coal Mine deviates far. In the low-pressure state, the coincidence degree is high, close to coincidence, and the average absolute error rate is 12.84%, which can meet the prediction requirements. It can be seen that the mathematical model of multiple linear regression has high reliability in predicting coal seam gas content in the Hancheng area. Because the average deviation is −1.2178 m^3/t, the predicted value W' is generally lower than the measured value W. Therefore, Equation (6) is modified by the average deviation of −1.2178 m^3/t. The modified mathematical model is:

$$W' = (11.031 + 0.148a + 0.084\Delta p - 7.919ARD + 2.263f)\ln K_1 + (4.724 - 0.339\Delta p + 4.725Q) + 1.2178 \tag{7}$$

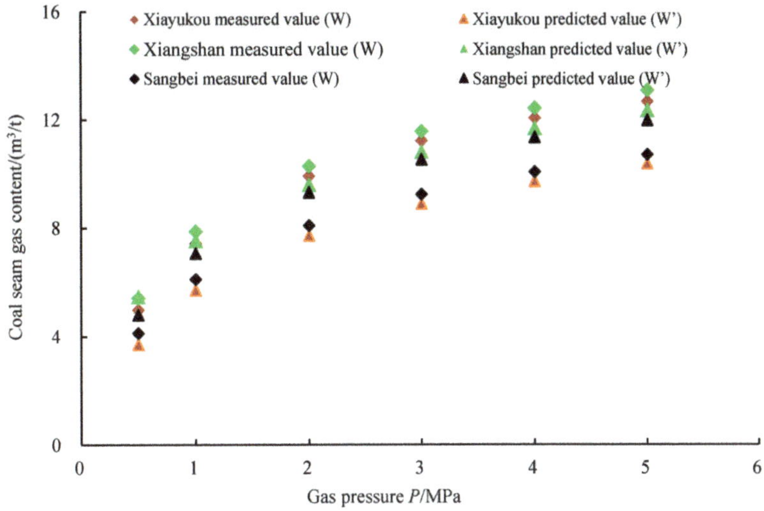

Figure 5. The predicted value W' of the regression model under different pressures is compared with the measured value W.

6. Conclusions

(1) Through the experimental study and analysis of coal samples in the Hancheng area of Shanxi, it is found that W and K_1 have a significant correlation. SPSS data analysis software is used for statistical analysis, and the main influencing factors of the regression coefficient C of the W–K_1 relationship model are a, ARD, Δp and f. Among them, a, Δp and f significantly affect positive C, and ARD significantly affects negative C. The main factors of regression coefficient D are Δp and Q, where Δp significantly affects negative D and Q significantly affects positive D. After testing, the multiple linear regression model has certain validity and reliability.

(2) When the gas pressure is 0.5, 1.0, 2.0, 3.0, 4.0 and 5.0 MPa, the maximum deviation between the predicted value W' of the mathematical model of coal seam gas content and the measured value W is -2.3189 m^3/t, the minimum deviation is 0.0711 m^3/t, and the average absolute error rate is 12.84%. It can meet the prediction requirements.

(3) The mathematical model of predicting coal seam gas content is only related to six parameters, such as a, ARD, Δp, f, Q and K_1, and has nothing to do with coal seam gas pressure. It avoids the problems of heavy task, long cycle, high cost and low efficiency in measuring coal seam gas pressure in a coal mine. The prediction model can quickly and accurately determine the gas content of a coal seam and provide a simple and feasible method for predicting the gas content of coal seams in the Hancheng area.

Author Contributions: Investigation, H.L., L.D., J.C., R.L. and B.W.; methodology, H.L., L.D. and J.C.; supervision, L.D., R.L. and B.W.; writing—original draft, H.L., L.D. and J.C.; writing—review and editing, H.L., L.D., J.C. and R.L.; funding acquisition, L.D., J.C., R.L. and B.W. All authors have read and agreed to the published version of the manuscript.

Funding: This research was funded by the National Natural Science Foundation of China (No. 51974358, No. 52104239 and No. 51874348), Natural Science Foundation of Chongqing (No. CSTB2022NSCQ-MSX1080, No. CSTB2022NSCQ-MSX0379 and No. cstc2020jcyj-msxmX1052), and Chongqing Science Fund for Distinguished Young Scholars (No. cstc2019jcyjjqX0019).

Data Availability Statement: All data and/or models used in the study appear in the submitted article.

Acknowledgments: We also would like to thank the anonymous reviewers for their valuable comments and suggestions that led to a substantially improved manuscript.

Conflicts of Interest: The authors declare no conflict of interest.

References

1. Dai, L.C.; Lei, H.Y.; Cheng, X.Y.; Li, R.F. Prediction of coal seam gas content based on the correlation between gas basic parameters and coal quality indexes. *Front. Energy Res.* **2023**, *10*, 1096539. [CrossRef]
2. Wang, F.K.; Zhao, X.S.; Liang, Y.P.; Li, X.L.; Chen, Y.L. Calculation model and rapid estimation method for coal seam gas content. *Processes* **2018**, *6*, 223. [CrossRef]
3. Wang, L.; Cheng, L.B.; Cheng, Y.P.; Liu, S.M.; Guo, P.K.; Jin, K.; Jiang, H.N. A new method for accurate and rapid measurement of underground coal seam gas content. *J. Nat. Gas Sci. Eng.* **2015**, *26*, 1388–1398. [CrossRef]
4. Chaffee, A.L.; Lay, G.; Marshall, M.; Jackson, W.R.; Fei, Y.; Verheyen, T.V.; Cassidy, P.J.; Scott, S.G. Structural characterisation of Middle Jurassic, high-volatile bituminous Walloon subgroup coals and correlation with the coal seam gas content. *Fuel* **2010**, *89*, 3241–3249. [CrossRef]
5. Yan, G.Q.; Wang, G.; Xin, L.; Du, W.Z.; Huang, Q.M. Direct fitting measurement of gas content in coalbed and selection of reasonable sampling time. *Int. J. Min. Sci. Technol.* **2017**, *27*, 299–305. [CrossRef]
6. Xu, H.J.; Ahmad, F.; Hu, B.L.; Sun, G.; Liu, H.H.; Ding, H.; Zhang, M.; Fang, H.H. Methodology for lost gas determination from exploratory coal cores and comparative evaluation of the accuracy of the direct method. *ACS Omega* **2021**, *6*, 19695–19704. [CrossRef]
7. Zhou, F.B.; Kang, J.H.; Wang, Y.P.; Zhang, R. Method of underground integrated automatic and accurate determination of coalbed gas content. *J. China Coal Soc.* **2022**, *47*, 2873–2882. [CrossRef]
8. Zhao, F.J.; Wen, Z.H.; Liu, M.J.; Wei, J.P.; Chen, M. Analysis content affected to seam gas pressure measured with indirect method. *Coal Sci. Technol.* **2010**, *38*, 52–54. [CrossRef]
9. Hou, X.W.; Liu, S.M.; Zhu, Y.M.; Yang, Y. Evaluation of gas contents for a multi-seam deep coalbed methane reservoir and their geological controls: In situ direct method versus indirect method. *Fuel* **2020**, *265*, 116917. [CrossRef]
10. Wang, Q.; Wang, Z.F.; Yue, J.W.; Wang, L.G.; Dong, J.X.; Tan, R.H. Experimental study on reasonable adsorption time in determination coalbed methane content. *Geofluids* **2022**, *2022*, 1–15. [CrossRef]
11. Creedy, D.P. Methods for the evaluation of seam gas content from measurements on coal samples. *Min. Sci. Technol.* **1986**, *3*, 141–160. [CrossRef]
12. Plaksin, M.S.; Kozyreva, E.N. Determining the Gas Content of Coal Beds. *Coal Chem.* **2021**, *64*, 144–147. [CrossRef]
13. Malinnikova, O.N.; Ul'yanova, E.V.; Kharchenko, A.V.; Pashichev, B.N. Influence of coal microstructure on gas content of the face area. *J. Min. Sci.* **2020**, *56*, 351–358. [CrossRef]
14. Li, D.; Peng, S.P.; Du, W.F.; Guo, Y.L. New method for predicting coal seam gas content. *Energy Sources Part A* **2019**, *41*, 1272–1284. [CrossRef]
15. Li, C.W.; Wang, Y.L.; Wang, Q.J.; Gao, X. Experimental study on accuracy of direct gas. *J. China Coal Soc.* **2020**, *45*, 189–196.
16. Zhang, J.F.; Xie, Y.D.; Yang, F.F.; Liu, G.L.; Zhang, J.J.; Miao, Z.Q.; Ma, D.Q. Study on prediction method of coal seam gas content based on principal component multiple regression. *Int. J. Oil Gas Sci. Eng.* **2020**, *2*, 1–7.
17. Chen, Y.Q.; Zheng, L.J.; Huang, J.; Zou, Z.; Li, C.H. Prediction of gas emission based on grey-generalized regression neural network. *IOP Conf. Ser. Earth Environ. Sci.* **2020**, *467*, 012056. [CrossRef]
18. Xu, G.; Wang, L.; Wang, H.T.; Wang, K.; Feng, Y.Z. Prediction method of coal seam gas content based on Grey theory and BP Neural Network. *Coal Technol.* **2019**, *38*, 82–85.
19. Wu, X.; Yang, Z.; Wu, D.D. Advanced computational methods for mitigating shock and vibration hazards in deep mines gas outburst prediction using SVM optimized by Grey relational analysis and APSO algorithm. *Shock Vib.* **2021**, *2021*, 5551320. [CrossRef]
20. Zhang, J.Q. Study on the gas content of coal seam based on the BP Neural Network. *Procedia Eng.* **2011**, *26*, 1554–1562.
21. Lv, Y.N.; Tang, D.Z.; Xu, H.; Tao, S. Productivity matching and quantitative prediction of coalbed methane wells based on BP neural network. *Sci. China Technol. Sci.* **2011**, *54*, 1281–1286.
22. Wang, L.; Li, J.H.; Zhang, W.B.; Li, Y. Research on the gas emission quantity prediction model of improved artificial bee colony algorithm and weighted least squares support vector machine (IABC-WLSSVM). *Appl. Bionics Biomech.* **2022**, *2022*, 4792988. [CrossRef] [PubMed]
23. Guo, J.H.; Zhang, Z.S.; Guo, G.S.; Xiao, H.; Zhu, L.Q.; Zhang, C.M.; Tang, X.; Zhou, X.Q.; Zhang, Y.N.; Wang, C. Evaluation of coalbed methane content by using Kernel Extreme Learning Machine and Geophysical Logging Data. *Geofluids* **2022**, *2022*, 1–28. [CrossRef]
24. Li, X.J. Experimental Research on the relationship between gas desorption index of drill cuttings and gas content. *Saf. Coal Mines* **2014**, *45*, 8–12.
25. Huang, H.X.; Sun, W.; Xiong, F.Y.; Chen, L.; Li, X.; Gao, T.; Jiang, Z.X.; Ji, W.; Wu, Y.J.; Han, J. A novel method to estimate subsurface shale gas capacities. *Fuel* **2018**, *232*, 341–350. [CrossRef]
26. Lei, H.Y.; Pang, J.L.; Chen, Y.T.; Cao, J. Experimental study on calibration of calculation formula of gas content in coal seam by indirect method. *Min. Saf. Environ. Prot.* **2018**, *45*, 37–40.
27. Ma, S.Y. Selection of characteristic particle size of drilling cuttings based on adsorption–desorption properties: Experiment and Simulation. *Adsorpt. Sci. Technol.* **2022**, *2022*, 1–19. [CrossRef]

28. Kong, S.L.; Cheng, L.B.; Wang, H.F.; Zhou, H.X. Determination and application on critical value of drilling cuttings gas desorption indices. *Coal Sci. Technol.* **2014**, *42*, 56–59+64.
29. Zhao, X.S.; Dong, Y.S.; Yue, C.P. Method for determining the sensitive index of coal and gas outburst prediction and its critical value. *Min. Saf. Environ. Prot.* **2007**, *34*, 28–30+52+89.
30. Lei, H.Y. Experimental study on rapid determination of critical value of drilling cutting gas desorption index. *Coal Sci. Technol.* **2019**, *47*, 129–134.
31. Zhu, T.H.; Zhang, Y.Z.; Xie, A.F.; Li, W.B.; Jiang, P.F. Upgrading and optimization of gas outburst parameter measuring instrument and its on-site comparative investigation and application. *Ind. Saf. Environ. Prot.* **2021**, *47*, 48–50.

processes

MDPI

Article

Study on the Deformation Mechanisms of the Surrounding Rock and Its Supporting Technology for Large Section Whole Coal Cavern Groups

Linlin Chen [1], Zhiliu Wang [2,*], Wen Wang [3] and Junxiang Zhang [2]

1 Gencun Coal Mine, Henan Dayou Energy Co., Ltd., Sanmenxia 472400, China
2 School of Civil Engineering and Architecture, Zhongyuan University of Technology, Zhengzhou 450007, China
3 School of Energy Science and Engineering, Henan Polytechnic University, Jiaozuo 454000, China
* Correspondence: 6855@zut.edu.cn

Abstract: The control difficulty of whole coal cavern groups is greatly increased due to the characteristics of soft rock with low strength, large sections, and the mutual influence of crossed cavern groups. The large section gas storage cavern group is taken as the research background. In this paper, the equivalent circle method is used to solve the loose circle of a rectangular roadway, and numerical calculation is used to obtain the deformation and stress distribution laws of the surrounding rock under the excavation conditions of large section whole coal cavern groups (WCCG). The deformation and failure mechanisms of the surrounding rock are revealed under the linkage impact between large section whole coal cavern groups. The stratified reinforcement ring concept of "long cable-bolt-grouting" (LBG) was proposed for the stability control of surrounding rock in the WCCG. On the roof of whole coal cavern groups, the supporting configuration of a high-strength bolt with a high pre-tightening force and the high-strength anchor with a high pre-tightening force were determined. On the two sides and floor of the WCCG, the grouting scheme was determined. These two supporting configurations in both the roof and sidewalls were applied to the large section gas storage cavern group. The results show that the surrounding rock presents asymmetric deformation and failure characteristics due to the large excavation area and complex structure. Tensile failure and mixed tensile-shear failure mainly occur in the shallow part of the surrounding rock, while shear failure mainly occurs in the deep part of the surrounding rock. The roof displacement curves show a symmetric distribution and saddle distribution in the low- and high-negative pressure caverns, respectively. The maximum displacements are on the left and right sides of the cavern roof. The range of the loose rings is 3.34 m and 2.54 m, respectively, on the roof and the two ribs. The stratified reinforcement ring support technology of LBG can effectively reduce the failure depth of surrounding rock, and the surrounding rock is in a stable state. The study can provide a theoretical basis for the layout of large section cavern groups and the stability control of surrounding rock.

Keywords: large section; cavern groups; stratified reinforcement ring; deformation of the surrounding rock

Citation: Chen, L.; Wang, Z.; Wang, W.; Zhang, J. Study on the Deformation Mechanisms of the Surrounding Rock and Its Supporting Technology for Large Section Whole Coal Cavern Groups. *Processes* **2023**, *11*, 891. https://doi.org/10.3390/pr11030891

Academic Editor: Zhongyang Luo

Received: 12 February 2023
Revised: 4 March 2023
Accepted: 13 March 2023
Published: 16 March 2023

1. Introduction

With the development of automation and intelligence in coal mine production, more and more large sections and super-large section cavern groups are required for large-scale mining and the transportation of machinery and equipment. The cavern group of the mine is relatively densely distributed. In deep geology, the rock mass is in a high-confining pressure state. It greatly increases the difficulty of the maintenance of the cavern group, and the serious deformation of the surrounding rock makes it easy to induce the instability of the surrounding rock [1–3]. In particular, the failure of one of the cavern groups can induce the linkage instability of the surrounding cavern, which will seriously affect the safety and stability of deep, large section caverns [4,5]. A large deformation disaster occurs in soft rock roadways and is usually characterized by high stress, large deformation, strong rheology, low strength, and difficult support. The deformation of soft rock often shows obvious nonlinear large deformation characteristics, and it is difficult to support. Therefore, the deformation mechanism and support technology of soft rock roadways with deep high stress in the large section whole coal cavern groups have become a hot topic in the research field of roadway support all over the world [6–8].

At present, many scholars have focused on these aspects of the research, including classification methods, surrounding rock instability mechanisms, and reinforcement control technology. In terms of classification methods, the cavern is divided on the basis of section area and span in the field of coal mining. According to the section area, it can be divided into small sections (<8 m^2), medium sections (8–12 m^2), large sections (12–20 m^2), and extra-large sections ($\geq$20 m^2) [9]. According to the span, it can be divided into a small section ($\leq$3 m), a medium section (3.1–4.0 m), a large section (4.1–5.0 m), and a super section ($\geq$5.1 m) [10]. The stability of the surrounding rock is also affected by the depth, excavation disturbance, and distance between the caverns. In terms of the failure and instability mechanisms of the cavern, the deep surrounding rock presents rheological instability and dynamic instability. He et al. [11] and Sun et al. [12] pointed out that interbedded shear slip deformation mechanisms and high stress dilatation deformation mechanisms are the main manifestations of asymmetric rock deformation and failure mechanisms. Numerical methods have been used to evaluate the safety of cavern groups and optimize the allowable pillar width [13–15]. Kang et al. [16,17] analyzed the rock rheological properties from the perspective of time and space scales and believed that the dynamic instability of deep surrounding rock was more frequent. Pan et al. [18] carried out a similar simulation test of explosion loading and obtained the dynamic failure process of the cavern under the action of an explosion load using digital speckle technology. Wang et al. [19] used a large-scale geomechanical model test system to study the deformation and failure response law of surrounding rock under the combined action of dynamic and static loads. By evaluating the stress and strain distribution, the stress and strain concentration can be easily determined [20]. The plastic zone can be assessed to determine whether elements yield [21]. Wu et al. used the true triaxial simulation technique to establish a model of the inclined strata and carry out high stress triaxial loading experiments. The deformation mechanism of high stress inclined rock masses is discussed in detail [22]. Zhao et al. [23] aimed to propose a dual-medium model, including equivalent continuous and discrete fracture media, to study the coupled seepage damage effect in fractured rock masses. However, it is difficult and counterintuitive to quantify the stability degree of the surrounding rock for large section whole coal cavern groups using these indices. The linkage impact between large section whole coal cavern groups is rarely studied.

Meanwhile, many scholars have made in-depth explorations into the reinforcement control technology of the surrounding rock. Wang et al. [23–26] proposed the asymmetric phased control method for large section caverns and verified the rationality of the method through field monitoring. Hou et al. [27,28] proposed methods of improving the stress state and mechanical properties of roadways surrounding rock and rationally selecting the support form of the roadway. New combined supporting system for weak floor reinforcement in deep underground coal mines is proposed by Yang and Kang et al. [29,30]. Jiao et al. [31] modified the traditional U-section steel group to effectively support and stabilize the loose and thick coal seam roadway. Li and Yao [32] proposed the key technologies of segmented roadway support after considering the time effect, local breakdown characteristics, and fracture development. Li et al. [33] developed a high-strength anchorage grouting support technology with a new type of high-strength hollow grouting bolt as the core. The problems, such as the serious weak crushing of surrounding rock and the frequent destruction of large section cavern supporting components, are solved. Singh et al. [34] studied the causes of the failure of two wire ropes used at high stress levels in two different coal mines in India. Rama et al. [35] conducted a detailed parameterized study of a roof bolt breaking line support (RBBLS) and proposed an empirical formula for the design of a RBBLS. An analytical solution for surrounding rock that took influences of water seepage, strain softening, dilatancy, and intermediate principals into account was given and a grouting measure was provided to improve roadway stability by Yuan et al. [36]. The deformation mechanical characteristics of weakly consolidated siltstone surrounding rock are described and control techniques are proposed [37]. The above scholars have achieved considerable achievements in the control of the surrounding rock in the cavern group. However, the failure mechanism and stability control theory of the surrounding rock in the whole coal cavern group need to be further perfected. With the increase of mining depth and cavern section, it is more urgent to study the deformation law and control technology under linkage impacts between large section whole coal cavern groups.

This paper aims for the stability of whole coal cavern groups with large sections and provides technical support for the stability of similar large section cave groups. From the background of practical engineering, the stress and deformation law of the surrounding rock of the WCCG under different section sizes and side pressure coefficients are studied by a numerical calculation method, and the deformation failure mechanism of the WCCG is obtained under linkage impact between large section whole coal cavern groups. The stratified reinforcement ring concept of LBG was proposed based on the obtained deformation mechanism under linkage impact between large section whole coal cavern groups. The surrounding rock control theory proposed can not only put forward a new control solution for rock cavern stability but also improve the technical aspects of the WCCG. It has important theoretical significance and practical value for the safe and stable production of working faces.

2. Study Area

The coal mine is located in the southeast region of Shanxi Province and northwest of Changzhi City. It is 8 km away from Tunliu County. The actual production capacity of the mine has reached 7.1 million tons. Coal seam No. 3 is mainly mined. The inclined angle of the No. 3 coal seam is from 0° to 7° in the east-west direction and from 0° to 22° in the north-south direction. The roof and floor are basically stable, and the geological structure is simple. According to the exploration data, the Taiyuan Formation, Shanxi Formation, and Xiashihezi Formation are weakly water-bearing. The hydrogeological conditions of the #3 coal seam are mainly simple type, and the local and regional geological conditions are intermediate type to complex type.

The gas pump station is located in the No. 91 mining area, which belongs to the whole coal cavern, including high- and low-negative pressure caverns. The relative positions of the two caverns are shown in Figure 1. A large gas extraction pump and pipeline need to be installed in the underground gas pumping station, which requires a large cavern size. The cavern area reached 50 m^2, and its support difficulty greatly increased. The section of the connection roadway is a rectangular section with a size of 6.4 × 6.15 m^2. The high-negative-pressure cavern is a rectangular chamber divided into two sections. The length of the roadway in the western section is 69.5 m, and the section size is 6.0 × 6.15 m^2. The eastern section of the cavern (small section roadway) is 25 m in length and 5.0 × 4.55 m^2 in size. The low-negative pressure cavern is a rectangular chamber divided into two sections. The length of the roadway in the western section is 60 m, and the section size is 7.4 × 6.75 m^2. The eastern section of the cavern (small section roadway) is 20 m in length and 6.0 × 4.55 m^2 in size. According to the exposure of surrounding rock in the cavern, the lithologic characteristics of the roof and floor of the cavern are shown in Table 1.

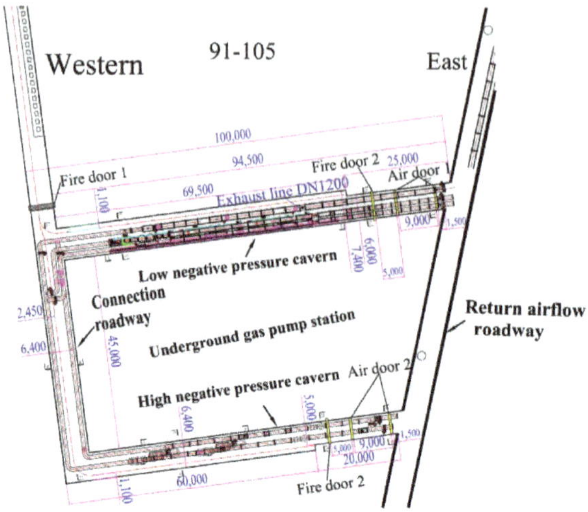

Figure 1. The relative positions of cavern groups.

Table 1. The lithologic characteristics of the roof and floor.

Name of Roof and Bottom	Name of Rock	Thickness/m	Lithological Characteristics
Upper roof	Mudstone	7	The rock color is gray-black and belongs to the thick layer of fine-grained sand structure. There are many argillaceous rocks, mainly composed of quartz and feldspar, followed by other dark minerals. The rock mass is argillaceous cement.
Immediate roof	Sandy mudstone	2.9	The color of the rock mass is black and belongs to a silty argillaceous structure. The shape of the fracture is flat and has scratches. The rock mass contains pyrite.
Immediate bottom	Sandy mudstone	3.95	The rock mass is black in color and blocky in shape. The rock mass contains plant fossils and argillaceous cement.
Lower bottom	Limestone	2.45	The rock mass is gray in color and blocky in shape. The material quality is not pure and contains mud.

3. Method and Material

3.1. Solve the Loose Circle of the Rectangular Roadway Based on the Equivalent Circular Method

Due to roadway excavation, surrounding rock stress will be redistributed. The original stress state is destroyed, and the stress is transferred to the depth. Failure occurs when the stress reaches or exceeds the strength of the surrounding rock. It is prone to roof collapses and slough accidents. Especially in the excavation of large section whole coal cavern group, each cavern group and intersection influence each other during excavation. Stress superposition leads to a high local stress concentration coefficient in large section caverns, which is more prone to deformation and failure.

The methods to solve the loose zone of a rectangular roadway mainly include the equivalent circle method, the complex function method, and the pressure arch method. In order to quantitatively solve the plastic range of the surrounding rock during excavation of a rectangular cavern, the equivalent circle method is adopted to theoretically calculate the influence range of the surrounding rock's loose circle caused by the excavation of a large section cavern [38]. Based on the loose circle of the circular roadway, the loose circle of the rectangular roadway is determined, and the plastic zone range of the rectangular roadway is finally determined:

(1) Construction of the mechanical model of the equivalent circle method expression

In order to calculate the stress state of the roadway, the circular roadway is taken as the theoretical analysis object, and the vertical stress is set as σ_v and the horizontal stress as σ_h. We can obtain:

$$\sigma_h = \lambda\sigma_v \tag{1}$$

where λ is the lateral pressure coefficient of the rock mass;

(2) Calculate the stress in the loose zone of the roadway surrounding the rock

For a circular roadway, the radial stress at any point is

$$\sigma_r = \sigma_v[\frac{1+\lambda}{2}(1-m) - \frac{1-\lambda}{2}(1+3m^2-4m)\cos(2\theta)], \tag{2}$$

The circumferential stress at any point is

$$\sigma_0 = \sigma_v[\frac{1+\lambda}{2}(1+m) + \frac{1-\lambda}{2}(1+3m^2)\cos(2\theta)], \tag{3}$$

The shear stress at any point is

$$\tau_{r0} = \sigma_v \cdot \frac{1-\lambda}{2}(1-3m^2+2m)\sin(2\theta), \tag{4}$$

where $m = {r_0}^2/{r_1}^2$

r_0 is the radius of the circular roadway, m;
r_1 is the radius of the loose ring, m;
θ is the polar angle of any point in the plastic zone of the surrounding rock, °;

(3) Determine the expression based on the equivalent circle method

The surrounding rock conforms to the Mohr–Coulom (M–C) constitutive by analyzing the physical and mechanical strength parameters of the surrounding rock. When the Mohr–Coulomb criterion is used to calculate the loosening range of the surrounding rock, its plastic condition is:

$$\sin\varphi_m = \frac{\sqrt{(\sigma_\theta-\sigma_r)^2+(2\tau_{r\theta})^2}}{\sigma_\theta+\sigma_r+2C_m\cot\varphi_m}(2\theta) \tag{5}$$

where C_m and φ_m are the cohesion and internal friction angles of the surrounding rock mass, respectively. σ_θ and τ_r are the circumferential stress and radial shear stress, respectively.

By substituting Equations (2–4) into Equation (5), we can obtain:

$$f_m = k_1 m^4 + k_2 m^3 + k_3 m^2 + k_4 m + k_5 \tag{6}$$

where $k_1 = 9(1-\lambda)$
$k_2 = 6(1-\lambda)(1+\lambda)\cos(2\theta) + 2(1-\lambda)$
$k_3 = 4(\sin^2\varphi_m - 3)(1-\lambda)^2\cos^2 2\theta + (1-\lambda)^2\cos(2\theta) + (1-6\lambda+\lambda^2)$
$k_4 = 4(1-\lambda)^2\cos(4\theta) + 2(1-\lambda)[(1+\lambda) - (1+\lambda+\frac{2C_m\cot\varphi_m}{\sigma_v})\cdot\sin^2\varphi_m]\cos(2\theta)$
$k_5 = (1+\lambda+\frac{2C_m\cot\varphi_m}{\sigma_v})^2\sin^2\varphi_m - (1-\lambda)^2$.

When θ is given, the relationship between the radius of the plastic loosening circle and the coefficient of lateral pressure of the roadway surrounding the rock can be obtained by Equation (6). Then the radius of the loose circle in the surrounding rock is calculated to provide the theoretical conditions for roadway support. When $\lambda = 1$ and $\sigma_h = \sigma_v$, we can substitute into Equation (6):

$$f_m = 4m^2 + (2 + \frac{2C_m \cot \varphi_m}{\sigma_v})^2 \sin^2 \varphi_m = 0 \quad (7)$$

Additionally the radius of the loosening ring can be obtained by Equation (7).

$$r_1 = r_0 \sqrt{\frac{\sigma_v}{(\sigma_v + C_m \cot \varphi_m) \sin \varphi_m}} \quad (8)$$

where r_0 is the radius of the circular roadway.

The stress change of the circular roadway can be obtained, as shown in Figure 2.

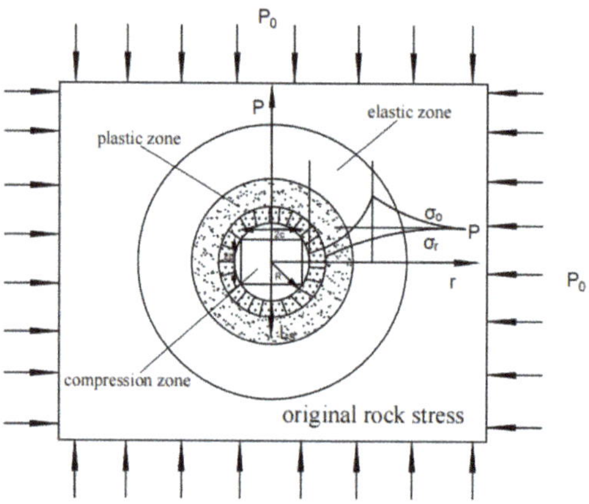

Figure 2. Stress distribution of the circular roadway.

At the interface of the elastic zone and the plastic zone, tangential stress σ_θ reaches its maximum value, and the stress at the interface gradually diffuses to the depth of the roadway (see Figure 1). When the elastic zone is exceeded, the surrounding rock stress gradually returns to its initial stress state.

According to the plastic zone of the circular roadway, the plastic zone range of rectangular roadway after excavation is also solved. Assuming that the width and height of the rectangular roadway are 2c and 2d, the radius of the circular roadway is solved,

$$r_0 = \sqrt{c^2 + d^2}. \quad (9)$$

The equivalent circular radius is substituted into r_0 to solve the radius range of the loose circle of the roof and two sides of the rectangular roadway. It provides the reliability basis for the design of roadway supporting parameters.

Therefore, the range of the loose circle is on the roof of the rectangular roadway:

$$l_s = r_1 - d \quad (10)$$

The range of loose rings on both sides:

$$l_e = r_1 - c \quad (11)$$

where d is half the height of the rectangular roadway, and c is half the width of the rectangular roadway.

Taking the cavern of the gas pump station as a case, the mechanical parameters of the roadway are $r_0 = 4.84$ m, $\varphi_m = 30°$, $C_m = 1$ MPa, and $\sigma_v = p_0 = 10$ MPa. By substituting into Equation (8), we obtain, $r_1 = 6.34$ m.

By substituting r_1 into Equations (9) and (10), we get the height of the roof loose ring, $l_s = 3.34$ m, and the loose ring of the two sides, $l_e = 2.54$ m. The range of loose circles provides a theoretical basis for future roadway support technology.

3.2. Numerical Calculation

According to the geological conditions, the direction of extension of the large destressing hole was in the x-axis (120 m), the axial direction of the cavern was in the y-axis (50 m), and the vertical direction of the coal and rock mass was in the z-axis (40 m), forming a numerical calculation model as shown in Figure 3. The M-C constitutive model was used in the calculation. The boundary conditions of the model are as follows: the upper boundary is free; the horizontal displacement of the boundary around the model is constrained; and the bottom boundary is fixed. By excavating the model, the influence law of stress distribution and plastic zone distribution of the surrounding rock is simulated. The mechanical parameters of rock strata are shown in Table 2.

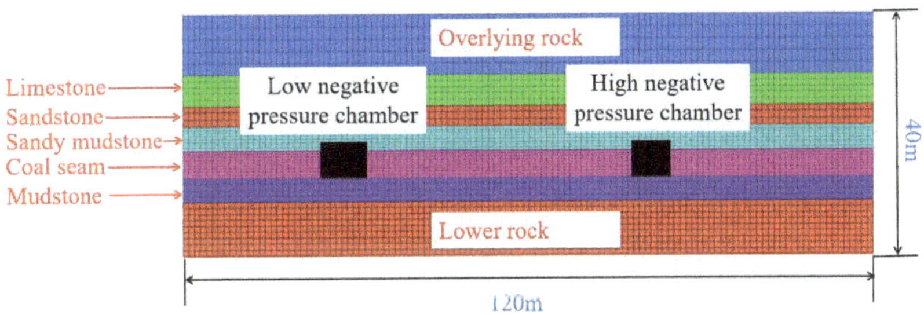

Figure 3. Numerical model diagram.

Table 2. Physical and mechanical properties of each stratum.

Lithology	Density/kg·m^{-3}	Bulk Modulus /GPa	Shear Modulus/GPa	Internal Friction Angle/°	Cohesion/MPa	Tensile Strength/MPa
Overlying strata	2570	4.85	3.4	32	4.5	3.9
Limestone	2590	5.02	3.6	36	5.2	4.3
Fine sandstone	2512	3.51	2.5	33	3.4	3.0
Sandy mudstone	2500	3.04	2.2	30	2.3	2.5
Coal seam	1300	2.32	1.4	29	1.5	1.8
Mudstone	2200	2.8	1.8	28	2.6	2.2
Underlying strata	2550	4.5	3.2	31	3.1	3.5

The cavern of the gas pump station studied in this paper includes a low- and high-negative pressure cavern. The numerical calculation model systematically studied the deformation and failure laws of the surrounding rock for the whole coal cavern group. The high- and low-pressure caverns are simulated according to the excavation sequence of the practical project. The excavation process is divided into five stages: the first stage is to excavate the connection roadway; the second stage is to excavate the western section of the high-negative pressure system cavern, and the length is 69.5 m; the third stage is to excavate the east section of the western section, and the length is 25 m; the fourth stage is to excavate the western section of the low-negative pressure system cavern, and the length is 60 m; and the fifth stage is to excavate the east section of the low-negative pressure system chamber, and the length is 20 m. It provides guidance for determining the supporting parameters of the whole coal cavern group.

4. Results and Discussion

4.1. Deformation Law of Surrounding Rock with Different Lateral Pressure Coefficients

The horizontal tectonic stress in the stratum has a great influence on the stability of the surrounding rock after excavation of the cavern. In this study, a low-negative pressure cavern was selected as the research object, and the influence of different lateral pressure coefficients ($\gamma = 0.5$, 1, 1.5, and 2) on the stability of the surrounding rock in the cavern was analyzed. The nephogram of maximum principal stress and the figure of the plastic zone were shown in Figures 4 and 5.

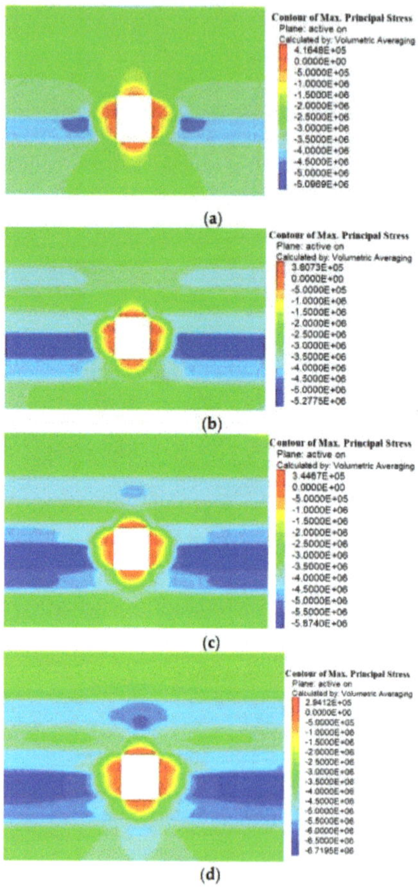

Figure 4. A cloud diagram of the maximum principal stress with different lateral pressure coefficients. (**a**) $\gamma = 0.5$; (**b**) $\gamma = 1.0$; (**c**) $\gamma = 1.5$; and (**d**) $\gamma = 2.0$.

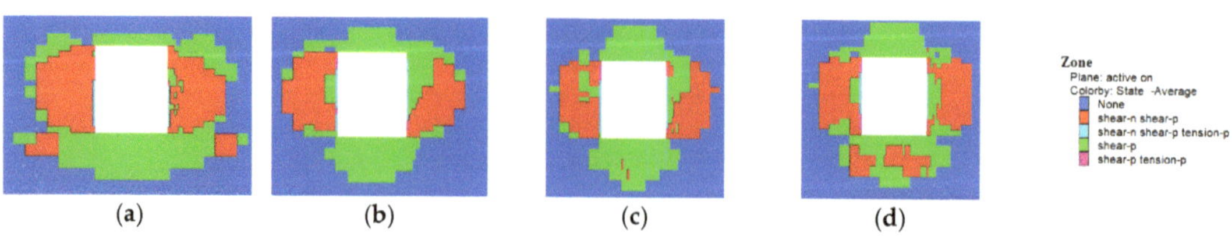

Figure 5. A cloud diagram of the plastic zone of the cavern with different lateral pressure coefficients. (**a**) $\gamma = 0.5$; (**b**) $\gamma = 1.0$; (**c**) $\gamma = 1.5$; and (**d**) $\gamma = 2.0$.

The peak value of the maximum principal stress in the cavern gradually increases from 5.09 MPa to 6.71 MPa as the lateral pressure coefficient γ increases (see Figures 4 and 5). The stress concentration area on both sides of the cavern increases laterally and then changes to longitudinal diffusion. When the lateral pressure coefficient $\gamma = 2.0$, the stress concentration area appears in the roof. The stress distribution of the vertical stress relative to the maximum principal stress changes little with the lateral pressure coefficient. The area of stress concentration near the cavern gradually increases, but the peak value of vertical stress gradually decreases. With the increasing lateral pressure coefficient, the plastic zone area of the cavern decreases. Additionally, under the influence of horizontal stress, the scope of the plastic zone gradually decreases in the two sides of the cavern, so as to gradually extend to the top and floor, and the distribution of the plastic zone changes from a chunky distribution to a thin distribution.

4.2. Stress Variation Characteristics for the Whole Coal Cavern Group

In order to analyze the distribution law of stress and plastic zone in the cavern groups with different heights, three sections were selected along the z-axis. The sections are located at h = 3 m (the middle position of the large roadway), h = 5 m (the roof position of the large roadway), and h = 8 m (the roof position of high- and low-negative pressure caverns) from the coal seam floor. The simulation results are shown in Figures 6–8.

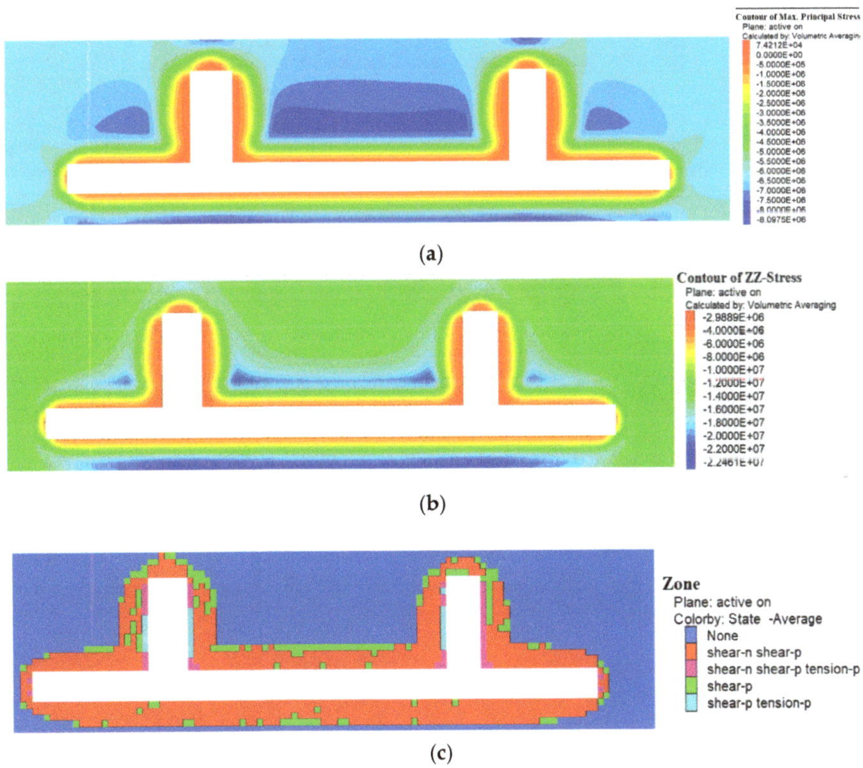

Figure 6. Surrounding rock variation characteristics for h = 3 m. (**a**) Maximum principal stress distribution; (**b**) vertical stress distribution; and (**c**) plastic differential distribution.

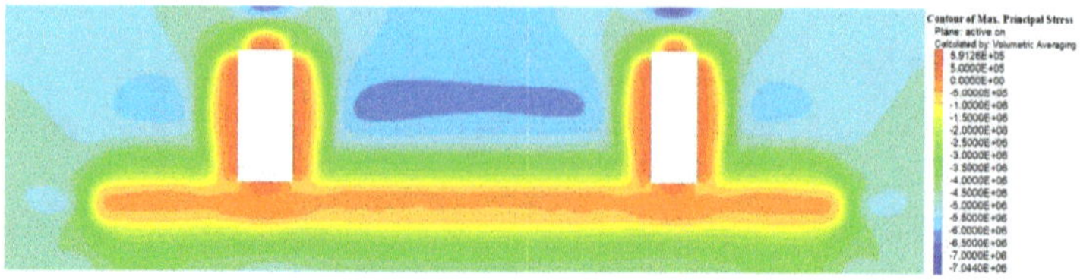

(a)

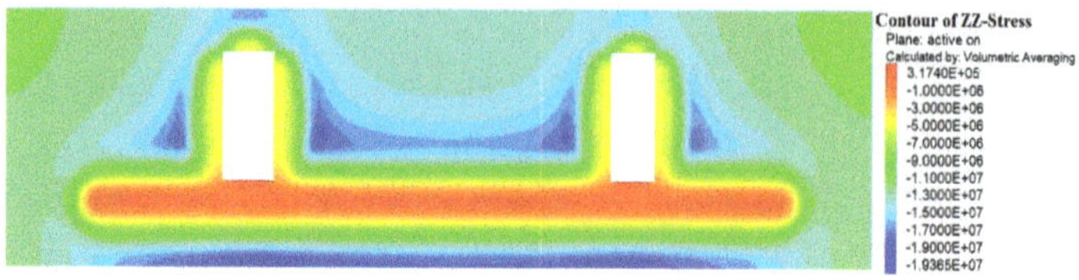

(b)

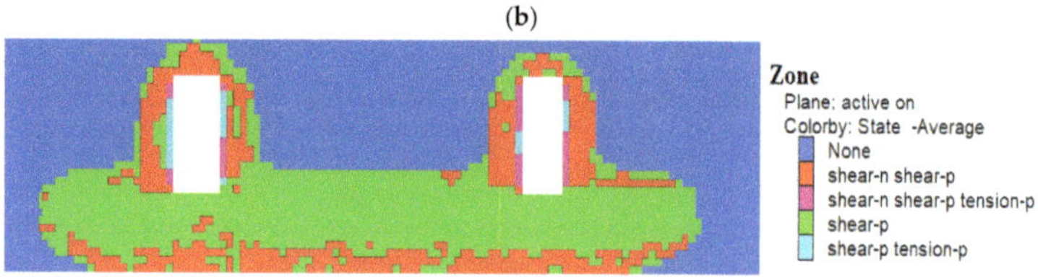

(c)

Figure 7. Surrounding rock variation characteristics for h = 5 m. (**a**) Maximum principal stress distribution; (**b**) vertical stress distribution; and (**c**) plastic differential distribution.

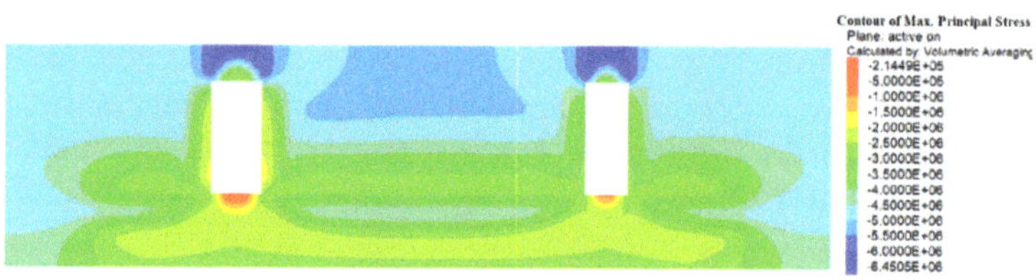

(a)

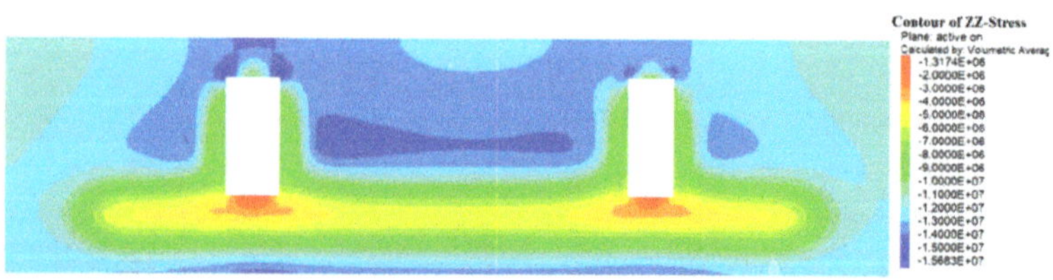

(b)

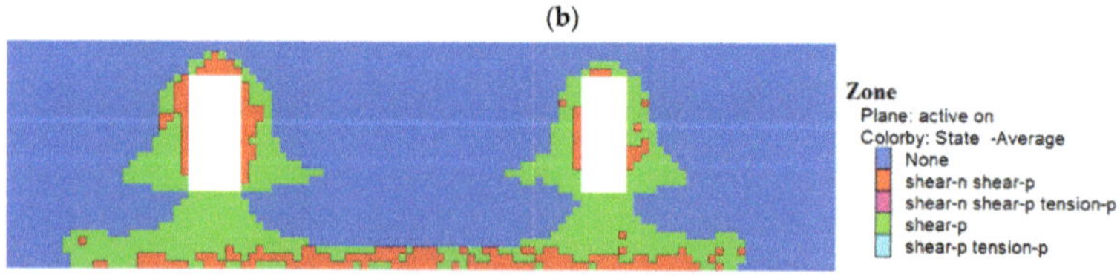

(c)

Figure 8. Surrounding rock variation characteristics for h = 8 m. (**a**) Maximum principal stress distribution; (**b**) vertical stress distribution; and (**c**) plastic differential distribution.

(1) The simulation result at h = 3 m

The height of the roadway near the underground pumping station is 3.5 m, and the surrounding rock variation in the position 3 m away from the coal seam floor is shown in Figure 6.

The distribution range of tensile stress generated by the excavation of the cavern is larger than that of the roadway (see Figure 6a). The maximum principal stress is larger in the area between the cavern groups, and the highest value is 8.09 MPa. The distribution law of vertical stress is roughly similar, but the maximum vertical stress value is 22 MPa (see Figure 6b). The larger the section of the cavern, the larger the scope of the plastic zone. The failure mode of the surrounding rock is mainly shear failure (see Figure 6c);

(2) The simulation result at h = 5 m

The surrounding rock variation in a position 5 m away from the coal seam floor is shown in Figure 7.

The peak values of the maximum principal stress and vertical stress decrease to 7.04 MPa and 19.3 MPa, respectively (see Figure 7a,b). The roof position of the roadway is obviously in the tensile stress zone, and the stress concentration area of the vertical stress begins to increase. The larger the section of the cavern, the greater the range of stress concentration. The larger the section of the cavern is, the larger the scope of the plastic zone is. The tensile failure occurs on the two sides of the cavern, the shear failure occurs in the rest of the scope, and the overall scope of the plastic zone increases gradually;

(3) The simulation result at h = 8 m

The surrounding rock variation in a position 8 m away from the coal seam floor, is shown in Figure 8.

The peak values of the maximum principal stress and vertical stress decrease to 6.45 MPa and 15.6 MPa, respectively (see Figure 8a,b). In this section, the stress near the cavern is not affected by the main roadway. The larger the section of the cavern, the greater the stress on the surrounding rock, and obvious stress asymmetry occurs. The plastic zone of the cavern continues to expand to the two sides as it moves away from the floor. The larger the section of the cavern, the larger the scope of the plastic zone. The range and distribution of the plastic zone obtained by numerical simulation are consistent with the theoretical results. These results can provide theoretical support for determining the supporting parameters, such as the length of the bolts and the cables;

(4) Stress distribution characteristics of cavern groups at different heights

Three sections were selected along the z-axis: the bottom of the cavern, the middle of the cavern, and the roof of the cavern. The stress curves for different heights were obtained, as shown in Figure 9.

The maximum principal stress of the roof attenuates sharply near the cavern and shows obvious stress concentration on the two sides of the cavern (see Figure 9). The variation trend of vertical stress is consistent with that of maximum principal stress. At the boundary of the model, the maximum principal stress and vertical stress increase and decrease sharply. In addition, the larger the section of the cavern, the greater the stress peak. In the middle of the cavern, the stress peak of the surrounding rock increases as a whole, and the distribution of the maximum principal stress is relatively stable among the cavern groups while the vertical stress increases gradually from high to low. Compared with the middle part of the cavern, the stress peak of surrounding rock near the bottom of the cavern gradually decreases, and the maximum and minimum principal stresses between the cavern groups show uniform distribution, but the influence range gradually decreases.

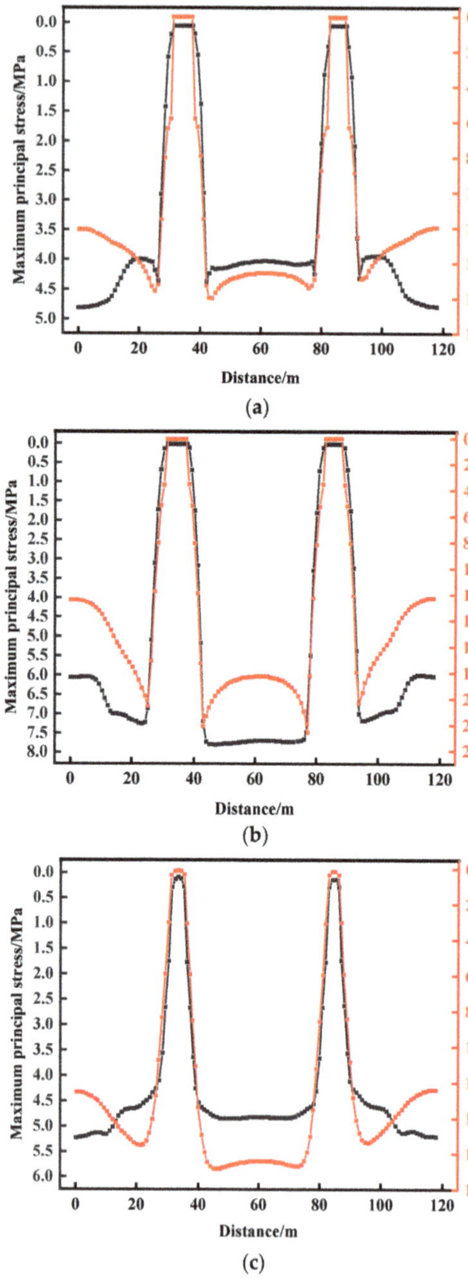

Figure 9. Stress variation characteristics of different section heights in the cavern group. (**a**) The roof position of the cavern; (**b**) the middle position of the cavern; and (**c**) the bottom position of the cavern.

4.3. Displacement Characteristics of the Whole Coal Cavern Group

The displacement cloud chart of the cavern group obtained by the numerical simulation is shown in Figure 10.

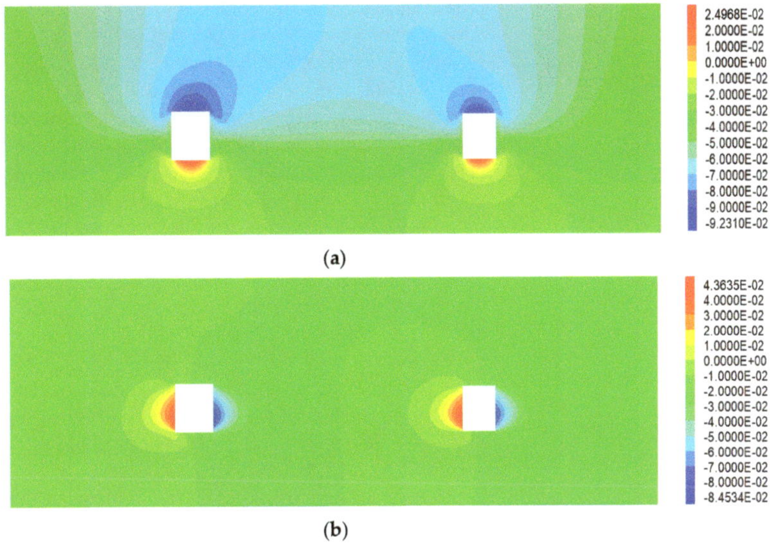

Figure 10. A cavern displacement cloud chart. (**a**) The vertical displacement cloud chart; and (**b**) the horizontal displacement cloud chart.

The larger the section of the cavern, the greater the displacement of the roof caused by the excavation of the cavern. The maximum displacements of low- and high-negative pressure caverns are 103.2 mm and 92.4 mm, respectively (see Figure 11). The maximum two-sided displacement is 160 mm and 125 mm, respectively. The roof displacement curve of the low-negative pressure cavern is asymmetrical. The roof displacement of the high-negative pressure cavern has a saddle-shaped distribution and reached a maximum in the cavern roof left and right, respectively. The two sides of the cavern have an asymmetrical arch distribution. It indicates that a two-sided displacement in the middle and upper parts of the cavern is larger.

4.4. Control Technology and Application for Large Section Whole Coal Cavern Groups

4.4.1. A Novel Stratified Reinforcement Ring Concept of the LBG

According to the stress distribution nephogram and curve of the cavern group, the surrounding rock of the roadway can be divided into internal and external bearing structures, as shown in Figure 12.

Different stress states in the broken zone, plastic zone, and elastic zone require different supporting strengths. The supporting effect should be coupled with the mechanical characteristics of the surrounding rock to effectively control the deformation and failure of the surrounding rock. A novel stratified reinforcement ring concept of "long cable-bolt-grouting" (LBG) was proposed to control the deformation of a whole coal cavern group based on the obtained deformation mechanism. The specific surrounding rock control mechanism is as follows:

(1) The surrounding rock of the cavern is divided into the fracture zone, the plastic zone, and the elastic zone from a shallow part to a deep part. The damage degree decreases gradually, and the support strength required to achieve stability also decreases gradually. The coordinated support of an anchor bolt, a grouting anchor cable, and a long anchor cable can meet this requirement. In the shallow part of the surrounding rock, the supporting density demand of a bolt, a grouted anchor cable, and a long anchor cable is small, while the supporting strength demand is large. In the deep surrounding rock, the support density demand decreases successively, and the sup-

port intensity decreases correspondingly. By partitioning the surrounding rock and adding reinforcement, a stable bearing circle that is consistent with the surrounding rock can be formed. For the fracture zone (0–2.5 m), plastic zone (3–8 m), and elastic zone (beyond 8 m), the coordinated support technology of an anchor bolt (3 m), a grouting anchor cable (8.3 m), and a long anchor cable (15 m) has been proposed and adopted to connect the broken zone, the plastic zone, and the elastic zone. Three kinds of supporting equipment are arranged at intervals and form three stable bearing rings of 0–3 m, 3–8 m, and 8–15 m from the shallow to deep layer of the surrounding rock;

(2) By using a high-strength and high-preload anchor cable support, the supporting function of an anchor bolt cable and the bearing capacity of surrounding rock can be fully demonstrated. In the early stages of excavation of a large section cavern, the surrounding rock's deformation speed is greater. A high-prestressed anchor bolt and a cable support can not only provide greater initial support resistance but also have good resistance-increasing performance, so that the rock strata in the broken zone, the plastic zone, and the elastic zone are locked as a whole. The internal friction angle and cohesion of the rock strata within the anchorage range can be improved so as to realize active and timely support and effectively limit the development of surrounding rock deformation and failure;

(3) Partition grouting reinforcement can improve the mechanical properties and stability of the surrounding rock. The cable support, with its high strength and high preload, can effectively control the deformation of the surrounding rock. However, primary and new fissures in the surrounding rock will expand to a certain extent. In order to achieve long-term stability of the surrounding rock, it is still necessary to improve the strength of the rock by grouting. Partition grouting reinforcement is mainly used to reinforce the fracture zone (0–2.5 m) and the plastic zone (3–8 m), respectively. Firstly, low-pressure grouting of a shallow hole is applied to the fracture zone, and then high-pressure grouting of the deep hole is applied to the surrounding rock in the plastic area. The grouting effect can be improved effectively through the combination of deep- and shallow-hole grouting. After grouting, the anchoring effect and supporting stiffness of an anchor bolt and an anchor cable are further strengthened, and the stability of the surrounding rock is further improved.

4.4.2. Determine the Surrounding Rock Control Parameters of the Whole Coal Cavern Group

The surrounding rock control parameters of the whole coal cavern group are determined based on the deformation and failure characteristics of the whole coal cavern group and the above surrounding rock control theory. On the roof of whole coal cavern groups, the supporting configuration of a high-strength bolt with a high pre-tightening force and a high-strength anchor with a high pre-tightening force were determined. On the two sides and floor of the whole coal cavern group, the grouting was determined. These two supporting configurations, in both the roof and two sides, were applied in the large section gas storage cavern group. The specific parameters are as follows:

(1) Roof support parameters of the cavern group

On the roof of whole coal cavern groups, the supporting configuration of a high-strength bolt with a high pre-tightening force and a high-strength anchor with a high pre-tightening force were determined.

In the roof and two sides of the cavern, the anchor cable specifications are Φ 22 $\times$ 9300 mm^2 and Φ 22 $\times$ 8300 mm^2, and the spacing and row layout are 1600 $\times$ 1600 mm^2 and 1500 $\times$ 1600 mm^2. The size of the bolt is Φ 22 $\times$ 2400 mm^2 with high-strength ribbed steel. The size of the laying metal mesh is 100 $\times$ 100 mm^2, and a WD 280-2.7 steel belt is adopted. The thickness of the concrete injection is 150 mm, and the roof is reinforced with a combined prestressed anchor cable. There are five combined anchor cables, each of which is prestressed with a dimension of 22 $\times$ 12,300 mm^2 and a spacing and row spacing of 2200 $\times$ 3200 mm^2. The supporting form of the roadway is shown in Figure 13;

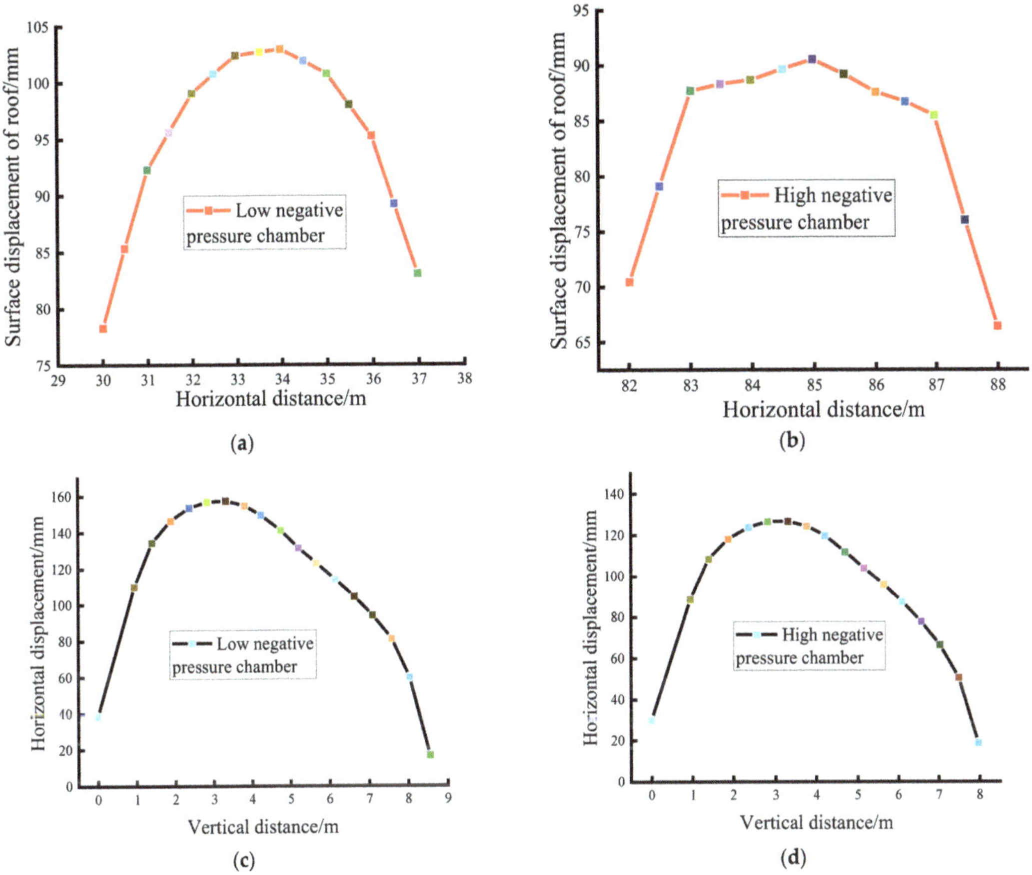

Figure 11. A cavern displacement diagram. (**a**) Roof displacement of a low-negative pressure cavern; (**b**) roof displacement of a high-negative pressure cavern; (**c**) two-sided displacement of a low-negative pressure cavern; and (**d**) two-sided displacement of a high-negative pressure cavern.

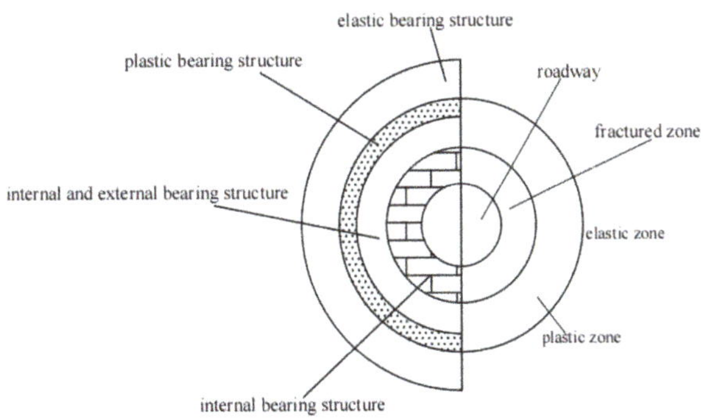

Figure 12. A schematic diagram of the inner and outer bearing structures of the surrounding rock.

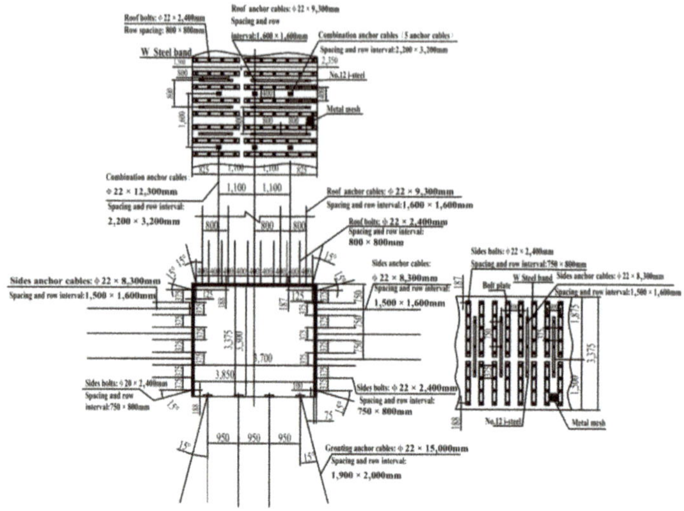

Figure 13. The roadway support layout plan.

(2) Grouting parameters

Three holes are arranged in each row of the cavern sections. The top row holes are 1.5 m away from the roof, the bottom row holes are 1.5 m away from the floor plate, and the middle holes are 3.3 m away from the floor plate. The holes are arranged in parallel. The diameter of the grouting hole is 42 mm, and the depth of the hole is 3 m. A seamless steel pipe is used as the grouting pipe. The hole sealing depth must be at least 1 m, and the hole sealing device is used to seal the hole. Shallow-hole grouting pressure is 2.0–4.0 MPa using a pneumatic single liquid grouting pump, and the diffusion radius is about 2 m.

5. Verification and Application

5.1. Numerical Simulation Results of LBG Control Technology

In order to study the control effect of the roadway surrounding rock with LBG technology and verify the rationality of the support scheme and parameters, a FLAC3D finite difference program was used to simulate the stress and deformation of the roadway surrounding rock. The stress and displacement nephograms are shown in Figures 14 and 15.

The stress and displacement nephograms (Figures 14 and 15) show that the LBG control technique can effectively improve the stress state of the surrounding rock in the chamber groups. The tensile stress of the surrounding rock is reduced, and surrounding rock control technology plays an active supporting effect. The maximum principal stress is reduced. The two-sided displacement and roof subsidence are reduced, and roadway deformation is completely controlled within the required range, which ensures the stability of the surrounding rock. It shows that LBG control technology is reasonable.

5.2. Engineering Practice

The supporting configurations in both the roof and sidewalls above were applied to the large section gas storage cavern group. In order to verify the rationality of the proposed surrounding rock control technology, a roof separation instrument and a borehole peeping instrument are arranged in the reinforcement section of the cavern to monitor the surface displacement change of the cavern. Eight measuring stations are arranged in the main body of the cavern (measuring stations 2–7) and the connecting crossover points of the cavern (measuring stations 1 and 8). The monitoring diagram and results are shown in

Figures 16 and 17. The field monitoring results and numerical simulation results on the maximum deformation of the surrounding rock are compared (Figures 18–20).

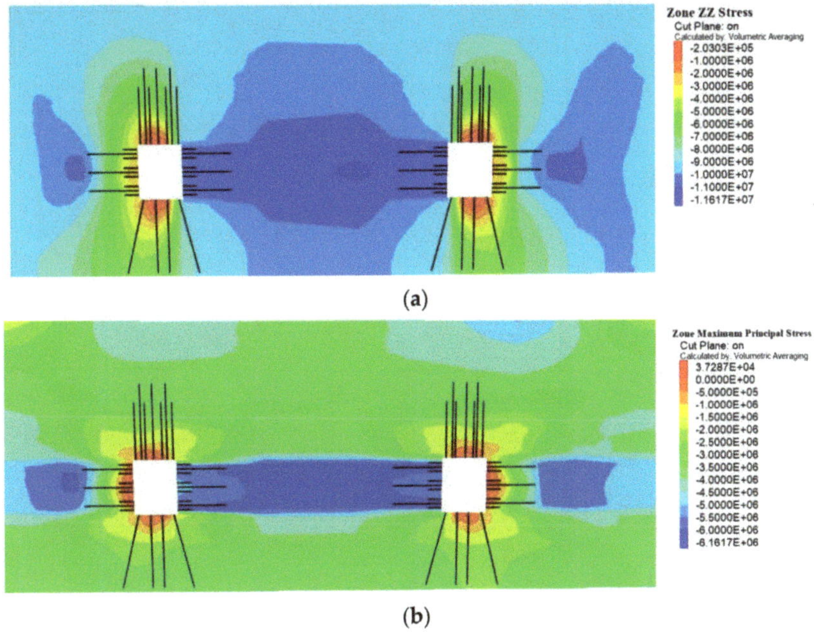

Figure 14. A stress cloud chart of LBG control technology. (**a**) A vertical stress cloud chart of LBG control technology; and (**b**) a maximum principal stress cloud chart of LBG control technology.

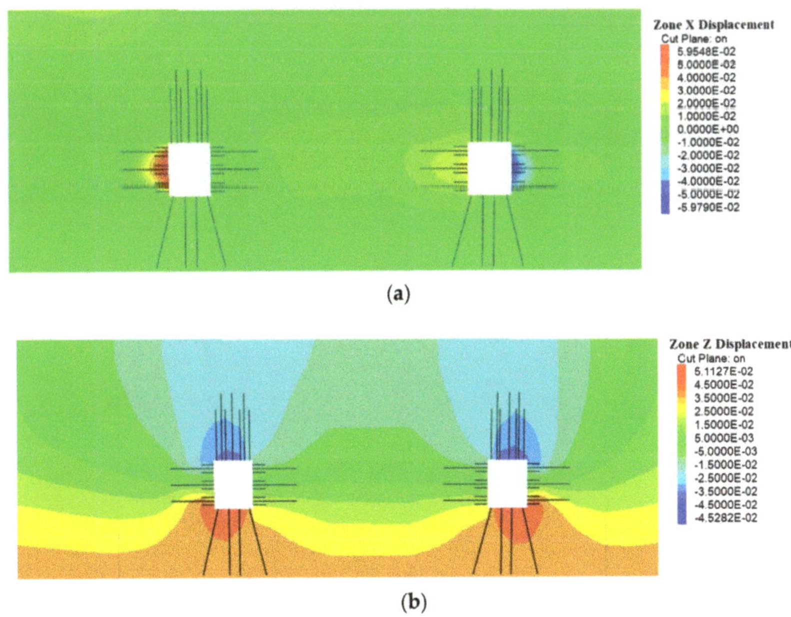

Figure 15. Displacement cloud chart of LBG control technology. (**a**) A horizontal displacement cloud chart of LBG control technology; and (**b**) a vertical displacement cloud chart of LBG control technology.

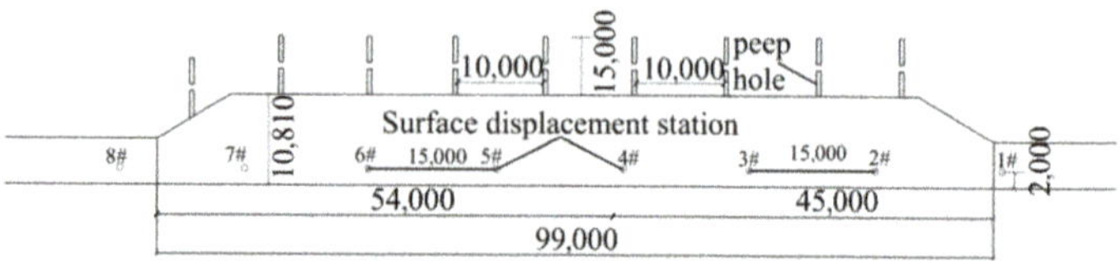

Figure 16. Layout of surface displacement measuring points and peep holes.

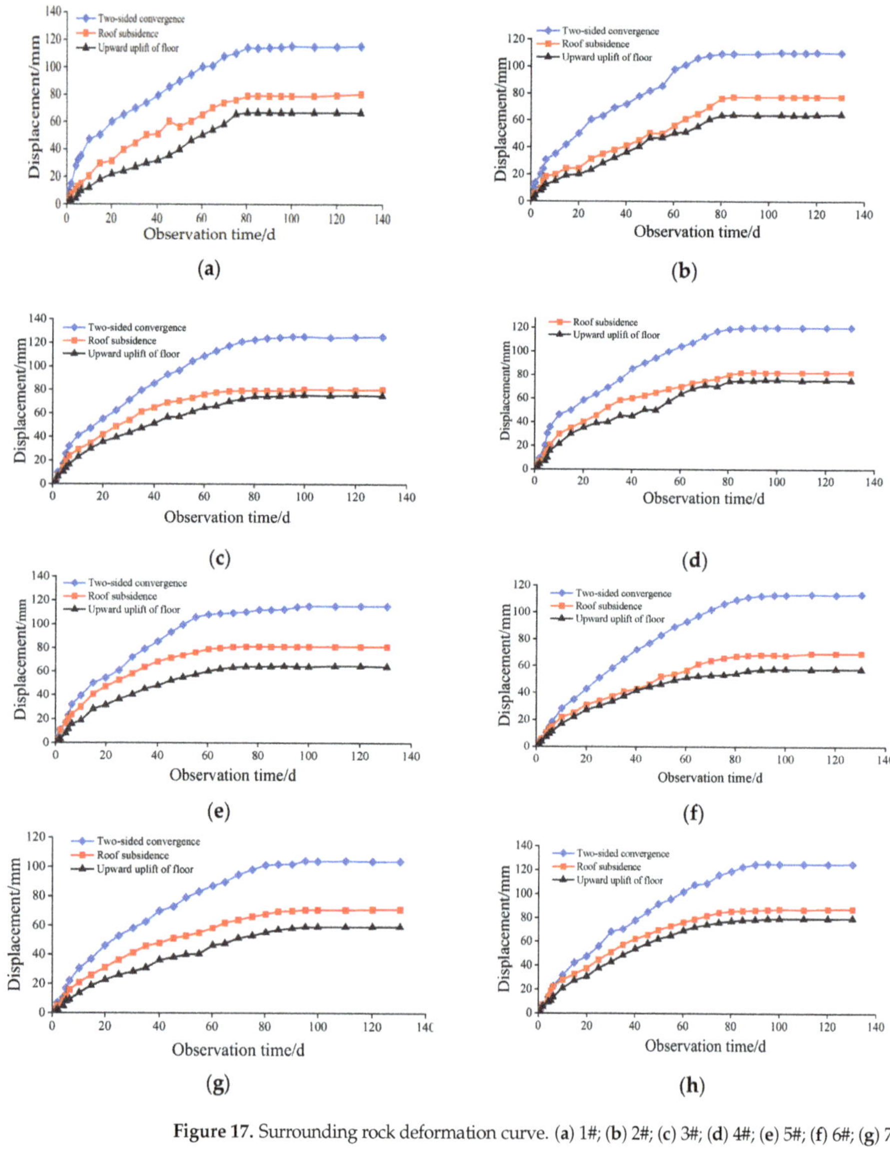

Figure 17. Surrounding rock deformation curve. (**a**) 1#; (**b**) 2#; (**c**) 3#; (**d**) 4#; (**e**) 5#; (**f**) 6#; (**g**) 7#; and (**h**) 8#.

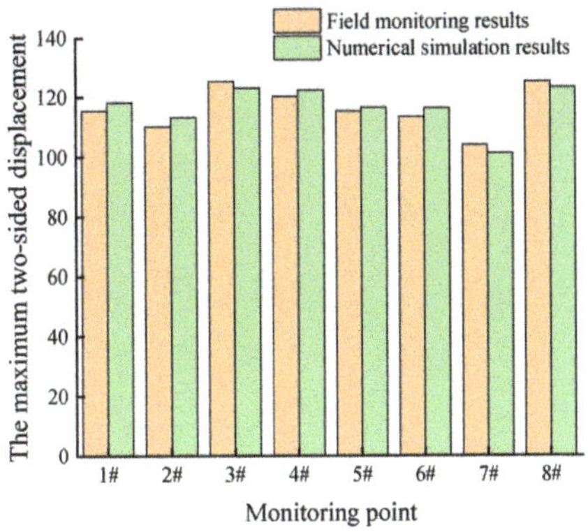

Figure 18. Numerical simulation results versus the field monitoring results of the maximum two-sided displacement.

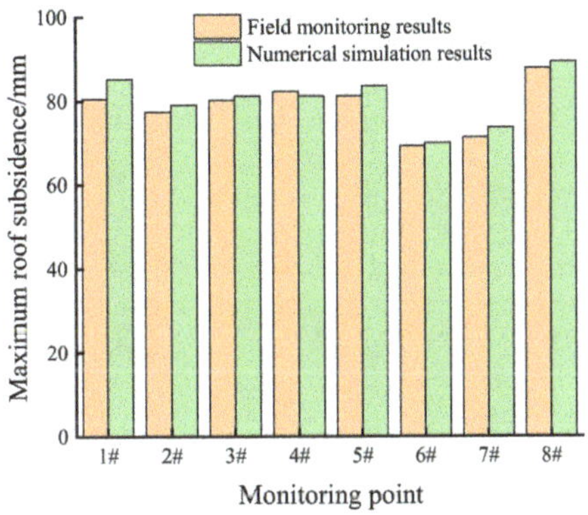

Figure 19. Numerical simulation results versus the field monitoring results of the maximum roof subsidence.

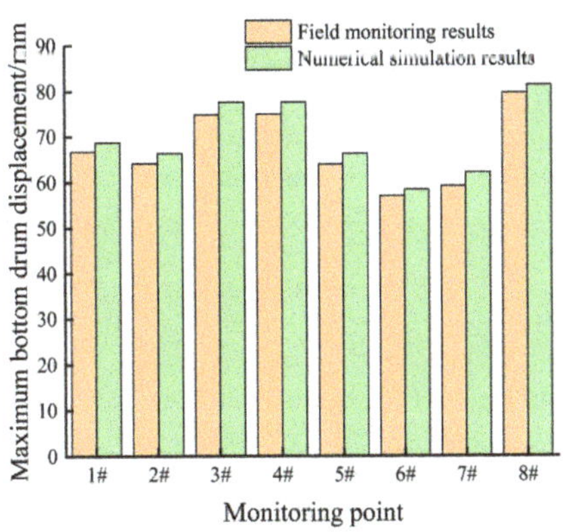

Figure 20. Numerical simulation results versus the field monitoring results of the maximum bottom drum displacement.

For measuring stations numbers 2–7 located in the main position of the cavern, the deformation of the surrounding rock in the cavern can be divided into three stages over time. They are the rapid deformation stage, the slow deformation stage, and the stable deformation stage. Within 10 d after excavation, the cavern is at the stage of rapid deformation. The maximum roof subsidence, bottom drum displacement, and two-sided displacement are 31 mm, 24 mm, and 46 mm, respectively. On the 15th day, the support technology of LBG was applied, and the excavation cavern began to increase slowly, which belongs to the slow deformation stage. Roof subsidence, bottom drum displacement, and two-sided displacement are 54 mm, 52 mm, and 79 mm, respectively. The deformation of the surrounding rock is almost constant after 74–95 d of excavation, which is the stable stage of deformation.

For the No. 1 and No. 8 measuring stations at the intersection and the connection of the roadway, the roof subsidence of the surrounding rock and the two-sided displacement are 85 mm, 91 mm, 125 mm, and 130 mm, respectively, and the bottom drum displacements are 77 mm and 79 mm, respectively. This is because measuring stations 1 and 8 are located in the passageway, and the roadway deformation is large due to the influence of concentrated stress.

Meanwhile, Figures 16–18 show the field-measured results are in good agreement with the numerical results. The monitoring results show that the deformation and failure of the surrounding rock in the cavern are obviously reduced by using a LBG control scheme. According to the peeping results of the drilling hole, the failure depth of the surrounding rock is only within 2 m of the surrounding rock, and there were no large tensile and shear failure areas. An anchor bolt and cable can play a better anchoring role. The novel support technology of a LBG can ensure the long-term stability of the cavern. There is no stratification within 2–15 m of the surface of the cavern.

6. Conclusions

In view of the problem of controlling the surrounding rock in large sections of whole coal cavern groups due to the influence of soft surrounding rock and a large section, this study proposes an innovative LBG support technology and obtains the following innovative conclusions, different from the perspective of improving the stress state of the surrounding rock:

(1) A mechanical model for WCCG with a rectangular roadway is established and solved based on an equivalent circular method. The theoretical calculation results showed that the heights of a loose ring in the roof and its two sides are 3.34 m and 2.54 m, respectively. It provides guidance for determining the supporting parameters of the whole coal cavern group;

(2) A numerical model under different section parameters is established for the linkage impact between whole coal cavern groups and the large section. With the increase of the lateral pressure coefficient, the stress concentration area on both sides of the caverns first increases laterally and then changes to a longitudinal diffusion. The plastic zone of the cavern tends to decrease with an increase in the lateral pressure coefficient. The range of the plastic zone decreases gradually on the two sides of the cavern, thus extending gradually to the top and bottom floors. The distribution of the plastic zone changed from stocky to tall. The stress and deformation mechanisms of the surrounding rock provide the basis for the control theory and parameters of the surrounding rock in the WCCG with a large section;

(3) The surrounding rock of the whole coal cavern group presents an asymmetric pressure relief form due to the excavation of the cavern. Far away from the surface of the cavern, the stress gradually increases, and the stress concentration appears in the depths. The stress concentration on the two sides is greater than that on the top and bottom. In the shallow part of the surrounding rock, tensile failure or tensile shear failure mainly occurs, and shear failure mainly occurs in the deep part of the surrounding rock. The bearing capacity of the surrounding rock is greatly reduced due to the excavation of the cavern. It is necessary to improve its stability through active, strong support;

(4) The stratified reinforcement ring concept of a LBG is proposed based on the obtained deformation mechanism to control surrounding rock deformation of the WCCG. The supporting technology of an anchor bolt (3 m), grouting anchor cable (8.3 m), and a long anchor cable (15 m) is proposed and adopted to connect the broken zone, the plastic zone, and the elastic zone in the gas storage cavern group with the large section. The roof subsidence and two-sided displacement are controlled at 140 mm and 120 mm, respectively, by using the supporting scheme of a LBG, and the surrounding rock deformation in the caverns is obviously reduced. The results of the numerical simulation and the field monitoring show that the LBG technology is effective for the stability control of the WCCG.

Author Contributions: Conceptualization, L.C. and Z.W.; methodology, Z.W.; software, L.C. and Z.W.; validation, L.C., W.W. and J.Z.; data curation, L.C. and Z.W.; supervision, W.W.; project administration, L.C., Z.W. and J.Z.; funding acquisition, Z.W. and J.Z. All authors have read and agreed to the published version of the manuscript.

Funding: The authors sincerely thank the following agents for their financial supports: the Scientific and Technological Project of Henan Province (222102320060), the Strength Improvement Plan of the Advantageous Disciplines of Zhongyuan University of Technology (SD202232), the Natural Science Foundation of China of Zhongyuan University of Technology (K2023QN013), the National Natural Science Foundation of China (52204264), and the Key Specialized Research and Development Breakthrough of Henan Province (222102320086).

Data Availability Statement: The data applied to support the results in this study are available from the corresponding author upon request.

Conflicts of Interest: The authors declare no conflict of interest.

References

1. Du, F.; Wang, K.; Wang, G.; Huang, Y.; Yi, L. The mechanism of damage in gas-bearing coal-rock combination bodies and gas seepage in coals. *Energy Sour. Part A Recovery Util. Environ. Eff.* **2021**, *43*, 1181–1201. [CrossRef]
2. Wang, W.; Wang, H.; Li, D.; Li, H.; Liu, Z. Strength and failure characteristics of naturaland water-saturated coal specimens under static and dynamic loads. *Shock. Vib.* **2018**, *2018*, 3526121.
3. Du, F.; Ma, J.; Guo, X.; Wang, T.; Dong, X.; Li, J.; He, S.; Nuerjuma, D. Rockburst mechanism and the law of energy accumulation and release in mining roadway: A case study. *Int. J. Coal Sci. Technol.* **2022**, *9*, 1–17. [CrossRef]
4. Zhang, J.; Li, B.; Liu, Y.; Li, P.; Fu, J.; Chen, L.; Ding, P. Dynamic multifield coupling model of gas drainage and a new remedy method for borehole leakage. *Acta Geotech.* **2022**, *17*, 4699–4715. [CrossRef]
5. Li, B.; Zhang, J.; Liu, Y.; Qu, L.; Liu, Q.; Sun, Y.; Xu, G. Interfacial porosity model and modification mechanism of broken coal grouting: A theoretical and experimental study. *Surf. Interfaces* **2022**, *33*, 102286. [CrossRef]
6. Bai, Q.; Tu, S.; Wang, F.; Zhang, C. Field and numerical investigations of gateroad system failure induced by hard roofs in a longwall top coal caving face. *Int. J. Coal Geol.* **2017**, *173*, 176–199. [CrossRef]
7. Chen, H.D.; Chen, X.J.; Wang, Z.F.; Li, Z.; An, F. Plastic zone range of a roadway considering the creep effect. *Sci. Rep.* **2020**, *10*, 20341. [CrossRef]
8. Zhang, Z.; Deng, M.; Bai, J.; Yan, S.; Yu, X. Stability control of gob-side entry retained under the gob with close distance coal seams. *Int. J. Min. Sci. Technol.* **2021**, *31*, 321–332. [CrossRef]
9. Yuan, L.; Xue, J.H.; Liu, Q.S.; Liu, B. Surrounding rock stability control theory and support technique in deep rock roadway for coal mine. *Meitan Xuebao/J. China Coal Soc.* **2011**, *36*, 535–543.
10. Kang, H. Sixty year'development and prospects of rock bolting technology for underground coal mine roadways in China. *J. China Univ. Min. Technol.* **2016**, *45*, 1071–1081.
11. He, M.; Gong, W.; Wang, J.; Qi, P.; Tao, Z.; Du, S.; Peng, Y. Development of a novel energy-absorbing bolt with extraordinarily large elongation and constant resistance. *Int. J. Rock Mech. Min. Sci.* **2014**, *67*, 29–42. [CrossRef]
12. Sun, X.M.; Zhang, G.F.; Cai, F.; Yu, S.B. Asymmetric deformation mechanism within inclinde rock strata induced by excavation in deep roadway and its controlling countermeasure. *Chin. J. Rock Mech. Eng.* **2009**, *28*, 1137–1143.
13. Deng, J.Q.; Yang, Q.; Liu, Y.R.; Pan, Y.W. Stability evaluation and failure analysis of rock salt gas storage caverns based on deformation reinforcement theory. *Comput Geotech.* **2015**, *68*, 147–160. [CrossRef]
14. Moghadam, S.N.; Nazokkar, K.; Chalaturnyk, R.J.; Mirzabozorg, H. Parametric assessment of salt cavern performance using a creep model describing dilatancy and failure. *Int. J. Rock Mech. Min.* **2015**, *79*, 250–267. [CrossRef]
15. Wang, T.; Yang, C.; Ma, H.; Li, Y.; Shi, X.; Li, J.; Daemen, J.J.K. Safety evaluation of salt cavern gas storage close to an old cavern. *Int. J. Rock Mech. Min.* **2016**, *83*, 95–106. [CrossRef]

16. Kang, H.P. Spatial scale analysis on coal mining and strata control technologies. *J. Min. Strat. Control Eng.* **2020**, *2*, 5–30.
17. Kang, H.P. Temporal scale analysis on coal mining and stratac ontroltechnologies. *J. Min. Strat. Control Eng.* **2021**, *3*, 013538.
18. Pan, Y.S.; Lv, X.F.; Li, Z.H.; Dai, S. Experimental study of dynamic failure process of roadway under high velocity impact loading. *Rock Soil Mech.* **2011**, *32*, 1281–1286.
19. Wang, S.W.; Liu, H.D.; Jiang, T. Large geomechanical model test on failure mechanism of rockburst tunnel under static and explosive loads. *Chin. J. Rock Mech. Eng.* **2014**, *33*, 2095–2100.
20. Shreedharan, S.; Kulatilake, P.H.S.W. Discontinuum-Equivalent continuum analysis of the stability of tunnels in a deep coal mine using the distinct element method. *Rock Mech. Rock Eng.* **2016**, *49*, 1903–1922. [CrossRef]
21. Deng, J.Q.; Yang, Q.; Liu, Y.R. Time-dependent behavior and stability evaluation of gas storage caverns in salt rock based on deformation reinforcement theory. *Tunn. Undergr. Sp. Tech.* **2014**, *42*, 277–292. [CrossRef]
22. Wu, H.; Jia, Q.; Wang, W.; Zhang, N.; Zhao, Y. Experimental test on nonuniform deformation in the tilted strata of a deep coal mine. *Sustainability* **2021**, *13*, 13280. [CrossRef]
23. Zhao, Y.; Liu, Q.; Zhang, C.; Liao, J.; Lin, H.; Wang, Y. Coupled seepage-damage effect in fractured rock masses: Model development and a case study. *Int. J. Rock Mech. Min. Sci.* **2021**, *144*, 104822. [CrossRef]
24. Wang, Q.; Zhan, H.J.; Jiang, B.; Huang, Y.B.; Zhang, P.; Xu, S.; Liu, B.H. Failure mechanism of deep large section cavern and anchor injection control method. *J. Min. Saf. Eng.* **2020**, *37*, 1094–1103.
25. Wang, Q.; Jiang, B.; Pan, R.; Li, S.-C.; He, M.-C.; Sun, H.-B.; Qin, Q.; Yu, H.-C.; Luan, Y.-C. Failure mechanism of surrounding rock with high stress and con-fined concrete support system. *Int. J. Rock Mech. Min. Sci.* **2018**, *102*, 89–100. [CrossRef]
26. Wang, Q.; Gao, H.; Yu, H.; Jiang, B.; Liu, B. Method for measuring rock mass characteristics and evaluating the grouting reinforced effect based on digital drilling. *J. Rock Mech. Rock Eng.* **2019**, *52*, 841–851. [CrossRef]
27. Hou, C.J. Effective approach for surrounding rock control in deep roadway. *J. China Univ. Min. Technol.* **2017**, *46*, 577–583.
28. He, M.C.; Li, G.F.; Ren, A.W.; Yang, J. Analysis of the stability of intersecting caverns in deep soft-rock roadway construction. *J. China Univ. Min. Technol.* **2008**, *37*, 167–170.
29. Yang, R.S.; Xue, J.H.; Guo, D.M.; Li, Y. Failure mechanism of surrounding rock of large section caverns in complex rock formations and its control. *J. China Coal Soc.* **2015**, *40*, 2234–2242.
30. Kang, Y.; Liu, Q.; Gong, G.; Wang, H. Application of a combined support system to the weak floor reinforcement in deep underground coal mine. *Int. J. Rock Mech. Min. Sci.* **2014**, *71*, 143–150. [CrossRef]
31. Jiao, Y.Y.; Song, L.; Wang, X.Z.; Adoko, A.C. Improvement of the U-shaped steel sets for supporting the roadways in loose thick coal seam. *Int. J. Rock Mech. Min. Sci.* **2013**, *60*, 19–25. [CrossRef]
32. Li, X.H.; Yao, Q.L.; Zhang, N. Fracture characteristics of a soft rock roadway: Staged and zoned control. *J. China Univ. Min. Technol.* **2009**, *38*, 618–623.
33. Li, S.C.; Wang, H.T.; Wang, Q.; Jiang, B.; Wang, F.Q.; Guo, N.B.; Liu, W.; Ren, Y.X. Failure mechanism of bolting support and high-strength bolt-grouting technology for deep and soft surrounding rock with high stress. *J. Cent. South Univ.* **2016**, *23*, 440–448. [CrossRef]
34. Singh, R.P.; Mousumi, M.; Verma, M.K. Studies on the failure behavior of wire rope used in underground coal mines. *Eng. Fail Anal.* **2016**, *70*, 290–304. [CrossRef]
35. Ram, S.; Kumar, D.; Singh, A.K.; Kumar, A.; Singh, R. Field and numerical modelling studies for an efficient placement of roof bolts as breaker line support. *Int. J. Rock Mech. Min. Sci.* **2017**, *93*, 152–162. [CrossRef]
36. Yuan, Z.; Zhao, J.; Li, S.; Jiang, Z.; Huang, F. A unified solution for surrounding rock of roadway considering seepage, dilatancy, strain-softening and intermediate principal stress. *Sustainability* **2022**, *14*, 8099. [CrossRef]
37. Yu, W.; Li, K.; Liu, Z.; An, B.; Wang, P.; Wu, H. Mechanical characteristics and deformation control of surrounding rock in weakly cemented siltstone. *Environ. Earth Sci.* **2021**, *80*, 337. [CrossRef]
38. Zheng, P.Q.; Liu, Y.Q.; Zhan, Z.W. Research on support optimization of rectangular roadway based on equivalent circle method. *Coal Technol.* **2022**, *41*, 10–13.

processes

Article

Study on Time-Dependent Failure Mechanisms and CBAG Differential Support Technology of Roadway in Steeply Inclined Coal Seam

Zhengzheng Xie [1], Jin Wang [1,*], Nong Zhang [1,2], Feng Guo [1], Zhe He [1], Zhe Xiang [1] and Chenghao Zhang [3]

1 State Key Laboratory of Coal Resources and Safe Mining, School of Mines, China University of Mining and Technology, Xuzhou 221116, China
2 School of Civil Engineering, Xuzhou University of Technology, Xuzhou 221018, China
3 Laboratory of Geotechnics, Department of Civil Engineering, Ghent University, Zwijnaarde, 9052 Gent, Belgium
* Correspondence: wang_jin@cumt.edu.cn; Tel.: +86-182-0520-3728

Abstract: In Sichuan Province, China, most coal seams that are mined are steeply inclined; their roadways' surrounding rocks are asymmetric, with non-equilibrium deformations and unstable anchorage structures, thus making major safety hazards highly likely. Using field observations and a universal distinct element code (UDEC) numerical simulation method, this paper analyzed the time-dependent failure of the ventilation roadway of Working Face 1961 of the Zhaojiaba Mine, revealing the preconditions for such damage and a bidirectional deterioration mechanism for the deformation as well as stress of surrounding rocks. Moreover, this paper built an anchorage mechanical model for the thick layer of the roadway roof and proposed a cross-boundary anchor-grouting (CBAG) differential support technique. Calculations proved that the new support was particularly effective in restraining the expansion of tension cracks, thus preventing the slipping and dislocation deformations of rock masses on the curved roof side. The feedback of engineering applications showed that the maximum development depths of cracks in the arc roof and straight inclined roof of the roadway 150 m behind the working face are only 1.5 m and 1.10 m, decreasing by 61.3% and 47.6%, respectively, compared with the primary support. The proposed technology offers an overall thick-layer bearing structure for the surrounding rocks of roadways, effectively restraining the non-equilibrium large deformations of roadways in steeply inclined coal seams.

Keywords: steeply inclined coal seam; time-dependent failure; CBAG; differential support; equivalent anchorage

Citation: Xie, Z.; Wang, J.; Zhang, N.; Guo, F.; He, Z.; Xiang, Z.; Zhang, C. Study on Time-Dependent Failure Mechanisms and CBAG Differential Support Technology of Roadway in Steeply Inclined Coal Seam. *Processes* **2023**, *11*, 866. https://doi.org/10.3390/pr11030866

Academic Editors: Feng Du, Aitao Zhou and Bo Li

Received: 27 February 2023
Revised: 11 March 2023
Accepted: 12 March 2023
Published: 14 March 2023

1. Introduction

China's proven reserves of steeply inclined coal seams account for about 20% of its total coal resources, but its annual mining volume only accounts for 4% of the total coal mining volume, showing huge mining potential of steeply inclined coal seams in the future [1]. Notably, in provinces such as Yunnan, Guizhou, and Sichuan in Southwest China, more than 50% of the coal seams are steeply inclined coal seams, and the mining of steeply inclined coal seams is crucial to the region's economy. Compared with flat coal seams or gently inclined coal seams, it is more difficult to control roadways in steeply inclined coal seams. Characterized by complex geological structures and unpredictable coal seam occurrences, such coal seams are prone to large deformations, destruction, and even roof collapses. A survey [2,3] shows that the repair rate of steeply inclined coal seam roadways exceeds 80%, and many roadways are scrapped due to improper maintenance. The main reason for this is that unreasonable roadway support parameters cause large deformations of surrounding rocks, severely hindering sustainable production. Therefore, the safety maintenance and control of roadways in steeply inclined coal seams have attracted more and more attention from experts and scholars.

The difficulties in controlling roadways in steeply inclined coal seams are that the self-stable performance of surrounding rocks is poor, and it is not easy to build a stable and reliable anchorage structure [4–6]. Three factors contribute to this situation: first, all of the occurrence angles of coal seams exceed 45°, and a coal–rock mass with a large dip angle will naturally be subject to slipping and dislocation deformations under the action of self-weight stress; second, most steeply inclined coal–rock masses have experienced multiple geological tectonic movements and have poor lithology as well as low strength; third, the thickness of a coal–rock seam varies greatly, the coal–rock interface is widely distributed, and the development of joints is obvious. However, the damage process and failure mechanisms of roadways in steeply inclined coal seams are still unclear, so the existing support system does not have a good control effect, and deformations as well as damage continue to occur.

Most roadways in underground coal mines are characterized by time-dependent deformations; that is, the deformations of surrounding rocks are due to the cumulative damage of coal–rock masses, which does not happen suddenly. Scholars study the time-dependent deformations of surrounding rocks mainly through numerical simulations, theoretical calculations, and field measurements [7,8]. The deformations and damage of roadway in steeply inclined coal seams are closely related to time; large deformations and damage of roadways often occur in later periods. Yang's analysis [9] shows that the gradual adjustment of surrounding rock stress is the main reason for large deformations of soft rock roadways with discrete elements. Sun [10] simulated the time-dependent deformation characteristics of roadways with FLAC 3D, and reduced the deformation speed of roadways by using jet grouting to impose constraints on surrounding rocks. Zhang [11] adopted concrete-filled steel tubular supports to passively control deep soft rock roadways and obtained deformation parameters of steel tubes via theoretical calculations to invert the deformation law of surrounding rocks. Tao [12] used a physical model to test the failure process of inclined layered soft rock roadways; through the real-time monitoring of the internal strain of the model during the failure process the failure mechanisms of inclined layered soft rock roadways are explored. On-site observation methods, such as borehole peeping, are the most intuitive means by which to study time-dependent failure of roadways, allowing clear observations of the number, angles, and openings of cracks in surrounding rocks for analyses of the evolution regularity of cracks in surrounding rocks of roadways over time [13–15].

In order to study the dynamic failure law and new control technologies of surrounding rocks of roadways in steeply inclined coal seams, this work studied the typical steeply inclined coal seam roadway of the Zhaojiaba Coal Mine in Sichuan Province, China. Methods such as numerical simulations, theoretical calculations, and field observations were employed to comprehensively analyze the time-dependent failure mechanisms of the roadway in the steeply inclined coal seam. A targeted CBAG differential support technology that has achieved a good control effect through field verification is proposed.

2. Asymmetric Deformation Characteristics of a Roadway in a Steeply Inclined Coal Seam

2.1. Basic Overview of the Roadway

Located in Guangyuan City, Sichuan Province, China, the Zhaojiaba Coal Mine, as shown in Figure 1a, is within the mining range of the Sichuan Coal Industry Group Limited Liability Company. The mine is a typical steeply inclined thin coal seam. At present, the main area undergoing mining is Mining Area 106, and the buried depth of the working face is 350–600 m. The roadway currently being excavated is the ventilation roadway of Working Face 1961, which is 15 m horizontally and 20 m vertically from the upper goaf. This roadway, a typical one in the steeply inclined coal seam, is excavated along two steeply inclined coal seams, i.e., 9# coal seam and 10# coal seam. The average dip angle of the two coal seams is 53°, the thickness of 9# coal seam is 1.1 m, and the thickness of 10# coal seam is 1.0 m, both of which are thin coal seams. The rock stratum between the two coal seams is carbonaceous mudstone with a thickness of 2.5 m. The ventilation roadway of

Working Face 1961 is 850 m long, and always crosses the two coal seams. The rock stratum histogram of the roadway is shown in Figure 1b.

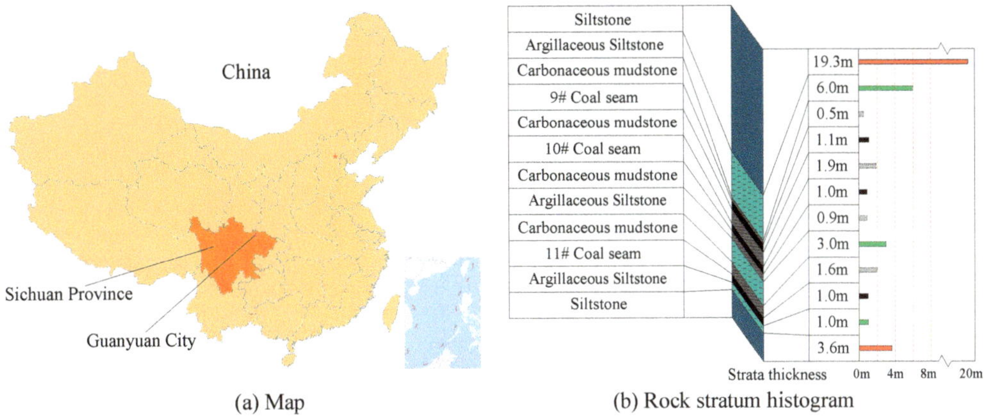

Figure 1. Coal mine location and coal seam histogram.

2.2. Primary Support Scheme

According to the engineering experience, the ventilation roadway of Working Face 1961 adopts the vertical wall irregular top section, with the roadway width being 4.00 m and the roadway height being 3.55 m, as shown in Figure 2. The coal–rock mass is cut in parallel to the immediate roof of 9# coal seam without damaging the rock mass structure of the immediate roof, and the roof on this side is a straight inclined roof; the coal–rock mass is then cut perpendicularly or obliquely to 9# coal seam and the mudstone interlayer to form an arc roof, and the roof on this side is curved.

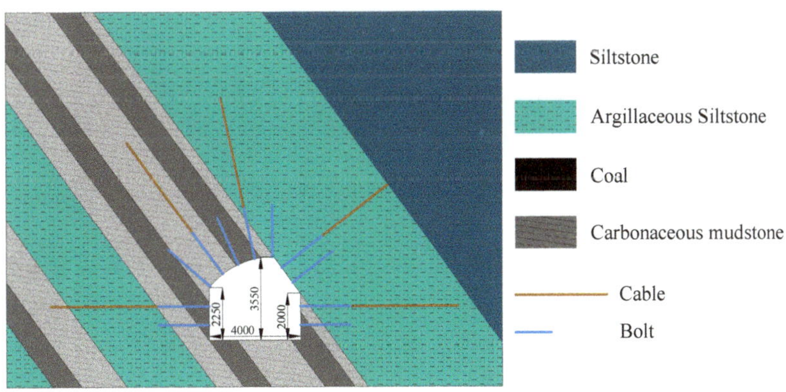

Figure 2. Irregular roadway section and primary support form.

As shown in Figure 2, the primary support of the ventilation roadway of Working Face 1961 is in the form of a bolt–anchor cable combined support, for which 11 bolts with a specification of $\Phi 20 \times 2200$ mm are installed in each row, with the bolt spacing and row spacing being 800 mm $\times$ 800 mm and the bolts being connected by reinforced ladder beams. Five anchor cables with a specification of $\Phi 17.8 \times 7000$ mm are installed in each row, with the anchor space and row spacing being 1600 mm $\times$ 1600 mm and the front as well as rear anchor cables of every two rows being supported by I-beams. The full section of the roadway is provided with a rhomboid metal mesh to protect its surface. Each bolt is matched with two K2335 anchoring agents, and each anchor cable is matched with five K2335 resin cartridges. On-site measurement feedback shows that the pre-tightening force of the bolt in the primary support is only 30 kN and that the pre-tightening force of the anchor cable is only 80 kN.

2.3. Asymmetric Deformation Characteristics of the Roadway

Figure 3 shows the deformation of the ventilation roadway of Working Face 1961 under the primary support, which is 100 m away from the excavation face. It can be seen from the figure that different positions of the roadway show differential deformation characteristics: ① The rock mass between the two rows of bolts on the straight inclined roof experiences a "bulge" deformation, with the surface layer being relatively loose and layered shedding occurring locally; ② the curved roof experiences quite evident deformation, with the coal–rock seam being displaced and deformed, the surface arc shape becoming flat, and the reinforced ladder beams at this position being generally bent and locally broken; and ③ the corner roadway zone on the arc roof side is "extruded" as a whole, while the floor experiences a slope-type deformation (the closer it is to the left side, the greater the floor heave). To sum up, roadway deformations are characterized by asymmetric deformations and failure. The starting point, A, of the asymmetric line is located at the intersection of the arc roof and the straight inclined roof, while the ending point, B, is located at the intersection of 10# coal seam and the mudstone interlayer. The surrounding rock deformation in the zone to the left of the straight line, AB, is significantly greater than in the zone to the right of the straight line, AB. The continued deformation in the left side area caused a continuous reduction in the section of the roadway, with a maximum reduction proportion of 37%. Furthermore, the anchor cables tended to break in some areas, greatly undermining mining safety.

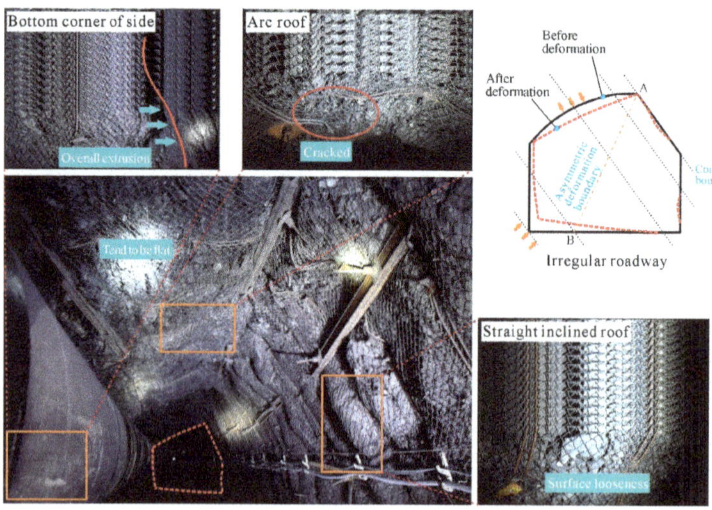

Figure 3. Failure status of the roadway under the primary support.

2.4. Evolution Characteristics of Cracks

In order to further evaluate the maintenance and control effects of the primary support, borehole peeping is carried out on the straight inclined roof and the arc roof at 4 m and 100 m away, respectively, from the excavation face to observe the development depths of initial cracks when the excavation begins, as well as to analyze the evolution of cracks in surrounding rocks after a period of roadway formation, as shown in Figure 4. The equipment used for the observation is a borehole peeping instrument, and the diameter and length of the borehole are 32 mm and 8 m, respectively.

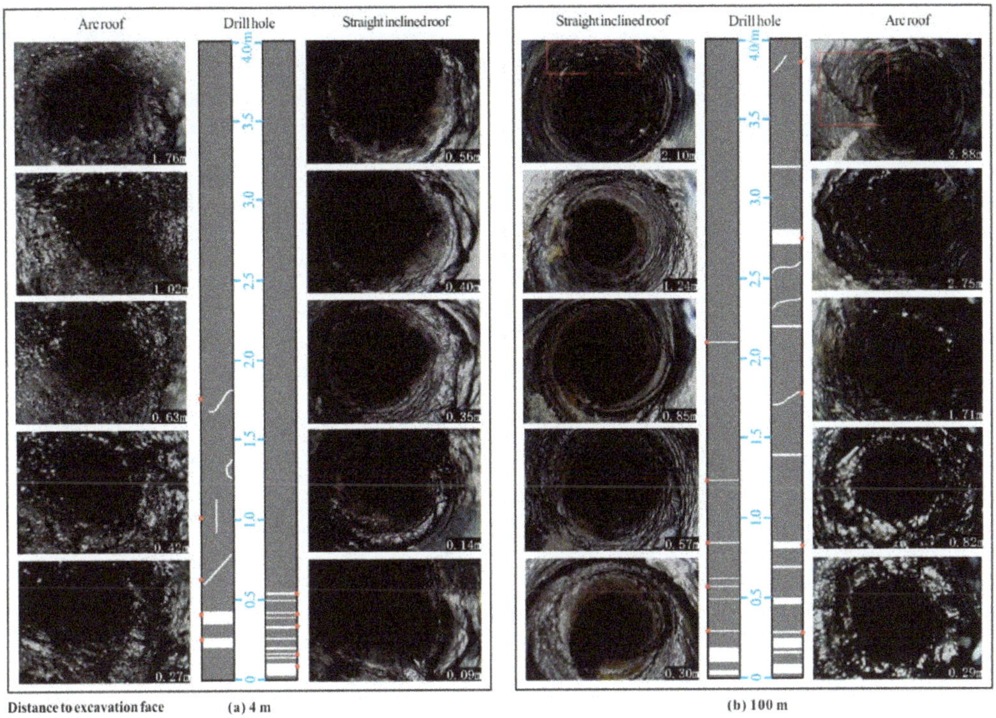

Figure 4. Distribution characteristics of cracks in the surrounding rocks of the roadway.

Figure 4a shows the distribution of cracks in the surrounding rocks when the roadway has just been excavated. At this time, the cracks' evolution depths at different roadway positions are obviously different. Cracks in the straight inclined roof are mainly within 0.5 m, and most of them are small annular cracks, with the maximum crack depth being only 0.56 m, while the coal–rock mass on the arc roof is of poor integrity, with the maximum crack depth being 1.76 m. In addition to annular cracks there are also inclined and longitudinal cracks in the hole. Compared with straight inclined roofs there are more cracks in the arc roof, and the cracks' openings are larger, undermining the arc roof's stability.

Figure 4b shows the distribution of cracks in the surrounding rocks 100 m away from the excavation face. Compared with when the excavation had just begun, the cracks on both sides of the surrounding rocks of the roadway expanded significantly at this time. The depths of the cracks in the straight inclined roof expand to 2.10 m, with a growth rate of 73.3%. It is additionally found from the figure that the cracks are mainly distributed within 1.5 m. The maximum development depths of the cracks in the arc roof reach 3.88 m, with an increase of 54.6%. The cracks and cracked zones are generally distributed within 3.0 m. When 100 m away from the excavation face the excavation stress tends to be stable, the cracks in the straight inclined roof expand to the bolt anchorage end area, and the cracks in the arc roof expand obviously beyond the bolt anchorage range, indicating that the shallow rock mass of the roadway as well as the bolt have experienced synchronous deformation and that the reinforcement effect of the bolt on the rock mass is very limited. Moreover, the expanding rate of the anchor cable is generally lower than 3.5%, and continuous rock mass deformation forces the anchor cable to break, such that the primary support can no longer meet the safety maintenance and control requirements of the roadway.

3. Time-Dependent Failure Mechanisms of a Roadway in a Steeply Inclined Coal Seam

3.1. Model Establishment and Scheme Design

3.1.1. Determination of Key Parameters

In order to analyze the differential damage and failure process of the roadway in the steeply inclined coal seam, the UDEC (universal distinct element code) numerical software is used to carry out a simulation study. The UDEC is a discontinuous mechanics method that uses discrete element method theory to provide accurate analyses for geotechnical engineering and mining engineering. It simulates the mechanical response of a jointed rock stratum under static or dynamic load, such as stress, displacement, and crack evolution [16,17]. The Trigon module in the software can realistically simulate the generation, expansion, and coalescence of cracks in brittle materials [18,19]. It was developed by Gao and Stead [20] on the basis of the Voronoi algorithm, which mainly divides a randomly generated polygonal mesh into triangular blocks. The resulting model has less dependence on a mesh, improves the problem of the post-peak strain hardening of rocks, and can realistically simulate fracturation as well as crack evolution in mining engineering.

In the discrete element model the mechanical parameters of the block and contact surface jointly determine the mechanical properties of a rock mass. The parameters of the block include density and the bulk as well as shear moduli, while those of the contact surface include normal stiffness, tangential stiffness, cohesion, and internal friction angle. The bulk modulus, K, and shear modulus, G, in the model are determined by the elastic modulus, E, and Poisson's ratio, ν; the specific conversion relationship (see Equations (1) and (2)) is as follows [21]:

$$K = \frac{E}{3(1-2\nu)} \tag{1}$$

$$G = \frac{E}{2(1+\nu)}, \tag{2}$$

In the Trigon module the elastic modulus of the triangular block depends on the normal stiffness and tangential stiffness of the contact surface of the block, and its calculation formula (see Equations (3) and (4)) is as follows [22]:

$$k_n = 10\left[\frac{K+\frac{4}{3}G}{\Delta Z_{\min}}\right], \tag{3}$$

$$k_s = 0.4k_n, \tag{4}$$

where k_n is the normal stiffness of the contact surface, k_s is the tangential stiffness of the contact surface, and $\Delta Z_{\min}$ is the minimum side length of the block.

Therefore, in order to determine the mechanical parameters of the model block and the contact surface, it is necessary to obtain those of the real rock stratum in combination with laboratory experiments to verify the rationality of the model parameters.

1. Determination of Rock Mass Parameters

The parameters of coal–rock samples, i.e., rock parameters, are obtained through experiments, while the parameters applied in UDEC are rock mass parameters; therefore, the conversion of mechanical parameters is required.

Zhang and Einstein [23] proposed the conversion formula between the elastic modulus of rock mass and the elastic modulus of rock, as shown in Equation (5):

$$\frac{E_m}{E_r} = 10^{0.0186RQD-1.91} \tag{5}$$

where E_r is the rock elastic modulus; E_m is the rock mass elastic modulus; and RQD is the rock quality index, which is obtained by peeping from a borehole in the ventilation roadway of Working Face 1961.

Singh and Seshagiri [24] found that there is a strong linear relationship between the nth power of the ratio of the rock mass elastic modulus to the rock elastic modulus and the ratio of the rock mass compressive strength to the rock compressive strength, as shown in Equation (6):

$$\frac{\sigma_{cm}}{\sigma_c} = \left(\frac{E_m}{E_r}\right)^j, \tag{6}$$

where σ_c is the rock compressive strength; σ_{cm} is the rock mass compressive strength; and the general value of j is 0.56.

The rock mass tensile strength is shown in Equation (7):

$$\sigma_{tm} = k\sigma_{cm}, \tag{7}$$

Hoek and Brown [25] proved that the value of k is generally 0.05–0.1, so its value is 0.1 this time.

The rock mechanical parameters are converted into rock mass mechanical parameters by calculation and filled into Table 1.

Table 1. Conversion of rock parameters and rock mass parameters.

Rock Stratum Lithology	Rock Mechanics Parameters		RQD	Rock Mass Mechanical Parameters		
	σ_c/MPa	E_r/GPa		σ_{cm}/MPa	E_m/GPa	σ_{tm}/MPa
Coal	7.9	1.4	63	1.2	0.3	0.1
Muddy siltstone	25.2	6.4	82	15.3	2.6	1.5
Carbonaceous mudstone	12.3	3.1	76	6.5	1.0	0.8
Siltstone	37.8	14.5	90	27.9	8.4	2.8
Fine-grained sandstone	40.5	16.2	95	33.7	11.7	3.4

2. Verification of Model Parameters

The micro-parameters of the contact surface are calculated with rock mass parameters according to Equations (1) to (4). A 5 × 10 m plane model is established by the UDEC to carry out uniaxial compression simulation experiments and obtain simulated mechanical parameters, such as the compressive strength and elastic modulus of each rock stratum. The obtained simulated values are compared with the experimental values. If the error between the simulated and experimental values is large, it is necessary to adjust the rock stratum parameters again for another simulation, and so on.

After many simulation adjustments, the microscopic parameters with an error of less than 5% are set as the final mechanical parameters of the model, as shown in Table 2.

Table 2. Mechanical parameters of each rock stratum in the model.

Rock Stratum Lithology	Block Parameters		Contact Surface Parameters				
	Density ρ/kg·m^{-3}	Elastic Modulus E_m/GPa	Normal Stiffness k_n/GPa·m^{-1}	Tangential Stiffness k_s/GPa·m^{-1}	Internal Friction Angle θ/°	Cohesion C/MPa	Tensile Strength/MPa
Coal	1350	0.3	16	7	10	1.0	0.1
Muddy siltstone	2180	2.6	53	21	12	4.1	1.5
Carbonaceous mudstone	2060	1.0	47	19	12	2.3	0.8
Siltstone	2300	8.4	109	44	16	8.7	2.8
Fine grained sandstone	2550	11.7	79	32	19	11.9	3.4

3.1.2. Model Establishment

The UDEC model is designed according to the geological occurrence of the ventilation roadway of Working Face 1961, and the model size is 50 m × 50 m. It can be seen from Figure 5 that the model is divided into twenty layers, including five lithologies, namely coal, argillaceous siltstone, carbonaceous mudstone, siltstone, and fine-grained sandstone, and that the dip angle of each rock stratum is set to 53°. The model adopts the mechanical parameters calibrated in Section 3.1.1, as shown in Table 2. The origin of the model is set at the lower-left corner, and the surrounding zone of the roadway (the red part in Figure 5) is the key observation zone. The rock mass joints in this zone are densified in order to observe the distribution of cracks in the surrounding rocks.

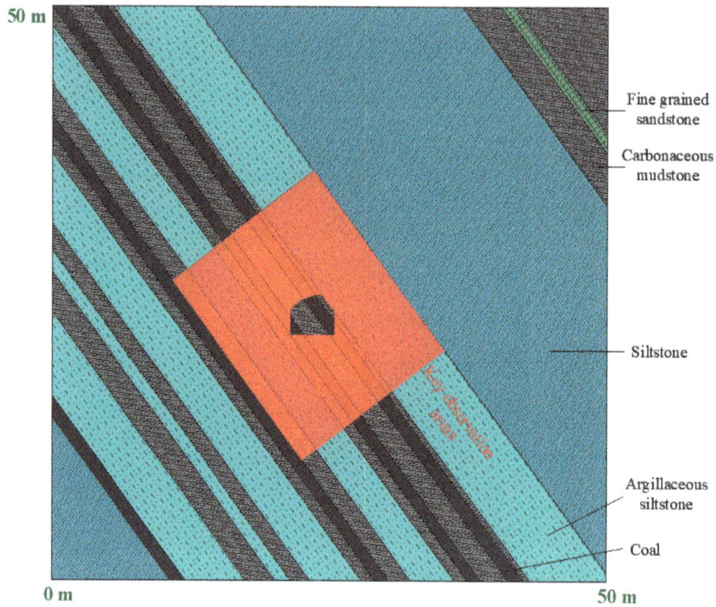

Figure 5. Model layout.

According to the geological data, the average burial depth of the ventilation roadway of Working Face 1961 is 450 m. If the lateral pressure coefficient is taken as 1, the vertical stress and horizontal stress applied by the model are both 11.76 MPa. The model is a plane strain model, the block is an elastic body, and the contact surface adopts the Mohr–Coulomb yield criterion. When the roadway module is deleted, this indicates that the excavation of the roadway has begun.

The cutting equipment for the ventilation roadway of Working Face 1961 is an EBZ260 TBM. Due to the limitation of the surrounding rocks around the roadway, the surrounding rock stress is reduced gradually rather than completely released at one time [26], so the Fish function "ZONK.FIS" (Itasca, 2012) in the UDEC is used to simulate the real process of stress release after excavation unloading of the roadway. The surface stress of the roadway is gradually released in ten stages, namely Stages ①, ②, ③, ④, ⑤, ⑥, ⑦, ⑧, ⑨, and ⑩. Of the surface stress, 10% is released in each stage until it drops to zero by Stage ⑩. The stress release coefficient is set as R, and the R of these ten stages is 0.1, 0.2, 0.3 … ~1.0, respectively, as shown in Figure 6. Four operation steps, namely four stress states, are selected to analyze the time-dependent failure law of the roadway. The operation steps are a (0.2 × 104), b (2.5 × 104), c (5.0 × 104), and d (10.0 × 104), and the stress release coefficients are 0.1, 0.7, 1.0, and 1.0, respectively.

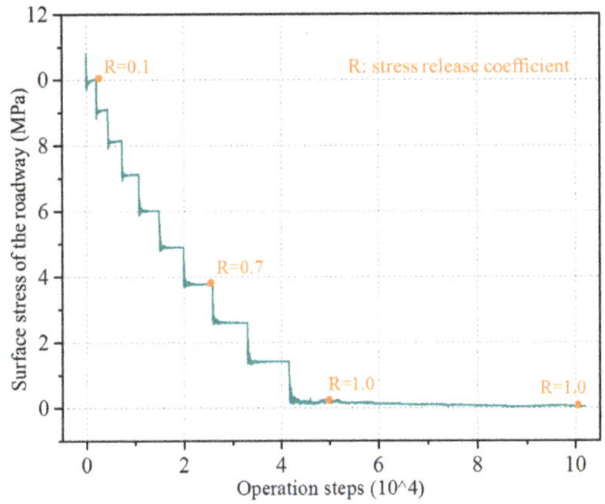

Figure 6. Surface stress evolution curve of the roadway.

3.2. *Time-Dependent Failure Process of a Roadway in a Steeply Inclined Coal Seam*

3.2.1. Stress Evolution Analysis

Figure 7 shows the nephogram of the maximum principal stress of the four stages in the stress release process of the roadway in the steeply inclined coal seam. It can be seen from the figure that after the excavation of the roadway is started the floor is damaged first, and its stress decreases accordingly. The nephogram of the maximum principal stress is in the shape of an "inverted triangle". The stress of the floor zone on the arc roof side decreases the most, with a relatively wide damage range, and the stress of the wall on this side also decreases in a small range. It is also found that stresses are concentrated in the argillaceous siltstone seam at the lower left of the roadway. With the gradual release of surrounding rock stress the damage degree of the surrounding rocks on the straight inclined roof side is small, and the degree of stress decreasing is not obvious; however, the damage to the surrounding rocks on the arc roof side becomes more and more serious, and the degree of stress decreases becomes increasingly obvious, which further causes the stresses to concentrate in the hard rock stratum at the lower left of the roadway. At this time, the roadway is at risk of being extruded from the lower left.

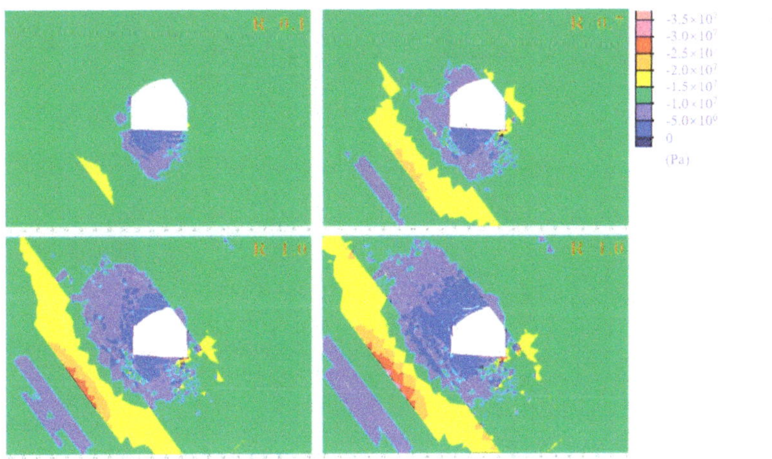

Figure 7. Nephogram of the maximum principal stress of the roadway during the stress release process.

3.2.2. Displacement Evolution Analysis

Figures 8 and 9 are the evolution diagrams of the surrounding rock displacement field of the roadway in the steeply inclined coal seam. It can be seen from Figure 8 that when the stress release coefficient, R, is 0.7, the coal–rock mass on the left side of the roadway floor is displaced upward, and the whole floor experiences a typical "slope-type" deformation with a large deformation on the left side and a small deformation on the right side. When the stress release coefficient, R, is 1.0, the coal–rock mass seam of the arc roof slips obviously and experiences a "falling" deformation, and the arc roof surface tends to be flat. It can be further concluded from Figure 9 that when the number of operation steps reaches 10 × 104 the maximum subsidence of the arc roof exceeds 300 mm, and the horizontal displacement at the left shoulder of the roadway is the largest, exceeding 200 mm. At the same time, the maximum floor heave of the roadway is located in the left corner of the roadway, and the deformation also exceeds 200 mm; however, the surrounding rock displacement on the straight inclined roof side is less than 50 mm. Therefore, the surrounding rocks of the roadway are steeply inclined coal seams that experience significant asymmetric differential deformations, the surrounding rocks on the arc roof side are a severely deteriorated zone, and the surrounding rocks on the straight inclined roof side is a slightly damaged zone.

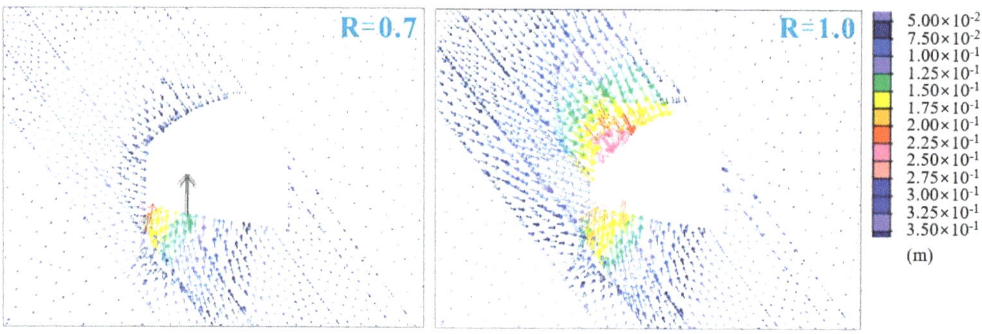

Figure 8. Vector diagram of the displacement field of the roadway.

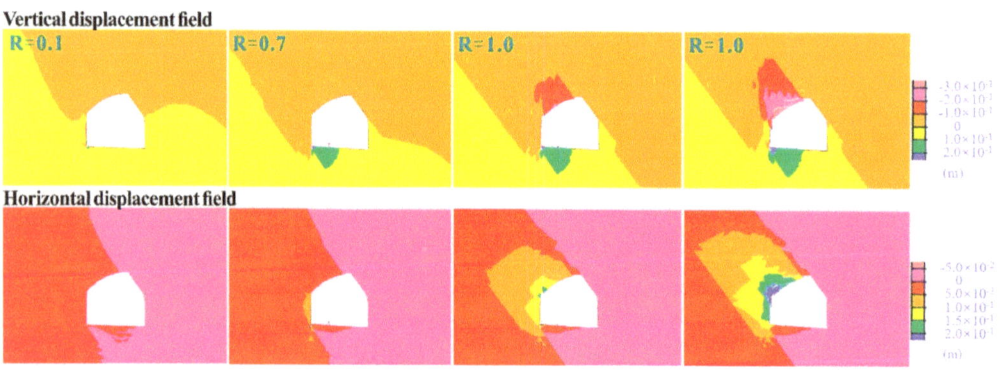

Figure 9. Nephogram of the displacement field of the surrounding rocks of the roadway.

3.2.3. Analysis of the Crack Evolution Law

Figure 10 shows the evolution characteristics of cracks in the surrounding rocks of the roadway in the steeply inclined coal seam. It can be seen from Figure 10a that the cracks in the surrounding rock are distributed in different zones, and that the number as well as evolution depths of the cracks in the surrounding rock zone on the arc roof side (Zone A) are significantly higher than those on the straight inclined roof side (Zone B).

Furthermore, Figure 10b shows that the cracks in the surrounding rocks are subject to the zonal evolution law. With the gradual release of surrounding rock stress, the cracks in Zone A increase significantly in stages. In particular, when R is 1 the number of cracks increases in a leaping manner; with the increase in calculation steps the number of cracks continues to increase, indicating that the deformation in Zone A has the property of long-term flowing deformation. However, the evolution trend of cracks in Zone B is significantly weaker than in Zone A, and the stability period is also significantly shortened. When the number of operation steps reaches 5×104, the number of cracks tends to stabilize. It can be found from the figure that the number of shear cracks in the two zones of the roadway is significantly larger than the number of tension cracks.

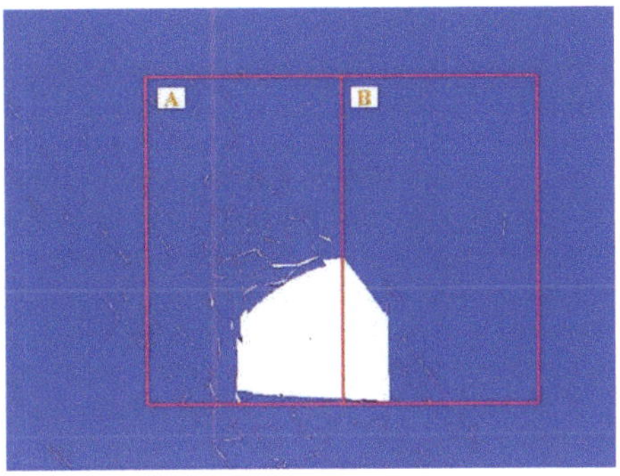

(a)

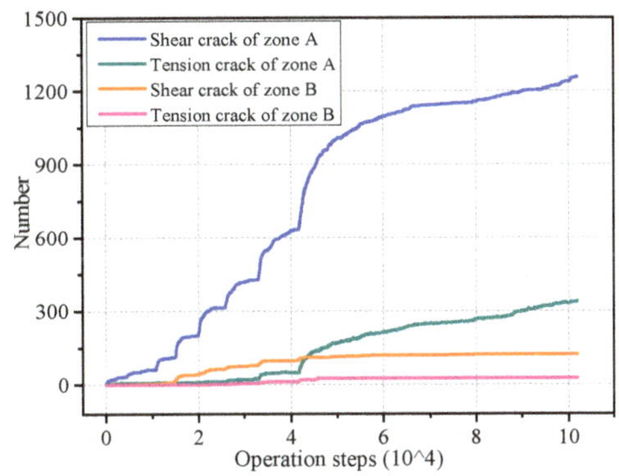

(b)

Figure 10. Differential evolution laws of cracks in surrounding rocks. (**a**) Zonal distribution characteristics of cracks; (**b**) time-dependent evolution curves of cracks.

3.2.4. Analysis of Time-Dependent Failure Mechanism

It can be concluded from the simulation results and field observations that the deformations of the surrounding rocks of the roadway in the steeply inclined coal seam are highly correlated with time and subject to the zonal differential evolution law, which is mainly reflected in the deterioration rate of the surrounding rocks on the arc roof side far exceeding that on the straight inclined roof side. The reasons for the time-dependent failure of the roadway are analyzed from the following two aspects:

(1) Fragmentation of rock mass in the floor corner on the arc roof side of the roadway in the steeply inclined coal seam is the precondition for the time-dependent failure of the surrounding rocks. It can be seen from Figure 8 that, after the excavation of the roadway is begun, the coal–rock mass in the floor corner zone first experiences an upward seam displacement deformation and that the damage degree at this position is the most serious, which provides deformation space for the later slipping failure of the surrounding rock on the arc roof side. The roadway floor also experiences a "slope-type" deformation with a large deformation on the left side and a small deformation on the right side. In order to realize the long-term maintenance and control of the roadway in the steeply inclined coal seam, it is necessary to control key parts of the floor on the arc roof side.

(2) Non-equilibrium deformations of the roadway in the steeply inclined coal seam are caused by surrounding rock stress. The occurrence of the steeply inclined coal–rock seam determines the deflection transfer of surrounding rock stress during the excavation process. Such asymmetric adjustment causes slipping and dislocation deformations of the coal–rock mass on the arc roof side along the seam, and the large-scale linkage of surrounding rocks tends to form a partial stress-bearing arch on the roadway, which intensifies the damage of the surrounding rocks on the peak stress side and structural instability damage. It can be seen from Figure 7 that the maximum principal stress is transferred to the hard rock stratum

deeper in the surrounding rocks, and the damaged as well as deteriorated parts further expand to the depth. The surrounding rock stress and the rock mass deformations have a bidirectional deterioration relationship, which leads to the failure of the bolt anchorage structure as well as the instability and collapse of the severely deteriorated zone; however, there is no large-scale stress concentration in the surrounding rock on the straight inclined roof side, and only a certain degree of damage occurs in the shallow part. Therefore, the floor angle of the roadway in the steeply inclined coal seam shall be timely controlled, and differential control measures shall be taken for surrounding rocks on both sides.

4. CBAG Differential Support Technology

4.1. Distribution Characteristics of Cracks in the Surrounding Rocks under the Combined Support of Short Bolts and Long Anchor Cables

The simulated roadway is controlled according to the original combined support of the mine in Figure 2 with the combination of 14 short bolts, with a length of 2.2 m, and 5 long anchor cables, with a length of 7.0 m. With the gradual release of the surface stress of the roadway in stages, the evolution characteristics of cracks in the surrounding rocks with the number of operation steps are shown in Figure 11. When the number of operation steps reaches 2.5 × 104, that is, R = 0.7, the number of cracks in the surrounding rocks in Zone A of the roadway is 257, while the number of cracks in the surrounding rocks in Zone B is only 43, indicating that the surface stress has not been completely released, such that the damage degree of the rock mass shows a great difference. It can be further seen from the crack distribution diagram that the maximum depth of crack evolution in Zone A is still in the bolt anchorage area, while the cracks in Zone B are only distributed on the shallow surface of the roadway. When the number of operation steps reaches 10 × 104, that is, R = 1.0, the number of cracks in the surrounding rocks in Zone A and Zone B is 794 and 117, respectively, and the cracks in different positions of the roadway increase to different extents. As shown in the lower-right figure, the cracks in Zone A expand significantly, and the evolution depth exceeds the bolt anchorage area, with the maximum crack depth reaching 3.5 m. At this time, the rock mass in the whole bolt anchorage area is seriously cracked, with the risk of overall collapse and instability.

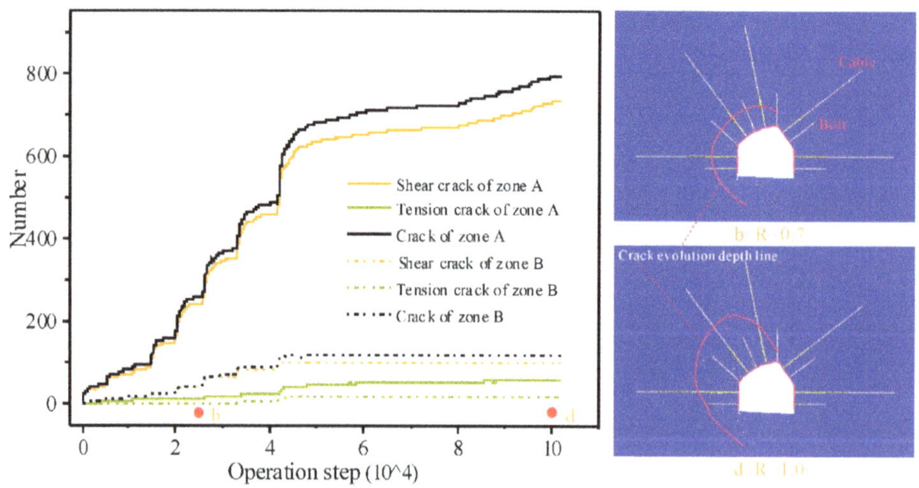

Figure 11. Evolution characteristics of cracks in the surrounding rocks under the primary support.

4.2. Thick-Layer Anchorage Principle and Differential Equivalent Reinforcement Technology

The number of cracks in the surrounding rocks in Zone A and Zone B of the roadway under the primary support of the mine is greatly reduced by 775 and 31, respectively, compared with that when there is no support, with a reduction rate of 49.4% and 20.9%, respectively. Although the primary support has a certain effect on restraining the development of cracks, it does not change the damage status of the seriously deteriorated zone of the roadway (Zone A). The anchor point in the bolt on this side is still located in the crack circle, and the bolt anchorage zone shows an overall movement trend; however, since the anchor cable is too long it has a limited reinforcement effect on the rock mass in this range, and the anchor cable is easy to break. Therefore, the bolt and anchor cable combined technology needs to be changed urgently.

It is assumed that the anchored rock beam of the foundation of the roadway roof is simplified as a simply supported beam, as shown in Figure 12. When the rock beam experiences pure bending deformation, according to the maximum tensile stress theory (the first strength theory), the maximum dangerous point of the roof is at the center of the anchored rock beam surface (i.e., Point C). At this time, the maximum tensile stress of the rock beam (see Equation (8)) is as follows:

$$\sigma_{\max} = \frac{M_{\max} y_{\max}}{I_Z}, \tag{8}$$

where $M_{\max} = \frac{qL^2}{8}$, $y_{\max} = \frac{h}{2}$, and $I_Z = \frac{bh^3}{12}$. The relationship (see Equation (9); k stands for constant) between the maximum tensile stress and the thickness of the rock beam is further obtained as follows:

$$\sigma_{\max} = \frac{3qL^2}{4bh^2} \Rightarrow k\frac{1}{h^2} \propto \frac{1}{h^2}, \tag{9}$$

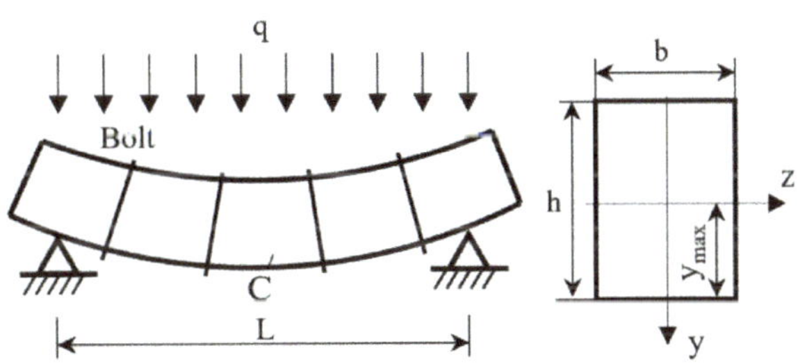

Figure 12. Mechanical model of a simply supported beam for an anchored rock beam.

It can be concluded from Equation (9) that the maximum tensile stress of an anchored rock beam is inversely proportional to the square of its thickness, indicating that the greater the thickness of an anchored rock beam the smaller the tensile stress value and the more stable the roof rock stratum. Therefore, in order to realize the long-term stable control of the roadway, the foundation anchorage thickness in the support system shall meet the cross-boundary requirements of the thick layer [27,28]; that is, the foundation bolt support shall span the surrounding rock crack circle as well as plastic circle, and the inner anchorage point shall be anchored into the elastic circle to form a thick reinforcement circle that is closely connected with the stable rock mass in the deep part. In order to improve the strength and stiffness of the anchorage bearing layer and resist the disturbance of mining stress as well as the long-term creep effect of soft rock, the bidirectional linkage control of the deep and shallow displacement is realized by increasing the thickness of the foundation anchorage layer and restricting the large displacement in the shallow part with the small displacement in the deep part, which is suitable for a roadway under any surrounding rock inclination angle.

According to the peep images and numerical simulation results of the surrounding rocks of the roadway, the initial cracks in the surrounding rocks of the roadway in the steeply inclined coal seam after excavation are characterized by an obvious asymmetric distribution. The crack depth in Zone A is generally greater than 1.5 m, while the crack depth in Zone B is generally less than 0.5 m, as shown in Figure 13a; however, the equal-length and equal-strength combined support form of the primary support is no longer suitable, and the key positions are not reinforced by deep anchorage, so the maintenance and control effects of the roadway are extremely poor. According to the cross-boundary anchorage principle, the bolt–anchorage thickness of the severely deteriorated Zone A shall be significantly greater than that of the slightly damaged Zone B. At the same time, due to the repeated staggered distribution of the steeply inclined coal–rock seams in Zone A, in order to ensure the long-term bearing of the anchored rock mass, the grouting method shall be adopted to modify the lithology of the surrounding rocks, especially the grouting reconstruction of the rock mass in the floor corner. Unequal-length differential anchor-grouting support technology is used to construct the overall thick-layer bearing structure of the full-section surrounding rocks of the roadway, that is, the equivalent reinforcement layer, to adapt to the long-term creep effect of the weak surrounding rocks, and then realize the long-term stable bearing of the roadway in the steeply inclined coal seam, as shown in Figure 13b.

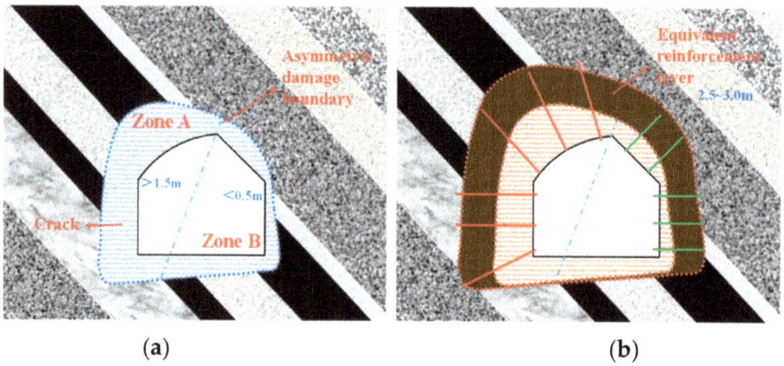

Figure 13. Asymmetric deterioration characteristics and equivalent reinforcement principle of the roadway in the steeply inclined coal seam. (**a**) Sketch of the asymmetric distribution of cracks; (**b**) equivalent reinforcement effect.

In view of different coal–rock properties, crack characteristics, and crack depths in each part of the roadway in the steeply sloping coal seam, two support forms, namely hollow grouting cables and flexible bolts, are used in Zone A and Zone B, respectively, to construct an equivalent reinforcement layer with a thickness of 2.5–3.0 m for the roadway in the steeply inclined coal seam. For the surrounding rocks in Zone A the anchor-grouting combined reinforcement method shall be adopted, and the foundation anchorage thickness shall exceed 4.0 m. For Zone B only the end anchorage is required, and the foundation anchorage thickness can be less than 3.5 m.

5. Engineering Case Analysis

5.1. Differential Support Scheme

Taking the intersection of the arc roof and the straight inclined roof of the irregular roadway as the dividing point, as shown in Figure 14, seven anchor cables with a length of 4300 mm and a diameter of 22 mm are installed on the left side of the roadway, with the red ones representing hollow grouting cables and the blue ones representing solid cables; four flexible bolts with a length of 3500 mm and a diameter of 22 mm are installed on the right side of the roadway. Both the anchor cables and flexible rods are matched with the 300 mm × 300 mm × 14 mm arched steel tray, and a reinforced ladder beam as well as a woven mesh are used to strengthen and protect the surface. The initial pre-tightening force of the anchor cable shall not be less than 120 kN, and the initial pre-tightening force of the flexible bolt shall not be less than 80 kN. The grouting material selection of Huainan Donghuaouke company KWJG-1 type mineral reinforced composite mortar, a product without shrinkage, micro-expansion, high liquidity, and pumping, does not contain chloride. A ZBYSB40/22-7.5 mining hydraulic grouting pump was selected for grouting pumping.

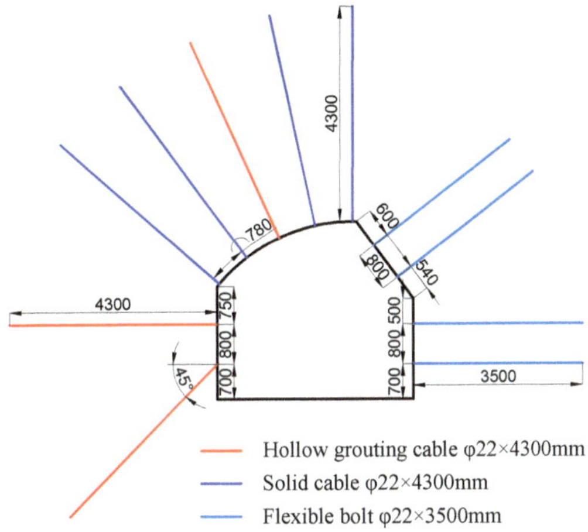

Figure 14. New support scheme.

It is required to note the following:① a long anchor cable with a size of Φ22 mm × 6500 mm shall also be installed for an arc roof area that is seriously cracked after the initial excavation; ② for serious floor heave, hollow grouting cables shall be installed at an inclination of 45° downward, and the deformations of the floor heave shall be suppressed by grouting coupling as well as the reinforcement of key parts of the floor corner.

5.2. Comparative Analysis of Simulation Effects of New Support

Figure 15 shows the characteristics and deformation of cracks in the surrounding rocks of the roadway under the new support when the number of operation steps reaches 10.0 × 104. Figure 15a shows that after the differential support is applied the number of cracks in the surrounding rocks of the roadway is significantly reduced, and that the distribution range is significantly narrowed. The above is specifically reflected in the following three aspects: ① the cracks on the arc roof side are all within the anchorage range, with the maximum evolution depth being reduced to 1.8 m, the cross-boundary support limiting the slipping as well as dislocation of the steeply inclined coal–rock mass, and the integrity of the surrounding rocks being significantly improved; ② the cracks on the straight inclined roof side are only distributed in the shallow part of the roadway, and the deformation is small; ③ and the number of floor cracks is also appropriately reduced, which is mainly because the 45° grouting anchor cable improves the shear capacity of the coal–rock mass in the floor corner, and the surrounding rock on the arc roof side is quite complete, which limits the deformations of the floor coal–rock mass as well as the development of cracks.

Figure 15b shows the evolution law of the displacement of the roadway's surrounding rocks with depths under different supports. The vertical line, EF, in the figure is the center line of the roadway. The distance between the horizontal line, GH, and the horizontal line, PQ, of the roadway floor is 2 m. A total of 40 measuring points are arranged vertically inward 5 m from points E, F, G, and H on the roadway's surface in order to obtain the deformation status of the roadway at different positions. The vertical displacement is monitored at each measuring point on the roof and floor, and the horizontal displacement is monitored at each measuring point on both sides. In addition, 40 measuring points are also arranged on the horizontal line, PQ, of the floor to observe the floor heave status.

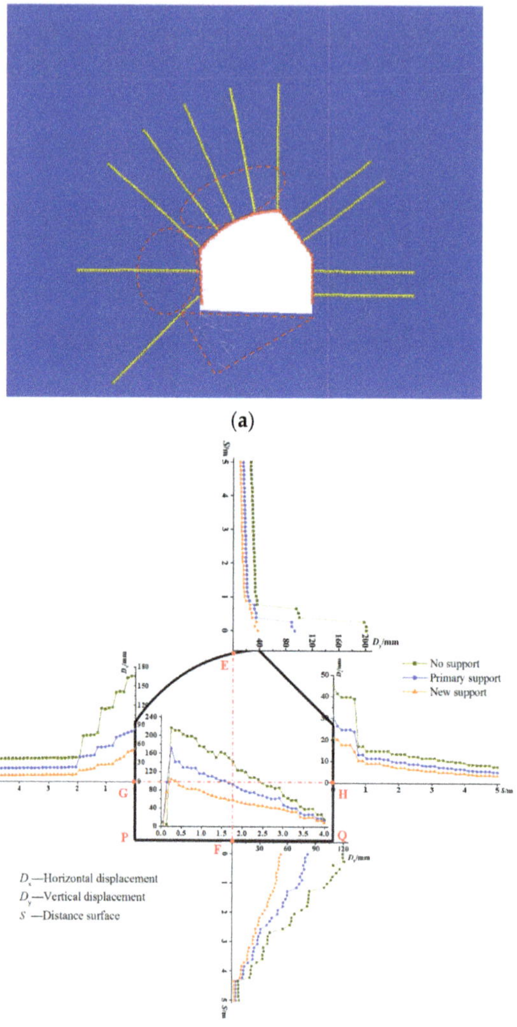

Figure 15. Maintenance and control effects of the roadway under the new support. (**a**) Distribution characteristics of cracks; (**b**) evolution law of the displacement of the roadway's surrounding rocks with depths under different supports.

It can be seen from the deformation status of the roadway at different positions that the control effects of different support forms are significantly different. It can be concluded from the subsidence data of the roof that deformation mainly occurs in shallow parts less than 1 m above the roof. The displacement of point E under no support, the primary support, and the new support is 205 mm, 98 mm, and 40 mm, respectively. The displacement under the new and primary supports decreases by 80.5% and 52.2%, respectively, compared with that under no support. It can be concluded from the displacement data of the wall on the arc roof side that deformations are mainly concentrated within 2 m of the wall. The displacement of point G under no support, the primary support, and the new support is 165 mm, 85 mm, and 50 mm, respectively. The displacement under the new and primary supports decreases by 69.7% and 48.5%, respectively, compared with that under no support.

It can be concluded from the displacement data of the wall on the straight inclined roof side that deformations are mainly concentrated within 0.8 m of the wall. The displacement of point H under no support, the primary support, and the new support are 45 mm, 31 mm, and 20 mm, respectively. The displacements under the new and primary supports decrease by 55.6% and 31.1%, respectively, compared with that under no support. It can be concluded from the displacement data of the floor that deformations are mainly concentrated within 4.2 m. The displacement of point F under no support, the primary support, and the new support is 120 mm, 81 mm, and 52 mm, respectively. The displacement under the new support and primary support decreases by 56.7% and 32.5%, respectively, compared with that under no support. It can be concluded from the displacement data of the horizontal line, PQ, of the floor that the maximum displacement point of the floor is 0.3 m away from point P as the origin, and it shows a single decreasing trend from 0.3 m to 4.0 m. The maximum floor heave under no support, the primary support, and the new support is 220 mm, 178 mm, and 102 mm, respectively. The maximum floor heave under the new support decreases by 53.6% and 19.1%, respectively, compared with that under the primary support and no support. It can be seen that the control effects of the new support at different positions are significantly better than those of the original support. The deformations under the new support decrease by 53.6–80.5% compared with those under no support, while the deformations under the primary support decrease by only 19.1–52.2% compared with those under no support.

5.3. Mine Pressure Monitoring and Analysis

The working load on anchor cables is monitored for a long time by the dynamometer clam between the tray and the nut, which has the functions of a digital display, historical data memory, and transmission. Figure 16 shows the evolution curve of the working load on anchor cables of the arc roof and the straight inclined roof under the new support. It can be seen from the figure that the evolution characteristics of the working load on anchor cables at different installation positions over time are quite different. It can be seen from Figure 16a that the stability period of the anchor cables is about 40 days, during which the load on the anchor cables is characterized by repeated oscillation up and down, with the maximum load being 168 kN and the minimum load being the pre-tightening force (123 kN), indicating that the coal–rock mass in the shallow part on the arc roof side is prone to damage and failure under compression, which will cause the tray to move and lead to a readjustment of the load on the anchor cables. After repeated mechanical action, balance is finally achieved and the load is stabilized, and the final load is about 161 kN. It can be seen from Figure 16b that the stability period of the flexible bolt is about 13 days, and its working load shows an evolutionary trend of a sharp increase at first, then a slow increase, and finally a stable bearing. The rock mass on this side has good integrity, and the lithology is relatively hard and flexible. There is no repeated oscillation or adjustment for the load on the flexible bolt. The initial pre-tightening force is 83 kN, and is finally stabilized at 165 kN. The characteristics of the load on the bolt–cable at different positions of the roadway also reflect differences in the properties of the rock mass, which confirms that the control methods also need to be differential.

Figure 17 shows the peeping images of the arc roof and straight inclined roof of the roadway. The cracks in the arc roof and the straight inclined roof are distributed within 1.50 m and 1.10 m, respectively. There are seven cracks and two cracked zones on the arc roof side, but only six small cracks on the straight inclined roof side. It can be concluded from the distribution of cracks in the surrounding rocks that the integrity of the rock mass is significantly improved, which is beneficial to the long-term stability of the roadway. Figure 18 is the site photo of the roadway after 150 m of excavation. The surfaces of the arc roof and the straight inclined roof of the roadway are very flat, and there are no deformations and damage similar to those that occur under the primary support.

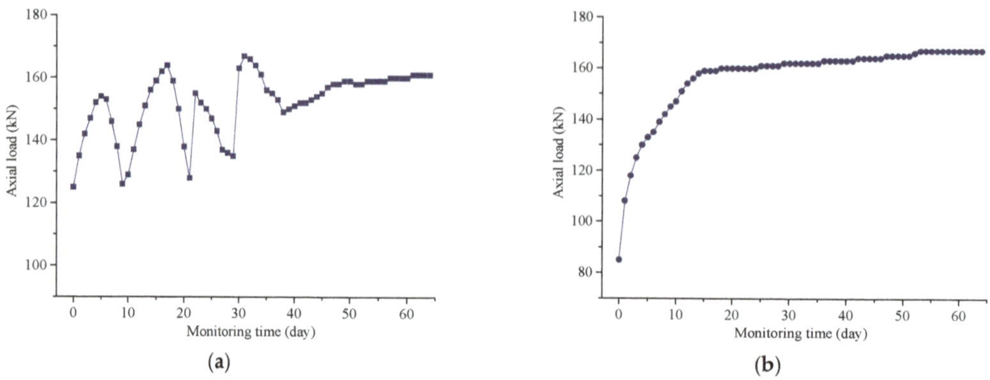

Figure 16. Curve of the working load on bolt/anchor cable under new support. (**a**) Arc roof; (**b**) Straight inclined roof.

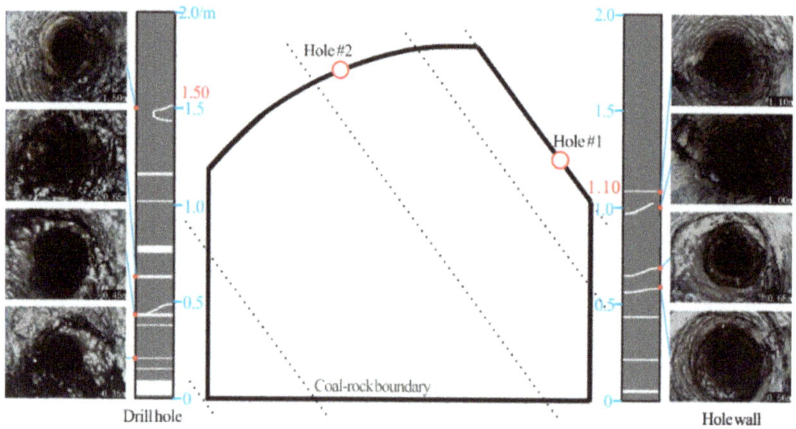

Figure 17. Peeping results of the surrounding rocks under the new support.

Figure 18. Maintenance and control effects of the roadway under the new support.

5.4. Discussion

5.4.1. Restraining Mechanism of Cross-Boundary Support on Cracks in the Surrounding Rocks

The selection of the roadway support in conditions of large deformations should consist in the common fulfillment of geometric, ventilation, and geomechanical criteria [29]. The new CBAG differential support technology improves the control effect of the irregular roadway. In order to further understand the difference in the control effect of the surrounding rocks under different support forms, Zone A, with the most obvious control effect, is selected for illustration. Figure 19 shows the evolution curves of cracks in Zone A of the irregular roadway under different support forms. It can be seen from Figure 19a that when the operation coefficient reaches 5 × 104, that is, the surface stress of the roadway is just released to zero, the number of shear cracks in Zone A under no support, the primary support, and the new support is 1005, 636, and 337, respectively. The number of shear cracks under the primary support and the new support decreases by 36.7% and 66.5%, respectively, compared to that under no support. At the same time, as the number of operation steps continues to increase, when the operation coefficient reaches 10 × 104 the number of shear cracks under no support, the primary support, and the new support increases by 23.1%, 15.1%, and 0.3%, respectively. It can be seen from Figure 19b that when the operation coefficient reaches 5 × 104 the number of tension cracks in the surrounding rocks of Zone A under no support, the primary support, and the new support is 171, 46, and 16, respectively. The number of tension cracks under the primary support and the new support decreases by 73.1% and 90.6%, respectively, compared with that under no support. When the operation coefficient reaches 10 × 104 the number of shear cracks under no support, the primary support, and the new support increases by 94.2%, 34.8%, and 0, respectively. It can be concluded from the data of shear cracks and tension cracks that as the number of operation steps continues to increase the evolution rate of tension cracks in the surrounding rocks under no support is significantly faster than that of shear cracks. At the same time, the CBAG new technology is used to reconstruct shallow surrounding rocks and bidirectional linkage control of deep and shallow rock masses, thus restraining the evolution of tension cracks and realizing the fundamental control of cracked surrounding rocks of irregular roadways. Therefore, restraining the expansion of tension cracks is the key to controlling the surrounding rocks of irregular roadways.

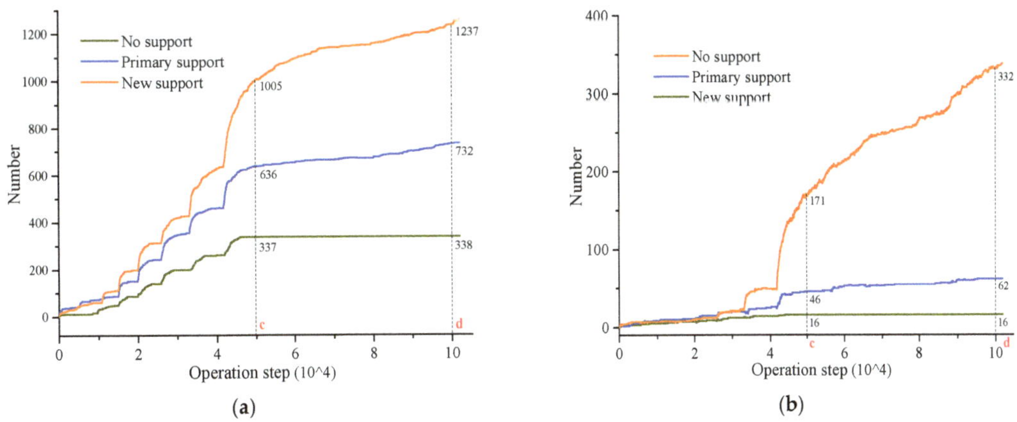

Figure 19. Evolution curves of cracks in Zone A of the roadway under different support forms. (**a**) Shear crack; (**b**) tension crack.

5.4.2. Discussion on CBAG New Support Structures and Grouting Sealing Method

Two new types of support products are adopted for the CBAG differential support, namely hollow grouting cables and flexible bolts, as shown in Figure 20. The hollow grouting cables, with a specification of 1 × 8–22–2000 MPa, are produced by the fully automatic production line of Shandong Yanxin Company. They are mainly composed of eight high-strength steel wires and stainless steel pipes, which overcome the disadvantages of the traditional welded grouting anchor cable structure, realize the high strength of the anchor cable bolt body, and also change the traditional method of plugging the hole at the tail with a rubber plug. They use a hard pipe to push and squeeze the hose to make the hose expand, and then move the plugging position from shallow inside to the hole of the complete surrounding rocks, that is, from State ① to State ②, which overcomes the disadvantage of poor plugging caused by the cracked shallow part. Another product is flexible bolts with a specification of 1 × 7–18–1860 MPa. They are composed of a steel strand bolt body and a tail threaded sleeve. They overcome the disadvantage that the length of the threaded steel bolt is limited by the roadway space and cannot meet requirements of cross-boundary support, and use the nut rotation installation method to realize the rapid installation of the steel strand bolt body.

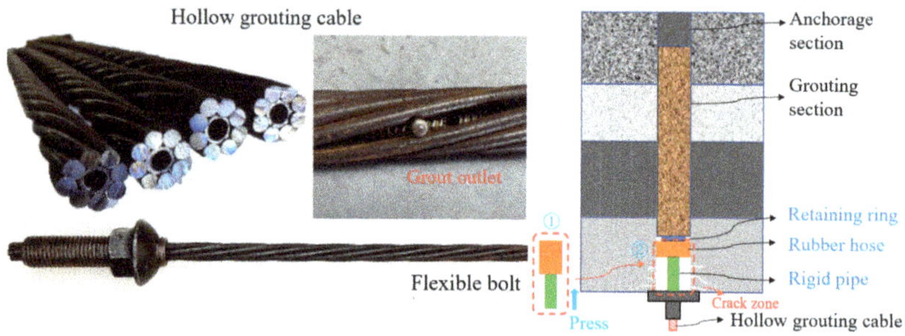

Figure 20. New support products and grouting sealing method.

6. Conclusions

(1) This work analyzes the asymmetric failure characteristics of the sliding and dislocation deformations of the coal–rock mass on the arc roof side and the loose deformations of the shallow rock mass on the straight inclined roof side of the roadway in a steeply inclined coal seam in Sichuan Province. The differential evolution regularities of cracks in the arc roof and the straight inclined roof during the initial and long-term excavation of the roadway are analyzed. The main reasons for the large deformations of the surrounding rocks are the coordinated deformations of the bolts, the shallow rock mass under the primary support, and the poor reinforcement of long anchor cables on the rock mass within this range.

(2) UDEC discrete element simulation was utilized to analyze the process stress evolution pattern. The result shows that stress release occurs first in the floor strata and then causes the gradual stress release of the surrounding rock on the entire arc roof side. Finally, the roadway shows a typical zonal failure characteristic of "slope-type" deformations for the floor, "dislocation falling" deformations for the arc roof, and micro-deformations for the straight inclined roof.

(3) Based on the non-equilibrium deformation characteristics of the roadway in the steeply inclined coal seam, this work proposes CBAG differential support technology for reinforcing key parts in the floor corner, thick-layer anchorage of the straight inclined roof and the arc roof, and grouting reconstruction. An equivalent thick-layer bearing structure was constructed for the full section of the surrounding rocks of the roadway. Using new hollow grouting cables and flexible bolts, this work anchored shallow grouting reconstructed rock mass into a deep rock stratum, such that a small displacement of deep rock mass can restrain a large displacement of shallow rock mass. In this way, it can adapt to the long-term creep effect of soft surrounding rocks of the roadway in the steeply inclined coal seam. Recommendation: The new grouted cable support should be adopted in all

roadways with similar conditions to the ventilation roadway in Working Face 1961 of the Zhaojiaba Mine.

(4) The analogue calculation shows that the new support greatly narrows the tensile stress zone of the surrounding rock of the roadway, effectively preventing slipping and dislocation deformation of the surrounding rock mass of the arc roof. On-site industrial testing shows that the depth of the maximum crack in the arc roof generally decreases by 61.3%, from 3.88 m to less than 1.50 m, and the development depth of cracks in the straight inclined roof decreases by 47.6%, showing a significant improvement in the surrounding rock control effect.

Author Contributions: Funding acquisition, Z.X. (Zhengzheng Xie) and N.Z.; Investigation, Z.X. (Zhengzheng Xie), J.W., N.Z., F.G., Z.H., Z.X. (Zhe Xiang) and C.Z.; Methodology, Z.X. (Zhengzheng Xie), J.W., F.G., Z.H. and Z.X. (Zhe Xiang); Supervision, N.Z. and C.Z.; Writing—Original Draft, Z.X. (Zhengzheng Xie), J.W. and F.G.; Writing—Review and Editing, N.Z., Z.H., Z.X. (Zhe Xiang) and C.Z. All authors have read and agreed to the published version of the manuscript.

Funding: This research was funded by the National Natural Science Foundation of China (52104104).

Institutional Review Board Statement: Not applicable.

Informed Consent Statement: Not applicable.

Data Availability Statement: The data are available from the corresponding author on reasonable request.

Acknowledgments: We would also like to thank the anonymous reviewers for their valuable comments and suggestions that lead to a substantially improved manuscript.

Conflicts of Interest: The authors declare that they have no conflict of interest.

References

1. Li, X.M.; Wang, Z.H.; Zhang, J.W. Stability of Roof Structure and Its Control in Steeply Inclined Coal Seams. *Int. J. Min. Sci. Technol.* **2017**, *27*, 359–364. [CrossRef]
2. He, S.Q.; Song, D.Z.; He, X.Q.; Chen, J.Q.; Ren, T.; Li, Z.L.; Qiu, L.M. Coupled Mechanism of Compression and Prying-Induced Rock Burst in Steeply Inclined Coal Seams and Principles for Its Prevention. *Tunn. Undergr. Space Technol.* **2020**, *98*, 103327. [CrossRef]
3. Deng, Y.H.; Wang, S.Q. Feasibility analysis of gob-side entry retaining on a working face in a steep coal seam. *Int. J. Min. Sci. Technol.* **2014**, *24*, 499–503. [CrossRef]
4. He, S.Q.; Song, D.Z.; Li, Z.L.; He, X.Q.; Chen, J.Q.; Li, D.H.; Tian, X.H. Precursor of Spatio-Temporal Evolution Law of MS and AE Activities for Rock Burst Warning in Steeply Inclined and Extremely Thick Coal Seams Under Caving Mining Conditions. *Rock Mech. Rock Eng.* **2019**, *52*, 2415–2435. [CrossRef]
5. Wu, G.J.; Jia, S.P.; Chen, W.Z.; Yuan, J.Q.; Yu, H.D.; Zhao, W.S. An Anchorage Experimental Study on Supporting a Roadway in Steeply Inclined Geological Formations. *Tunn. Undergr. Space Technol.* **2018**, *82*, 125–134. [CrossRef]
6. Wu, G.J.; Chen, W.Z.; Jia, S.P.; Tan, X.J.; Zheng, P.Q.; Tian, H.M.; Rong, C. Deformation Characteristics of a Roadway in Steeply Inclined Formations and Its Improved Support. *Int. J. Rock Mech. Min. Sci.* **2020**, *130*, 104324. [CrossRef]
7. Xue, Y.; Gao, F.; Liu, X.G.; Liang, X. Permeability and Pressure Distribution Characteristics of the Roadway Surrounding Rock in the Damaged Zone of an Excavation. *Int. J. Min. Sci. Technol.* **2017**, *27*, 211–219. [CrossRef]
8. Chen, D.X.; Sun, C.; Wang, L.G. Collapse Behavior and Control of Hard Roofs in Steeply Inclined Coal Seams. *Bull. Eng. Geol. Environ.* **2021**, *80*, 1489–1505. [CrossRef]
9. Yang, S.Q.; Chen, M.; Jing, H.W.; Chen, K.F.; Meng, B. A Case Study on Large Deformation Failure Mechanism of Deep Soft Rock Roadway in Xin'An Coal Mine, China. *Eng. Geol.* **2017**, *217*, 89–101. [CrossRef]
10. Sun, Y.T.; Li, G.C.; Zhang, J.F.; Qian, D.Y. Experimental and Numerical Investigation on a Novel Support System for Controlling Roadway Deformation in Underground Coal Mines. *Energy Sci. Eng.* **2020**, *8*, 490–500. [CrossRef]
11. Zhang, J.P.; Liu, L.M.; Cao, J.Z.; Yan, X.; Zhang, F.T. Mechanism and Application of Concrete-Filled Steel Tubular Support in Deep and High Stress Roadway. *Constr. Build. Mater.* **2018**, *186*, 233–246. [CrossRef]
12. Tao, Z.; Zhu, C.; Zheng, X.; Wang, D.; Liu, Y.; He, M.; Wang, Y. Failure mechanisms of soft rock roadways in steeply inclined layered rock formations. *Geomat. Nat. Hazards Risk* **2018**, *9*, 1186–1206. [CrossRef]
13. Xie, Z.Z.; Zhang, N.; Feng, X.W.; Liang, D.X.; Wei, Q.; Weng, M.Y. Investigation on the Evolution and Control of Surrounding Rock Fracture under Different Supporting Conditions in Deep Roadway during Excavation Period. *Int. J. Rock Mech. Min. Sci.* **2019**, *123*, 104122. [CrossRef]
14. An, Y.P.; Zhang, N.; Zhao, Y.M.; Xie, Z.Z. Field and Numerical Investigation on Roof Failure and Fracture Control of Thick Coal Seam Roadway. *Eng. Fail. Anal.* **2021**, *128*, 105594. [CrossRef]

15. Das, A.J.; Mandal, P.K.; Bhattacharjee, R.; Tiwari, S.; Kushwaha, A.; Roy, L.B. Evaluation of Stability of Underground Workings for Exploitation of an Inclined Coal Seam by the Ubiquitous Joint Model. *Int. J. Rock Mech. Min. Sci.* **2017**, *93*, 101–114. [CrossRef]
16. Zhang, S.; Zhang, D.S.; Wang, H.Z.; Liang, S.S. Discrete Element Simulation of the Control Technology of Large Section Roadway along a Fault to Drivage under Strong Mining. *J. Geophys. Eng.* **2018**, *15*, 2642–2657. [CrossRef]
17. Ross, C.; Conover, D.; Baine, J. Highwall Mining of Thick, Steeply Dipping Coal-a Case Study in Geotechnical Design and Recovery Optimization. *Int. J. Min. Sci. Technol.* **2019**, *29*, 777–780. [CrossRef]
18. Gao, F.Q.; Stead, D.; Coggan, J. Evaluation of Coal Longwall Caving Characteristics Using an Innovative UDEC Trigon Approach. *Comput. Geotech.* **2014**, *55*, 448–460. [CrossRef]
19. Suner, M.C.; Tulu, I.B. Examining the Effect of Natural Fractures on Stone Mine Pillar Strength Through Synthetic Rock Mass Approach. *Min. Met. Explor.* **2022**, *39*, 1863–1871. [CrossRef]
20. Gao, F.Q.; Stead, D. The Application of a Modified Voronoi Logic to Brittle Fracture Modelling at the Laboratory and Field Scale. *Int. J. Rock Mech. Min. Sci.* **2014**, *68*, 1–14. [CrossRef]
21. Li, X.H.; Ju, M.H.; Yao, Q.L.; Zhou, J.; Chong, Z.H. Numerical Investigation of the Effect of the Location of Critical Rock Block Fracture on Crack Evolution in a Gob-Side Filling Wall. *Rock Mech. Rock Eng.* **2016**, *49*, 1041–1058. [CrossRef]
22. Kazerani, T.; Zhao, J. Micromechanical Parameters in Bonded Particle Method for Modelling of Brittle Material Failure. *Int. J. Numer. Anal. Methods Geomech.* **2010**, *34*, 1877–1895. [CrossRef]
23. Zhang, L.Y.; Einstein, H.H. Using RQD to Estimate the Deformation Modulus of Rock Masses. *Int. J. Rock Mech. Min. Sci.* **2004**, *41*, 337–341. [CrossRef]
24. Singh, M.; Rao, K.S. Empirical Methods to Estimate the Strength of Jointed Rock Masses. *Eng. Geol.* **2005**, *77*, 127–137. [CrossRef]
25. Hoek, E.; Brown, E.T. Practical estimates of rockmass strength. *Int. J. Rock Mech. Min. Sci.* **1997**, *34*, 1165–1186. [CrossRef]
26. Li, S.C.; Wang, Q.; Wang, H.T.; Jiang, B.; Wang, D.C.; Zhang, B.; Li, Y.; Ruan, G.Q. Model Test Study on Surrounding Rock Deformation and Failure Mechanisms of Deep Roadways with Thick Top Coal. *Tunn. Undergr. Space Technol.* **2015**, *47*, 52–63. [CrossRef]
27. Xie, Z.Z.; Zhang, N.; Qian, D.Y.; Han, C.L.; An, Y.P.; Wang, Y. Rapid Excavation and Stability Control of Deep Roadways for an Underground Coal Mine with High Production in Inner Mongolia. *Sustainability* **2018**, *10*, 1160. [CrossRef]
28. Yuan, Y.X.; Han, C.L.; Zhang, N.; Feng, X.W.; Wang, P.; Song, K.; Wei, M. Zonal Disintegration Characteristics of Roadway Roof Under Strong Mining Conditions and Mechanism of Thick Anchored and Trans-Boundary Supporting. *Rock Mech. Rock Eng.* **2022**, *55*, 297–315. [CrossRef]
29. Skrzypkowski, K.; Zagórski, K.; Zagórska, A.; Apel, D.B.; Wang, J.; Xu, H.; Guo, L. Choice of the arch yielding support for the preparatory roadway located near the fault. *Energies* **2022**, *15*, 3774. [CrossRef]

processes

MDPI

Article

Study on the Variation Laws and Fractal Characteristics of Acoustic Emission during Coal Spontaneous Combustion

Jueli Yin [1], Linchao Shi [2], Zhen Liu [1,*], Wei Lu [2], Xingsong Pan [3], Zedong Zhuang [2], Lei Jiao [4] and Biao Kong [2,*]

[1] College of Finance and Economics, Shandong University of Science and Technology, Taian 271000, China
[2] College of Safety and Environmental Engineering, Shandong University of Science and Technology, Qingdao 266590, China
[3] Dongtan Coal Mine, Yanzhou Coal Mining Company Ltd., Jining 272000, China
[4] Qingdao Sinotrans Logistics Co., Ltd., Qingdao 266000, China
* Correspondence: jinanlanchuan@163.com (Z.L.); kongbiao8807@163.com (B.K.)

Abstract: Acoustic emission (AE) technology has the advantage of online localization to study the change laws of AE in the process of coal spontaneous combustion and to reveal the generation mechanisms of AE signal during the process of heating and rupture of coal body from a microscopic perspective. This paper first constructs a large-scale coal spontaneous combustion AE test system and conducts experimental tests on the AE signal in the process of coal spontaneous combustion. The results show that with the increase of temperature in the process of coal spontaneous combustion, the AE signal shows a trend of increasing fluctuations. Low-temperature nitrogen adsorption experiments studied the pore structure of coal spontaneous combustion, and the results showed a correspondence between the development of pores and the temperature of coal spontaneous combustion. Further, through the analysis of the evolution of the pore structure of coal through Fourier transform and fractal theory, it is found that the high-frequency main frequency AE signal and average frequency are continuously enhanced with the increase of temperature. The fractal dimension of the pore structure and the fractal dimension of the AE count of the coal body first rise and then decline. The mechanism of coal spontaneous combustion AE of coal is revealed, and the pore development caused by thermal stress when coal heats up is the main source of AE signal generation. The research in this paper is of great significance for applying AE technology to detect the position of coal spontaneous combustion.

Keywords: coal spontaneous combustion; thermal rupture; AE; time frequency; fractal

Citation: Yin, J.; Shi, L.; Liu, Z.; Lu, W.; Pan, X.; Zhuang, Z.; Jiao, L.; Kong, B. Study on the Variation Laws and Fractal Characteristics of Acoustic Emission during Coal Spontaneous Combustion. *Processes* **2023**, *11*, 786. https://doi.org/10.3390/pr11030786

Academic Editors: Feng Du, Aitao Zhou and Bo Li

Received: 7 February 2023
Revised: 1 March 2023
Accepted: 2 March 2023
Published: 7 March 2023

1. Introduction

Coal spontaneous combustion is serious in China, resulting in a massive waste of resources. Coal spontaneous combustion monitoring and early warning technologies mainly include temperature detection methods, infrared detection methods, gas detection methods, magnetic force detection methods, transient electromagnetic methods, resistivity detection methods, etc. [1]. In recent years, bundle tube monitoring, distributed optical fiber temperature measurement, and wireless temporary network temperature measurement technology have also been widely used in coal spontaneous combustion disaster monitoring and early warning. However, these monitoring and early warning methods still have certain drawbacks [2]. Bundle tube monitoring temperature measurement and distributed fiber optic temperature measurement are subject to large environmental factors in practical applications, and the data is easily distorted in complex mine environments [3–5]. Wireless self-assembling network temperature measurement systems will not be affected by problems such as pipelines being damaged. Due to the complexity of the mining area environment, the wireless signal transmission is unstable, and difficult to achieve efficient monitoring [6].

AE detection technology based on coal heating and the heating process fracture to generate the elastic waves phenomenon for monitoring and early warning of coal spontaneous combustion is a non-contact, non-destructive testing technology. In recent years, many scholars have linked AE with the prediction of coal body thermodynamic disasters, realized non-contact detection of coal body thermodynamic disasters in coal mines through AE technology, and provided precursor information of coal body thermodynamic disasters through the characteristics of AE activities.

At present, scholars at home and abroad have conducted a large number of studies on the AE law of coal rock. Ganne et al. studied the AE characteristics of hard and brittle rocks and established the relationship between micro-rupture and AE parameters [7]. Ishida et al. studied the distribution characteristics of AE sources during misalignment tests [8]. Han et al. investigated the AE characteristics of different coal samples during the process under uniaxial compression and found that there were significant differences in the AE characteristics of coal with different strengths during the compression process [9]. Kong et al. [10–12] studied the AE nonlinear and fractal characteristics of AE under multiple heating and loading damage conditions of coal rocks. Chen, under laboratory conditions, analyzed the acoustic emission count, dissipation energy, and fracturing point distribution of different deformation stages of coal [13]. Zhang studied the AE signal pattern and multiple fractal characteristics of coal deformation damage under different conditions [14]. The AE response characteristics of anthracite coal were studied through ultrasonic experiments [15]. In uniaxial compression experiments, it was found that when the density was higher than a specific value, the wave velocity of the coal sample increased with the increase of density, and the distribution of AE events over time was approximately normally distributed. Li et al. studied the AE characteristics of different coal-thick coal-rock assemblies during the rupture process and found that the peak count of AE was negatively correlated with coal thickness, and the cumulative count of AE was positively correlated with coal thickness [16].

Scholars at home and abroad have also done a lot of research on the changing laws of the pore structure of coal bodies. They studied the pore structure and fractal characteristics of different coals and found that the pores and fractures in coal become simpler and simpler during the crushing of coal samples, which is more favorable for gas storage and transport [17,18]. Li et al. conducted uniaxial compression tests on coal samples with prefabricated cracks at different angles and found that the AE response showed significantly different characteristics at different stages [19]. Zhao et al. analyzed the influence of the question on the transformation of coal pore structure and concluded that the higher the temperature, the greater the pore fractal dimension of the large pore in the coal [20]. Yi et al. analyzed the pore evolution of hydrochloric acid-treated coal samples by the low-temperature gas adsorption (LTGA) method and found that fractal dimension D2 was significantly positively correlated with ash content and increased with the increase of hydrochloric acid concentration [21]. Xu et al. studied the effect of microwave-assisted oxidant stimulation based on low-temperature nitrogen adsorption on the pore structure and fractal characteristics of bituminous coal. They found that microwave-assisted oxidation expanded the internal pore space and promoted pore expansion [22].

The above studies show that during the heating process, the structure of the coal body will change significantly with the increase in temperature. At different temperatures, it will show different characteristics, mainly changes in the structure of pore fractures [23–30]. There are many similarities between the damage change of coal in the process of heating and the damage change at the time of load deformation and rupture, and both processes are accompanied by the generation of AE signal, which provides conditions for the monitoring and early warning of hidden fire sources under coal mines through AE technology. The fractal theory is primarily used in the study of uniaxial compression and thermal rupture damage of coal rocks, and in recent years, the research on AE of coal bodies has gradually increased, but these studies have less analysis of the law and generation mechanism of coal spontaneous combustion AE signal. The passive monitoring technology of AE can realize the source finding and locating of underground fire sources and improve the efficiency of coal spontaneous combustion monitoring and early warning, and there is less research on the monitoring and early warning technology of the whole process of coal spontaneous combustion, which needs further research exploration in theory.

Based on the advantages of the acoustic method of temperature measurement, the application of AE technology for coal spontaneous combustion monitoring to achieve underground hidden fire source early warning has a good prospect. In contrast, the characteristics of AE signal changes during coal warming have not been studied in depth, and the mechanism of AE generation during coal warming has not been revealed. Based on this, this paper constructs a large-scale coal spontaneous combustion AE test system and a low-temperature nitrogen adsorption experimental platform, studies the thermal rupture and AE signal laws during the spontaneous combustion and heating of coal, analyzes the AE count and energy changes during the coal spontaneous combustion, reveals the temporal and frequency characteristics of AE signal by using Fourier transform, analyzes the fractal characteristics of low-temperature nitrogen adsorption and AE counts by fractal theory, and discusses the mechanism of the pores generated by thermal damage of coal bodies during coal spontaneous combustion heating process as the source of AE signal. The counting, energy, and spectral laws of the AE signal in the coal spontaneous combustion process are studied, and the fractal laws of BET of coal after different temperature treatments are analyzed. The nonlinear characteristics between AE and the pore structure of coal are revealed. This study lays a foundation for further exploring the heating coal body's AE signal generation mechanism and improving the monitoring and early warning of the spontaneous combustion fire of the AE signal coal.

2. Experimental System and Experimental Protocol

2.1. Large-Scale Coal Spontaneous Combustion AE Experiment

2.1.1. Preparation and Analysis of Coal Samples

In this paper, the main objective is to investigate the characteristic patterns and differences of AE signals generated at different temperature stages during the warming and combustion of coal. In addition, we studied the mechanism of AE generation, which provides a theoretical basis for AE signal inversion of temperature and monitoring of coal spontaneous combustion. Based on this, bituminous coal (DT) sample from the Dongtan coal mine in Shandong, China, was selected for the large-scale coal spontaneous combustion AE experiment, with a weight of 3 tons. In order to ensure that the coal pile has better heat storage conditions, some of the coal blocks were crushed below 30 mm. The volatile fraction (V_{daf}) and five elemental contents of C, H, O, N, and S of DT coal samples at room temperature were analyzed by a 5E-MAG6700 fully automatic industrial analyzer, and the volatile fraction and elemental composition of coal samples were determined according to the national standard GB/T 212-2008 as shown in Table 1.

Table 1. Volatile and elemental analysis of coal samples.

DT Coal Sample (Room Temperature)					
V_{daf}, wt%	C_{daf}/%	H_{daf}/%	O_{daf}/%	N_{daf}/%	$S_{t.d}$/%
40.96	81.16	6.11	0.33	1.16	11.20

2.1.2. Experimental System

The large-scale coal spontaneous combustion AE test system is mainly composed of a furnace body, an AE monitoring system, a hot air input system, and a thermocouple temperature measurement system. Among them, the furnace body is cylindrical, the surface is a metal shell, and the interior is composed of a refractory brick layer, polyurethane thermal insulation layer, water sandwich layer, etc. The hot air input system consists of an air pump, a heating tank, a metal hose, and a temperature control box. The thermocouple temperature measurement system is composed of K type thermocouple and a temperature recorder. The AE monitoring system is composed of an AE probe, AE host, preamplifier, and waveguide rod, where the AE sensor operates in the frequency range of 10 KHz–2 MHz. Figure 1 shows a large-scale coal spontaneous combustion AE test system diagram.

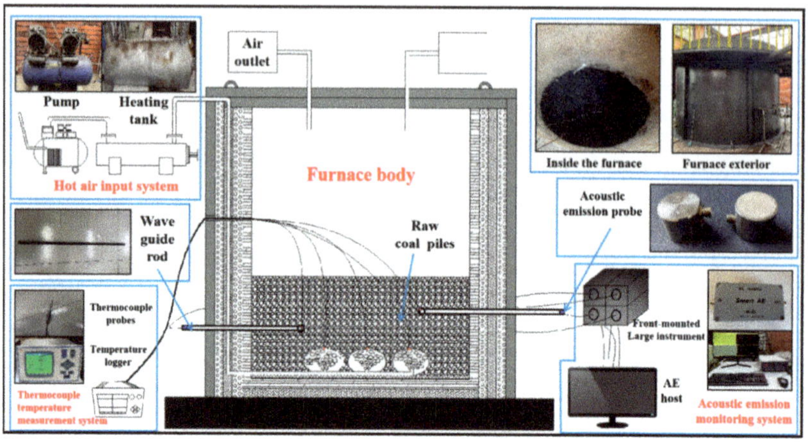

Figure 1. Large-scale coal spontaneous combustion AE test system diagram.

2.1.3. Experimental Methods and Steps

The large-scale coal spontaneous combustion AE experiment is divided into three steps:

(1) First, the 4 waveguide rods are extended into the furnace through the pores, and the horizontal distance of each waveguide rod from the nearest vent is 10 cm, 20 cm, 20 cm, and 40 cm. The distance from the bottom of the furnace is 37 cm, 35 cm, 15 cm, and 35 cm. The 4 waveguide rods are defined as channels 1–4, and 1 thermocouple probe is arranged at the top of the 4 waveguide rods and 3 outlets to measure the temperature of the surrounding coal body.

(2) After the arrangement is completed, 3 tons of raw coal are poured into the inside of the furnace body, the top cover of the furnace body is covered, and the top cover is sealed with glue, the ventilation hole is reserved, and the air is pressed into the heating tank by the air pump, and the heated air enters the bottom of the furnace body through the metal hose, flows out through 3 outlets, and continuously heats the coal body inside the furnace body, of which the ventilation pipe is located in the middle of the furnace body close to the bottom of the furnace body.

(3) The top of the waveguide rod extends into the inside of the furnace, and the tail end is connected to the AE probe. The sound wave can be transmitted to the AE probe by low loss through the waveguide rod to be monitored and recorded, and the waveform is finally saved to the AE host.

The noise reduction process is performed before the experiment, and the threshold value is set in advance. In order to avoid noise triggering, the threshold value is set relatively high, and the signal below the threshold value is not processed in any way. If the waveform below the threshold value is also found to be valuable, then the relatively weak signal can be analyzed by resetting the threshold value. During this experiment, we try to keep it quiet to avoid generating too much noise, and the threshold is set to 5 mV. When extracting the data, the part of the signal is masked out by the set threshold value. The preamplifier used in this experiment has a multiple of 40 dB and a sampling frequency of 200 kHz. During the heating process, the thermocouple temperature measurement system continuously records the internal temperature of the furnace body and saves the temperature data to the temperature recorder.

2.2. Coal Sample Low-Temperature Nitrogen Adsorption Experiment with Different Temperature Treatment

During the experiment, after transporting the packaged fresh coal samples to the laboratory, the whole coal sample is knocked out with a geological hammer to take a small coal sample with a length of no more than 1 cm in width and height. The small coal samples were put into a programmed heating furnace and treated with a temperature constant temperature of 30 °C, 100 °C, 200 °C and 300 °C for 3 h and cooled freely. For 4 groups of experimental samples, the automatic specific surface area and porosity analyzer were selected for low-temperature nitrogen adsorption experiments. The coal samples treated at different temperatures were screened, and the pulverized coal with specifications 40–80 was obtained, and then the sieved pulverized coal was put into the dry sample tube for degassing because the process of thermal expansion and rupture of the coal rock mass was irreversible [31,32]. Thus, in order to avoid excessive deregulation temperature to change the coal pore fracture structure, the degassing temperature of the coal sample was set below the treatment temperature. The pore fracture structure of the coal body would not be affected at this time. Since degassing requires the removal of moisture and air, it is difficult to remove moisture at room temperature, so the samples treated at 30 °C are degassed in a vacuum for 48 h at 30 °C, while the samples treated at 100 °C, 200 °C and 300 °C are degassed at 100 °C vacuum for 12 h.

3. Study on the Thermal Damage Law of Coal Body during Spontaneous Combustion and Heating Process of Coal

3.1. AE Counting and Energy Analysis during Spontaneous Combustion and Heating of Coal

Through the two aspects of AE characteristic parameters and spectrum, the data obtained by the large-scale coal spontaneous combustion AE test system are analyzed. The characteristic parameters are selected for counting and energy. The current research on coal spontaneous combustion shows that the occurrence and development of coal spontaneous combustion have stage characteristics in different temperature bands [33]. Based on this, the experimental process in Figures 2 and 3 is divided according to the temperature stages, and the change law of the AE signal is analyzed. The third stage of the sample has entered the high-temperature state after rapid warming, and the subsequent warming is slower [10]. Figure 2 shows the law of the change of AE counts with time during the heating process of coal.

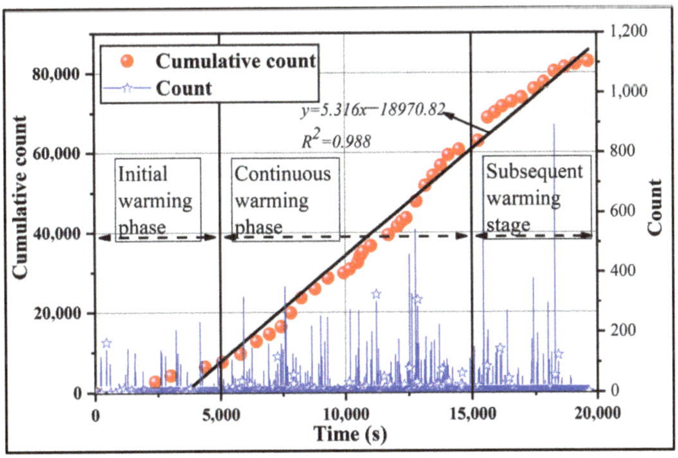

Figure 2. AE count and cumulative count plot.

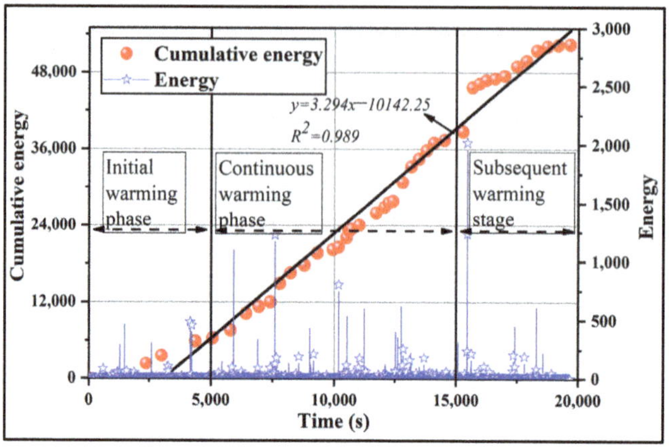

Figure 3. AE energy and cumulative energy map.

As can be obtained from Figure 2, with the progress of the experiment, there is a gradual increase in the AE count in the initial heating stage of 0~5000 s due to the low temperature. The coal body has not obtained enough energy, and the thermal stress is not enough to destroy the coal body itself. Only more fine cracks are generated. Therefore, the integrity of the coal sample is better, and the AE signal is less. In the continuous heating stage of 5000~15,000 s, the coal body has a higher temperature. With the increase in temperature, the thermal stress of the coal body is enough to destroy the weaker structure, and the internal pore system of the coal body is reconstructed. The native cracks and new cracks inside the coal sample expand and develop at this stage. Hence strong AE signals are generated. In the subsequent heating stage of 15,000~20,000 s, the surface temperature of the coal sample continues to rise. More and more particles beyond the yield stress are inside the coal sample, so the destruction of pore structure and crack expansion are also increasing, the integrity of the coal body is seriously damaged, and the AE signal is further enhanced.

The cumulative count curve of AE is fitted to obtain a univariate linear regression curve with a slope of 5.316 and an apparent linear law, indicating that with the increase of temperature during the heating of the coal body, the AE signal continues to increase, and the level of AE count continues to increase.

As can be seen from Figure 3, with the progress of the experiment, the AE signal intensity shows a significantly enhanced trend. It has good consistency with the AE cumulative count diagram. In the initial heating stage of 0~5000 s, the coal body, due to the low temperature and the internal functional group of the coal body, failed to achieve the required activation energy. At this time, the redox reaction is weak, the coal body structure does not change significantly, and the AE energy is weak. In the continuous heating stage of 5000~15,000 s, the coal body has a higher temperature, the redox reaction of the coal body itself is more active, and the structure of the coal body is slightly damaged, so the AE signal energy is strong. In the subsequent heating stage of 15,000~20,000 s, the surface temperature of the coal sample continues to rise, the coal body obtains sufficient activation energy, and the internal structure is continuously decomposed and converted so that the coal structure is damaged. The AE energy continues to increase.

The cumulative energy curve of AE is fitted to obtain a univariate linear regression curve with a slope of 3.294 and an apparent linear law, indicating that with the increase of temperature during the heating of coal, the level of AE energy continues to increase, and the AE signal continues to increase.

In order to obtain the difference between the AE signal data at different temperatures, the AE counts and energy data of 5 min before and after 30 °C, 50 °C, 100 °C, 200 °C, and 300 °C are intercepted and plotted in Figures 4 and 5 below.

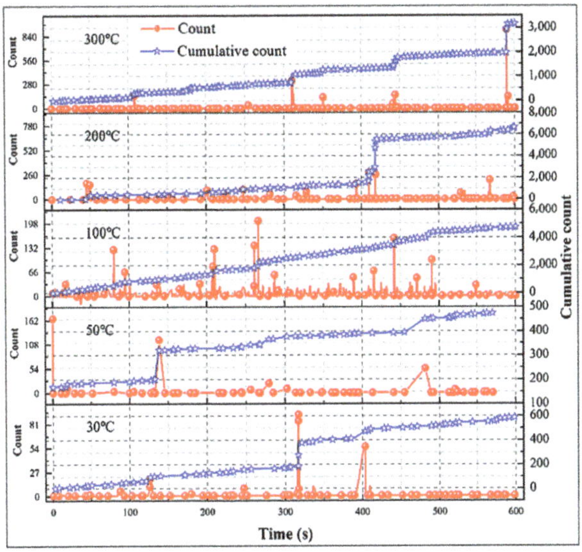

Figure 4. 600 s AE count of different temperature stages in the spontaneous combustion process of coal.

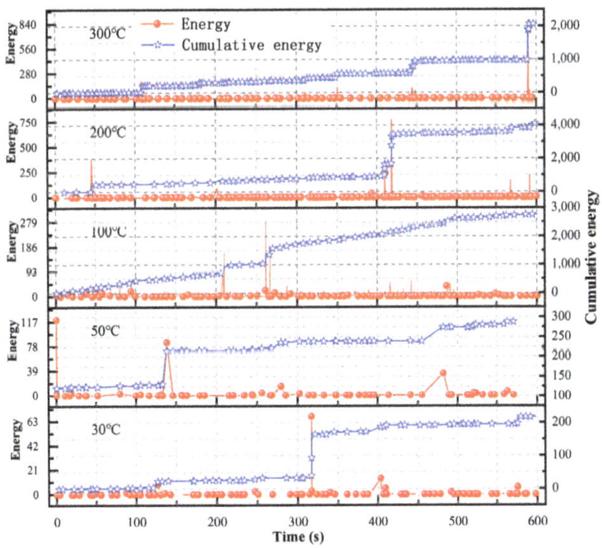

Figure 5. 600 s AE energy map of different temperature stages in the process of coal spontaneous combustion.

It can be seen from Figure 5 that there are fewer overall AE events at 30 °C and 50 °C. More AE signals at low counting points are generated, indicating that due to the lower temperature in the temperature band, the internal thermal expansion of the coal body is weaker, and the AE events are fewer. At 100 °C, there are more AE events, and a higher count of AE signals appear, indicating that in this temperature segment, a large amount of crystalline water disappears by heat evaporation of the coal body, resulting in more AE signals, and more pores have appeared on the surface of the coal sample, and the pore cracks are gradually increasing with the increase of temperature. At 200 °C, the internal moisture of the coal body is tiny, and the thermal rupture phenomenon begins to appear. Hence, a robust AE signal appears, and the cracks on the surface of the coal sample expand and develop in this temperature segment. At 300 °C, the AE event of the coal body is more active, but the AE signal is also stronger, the thermal rupture phenomenon in the temperature band is completed, the thermal decomposition begins to appear, and the integrity of the coal sample at this time is seriously damaged, and it can be seen from the AE count data that the increase in temperature has a promoting effect on the generation of AE signals.

As shown in Figure 5, at 30 °C and 50 °C, the overall performance is relatively calm, a relatively small number of high-energy events occurred, and the overall energy amount value is low. At 100 °C, the coal body began to surge, the AE activity was active, and the AE signal at this time mainly came from the expansion and closure of the native fissure during the thermal expansion of the coal sample. At 200 °C, the AE activity is more active, with more high-energy events. The AE signal at this time mainly comes from the thermal expansion and rupture of the coal body, resulting in many pore fractures. At 300 °C, the AE activity is still relatively active. However, the intensity is slightly lower, indicating that the hot expansion material inside the coal body at this time has been completely expanded. A large number of cracks appear on the surface of the coal body, and the AE signal is mainly caused by the oxidative decomposition of the coal body by heat.

In order to explore the variation law of the maximum count and maximum energy in different temperature bands, the maximum values of the count and energy in the five temperature bands are selected and plotted, as shown in Figure 6 below.

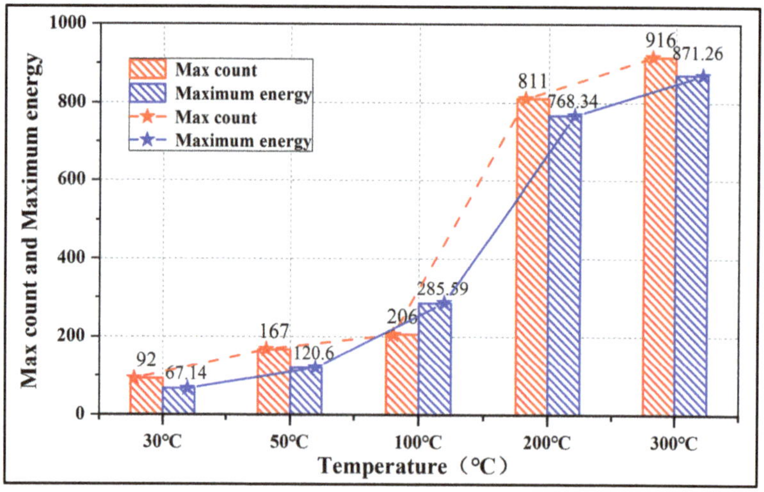

Figure 6. Maximum count and maximum energy of AE in different temperature segments.

Figure 6 is a graph of the maximum count and maximum energy change in the five temperature bands, and it can be seen from the figure that with the increase in temperature, the maximum count of AE shows a gradual increase trend, indicating that the temperature has a promoting effect on the AE signal. The maximum increase in the 30 °C–100 °C section is slight, indicating that the AE signal of the coal body is small and the change of the coal structure is small during this stage. In the temperature range of 100 °C to 200 °C, the maximum count of coal bodies increased significantly, indicating that during this time period, the rupture state of coal bodies changed, the AE signal was more active, and the internal structure of coal bodies changed greatly. There is still a slow growth trend in the maximum count of coal bodies at 300 °C, indicating that the coal body is in the same rupture state at 200 °C–300 °C. The maximum energy is in the temperature range of 30 °C to 100 °C. There is a slow growth trend, indicating that the coal structure changes little at this stage. In the temperature range of 100 °C to 200 °C, the maximum energy growth of the coal body is large, indicating that in this temperature range, the rupture mode of the coal body changes so that the energy is greater. At 300 °C, the maximum energy of the coal body still increases compared with 200 °C, indicating that the structural evolution of the coal body is continuing.

3.2. Time-Frequency Analysis of AE Signals during Coal Spontaneous Combustion Heating

The measured data are analyzed using a fast Fourier transform (FFT), and the following is an introduction to the principle of FFT [34,35]:

$$x(k) = \sum_{n=0}^{N-1} x(n) e^{-2j\pi nk/N} \quad (1)$$

The inverse transformation is:

$$x(k) = \frac{1}{\mathrm{N}} \sum_{n=0}^{N-1} x(k) e^{2j\pi nk/N} \quad (2)$$

where: $x(k)$ is the kth value of the discrete spectrum, $k = 0, 1, \ldots, N - 1$; $x(n)$ is the nth value of time domain sampling, $n = 0, 1, \ldots, N - 1$.

The original waveform signals emitted by sound emissions from five different temperature bands during the heating process are extracted. Fast Fourier transforms (FFT) are performed to obtain a two-dimensional spectrogram.

The four small plots in the four typical power spectra of Figure 7 are interpreted for the four states. The main frequency of AE (peak frequency) corresponds to the maximum amplitude in the two-dimensional spectrogram. The main frequency and secondary frequency not in the same main peak is called a double peak, and the same main peak is called a single peak. When the frequency of the second frequency is higher than the main frequency, it is called high frequency, and when the frequency of the second frequency is lower than the main frequency, it is called low frequency. In this paper, eight AE events are selected around the five temperatures, and the probability of occurrence of each type of frequency is obtained by the fast Fourier transform, as shown in Table 2.

Table 2. Power spectrum probability table for different temperature types.

Waveform Type	Single Peak High Frequency	Single Peak Low Frequency	Bimodal High Frequency	Bimodal Low Frequency
30 °C	10%	30%	40%	20%
50 °C	10%	20%	70%	0%
100 °C	10%	20%	20%	50%
200 °C	10%	0%	60%	30%
300 °C	0%	10%	80%	10%

From Table 2 above, it can be seen that the number of occurrences of high bimodal frequency and bimodal low frequency measured in the experiment is more. The single peak high frequency and single peak low frequency are less. Fourier transform through MATLAB was used to obtain the main frequency amplitude of five temperatures. The AE signal at 30 °C–100 °C is mainly low frequency, the amplitude is slightly increased, and at 200 °C, it is mainly based on low frequency and high amplitude. At 300 °C, the AE signal is mainly based on high frequency and low amplitude. The difference between the main frequency and the amplitude category at different temperatures is large. It shows that with the increase of temperature, the probability of higher frequency main frequency AE signal will also increase, and there will be almost no high-frequency main frequency AE event before 50 °C, and when the temperature exceeds 50 °C, the high-frequency AE events begin to appear. Thus, the AE frequency provides certain conditions for coal spontaneous combustion monitoring and early warning.

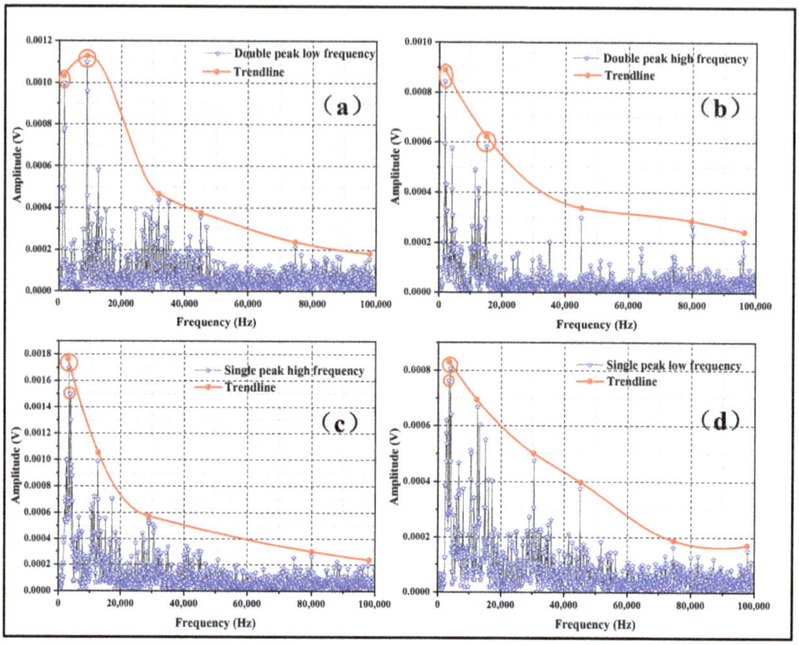

Figure 7. Spectrograms of four typical power articles. (**a**) double peak low frequency; (**b**) double peak high frequency; (**c**) single peak high frequency; (**d**) single peak low frequency.

The average frequency of the main frequency of the AE signal of five coal samples is calculated, and the average frequency fitting curve of coal at different temperatures is plotted, as shown in Figure 8. The fitting curve of the main frequency average frequency of coal at different temperatures shows that with the increase of temperature, the average frequency of the main frequency of AE gradually increases, so it can be known that the main frequency of AE shows a positive correlation with temperature, and this law can be used as the basis for judging the temperature change of the goaf area under the coal mine.

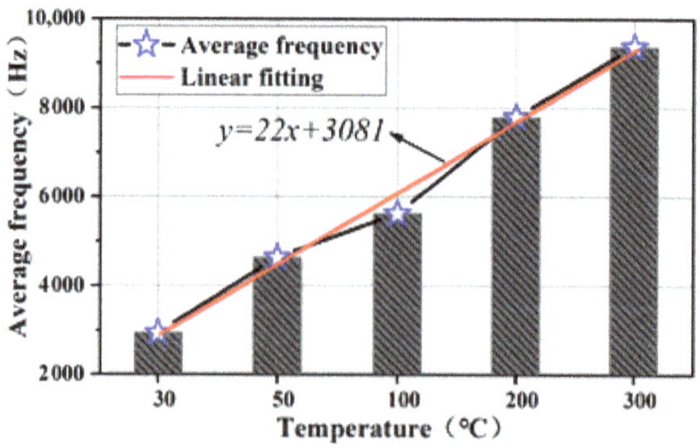

Figure 8. The average frequency of the main frequency of coal under different temperature conditions.

4. Study on the Fractal Law of Coal Spontaneous Combustion Heating Process

4.1. Fractal Law of Pore Structure of Coal Body after Treatment at Different Temperatures

In this paper, the fractal dimension "A = D − 3" is calculated by using the FHH model and capillary force. The nitrogen adsorption curve is divided into two parts at P/P0 = 0.5; the fractal dimension calculated from the P/P0 < 0.5 curve part is recorded as D2 as the fractal dimension of the pore surface. The fractal dimension calculated from the P/P0 > 0.5 curve part is the fractal dimension of the pore structure is recorded as D1. The fractal calculation of the coal samples treated at four different temperatures is performed, and the obtained data is plotted in Figure 9 below.

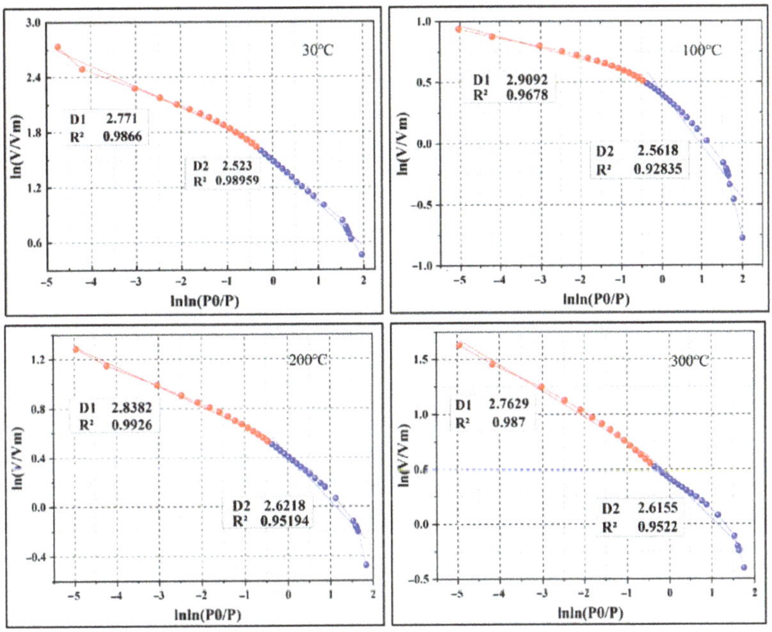

Figure 9. Fractal dimension fitting curve after different temperature treatments.

D1 and D2 are obtained from the fractal dimension fitting curve of Figure 9, where D1 is the fractal dimension of the high-pressure segment, that is, the fractal dimension of the pore structure. D2 is the fractal dimension of the low-pressure segment, that is, the fractal dimension of the pore surface. In Table 3, fractal dimensions of different pore sizes are obtained by different temperature treatments.

Table 3. Fractal dimensions after different temperature treatments.

	30 °C	100 °C	200 °C	300 °C
D1	2.771	2.909	2.838	2.763
D2	2.523	2.562	2.622	2.616

It can be seen from the table that with the increase of temperature, the fractal dimension of pore structures D1 and D2 are the same, in the temperature range from 30 °C to 100 °C; as the temperature rises due to thermal evaporation, the coal structure gradually becomes more complex, the pore structure continues to increase, and D1 increases. In the temperature section of 100~300 °C, the internal moisture of the coal body has been volatilized in large quantities, and the thermal expansion phenomenon of coal has begun to appear, resulting in the squeezing and closing of some pores and the reduction of D1. Coal pore surface fractal dimension D2 generally showed an upward trend, indicating that the coal surface pore fracture structure tends to be complex. With the temperature of 30 °C to 200 °C, the surface moisture and volatile substances of the coal body are heated and evaporated so that the number of pores on the surface of the coal increases, and the coal surface structure is more complex. When the temperature reaches 200 °C, D2 shows a downward trend indicating that the coal surface structure tends to be flattened at this temperature. The surface part of the pore cracks is squeezed and closed due to heat expansion, resulting in a decrease in the complexity of the coal surface structure. D2 is reduced accordingly.

4.2. Fractal Law of Large-Scale Coal Spontaneous Combustion AE Count

4.2.1. Fractal Theory and the Determination of Phase Space Dimensions

In this paper, the associative dimension characteristics of AE counts are introduced, combined with the G-P algorithm [36], MATLAB software is used to calculate the bi-logarithmic relationship under different experimental conditions [37], and the count of AEs is used as the basic parameter to determine a series set with a capacity of n:

$$X = \{x_1, x_2, \ldots x_n\} \tag{3}$$

The first m number in Equation (3) constitutes the phase space (m < n) of the m dimension,

$$X_1 = \{x_1, x_2, \ldots x_m\} \tag{4}$$

The following data bits are sequentially extended by 1 to obtain N = n − m + 1 vector,

$$X_2 = \{x_2, x_3, \ldots x_{m+1}\}, X_3 = \{x_3, x_4, \ldots x_{m+2}\} \tag{5}$$

The corresponding associated dimensions are:

$$C(r(k)) = \frac{1}{N^2}\sum_{i=1}^{N}\sum_{i=1}^{N} h[r(k) - |X_i - X_j|] \tag{6}$$

where: h is the Riverside function; $r(k)$ is a given scale function. Each determined scale $r(k)$ corresponds to it. Each (lgr, lgC(r)) coordinate point is linearly fitted, and the slope after fitting is the associated dimension of the AE count. m is the phase space dimension, the size of m has a specific influence on the associated dimension, and the value of m is 1, 2, 3, 4, 5 ... 20. With the increase of m value, the correlation dimension gradually tends to be saturated and stable. At m = 10, the associative dimension gradually tends to saturate and stabilize, so the phase space dimension m = 10 is determined.

4.2.2. Determination of Fractal Features of AE Counts

According to the introduction of literature [38–46], when the scale coefficient $k \leq 0.1$, the fractal characteristics of the AE sequence are not prominent. This paper takes the scale factor k > 0.1 to calculate the associated dimension in the process of aqueous coal sample rupture. Moreover, the slope of the regression function is the associated fractal dimension. Figure 10 is a curve plot obtained by selecting 5 points with good lnr and lnC value conditions processed by MATLAB software, and after performing univariate linear regression analysis. It is found that the correlation coefficients R^2 of the AE counts of 30 °C, 100 °C, 200 °C, and 300 °C are 0.96, 0.97, 0.99, and 0.98, respectively, and the correlation coefficient R^2 is more significant than 0.90. The correlation between lnr and lnC is relatively high, indicating that the correlation between AE counts at different temperatures is high, and the AE count sequences of different temperatures during the spontaneous combustion of large-scale coals have obvious fractal characteristics.

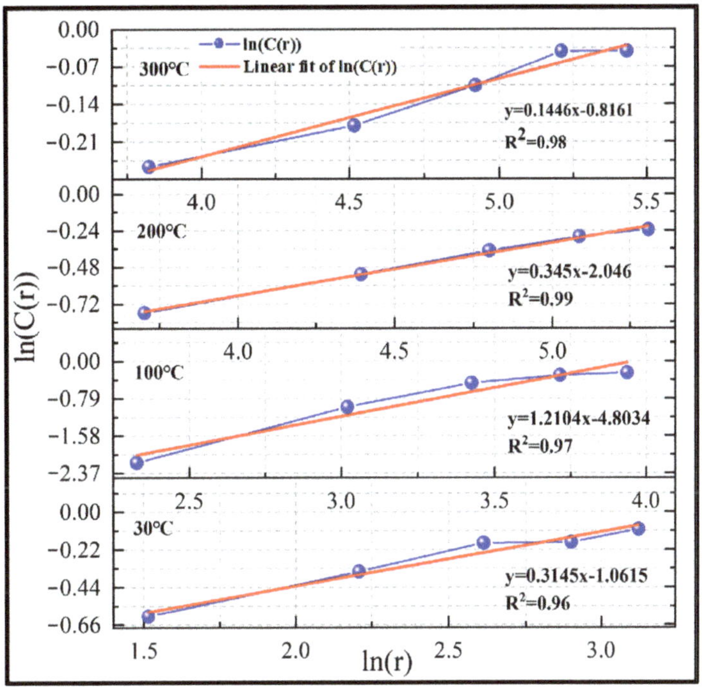

Figure 10. Fit curve plot of double logarithmic relationship at different temperatures.

4.2.3. Fractal Characteristics of AE Counting

According to the fractal characteristics of the AE count determined above, the fractal dimension change of the large-scale coal spontaneous combustion AE count is plotted, as shown in Figure 11.

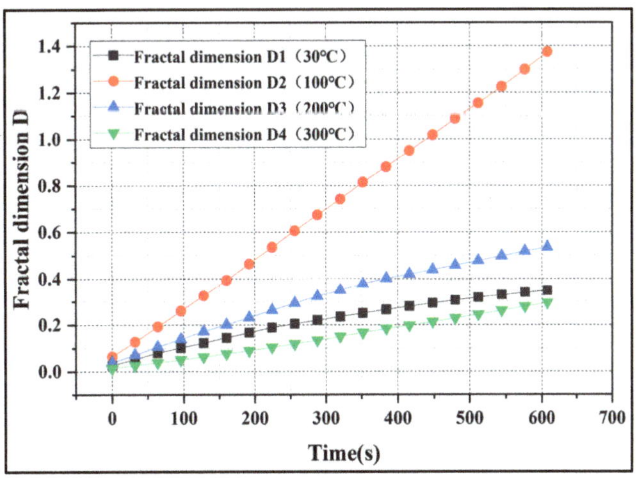

Figure 11. Fractal dimension change of AE counts at different temperatures.

The fitted slopes of the different fractal dimension change curves were 0.999, 0.987, 0.981, and 0.998, respectively. According to the fractal dimension change chart of the

AE count within 300 s before and after the temperature points of 30 °C, 100 °C, 200 °C, and 300 °C in the large-scale coal spontaneous combustion heating process, it can be seen that the fractal dimension of the sound emission count of the four temperature gradients continues to increase with the increase of time in the 300 s before and after the temperature, the linear law is very obvious, the slope is stable, and it is greater than 0.950, indicating that the AE events of the coal body continue to increase during the heating process, and the AE signal is continuously enhanced. The fractal dimension of the AE count at 30 °C increased from 0.0258 to 0.3487, the fractal dimension of AE at 100 °C increased from 0.0638 to 1.3735, the fractal dimension of the AE count at 200 °C increased from 0.0377 to 0.5630. The fractal dimension of AE at 300 °C increased from 0.0130 to 0.2937, indicating that the fractal dimension of the AE count was greatest at 100 °C and the smallest at 300 °C, showing a trend of first growing and then declining from 30 °C to 300 °C.

Figure 12 is the selection of m = 4, 8, 12, 16, 20 of the AE fractal dimensions with the temperature of the trend plot. The plot presents an inverted V-shaped feature; the fractal dimension of the AE count with the increase of temperature first shows an increase. Then it shows a downward trend; the m value is different, the change situation is the same, and the larger the m value, the greater the fractal dimension. According to the literature, the changing trend of fractal dimension can reflect the failure process of the coal body and the timing change of the AE signal. The larger the fractal dimension, the more disordered the AE event. Therefore, in the process of coal spontaneous combustion and heating of large-scale coal, the AE event of coal spontaneous combustion developed from orderly to disorderly with the increase of temperature and then from disorder to order. It further shows the increasing level of acoustic emission count and the increasing acoustic emission signal during the warming process of coal spontaneous combustion. The acoustic emission events show a change process from order to disorder and then from disorder to order.

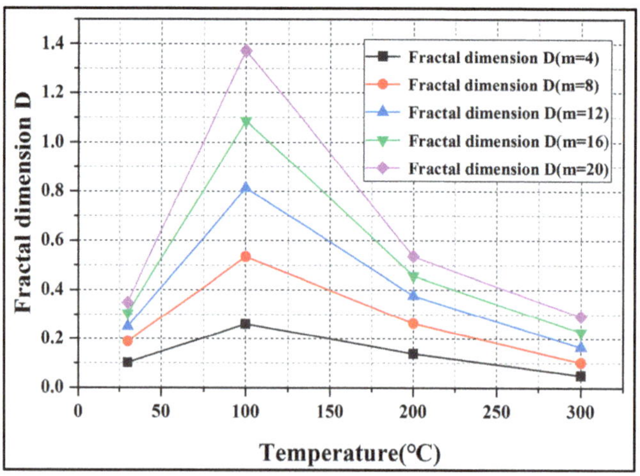

Figure 12. Variation of fractal dimension of AE counts with different m values.

The above trend of fractal dimension of AE count shows that there is a certain regularity in the sound emission count during the spontaneous combustion heating of coal. The fractal dimension level of the AE count near a specific temperature is continuously improved, which further indicates that the sound emission count level of the coal spontaneous combustion heating process is continuously enhanced, and the AE signal is continuously enhanced. The AE event shows a change process from order to disorder and then from disorder to order.

4.3. Discussion

In this paper, by constructing a large-scale AE test system, we tested the change law of the AE signal during the warming process of coal. The obtained data were Fourier transformed to systematically test the AE signal generated during the warming process of coal and comprehensively compare and analyze the intensity, energy scale, and timing characteristics of the AE signal, which showed a gradual increase as the experiment proceeded. At the beginning of the experiment, the coal did not obtain enough energy, and the thermal stress was not enough to destroy the coal itself, and only more small cracks were produced; at the middle of the experiment, the thermal stress of the coal itself was enough to destroy the weaker structure, and the internal pore system of the coal was reconstructed, and at this stage, the primary cracks and new cracks inside the coal sample expanded and developed so more strong acoustics. At the later stage of the experiment, the surface temperature of the coal sample was increasing, so the destruction of the pore structure and crack expansion were also increasing, and the integrity of the coal body was seriously damaged. This indicates that the AE signal can effectively respond to the damage and crack evolution of the coal body and can realize the monitoring and early warning of coal spontaneous combustion. The passive monitoring of AE, combined with the corresponding active detection of the fire source, can effectively improve the efficiency of coal spontaneous combustion monitoring and early warning and realize the joint detection of coal spontaneous combustion fire, which provides theoretical guidance for the subsequent research.

In order to study the AE signal to realize the monitoring and early warning of coal spontaneous combustion, this paper analyzes the time-frequency characteristics of the AE signal in the process of coal spontaneous combustion and the change law of the main frequency of the AE signal in the process of coal spontaneous combustion is obtained by Fourier transform calculation. In addition, the characteristic frequency spectrum is divided into four types, and the fitted curve of the main frequency average frequency of coal under different temperature conditions is analyzed and obtained, which can effectively reflect the phase change of the coal spontaneous combustion process by the AE signal. Chai and Ambrosio [47,48] studied the structural health detection of objects by the peak frequency and center of mass frequency of AE signal, and the research has been applied in nondestructive testing, welding, and microseismic, which can effectively respond to the changes of AE signals. Therefore, it is important to deeply explore the spectral characteristics of AE signal during coal spontaneous combustion through peak frequency and center-of-mass frequency to determine the correlation between temperature and AE signal during coal spontaneous combustion, and it is needed to further explore the correlation between damage rupture and AE signal during coal spontaneous combustion so as to realize the efficiency of coal spontaneous combustion monitoring and early warning.

The fractal dimension law of the pore structure of low-temperature nitrogen-adsorbed coal during the warming process and the fractal dimension law of AE counting the two fractal dimension laws are obvious; the fractal dimension law of the pore structure of low-temperature nitrogen-adsorbed coal is similar to the fractal dimension law of AE counting, and the two fractal dimension change trends are the same. In order to analyze the characteristics of pore structure changes during the warming process of coal synchronously, our team has characterized the pore structure changes at different temperatures by SEM experiments in the previous study. The integrity is good, and the pore structure is single under the room temperature condition (30 °C); after treatment at 100 °C, the integrity of the coal sample is slightly damaged, and the number of surface pores increases. At 200 °C, the integrity of the coal sample is damaged, and the number of surface pores increases. The integrity of the coal sample was damaged after treatment at 200 °C, and a large number of regional pore clusters appeared on the surface, and the pore size increased significantly compared with that at room temperature. At 300 °C, the integrity of the coal sample was severely damaged, and some of the fissures were connected with each other, and the connectivity was good. Based on this, the fractal dimension law of coal pore structure and the fractal dimension law of AE counting were investigated by low-temperature nitrogen adsorption experiments. The results of the study showed that from 30 °C to 100 °C, the fractal dimension D1 of the pore structure of the coal body and the fractal dimension of the AE count showed a significant upward trend, combined with the analysis of the AE count and the cumulative count of AE, the coal body was heated and ruptured by heating during the heating process, and the nitrogen adsorption of the coal body continued to increase. From 100 °C to 300 °C, the fractal dimension D1 of the pore structure of the coal body and the fractal dimension of the AE count showed a downward trend, which was due to the thermal expansion phenomenon caused by the evaporation of a large amount of water in the coal body, resulting in the extrusion and closure of the pores of the coal body, and the growth rate of nitrogen adsorption decreased. Similarly, the fractal dimension D2 of the pore surface of the coal body has the same change trend at 30 °C–200 °C and 200 °C–300 °C, which further indicates that the pore structure of the coal body produces thermal rupture with the heating of the coal body. The thermal expansion phenomenon leads to the closure of some pores of the coal body. With the increase in temperature, the level of AE count continues to increase. The fractal dimension of the AE count rises first and then declines, indicating that with the increase in temperature, the AE signal continues to increase, and the AE signal shows a change law from order to disorder and then from disorder to order.

The experiment verifies the law that the pore structure of the coal body continues to generate and gradually increases during the heating process of the coal body. The fractal law of the low-temperature nitrogen adsorption curve and the fractal law of the AE count are analyzed, which further illustrates that the pore structure of the coal body in the process of heating up is the root cause of the AE signal of the coal body. The number of pores and the pore size of the coal body in spontaneous combustion and heating is increasing, which in turn leads to the continuous enhancement of the AE signal. Therefore, AE signals can be used for monitoring and early warning of coal spontaneous combustion.

5. Conclusions

1. With the increase in time and temperature, the count and energy of AE continue to increase, and the maximum count and maximum energy of AE continue to increase. The cumulative count of AE and the cumulative energy of AE continue to increase; it shows that the integrity of coal samples is destroyed during the heating process, the pore structure is constantly complex, the number of pores is increasing, the thermal damage rupture is increasing, and the AE signal released by the coal body is continuously enhanced.
2. The probability of the occurrence of high-frequency main frequency AE signals increases with the increase in temperature. The average frequency of the main frequency also increases. The main frequency of the acoustic transmission signal and its average frequency show a positive correlation with temperature; it shows that with the increase in temperature, the thermal damage of the coal body increases, and the AE signal continues to increase.
3. During the spontaneous combustion heating process of the coal body, with the increase of time and the increase of temperature, the fractal dimension of the low-temperature nitrogen adsorption curve and the fractal dimension of the AE count appear to rise first and then decline. It shows that due to the thermal expansion of the coal body in the process of heating up, the growth of the number of pores in the coal body rises first and then declines. The AE signal intensity of the coal body rises first and then declines, and the AE signal from order to disorder, from disorder to order, shows an increasing trend.
4. The law that the pore structure of the coal body is continuously generated and gradually enhanced during the heating process of the coal body further illustrated that the AE signal in the process of heating and heating of the coal body comes from the generation of the pore structure. The number of pores in the spontaneous combustion and heating process of the coal body continues to increase, and the

complexity increases, which in turn leads to the continuous enhancement of the AE signal. Therefore, the AE signal provides favorable conditions for monitoring and early warning of coal spontaneous combustion.

Author Contributions: Writing—original draft, J.Y., B.K. and L.S.; Writing—review & editing, Z.L. and W.L.; Formal analysis, X.P.; Supervision, Z.Z.; Data curation, L.J. All authors have read and agreed to the published version of the manuscript.

Funding: This research was jointly supported by the National Natural Science Foundation of China (51904172), the China Postdoctoral Science Foundation (2020M682209), the Natural Science Foundation of Shandong Province (ZR2019QEE041).

Data Availability Statement: The data used and/or analyzed during the current study are available from the corresponding author on reasonable request.

Conflicts of Interest: No potential conflict of interest was reported by the author(s).

References

1. Wu, L.X. The responsibility and development of coal under the double carbon target. *Coal Econ. Res.* **2021**, *41*, 1. [CrossRef]
2. Liu, C.; Xie, J.; Xin, L. A review of the theory and technology of spontaneous combustion prediction and forecasting of coal. *Min. Saf. Environ. Prot.* **2019**, *46*, 92–95+99.
3. Lu, W.; Wen, J.M.; Wang, D.; Zhang, W.R.; Hu, X.M.; Wu, H.; Kong, B. Study on gas transport characteristics of positive pressure beam tube system and its application to coal spontaneous combustion early warning. *J. Clean. Prod.* **2021**, *316*, 128342. [CrossRef]
4. Kong, B.; Li, J.H.; Lu, W.; Fu, W.; Song, H.Z.; Liu, J.F. Research on a Spontaneous Combustion Prevention System in Deep Mine: A Case Study of Dongtan Coal Mine. *Combust. Sci. Technol.* **2022**. [CrossRef]
5. Zhang, W.T.; Hou, Y.H.; Peng, D.H.; Fan, C. Distributed fiber optic temperature measurement and fire warning system in coal mines. In Proceedings of the 23rd National Conference on Coal Mine Automation and Informatization and the 5th China High Level Forum on Coal Mine Informatization and Automation, Beijing, China, 6–7 December 2013; pp. 74–80.
6. Zhang, X.H.; Liu, Q.; Zheng, X.Z.; Wang, W.F. Analysis of ZigBee-based wireless self-assembled temperature measurement system in mining area. *Coal Eng.* **2012**, *9*, 122–124.
7. Ganne, P.; Vervoort, A.; Wevers, M. Quantification of pre-peak brittle damage: Correlation between acoustic emission and observed micro-fracturing. *Int. J. Rock Mech. Min. Sci.* **2007**, *44*, 720–729. [CrossRef]
8. Ishida, T.; Kanagawa, T.; Kanaori, Y. Source distribution of acoustic emissions during an in-situ direct shear test: Implications for an analog model of seismogenic faulting in an inhomogeneous rock mass. *Eng. Geol.* **2010**, *110*, 66–76. [CrossRef]
9. Han, J.; Han, S.Z.; Ma, S.W.; Zhang, L.Y.; Cao, C. Study on acoustic emission characteristics of coal bodies of different strengths. *Chin. J. Undergr. Space Eng.* **2021**, *17*, 739–747.
10. Kong, B.; Zhuang, Z.D.; Zhang, X.Y.; Jia, S.; Lu, W.; Zhang, X.Y.; Zhang, W.R. A study on fractal characteristics of acoustic emission under multiple heating and loading damage conditions. *J. Appl. Geophys.* **2022**, *197*, 104532. [CrossRef]
11. Kong, B.; Zhong, J.H.; Hu, X.M.; Xin, L.; Zhang, B.; Zhang, X.L.; Zhuang, Z.D. Study on the change pattern of acoustic emission signal and generation mechanism during coal heating and combustion process. *Coal Sci. Technol.* **2023**, 1–8. [CrossRef]
12. Kong, B.; Zhu, S.X.; Zhang, W.R.; Sun, X.L.; Lu, W.; Ma, Y.K. Mechanical deterioration and thermal deformations of high-temperature-treated coal with evaluations by EMR. Geomechanics and Engineering. *Geomech. Eng.* **2023**, *32*, 233–244. [CrossRef]
13. Chen, C. Laboratory Study of Deformational Characteristics and Acoustic Emission Properties of Coal with Different Strengths under Uniaxial Compression. *Minerals* **2021**, *11*, 1070.
14. Zhang, R.; Liu, J.; Sa, Z.; Wang, Z.; Lu, S.; Wang, C. Experimental investigation on multi-fractal characteristics of acoustic emission of coal samples subjected to true triaxial loading-unloading. *Fractals* **2020**, *28*, 2050092. [CrossRef]
15. Wang, H.Y.; Wang, G.D.; Zhang, G.J.; Du, F.; Ma, J. Acoustic Emission Response Characteristics of Anthracitic Coal under Uniaxial Compression. *Shock. Vib.* **2020**, 1–12. [CrossRef]
16. Li, H.G.; Li, H.M.; Gao, B.B. Study on acoustic emission characteristics of rupture process of different coal-thick coal-rock assemblies. *J. Henan Polytech. Univ. (Nat. Sci. Ed.)* **2021**, *40*, 30–37.
17. Cai, Y.; Liu, D.; Yao, Y.; Li, J.; Liu, J. Fractal Characteristics of Coal Pores Based on Classic Geometry and Thermodynamics Models. *Acta Geol. Sin.* **2011**, *85*, 1150–1162. [CrossRef]
18. Guo, H.; Yuan, L.; Cheng, Y.; Wang, K.; Xu, C. Experimental investigation on coal pore and fracture characteristics based on fractal theory. *Powder Technol.* **2019**, *346*, 341–349. [CrossRef]
19. Li, Z.H.; Tian, H.; Niu, Y.; Wang, E.Y.; Zhang, X.; He, S.; Wang, F.Z.; Zheng, A.Q. Study on the acoustic and thermal response characteristics of coal samples with various prefabricated crack angles during loaded failure under uniaxial compression. *J. Appl. Geophys.* **2022**, *200*, 104618. [CrossRef]
20. Zhao, S.; Chen, X.J.; Li, X.J.; Qi, L.L.; Zhang, G.X. Experimental analysis of the effect of temperature on coal pore structure transformation. *Fuel* **2021**, *305*, 121613. [CrossRef]

21. Yi, M.H.; Cheng, Y.P.; Wang, C.H.; Wang, Z.Y.; Hu, B.; He, X.X. Effects of composition changes of coal treated with hydrochloric acid on pore structure and fractal characteristics. *Fuel* **2021**, *294*, 120506. [CrossRef]
22. Xu, C.P.; Li, H.; Lu, Y.; Liu, T.; Lu, J.X.; Shi, S.L.; Ye, Q.; Jia, Z.Z.; Wang, Z. Influence of microwave-assisted oxidant stimulation on pore structure and fractal characteristics of bituminous coal based on low-temperature nitrogen adsorption. *Fuel* **2022**, *327*, 125173. [CrossRef]
23. Jiang, L.; Chen, Z.; Ali, S.F. Thermal-hydro-chemical-mechanical alteration of coal pores in underground coal gasification. *Fuel* **2020**, *262*, 116543. [CrossRef]
24. Cheng, Q.Y.; Huang, B.X.; Li, Z.H. Research status of porosity and fracture of coal. *Coal Eng.* **2011**, *12*, 91–93.
25. Li, H.; Shi, S.L.; Lu, J.X.; Ye, Q.; Lu, Y.; Zhu, X.N. Pore structure and multifractal analysis of coal subjected to microwave heating. *Powder Technol.* **2019**, *346*, 97–108. [CrossRef]
26. Kong, B.; Wang, E.Y.; Li, Z.H.; Lu, W. Study on the feature of electromagnetic radiation under coal oxidation and temperature rise based on multifractal theory. *Fractals* **2019**, *27*, 1950038. [CrossRef]
27. Yang, S.Q. Study on the Evolution of Microstructure and Mechanical Response Law of Oil Shale under High Temperature Real-Time Action. Ph.D. Thesis, Taiyuan University of Technology, Taiyuan, China, 2021.
28. Li, J.H.; Xue, C.Z.; Han, Q. Study on the evolution characteristics of pore fractures in coal under different thermal fracture temperatures. *Coal Mine Saf.* **2020**, *51*, 22–25+29.
29. Lu, W.; Zhuang, Z.; Zhang, W.; Zhang, C.; Song, S.; Wang, R.; Kong, B. Study on the Pore and Crack Change Characteristics of Bituminous Coal and Anthracite after Different Temperature Gradient Baking. *Energy Fuels* **2021**, *35*, 19448–19463. [CrossRef]
30. Gan, Q.; Xu, J.; Peng, S.; Yan, F.; Wang, R.; Cai, G. Effects of heating temperature on pore structure evolution of briquette coals. *Fuel* **2021**, *296*, 120651. [CrossRef]
31. Gbadamosi, A.R.; Onifade, M.; Genc, B.; Rupprecht, S. Analysis of spontaneous combustion liability indices and coal recording standards/basis. *Int. J. Min. Sci. Technol.* **2020**, *30*, 723–736. [CrossRef]
32. Zhao, Y.S.; Meng, Q.R.; Kang, T.H. Micro-CT experimental technology and meso-investigation on thermal fracturing characteristics of granite. *Rock Mech. Eng.* **2008**, *27*, 28–34.
33. Kong, B.; Wang, E.; Li, Z.; Wang, X.; Chen, L.; Kong, X. Nonlinear characteristics of AEs during the deformation and fracture of sandstone subjected to thermal treatment. *Int. J. Rock Mech. Min. Sci.* **2016**, *90*, 43–52. [CrossRef]
34. Zhao, K.; Yang, D.X.; Gong, C.; Zhuo, Y.L.; Wang, X.J.; Zhong, W. Evaluation of internal microcrack evolution in red sandstone based on time–frequency domain characteristics of acoustic emission signals. *Constr. Build. Mater.* **2020**, *260*, 120435. [CrossRef]
35. Jia, X.N.; Jiang, Q.G.; He, M.C.; Ren, F.Q.; Du, S. Characteristic analysis of acoustic radio frequency spectroscopy of rhesus rock burst experiment in Laizhou granite. *J. Undergr. Space Eng.* **2018**, *14*, 51–57.
36. Song, X.X.; Tang, Y.G.; Li, W.; Wang, S.Q.; Yang, M.M. Fractal characteristics of structural coal adsorption holes in Zhongliang-shannan Mine. *J. China Coal Soc.* **2013**, *38*, 134–139.
37. Ji, H.G.; Wang, J.C.; Shan, X.Y.; Cai, M.F. Fractal characteristics of acoustic emission process of concrete materials and their application in fracture analysis. *Chin. J. Rock Mech. Eng.* **2001**, *6*, 801–804.
38. Tan, J.N.; Wang, B.; Feng, T.; Ning, Y.; Liu, B.B.; Zhao, F.J. acoustic emission characteristics of anchored sandstone under uniaxial compression and its connection with rock burst. *J. Cent. South Univ. (Nat. Sci. Ed.)* **2021**, *52*, 2828–2838.
39. Ding, X.; Xiao, X.C.; Lv, X.F.; Wu, D.; Wang, L.; Fan, Y.F. Study on fractal characteristics and acoustic emission law of coal body rupture. *J. China Coal Soc.* **2018**, *43*, 3080–3087.
40. Gao, B.; Li, H.; Li, H. Study on Acoustic Emission and Fractal Characteristics of Different Damage Types of Rock. *J. Undergr. Space Eng.* **2015**, *11*, 358–363.
41. Gao, B.B.; Li, H.G.; Li, L.; Wang, X.L.; Yu, S.J. acoustic emission and fractal characteristics of soft hard coals in the same group. *Chin. J. Rock Mech. Eng.* **2014**, *33*, 3498–3504.
42. Jin, P.J.; Wang, E.Y.; Song, D.Z. Study on Correlation of Acoustic Emission and Plastic Strain Based on Coal-rock Damage Theory. *Geomech. Eng.* **2017**, *12*, 627–637. [CrossRef]
43. Li, H.R.; Shen, R.X.; Qiao, Y.F.; He, M.C. acoustic emission signal characteristics and its critical slowing down phenomenon during the loading process of water-bearing sandstone. *J. Appl. Geophys.* **2021**, *194*, 104458. [CrossRef]
44. Yang, X.H.; Ren, T.; He, X.Q. Experimental study of coal burst risk prediction using fractal dimension analysis of acoustic emission spatial distribution. *J. Appl. Geophys.* **2020**, *177*, 104025. [CrossRef]
45. Wang, C.L.; Hou Xl Liao, Z.F.; Chen, Z.; Lu, Z.J. Experimental investigation of predicting coal failure using acoustic emission energy and load-unload response ratio theory. *J. Appl. Geophys.* **2019**, *161*, 76–83. [CrossRef]
46. Liu, K.; Ostadhassan, M. Multi-scale fractal analysis of pores in shale rocks. *J. Appl. Geophys.* **2017**, *140*, 1–10. [CrossRef]
47. Chai, M.Y.; Hou, X.L.; Zhang, Z.X.; Duan, Q. Identification and prediction of fatigue crack growth under different stress ratios using acoustic emission data. *Int. J. Fatigue* **2022**, *160*, 106860. [CrossRef]
48. Ambrosio, D.; Dessein, G.; Wagner, V.; Yahiaoui, M.; Paris, J.-Y.; Fazzini, M.; Cahuc, O. On the potential applications of acoustic emission in friction stir welding. *J. Manuf. Processes* **2022**, *75*, 461–475. [CrossRef]

processes

MDPI

Article

Investigation of Formation Process and Intensity of Coal and Gas Outburst Shockwave

Dongling Sun [1,2], Jie Cao [1,2], Linchao Dai [1,2,*], Rifu Li [1,2] and Yanbao Liu [1,2]

1 State Key Laboratory of the Gas Disaster Detecting, Preventing and Emergency Controlling, Chongqing 400037, China

2 China Coal Technology and Engineering Group, Chongqing Research Institute, Chongqing 400037, China

* Correspondence: 20212001047g@cqu.edu.cn; Tel.: +86-023-6523-9611

Abstract: The shock wave of a coal and gas outburst is a high-pressure and high-speed impact airflow formed rapidly after the outburst. The propagation destroys the ventilation facilities and causes the destruction of the ventilation system. The theoretical research on the outburst shock wave is of great significance. In order to deeply understand the formation mechanism of the outburst shock wave, this paper draws on the shock wave theory to theoretically analyze the microscopic formation process of the outburst shock wave. The main difference between the formation process of a coal and gas outburst shock wave and the formation process of a general shock wave is that the outburst shock wave has a solid–gas flow zone in the high-pressure zone. The calculation formulas of pressure, density, temperature and other parameters before and after the outburst shock wave are derived. After the outburst shock wave passes through, the pressure, temperature and density of the roadway air will change suddenly. The relationship expression between outburst gas pressure and outburst shock wave intensity is derived, which can reflect the role of pulverized coal in the formation process of a shock wave. In order to facilitate the understanding and calculation, the concept of equivalent sound velocity of coal-gas flow is proposed, and the direct calculation of the impact strength of a coal and gas outburst is attempted. This paper is helpful to improve the understanding of the essence of a coal and gas outburst shock wave. It is also of great significance to outburst disaster relief.

Keywords: coal and gas outburst; outburst the shock wave; gas–solid flow; initial gas pressure

Citation: Sun, D.; Cao, J.; Dai, L.; Li, R.; Liu, Y. Investigation of Formation Process and Intensity of Coal and Gas Outburst Shockwave. *Processes* **2023**, *11*, 659. https://doi.org/10.3390/pr11030659

Academic Editors: Feng Du, Aitao Zhou and Bo Li

Received: 9 February 2023
Revised: 21 February 2023
Accepted: 21 February 2023
Published: 22 February 2023

1. Introduction

In the process of coal and gas outbursts, the direct damage to underground personnel and equipment in coal mines is the impact of airflow and pulverized coal that rushes with high-speed movement into the working space [1–3]. The damage of the impact airflow can be divided into high-pressure airflow shock wave overpressure damage and suffocation damage, among which the shock wave damage is the most direct and fastest [4–6]. The former Soviet Union scientist Savonko [7] studied the influence of roadway section reduction and expansion on the pressure of outburst shock waves and obtained the attenuation coefficient of air shock waves during movement. Zhang [8] pointed out that the outburst shock wave belongs to the weak shock wave and constructed the relationship between the propagation attenuation of the outburst shock wave and the initial energy, propagation distance and friction resistance of the roadway. Sun et al. [9] studied the phenomenon of gas accumulation in the roadway after the outburst airflow by using the outburst test device. According to the theory of aerodynamics, Cheng et al. [10,11] established the mathematical model of motion and dynamics of outburst shock wave propagation, and established the relationship between shock wave overpressure, impact airflow velocity and propagation distance, coal seam gas pressure, etc. Based on theoretical analysis and numerical simulation, Tang [12] studied the influence of gas pressure on outburst energy, outburst intensity and gas emission. Miao [13] studied the expansion characteristics of the outburst impact airflow and established the relationship between the expansion characteristics and the expansion failure strength. Hu [14] conducted a numerical study on the impact dynamic effect of the impact airflow on the road header and the outburst prevention door. Zhou et al. [15–17] studied the formation and propagation process of outburst shock waves and gas flow in different types of roadways. They analyzed the best roadway form to prevent the spread of outburst shock, studied the influence of gas outburst shock waves on airflow disturbance in mine roadways, and pointed out that the airflow disturbance caused by outburst shock waves was mainly controlled by shock wave overpressure and main fan wind pressure.

Although there have been many studies on the formation and development of outburst shock waves, most of them rely on physical experiments or numerical simulations. Unfortunately, the theoretical analysis of the underlying mechanism is not as comprehensive. Many of the expressions used in these studies are based on aerodynamic theory alone and do not take into account the effects of outburst pulverized coal on a shock wave's formation. It is difficult to reveal the theoretical mechanism in the formation and development of outburst shock waves.

2. The Relationship between Outburst Shock Wave and General Shock Wave

2.1. The Formation of General Shock Waves

In gases, liquids or solids, if the pressure, density and temperature change (nonlinear change) on a cross section, there is a shock wave inside the material [18–21]. The reason is generally that there is a wave source in the object whose moving speed exceeds the wave velocity. Shock waves are very common in nature and human society and may be called differently in different situations. For example, the shock wave formed during an explosion can be an explosive wave. The shock wave generated by a flying object when the propagation speed in the air exceeds the speed of sound is generally called sonic explosion. The water hammer effect formed by the sudden closure of water flow in a hard water pipe is also a shock wave. In aerodynamics, a strong shock wave is generally called a shock wave.

The formation process of the shock wave is shown in Figure 1. There is an ideal gas with pressure p_1, density ρ_1 and temperature T_1 in a straight round tube. There is a piston on the left side of the ideal gas, and the piston and gas are static at the initial time, as shown in Figure 1a. In the parameter curve, only the change of pressure p is shown to represent the change of the gas state, so as to avoid repeating. In a very short period of time $\Delta t'$, the piston starts moving to the right at speed $\Delta t'$ and squeezing the gas in the tube, as shown in Figure 1b. A weak compression wave is formed at the intersection of the gas and the tube, and the compression wave head A_1-A_1 pushes to the right at the sound velocity c_1. The sound velocity c_1 is the local sound velocity of stationary gas in the round tube. With the wave head as the boundary, the air parameters on the right side are the initial state in the tube, and the air parameters on the left side are pressure $p_1 + \Delta p'$, density $\rho_1 + \Delta\rho'$ and temperature $T_1 + \Delta T'$. In addition, the velocity increases from 0 to $\Delta v'$, and the increments of the four parameters are all positive. The piston continues to move forward, and the piston speed reaches $\Delta t''$ after time $\Delta v''$. The gas behind the wave head formed in the previous stage is further compressed, and a new compression wave will be formed behind it. The compressed wave head A_2-A_2 pushes the gas to the right at the sound velocity $c_1{}'$ relative to wave A_1-A_1, as shown in Figure 1c. The sound velocity c_2 is the local sound velocity of the gas behind A_2-A_2 in the round roadway. The absolute velocity of A_2-A_2 is $c_1{}' + \Delta v'$. With the wave head A_2-A_2 as the boundary, the air parameters on the right side are pressure $p_1 + \Delta p'$, density $\rho_1 + \Delta\rho'$ and temperature $T_1 + \Delta T'$. The air parameters on the left side are pressure $p_1 + \Delta p''$, density $\rho_1 + \Delta\rho''$ and temperature $T_1 + \Delta T''$. The velocity is increased from $\Delta v'$ to $\Delta v''$. At this time, there are two wave heads, A_1-A_1 and A_2-A_2, in the round roadway. As time goes on, the piston will form a series of wave heads in turn as it moves to the right.

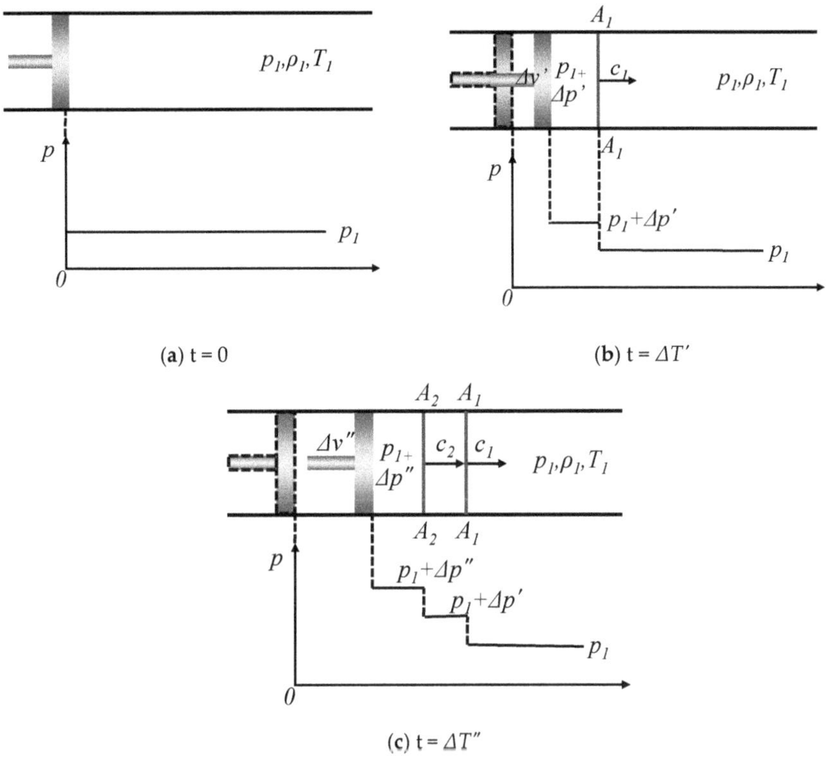

Figure 1. Schematic diagram of shock wave formation.

The formula of sound velocity in gas is:

$$a = \sqrt{kRT} \tag{1}$$

In the formula, k is the specific heat ratio of air, T is the thermodynamic temperature, R is the universal gas constant of air and the universal gas constant is $R = 8.31$ J/(K·mol). The air-specific heat ratio is also constant $k = 1.4$ in non-extreme cases. It is easy to see that the sound velocity is only related to the gas temperature. The sound velocity is different at different temperatures in the flow field.

The moving speed of A_1-A_1 is:

$$v_1 = c_1 = \sqrt{kRT_1} \tag{2}$$

The moving speed of A_2-A_2 is:

$$v_2 = c_2 + \Delta v = \sqrt{kR(T_1 + \Delta T')} + \Delta v \tag{3}$$

Obviously, the wave head A_2-A_2 will eventually catch up with the wave head A_1-A_1 and merge with it after a certain period of time. By analogy, A_3-A_3 and A_4-A_4 after the wave head A_2-A_2 will catch up and merge with the initial wave head A_1-A_1. Since the air compression wave is also a mechanical wave, the strength of the compression wave obeys the superposition principle, and the strength of the wave at the compression wave is strengthened. Due to the continuous advancement of the piston, new wave heads are continuously generated, and the compression wave is continuously strengthened. Finally, a compression wave with great intensity is formed, which is generally called a shock wave. The continuously strengthened wave head is called the wave front. After the shock wave front passes through, the gas parameters change from the initial p_1, ρ_1 and T_1 to p_2, ρ_2 and T_2 (the values are calculated in the later chapters), and the gas velocity behind it changes from 0 to v_f, which is called the accompanying velocity. It should be emphasized that the adjoint velocity is the velocity of the gas after the wave, and the velocity of the shock wave front is the propagation velocity of the mechanical wave. The two are completely independent concepts. The shock wave front velocity is greater than the accompanying velocity.

In summary, the shock wave is a strong disturbance wave. Its propagation velocity and wave front velocity are greater than the sound velocity. The gas state parameters of the shock wave change abruptly (pressure, temperature and density are all increased). The so-called mutation is because the gas state parameters have a large parameter gradient at the wave front and are macroscopically discontinuous. The greater the parameter gradient is, the more obvious the viscous dissipation effect of the gas is at the time of sudden change. The result is that the gas inertial force and viscous dissipation reach equilibrium, as shown in Figure 2a. The spatial thickness involved in this equilibrium stage is several molecular average free paths. The calculation formula of the average free path λ of gas molecules:

$$\lambda = \frac{kT}{\sqrt{2}\pi d_0^2 p} \tag{4}$$

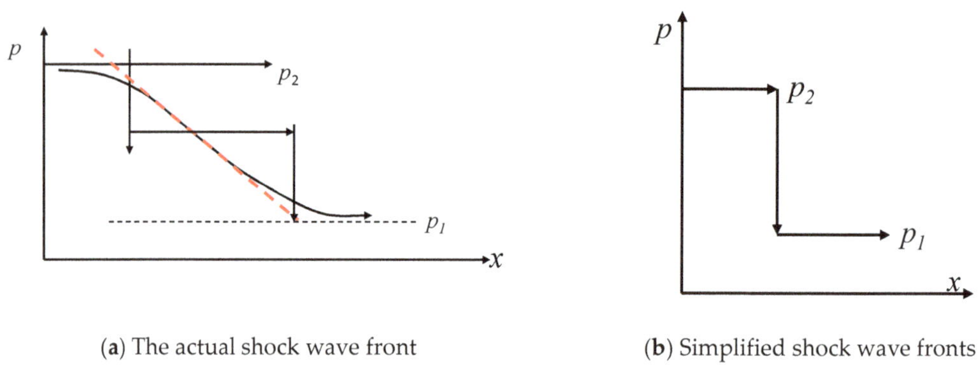

Figure 2. Wave front of a shockwave.

In the formula, k is the Boltzmann constant, $k = 1.38 \times 10^{-23}$ J/K; T is the absolute temperature, K, in the standard state, $T = 273.15$ K; d_0 is the molecular effective diameter, m; the effective average diameter of air molecules is 3.7×10^{-10} m; p is gas pressure, Pa; the atmospheric pressure is 1×10^5 Pa; and the average free path of air molecules in the standard state is 5.9×10^{-8} m. This length can be ignored on the macro scale. Therefore, the shock wave front can be completely described as a vertical plane, as shown in Figure 2b.

The shock wave front discussed above is perpendicular to the propagation direction of the wave front, which is called a normal shock wave. Because the axis direction of the roadway is perpendicular to the shock wave of a coal and gas outburst in the roadway, which is similar to the normal shock wave, the shock waves described in this paper are all normal shock waves.

2.2. Coal and Gas Outburst Formation Characteristics

After the formation of underground production and ventilation systems in a coal mine, although there are factors such as altitude and ventilation system, compared with the gas pressure in the coal seam (several atmospheric pressures), a strong gas pressure gradient is formed between the air pressure in the roadway and the gas pressure in the coal seam.

Under certain combination conditions of gas pressure, ground stress and physical and mechanical properties of coal, a coal and gas outburst will occur.

The simplified coal and gas outburst scene is shown in Figure 3. Under the action of ground stress and gas pressure, the outburst is excited, and a large number of broken pulverized coal and gas migrate from the coal seam to the roadway at a high speed. From left to right, it can be divided into the stable coal seam area, coal-gas flow area, air compression area and roadway unaffected area. The location of these areas will change with the change of the outburst influence stage, and the interior of each area is also uneven (except for the temporarily unaffected area of the roadway).

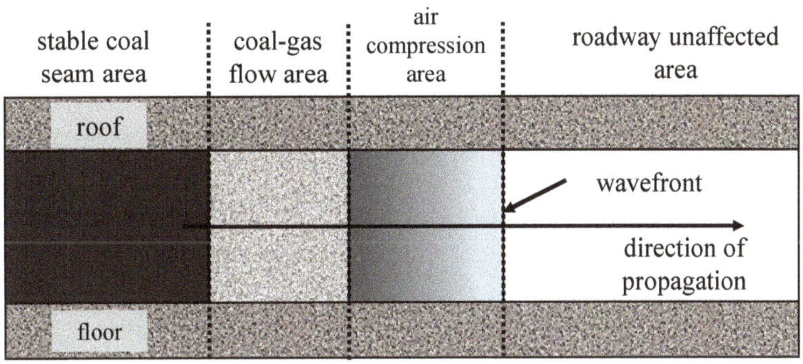

Figure 3. Schematic diagram of a coal-gas outburst.

After the failure of the weak layer at the outburst mouth disappears, the pulverized coal-gas in the high-pressure space of the coal seam are released and freely migrate to the roadway space. The coal-gas flow will first form in the coal seam. The initial coal-gas flow first compresses the roadway air at the outburst port, and as the coal-gas flow moves, its interface with the air continues to move forward (right side). The compressed roadway air moves forward and compresses the more forward roadway air. The front of the compressed roadway air is called the outburst wave front. The range between the wave front and the coal-gas flow is the air compression zone. The roadway in front of the wave front has not been disturbed by the outburst, which is called the temporarily unaffected area. The air parameters in the area are stable and maintain initial parameters. The moving speed of the wave front is the development speed of the influence range of a coal and gas outburst. From the above discussion, it can be known that the forefront of the impact of a coal and gas outburst is the air compression zone, not coal-gas flow.

3. Coal and Gas Outburst Shock Wave Theory

3.1. Overview of Shock Tube

Shock wave is an important research topic in many fields of nature and laboratory. Its shock wave theory plays an important role in aerospace, aviation, explosion engineering and other fields. A shock tube can form stable and controllable shock waves in laboratory research, which is an important instrument for shock wave-related research. The shock wave process formed in coal and gas outbursts is highly similar to the shock wave process formed by a shock tube. Therefore, this section takes shock wave theory as the starting point to study the shock waves of coal and gas outbursts.

The world's first shock tube was born in France in 1861. The chemist P. Vieille used it to obtain a moving shock wave with a velocity of 600 m/s in the process of studying the detonation problem in combustion. With the development of shock tube technology and the increase in social development needs, a shock tube is of great significance in many fields. In addition to the basic theories of physics and chemistry, it has been widely used in the fields of electromagnetic fluid mechanics, pneumatic lasers, anti-explosion processes and combustion [22–26]. The shock tube is widely used. The main reason is that the shock tube itself has many advantages, such as a simple structure and strong controllability of shock wave parameters.

The shock tube in the laboratory can be in a variety of forms, but its core basic forms are consistent. The basic shock tube form is called the shock tube later. The shape of this shock tube is a straight tube with an equal cross-section, which is divided into two parts: a high-pressure section and low-pressure section, which are isolated by a bursting disc, as shown in Figure 4. When a shock wave is needed, the bursting disc breaks and the high-pressure gas in the high-pressure section suddenly rushes into the low-pressure section to form a shock wave. By changing the gas species in the high- and low-pressure sections of the shock tube and the gas pressure in the high- and low-pressure sections, the shock tube can form different shock waves [27–30].

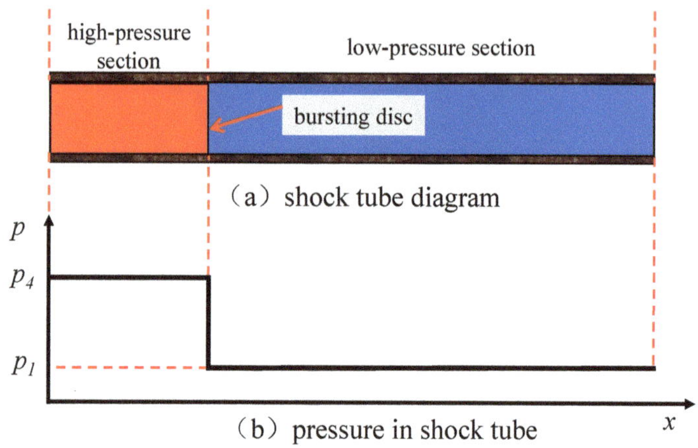

Figure 4. Initial state of the shock tube.

In the shock tube experiment, when the pressure difference on both sides of the bursting disc exceeds its critical value, the bursting disc ruptures. Due to the huge pressure difference on the two sides of the bursting disc, an incident shock wave is generated from the bursting disc position to the right side, and a sparse wave is generated to the left side, as shown in Figure 5. In order to facilitate the later description, the original high-pressure section of the shock tube is called zone 4, and the low-pressure section is called zone 1. When the incident shock wave is formed, its power is the high-pressure gas on the left side, but after the incident shock wave begins to move, it begins to cause a sudden change in the parameters of the low-pressure section at the junction with it. The shock wave front is transferred to the original low-pressure section gas, and the gas parameters (including pressure) of the low-pressure section of this part are surged to form zone 2. With the passage of time, zone 2 gradually moves to zone 1, and the moving speed is the shock wave front speed. The influence range of the sparse wave moving to the left is called zone 3, and the gas parameters (including pressure) in zone 3 become smaller. One difference between rarefaction waves and shock waves is that rarefaction waves do not cause sudden changes in gas parameters, so rarefaction waves are generally not ignored as a line in the vertical direction in space. However, the spatial length of the sparse wave is very short compared with zone 3, and they are all formed by the original high-pressure gas, so this zone is merged into zone 3 and its parameters are regarded as zone 3. Therefore, after the rupture of the bursting disc, the shock tube is divided into zone 4, zone 3, zone 2 and zone 1 from left to right. The pressure in zone 4 is the largest, the pressure in zone 1 is the smallest, zone 3 and zone 2 are equal and between zone 4 and zone 1. Before the incident shock wave or rarefaction wave reaches the two ends of the shock tube, the range of zone 2 and zone 3 gradually increases, and the range of zone 1 and zone 4 gradually decreases. It is easy to see that the gas in zone 1 and zone 2 is the original low-pressure gas, and the gas in zone 3 and zone 4 is the original high-pressure gas. Zone 3's position spans both sides of the original bursting disc position.

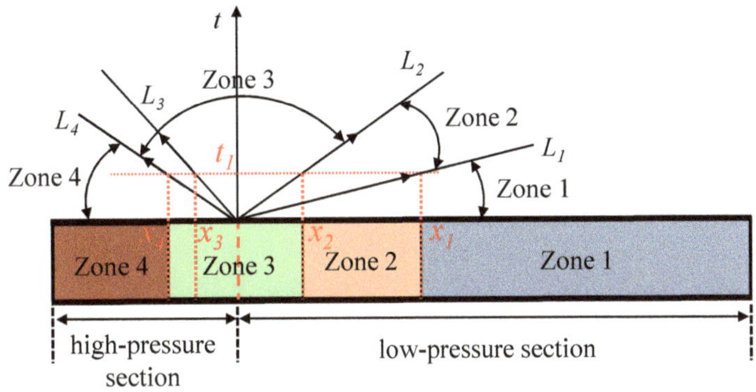

Figure 5. Region division and wave system diagram in shock tube.

The upper part of Figure 5 shows the shock tube wave diagram. In the figure, the ordinate represents the time, and the abscissa represents the spatial position of the shock tube axis. L_1, L_2, L_3 and L_4 represent the shock wave front, the original high-pressure and low-pressure gas interface, the sparse wave right end and the sparse wave left end, respectively. At time t_1, these three sections move to x_1, x_2, x_3 and x_4, respectively.

In Figure 6, the two figures are the pressure curves in the shock tube at the initial time and $t = t_1$ time, respectively.

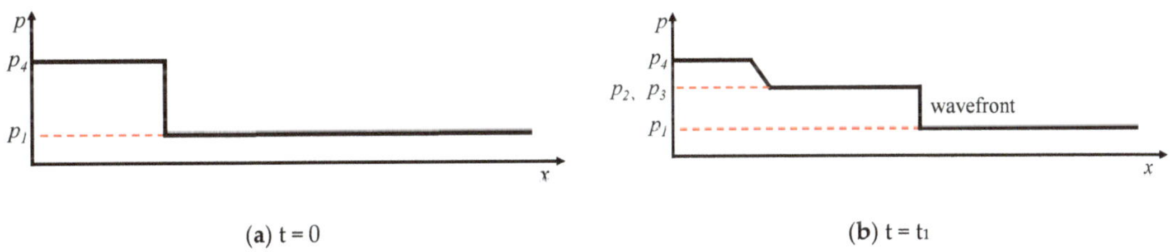

Figure 6. Pressure curve in shock tube.

When the shock wave theory is applied to the theoretical analysis of a coal and gas outburst, zone 1 is the area where the roadway has not been affected, zone 2 is the roadway air affected by the shock wave, zone 3 is the coal-gas flow area and zone 4 is the outburst area in the coal seam.

3.2. Derivation of Outburst Shock Wave Parameters

Although most of the roadways in coal mines are curved, most of the roadways are also straight in a small range, and most of the coal and gas outbursts occur in the tunneling roadways; these roadways are also straight, which is similar to the geometry of the shock tube. Additionally, the space of the coal mine roadway is much larger than that of the ordinary shock tube. From this aspect, the coal mine roadway is more in line with the assumption of the ideal shock tube.

Some idealized assumptions are needed in the theoretical derivation of coal and gas outburst flow and fluid in shock tubes. Based on these assumptions, complex practical phenomena, in reality, can be transformed into strict formulas. Facts have proved that these assumptions are reasonable, and the calculated results only slightly deviate from the actual gas flow. The main assumption is that the gas in the shock tube is an ideal gas without viscosity. The shock tube wall is rigid. There is no heat exchange between the gas and the shock tube wall. The gas flow in the shock tube is a one-dimensional flow. The rupture of the bursting disc ends instantaneously and completely. In sparse waves, the gas is isentropic. The shock tube that satisfies the above assumptions is also called an ideal shock tube.

As shown in Figure 7a, the incident shock wave front passes through the gas in zone 1 with v_s, and then the gas in this zone suddenly changes to the gas in zone 2. The gas changes from the stationary state to v_b, and v_b is the adjoint velocity. In order to facilitate the analysis, the coordinate transformation is carried out. The coordinate system is established on the shock wave front, and the incident shock wave is transformed into a stationary state to become a stationary shock wave. At this time, the gas in zone 1 enters the wave front at the speed of $v_1 = v_s$ (facilitate the calculation of the retention value size and ignore the speed direction, the same below). The gas after the parameter mutation moves to the left at the speed $v_2 = v_s - v_b$, as shown in Figure 7b. Gas thermodynamic static parameters such as pressure, density and temperature are independent of coordinate transformation.

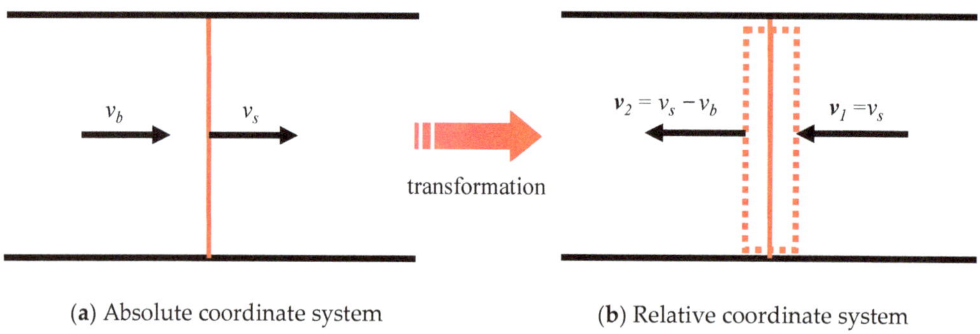

Figure 7. Transformation of coordinate system of wave front.

In fluid mechanics, Mach number M is an important parameter to describe the flow state. It is defined as the ratio of the velocity v of a point in the flow field to the local sound velocity c of the point:

$$M = \frac{v}{c} \tag{5}$$

Similar to the concept of local sound speed, the Mach number describes the local properties in the flow field. If the local sound speed of two points in the flow field is different, even if the two points have the same fluid velocity, the Mach number is also different.

The gas Mach number M before and after the wave front after the coordinate transformation:

$$M_1 = \frac{v_1}{c_1} = \frac{v_s}{c_1} \tag{6}$$

$$M_2 = \frac{v_2}{c_2} = \frac{v_s - v_b}{c_2} \tag{7}$$

In the formula, the subscript of 1 is the gas parameter on the right side of the shock wave, and the subscript of 2 is the gas parameter on the left side of the shock wave.

In Figure 7b, the gas in the imaginary frame is used as the control body, because the standing shock wave is a parameter-stable structure and obeys the mass conservation equation:

$$\oiint_S \rho \vec{V} \cdot dS = 0 \tag{8}$$

Only the front and back sides of the control body boundary S have fluid in and out. A_1 is the area of the inlet surface of the standing shock wave, and A_2 is the area of the outlet surface of the standing shock wave. The two values are equivalent and the direction is opposite, so $S = A_1 - A_2$, the above formula can be changed into:

$$A_1\rho_1 v_s - A_2\rho_2[(v_s - v_b)] = 0 \tag{9}$$

The formula can be simplified to:

$$\rho_1 v_1 = \rho_2 v_2 \tag{10}$$

The momentum conservation is obeyed in the control body, and the impulse of the pressure resultant force is equal to the difference between the momentum of the outflow fluid and the momentum of the inflow fluid.

$$\oiint_S (\rho\vec{V}\cdot d\vec{S})\vec{V} = \oiint_S p dS \tag{11}$$

The above formula can be changed to:

$$p_1 A_1 - p_2 A_2 = m v_2 - m v_1 \tag{12}$$

In the formula, m denotes the mass flow rate of gas in and out of the control body, $m = \rho v A$. The above formula can be transformed into:

$$p_1 + \rho_1 v_1^2 = p_2 + \rho_2 v_2^2 \tag{13}$$

Similarly, the energy equation is expressed as:

$$\oiint_S \rho\left(e + \frac{V^2}{2}\right)\vec{V}\cdot ds = -\oiint_S p\vec{V}\cdot d\vec{S} \tag{14}$$

The above formula can be transformed into:

$$h_1 + \frac{v_1^2}{2} = h_2 + \frac{v_2^2}{2} \tag{15}$$

In the formula, $\frac{{v_1}^2}{2}$ represents the kinetic energy of the airflow, which is a macroscopic parameter. Enthalpy h represents the sum of gas internal energy and pressure potential energy, which is a microscopic parameter.

The momentum Equation (13) is divided by the continuity Equation (10) to obtain:

$${v_1}^2 - {v_2}^2 = (p_2 - p_1)\left(\frac{1}{\rho_1} + \frac{1}{\rho_2}\right) \tag{16}$$

According to:

$$h = \frac{k}{k-1}\frac{p}{\rho} \tag{17}$$

The energy Equation (15) can be transformed into:

$${v_1}^2 - {v_2}^2 = \frac{2k}{k-1}\left(\frac{p_2}{\rho_2} - \frac{p_1}{\rho_1}\right) \tag{18}$$

Combining Equation (16) with Equation (18):

$$\frac{\rho_2}{\rho_1} = \frac{1 + \frac{k+1}{k-1}\frac{p_2}{p_1}}{\frac{k+1}{k-1} + \frac{p_2}{p_1}} \tag{19}$$

This function represents the relationship between the density ratio and the pressure ratio before and after the normal shock wave, also known as the shock adiabatic relation. In the experiment, the pressure before and after the shock wave can be directly measured by the pressure sensor, and the density cannot be directly measured, so this formula can be used for density calculation.

The isentropic change of gas is a reversible adiabatic process. The relationship between pressure change and density change is:

$$\frac{\rho_2}{\rho_1} = (\frac{p_2}{p_1})^{1/k} \tag{20}$$

The density changes in the isentropic process and the shock process are compared with $k = 1.4$, as shown in Figure 8. It can be seen from the diagram that when the pressure is relatively small, the density ratio of the isentropic process and the shock wave process is not much different. As the pressure ratio increases, the density ratio in the shock wave mutation process becomes significantly smaller. The difference between the shock wave process [31] and the isentropic process curve shows that the shock wave process is entropy-increasing, that is, an irreversible process. In addition, as the pressure ratio further increases, the density ratio of the shock wave process has a limit, and the limit value is:

$$\frac{\rho_2}{\rho_1} = \frac{k+1}{k-1} \tag{21}$$

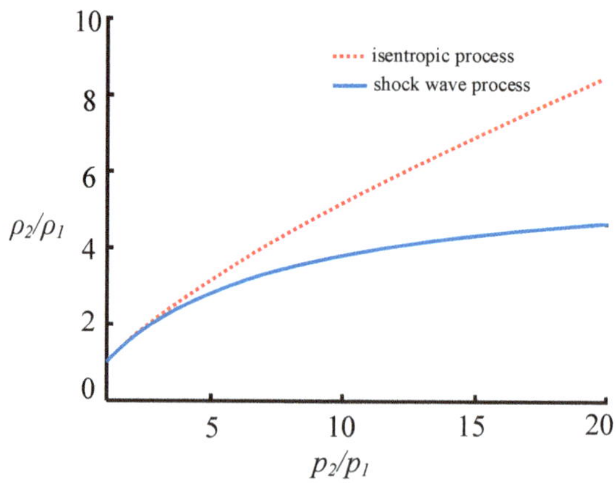

Figure 8. Comparison of density changes in isentropic and shock processes.

The calculated limit density ratio is 6. This shows that no matter how strong the shock wave is, the maximum pressure behind the shock wave can reach 6 times the pressure before the shock wave.

Equations (9) and (13) are combined to obtain:

$$v_s = \sqrt{\frac{\rho_2}{\rho_1} \cdot \frac{p_2 - p_1}{\rho_2 - \rho_1}} \tag{22}$$

Combined with the shock adiabatic relationship (19) and the sound velocity relationship, we can get:

$$v_s = \sqrt{\frac{\rho_2}{\rho_1} \cdot \frac{p_2 - p_1}{\rho_2 - \rho_1}} = c_1\sqrt{1 + \frac{k+1}{2k}(\frac{p_2}{p_1} - 1)} \tag{23}$$

In the formula, c_1 is the sound speed of the gas before the disturbance (before the shock wave). Since the shock wave pressure p_2 is always greater than the front atmospheric pressure p_1, $\frac{p_2}{p_1} > 1$, according to the above formula, it is easy to see that v_s is greater than c_1. That is, the velocity of the shock wave front relative to the front gas is theoretically calculated to be supersonic. The greater the shock wave strength, the greater the shock wave propagation speed. When the shock wave strength is very weak $\frac{p_2}{p_1} \to 1$, the shock wave velocity v_s is infinitely equal to the unperturbed gas sound velocity.

Defined by the shock wave front Mach number:

$$M_1 = \frac{v_s}{c_1} \tag{24}$$

Derived from Equations (23) and (24):

$$M_1 = \sqrt{1 + \frac{k+1}{2k}(\frac{p_2}{p_1} - 1)} \tag{25}$$

The shock wave front Mach number can be obtained from the pressure difference before and after the shock wave. The shock Mach number can be obtained by Equation (24) and Equation (25). However, in Equation (24), the velocity of the wave front needs to be measured, while in Equation (25), the pressure before and after the wave front needs to be measured. The difficulty and accuracy of pressure measurement are higher than that of velocity measurement, so the shock Mach number is generally calculated directly by Equation (24).

In the calculation of shock wave-related parameters, Mach number M_1 can be considered an earlier determined parameter. Taking the Mach number as the known number, other parameters can be calculated and deduced.

The inverse operation of Equation (25) is obtained:

$$\frac{p_2}{p_1} = \frac{2k}{k+1}{M_1}^2 - \frac{k-1}{k+1} \tag{26}$$

Using this formula, the pressure behind the shock wave can be calculated by the shock wave Mach number.

The local thermodynamic parameters of a point in the flowing gas in fluid mechanics are called static parameters. If the isentropic velocity of the fluid at this point is reduced to zero, it is called the stagnation state, and then the parameter value corresponding to the static parameter becomes the stagnation parameter. Static parameters are real state parameters, and stagnation parameters are parameters based on theoretical assumptions. The sign of the stagnation parameter is generally 0. The stagnation parameters include stagnation enthalpy h_0, stagnation temperature T_0, stagnation pressure p_0, stagnation sound speed c_0 and stagnation density ρ_0. These parameters are also called total enthalpy, total temperature, total pressure and total density.

According to the Bernoulli equation, any two points on a streamline satisfy the formula:

$$h_1 + \frac{{v_1}^2}{2} = h_2 + \frac{{v_2}^2}{2} = const \tag{27}$$

According to the definition of stagnation state, $v_2 = 0$, then:

$$h_0 = h + \frac{1}{2}v^2 \tag{28}$$

It can be seen from the Equation (28) that the total enthalpy is the sum of the enthalpy related to the temperature of each point and the enthalpy related to the dynamic pressure of the point. It essentially represents the total energy per unit mass of gas. The Equation (27)

From the above theory of total temperature, the shock wave process is an adiabatic process, so the total temperature before and after the wave front remains unchanged, so:

$$T_1(1+\frac{k-1}{2}{M_1}^2) = T_2(1+\frac{k-1}{2}{M_2}^2) \tag{41}$$

Combining Equation (40) with Equation (41), we can obtain:

$$\frac{T_2}{T_1} = \frac{(1+\frac{k-1}{2}{M_1}^2)(\frac{2k}{k-1}{M_1}^2-1)}{\frac{(k+1)^2}{2(k-1)}{M_1}^2} \tag{42}$$

From the general gas state equation and Equations (26) and (42), it is concluded that:

$$\frac{\rho_2}{\rho_1} = \frac{p_2}{p_1}\frac{T_1}{T_2} = \frac{k+1}{2}(\frac{{M_1}^2}{1+\frac{k-1}{2}{M_1}^2}) \tag{43}$$

By the continuity Equation (10):

$$\frac{v_1}{v_2} = \frac{\rho_2}{\rho_1} = \frac{2+(k-1)M^2}{(k+1)M^2} \tag{44}$$

In the continuity equation, v_2 is expressed as $v_2 = (v_s - v_b)$. Combining Equation (43) and the sound velocity formula, the expression of the accompanying velocity can be obtained:

$$v_b = c_1\frac{2}{k+1}\frac{M-1}{M} \tag{45}$$

According to the second law of thermodynamics and the definition of entropy:

$$ds = \frac{\delta q}{T} = \frac{de + p\cdot d(\frac{1}{\rho})}{T} = c_V\cdot\frac{dT}{T} + R\frac{d(\frac{1}{\rho})}{(\frac{1}{\rho})} \tag{46}$$

Integral on both sides:

$$s_2 - s_1 = c_V\ln(\frac{T_2}{T_1}) + R\ln(\frac{\rho_1}{\rho_2}) = c_V\ln\left[\frac{(\frac{T_2}{T_1})}{(\frac{\rho_2}{\rho_1})^{k-1}}\right] \tag{47}$$

Substituting the shock temperature ratio Equation (42) and the density ratio Equation (43), we obtain:

$$s_2 - s_1 = R\ln\left[\frac{2+(k-1){M_1}^2}{(k+1){M_1}^2}\right]^{\frac{k}{k-1}} + R\ln\left[\frac{2k{M_1}^2-(k-1)}{(k+1)}\right]^{\frac{1}{k-1}} \tag{48}$$

$M_1 > 1$ in the process of shock wave is substituted into the above formula:

$$s_2 - s_1 \geq 0 \tag{49}$$

Therefore, the entropy value increases after the shock wave of a coal and gas outburst, which is an irreversible process.

3.3. The Attenuation of Outburst Shock Wave Intensity

According to the ideal shock wave theory, when the shock wave source does not attenuate, the ideal gas shock wave front will not weaken with the increase of propagation distance in the process of propagation in the roadway. Even if the shock wave source can only maintain a short time, the propagation of the shock wave in the tunnel may not be affected in a short time, and the specific principle is no longer described. However, the non-attenuation of shock wave is not consistent with the traditional understanding and does not conform to the fact. In the process of real shock wave propagation, due to the influence of viscosity in the moving gas and heat conduction between the roadway, the intensity (or propagation speed) of the shock wave gradually decays, but the attenuation speed is slow [32,33]. In summary, the attenuation of shock waves in an underground ventilation system mainly depends on the bending of facilities such as anti-burst doors or the roadway itself. Some other studies say that there is a linear relationship between the attenuation of the prominent shock wave and the propagation distance of the shock wave, which is not consistent with the facts.

4. Relationship between Gas Pressure and Outburst Shock Wave Intensity

The above part is the analysis of the shock wave propagation process after the occurrence of a coal and gas outburst and does not involve the analysis of the influence of the initial gas pressure of the coal seam on the shock wave intensity. The following will be combined with the above theory for further analysis to discuss the relationship between gas pressure and shock wave formation strength.

As shown in Figure 9, the region division and pressure magnitude diagram of the shock tube at a certain time are shown. There is a left-lateral sparse wave between zone 3 and zone 4. The high-pressure coal-gas flow in zone 4 accelerates to zone 3 through expansion, and its flow is an isentropic flow. The gas flow in zone 3 and zone 4 on the same streamline conforms to the formula:

$$u_3 + \frac{2}{k_3 - 1}c_3 = u_4 + \frac{2}{k_4 - 1}c_4 \tag{50}$$

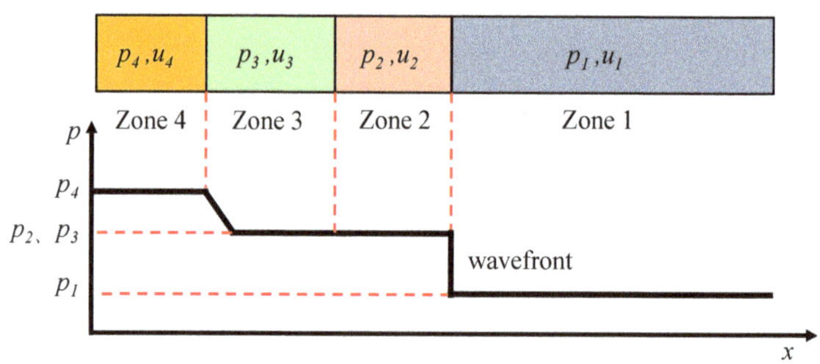

Figure 9. Pressure distribution in shock tube.

In the formula, u_3 and u_4 represent the fluid flow velocity in zone 3 and zone 4, respectively, and zone 4 is the static region, so $u_4 = 0$. k_3 and k_4 represent the specific heat ratio of fluid in zone 3 and zone 4, respectively. Zone 3 and zone 4 are essentially the same substance, so $k_3 = k_4$. Therefore, the above formula can be rewritten as:

$$u_3 + \frac{2}{k_4 - 1}c_3 = \frac{2}{k_4 - 1}c_4 \tag{51}$$

The isentropic flow satisfies the following equation:

$$\frac{p_4}{p_3} = \left(\frac{c_4}{c_3}\right)^{\frac{2k_4}{k_4-1}} \tag{52}$$

The contact surface of zone 2 and zone 3 meets the compatibility relationship:

$$u_2 = u_3, p_2 = p_3 \tag{53}$$

The pressure relationship between zone 4 and zone 1 satisfies:

$$\frac{p_4}{p_1} = \frac{p_4}{p_3}\frac{p_3}{p_1} = \frac{p_4}{p_3}\frac{p_2}{p_1} \tag{54}$$

Using the above formula can be obtained:

$$\frac{p_4}{p_1} = \frac{2k_1M^2 - (k_1 - 1)}{k_1 + 1}\left\{1 - \left(\frac{k_4 - 1}{k_1 + 1}\right)\left(\frac{c_1}{c_4}\right)\left(M - \frac{1}{M}\right)\right\}^{-\left(\frac{2k_4}{k_4 - 1}\right)} \tag{55}$$

This formula represents the relationship between the pressure ratio p_4/p_1 and the shock Mach number M (shock wave intensity). The parameters k_1 and c_1 in the formula are roadway air parameters, which are easy to obtain. k_4 and c_4 represent the equivalent specific heat ratio and equivalent sound velocity of coal-gas flow. They have no known fixed values and can only be obtained through experiments. Because there are many distribution positions of parameter k_4 in Equation (55), it is difficult to unify it. Therefore, this paper sets the parameter as 1.4. The relationship between the theoretical pressure ratio p_4/p_1 and the experimental results is calculated by changing the equivalent sound velocity. Figure 10 shows the ratio of initial gas pressure to atmospheric pressure p_4/p_1 when the ratio of standard atmospheric sound velocity to the equivalent sound velocity of coal-gas flow c_1/c_4 takes different values at Mach number M = 1.5. The inverse calculation of Equation (55) can directly calculate the outburst shock wave M from the initial gas pressure of the coal seam. If the initial gas pressure p_4 of the coal seam and the intensity of the outburst shock wave are known, the equivalent sound velocity of the coal-gas flow under this condition can be calculated.

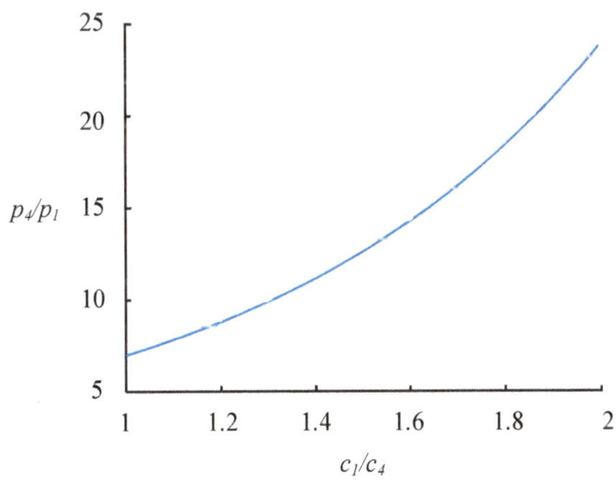

Figure 10. The relationship between equivalent sound ratio and pressure ratio.

It can be seen from Figure 10 that the pressure ratio is very sensitive to the equivalent sound velocity of the coal-gas flow. The pressure ratio is 7 when c_1/c_4 is the same sound velocity in zone 4 and zone 1. When c_1/c_4= 1.5, the pressure ratio is 12.6. When c_1/c_4= 2, the pressure ratio is 24.

5. Conclusions

(1) When outburst occurs, a large amount of broken coal and gas migrates from the coal seam to the roadway under the action of ground stress and gas pressure. It can be divided into stable coal seam area, coal-gas flow area, air compression area and roadway unaffected area. The location of these zones will change with the development of the outburst, and the internal of each region is uneven. The outburst

energy accumulates continuously at the interface between the air compression zone and the unaffected zone of the roadway, forming an outburst wave front, and its formation and propagation satisfies the aerodynamic theory;

(2) Based on the shock wave theory, the shock wave of a coal and gas outburst is studied, and the theoretical model of the shock wave of a coal and gas outburst is established. The propagation velocity of an outburst shock wave is greater than the speed of sound, and much larger than the velocity of airflow. After the outburst shock wave passes through, the pressure, temperature and density of the roadway air will change suddenly. Due to the viscosity of the moving gas and the heat conduction between the roadway and other factors, the strength of the shock wave gradually decays, and the attenuation speed is slow in the straight roadway. The attenuation of a shock wave in an underground ventilation system mainly depends on the bending of facilities such as anti-burst doors or the roadway itself;

(3) According to the shock wave theory, the expressions of outburst gas pressure and outburst shock wave intensity are derived. In order to facilitate understanding and calculation, the concept of the equivalent sound velocity of coal-gas flow is proposed. Under the condition of determined outburst intensity, the initial gas pressure is very sensitive to the equivalent sound velocity. According to the initial gas pressure and the intensity of the outburst shock wave, the equivalent sound velocity can be calculated. Or, under the condition of the known equivalent sound velocity, the intensity of the outburst shock wave can be directly calculated according to the initial gas pressure.

Author Contributions: Investigation, D.S., J.C., L.D., R.L. and Y.L.; Methodology, D.S., J.C. and L.D.; Supervision, R.L. and Y.L.; Writing—original draft, D.S., J.C. and L.D.; Writing—review and editing, D.S., J.C., L.D. and R.L; funding acquisition, D.S., J.C., and L.D. All authors have read and agreed to the published version of the manuscript.

Funding: This research was funded by the National Natural Science Foundation of China (No. 51974358, No. 52104239, No. 51774319), Natural Science Foundation of Chongqing (No. CSTB2022NSCQ-MSX1080, No. CSTB2022NSCQ-MSX0379, No. cstc2021jcyj-msxmX0564).

Data Availability Statement: The data are available from the corresponding author on reasonable request.

Acknowledgments: We also would like to thank the anonymous reviewers for their valuable comments and suggestions that lead to a substantially improved manuscript.

Conflicts of Interest: The authors declare no conflict of interest.

References

1. Yuan, L. Deep coal and gas mining strategic thinking in China. *J. China Coal Soc.* **2016**, *41*, 1–6.
2. Wang, K.; Du, F. Coal-gas compound dynamic disasters in China: A review. *Process Saf. Environ. Prot.* **2020**, *133*, 1–17. [CrossRef]
3. Du, F.; Wang, K.; Zhang, X.; Xin, C.P.; Shu, L.Y.; Wang, G.D. Experimental Study of Coal-gas Outburst: Insights from Coal–Rock Structure, Gas Pressure and Adsorptivity. *Nat. Resour. Res.* **2020**, *29*, 1–13. [CrossRef]
4. Ding, Y.; Yue, Z.Q. Outburst characteristics of CO_2 gas-coal mixture with tunnel outburst simulator in deep mining. *Geomech. Geophys. Geo-Energy Geo-Resour.* **2022**, *8*, 1–23. [CrossRef]
5. Zhang, C.L.; Wang, Y.B.; Wang, E.Y.; Zhou, X.F.; Wang, P.Z.; Zeng, W. Influence of coal seam gas pressure on the propagation mechanism of outburst two-phase flow in visual roadway. *Fuel* **2022**, *322*, 124296. [CrossRef]
6. Zhao, X.S.; Cao, J.; Wang, B.; Yang, X.L. Experiment Study of Outburst Pulverized Coal-Gas Two-Phase Flow and Characteristic Analysis of Outburst Wave. *Geofluids* **2021**, *2021*, 8186230. [CrossRef]
7. Savonko, C.K.; Jilin, A.A., II. *Underground Air Shock Wave*; Metallurgical Industry Press: Beijing, China, 1979.
8. Zhang, J.F.; Wang, K.; Wei, C.P. Formation and Propagation of Shock Waves During Coal and Gas Outburst. *J. Min. Saf. Eng.* **2010**, *27*, 67–71.
9. Sun, D.L.; Cao, J.; Miao, F.T.; Sun, H.T.; Zhao, X.S.; Dai, L.C.; Wang, B. Migration law of outburst coal and gas in roadway. *J. China Coal Soc.* **2018**, *43*, 2773–2779. [CrossRef]
10. Cheng, W.Y.; Liu, X.Y.; Wang, K.J.; Li, X. Study on regulation about shock-wave-front propagating for coal and gas outburst. *J. China Coal Soc.* **2004**, *29*, 57–60.
11. Zhang, S.Z.; Zhang, R.L. Research on Injuries Due to Shock Wave and Gas Flow from Coal and Gas Outburst. *China Saf. Sci. J.* **2012**, *22*, 62–66. [CrossRef]

12. Tang, J.P.; Yu, N.; Chen, S. Impact of the gas pressure on the jet outburst energy of coal and gas. *J. Saf. Environ.* **2017**, *17*, 943–948. [CrossRef]
13. Miao, F.T.; Sun, D.L.; Hu, Q.T. The formation mechanism of shock waves in the coal and gas outburst process. *J. China Coal Soc.* **2013**, *38*, 367–372. [CrossRef]
14. Hu, W.J. *Theoretical and Experimental Research for Gas-Pulverized Coal Impactive Dynamic Effect after Outburst*; University of Mining and Technology: Beijing, China, 2013.
15. Zhou, A.T.; Wang, K.; Wu, Z.Q. Propagation law of shock waves and gas flow in cross roadway caused by coal and gas outburst. *Int. J. Min. Sci. Technol.* **2014**, *24*, 23–29. [CrossRef]
16. Zhou, A.T.; Wang, K. Airflow stabilization in airways induced by gas flows following an outburst. *J. Nat. Gas Sci. Eng.* **2016**, *35*, 720–725. [CrossRef]
17. Zhou, A.T.; Zhang, M.; Wang, K.; Elsworth, D.; Wang, J.W.; Fan, L.P. Airflow disturbance induced by coal mine outburst shock waves: A case study of a gas outburst disaster in China. *Int. J. Rock Mech. Min. Sci.* **2020**, *128*, 104262. [CrossRef]
18. Li, W.X. *One-Dimensional Unsteady Flow and Shock Wave*; National Defense Industry Press: Beijing, China, 2003.
19. Tan, H. *Physical Guidance of Experimental Shock Wave*; National Defense Industry Press: Beijing, China, 2007.
20. Cheng, B.Q. *Gas Dynamics*; National Defense Industry Press: Beijing, China, 2009.
21. John, D.; Anderson, J. *Fundamentals of Aerodynamics*; University of Maryland: College Park, MD, USA, 2009.
22. Wang, G.F. *A Shock Tube Study of Methane Combustion Chemical Kinetics Mechanism*; University of Science and Technology of China: Hefei, China, 2008.
23. Cui, H.T.; Liu, Q.M. Dynamic calibration of shock wave pressure measurement system. *Exp. Meas. Fluid Mech.* **2004**, *18*, 92–96.
24. Zhang, D.Y. The Application of Shock tube in Testing and Calibrating the Performance of Pressure Sensor. *J. Astronaut. Metrol. Meas.* **2004**, *24*, 24–27.
25. Xu, L.; Sun, J.S. SPH Simulation of One-dimensional Shock Tube Problems. *Chin. J. Comput. Phys.* **2003**, *20*, 153–156. [CrossRef]
26. Li, J.P.; Feng, H.; Jiang, Z.L.; Yu, H.R. Numerical computation on the tailored shock Mach numbers for a hydrogen-oxygen detonation shock tube. *Acta Aerodyn. Sin.* **2008**, *112*, 291–296.
27. Davidson, D.; Hanson, R. Interpreting shock tube ignition data. *Int. J. Chem. Kinet.* **2004**, *36*, 510–523. [CrossRef]
28. Gauthier, B.; Davidson, D.F.; Hanson, R.K. Shock tube determination of ignition delay times in full-blend and surrogate fuel mixtures. *Combust. Flame* **2004**, *139*, 300–311. [CrossRef]
29. Duff, R.E. Shock-tube performance at low initial pressure. *Phys. Fluids* **1959**, *2*, 207–216. [CrossRef]
30. Chen, Q. *Theory and Experimental Technique of Shock Tube Flow*; China University of Science and Technology Press: Hefei, China, 1979.
31. Chen, F.; Quan, X.B.; Song, Y.P. *Aerodynamic Basis*; Harbin Institute of Technology Press: Harbin, China, 2015.
32. Mirels, H. *Attenuation in a Shock Tube due to Unsteady-Boundary-Layer Action*; National Advisory Committee for Aeronautics Collection: Washington, DC, USA, 1957.
33. Kitagawa, K.; Yasuhara, M.; Takayama, K. Attenuation of shock waves propagating in polyurethane foams. *Shock Waves* **2006**, *15*, 437–445. [CrossRef]

processes

Article

Overlying Strata Dynamic Movement Law and Prediction Method Caused by Longwall Coal-Mining: A Case Study

Guosheng Xu [1,*], Dehai Li [2], Yanbin Zhang [3] and Huigui Li [1]

1 School of Mining Engineering, Guizhou University of Engineering Science, Bijie 551700, China
2 Ming Research Institute, Henan Polytechnic University, Jiaozuo 454000, China
3 School of Continuing Education, Henan Polytechnic University, Jiaozuo 454000, China
* Correspondence: teesn286@gues.edu.cn; Tel.: +86-15761440314

Abstract: The surface subsidence caused by underground mining is a spatiotemporal process. The impact of mining on surface structures (houses, highways, railways, dikes, etc.) and structures in rock strata (shafts, roadways, chambers, etc.) is a dynamic process. It is necessary to study the dynamic movement law of the surface and overlying strata in the mining process of the working face to predict the extent of the impact of mining on the aforementioned structures. It provides a reference for pre-reinforcement and post-mining treatment. This paper studies the variation of surface dynamic movement based on the survey line above the working face of Peigou Coal Mine. The numerical simulation model of the overlying strata dynamic movement is established to study the dynamic movement law of rock strata with different depths, and the fitting function of surface and overlying strata dynamic movement is determined. Finally, the subsidence velocity prediction function of the major section of the surface and overlying strata in the Peigou Coal Mine is established. The accuracy of this prediction function is demonstrated by contrasting the subsidence and subsidence velocity curves of the surface subsidence basin survey line with the numerical model. In this paper, a numerical simulation method for the dynamic movement of the surface and overlying strata and a function for predicting the subsidence velocity on the strike major section are established, which provides an important theoretical reference for the dynamic protection of the structures on the surface and in the overlying strata.

Keywords: surface subsidence; overlying strata movement; maximum subsidence velocity; log distance of maximum subsidence velocity; dynamic movement

Citation: Xu, G.; Li, D.; Zhang, Y.; Li, H. Overlying Strata Dynamic Movement Law and Prediction Method Caused by Longwall Coal-Mining: A Case Study. *Processes* **2023**, *11*, 428. https://doi.org/10.3390/pr11020428

Academic Editors: Feng Du and Aitao Zhou

Received: 2 January 2023
Revised: 25 January 2023
Accepted: 27 January 2023
Published: 31 January 2023

1. Introduction

The surface subsidence caused by underground mining is a spatiotemporal process. In the mining process, the impact of mining varies for various areas of the surface, and its influence is dominated by the relative position of the working face and the surface area. In production practice, it is nowhere near sufficient to solve the actual problems on site according to the final subsidence law [1,2]. It is often necessary to study the dynamic subsidence and master the variation of subsidence velocity of surface and overlying strata in order to judge the intensity and location of surface and overlying strata movement and deformation, as well as to protect and repair surface structures (houses, dams, roads, and railways, etc.) and structures in overlying strata (chambers, roadways, etc.) [3–5].

The dynamic movement of the surface and overlying strata caused by coal mining has been deeply studied in China and other countries. Considering the changes in stress-strain state of rock mass and coal mass around the goaf and the working face in the process of mining [6–10], there is a close relationship between surface movement and overlying strata movement. Shu [11,12] proposed a theoretical model of the relationship between sub-surface and surface subsidence movements, which can be used to predict sub-surface subsidence components at any point within the zone influenced by the extraction of a panel according to the subsidence components obtained from field measurements. Based on the subsidence data in the Southern Coalfield of the Sydney Coal Basin, the proposed prediction is examined. Amar [2] studied the behavior of dynamic active and residual subsidence for a few panels of Jharia coalfield, and found that the subsidence and slope were linearly related to time. Compressive and tensile strains showed a typical fluctuating characteristic behavior, and the rate of mining being a key and controlling parameter for the rate of subsidence and its inter-relationship was developed. The time effect is studied in the coalfield of Asturias, Northern Spain, with the aim of predicting subsidence phenomena and characterizing the trough in the different intermediate stages of the process of excavation and subsidence. In addition, the subsidence is predicted following the models of Knothe and Sroka–Schober and a new time function based on the normal distribution function [13].

According to the difference of surface subsidence velocity in the process of surface dynamic subsidence, Huang [14] divided the whole process of surface subsidence into three stages: Subsidence development, full subsidence, and subsidence attenuation. The development and variation laws of surface subsidence deformation in these three stages were respectively analyzed, and a new concept of "deformation velocity in surface subsidence" was also put forward. Chudek [15] analyzed the impact of overlying strata lithology on the time influence parameters of surface movement and determined the relationship between these parameters and mining depth according to a large amount of survey data. Knothe [16], a foreign scholar, introduced the time function into the prediction of dynamic surface subsidence. Numerous scholars have conducted extensive studies based on Knothe time function, and the findings have been able to accurately predict the entire process of surface subsidence. Scholars have worked to improve the Knothe time function in light of its drawbacks and defects, and they have provided the ideal distribution form of time function to make up for its shortcomings. Based on the static prediction model of probability integral, Zhang [17–19] discussed the prediction method of dynamic subsidence for the inclined main section using the "optimized segmented Knothe time function". The outcome showed that regardless of which direction the mining begins from or whether the prediction direction is uphill or downhill, the results of dynamic subsidence and deformation of the inclined major section are always consistent with the rules of the theory. Guo [20] analyzed the dynamic surface movement characteristics and parameters affected by fully mechanized mining under thick hydraulic collapse, based on the data from the monitoring station. The results indicate that the surface movement is intensive, the surface subsidence velocity is rapid, and the surface deformation and damages are serious in this mining area with thick loess. Deng [21] used the relationship between the maximum subsidence velocity and the relative position of the working face to calculate the prediction formula of the subsidence velocity at any point and time during the mining process. For monitoring the movement of overlying strata, vertical boreholes were drilled from the ground into the rock stratum, Wang [22] adopted the distributed optical fiber sensor monitoring technology and multipoint borehole extensometer methods to the movement data of the strata. Guo [23] conducted field monitoring of overburden displacement, stress, and water pressure changes at the longwall panel in Anhui, China. A three-dimensional annular-shaped overlying zone along the perimeter of the longwall panel is identified for optimal methane drainage during mining. The above researches mainly studied the process and stage of surface subsidence induced by underground longwall coal mining, and obtained the prediction model of surface dynamic subsidence. However, for the protection of surface and underground structures, the distribution characteristics and prediction of subsidence velocity are crucial. In the existing researches, the impact of mining extraction area on surface dynamic movement characteristics is not considered, which makes it impossible to effectively predict the whole process of the working face mining. Moreover, due to the difficulty of overlying the strata field monitoring experiment, the research on overlying strata movement is mainly a static movement, and the relationship between the dynamic movement of the ground surface and the dynamic movement of the overlying strata is not established. As a result, there are few research results on the dynamic movement law and prediction theory. This paper first examines the relationship of movement and deformation between surface and overlying strata, and then analyzes the variation characteristics of subsidence value and subsidence velocity at the maximum subsidence point along the survey line with different goaf areas. The dynamic relationship between the surface dynamic movement parameters (maximum subsidence velocity and lag distance of maximum subsidence velocity) and the advancing distance of the working face is analyzed. The numerical simulation model of the dynamic movement of overlying strata is established, and the rock mechanics and dynamic excavation parameters are determined. When comparing the subsidence and subsidence velocity curves of the surface survey line and monitoring the line of the numerical model, it is considered that the parameters of the numerical model are accurate and can provide appropriate data for analyzing the movement of overlying strata. The variation functions and lag distance of the maximum subsidence velocity at different cover depths of rock strata during the advancement of the working face are determined based on numerical simulation results. The subsidence velocity prediction function on the major section of the surface and overlying strata subsidence basin is obtained based on the distribution function of surface subsidence velocity with supercritical extraction on the major section and the dynamic subsidence prediction method of the major strike section.

In this paper, based on the surface survey data above the panel in Peigou Coal Mine, the variation characteristics of surface dynamic movement parameters, the maximum subsidence velocity (MSV), and the lag distance of maximum subsidence velocity (LDMSV) are obtained during different advancing distances of the working face. To explore the overlying strata movement law, a numerical simulation model is established, and variation functions of dynamic movement parameters for rock strata with different depths are obtained based on the results of the numerical simulation. The prediction model of subsidence velocity at any time and any position on the strike major section of the subsidence basin is established through the dynamic description of surface and overlying strata movement. The study's findings not only provide guidance for surface and overlying strata structures protection and treatment technologies, but also further enrich the body of knowledge regarding the dynamic movement of the surface and overlying strata.

2. Relationship between Surface and Overlying Strata Movement and Deformation

In the past, the overlying strata movement mainly focused on the strata pressure around the working face, the fluid migration of the overlying strata aquifer, the mining induced fracture evolution, and relieved methane delivery. These factors form an important basis for addressing a number of safety and environmental issues, such as ground pressure accidents, water and gas disasters, surface subsidence, etc. [24]. Studies on surface subsidence focus on environmental, facility, life, and ecological impacts. However, there is a link between overlying strata and surface subsidence, and mining excavation will inevitably cause the movement of the overlying strata. Considering the key layer as the main body of rock strata movement research, mechanical methods are used to explain changes in the structural form, stress field, and mining-induced fracture field of rock strata. Therefore, there is a relatively unified understanding and complete mechanical description of ground pressure, mining subsidence, and water and gas migration in mining-induced rock strata [25].

The relationship between surface subsidence and rock strata movement was studied by similarity simulation and numerical simulation tests. Adhikary [26] established a practical hydrogeological model based on the large coal seam sub-critical longwall panel in Australia, and stated that the rock collapse and fracture zones are trapezoidal large structures similar to the "inverted funnel" arch. Zuo [27] used the mechanical model to explain the complex mechanical relationship between strata movement and surface subsidence based on the hyperbola-like shape of the broken boundary of strata movement. The boundary of surface subsidence and strata movement was unified as a whole, and an overall strata movement model that can describe the movement of overlying strata to surface subsidence was established. The movement and deformation of overlying strata caused by coal mining are a mechanical phenomenon in rock mass. Due to the presence of rock mass and the complexity of mining boundary conditions, it is difficult to develop a complete mathematical model and its parameters in order to express the overlying strata damage and movement deformation caused by coal mining. According to the laws of strata subsidence and displacement at different buried levels [28,29], as shown in Figure 1, it is assumed that the subsidence curve of different strata considers the projection line of coal wall as the inflection point, and the relationship between overlying strata and surface movement and deformation lies in the following:

(1) The formation of goaf space will cause the bending deformation of the lower rock strata. After the upper rock strata lose support, their weight and the load of the overlying strata will be transferred to the top of the coal pillar, forming an incremental stress (called abutment pressure) that is greater than the original stress and acting on the lower coal and rock mass to cause compression deformation. The subsidence value of the strata outside of the inflection point is caused by the compression deformation of overlying strata above the coal pillar, which accumulates layer by layer from the coal seam to the ground surface, and shows a gradual increasing trend as the distance between the strata and the coal seam increases.

(2) Although the fractured rock above the goaf is recomputed by the upper pressure, the rock strata subsidence coefficient from the goaf to the surface gradually decreases due to the plastic expansion of the caved zone and fractured zone above the goaf and the elastic expansion effect of the continuous deformation zone [25]. Due to the weakening of the mining influence, the subsidence coefficient decreases more slowly over time.

(3) Considering the cantilever effect of the strata at the side of the coal wall, the inflection point of the subsidence curve (the subsidence value is 0.5 W_0) is projected onto the calculation boundary on the coal seam. The distance between the actual mining boundary and the calculation boundary becomes the offset of inflection point. The surface and the strata outside of the inflection point produce tensile deformation in the horizontal direction, and the position of the maximum displacement is located at the inflection point. On the other hand, compressive deformation of the surface and upper strata occurs in the horizontal direction within the inflection points on both sides. However, for the lower strata, they are compressed both horizontally and vertically outside of the inflection point. Within the inflection point, they are subject to tensile deformation in the horizontal direction due to the accumulative effect of broken rock mass. From the perspective of rock strata mining influence range, a subsidence basin that is significantly larger than the goaf area appears on the surface with the expansion of the damage range of the rock strata.

In summary, due to the complex mechanical behavior of the mining-induced rock mass, rock strata movement is closely related to the surface movement. The surface subsidence basin is the external manifestation of the rock strata movement, and the rock strata movement is the fundamental reason for the surface subsidence. Due to the complexity of the occurrence of the rock mass itself and the mechanical boundary condition, it is impossible to fix many in-situ monitoring points for the movement of boreholes in the overlying strata as the surface survey stations [30,31]. Therefore, based on the measured surface data, this paper establishes a numerical simulation model of overlying strata dynamic movement and studies its law. Combined with the aforementioned information, the dynamic movement prediction model of the surface and overlying strata is established.

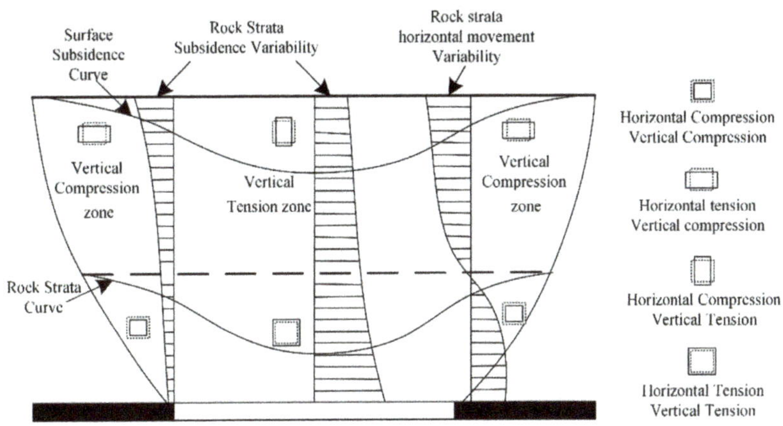

Figure 1. Movement and deformation characteristics of the overlying strata.

3. Surface Dynamic Movement Characteristics

3.1. Geological Setting and Survey Lines Plan

The Peigou Coal Mine is located in Zhengzhou mining area of the Henan Province, China (Figure 2a,b). The panel 31,071 has a width of 130 m, a strike length of 1100 m (Figure 2c), an average mining height of 7.5 m, an average incline angle of 15°, and an average burial depth of 300 m. The mining method is a fully mechanized top-coal caving. The roofs of the mining coal seam are mainly fine sandstone and sandy mudstone. The sandstone fissure aquifer of lower Shanxi formation is the upper main aquifer in the local coal mine, and this aquifer is widely distributed in the entire coal mine with an average thickness of 37.13 m. The distance between the bottom interface of this aquifer and the coal seam is approximately 6 m. The aquifer consists of sandstones with medium and coarse grain sizes. Moreover, fractures have been found in certain local regions of the aquifer. In general, the aquifer in Shanxi formation exhibits poor permeability and a weak water yield property. However, there is a reservoir called Modongwang Up Neath (panel 31071), where the risk of reservoir water inrush is high. In order to accurately evaluate the impact of mining damage on Modongwang reservoir and its dam, a surface movement survey station is established on the surface above the panel. Before mining, movement survey lines in the strike and the inclined directions are arranged on the surface above the panel (Figure 2c).

3.2. Subsidence and Subsidence Velocity Curves of Survey Station with Maximum Subsidence

Figure 3 shows the variation curves of the subsidence velocity and subsidence value at the survey station QZ in the surface strike survey line. Due to survey error, surface subsidence also occurs when the working face is distant. Therefore, this paper mainly considers the increase in subsidence velocity as the basis for judging the beginning of the mining influence. When the distance between the working face and the survey station is about 54 m, this station begins to subside. It can be determined that the working face's leading distance is 54 m, and the advance effect angle is equal to 79.27°. When the working face passes over this station, the subsidence velocity increases gradually, and then at a distance of 76 m, the subsidence velocity reaches its maximum value of 38.36 mm/d. The surface position subsides most violently at this point. Then, the subsidence velocity gradually decreases, and the subsidence curve of the maximum subsidence position essentially flattens out. When the working face passes over this station by roughly 296 m, the subsidence velocity of this station is 1.67 mm/d, the active phase of the station subsidence ends, and the recession phase starts. Then, the subsidence velocity of this station slows down, and its subsidence value becomes stable at 2140 mm.

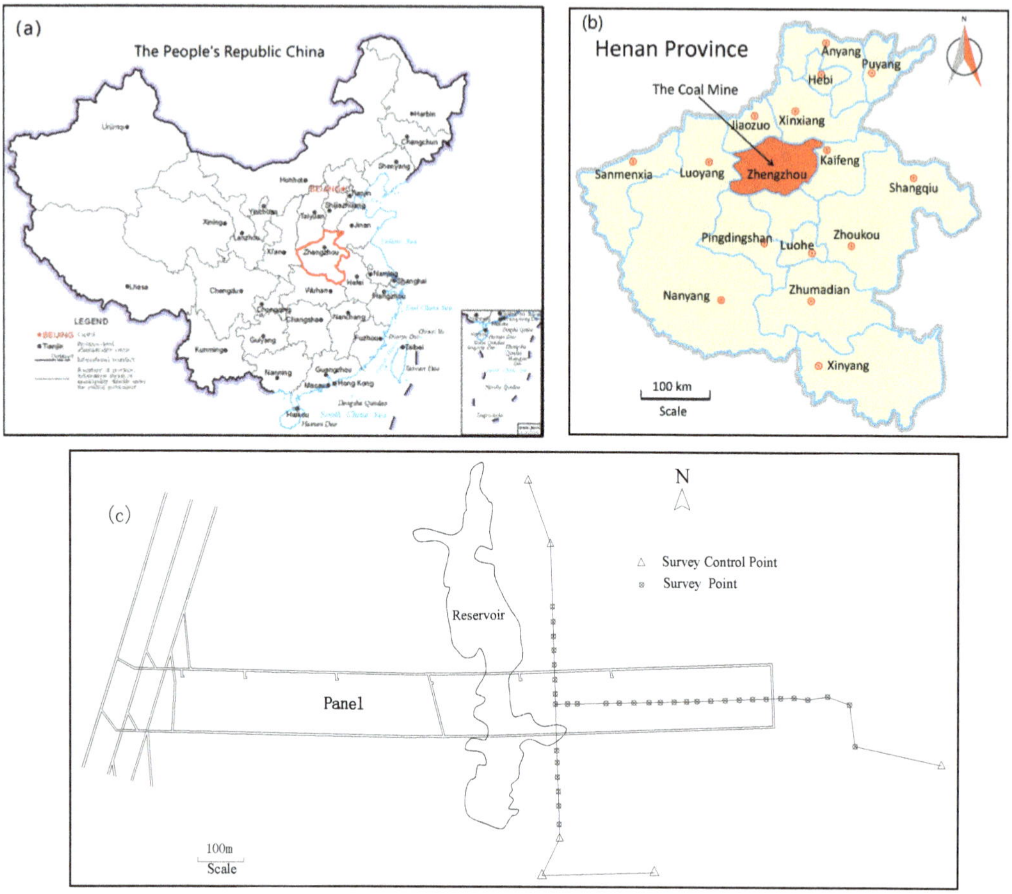

Figure 2. Location of survey lines, (**a**) map of the People's Republic of China, (**b**) location of the Peigou Coal Mine, (**c**) position relationship between the panel and survey lines.

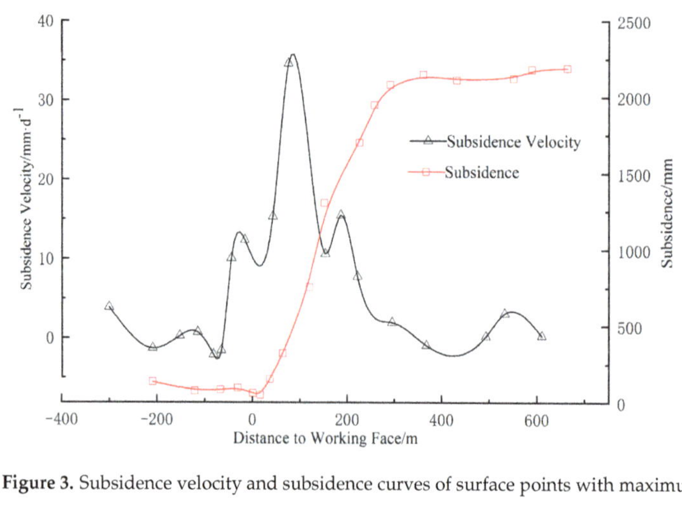

Figure 3. Subsidence velocity and subsidence curves of surface points with maximum subsidence.

3.3. Variation of Surface Dynamic Movement Parameters in the Working Face Mining Process

1. The MSV of surface subsidence basin

To express the influence induced by underground mining to the surface with different mining area sizes, subsidence velocity is considered as an index to measure the severity of mining influence. The relationship between the maximum subsidence velocity on the major section of the surface subsidence basin and the advancing position of the working face is depicted in Figure 4.

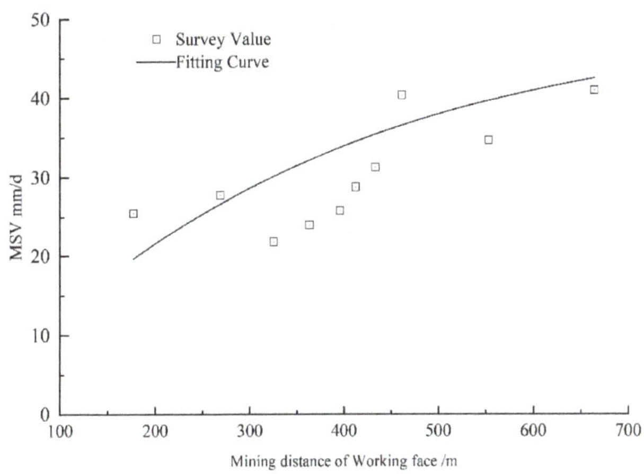

Figure 4. Maximum subsidence velocity (MSV) curve of the surface survey line.

Based on the Knothe time function of the surface dynamic subsidence, the MSV(V_{max}) function formula is obtained by curve fitting:

$$V_{max} = 50.40(1 - e^{-0.0028d}) \quad (1)$$

In Equation (1), d is the mining distance of the working face, m.

It can be seen from Figure 4 that with the advancement in the working face, the goaf area increases, and the influence of underground mining on the surface intensifies. As a result, the MSV on the major section of the surface trend gradually increases. When the mining distance of the working face reaches 663.8 m, the V_{max} (MSV) of surface survey station S9 increases from 0 to 45.43 mm/d. However, when the mining distance exceeds 400 m (i.e., 1.4 times the mining depth), it reaches supercritical mining in the strike direction, and the increase amplitude of the MSV station gradually decreases, reaching a stable value of 50.40 mm/d.

2. Variation of LDMSV

On the surface subsidence velocity curve, the position of the MSV always lags behind the working face by a certain distance. This phenomenon is called the lag phenomenon of MSV. Knowing the LDMSV of the surface is useful in identifying the area where the surface moves violently during the working face mining process and the time when the maximum subsidence velocity occurs, which is significant as a guide for the protection of surface buildings. It is well known that when the working face starts mining from the open-off cut, each station on the surface undergoes a process from subcritical mining to supercritical mining, and LDMSV will be a dynamic process. Therefore, studying the variation of LDMSV can dynamically determine the area with the most violent surface movement in the mining process.

To describe the variation of LDMSV on the strike major section of the surface subsidence basin during the mining process, the nonlinear functional relationship between the LDMSV(L) and the mining distance is obtained by curve fitting:

$$L = 95.60(1 - e^{-0.11d}) \tag{2}$$

Figure 5 illustrates the relationship between the LDMSV and the mining distance as an exponential function. The LDMSV significantly increases with the mining distance before the latter is advanced to 400 m. When the working face reaches supercritical mining, the LDMSV gradually flattens, increases to a certain extent, and then essentially does not increase, maintaining stability at 95.50 m; namely, the lag angle of MSV is 71.47°. It shows that in the mining process in this coal mine, the proportion of mudstone and sandstone in the overlying strata is large, and the overlying strata structure is weak. As the goaf area increases, the strata movement is quickly transmitted to the surface, resulting in a large lag angle for the MSV on the surface.

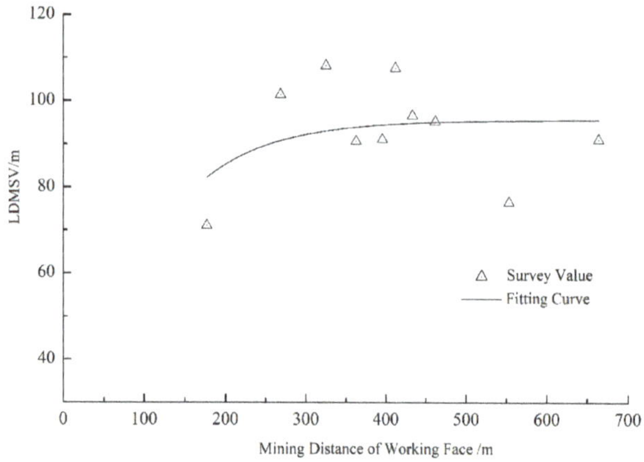

Figure 5. Dynamic curve of lag distance of maximum subsidence velocity (LDMSV).

4. Numerical Simulation of Dynamic Movement of Overlying Strata

4.1. Establishment of Rock Strata Movement Numerical Model

The strike section of the panel is selected as the calculation object in the numerical model. The dimensions of the model are 350 × 1000 m, the mining distance is 440 m, and the coal pillars of the numerical model are 275 m and 285 m, which can ensure that the left and right boundaries of the model are outside of the range of the influence angle of strata movement and avoid the influence of boundary conditions on the movement caused by the excavation. As for the boundary conditions of the model, the left and right sides of the model have roller constraints, the bottom has full constraints, and the top has free boundary conditions. The input parameters of the numerical model rock mass are repeatedly checked by adjusting the geological strength index (GSI) [32–34], in order that the numerical model can truly reflect the strata movement. These parameters are obtained by the inversion method. The physical and mechanical parameters of coal and rock mass and the parameters related to Hoek-Brown strength criterion are shown in Table 1.

Table 1. Mechanical property parameter of rock and coal in numerical model.

Position	lithology	Density/ (kg·m^{-3})	Elastic Modulus/GPa	Cohesion/ MPa	Internal Friction Angle/°	Uniaxial Compressive Strength/ MPa	Tensile Strength/ MPa	GSI	m_i
Roof	Mudstone	2300	2.97	1.13	21.56	35	0.07	38	5
	Sandy Mudstone	2300	3.98	1.78	23.49	50	0.09	40	6
	Sandstone Interbedding	2400	4.09	1.63	25.26	42	0.07	42	7
	Medium Grained Sandstone	2550	6.91	4.14	30.53	85	0.11	45	12
	Fine Grained sandstone	2550	7.50	4.87	29.04	106	0.16	45	10
Coal Seam	Coal	1400	1.58	0.88	24.60	25	0.01	30	10
Floor	Mudstone	2300	2.97	1.13	21.56	35	0.07	38	5
	Medium Grained Sandstone	2300	6.91	4.14	30.53	85	0.11	45	12
	Limestone	2500	5.71	3.82	27.00	92	0.12	41	9

4.2. Verification of the Numerical Model of Strata Movement

The rationality of the numerical model directly determines the accuracy of the numerical simulation analysis. In this paper, the accuracy of the numerical model is assessed through the comparative analysis of the survey data of the surface movement survey line, the theoretical calculation value of the abutment pressure, and the theoretical calculation value of the stress-recovery distance in the goaf with the numerical simulation.

(1) The uniaxial compressive strength of intact rock sample σci is 25 MPa, the GSI is 30, and the coal mechanical parameters are obtained by Willson's theoretical formula [35].

As shown in Figure 6, the peak abutment pressure of the coal seam is 17.50 MPa, the stress concentration factor K is 2.49, the distance between the peak abutment pressure and the coal wall is 16.21 m, and the influence distance before the peak abutment pressure is 65.86 m. According to the numerical simulation, the abutment pressure peak value is located 15 m near the front of the coal wall, and its peak value is 18.52 MPa. The errors of peak stress value and peak position are 5.51% and 8.06%, respectively. Moreover, according to the fact that the ground pressure of the headentry and tailentry of the panel is apparent in the area of 14~17 m, the correctness of the numerical model for the calculation value of abutment pressure is verified.

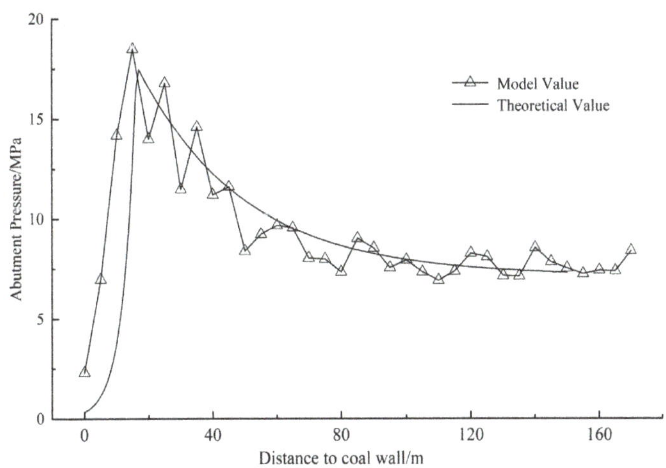

Figure 6. The distribution of abutment pressure ahead of the working face.

(2) According to the numerical model stress monitoring in the goaf, it is found that the stress in the goaf at 101.5 m behind the coal wall reaches the original rock stress, and the corresponding stress-recovery distance in the goaf is 101.5 m (i.e., 0.6 times the mining depth), conforming to the laws of domestic and foreign experience, which is the mining depth of 0.3~0.4 times the stress-recovery distance in the goaf.

(3) The parameters of the surface probability integral method are obtained based on research on the strike survey line over the panel: The subsidence coefficient q = 0.80, the tangent of influence angle tan β = 1.88, and the offset of inflection point is 0.15 H, where H is the average depth of coal seam.

The subsidence data of the surface are extracted with the mining distance of 440 m, and the numerical model subsidence curve is obtained and used for comparison with the data calculated by the probability integral method. The subsidence curves of the measured data and numerical model are shown in Figure 7. It can be surveyed that the subsidence coefficient q is 0.80 with supercritical mining. The maximum subsidence obtained by the numerical model is 5.78 m, i.e., the subsidence coefficient is 0.77, and thus the error rate is 3.66%. It can be considered that the measured results and the numerical model results are consistent in terms of accuracy and distribution law. This shows that the numerical model can accurately inverse the rock strata movement and can be used as an effective means to study the law of overlying strata and surface movement [36].

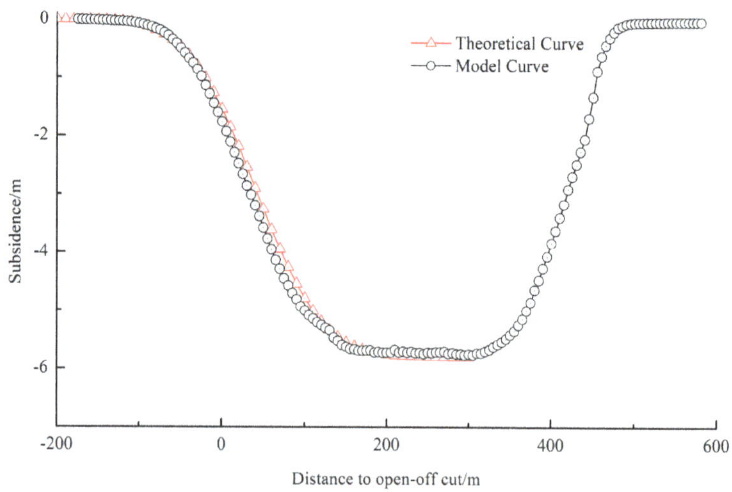

Figure 7. Contrast subsidence curves between the survey data and the numerical model.

4.3. Determination of Dynamic Excavation Parameters for the Numerical Model

In the discrete element numerical simulation software, the calculation of the model is mainly controlled by two different methods [37]. The first method is by setting the number of the software's operation cycle steps, and stopping the operation before moving on to the next stage. Moreover, the operation will stop or move on to the next stage if the maximum unbalance force ratio (the ratio of the maximum unbalanced force to the initial unbalanced force) is less than the default value even if the number of calculation steps has not reached the set cycle steps. The second method is by setting the maximum unbalanced force ratio (MUFR). When the maximum unbalanced force ratio calculated by the software is less than the aforementioned set value, the operation will stop or move on to the next stage. The calculation principle is shown in Figure 8.

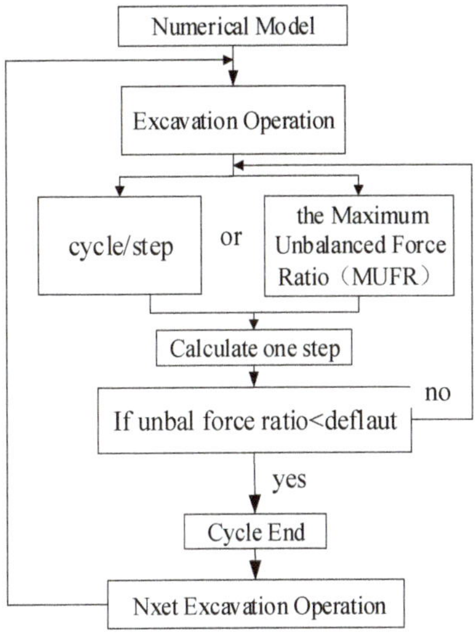

Figure 8. The procedure of continuous excavation in the numerical modeling.

For the dynamic representation of coal seam excavation, a certain excavation step is generally set, and a certain cycle step is given, or the value of MUFR is changed. When the calculation is completed, the model enters the next excavation operation. However, in the continuous excavation, the deformation and failure of the overlying strata are a constantly changing process. It is necessary to set appropriate excavation steps and cycle steps (or MUFR) to accurately simulate an actual mining operation. Therefore, it is necessary to check the parameters used in the numerical simulation through relevant data measured on site, in order that the numerical simulation can be used as a reliable means to study the movement of the surface and overlying strata.

(1) Determination of excavation step

Kbj-60 III rock pressure observation system is installed in Peigou Coal Mine to accurately grasp the ground stress and other data of roof strata. During the mining, the monitoring system continuously logs the stress of the hydraulic support for the underground working face in the support extension. The extension data are gathered by specialized personnel and uploaded to a specialized computer software. Based on the statistical data of the mine pressure monitoring system during the mining, it is known that the average periodic stress step distance of the panel is 18.5 m, and the numerical simulation excavation step distance is determined to be 20 m.

(2) Setting of MUFR

According to the schematic diagram of the dynamic calculation scheme in Figure 9, an initial value is first set, and the numerical simulation calculation is compared with the measured surface subsidence curve. Then, the maximum unbalanced force ratio is continuously adjusted in order that the numerical simulation results are consistent with the field-measured results. It is found that when the excavation step is 20 m, the maximum unbalanced force ratio is 4.8 e^{-5}. After the numerical simulation of dynamic excavation, the dynamic subsidence data of the surface during the mining process are obtained, and the subsidence curve of survey station Z8 and the corresponding numerical model positions are drawn, as shown in Figure 10. The subsidence curves of the survey stations on the surface and the corresponding points in the numerical model are very similar. The relationship curve between the survey station subsidence and mining distance presents an asymmetric "s" shape with a small front end and a large rear end. The two curves can show three stages of dynamic development of the surface.

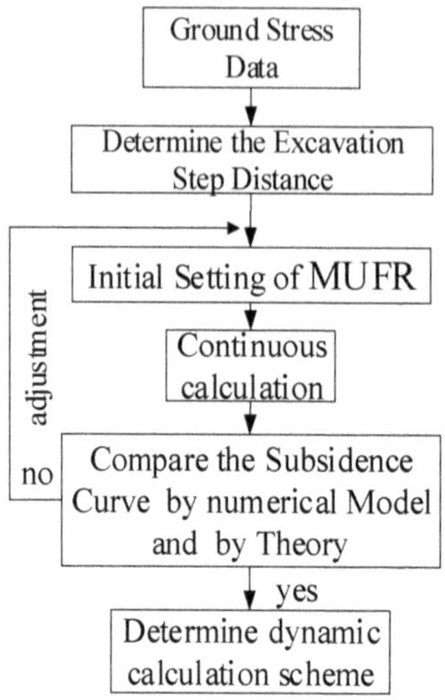

Figure 9. The dynamic calculation scheme.

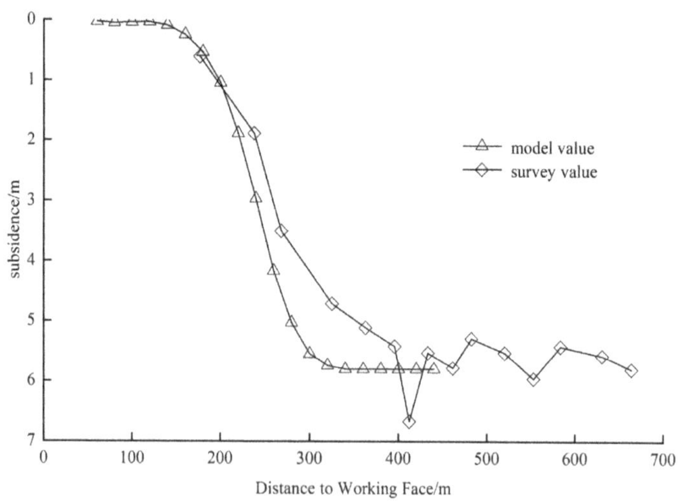

Figure 10. The subsidence velocity between the numerical modeling and the survey data.

Figure 11 illustrates a comparison of measured data and numerical simulation results of subsidence velocity. It can be noticed that when two adjacent survey stations with large measurement error are excluded with a mining distance of about 400 m, the curve of subsidence velocity obtained by measurement is similar to the one obtained from the numerical model in curve shape, the MSV on the curve are 124.32 mm/d and 132.78 mm/d, and the relative error rate is 6.4%. This demonstrates that the numerical model can accurately calculate the surface dynamic movement and can therefore be used to study the dynamic movement of the surface and overlying strata.

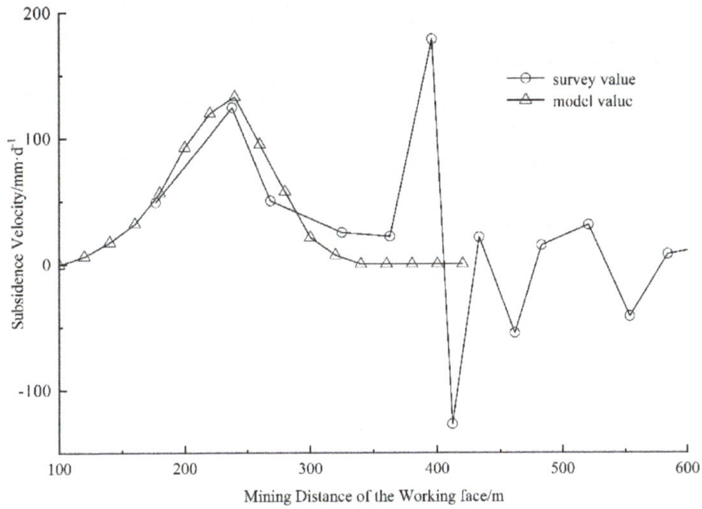

Figure 11. Surface subsidence velocity between the numerical modeling and the survey data.

4.4. Dynamic Movement Characteristics of Overlying Strata

(1) Variation of overlying strata MSV

Considering the surface subsidence curve of the numerical model as an example, the MSV of the surface is fitted according to the analysis method of the surface dynamic movement, as shown in Figure 4. The functional relationship between the surface MSV and the mining distance is obtained:

$$V_{\max} = 137.96(1 - e^{-0.0083d}) \tag{3}$$

The subsidence data of rock stratus are then selected to study the dynamic movement, which are 250 m, 210 m, 170 m, 130 m, 90 m, 50 m, and 10 m away from the coal seam. Due to its discreteness, the MSV of rock strata under different mining distances is fitted and defined by referral to the form of the relationship function between the MSV of the surface and the mining distance (Equation (3)):

$$V^i_{\max} = V'(1 - e^{-bd}) \tag{4}$$

In Equation (4), $V^i{}_{\max}$ is the MSV of the strata with the i depth, mm/d; V 'is the stable value of the MSV, mm/d; and b is the growth coefficient.

By substituting the MSV of different depths of strata in the mining process into Equation (4), the parameter of the dynamic change function of the MSV of strata is obtained as shown in Table 2.

Table 2. The MSV parameter of the overlying strata.

Distance to Coal Seam (m)	300	250	210	170	130	90	50	30
V' (mm/d)	137.96	139.04	141.48	162.45	165.5	188.41	231.3	261.58
b	0.0083	0.0085	0.0093	0.0083	0.0093	0.0095	0.0127	0.01

The MSV of different depths of strata gradually increases with the mining distance. When the working face advances to a certain distance, the MSV of strata tends to have a fixed value. It can be seen from Table 2 that the MSV of the rock strata gradually increases with the distance, which is between the rock strata and the coal seam. The MSV of the surface increases from 137.96 to 261.58 mm/d when the distance from the coal seam is 30 m. To express the relationship between the MSV after the rock strata is stabilized and the distance between the rock strata and the coal seam, the MSV of the rock strata in Table 2 is nonlinearly fitted. In addition, the fitting formula (Equation (5)) is obtained, as well as the variation curve of the MSV at various distances between the rock stratus and the coal seam (Figure 5):

$$V'_{\max}(z) = V_0(\frac{H_0 - z}{H_0})^{-0.30} \tag{5}$$

In Equation (5), z is the depth of the rock strata, m; H_0 is the depth of the coal seam, m; and V_0 is the MSV of the surface, mm/d, i.e., 132.75 mm/d.

It can be seen from Table 2 that the growth parameter B in the fitting parameter of the MSV of the rock strata increases with the decrease in the distance from the coal seam, indicating that when the MSV of the rock strata close to the coal seam is stable, the mining distance is small. According to the curve fitting of the above data, the relationship between the growth parameter b and the depth of rock strata z is obtained:

$$b(z) = 0.008(\frac{H_0 - z}{H_0})^{-0.12} \tag{6}$$

Substituting Equations (5) and (6) into Equation (4), the relationship between the MSV of rock strata and the mining distance is obtained:

$$V^i = V_0(\frac{H_0 - z}{H_0})^{-0.30}[1 - e^{-0.008(\frac{H_0 - z}{H_0})^{-0.12}d}] \tag{7}$$

(2) Variation of overlying strata LDMSV

Due to the different distances between the rock strata and the coal seam, the mining-induced impact and the dynamic movement of the strata are also different. This paper selects the numerical calculation data with strata depths of 250 m, 210 m, 170 m, 130 m, 90 m, 50 m, and 10 m in order to study the variation of the LDMSV of rock strata in the mining process.

Similar to the form of Equation (4), L' is defined as the LDMSV of rock strata, and the parameter c represents the growth trend of the LDMSV with the mining distance, as shown in Table 3.

Table 3. The LDMSV parameter of the overlying strata.

Distance to Coal Seam (m)	300	250	210	170	130	90	50	30
L' (mm/d)	79.51	76.53	59.92	55.13	55.13	49.89	45.68	40.36
c	0.014	0.015	0.014	0.013	0.013	0.014	0.011	0.013

When the distance between the rock strata and the coal seam gradually decreases, L' gradually decreases; namely, from 79.51 to 40.36 m when the distance between the rock strata and the coal seam is 30 m. However, L' of the rock strata has little change with the growth parameter c, which is between 0.011 and 0.015, and its average value can be considered as 0.013. Similarly, the stable value $L'_{\max}$ of the lag distance from the MSV of the rock strata can be determined as shown in Equation (8). Therefore, the calculation formula of L' with different mining depths can be obtained (Equation (9)):

$$L'_{\max}(z) = 41.84\frac{H_0 - z}{H_0} + 36.32 \tag{8}$$

$$L^i = (41.84\frac{H_0 - z}{H_0} + 36.32)(1 - e^{-0.013d}) \tag{9}$$

5. Calculation Method of Subsidence Velocity

5.1. Calculation Method of Subsidence Velocity in Strike Major Section of Surface Subsidence Basin

According to the variation of the subsidence velocity of the survey stations, the study in [13] shows that the distribution curve of the surface subsidence velocity is similar to the "quadratic curve" distribution. If the projection position on the surface at a certain time is considered as the coordinate origin, the direction of the working face on the strike is the positive direction of the x axis, and the subsidence velocity of the station is the y axis. Therefore, the relative relationship between the subsidence velocity curve of the ground surface and the working face position can be expressed. The subsidence velocity formula at one position on the strike major section is:

$$V(x) = \frac{V_{\max}}{1 + (\frac{x+L}{a})^2} \tag{10}$$

In Equation (10), x is the distance to the working face, a is the morphological parameter, which indicates the steepness of the curve.

For the solution of the shape parameter a, this paper assumes that the a calculation point on the surface starts when it is significantly distant from the working face, and its subsidence tends to have a stable value when the working face pushes over significantly away from the point. The following results are obtained:

$$\Delta W = \int_{+\infty}^{-\infty} \frac{V_{\max}}{1 + (\frac{x+L}{a})^2} d(-\frac{x}{c}) \tag{11}$$

Then:

$$a = \frac{W_{\max}}{V_{\max}} \cdot \frac{c}{\pi} \tag{12}$$

In Equation (12), $W_{\max}$ is the maximum surface subsidence and can be obtained by $W_{\max} = qm\sqrt[3]{n_1 n_2}\cos\alpha$; m is the thinness of coal seam, n_1 and n_2 are respectively equal to the ratio of mining size and coal seam burial depth on the strike and incline direction, and c is the average mining velocity of the working face, m/d. Moreover, $V_{\max}$ can be calculated according to Equation (1).

From Equation (12), it can be seen that when the mining velocity of the working face is constant, the value of the shape parameter a changes with the change in the maximum subsidence and maximum subsidence velocity values. Moreover, it can be judged that before reaching the supercritical mining, the shape parameter a and parameter L of the subsidence velocity curve will change with the increase in the goaf area, and the change law can be obtained from the relationship between the MSV, LDMSV, and mining distance. After a mining distance of 400 m, the maximum subsidence $W_{\max}$ no longer increases due to supercritical mining in the strike direction of the working face, while the MSV of Equation (1) increases, but the increasing range gradually decreases. Therefore, it can be considered that the parameter a of the subsidence velocity curve increases slightly after supercritical mining, and the shape of subsidence velocity curve becomes steeper.

In summary, the variation of the subsidence velocity curve with the mining process can be described as follows: With the increase in the goaf area, the MSV gradually increases, the subsidence velocity curve shape gradually becomes steep, and the LDMSV gradually increases. When the working face reaches supercritical mining; namely, the working face mining distance exceeds 400 m, the subsidence velocity curve of the strike major section maintains a certain lag distance with the working face in a fixed form and moves forward with the mining. Figure 12 shows the relative position of the subsidence and subsidence velocity curves of the strike major section of the surface, when the working face mining distance is 200m, 400m, 600m, and 800 m, respectively.

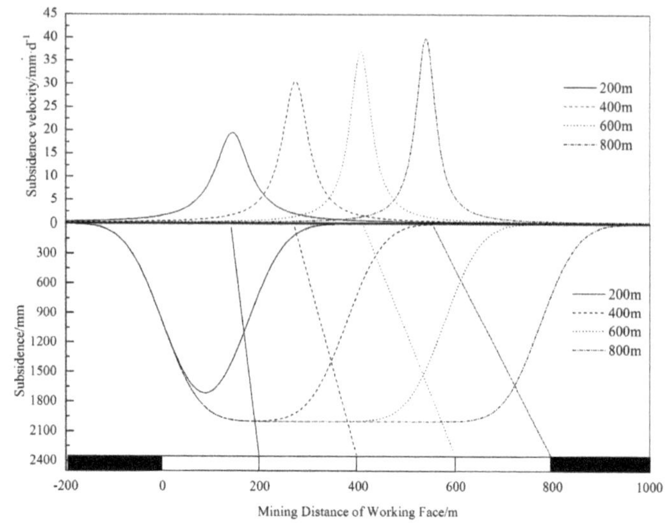

Figure 12. Surface subsidence and subsidence velocity relative locations during mining.

The mining and geology parameters of the panel of Peigou Coal Mine are brought into Equations (1), (2), and (12), and the formula for calculating the subsidence velocity at any position on the strike major section at any time of the working face is obtained by bringing in Equation (10):

$$V(x,d) = \frac{50.40(1-e^{-0.0028d})}{1+\left(\frac{x+95.60(1-e^{-0.11d})}{qm\sqrt[3]{n_1 n_2}\cos\alpha\cdot c}\cdot 50.40(1-e^{-0.0028d})\cdot\pi\right)^2} \quad (13)$$

When the working face mining distance is equal to 360 m, the subsidence velocity calculation points on the strike major section of subsidence basin at different mining distances are obtained according to Equation (13) and compared with the measured values, as shown in Figure 13.

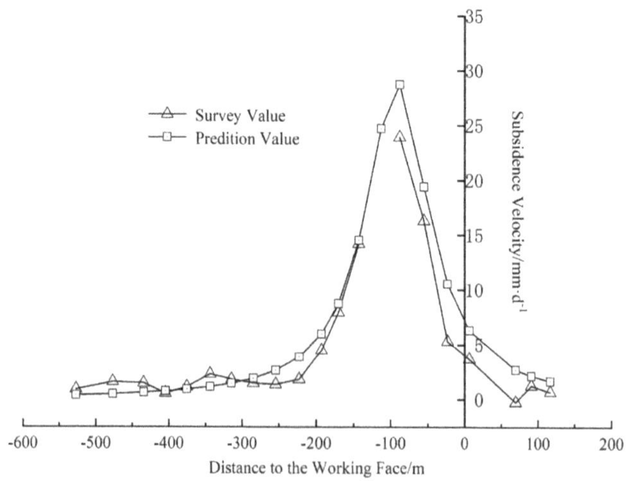

Figure 13. Surface subsidence velocity curve while d = 360 m.

It can be seen from Figure 13 that when the mining distance of the working face is 360 m, the deviation between the measured and calculated values of the subsidence velocity at each point is small, and the average error is 1.57 mm/d, indicating that the predicted results of dynamic subsidence velocity can meet engineering needs.

5.2. Prediction Model of Overlying Strata Dynamic Movement

For the correction of the LDMSV of rock strata, a correction factor a_L can be added before Equation (9), and its value is equal to $Ls/L'(z) \mid z = H_0$. Therefore, the calculation formula of LDMSV of overlying strata based on the measured LDMSV of the surface is obtained:

$$L^i = a_L(41.84\frac{H_0 - z}{H_0} + 36.32)(1 - e^{-0.013d}) \quad (14)$$

Assuming that the subsidence velocity distribution curve on the major section of the surface and overlying strata is similar to the form of a quadratic curve during the working face mining, referring to the subsidence velocity function of the strike major section of the surface (Equation (10)), the subsidence velocity formula of the overlying strata is:

$$V^i(x) = \frac{V^i{}_{\max}}{1 + (\frac{x+L^i}{a^i})^2} \quad (15)$$

According to Equation (11), the parameter a of the subsidence velocity curve will change with the increase in mining distance. When the working face moves from subcritical mining to supercritical mining, the shape of the subsidence velocity curve becomes steeper. The shape parameter a^i of the distribution function of rock strata subsidence velocity is mainly calculated according to the maximum subsidence value $W^i{}_{MAX}$, the $V^i{}_{\max}$ (MSV), and the working face mining velocity c of the rock strike under this level. The maximum subsidence velocity $V^i{}_{\max}$ and the rock strata can be referred to Equation (5), and the maximum subsidence $W^i{}_{MAX}$ of the rock strata can be determined according to the mining distance and the distance between the rock strata and the coal seam; namely:

$$W^i_{\max} = Mq_i \sqrt[\beta]{n^i_1 n^i_2} \cos\alpha \quad (16)$$

In Equation (16), for rock strata i, the mining coefficients $n^i{}_1$ and $n^i{}_2$ of the working face are equal to $\eta D/(H_0\text{-}z)$, where η is a coefficient generally considered as 0.7~0.9, D is the penal length of incline or strike (m), and z is the depth of rock strata i (m).

Therefore, Equations (7), (14), and (16) are combined with Equation (12) and brought into Equation (15) to obtain the calculation formula of subsidence velocity of the surface and overlying strata on the strike major section at different mining distances:

$$V(x,d,z) = \frac{V_S(\frac{H_0-z}{H_0})^{-0.30}[1 - e^{-0.008(\frac{H_0-z}{H_0})^{-0.12}d}]}{1 + \left\{\frac{x+a_L(41.84\frac{H_0-z}{H_0}+36.32)(1-e^{-0.013d})}{Mq_i\sqrt[3]{n^i_1 n^i_2}\cos\alpha\cdot c} \cdot V_S(\frac{H_0-z}{H_0})^{-0.30}[1 - e^{-0.008(\frac{H_0-z}{H_0})^{-0.12}d}] \cdot \pi\right\}^2} \quad (17)$$

According to the above analysis of the MSV and the LDMSV on the strike major section of the surface in Peigou Coal Mine, it is known that in Equation (17), V_0 is equal to 50.40 mm/d, and the correction coefficient a_L of the LDMSV is equal to 1.22. Bringing the above parameters into Equation (17) results in the calculation formula of the subsidence velocity of the surface and overlying strata. When the mining velocity is 2.3 m/d and the mining time is selected as 20d, 40 d, 60 d, 100 d, and 140 d, the subsidence velocity distribution curves of the rock strata with the depths of 100 and 200 m, respectively, are shown in Figures 5–13.

It can be seen from Figure 14 that the subsidence velocity of rock strata shows the following laws: First, the peak value on the subsidence velocity curve gradually increases with the increase in mining distance, and when the working face mining reaches supercritical mining, the increase amplitude of the MSV gradually decreases, finally stabilizing at a certain value. In addition, the shape of the subsidence velocity curve of the rock strata becomes steeper with the increase in mining distance, and finally tends to have a stable shape. Finally, the LDMSV of the subsidence velocity curve increases gradually with the increase in mining distance. In general, the peak value of the rock subsidence velocity curve gradually increases, and the curve shape gradually becomes steep during the mining. When the mining distance reaches a certain value, the peak value and shape of the rock strata subsidence velocity curve tend to be stable and maintain a certain lag distance from the working face, moving forward with the mining of the working face. By comparing Figure 14a,b, it can be found that with the same mining distance, the mining-induced influence of the strata close to the coal seam is greater, resulting in a larger MSV of the rock strata, a smaller LDMSV, and a steeper shape of the subsidence velocity curve.

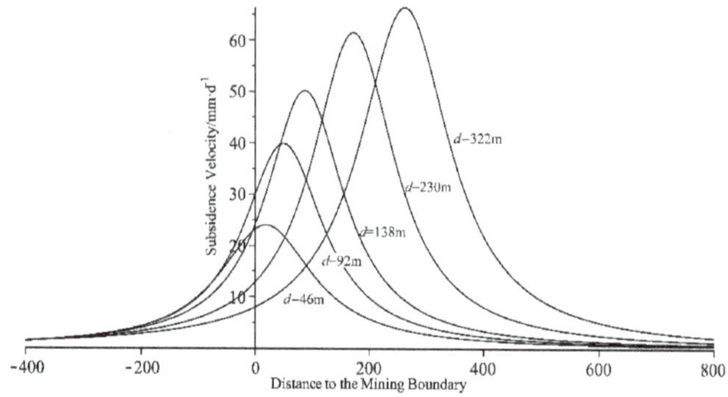

(**a**). z = 100 m

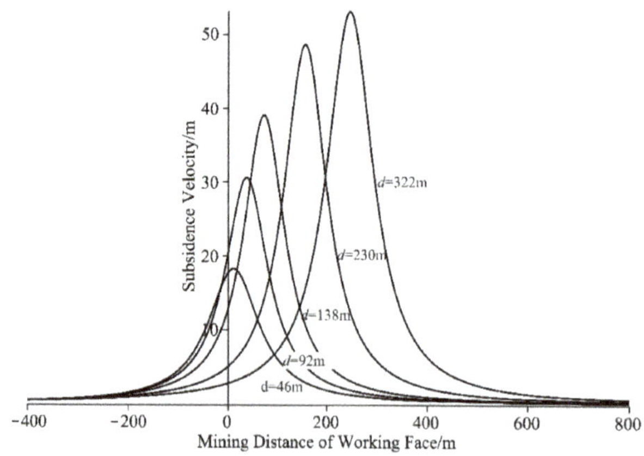

(**b**). z = 200 m

Figure 14. The time-dependent subsidence velocity of strata in different buried depths with mining.

The dynamic impact of mining on structures (shafts, roadways, and other structures) in overburden can be predicted and evaluated according to the variation rule of overlying strata dynamic subsidence and movement. During the mining process, the position and time of the most severe rock strata movement can be determined to reinforce, prevent in advance, and ensure the safe operation of underground mining.

6. Conclusions and Discussion

In this paper, the dynamic movement law and prediction method of the surface and overburden are studied with the stratum structure in Peigou Coal Mine. Since the dynamic movement in the overlying strata is not easy to be observed, this paper innovatively put forward a numerical simulation method to study the dynamic movement, which is used to establish the relationship between surface and overlying strata dynamic movement parameters. Therefore, the dynamic movement law and prediction method of overlying strata and surface are obtained. The research results can be used to predict the dynamic subsidence impact on the structures on the surface and in the overlying strata during the mining process, in order to obtain the mining time or position of working face, which induces the strongest impact on the these structures, and provides a reference for the protection of the structures. The key conclusions from this study are as follows:

(1) In order to study the movement of overlying strata, a numerical simulation model was constructed, and the accuracy of the numerical model was demonstrated by verifying the surface subsidence data, the law of ground pressure, and the recovery distance of the goaf. Moreover, the dynamic excavation parameters of the numerical simulation were determined, and a simulation method of dynamic movement of overlying strata was provided.

(2) By analyzing the simulation results of the numerical model, the dynamic movement characteristics of the rock mass with different depths during the mining of the working face were obtained, including the variation laws of the MSV and LDMSV. The dynamic movement parameter functions of rock strata with different depths were determined by curve fitting.

(3) According to the variation law of dynamic movement parameters of surface and rock mass during mining, the prediction formula of subsidence velocity at different positions and mining distances on the strike major section was established. Considering the mining distance of the working face as 360 m and comparing the predicted and measured values of the surface subsidence velocity, the deviation between the measured and calculated values of the subsidence velocity at each point on the surface strike section was small, and the average error was 1.57 mm/d, indicating that the predicted results of dynamic movement can meet the needs of the project.

(4) At the same depth, the subsidence velocity of rock strata is distributed in the form of a quadratic curve. During the mining process of the working face, the MSV of rock strata subsidence velocity curve gradually increases, and the shape gradually becomes steep. When supercritical mining is reached, the MSV and shape of the rock strata subsidence velocity curve tend to be stable and maintain a certain lag distance from the working face, moving forward with the mining of the working face. Under the condition of the same mining distance of the working face, the MSV of the subsidence velocity curve of the rock strata close to the coal seam is larger, the LDMSV is smaller, and the shape of the subsidence curve is relatively steeper. The position and time of the most severe rock strata movement can be determined to reinforce and prevent in advance and ensure the safe operation of underground mining.

Due to the complexity and diversity of stratum structure, the strata with different structures show different movement laws induced by underground mining, but entire or partial overburden strata movement after mining is controlled by strata structures of "key stratum". Due to the different locations of the key strata in the overlying strata, the propagation law of the movement and deformation in the rock stratum is different. Another important factor influencing the dynamic movement of overlying strata is the bulking factor of the roof strata in the goaf [38–43]. The stratum immediately above the seam bends downward and later breaks into different size blocks depending on their lithological nature. This phenomenon continues upward until the lowest uncaved stratum receives the support of the caved and bulked rock pile. During this process, the strata are broken into blocks by vertical and/or sub-vertical fractures and horizontal cracks due to the vertical bed separation. Vertical and/or sub-vertical fractures andhorizontal crackswill be re-compacted by the stress induced by the weight of overlying strata, and the surface and rock stratum dynamic movement is closely related to the re-compaction of goaf. The above two aspects are the focus of further research.

Author Contributions: Writing—original draft preparation, G.X.; project administration, D.L.; resources, Y.Z.; data curation, H.L. All authors have read and agreed to the published version of the manuscript.

Funding: This research was funded by the regional first-class discipline of Ecology in Guizhou province (XKTJ [2020]22), National Natural Science Foundation of China (Grant No. 51974105), and Fund Project of Bijie Science and Technology Bureau (G [2019]1).

Data Availability Statement: Not applicable.

Acknowledgments: The authors would like to acknowledge the support of the regional first-class discipline of Ecology in Guizhou province (XKTJ [2020] 22), National Natural Science Foundation of China (Grant No. 51974105), and Fund Project of Bijie Science and Technology Bureau (G [2019]1).

Conflicts of Interest: The authors declare no conflict of interest.

References

1. Tang, J.; Wang, J.A.; Wang, L. Dynamic laws and characteristics of surface movement induced by mining under thin alluvium. *Rock Soil Mech.* **2014**, *35*, 2958–2968. (In Chinese)
2. Amar, P.; Ajay, K.; Kalendra, S. Dynamic Subsidence Characteristics in Jharia Coalfield, India. *Geotech. Geol. Eng.* **2014**, *32*, 627–635.
3. Hou, D.F.; Li, D.H.; Xu, G. H Impact of mining thickness on dynamic subsidence characteristics in condition of mining under thick un-consolidated layers. *Coal Sci. Technol.* **2016**, *44*, 191–196. (In Chinese)
4. Xu, G.S.; Li, D.H.; Hou, D.F. Measurement and pre-diction of surface dynamic movement and deformation under thick loose layer mining. *Rock Soil Mech.* **2016**, *37*, 2056–2062. (In Chinese)
5. Li, Q.S.; Guo, J.T.; Dai, H.Y. Surface dynamic subsidence prediction method based on mining sufficiency degree. *J. China Coal Soc.* **2020**, *45*, 160–167. (In Chinese)
6. Dychkovskyi, R.; Shavarskyi, I.; Saik, P.; Lozynskyi, V.; Falshtynskyi, V.; Cabana, E. Research into stress-strain state of the rock mass condition in the process of the operation of double-unit longwalls. *Min. Miner. Depos.* **2020**, *14*, 85–94. [CrossRef]
7. Vu, T.T. Solutions to prevent face spall and roof falling in fully mechanized longwall at underground mines, Vietnam. *Min. Miner. Depos.* **2022**, *16*, 127–134. [CrossRef]
8. Yuan, S.; Sun, B.; Han, G.; Duan, W.; Wang, Z. Application and Prospect of Curtain Grouting Technology in Mine Water Safety Management in China: A Review. *Water* **2022**, *14*, 4093. [CrossRef]
9. Yuan, S.; Han, G. Combined Drilling Methods to Install Grout Curtains in a Deep Underground Mine: A Case Study in Southwest China. *Mine Water Environ.* **2020**, *39*, 902–909. [CrossRef]
10. Fan, K.F.; Li, W.P.; Wang, Q.Q.; Chen, W.C.; Yang, Y.R. Distribution characteristic of mining-induced horizontal fracture in longwall panel: Field investigation and prediction model. *Eng. Fail. Anal.* **2022**, *142*, 106778. [CrossRef]
11. Shu, D.M.; Bhattacharyya, A.K. Prediction of sub-surface subsidence movements due to underground coal mining. *Geotech. Geol. Eng.* **1993**, *11*, 221–234.
12. Shu, D.M.; Bhattacharyya, A.K. Relationship between sub-surface and surface subsidence—A theoretical model. *Min. Sci. Technol.* **1990**, *11*, 307–319. [CrossRef]
13. Gonzalez-Nicieza, C.; Alvarez-Fernandez, M.I.; Menendez-Diaz, A.; Alvarez-Vigil, A.E. The influence of time on subsidence in the central Austrian coalfield. *Bull. Eng. Geol. Environ.* **2007**, *66*, 319–329. [CrossRef]
14. Huang, L.T.; Wang, J.Z. Study on the three stages and deformation velocity of dynamic surface subsidence deformation. *J. China Coal Soc.* **2006**, *31*, 420–424. (In Chinese)
15. Chuedk, M.; Li, D.H. Effects of the progress of the working face front upon the rate of deformations of the surface area. *J. Jiaozuo Min. Inst.* **1993**, *1*, 64–74. (In Chinese)
16. Cui, X.M.; Wang, J.C.; Liu, Y.S. Prediction of progressive surface subsidence above longwall coal mining using a time function. *Int. J. Rock Mech. Min. Sci.* **2001**, *38*, 1057–1063. [CrossRef]
17. Zhang, B.; Cui, X.M.; Zhao, Y.L. Prediction model and algorithm for dynamic subsidence of inclined main section. *J. China Coal Soc.* **2021**, *46*, 1864–1873. (In Chinese)
18. Zhang, B.; Cui, X.M.; Zhao, Y.; Li, C. Parameter calculation method for optimized segmented Knothe time function. *J. China Coal Soc.* **2018**, *43*, 3379–3386. (In Chinese)
19. Zhang, B.; Cui, X.M. Optimization of segmented Knothe time function model for dynamic prediction of mining subsidence. *Rock Soil Mech.* **2017**, *38*, 541–548. (In Chinese)
20. Guo, W.B.; Huang, C.F.; Chen, J.J. The dynamic surface movement characteristics of fully mechanized caving mining under thick hydrous collapsed loess. *J. China Coal Soc.* **2010**, *35*, 38–43. (In Chinese)

21. Deng, K.Z.; Wang, H.Z.; Xing, A.S. On predicting the surface subsidence velocity in undermining process. *J. China Min. Inst.* **1983**, *12*, 71–82. (In Chinese)
22. Wang, X.Z.; Xie, J.L.; Zhu, W.B.; Xu, J.L. The field monitoring experiment of the high-level key stratum movement in coal mining based on collaborative DOFS and MPBX. *Sci. Rep.* **2022**, *12*, 665. [CrossRef] [PubMed]
23. Guo, H.; Yuan, L.; Shen, B.T.; Qu, Q.D.; Xue, J.H. Mining-induced strata stress changes, fractures and gas flow dynamics in multi-seam longwall mining. *Int. J. Rock Mech. Min. Sci.* **2012**, *54*, 129–139. [CrossRef]
24. Xu, J.L.; Qin, W.; Xuan, D.Y.; Zhu, W.B. Influencing factors of accumulative effect of overburden strata expansion induced by stress relief. *J. China Coal Soc.* **2022**, *47*, 115–127. (In Chinese)
25. Qian, M.G.; Miao, X.X.; Xu, J.L. Theoretical study of key stratum in ground control. *J. China Coal Soc.* **1996**, *21*, 225–230.
26. Adhikary, D.P.; Guo, H. Modelling of Longwall Mining-Induced Strata Permeability Change. *Rock Mech. Rock Eng.* **2015**, *48*, 345–359. [CrossRef]
27. Zuo, J.P.; Wu, H.S.; Sun, Y.J. Investigation on the inner and outer analogous hyperbola model (AHM) of strata movement. *J. China Coal Soc.* **2021**, *46*, 333–343. (In Chinese)
28. Sui, W.H.; DI, Q.S. Study on interaction between soil mass deformation and fore pressure during subsidence by mining. *J. Eng. Geol.* **1999**, *7*, 303–309.
29. Zhou, D.W. The Synergy Mechanism between Rock Mass and Soil in Mining Subsidence and Its Prediction. Doctor Dissertation, China University of Mining and Technology, Xuzhou, China, 2015. (In Chinese).
30. Ju, J.Y.; Xu, J.L.; Liu, Y.J. Key strata movement monitoring during underground coal mining and its 5-stage movement law inversion: A case study in Hongqinghe Mine. *J. China Coal Soc.* **2022**, *47*, 611–622. (In Chinese)
31. Su, Y.H.; Feng, L.Z.; Li, Z.Y. Quantification of elements for geological strength index in Hoek—Brown criterion. *Chin. J. Rock Mech. Eng.* **2009**, *28*, 679–686. (In Chinese)
32. Chen, X.X.; Xie, W.B.; Jin, S.G. Determination of mechanics parameters of mining induced rock mass for numerical simulation. *J. Min. Saf. Eng.* **2006**, *23*, 341–345. (In Chinese)
33. Xu, G.S.; Guan, J.F.; Li, H.G. Modeling methods and verification of rock strata movement in coal mine. *Coal Sci. Technol.* **2018**, *46*, 8–13. (In Chinese) [CrossRef]
34. Peng, S.S. Topical areas of research needs in ground control—A state of the art review on coal mine ground control. *Int. J. Min. Sci. Technol.* **2015**, *25*, 1–6. [CrossRef]
35. Wilson, A.H. The Stability of Underground Workings in the Soft Rocks of the Coal Measures. *Int. J. Min. Eng.* **1983**, *1*, 91–187. [CrossRef]
36. Xu, G.S.; Li, H.G.; Li, D.H. Method to Calculate Mining-Induced Fracture Based on the Movement and Deformation of Overburden Strata. *Shock. Vib.* **2021**, *2021*, 9965466. [CrossRef]
37. Itasca Consulting Group, Inc. *3DEC User Manual*; Itasca Consulting Group, Inc.: Minneapolis, MN, USA, 2005.
38. Cui, X.M.; Che, Y.H.; Malinowska, A.; Zhao, Y.L.; Li, P.X.; Hu, Q.F.; Kang, X.L.; Bai, Z.H. Method and problems for subsidence prediction in entire process induced by underground mining. *J. China Coal Soc.* **2022**, *47*, 2170–2181. (In Chinese)
39. Smoliński, A.; Malashkevych, D.; Petlovanyi, M.; Rysbekov, K.; Lozynskyi, V.; Sai, K. Research into Impact of Leaving Waste Rocks in the Mined-Out Space on the Geomechanical State of the Rock Mass Surrounding the Longwall Face. *Energies* **2022**, *15*, 9522. [CrossRef]
40. Malashkevych, D.; Petlovanyi, M.; Sai, K.; Zubko, S. Research into the coal quality with a new selective mining technology of the waste rock accumulation in the mined-out area. *Min. Miner. Depos.* **2022**, *16*, 103–114. [CrossRef]
41. Palchik, V. Bulking factors and extents of caved zones in weathered overburden of shallow abandoned underground workings. *Int. J. Rock Mech. Min. Sci.* **2015**, *79*, 227–240. [CrossRef]
42. Yavuz, H. An estimation method for cover pressure re-establishment distance and pressure distribution in the goaf of longwall mines. *Int. J. Rock Mech. Min. Sci.* **2004**, *41*, 193–205. [CrossRef]
43. He, J.R.; Malinowska, A. Significance of the uncertainty level for the modeling of ground deformation ranges. *Int. J. Rock Mech. Min. Sci.* **2016**, *83*, 140–148.

processes

MDPI

Article

Experimental Research on the Influence of Short-Term Noise Exposure on Miners' Physiology

Jing Li, Zhongjie Cai, Huiyan Liu * and Yanli Xin

School of Emergency Management and Safety Engineering, China University of Mining and Technology, Beijing 100083, China
* Correspondence: zqt2110101018@student.cumtb.edu.cn

Abstract: Coal mine noise affects human physiology, psychology, and behavior. It causes errors at work and increases accidents. In this study, we built a coal mine noise simulation experiment system. The system not only included an experimental environment simulation system and a physiological indicator test system, but it also added a miners' working simulation system. This paper aimed to study the effect of different short-term (25 min) noise levels (60 dB, 70 dB, 80 dB, 90 dB, and 100 dB) on human physiology (skin conductivity and heart rate). Critical analysis showed that the stronger the noise intensity is, the shorter the contact time it takes for physiological indicators to present significant changes, and by setting different noises and measuring the skin conductivity and heart rate of human body, it was concluded that the noise level should be reduced to 90 dB to reduce accidents of miners.

Keywords: coal mine safety; occupational noise; physiology indicator; human health

Citation: Li, J.; Cai, Z.; Liu, H.; Xin, Y. Experimental Research on the Influence of Short-Term Noise Exposure on Miners' Physiology. *Processes* **2023**, *11*, 425. https://doi.org/10.3390/pr11020425

Academic Editor: Carlos Sierra Fernández

Received: 11 January 2023
Revised: 23 January 2023
Accepted: 23 January 2023
Published: 31 January 2023

1. Introduction

All countries in the world regard coal as an important energy source. Coal consumption accounts for 60% of the total energy consumption in China. However, coal mine accidents in China represent more than 70% of the world's [1]. As we all know, coal mining is recognized as one of the most dangerous industries [2,3]. The national statistics of coal mine accidents and injuries can show the danger of coal mine operations [4]. In addition, mining companies may even go bankrupt or suffer heavy losses due to accidents [5]. Human factors are related to industrial accidents and heavy losses [6–9]. Statistics show that major coal mine accidents in China were caused by unsafe human behavior [10]. The unsafe behavior of miners is related to various environmental factors (such as high temperature, low light, and high dust density), and noise is one of them [11].

Modern coal mines often use noisy mechanical equipment, but thick walls, narrow roadways, and closed working spaces make it difficult for noise to disperse or be absorbed. Usually, the noise level in coal mines exceeds the maximum noise value (85 dB/A)(work 5 d a week, 8 h a day, the steady-state noise limit is 85 dB (A)) in the "Occupational Exposure Limits for Hazardous Agents in The Workplace-Physical Factors" (GBZ2.2-2007). In such a high-level noise environment, human psychology and physiology change. Human psychological and physiological systems are interrelated and interact with each other. If a person is in poor physical condition, it affects his mental state. On the other hand, poor mental condition affects the disorder of the physiological system to a large extent. In addition, the external environment affects the overall state of the psychological and physiological system and ultimately affects behavior and operation [12]. Noise can easily lead to the formation of pressure, thus damaging personal physical and mental health, affecting work efficiency, and increasing the probability of accidents.

Previous studies have shown that noise affects human physiology, psychology, and behavior. Physiologically, studies [13–16] opined that noise could cause human hearing loss and affect verbal communication. Liu et al. [17] expressed that noise affects miners' blood tension and hearing. Babisch et al. [18] stated that the incidence rate of cardiovascular diseases increased significantly with the increase of noise. Additionally, noise affects human physiological indicators. Alanalyn N et al. [19] used the change in pupil dilation response of participants caused by each type of noise as an assessment of physiological stress response to noise stimulation. Lusk, SL et al. [20,21] found that heart rate was significantly positively associated with noise exposure. Sajad Zare et al. [22] found that increasing noise SPL resulted in increased serum cortisol concentrations; S, H. Park et al. [23] found that the noise level was significantly associated with skin conductivity and RR changes but not to heart rate. Jing Guoxun et al. [24] found that RRI and HRV decreased, sympathetic nervous system activity increased, and HR increased with increasing noise level. Psychologically, Mohapatra, H et al. [25] believed that the human annoyance increased with increasing noise levels. Wang, FS et al. [26] found that noise induces tension, depression, boredom, and anger, which bring about unpleasant moods and reduce accuracy in judging the degree of environmental safety, causing a breach of regulations. In terms of safety behavior, C. Clark et al. [27] pointed out that exposure to noise increases children's noise problems, and reading and understanding ability also depends on exposure to noise. Jing et al. [28] said that the behavioral reliability of fully mechanized coal mining operators is high with a low accident occurrence rate under the noise of 50~70 dB. Under 70~90 dB, their behavioral reliability is with potential accident risks. The behavioral reliability is low when noise is 90~110 dB, under which accidents may easily take place; Li et al. [29] opined that increased levels of environmental noise affect indicators of safe behavior capacity, and the test indicators of attention, fatigue, and reaction significantly change when the noise exceeds 70–80 dB.

The above research shows that a person's physical and mental health is affected by noise, which can easily lead to unsafe behaviors of miners. However, previous studies have had two limitations. First, some focused on the analysis of the occupational hazard of noise and failed to study the impact of noise on humans from an objective perspective. Second, they were unable to research whether and how short-term noise exposure under different noise levels affects human physiological indicators. To make up for the lack of previous studies, we built a coal mine noise simulation system and studied how short-term noise exposure (0–5 min, 10–15 min, 20–25 min) affects human indicators at five noise levels (60 dB, 70 dB, 80 dB, 90 dB, 100 dB). At the same time, we hope to provide new ideas to take noise protection measures (especially shifting systems for workers), prevent unsafe behaviors, and reduce accidents.

2. Methods

2.1. Selecting Physiological Indicators

In this quantitative research, skin conductivity and heart rate were selected as research indicators according to the experimental conditions and the relevant literature [21,23].

Skin conductivity (SC) is the potential difference between two points on a person's complexion. It is closely related to human emotions. When people are nervous or eager, caring nerves are activated, perspiration discharge increases, complexion freedom fighters decrease, and skin conductivity increases. Psychologists believe that skin conductivity can be used as a physiological indicator for measuring emotions. This is because the skin's electrical conductivity is controlled by the autonomic nervous system [30].

Pulse is the number of beats per minute. It is closely related to stress. When a human is under pressure, their blood flow becomes faster, which increases the heart rate.

2.2. Establishing an Experimental System

Our experiment system consisted of a simulation system and a physiological indicator test system.

(1) Miners' working intensity simulation system

According to studies [31], the labor intensity of miners is moderate or high. Reflections on West German coal mines' labor intensity [32] defined that the metabolic value of medium labor intensity ranges from 232~348 W, and the high labor intensity is above 348 W. Thus, we put the above data into the metabolism formula established by Pandolf [33] (Equation (1)) and calculated that the speed of the treadmill should be 5 km/h. In the experiment, as shown in Figure 1, the work intensity of underground miners was simulated by letting the subjects walk on a treadmill.

$$M_W = 1.5W + 2.0(W+L)\left(\frac{L}{W}\right)^2 + \eta(W+L)\left(1.5V^2 + 0.35VG\right) \quad (1)$$

where M_W is metabolic rate (watts); W is subject weight (kg); L is load carried (kg); η is terrain factor (η = 1.0 for treadmill); V is speed of walking (m·s^{-1}); G is grade (%).

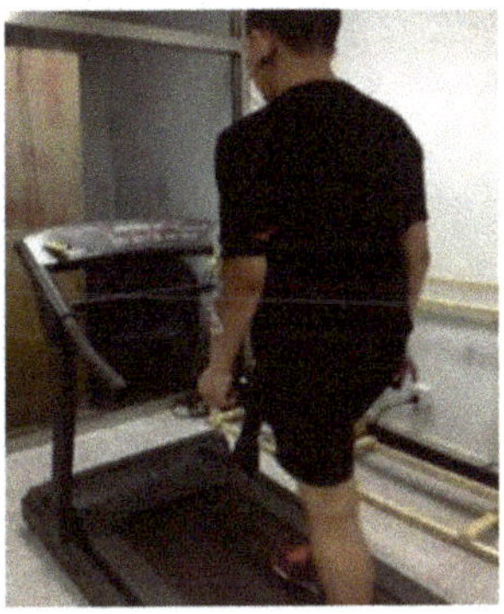

Figure 1. A treadmill used to simulate the intensity of a miner's working.

(2) Coal mine noise simulation experiment system

In this experiment, we recorded the real coal mine noise as the noise source and controlled noise levels through a speaker and a sound level meter (see Table 1). The experiment temperature was at 25 °C, the humidity 50%, and the wind speed was 0.25 m/s. These can make sure experimental results are free from any changes in temperature, humidity, and wind speed [34].

Table 1. Noise test equipment.

Noise System Device Name	Equipment Model
Computer	Dell–Inspiron 7557
Louder speaker box	JBL–Charge4
Sound meter	TES1350A

Our experiment originally set 8 noise levels, a control group (50 dB), and seven experimental groups (60 dB, 70 dB, 80 dB, 90 dB, 100 dB, 110 dB, and 120 dB). First, we tested the calm state in the laboratory and found the noise level was 50 dB; thus, we set the control group as 50 dB. Second, we set 80 dB, 90 dB, 100 dB, 110 dB, and 120 dB as experimental groups because our field investigation of Shanxi Province, China found that the range of coal mine noise is mainly 80 dB~120 dB. Thirdly, we added 60 dB and 70 dB because we believe it helps to discuss the general effects on miner physiology and would improve the credibility of the experiment's credibility. Unfortunately, 7 subjects of 12 showed serious nausea, felt dizzy, or vomited after 5 min under 100 dB. For safety concerns, the 110 dB and 120 dB tests were canceled, and the test of 100 dB was 0–5 min.

(3) Physiological indicator test system

The physiological indicator test system consisted of an ErgoLAB human–machine environment synchronization platform and wireless physiological sensors. First, the skin conductivity, heart rate, and other physiological indicators of the subjects were measured by wireless physiological sensors (see Table 2 and Figure 2). Then, the data were synchronously collected and stored by the ErgoLAB human–machine environment platform.

Table 2. Test indicators and equipment.

Test Indicator	Equipment Name	Equipment Model	Manufacturer
Heart rate	ErgoLAB wireless blood volume pulse sensor	ErgoLAB-PPG/BVP	KINGFAR, Beijing, China
Skin conductance	ErgoLAB wireless skin conductance sensor	ErgoLAB-EDA	

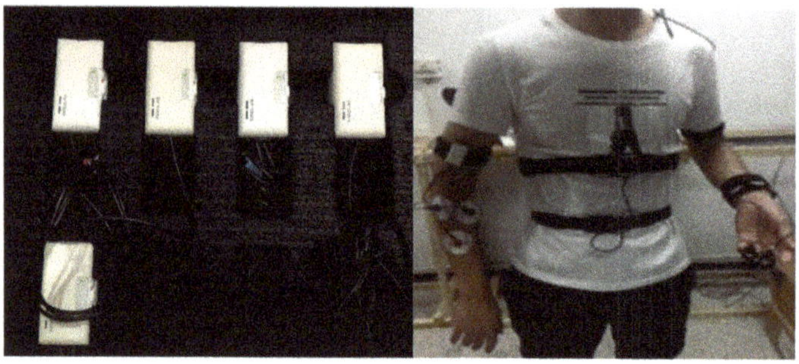

Figure 2. ErgoLAB wireless physiological sensors and participants wearing wireless physiological sensors.

2.3. Selecting Subjects

Based on this study [35], researchers selected 12 participants. Considering the particularity of the miners, the volunteers were men. They were (22.5 ± 2.5) years old and weighed (56 ± 10) kg, which avoids, to a certain extent, the individual differences due to age, body type, etc. In addition, participants were healthy. Volunteers should have voluntarily participated in the study without compulsion. A total of 12 subjects were numbered sequentially. They were not allowed to take any medicine during the test to make sure not to overstimulate their nervous system. They kept an adequate sleep of no less than 8 h [36] before the 24 h of this experiment and wore comfortable clothes.

2.4. Experimental Process

The experiment was divided into two stages, before the experiment and while experimenting.

Preparation before the experiment was as follows: (a) We contacted the subjects, explained what they need to pay attention to, and taught them how to use instruments. (b) We set the experimental environment, adjusted the laboratory temperature, humidity, and wind speed, and debugged tools.

During the experiment, except for researchers, volunteers, and recorders, other personnel were temporarily evacuated from the experimental site. The proper experimental steps were as follows: (a) 2 subjects wore the EDA wireless skin sensors and PPG/BVP wireless blood volume pulse sensors with the help of the researcher. (b) The volunteer adapted to a quiet test environment (50 dB) for 30 min before the test to ensure they were in perfect condition. (c) After 30 min, the researcher measured volunteers' physiological indicators in a quiet environment (50 dB) for about 3 min. At the same time, two subjects were asked to continuously watch the indicators displayed on the computer screen within three minutes. (d) The researcher changed the noise level to 60 dB and collected the subjects' physiological indicators when subjects were watching the computer screen (as a working condition). This step continued for 30 min. (e) After the first test, we removed the sensors, debugged them, and asked the 2 subjects of the second group to wear the sensors. We conducted the second group by repeating (a)–(d). We all together performed 6 tests and tested the indicators of 12 subjects. (f) After all the 12 tests, we saved the test results. (g) We turned off the ErgoLAB human-machine environment platform, debugged all the instruments, and put everything in the experimental environment in its right place. (h) We conducted 60 dB/70 dB/80 dB/90 dB/100 dB tests at the same period (8.00–11:30 and 2:00–5:30) in the following 4 days. The experiment process was the same as (a)–(g).

It is very noteworthy that we performed the following things to make sure the results were credible: (a) The tests of the 5 experimental groups were carried out separately. That is to say, we tested indicators on only one noise level in 1 d to avoid noise adaptability. (b) We conducted a control group test before the tests of each experimental group. This minimized the possible errors caused by subjects' different conditions every day. (c) The test time was between 8:00–11:30 and 14:00–17:30 every day, and the experiment lasted for 5 days. (d) Each subject placed the wireless physiological sensors in the same place.

2.5. Analytical Method

In this paper, noise levels and exposure time were independent variables; skin conductivity and heart rate were dependent variables. A bar graph was used to show the distribution of all indicators and the variation trend of their mean values. SPSS26.0 was used to perform paired sample *t*-test analysis. It could determine when the hands were statistically significant and when they were not under different noise levels. We checked the normality of the physiological data before the paired sample *t*-test because we can perform a paired sample *t*-test only when statistics show normal distribution.

3. Results

3.1. Skin Conductivity

Figure 3 shows how skin conductivity changed with time under five noise levels. We can see that overall skin conductivity increased exposure time. Skin conductivity under all five noise levels changed relatively rapidly at a particular time, but the specific time when the rapid changes occurred was somehow different. When the noise was at 60 dB, skin conductivity sharply rose in 20–25 min. At 70 dB and 80 dB, skin conductivity did not sharply rise until 10–15 min. At 90 dB and 100 dB, skin conductivity sharply increased in the first 0–5 min. We further determined the statistical significance of the above changes through the paired sample *t*-test results to speculate the reasons for the trend change.

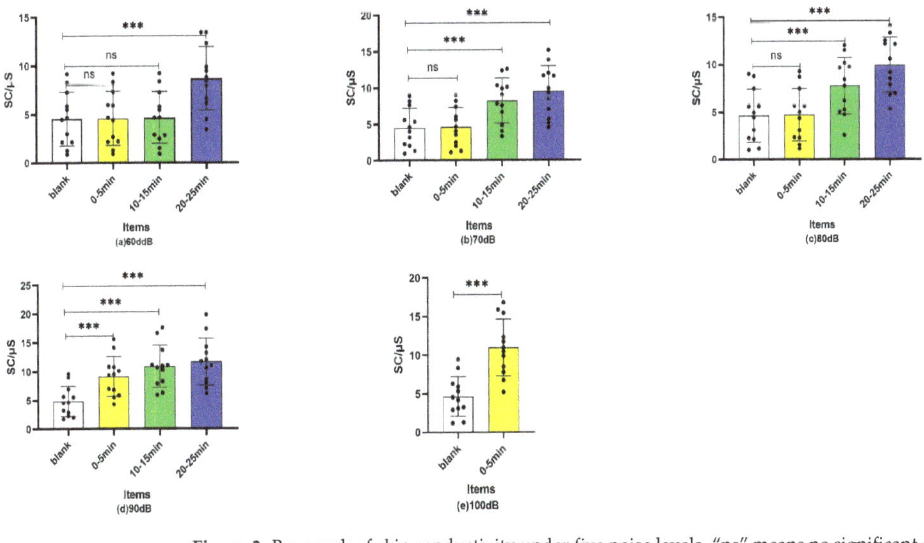

Figure 3. Bar graph of skin conductivity under five noise levels. "ns" means no significant difference, "***" means have significant differences.

Table 3 shows the normality test results of skin conductivity. The null hypotheses H0 are the skin conductivity of the control group, 0–5 min at 60 dB/70 dB/80 dB/90 dB/100 dB, 10–15 min at 60 dB/70 dB/80 dB/90 dB and 20–25 min at 60 dB/70 dB/80 dB/90 dB, and they were normally distributed. We can see that under the test standard $\alpha = 0.05$, the p-value of the S-W test for skin conductivity in different noise levels and their different test periods was greater than 0.05; thus, the null hypotheses H0 were accepted. That is to say, the distribution of skin conductivity was normal, and we were able to perform the paired sample t-test.

Table 3. Normality test results of skin conductivity.

Items	Shapiro-Wilk			Decision-Making
	Statistics	df	Significant p	
Control	0.947	12	0.598	Accept H_0
t0–5/60 dB	0.938	12	0.478	Accept H_0
t10–15/60 dB	0.955	12	0.703	Accept H_0
t20–25/60 dB	0.956	12	0.732	Accept H_0
Control	0.940	12	0.501	Accept H_0
t0–5/70 dB	0.949	12	0.621	Accept H_0
t10–15/70 dB	0.951	12	0.646	Accept H_0
t20–25/70 dB	0.957	12	0.740	Accept H_0
Control	0.933	12	0.414	Accept H_0
t0–5/80 dB	0.938	12	0.470	Accept H_0
t10–15/80 dB	0.954	12	0.694	Accept H_0
t20–25/80 dB	0.941	12	0.508	Accept H_0
Control	0.921	12	0.294	Accept H_0
t0–5/90 dB	0.950	12	0.638	Accept H_0
t10–15/90 dB	0.929	12	0.371	Accept H_0
t20–25/90 dB	0.954	12	0.700	Accept H_0
Control	0.957	12	0.736	Accept H_0
t0–5/100 dB	0.960	12	0.786	Accept H_0

Table 4 shows the results of the paired sample t-test analysis of skin conductivity. The null hypotheses H0 were that there was no significant difference between the skin conductivity of 0–5 min/10–15 min/10–15 min at 60 dB/70 dB/80 dB/90 dB/100 dB and that of the control group. We can see that the higher the noise levels were, the shorter the contact time it took for skin conductivity to present significant changes. In detail, under the test standard $\alpha = 0.05$, $p < 0.05$ was in 20–25 min at 60 dB; $p < 0.05$ was in 10–15 min at both 70 dB and 80 dB; $p < 0.05$ was in the first 0–5 min at both 90 dB and 100 dB. This shows that the changes in the box plot in Figure 3 were statistically significant. Thus, it can be imagined that the increase in noise intensity and exposure time makes the human body easily nervous, anxious, and mentally sweating, resulting in a decrease in skin resistance and an increase in skin conductivity.

3.2. Heart Rate

Figure 4 shows how heart rate changed with time under five noise levels. We can see that, overall, heart rate increased with the rise of exposure time, but this trend was not as strong as skin conductivity. The heart rates under all five noise levels changed relatively rapidly at a particular time, but the specific time when the rapid change occurred was somehow different. When noise was at 60 dB, the heart rate slightly increased, but there was no drastic change within 25 min. At 70 dB, the heart rate sharply rose in 20–25 min. At 80 dB and 90 dB, heart rate drastically grew in 10–15 min. At 100 dB, the heart rate sharply grew in the first 0–5 min. We further determined the statistical significance of the above changes through the paired sample t-test results to speculate the reasons for the trend change.

Table 4. Paired sample *t*-test results of skin conductivity.

Matching Design	Pairing Difference					t	Degree of Freedom	Significant *p* (Two-Tailed)	Decision-Making
	Average Value (E)	Standard Deviation	Standard Error Mean	95% Confidence Interval for the Difference					
				Lower Limit	Upper Limit				
Control-t0~5/60 dB	−0.04167	0.15976	0.04612	−0.14318	0.05984	−0.903	11	0.386	Accept H_0
Control-t10~15/60 dB	−0.11417	0.23956	0.06916	−0.26638	0.03804	−1.651	11	0.127	Accept H_0
Control-t20~25/60 dB	−4.14667	1.18646	0.34250	−0.490051	−3.39283	−12.107	11	0.000	Reject H_0
Control-t0~5/70 dB	−0.11500	0.22359	0.06454	−0.25706	0.02706	−1.782	11	0.102	Accept H_0
Control-t10~15/70 dB	−3.72250	1.10470	0.31890	−4.42439	−11.673	−11.673	11	0.000	Reject H_0
Control-t20~25/70 dB	−5.03833	1.74591	0.50400	−6.14763	−9.997	−9.997	11	0.000	Reject H_0
Control-t0~5/80 dB	−0.08583	0.14817	0.04277	−0.17997	0.00831	−2.007	11	0.070	Accept H_0
Control-t10~15/80 dB	−3.11500	0.87780	0.25340	−3.67273	−2.55727	−12.293	11	0.000	Reject H_0
Control-t20~25/80 dB	−5.31000	0.27414	0.27414	−5.91337	−4.70663	−19.293	11	0.000	Reject H_0
Control-t0~5/90 dB	−4.31000	1.33697	0.38595	−5.15947	−3.46053	−11.167	11	0.000	Reject H_0
Control-t10~15/90 dB	−6.04167	1.47611	0.42612	−6.97954	−5.10379	−14.178	11	0.000	Reject H_0
Control-t20~25/90 dB	−6.82917	1.83239	0.52897	−7.99341	−5.66492	−12.910	11	0.000	Reject H_0
Control-t0~5/100 dB	−6.32333	1.64988	0.47628	−7.37162	−5.27505	−13.667	11	0.000	Reject H_0

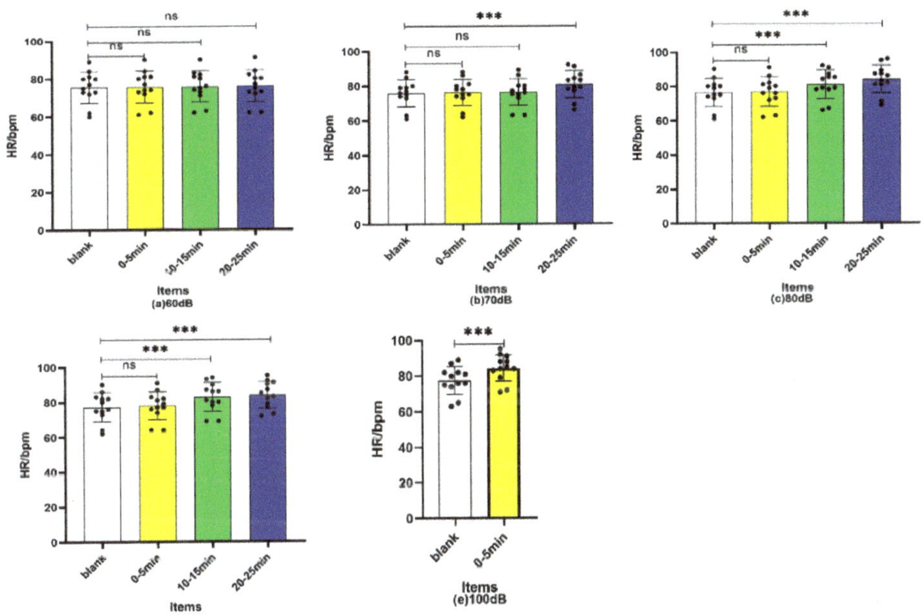

Figure 4. Bar graph of heart rate under five noise levels. "ns" means no significant difference, "***" means have significant differences.

Table 5 shows the normality test results of heart rate. The null hypotheses H0 are the heart rate of the control group, 0–5 min at 60 dB/70 dB/80 dB/90 dB/100 dB, 10–15 min at 60 dB/70 dB/80 dB/90 dB, and 20–25 min at 60 dB/70 dB/80 dB/90 dB, and they were normally distributed. We can see that, under the test standard $\alpha = 0.05$, the *p*-value of the S-W test for heart rate in different noise levels and their different test periods was greater than 0.05; thus, the null hypothesis H0 was accepted. That is to say, the distribution of heart rate was normal, and we were able to perform the paired sample *t*-test.

Table 6 shows the results of the paired sample *t*-test analysis of heart rate. The null hypotheses H0 were that there was no significant difference between the heart rate of 0–5 min/10–15 min/10–15 min at 60 dB/70 dB/80 dB/90 dB/100 dB and that of the control group. We can see that the higher the noise levels were, the shorter the contact time it took for the heart rate to present significant changes. In detail, under the test standard $\alpha = 0.05$, when the noise was at 60 dB, the *p*-value was greater than 0.05 within 25 min. At

70 dB, $p < 0.05$ was in 20–25 min. At 80 dB and 90 dB, $p < 0.05$ was 10–15 min in both. At 100 dB, $p < 0.05$ was in the first 0–5 min. Thus, the changes in the box plot in Figure 4 were statistically significant. Thus, it can be imagined that the increase in noise intensity and exposure time increases the psychological load on humans, making it easy to be nervous and impetuous and resulting in an increase in heart rate.

Table 5. Normality test results of heart rate.

Items	Shapiro-Wilk			Decision-Making
	Statistics	df	Significant p	
Control	0.949	12	0.626	Accept H_0
t0–5/60 dB	0.950	12	0.631	Accept H_0
t10–15/60 dB	0.962	12	0.806	Accept H_0
t20–25/60 dB	0.954	12	0.689	Accept H_0
Control	0.931	12	0.392	Accept H_0
t0–5/70 dB	0.946	12	0.574	Accept H_0
t10–15/70 dB	0.927	12	0.349	Accept H_0
t20–25/70 dB	0.951	12	0.648	Accept H_0
Control	0.943	12	0.537	Accept H_0
t0–5/80 dB	0.959	12	0.773	Accept H_0
t10–15/80 dB	0.918	12	0.267	Accept H_0
t20–25/80 dB	0.938	12	0.470	Accept H_0
Control	0.947	12	0.590	Accept H_0
t0–5/90 dB	0.940	12	0.499	Accept H_0
t10–15/90 dB	0.932	12	0.401	Accept H_0
t20–25/90 dB	0.960	12	0.778	Accept H_0
Control	0.941	12	0.505	Accept H_0
t0–5/100 dB	0.937	12	0.456	Accept H_0

Table 6. Paired sample *t*-test analysis results of heart rate.

Matching Design	Pairing Difference					t	Degree of Freedom	Significant p (Two-Tailed)	Decision-Making
	Average Value (E)	Standard Deviation	Standard Error Mean	95%Confidence Interval for the Difference					
				Lower Limit	Upper Limit				
Control-t0~5/60 dB	−0.25000	0.75378	0.21760	−0.72893	0.22893	−1.149	11	0.275	Accept H_0
Control-t10~15/60 dB	−0.33333	1.07309	0.30977	−1.01514	0.34847	−1.706	11	0.305	Accept H_0
Control-t0~5/70 dB	−0.34323	0.88763	0.25624	−0.89730	0.23064	−1.301	11	0.220	Accept H_0
Control-t10~15/70 dB	−0.41637	0.99620	0.28758	−1.04963	−0.21629	−1.449	11	0.175	Accept H_0
Control-t0~5/80 dB	−0.41667	1.08362	0.31282	−1.10517	0.27184	−1.332	11	0.210	Accept H_0
Control-t10~15/80 dB	−4.58333	1.78164	0.51432	−5.71533	−3.45133	−8.912	11	0.000	Reject H_0
Control-t0~5/90 dB	−0.58333	1.08362	0.31282	−1.27184	−0.10517	−1.865	11	0.089	Accept H_0
Control-t10~15/90 dB	−5.41667	1.67649	0.48396	−6.48186	−4.35148	−11.192	11	0.000	Reject H_0
Control-t0~5/100 dB	−6.83333	2.08167	0.60093	−8.15596	−5.51071	−11.371	11	0.000	Reject H_0

4. Discussion

In the high-risk coal mine environment, due to the particularity of their work, miners are under much greater pressure than ordinary workers, and the potential unintended consequences of accidents caused by stress disorders are even greater. As external influences and internal psychological conditions are easy to damage human performance, whether high-risk coal mine workers can take effective measures to mitigate and adjust their emotions and attitudes under noise conditions in a timely manner is crucial to recover losses, reduce losses, and control the scale of disasters. In risk assessment, environmental factors must be scientifically assessed.

Previous studies explored the impact of noise on miners through cognitive level data of attention, reaction time, and mental fatigue level. For a more in-depth study, we analyzed the impact of noise on miners based on data produced, making use of the principles of neuroscience. Through studying whether and how short-term noise exposure affects human physiological indicators, we compared when dramatic changes occurred in skin conductivity and heart rate under different noise levels and found that: (a) When the noise was at 60 dB, skin conductivity did not present significant growth until 20–25 min, while heart rate showed no significant change within 25 min. (b) When noise was at 70 dB, skin conductivity significantly increased in 10–15 min, while heart rate significantly rose in 20–25 min. (c) When noise was at 80 dB, skin conductivity and heart rate both presented as significantly increased in 10–15 min. (d) When the noise was at 90 dB, skin conductivity significantly changed in 0–5 min, while heart rate significantly changed in 10–15 min. (e) When the noise level was at 100 dB, skin conductivity and heart rate presented significant changes in 0–5 min. Relevant studies [21,23] also found that human heart rate, heart rate variability, skin conductivity, and other physiological indicators all change in a noisy environment. The center rate, heart rate variability, skin conductivity, and respiratory rate can reflect human comfort. Physiological indicators are widely used in the study of environmental factors.

In short, the noise intensity standard is not absolute, and it cannot be said that a noise environment lower than the noise intensity standard will not affect humans. An intensity of 60 dB~80 dB affects skin conductivity and heart rate relatively short-term, although this range is not beyond what is required by occupational health standards. Moreover, sometimes time spent working in a noisy environment is not long, but it may still have an impact on the human body. Additionally, noise protection is necessary despite shorter working hours or an environment of low-level noise(60 dB~80 dB), and the noise level should preferably be kept no higher than 90 dB in coal mines. We hope that this study can provide scientific basis for formulating noise occupational health standards and preventing accidents.

Considering the experimental operation, the researchers just selected the mining college students as volunteers. By reason of the influence of the cultural level, physiological and psychological quality, work experience, and other factors, students' reactions to coal mine accidents may be different from miners' reactions, and the changes of physiological indicators may also be different under different noise levels. In future research, the experiment can be further expanded to make the selected research volunteers closer to the actual situation of miners. In real coal mines, the walls have noise-absorptive and reflective effects, and this experiment did not consider this. However, it still can be proven that the collected coal mine noise is an adequate stressor. This aspect needs further research to use virtual reality technology to simulate a natural coal mine noise environment.

This research can also be further explored later. For example, in real accidents, individuals may suffer from physical injury and irritable psychological and irrational factors, which may lead to accidents. The possibility and extent of operational errors, inattention, and unreasonable behaviors caused by physiological and psychological changes of volunteers under different noise levels were not taken into account.

5. Conclusions

Based on human factor engineering, we selected two physiological indicators and studied how five noise levels affect the indicators in a short time. The results are as follows:

a. Skin conductivity and heart rate presented an overall increasing trend, with the increase of noise exposure time under five different noise levels, although the noise exposure time was short. The higher the noise level, the shorter the contact time required for significant changes in skin conductivity and heart rate.
b. An intensity of 60 dB~80 dB affects skin conductivity and heart rate in a relatively short-term aspect, although this range is not beyond what is required by occupational health standards. Thus, it is necessary to properly carry out safety regulations for noise prevention.
c. When the noise reached 90 dB or above, skin conductivity or heart rate significantly changed in the first 0–5 min. Without protection, it is necessary to control noise levels below 90 dB and reduce the long-term exposure of miners in such an environment.

d We compared how much time it takes for skin conductivity and heart rate to present significant changes. We found that skin conductivity needs relatively less time to show substantial change under the five noise levels.

To conclude, from the perspective of accident prevention, we suggest coal mines pay attention to the noise protection of miners even though miners are in an environment of low-level noise. Additionally, noise protection is necessary despite shorter working hours. Notably, the noise level should preferably be kept no higher than 90 dB in coal mines without protection. These can effectively reduce accidents.

Author Contributions: J.L. designed the study. J.L. and Z.C. wrote the manuscript. H.L. performed data collection and data analysis. Y.X. performed the reviewing and editing of the manuscript. All authors have read and agreed to the published version of the manuscript.

Funding: This work was supported by the 2 Batch of 2022 MOE of PRC Industry University Collaborative Education Program (Program No. 220705329280548, Kingfar-CES "Human Factors and Ergonomics" Program) and the Fundamental Research Funds for the Central Universities (2022YJSAQ20). The funding body played no role in the design of the study and collection, analysis, and interpretation of data and in writing the manuscript.

Institutional Review Board Statement: All research procedures were approved prior to the commencement of the study by the China University of mining and technology (Beijing). All participants signed an informed consent form. Our research received ethics approval from China University of Mining and Technology (Beijing) and it conformed to the ethics guidelines of the Declaration of Helsinki.

Informed Consent Statement: Not applicable.

Data Availability Statement: The datasets used and/or analysed during the current study are available from the corresponding author upon reasonable request.

Conflicts of Interest: The authors declare that they have no competing interest.

References

1. Liu, Q.; Li, X.; Hassall, M. Evolutionary game analysis and stability control scenarios of coal mine safety inspection system in China based on system dynamics. *Saf. Sci.* **2015**, *80*, 13–22. [CrossRef]
2. Amponsah-Tawiah, K.; Jain, A.; Leka, S.; Hollis, D.; Cox, T. Examining psychosocial and physical hazards in the Ghanaian mining industry and their implications for employees' safety experience. *J. Saf. Res.* **2013**, *45*, 75–84. [CrossRef]
3. Gyekye, S.A. Causal attributions of Ghanaian industrial workers for accident occurrence: Miners and non-miners' perspective. *J. Saf. Res.* **2003**, *34*, 533–538. [CrossRef]
4. Paul, P.S. Predictors of work injury in underground mines-an application of a logistic regression model. *Min. Sci. Technol.* **2009**, *19*, 282–289. [CrossRef]
5. Mahdevari, S.; Shahriar, K.; Esfahanipour, A. Human health and safety risks management in underground coal mines using fuzzy TOPSIS. *Sci. Total Environ.* **2014**, *488–489*, 85–99. [CrossRef] [PubMed]
6. Griffith, C.D.; Mahadevan, S. Human reliability under sleep deprivation: Derivation of performance shaping factor multipliers from empirical data. *Reliab. Eng. Syst. Saf.* **2015**, *144*, 23–34. [CrossRef]
7. Jeong, K.; Choi, B.; Moon, J.; Hyun, D.; Lee, J.; Kim, I.; Kim, G.; Kang, S. Risk assessment on abnormal accidents from human errors during decommissioning of nuclear facilities. *Ann. Nucl. Energy* **2016**, *87*, 1–6. [CrossRef]
8. Moura, R.; Beer, M.; Patelli, E.; Lewis, J.; Knoll, F. Learning from major accidents to improve system design. *Saf. Sci.* **2016**, *84*, 37–45. [CrossRef]
9. Park, J. Scrutinizing inter-relations between performance influencing factors and the performance of human operators pertaining to the emergency tasks of nuclear power plant—An explanatory study. *Ann. Nucl. Energy* **2011**, *38*, 2521–2532. [CrossRef]
10. Zhang, J.S.; Gui, F.; Chen, D.W.; Zhou, X.W.; Liu, Z.M.; Coal Industry Publ. H. (Eds.) The design and experiment for mining unsafe-behavior prevention. In Proceedings of the 3rd International Symposium on Modern Mining and Safety Technology, Beijing, China, 4–6 August 2008; Liaoning Technical University: Fuxin, China, 2008.
11. Feng, X.W. (Ed.) The application of ergonomics on mining safety management. In Proceedings of the 3rd International Symposium on Safety Science and Technology (2002 ISSST), Tai'an, China, 10–13 October 2002.
12. Cheng, G.Y.; Chen, S.J.; Qi, J.L.; Cheng, Y. (Eds.) Influence of underground noise to people's unsafe behavior in coal mines. In Proceedings of the 2nd International Symposium on Mine Safety Science and Engineering, Beijing, China, 21–23 September 2013.
13. Farooqi, Z.U.R.; Sabir, M.; Latif, J.; Aslam, Z.; Ahmad, H.R.; Ahmad, I.; Imran, M.; Ilić, P. Assessment of noise pollution and its effects on human health in industrial hub of Pakistan. *Environ. Sci. Pollut. Res. Int.* **2020**, *27*, 2819–2828. [CrossRef]

14. Xin, J.; Shi, Z.; Zhang, M. The role of noise temporal structure in noise-induced hearing loss among manufacturing workers. *Saf. Health Work* **2022**, *13*, S234. [CrossRef]
15. Kovalchik, P.G.; Matetic, R.J.; Smith, A.K.; Bealko, S.B. Application of Prevention through Design for hearing loss in the mining industry. *J. Safety Res.* **2008**, *39*, 251–254. [CrossRef] [PubMed]
16. Wouters, N.L.; Kaanen, C.I.; Ouden, P.J.D.; Schilthuis, H.; Böhringer, S.; Sorgdrager, B.; Ajayi, R.; De Laat, J.A.P.M. Noise Exposure and Hearing Loss among Brewery Workers in Lagos, Nigeria. *Int. J. Environ. Res. Public Health* **2020**, *17*, 2880. [CrossRef] [PubMed]
17. Liu, J.; Xu, M.; Ding, L.; Zhang, H.; Pan, L.; Liu, Q.; Ding, E.; Zhao, Q.; Wang, B.; Han, L.; et al. Prevalence of hypertension and noise-induced hearing loss in Chinese coal miners. *J. Thorac. Dis.* **2016**, *8*, 422–429. [CrossRef]
18. Babisch, W. Cardiovascular effects of noise. *Noise Health* **2011**, *13*, 201–204. [CrossRef]
19. Pinaula-Toves, A.N.; Nelson, P.; Gianakas, S.P.; Sullivan, J.; Winn, M.B. Relationship of noise history to perceived loudness and physiological arousal from noise. *J. Acoust. Soc. Am.* **2022**, *151*, A256. [CrossRef]
20. Rahmani, R.; Aliabadi, M.; Golmohammadi, R.; Babamiri, M.; Farhadian, M. Body physiological responses of city bus drivers subjected to noise and vibration exposure in working environment. *Heliyon* **2022**, *8*, e10329. [CrossRef]
21. Dai, C.Z.; Lian, Z.W. The effects of sound loudness on subjective feeling, sympathovagal balance and brain activity. *Indoor Built Environ.* **2018**, *27*, 1287–1300. [CrossRef]
22. Zare, S.; Baneshi, M.R.; Hemmatjo, R.; Ahmadi, S.; Omidvar, M.; Dehaghi, B.F. The Effect of Occupational Noise Exposure on Serum Cortisol Concentration of Night-shift Industrial Workers: A Field Study. *Saf. Health Work* **2019**, *10*, 109–113. [CrossRef]
23. Park, S.H.; Lee, P.J. Effects of floor impact noise on psychophysiological responses. *Build. Environ.* **2017**, *116*, 173–181. [CrossRef]
24. Guoxun, J.; Min, W. Effects of noise on cognitive performance of workers based on physiological indicators. *Saf. Coal Mines* **2021**, *52*, 243–247.
25. Mohapatra, H.; Goswami, S. Assessment and analysis of noise levels in and around lb river coalfield, Orissa, India. *J. Environ. Biol.* **2012**, *33*, 649–655. [PubMed]
26. Wang, F.S.; Guo, L.W.; Zhang, Y. (Eds.) The Analysis and Countermeasures of Miners' Unsafe Behaviors in Operating Environment. In *Chinese Seminar on the Principles of Safety Science and Technology*; Beijing, China, 2010.
27. Clark, C.; Head, J.; Stansfeld, S.A. Longitudinal effects of aircraft noise exposure on children's health and cognition: A six-year follow-up of the UK RANCH cohort. *J. Environ. Psychol.* **2013**, *35*, 1–9. [CrossRef]
28. Wang, Y.S.; Jing, G.X.; Guo, S.S.; Zhou, F. Monte Carlo Method-Based Behavioral Reliability Analysis of Fully-Mechanized Coal Mining Operators in Underground Noise Environment. *Teh. Vjesn. Tech. Gaz.* **2021**, *28*, 178–184.
29. Li, J.; Qin, Y.; Yang, L.; Wang, Z.; Han, K.; Guan, C. A simulation experiment study to examine the effects of noise on miners' safety behavior in underground coal mines. *BMC Public Health* **2021**, *21*, 324. [CrossRef]
30. Scheirer, J.; Fernandez, R.; Klein, J.; Picard, R.W. Frustrating the user on purpose: A step toward building an affective computer. *Interact. Comput.* **2002**, *14*, 93–118. [CrossRef]
31. Peng, X.S. *Study on Coupling Relation of Man-Environment under Complicated Conditions of Fully Mechanized Excavation FaceExperimental Study*; Henan Polytechnic University: Jiaozuo, China, 2011.
32. Liu, H.Q.; Gao, L.Y.; You, B.; Wu, S.X.; Mi, L.H.; Chen, F.; Zhu, K.Y. Experiment on factors affecting thermal comfort of gas-cooled clothing. *J. Xi'an Univ. Sci. Technol.* **2018**, *38*, 910–918.
33. Pandolf, K.B.; Givoni, B.; Goldman, R.F. Predicting energy expenditure with loads while standing or walking very slowly. *J. Appl. Physiol.* **1977**, *43*, 577–581. [CrossRef]
34. Masterson, E.A.; Bushnell, P.T.; Themann, C.L.; Morata, T.C. Hearing Impairment Among Noise-Exposed Workers—United States, 2003–2012. *MMWR Morb. Mortal. Wkly. Rep.* **2016**, *65*, 389–394. [CrossRef]
35. Mu, Z.; Hu, J.; Min, J. Driver fatigue detection system using electroencephalography signals based on combined entropy features. *Appl. Sci.* **2017**, *7*, 150. [CrossRef]
36. Paruthi, S.; Brooks, L.J.; D'Ambrosio, C.; Hall, W.A.; Kotagal, S.; Lloyd, R.M.; Malow, B.A.; Maski, K.; Nichols, C.; Quan, S.F.; et al. Recommended Amount of Sleep for Pediatric Populations: A Consensus Statement of the American Academy of Sleep Medicine. *J. Clin. Sleep Med.* **2016**, *12*, 785–786. [CrossRef] [PubMed]

processes

Article

Study on the Shadow Effect of the Stress Field around a Deep-Hole Hydraulic-Fracturing Top-Cutting Borehole and Process Optimization

Shuanlin Wang [1,2] and Jianqiao Luo [3,*]

1 School of Mining Engineering, Xi'an University of Science and Technology, Xi'an 710054, China
2 Safety Branch, China Coal Research Institute, Beijing 100013, China
3 School of Energy and Mining Engineering, China University of Mining Science and Technology (Beijing), Beijing 100083, China
* Correspondence: luojianqiao57@163.com

Abstract: The clean utilization and green development of coal resources have become a research focus in recent years. Underground hydraulic fracturing technology in coal mines has been widely used in roof pressure relief, top coal pre-splitting, gas drainage, roadway pressure relief and goaf disaster prevention. Different in situ stress types cause great differences in the stress field around the boreholes, the critical pressure of the fracture initiation, and the direction of the fracture expansion trend; in addition, the stress shadow effect generated by the superposition of stress fields between boreholes relatively close together has a mutual coupling effect on the evolution of the stress field, the development of the plastic zone, and the crack propagation of the rock mass. Therefore, an effective method to solve the problem is to establish a mechanical model of hydraulic fracturing in boreholes for theoretical calculation, determine the influence mechanism of the crack shadow effect, and design a numerical simulation experiment of the equivalent stress fluid–solid coupling of hydraulic fracturing under different pore diameters and spacings. In addition, combining rock mechanics and fracture mechanics to analyze the influence of the shadow effect of the stress field between cracks on the evolution of the equivalent stress and the plastic zone is one of the important advances in this paper. Considering the engineering background of the site, the geological conditions and the requirements of general regulations, it is considered that the parameter selection of roof fracturing hydraulic fracturing technology in the Yushen mining area is more suitable when 0.12 m hole diameter and 3.5 m hole spacing are selected.

Keywords: prescribed hydraulic fracturing; fracture mechanics; fluid–solid coupling; stress shadow effect

Citation: Wang, S.; Luo, J. Study on the Shadow Effect of the Stress Field around a Deep-Hole Hydraulic-Fracturing Top-Cutting Borehole and Process Optimization. *Processes* **2023**, *11*, 367. https://doi.org/10.3390/pr11020367

Academic Editors: Feng Du, Aitao Zhou and Bo Li

Received: 15 December 2022
Revised: 6 January 2023
Accepted: 10 January 2023
Published: 24 January 2023

1. Introduction

In recent years, the research into the fluid–solid coupling of hydraulic fracturing has attracted the attention of experts and scholars in many fields around the world. In 1925, Terzaghi [1] regarded the consolidation of deformable saturated porous media as a coupling problem of flow boundary–solid boundary deformation. On the basis that the flow field satisfied Darcy's law, taking the total stress and pore fluid pressure of the saturated rock and soil as the state variables, a relatively complete three-dimensional consolidation theory was established. Barenblatt [2,3] (1959) introduced the definition of the field near the crack boundary and analyzed the linear elasticity by using the relevant provisions and assumptions in elasticity.

Therefore, the rational development of some resources with special mining difficulties in the past two decades has also become the global focus of scientific mining researchers. The research and development of safe and effective mining technology for near-vertical coal seams, deep high-thermal coal seams, short-distance thin coal seams, and deep strong-impact-prone coal seams has entered a new stage. In the 1990s, the hydraulic fracturing process carried out by the American Petroleum System for shale gas development and oil penetration and production was introduced into the mining industry and successfully applied to the seam cutting of the hard roof plate of the mine, the pressure relief of the roadway, and the treatment of the weakening of the top coal. In order to determine that the layout plan of the prescribed hydraulic-fracturing (PHF) boreholes can ensure that the fractures can be successfully expanded and the roof can be cut without premature penetration leading to early pressure reduction, it is important and difficult to reasonably design the spacing of the hydraulic-fracturing boreholes.

On this basis, Hutchinson [4], Sneddon [5], Green [6], Warpinski et al. [7], Wang [8], and Nagel et al. [9] have made some progress in determining the spatial distribution characteristics of stress shadowing. Considering the effects of crack spacing, rock mechanics parameters, and in situ stress in three-dimensional conditions, it was found that the influence range of the increase in the minimum horizontal stress was large and extended in the direction of the crack height, but it had little effect on the crack tip region. Zhao et al. [10] studied the three-dimensional propagation mechanism of the hydraulic fracture in a borehole with multiple angles and different spacings in a three-dimensional rock mass and concluded that the increase in the intermediate principal stress suppressed the propagation of the fracture perpendicular to the direction of the intermediate principal stress; the propagation track of the crack in the direction of the intermediate principal stress became straighter. The change in the intermediate principal stress did not change the main direction of the crack propagation (the direction of the maximum principal stress).

In terms of the development of multi-physical field numerical simulation experimental software, in the past two decades, scholars at home and abroad have mainly carried out engineering scale research on the basis of binary numerical models to explore the law of interaction between porous hydraulic fracturing and fracture propagation. Wang [11], Liu et al. [12], and Michael [13] found in experiments that when designing the perforation spacing and aperture, the stress shadow effect could be weakened by reasonably optimizing the setting of large or nonuniform perforation spacing, so as to make the spatial morphology of the hydraulic fracture propagation more uniform.

To summarize, it is of new significance to study the influence of the borehole size and borehole spacing on the shadow effect between the hydraulic fracturing fractures within the three-dimensional rock mass scale model. Building a three-dimensional mechanical model and grid division of three-dimensional multi-cluster perforated rock materials can more realistically simulate the propagation of fracturing fractures in order to then design the reasonable arrangement of hole spacing and aperture, so that the spatial form of fracture propagation is more conducive to engineering applications. It has been planned to obtain the optimal parameters applicable to roof hydraulic fracturing drilling in the Yushen mining area under the influence of special conditions, such as shadow effect, by means of mechanical model establishment, theoretical calculation and analysis, multi physical field coupling numerical simulation experiments and mutual verification.

2. Simulation Engineering Background and Boundary Condition Setting

Geological Condition Background and Working Face Overview

Hydraulic fracturing and roof cutting are widely used to prevent rock burst or strong ground pressure in deep buried or deep mines from 400 m to 1000 m. With the increase in the buried depth and the change in the importance of the tectonic stress in the in situ stress of the unit coal body, the type of in situ stress changes, as shown in Figure 1a. Therefore, the research background of the shadow effect of deep-hole hydraulic-fracturing jet-fracturing boreholes was set as the type III geostress type with vertical stress as the maximum principal stress, as shown in Figure 1e; in the absence of cutting-induced fractures and the influence of the stress shadow effect between adjacent boreholes, the dominant trend of fracture space expansion gradually moves to the minimum horizontal principal stress σ_3 direction deflection. In most cases, deep-hole hydraulic fracturing is conducted from the leading position of the roadway to the hard roof above the stope, see Figure 1b. The overall development trend of fracturing fractures is perpendicular to the advancing direction of the working face and along the vertical stress direction. This fracture development mode has natural advantages for cutting the hard roof along the mining fracture direction, shortening the fracture step, and weakening the leading support pressure of the stope. Therefore, it is particularly important to reasonably optimize the spacing and aperture of the boreholes to reduce the disturbance of the stress shadow effect on the crack propagation and to avoid the influence of water leakage from the adjacent boreholes on the effect of roof cutting.

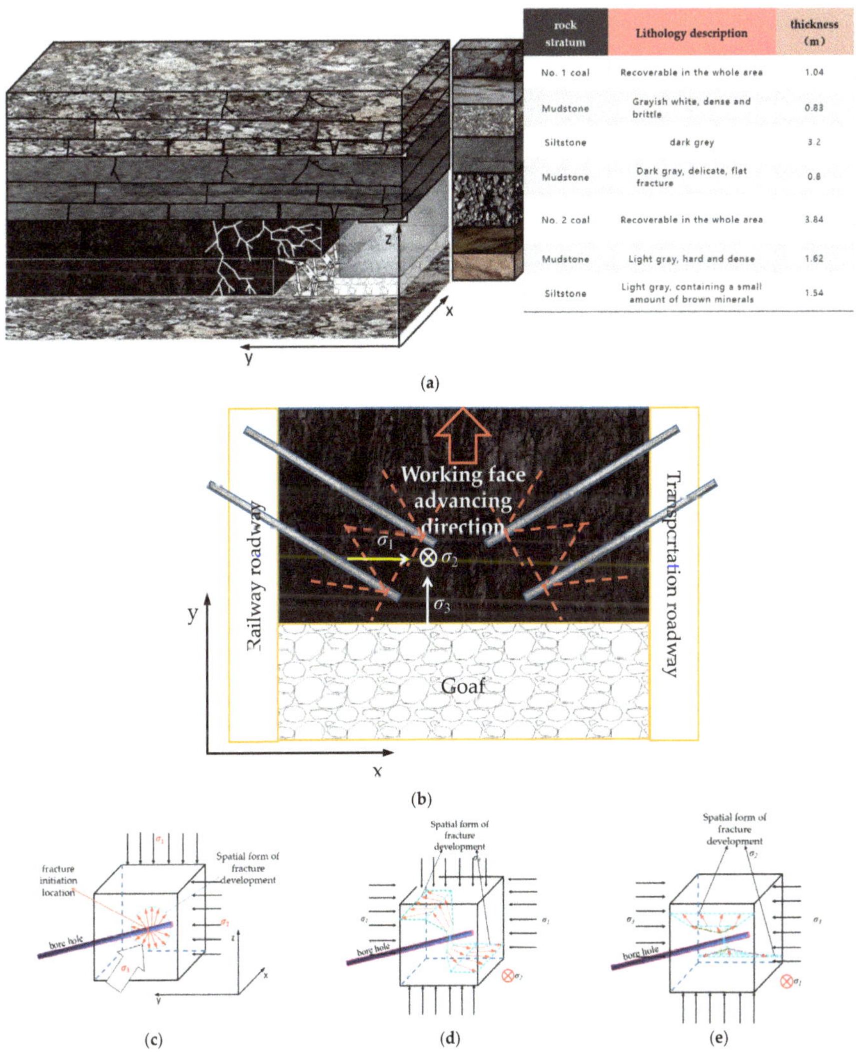

Figure 1. Relationship between the spatial development pattern of the deep-hole hydraulic-fracturing fractures and the evolution of the in situ stress types. (**a**) Comprehensive geological histogram. (**b**) The spatial relationship between the layout of the long-hole hydraulic-fracturing drilling site and the stope roof in the advance roadway (σ_V is the vertical maximum principal stress; σ_H is the maximum horizontal principal stress; and σ_h is the minimum horizontal principal stress); the red dotted line is the possible jet direction when the high-pressure water-jet drill pipe is used to induce the kerf. (**c**) Type I geostress. (**d**) Type II geostress. (**e**) Type III geostress.

We selected the geological conditions in the Xiangning mining area of Hedong Coalfield, Shanxi Province, where the buried depth of the coal seam was 555 m~631 m, which was the Shennan'ao coal mine. In order to facilitate a unified calculation, the buried depth parameter was determined to be 600 m, the working face length was 180 m, and the advancing length was 1286 m. The hard sandstone on the roof was treated to weaken the rock pressure and high-energy dynamic load test pieces, and the multilayer mudstone and fine sandstone on the roof were subject to hydraulic fracturing.

As shown in Figure 1c–e, there were three distinct development stages in the spatial form of hydraulic-fracturing fracture propagation under different in situ stress types: 1. the crack initiation stage along the prefabricated fracture or the jet nozzle; 2. the deflection stage dominated by the ground stress; and 3. the continuous expansion along the direction of the maximum principal stress and perpendicular to the direction of the minimum principal stress.

An excessive size of a three-dimensional fluid–structure coupling experimental model leads to non-convergence; however, too small a size makes it difficult to meet the actual rock-block size at the field. Considering this, a similar model of a normal-cube rock mass with a side length of 10 m was established. The boundary conditions were selected to assign parameters to the model through the measured in situ stress state of the basic roof rock in the Xiangning mining area; by comparing the convergence effects of three adaptive mesh generation methods (coarsening, thinning, and refining), it was determined that the refined mesh better considered the above experimental purposes.

The experimental scheme of borehole spacings of 3 m, 3.5 m, and 4 m was intended to further study the common situation of the borehole spacing for hydraulic fracturing in coal mines in "Detailed rules for prevention and control of rock burst in coal mines" published by the Ministry of Coal Industry of the People's Republic of China.

The pore diameters of the high-pressure fracturing drills were taken to be 0.08 m, 0.10 m, and 0.12 m, which were designed according to the three most commonly used boreholes for hydraulic fracturing in coal mines.

Based on the above two considerations, an experimental scheme was designed for the research object of this manuscript: first, the scheme most suitable for the studied engineering background geological conditions was selected by comparing the fluid–structure coupling boundary surface stress and flow velocity of three different hole diameters; then, the influence of the stress shadow effect on the fracturing effect under 3.0 m, 3.5 m, and 4.0 m hole spacing was compared on the basis of this aperture.

Due to the self-adaptive mesh generation function developed in the multi-physical field coupling numerical simulation software COMSOL 5.6, the mesh at the boundary of the borehole wall of the hydraulic fracturing jet could automatically adapt to the transition state of laminar flow and turbulence, better converge the calculation of the surface velocity and the equivalent stress at the fluid–solid coupling boundary in the turbulent state, where the internal flow velocity exceeded 2300 Reynolds number, and densify and set smooth grids at fewer positions at the borehole wall boundary, as shown in Figure 2. In the practical application of underground hydraulic-fracturing engineering in coal mines, the most common drill pipe sizes are 0.08, 0.10, and 0.12 m. These three sizes of drill pipes can simultaneously meet the rapid and high-pressure passage of the large-flow fracturing fluid and have no destructive impact on the coal pillar in the cross roadway. Three drilling sizes with different diameters of 0.08 m, 0.1 m, and 0.12 m were set, with a hole depth of 3 m, 5 m, and 8 m (5% of different hole spacings) × five × 5 m sandstone-rock material test block. The density was ρ 2460 kg/m^3; the Young's modulus E was 4.2 Gpa; Poisson's ratio was μ 0.75; the vertical stress σ_V was 20 Mpa; the horizontal maximum principal stress σ_H was 18.5 Mpa; and the horizontal minimum principal stress σ_h was 15.5 Mpa. First, we determined the optimal aperture; then, we conducted the stress shadow effect experiment under a fixed aperture size.

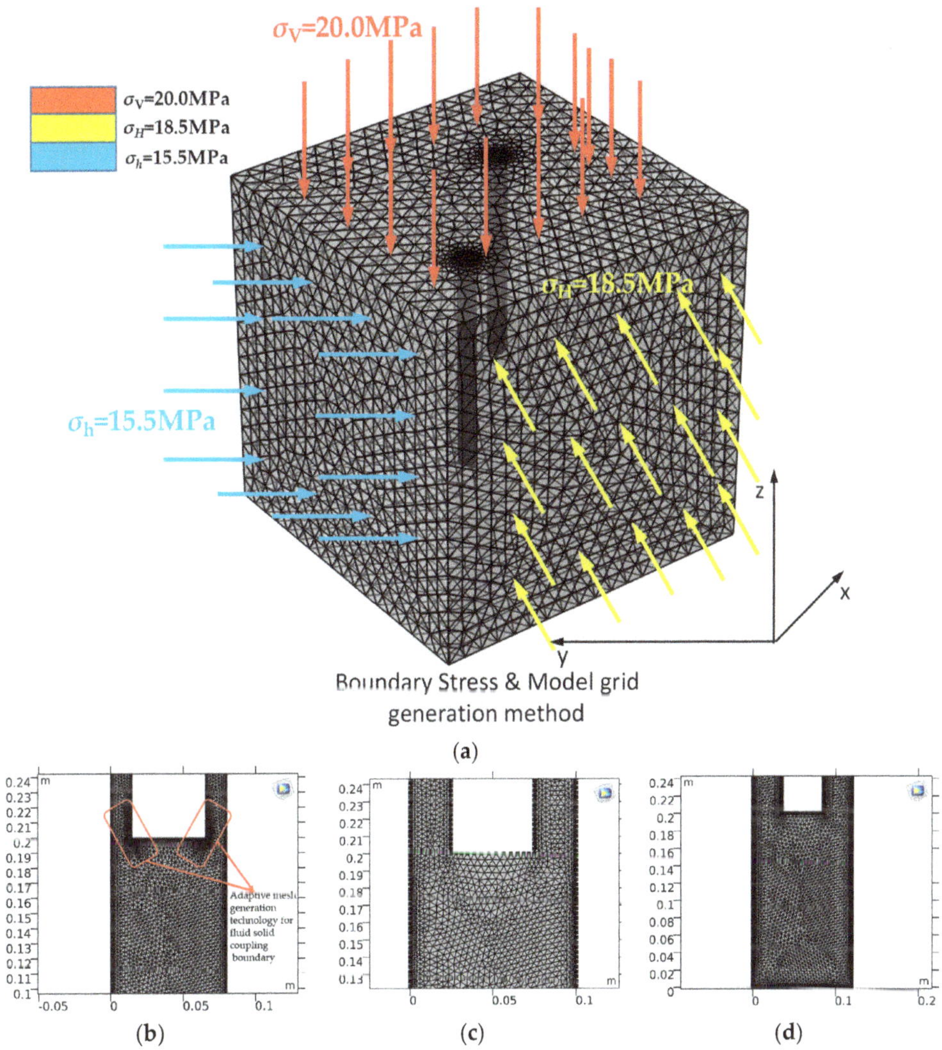

Figure 2. Boundary conditions and mesh generation. (**a**) Boundary load setting method of type I in situ stress rock specimen. (**b**) Grid of 0.08 m borehole. (**c**) Grid of 0.10 m borehole. (**d**) Grid of 0.12 m borehole.

In the fluid solid coupling simulation experiment, for the fluid mechanics calculation part, the water inlet (the starting point of the fracturing fluid inflow), the wall (the part where the fluid contacts the solid) and the water outlet (the end point of the fracturing fluid outflow) are usually set. The water outlet is free in the jet stage, closed in the fracturing stage, and the model is pressurized.

Figure 1b shows the development trend of fracture propagation direction induced by different geostress types, while Figure 2 shows the experimental model to be consistent with the standard specimen of rock mechanics experiment in the laboratory and the drilling hole layout.

The flow field inside the borehole was determined by the inlet flow velocity, outlet pressure, and wall equation, where the inlet flow velocity was 3×10^2 m/s, the outlet was open and free, and the orifice pressure was 0 Pa. The flow field wall was smooth without sliding. The total pressure of the internal flow field was 35 Mpa under the pressure holding state.

Considering that the COMSOL multi physical field simulation software is based on the calculation principle developed by finite element method, the excellent simulation on the boundary of the two physical fields and the relatively accurate calculation of the corresponding stress field and velocity field are highlighted when the hydraulic fracturing process is simulated. However, there is also a defect that the fracture expansion cannot be visualized. Therefore, this manuscript illustrates the scientific problems of the research object by comparing its equivalent force field and surface velocity field, and looks forward to fitting the reflection of crack propagation mode in the laboratory with the simulation results in this manuscript in the future research work.

In consideration of both convergence and experimental effect, it is considered that the finer adaptive mesh generation method is most suitable for the experimental study of fluid structure coupling mechanical model established in this manuscript.

Based on the above boundary conditions and the multi-physical field fluid–solid coupling parameter settings, the mesh division and model size are shown in Figure 2.

The turbulent governing equation of the transient solver inside the borehole was as follows:

$$\begin{aligned} &\rho\tfrac{\partial u}{\partial t} + \rho(u\cdot\nabla)u = \nabla\cdot[-p\mathrm{I}+\kappa]+F \\ &\rho\nabla\cdot u = 0 \end{aligned} \quad (1)$$

$$\kappa = (\mu+\mu_T)(\nabla u_2 + (\nabla u_2)^T) \quad (2)$$

$$\rho\frac{\partial k}{\partial t} + \rho(u\cdot\nabla)K = \nabla\cdot[(\mu+\frac{\mu_T}{\sigma_K})\nabla K] + P_k - \rho\varepsilon \quad (3)$$

$$\begin{aligned} &\rho\tfrac{\partial \varepsilon}{\partial t} + \rho(u\cdot\nabla)\varepsilon = \nabla\cdot[(\mu+\tfrac{\mu_T}{\sigma_\varepsilon})\nabla\varepsilon] + C_{\varepsilon 1}\tfrac{\varepsilon}{k_2}P_k - C_{\varepsilon 2}\rho\tfrac{\varepsilon^2}{k_2} \\ &\varepsilon = ep \end{aligned} \quad (4)$$

$$\mu_T = \rho C_\mu\frac{k^2}{\varepsilon}, P_k = \mu_T[\nabla u : (\nabla u + (\nabla u)^T)] \quad (5)$$

where u is the velocity field component, m/s; P is the pressure, MPa; F is the boundary pressure of the solid mechanics, MPa; K (K is an capital English letter) is the turbulent flow energy N·m; ε is the turbulent dissipation rate; ρ is the fluid density, kg·m^3; $\bigtriangledown$ is the symbol of the nabrakin operator; κ (kappa 'κ' is a Greek alphabet) is the directional gradient matrix, μ is the dynamic viscosity coefficient, n·s/m^3; $C_{\varepsilon 1}$ is the compressibility of the fluid; here, the incompressible flow was selected, and the parameter was 1; and k (k is an italic lowercase English letter) is the turbulence auxiliary variable constant.

The solid boundary coupling governing equation is:

$$\rho\frac{\partial^2 u}{\partial t^2} = \nabla\cdot s + \mathrm{F_V} \quad (6)$$

where s is the displacement of the solid boundary, m; and $\mathrm{F_V}$ is the normal stress at the solid boundary, MPa.

3. Internal Flow Field and Boundary Surface Pressure of Solid Mechanics in Different Boreholes

Based on the above flow-field transient control equations and solid-mechanics boundary equations, the transient solver was used to solve them. The borehole surface velocity and fluid pressure obtained after convergence and equilibrium are shown in Figures 3–5.

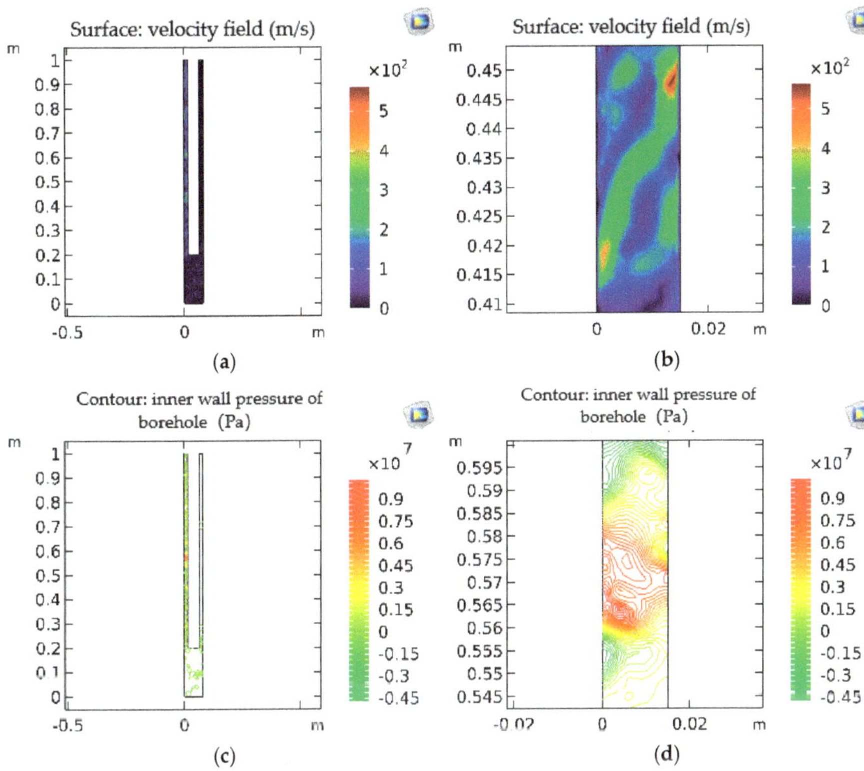

Figure 3. Fluid–solid coupling boundary state of 0.08 m borehole. (**a**) Surface velocity of inner wall of 0.08 m borehole. (**b**) Surface velocity of inner wall of 0.08 m borehole (velocity peak position). (**c**) Hole wall pressure of 0.08 m hole diameter. (**d**) Hole wall pressure of 0.08 m hole diameter (stress peak position).

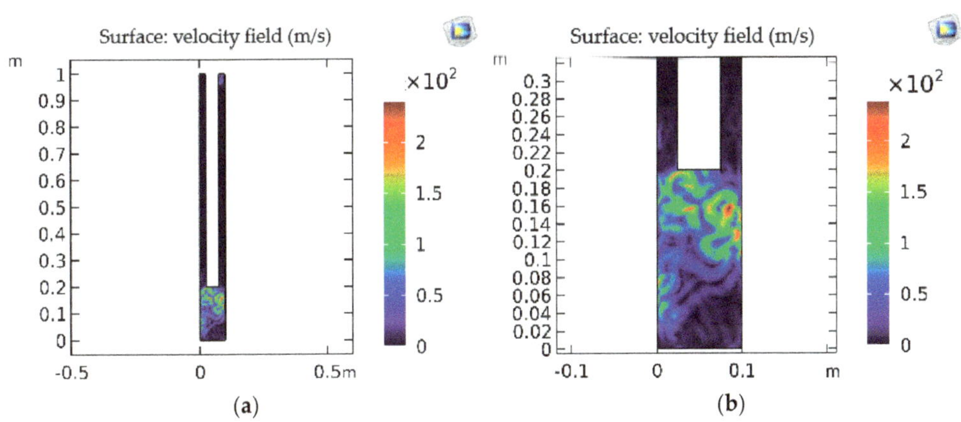

Figure 4. *Cont.*

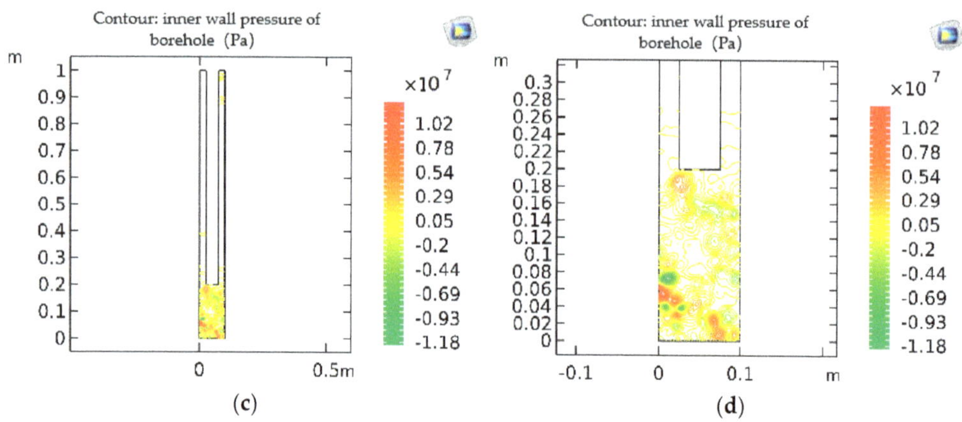

Figure 4. Fluid–solid coupling boundary state of 0.10 m borehole. (**a**) Surface velocity of inner wall of 0.10 m borehole. (**b**) Surface velocity of inner wall of 0.10 m borehole (velocity peak position). (**c**) Hole wall pressure of 0.10 m hole diameter. (**d**) Hole wall pressure of 0.10 m hole diameter (stress peak position).

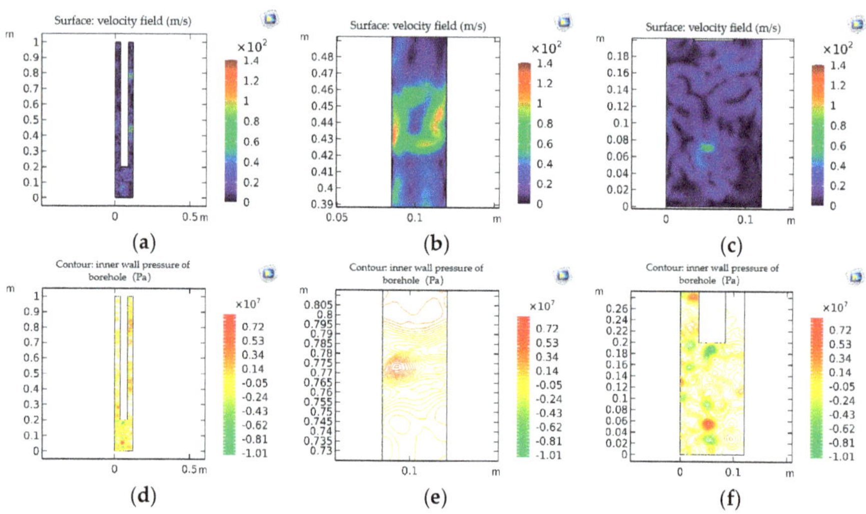

Figure 5. Fluid–solid coupling boundary state of 0.12 m borehole. (**a**) Surface velocity of inner wall of 0.12 m borehole. (**b**) Surface velocity of inner wall of 0.12 m borehole (velocity peak position 1). (**c**) Surface velocity of inner wall of 0.12 m borehole (velocity peak position 2). (**d**) Hole wall pressure of 0.12 m hole diameter. (**e**) Hole wall pressure of 0.12 m hole diameter (stress peak position 1). (**f**) Hole wall pressure of 0.12 m hole diameter (stress peak position 2).

According to the analysis of the flow velocity–stress state of the inner wall of the 0.08 m borehole as shown in Figure 3, under the abovementioned computational fluid dynamics (CFD) flow-field parameter setting conditions, the peak flow velocity in the 0.08 m borehole reached about 1.47 Mach number, which met the turbulence state, and the peak velocity appeared at the slot position of the borehole wall. The peak pressure on the surface of the inner wall of the hole reached about 9 MPa, and the position of the peak pressure was basically the same as the position of the peak flow rate. This phenomenon indicated that the crack initiation position of the 0.08 m hole was more likely to occur at the wall of the hole but less likely to occur at the fracturing target rock layer at the bottom of the hole. Therefore, it is not recommended to select the 0.08 m hole for drilling and fracturing hard-rock layers under the working conditions assumed in this paper.

As shown in Figure 4, the flow velocity and boundary surface pressure of the internal flow field of the 0.10 m borehole were analyzed. The peak position of the flow velocity mainly occurred in the area between the jet nozzle and the bottom of the borehole. The peak flow velocity reached about 2×10^2 m/s. The peak position of the surface pressure also mainly appeared at the hole bottom, and the peak stress was about 10.2 MPa. Compared with the 0.08 m hole diameter, the peak flow velocity and peak pressure were reduced to some extent, but the flow velocity and pressure in the general area were increased, and the area of the peak position was also significantly increased, mainly at the bottom of the hole, which may weaken the roof by causing multiple cracks.

Figure 5 shows the internal flow velocity and fluid–solid coupling boundary pressure state of the 0.12 m borehole. When the 0.12 m borehole was selected for hydraulic fracturing, the peak flow velocity reached 1.4×10^2 m/s, which was widely distributed at the slit between the hole bottom, the hole orifice, and the hole wall. The surface peak pressure was about 7 MPa, which was distributed throughout the hole wall. Through comprehensive analysis of Figures 3–5, with the gradual increase in the borehole diameter, the peak flow velocity and peak surface pressure of the flow field weakened to a certain extent but remained on the same order of magnitude as a whole, while the distribution of the peak area increased significantly, which was not advantageous for the pressure of hard but thin rock layers. However, during the fracturing operation of hard roof plates in deep mines, the target fracturing rock layers are generally thicker basic roof layers. Therefore, in these three schemes, the selection of the 0.12 m borehole diameter for the next step of the stress-shadow effect experiment comparison is of more practical significance for the fracturing optimization under the hypothetical conditions in this paper. When the borehole spacing is small, a stress shadow effect appears, and the influence range of the plastic zone overlaps, resulting in the rock between the boreholes being relatively broken, and the stress in the plastic zone decreases. When the two boreholes move further away, the cracks around the boreholes are developed and connected, but the stress shadow effect does not appear; hence, the stress field value between the boreholes increases.

4. Contrast Experiment of the Stress Shadow Effect between Different Borehole Spacings

The study of the stress shadow effect between natural primary cracks or artificial cracks was initially carried out for fractures created during shale gas extraction [14,15]; in Yu et al. [16], an analytical equation was given for the redistribution of the induced stress components between the stress shadows [17–19]:

$$
\begin{aligned}
\Delta\sigma_{xx,induced} &= \frac{\int_{-A_c}^{A_c} (\sigma_{xx,induced} {l_1}^2 + \sigma_{yy,induced} {m_1}^2 + 2\tau_{xy,induced} {l_1}^2 {m_1}^2) dx}{2A_c} \\
\Delta\sigma_{yy,induced} &= \frac{\int_{-A_c}^{A_c} (\sigma_{xx,induced} {l_2}^2 + \sigma_{yy,induced} {m_2}^2 + 2\tau_{xy,induced} {l_2}^2 {m_2}^2) dx}{2A_c}
\end{aligned}
\quad (7)
$$

where x is the average induced stress along the tangent direction and normal direction of the crack, A_c is the half length of the crack, taken as 0.5 m; m_i is the distance between end cracks, m; l_i is the crack width, m; $\sigma_{yy,induced}$, $\sigma_{xx,induced}$, $\tau_{xy,induced}$ is the stress tensor, respectively, MPa; $\Delta\sigma_{yy,induced}$, $\Delta\sigma_{xx,induced}$, $\Delta\tau_{xy,induced}$ is the increment in the stress tensor, respectively, MPa [20–23].

Figure 6a,b shows the influence range of the von Mises equivalent stress in the rock mass caused by 0.12 m borehole fracturing with 3.0 m spacing; Figure 6c,d shows the influence range of the von Mises equivalent stress in the rock mass caused by 0.12 m borehole fracturing with 3.5 m spacing, and Figure 6e,f shows the influence range of the von Mises equivalent stress in the rock mass caused by 0.12 m borehole fracturing with 4.0 m spacing. Through comparative analysis, with the increase in the borehole spacing, the damage range around the borehole also increased, especially the damage range at the orifice, and the overall average value of the equivalent stress around the borehole also increased at the spacing of 4.0 m. Therefore, in order to avoid the influence of the stress shadow effect on the spatial shape of the fracture development and the expansion and the pressure relief in advancing during the actual hydraulic-fracturing rock-breaking construction, it is necessary to ensure that the influence range of the equivalent stress of the fracture is at the overlapping critical edge as far as possible.

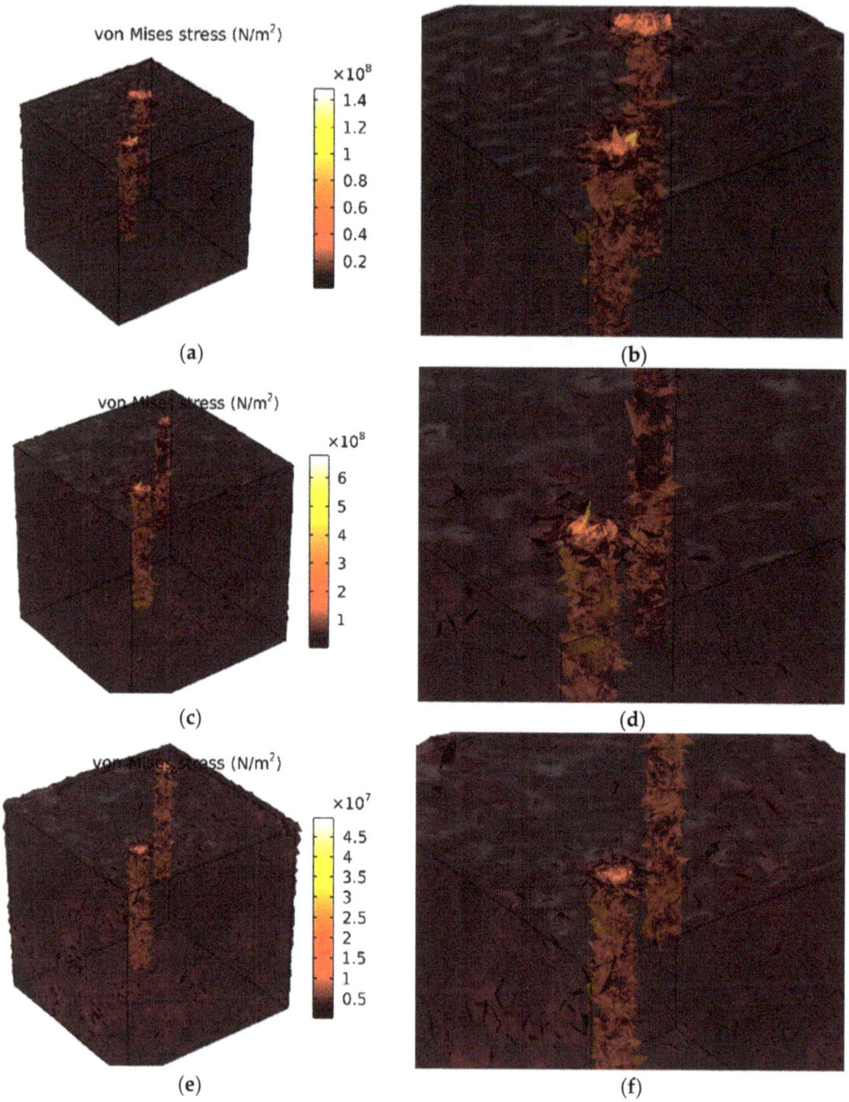

Figure 6. Stress shadow effect represented by an equal-effect stress field with different hole spacings. (**a**) Equivalent stress field of 0.12 m borehole with 3 m spacing. (**b**) Equivalent stress field of 0.12 m borehole with 3 m spacing (details). (**c**) Equivalent stress field of 0.12 m borehole with 3.5 m spacing. (**d**) Equivalent stress field of 0.12 m borehole with 3.5 m spacing (details). (**e**) Equivalent stress field of 0.12 m borehole with 4 m spacing. (**f**) Equivalent stress field of 0.12 m borehole with 4 m spacing (details).

With the increase in the buried depth, the spatial shape trend of the fracture surface development and the expansion of the hydraulic-fracturing borehole wall changes significantly. Under the condition of shallow-buried in situ stress, where the horizontal maximum principal stress is the maximum principal stress, and the vertical stress is the minimum principal stress, the fracture surface tends to expand along the water square direction to weaken the thickness of the fractured rock stratum. Under the in situ stress condition of deep mines, where the vertical stress is the maximum principal stress, and the horizontal stress is the minimum principal stress, the fracture plane tends to shorten the fracture step of the fractured rock layer along the vertical direction. Therefore, it was concluded that under different in situ stress types and different purposes of cutting hard-rock layers, appropriate drilling orientations and induction methods should be selected to optimize and control the spatial morphology of the fracture plane.

5. Discussion and Conclusions

Based on the established rock-mechanics multi-physical field fluid–solid coupling hydraulic-fracturing mechanical model and the assumptions and boundary conditions in computational fluid mechanics and rock mechanics, the turbulent control equation of the internal flow field in the borehole, the boundary laminar-flow control equation, and the fluid–solid coupling boundary conditions were obtained, and the rock-mass model borehole surface was substituted as the load boundary condition to simulate the impact of the stress shadow effect on the fracturing effect under different borehole spacings. At the same time, the borehole diameter and spacing suitable for the geological conditions of the example were compared.

In practice, the conclusions of this paper are the following:

1. Through the comparison of the fluid–solid coupling numerical simulation experiments, it was concluded that hydraulic fracturing is more effective for roof presplitting and pressure relief when the 0.12 m diameter borehole spacing is 3.5 m, considering the spatial development scale of the fracture network and fracturing efficiency (large-flow borehole).
2. The coupling relationship between the traditional hydraulic-fracturing fracture propagation model, the surface flow field velocity–stress model, and the solid material is difficult to express with the three-dimensional model. In view of the improvement of the above conditions, the displacement and stress constitutive relationship of the solid boundary were selected as appropriate functions to reconcile, and the boundary solution formula suitable for the three-dimensional shape of the borehole was obtained.
3. Further, deep-seated mechanism research should also be carried out on the flow field conditions of the cutting effect of the hydraulic fracturing technology on rock materials. The impact of the inlet flow velocity and pressure, outlet aperture, outlet pressure and flow, and the seepage state of the whole section of the borehole under the pressure-holding state on fracture propagation will be reflected in our further research.

Author Contributions: Conceptualization, S.W.; methodology, S.W.; software; J.L.; validation, S.W.; formal analysis, S.W.; investigation, S.W.; resources, S.W.; data curation, S.W.; writing—original draft preparation, S.W.; writing—review and editing, J.L.; visualization, J.L.; supervision, S.W.; project administration, S.W. All authors have read and agreed to the published version of the manuscript.

Funding: This research received no external funding.

Institutional Review Board Statement: Not applicable.

Informed Consent Statement: Informed consent was obtained from all subjects involved in the study.

Data Availability Statement: The experimental and analytical calculation data used to support the findings of this study are included within the article.

Conflicts of Interest: The authors declare that there are no conflicts of interest regarding the publication of this paper.

References

1. Terzaghi, K.; Peck, R.B.; Mesri, G. *Soil Mechanics in Engineering Practice*; Wiley: New York, NY, USA, 1996.
2. Barenblatt, G.I. The formation of equilibrium cracks during brittle fracture. General ideas and hypotheses. Axially-symmetric cracks. *J. Appl. Math. Mech.* **1959**, *23*, 622–636. [CrossRef]
3. Barenblatt, G.I. Equilibrium cracks formed during brittle fracture rectilinear cracks in plane plates. *J. Appl. Math. Mech.* **1959**, *23*, 1009–1029. [CrossRef]
4. Hutchinson, J.W. Singular behavior at the end of a tense crack in a hardening material. *J. Mech. Phys. Solids* **1968**, *16*, 13–31. [CrossRef]
5. Sneddon, I.N.; Elliot, H.A. The opening of a Griffith crack under internal pressure. *Q. Appl. Math.* **1946**, *4*, 262–267. [CrossRef]
6. Sneddon, I.N. The Distribution of Stress in the Neighbourhood of a Crack in an Elastic Solid. *Proced. R. Soc. Lond.* **1946**, *187*, 229–260.
7. Warpinski, N.R.; Branagan, P.T. Alternate stress fracturing. *J. Pet. Technol.* **1989**, *41*, 990–997. [CrossRef]
8. Wang, B.; Zhou, F.; Hu, J.; Gao, L.; Yuan, L.; Wang, Y. Fracture Interaction during Temporarily Plugging Staged Fracturing. *J. Pet. Sci. Eng.* **2018**, *4*, 8–13.
9. Nagel, N.B.; Sanchez Nagel, M. Stress shadowing and microseismic events: A numerical evaluation. In Proceedings of the SPE Annual Technical Conference and Exhibition, Denver, CO, USA, 30 October–2 November 2011.
10. Zhao, K.; Zhang, Z.; Li, W.; Wang, X.; Yi, K.; Sun, Z. Study on three-dimensional propagation of hydraulic fracture in borehole initiation based on xsite. *J. Geotech. Eng.* **2021**, *43*, 1483–1491.
11. Wang, H. Poro-elasto-plastic modeling of complex hydraulic fracture propagation: Simultaneous multi-fracturing and producing well interference. *Acta Mech.* **2016**, *227*, 507–525. [CrossRef]
12. Liu, C.; Li, M.; Hao, L.; Hu, H. Numerical simulation of hydraulic fracture propagation in heterogeneous unconventional reservoir. *AIP Conf. Proc.* **2017**, *1890*, 020010.
13. Michael, A. Hydraulic Fracturing Optimization: Experimental Investigation of Multiple Fracture Growth Homogeneity via Perforation Cluster Distribution. Ph.D. Thesis, The University of Texas at Austin, Austin, TX, USA, 2016.
14. Olson, J.E. Multi fracture propagation modeling: Applications to hydraulic fracture in shale and tight gas sands. In Proceedings of the 42nd American Rock Mechanics Association, San Francisco, CA, USA, 29 June–2 July 2008.
15. Sepehri, J.; Solid, M.Y.; Morse, S.M. Application of extended finite elementmethod to simulate hydraulic fracture propagation from orientated performances. In Proceedings of the SPE Hydraulic Fracturing Technology Conference, The Woodlands, TX, USA, 3–5 February 2015.
16. Yu, Y.J.; Zhu, W.C.; Li, L.C.; Wei, C.H.; Dai, F.; Liu, S.Y.; Wang, W.D. Theoretical analysis of stress shadow effect of mutual interference of hydraulic fracturing fractures. *J. Rock Mech. Eng.* **2017**, *36*, 2926–2939. [CrossRef]
17. Zhang, Z.L.; Shu, C.; Khalid, M.S.U.; Chen, Z.; Yuan, Z.Y.; Liu, W. Investigations on the hydroelastic slamming of deformable wedges by using the smoothed particle element method. *J. Fluids Struct.* **2022**, *114*, 103732. [CrossRef]
18. Long, T.; Zhang, Z.L.; Liu, M.B. A multi-resolution technique integrated with the smoothed particle element method (SPEM) for modeling fluid-structure interaction problems with free surfaces. *Sci. China Phys. Mech. Astron.* **2021**, *64*, 284711. [CrossRef]
19. Wang, X.; Li, J.C.; Xiao, B.Z.; Yue, L. Propagation characteristics and prediction of blast-induced vibration on closely spaced rock tunnels. *Tunn. Undergr. Space Technol.* **2022**, *123*, 104416. [CrossRef]
20. Abe, A.; Kim, T.W.; Horne, R.N. Laboratory hydraulic stimulation experiments to investigate the interaction between newly formed and preexisting fractures. *Int. J. Rock Mech. Min. Sci.* **2021**, *141*, 104665. [CrossRef]
21. Li, B.; Zhang, J.; Liu, Y.; Qu, L.; Liu, Q.; Sun, Y.; Xu, G. Interfacial porosity model and modification mechanism of broken coal grouting: A theoretical and experimental study. *Surf. Interfaces* **2022**, *33*, 102286. [CrossRef]
22. Zhang, J.; Li, B.; Liu, Y.; Li, P.; Fu, J.; Chen, L.; Ding, P. Dynamic multifield coupling model of gas drainage and a new remedy method for borehole leakage. *Acta Geotech.* **2022**, *17*, 4699–4715. [CrossRef]
23. Li, B.; Shi, Z.; Li, L.; Zhang, J.; Huang, L.; He, Y. Simulation study on the deflection and expansion of hydraulic fractures in coal-rock complexes. *Energy Rep.* **2022**, *8*, 9958–9968. [CrossRef]

processes

MDPI

Article

Vibration Response of the Interfaces in Multi-Layer Combined Coal and Rock Mass under Impact Load

Feng Li, Guanghao Wang *, Guangyou Xiang, Jia Tang, Baorui Ren and Zhibang Chen

School of Emergency Management and Safety Engineering, China University of Mining and Technology, Beijing 100083, China
* Correspondence: zqt2010102037@student.cumtb.edu.cn; Tel.: +86-177-660-77-309

Abstract: The stress wave generated by impact or dynamic load will produce significant reflection and transmission at the rock coal or rock interface during the propagation process. This will produce dynamic effects such as dynamic tensile, stress superposition and mutation. These dynamic effects will lead to obvious vibration at the interfaces, which is a key factor leading to dynamic damage and the failure of coal and rock mass. In the process of underground engineering excavation, the dynamic damage of a series of layered rock masses is one of the important factors causing geological disasters. Based on the two–dimensional similar material simulation experiment, the coal and rock mass combined of five layers of fine sandstone, medium sandstone, coal, coarse sandstone and mudstone was taken as the research object, and single and multi-point excitation (synchronous/step-by-step) were used to test the time–history vibration curves of rock–coal and rock–rock interfaces under impact load. It was concluded that the change of extreme value of vibration amplitude presented two stages: first increase, and then attenuation. Most of them required 2.25 cycles to reach the peak value, and the dynamic attenuation of amplitude conformed to the law of exponential. Based on Fast Fourier transform (FFT), the spectrum structures of the amplitude–frequency of interface vibration were studied, and the two predominant frequencies were 48.9~53.7 Hz and 92.4 Hz, respectively. Based on the Hilbert-Huang transform and energy equation, 5~7 vibration modes (IMF) were obtained by decomposing the time–history curves. The three modes, IMF1, IMF2, and IMF3, contained high energy and were effective vibration modes. IMF2 accounted for the highest proportion and was the main vibration mode whose predominant frequencies were concentrated in 45.6~50.2 Hz. Therefore, IMF2 played a decisive role in the whole vibration process and had an important impact on the dynamic response, damage and failure of coal and rock mass. In real conditions, the actual predominant frequencies can be converted according to the size and mechanical properties of the coal and rock mass, and the vibration response characteristics of the interfaces between coal and rock mass under impact load were preliminarily revealed. This study can provide reference for monitoring and early warning of coal and rock dynamic disasters, prevention and control of coal and gas outburst and technical development.

Keywords: impact load; multi-layer combined coal and rock mass; interface vibration; effective vibration mode; predominant frequency

Citation: Li, F.; Wang, G.; Xiang, G.; Tang, J.; Ren, B.; Chen, Z. Vibration Response of the Interfaces in Multi-Layer Combined Coal and Rock Mass under Impact Load. *Processes* **2023**, *11*, 306. https://doi.org/10.3390/pr11020306

Academic Editor: Yidong Cai

Received: 6 November 2022
Revised: 29 December 2022
Accepted: 5 January 2023
Published: 17 January 2023

1. Introduction

The sedimentary process and the arrangement of mineral particles during the formation of coal measure rock mass give an obvious bedding structure and show layered structure macroscopically [1–3]. This layered rock mass widely exists in mining, construction, transportation and water conservancy projects [4–6], and is prone to vibration, damage and even failure under the action of external dynamic disturbances, such as blasting, roof and floor breaking, rock instability and so on [7,8]. For example, the roof strata of working face in mines, especially in the early stage of mining, is subjected to the loading/unloading action of dynamic disturbance and the appearance of mineral pressure is violent, which leads to obvious mine earthquakes and even roof collapse. Therefore, in the process of underground engineering excavation, the dynamic damage of a series of layered rock masses is one of the most important factors causing geological disasters.

At present, scholars in China and other countries have mainly studied the static mechanical properties of single, two-layer or three-layer combined coal and rock mass. Zuo et al. demonstrated that both the breaking strength and elastic modulus of the coal–rock assemblage increased to some extent compared with the single coal sample, with the rock–coal–rock assemblage having the largest breaking strength and elastic modulus [9]. Li et al. demonstrated that more than one peak was observed in the spectrum of green sandstone with repeated impact loads, and the relative weight of peak frequencies increased in the low frequency range [10]. Mu et al. demonstrated that the damage state of coal rock specimens was determined by the post peak stiffness of the coal in relation to the rock stiffness [11]. Liu et al. demonstrated that the dynamic strength and fragmentation dimension of the combined coal rock specimens increased with increasing stress wave energy and showed a trend of first increasing and then decreasing with static load [12]. However, the research on dynamic response, the vibration characteristics and dynamic damage of coal and rock mass, especially multi-layer combined coal and rock mass, are still in the stage of experimental exploration. The stress wave generated by impact or dynamic load has significant reflection and transmission at the rock–coal or rock–rock interfaces in the process of propagation, resulting in dynamic effects such as dynamic tensile, stress superposition and mutation. These dynamic effects lead to obvious vibration at the interfaces, which is a key factor leading to dynamic damage and failure of coal and rock mass [13–16]. Meanwhile, Li et al. demonstrated that the initial location of damage and failure can be determined by the lowest frequency effective vibration mode [17]. These articles illustrate that the study of coal rock vibration signals can help to initially reveal the vibration response characteristics of the coal rock interface under impact loading. Unlike previous studies on the static response characteristics of combined coal and rock mass below three layers, this paper conducts a study on the dynamic response characteristics of five-layer combined coal and rock mass.

Based on the two-dimensional similar material simulation experiment, the coal and rock mass combined by five layers of fine sandstone, medium sandstone, coal, coarse sandstone and mudstone was taken as the research object, single and multi-point excitation (synchronous/step-by-step) were used to test the time–history vibration curves of rock–coal and rock–rock interfaces under impact load, and the dynamic attenuation law of amplitude was obtained. Based on Fast Fourier transform (FFT), the spectrum structures of amplitude–frequency of interfaces vibration were studied. Based on Hilbert-Huang transform and energy equation, the time–history curves were decomposed by Ensemble Empirical Mode Decomposition (EEMD) and the vibration modes were obtained. According to the distribution laws of marginal spectrum and energy proportion, the most effective and main vibration modes and the predominant frequencies were obtained. The vibration response characteristics of the interfaces between coal and rock mass under impact load were preliminarily revealed.

2. Experimental Design

2.1. Experimental Model

The two-dimensional similar material simulation experiment platform was often used to simulate the deformation, damage and failure laws of overlying strata in the process of mining [18,19]. The size of similar material simulation experiment platform was 1800 × 160 × 1100 mm in the lab. The multi-layer combined coal and rock mass was composed of five layers, each with a thickness of 200 mm; from top to bottom, they were fine sandstone, medium sandstone, coal, coarse sandstone and mudstone. The similarity density constant of rock mass is 1.6, and that of coal is 1; the composition and ratio of similar materials of each layer in coal and rock mass can be seen in Table 1. The vibration amplitude was measured by magnetoelectric speed sensors (2D001) with a size of 63 × 63 × 63 mm.

The distributed network dynamic signal test system (DH5981) was used for data acquisition and analysis, as shown in Figure 1.

Table 1. The composition and ratio of similar materials of each layer.

Serial Number	Layers of Coal and Rock Mass	Raw Material Compressive Strength (MPa)	Similar Material Compressive Strength (MPa)	Fine Sand (Kg)	Cement (Kg)	Gypsum (Kg)	Water (Kg)
1	Fine sandstone	120	0.75	72.98	7.30	17.03	9.73
2	Medium sandstone	80.1	0.50	81.09	8.11	8.11	9.73
3	Coal	10	0.063	86.49	7.57	3.24	9.73
4	Coarse sandstone	61.5	0.384	83.40	6.95	6.95	9.73
5	Mudstone	27.4	0.171	88.46	6.32	6.32	10.11

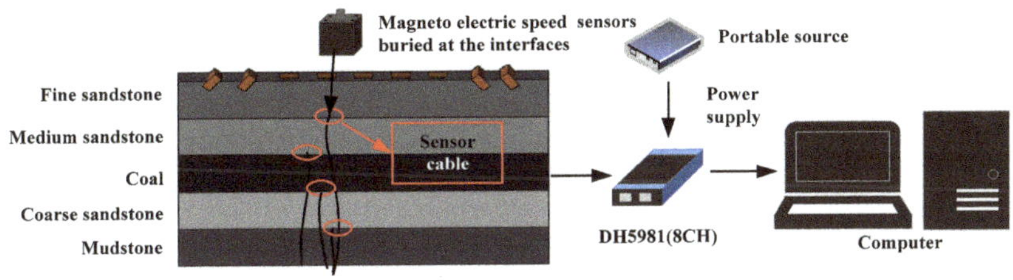

Figure 1. Experimental model and data acquisition system.

2.2. Action Points and Sensors Distribution

Four sensors (No. 1–4) were distributed at the interfaces of each layer of combined coal and rock mass, and each sensor was at the midpoint of each interface. The lower surface of the sensor (signal receiving surface) coincided with each interface of coal and rock mass. An action point was set every 100 mm on the upper surface of the fine sandstone for percussion excitation. An iron block of 80 × 80 mm was placed at action points to prevent damage to fine sandstone during excitation; a total of five vertical action points (1–5#) and four inclined action points with an angle of 45° to the horizontal direction (6–9#) were arranged, as shown in Figure 2.

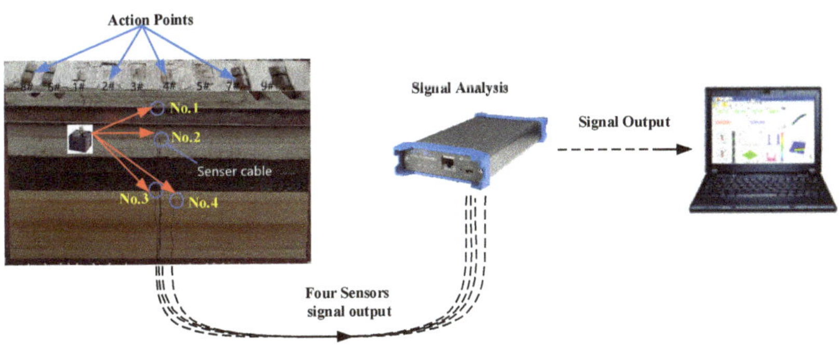

Figure 2. Action points and sensors distribution in combined coal and rock mass.

2.3. Experimental Process

Single point and multi-point excitation (synchronous/step-by-step) were used in the experiment, and the specific excitation steps were as follows:

(1) the action point (1~9#) was excited respectively; (2) two action points were executed synchronously or step-by-step as follows: 1# and 2#, 1# and 5#, 2# and 4# (excitation step-by-step), 6# and 7#, 6# and 8#, 7# and 8#, 8# and 9#; (3) four action points (6~9#) and five action points (1~5#) were excited synchronously and respectively.

Before the experiment, the data acquisition instrument needed to be zeroed first, and when the impact actions caused by the excitation were balanced, the data acquisition would be terminated and saved.

3. Experimental Results

3.1. Amplitude Variation of Interface Vibration under Impact Load

3.1.1. Amplitude Variation of Interface Vibration under Single Point Excitation

When a single point was excited, the dynamic variation laws of the vibration amplitude of each interface obtained from the four sensors were similar; the amplitude time-history curves of 2#, 4# and 6# action points excited separately are shown in Figures 3–5. It can be seen from Figure 3 that when the 2# action point was excited alone, the amplitude curve (2#-1 in Figure 3) of No. 1 reached the first extreme value $s_1 = -0.005$ mm when $t_1 = 15$ ms and reached a peak $s_6 = 0.014$ mm when $t_6 = 51$ ms. When 4# action point was excited alone (Figure 4), the amplitude curve of No. 1 (2#-1 in Figure 4) reached the first extreme value $s_1 = -0.004$ mm when $t_1 = 12$ ms and the peak $s_6 = 0.015$ mm when $t_6 = 49$ ms. Similarly, when 6# action point was excited alone (Figure 5), the amplitude curve of No. 1 (6#-1 in Figure 5) reached the first extreme value $s_1 = -0.001$ mm when $t_1 = 7$ ms and the peak value $s_6 = 0.005$ mm when $t_6 = 18$ ms. The extreme points of amplitude curves of single point excitation (2#, 4# and 6#) and their corresponding time can be seen in Table 2. The extreme values of the amplitude curves of these three action points showed two stages as follows: first increasing and then decreasing; the vibration curves of 2# and 4# action points reached the peak after 2.25 cycles, and that of 6# action point the peak after 1.25 cycles.

Table 2. The extreme points of amplitude curves of 2#, 4# and 6# action points.

Extreme Value (mm)/Time (ms)	2#	4#	6#
s_1/t_1	0.005/15	0.004/12	0.001/7
s_2/t_2	0.014/18	0.009/15	0.002/13
s_3/t_3	0.011/21	0.007/22	0.005/18
s_4/t_4	0.010/33	0.012/29	0.005/25
s_5/t_5	0.011/42	0.013/38	0.005/33
s_6/t_6	0.014/51	0.015/49	0.005/43
s_7/t_7	0.014/61	0.015/58	0.005/52
s_8/t_8	0.013/70	0.014/67	0.004/62
s_9/t_9	0.011/80	0.011/78	0.003/72
s_{10}/t_{10}	0.008/90	0.008/86	0.003/82
s_{11}/t_{11}	0.006/99	0.007/97	0.002/92
s_{12}/t_{12}	0.004/109	0.004/117	0.002/101

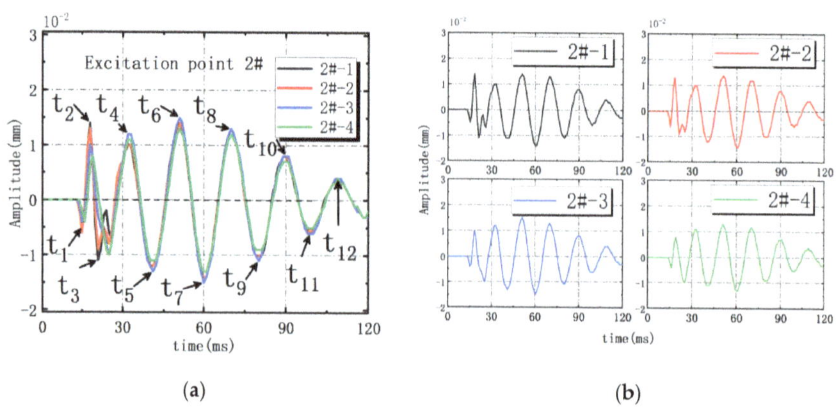

Figure 3. Amplitude time–history curves under 2# action point excited alone. (**a**) Overlapping display of the four sensor signals; (**b**) Separate display of the four sensor signals.

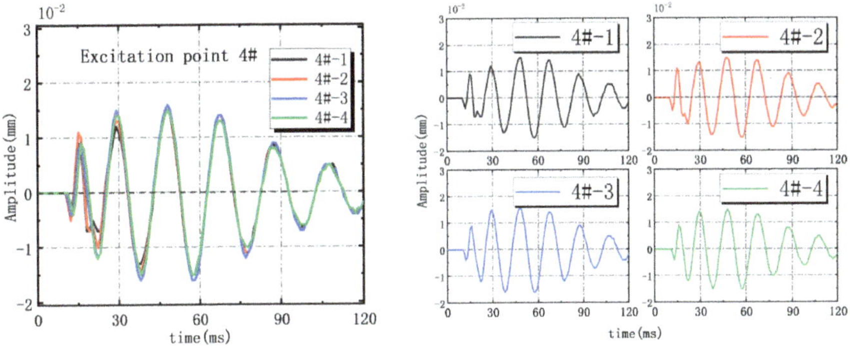

(a) (b)

Figure 4. Amplitude time–history curves under 4# action point excited alone. (**a**) Overlapping display of the four sensor signals; (**b**) Separate display of the four sensor signals.

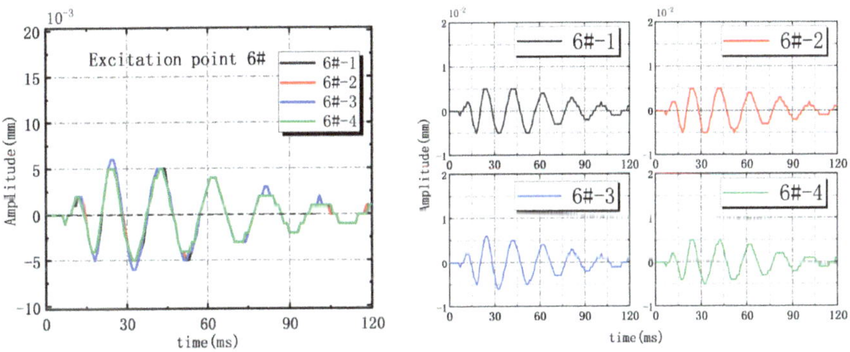

(a) (b)

Figure 5. Amplitude time-history curves under 6# action point excited alone. (**a**) Overlapping display of the four sensor signals; (**b**) Separate display of the four sensor signals.

3.1.2. Amplitude Variation of Interface Vibration under Multi-Point Excitation

When 2# and 4# action points were excited step-by-step, the curve of No. 1 (24#-1 in Figure 6) reached the first extreme value $s_1 = -0.01$ mm when $t_1 = 11$ ms, the peak $s_2 = 0.029$ mm when $t_2 = 14$ ms, and the second peak $s_6 = 0.021$ mm when $t_6 = 46$ ms. Compared with No. 3 and No. 4, No. 1 and No. 2 were close to the action points, and there were complex micro vibrations during step-by-step excitation (24#-1 and 24#-2 in Figure 6), but the whole dynamic change of all measuring points were similar. When 6# and 7# action points were excited synchronously, the curve of No. 1 (67#-1 in Figure 7) reached the first extreme value $s_1 = -0.001$ mm when $t_1 = 12$ ms, the peak $s_6 = 0.016$ mm when $t_6 = 48$ ms, and the whole dynamic change of all measuring points were also similar. The extreme values of the amplitude curves of these two groups of action points showed two stages as follows: first increasing and then decreasing, and the vibration curves reached the peak after 2.25 cycles. The extreme points of amplitude curves of multi-point excitation (2# and 4#, 6# and 7#) and their corresponding time could be seen in Table 3.

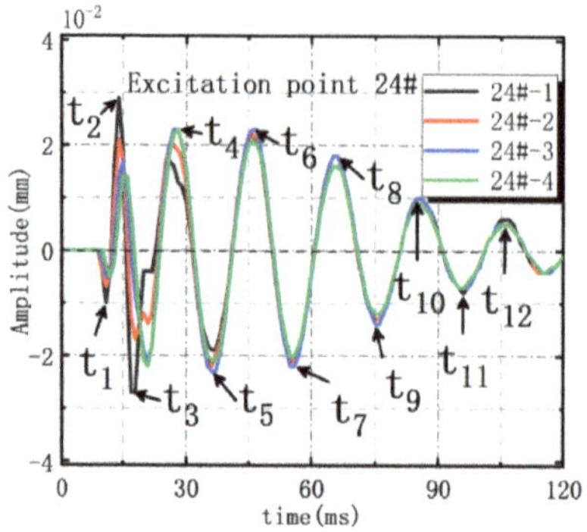

(**a**)

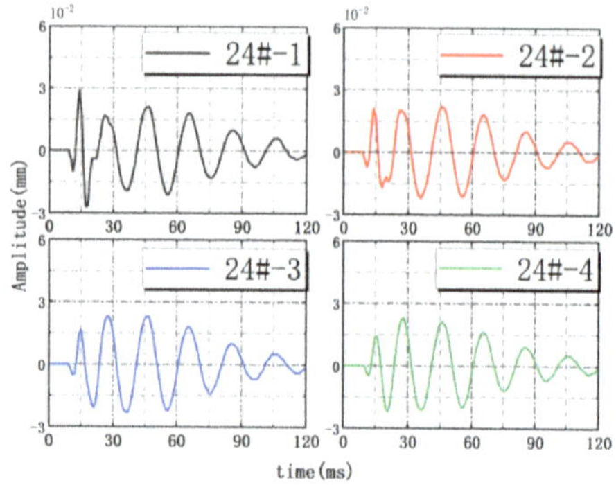

(**b**)

Figure 6. Amplitude time-history curves under 2# and 4# action points excited step by step. (**a**) Overlapping display of the four sensor signals; (**b**) Separate display of the four sensor signals.

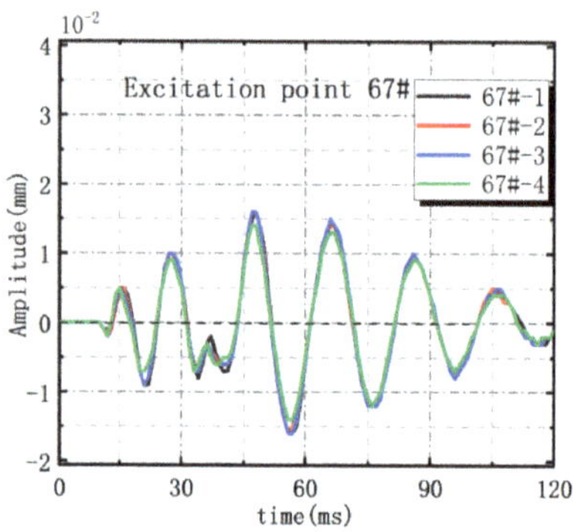

(**a**)

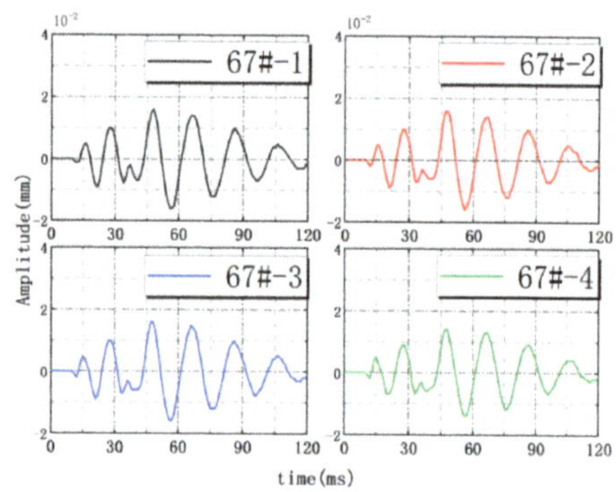

(**b**)

Figure 7. Amplitude time-history curves under 6# and 7# action points excited synchronously. (**a**) Overlapping display of the four sensor signals; (**b**) Separate display of the four sensor signals.

Table 3. The extreme points of amplitude curves of two groups of action points (2# and 4#, 6# and 7#).

Extreme Value (mm)/Time (ms)	2# and 4#	6# and 7#
s_1/t_1	0.010/11	0.001/12
s_2/t_2	0.029/14	0.004/17
s_3/t_3	0.027/18	0.009/22
s_4/t_4	0.016/27	0.010/28
s_5/t_5	0.019/37	0.008/34
s_6/t_6	0.021/47	0.016/48
s_7/t_7	0.021/56	0.016/57
s_8/t_8	0.018/66	0.014/67
s_9/t_9	0.013/76	0.012/77
s_{10}/t_{10}	0.010/86	0.010/86
s_{11}/t_{11}	0.008/96	0.007/97
s_{12}/t_{12}	0.006/107	0.005/106

3.2. Amplitude Attenuation Law of Interface Vibration under Impact Load

3.2.1. Dynamic Attenuation Law of Amplitude under Single Point Excitation

The analysis results of the amplitude extreme value under single point excitation showed that the dynamic changes were similar. The extreme values distribution of the test curves when 6# action point was excited alone can be seen in Table 4, and the curve-fitting results are shown in Figure 8. It could be concluded that the dynamic attenuation of amplitude under single point excitation conformed to the law of exponential variation y = y0 + Aexp (x/k), and the fitting degree was as high as 0.99.

Table 4. Amplitude extreme points of No. 1–4 under 2# action point excitation.

Extreme Point	Time (ms)	No. 1 (mm)	Time (ms)	No. 2 (mm)	Time (ms)	No. 3 (mm)	Time (ms)
T_1	61	0.014	61	0.014	60	0.014	61
T_2	70	0.013	70	0.012	70	0.013	70
T_3	80	0.011	80	0.01	80	0.011	80
T_4	90	0.008	90	0.008	90	0.008	90
T_5	99	0.006	99	0.006	99	0.006	99

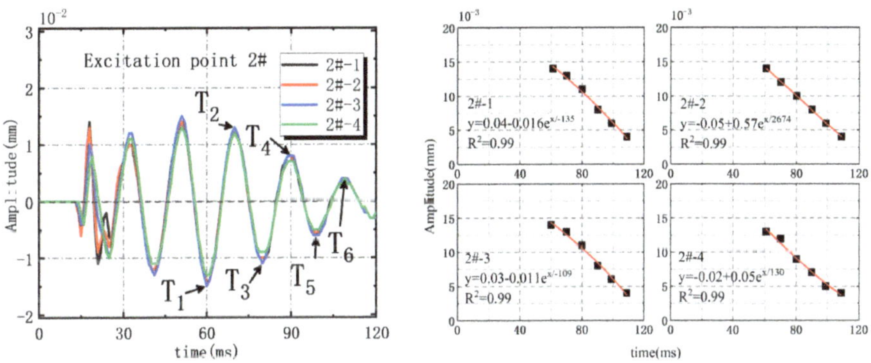

(**a**) (**b**)

Figure 8. Amplitude attenuation law under 2# action point excitation. (**a**) Overlapping display of the four sensor signals; (**b**) Curve-fitting results of curve decay extreme value.

3.2.2. Dynamic Attenuation Law of Amplitude under Multi-Point Excitation

The analysis results of the amplitude extreme value under multi–point excitation showed that the dynamic changes were similar. The extreme values distribution of the test curves when 6# and 7# action points were excited synchronously can be seen in Table 5, and the curve–fitting results are shown in Figure 9. It could be concluded that the dynamic attenuation of amplitude under single point excitation also conformed to the law of exponential variation y = y0 + Aexp (x/k), and the fitting degree was as high as 0.99.

Table 5. Amplitude extreme points of No. 1–4 under 6# and 7# action points excited synchronously.

Extreme Point	Time (ms)	No. 1 (mm)	Time (ms)	No. 2 (mm)	Time (ms)	No. 3 (mm)	Time (ms)
T_1	57	0.016	56	0.016	57	0.016	57
T_2	67	0.014	67	0.014	66	0.015	67
T_3	77	0.012	77	0.012	77	0.012	76
T_4	86	0.01	86	0.01	86	0.01	87
T_5	97	0.007	97	0.007	96	0.008	96

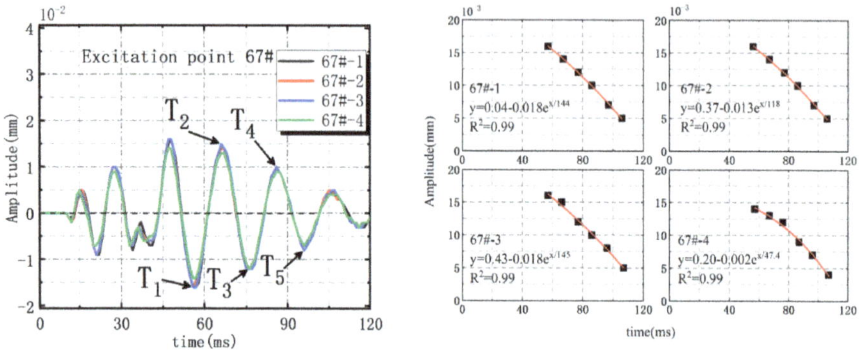

(a) (b)

Figure 9. Amplitude attenuation law under 6# and 7# action points excited synchronously. (**a**) Overlapping display of the four sensor signals; (**b**) Curve-fitting results of curve decay extreme value.

4. Amplitude-Frequency Distribution of Interface Vibration under Impact Load

4.1. Amplitude-Frequency Distribution of Interface Vibration under Single Point Excitation

The amplitude–frequency distributions when the 2#, 4# and 6# action points were excited separately are shown in Figures 10–12. When the 2# action point was excited, the amplitude of No. 1 (2#-1 in Figure 10) reached the peak 7.2×10^{-3} mm at P_1 = 50 Hz. When the 4# action point was excited, the amplitude of No. 1 (4#-1 in Figure 11) reached the peak 7.5×10^{-3} mm at P_1 = 50 Hz. When the 6# action point was excited, the amplitude of No. 1 (6#-1 in Figure 12) reached the peak 1.6×10^{-3} mm at P_1 = 53.7 Hz. The amplitude variations of No. 2~4 were all similar to that of No. 1.

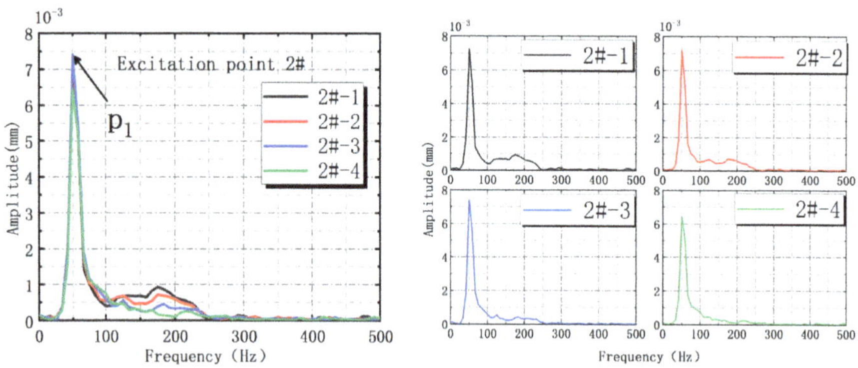

(a) (b)

Figure 10. Amplitude–frequency distribution of interface vibration under 2# action point excitation. (**a**) Overlapping display of the four sensor signals; (**b**) Separate display of the four sensor signals.

4.2. Amplitude-Frequency Distribution under Multi-Point Excitation Step-by-Step

The amplitude–frequency distributions when 2# and 4# action points were excited step-by-step are shown in Figure 13. The amplitude of No. 1 (24#-1 in Figure 13) reached the peak 5.4×10^{-3} mm at P_1 = 52.1 Hz, compared with single point excitation, the vibration complexity of multi-point excitation step-by-step was relatively high, and the amplitude variations of No. 2~4 were all similar to that of No. 1.

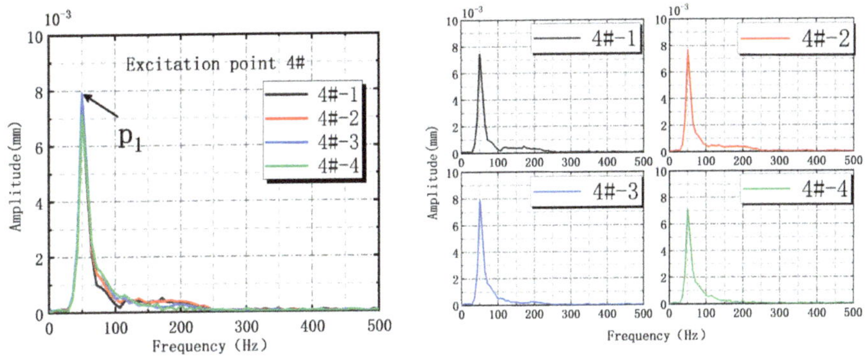

(**a**) (**b**)

Figure 11. Amplitude–frequency distribution of interface vibration under 4# action point excitation. (**a**) Overlapping display of the four sensor signals; (**b**) Separate display of the four sensor signals.

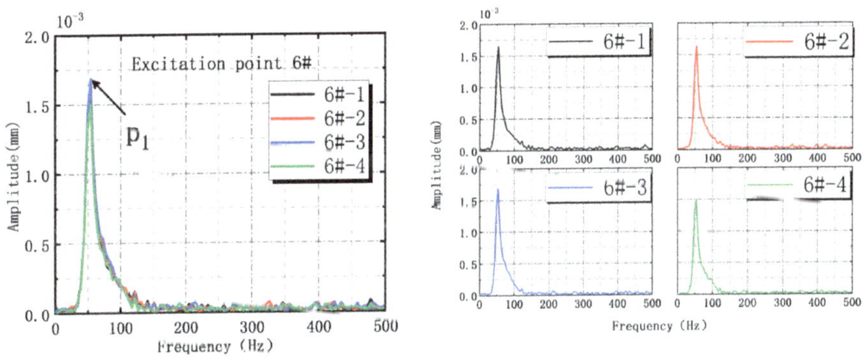

(**a**) (**b**)

Figure 12. Amplitude–frequency distribution of interface vibration under 6# action point excitation. (**a**) Overlapping display of the four sensor signals; (**b**) Separate display of the four sensor signals.

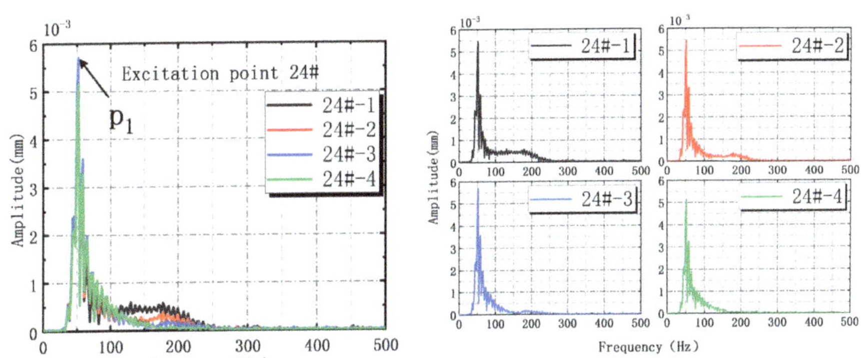

(**a**) (**b**)

Figure 13. Amplitude–frequency distribution under 2# and 4# action points excited step-by-step. (**a**) Overlapping display of the four sensor signals; (**b**) Separate display of the four sensor signals.

4.3. *Amplitude-Frequency Distribution under Multi-Point Synchronous Excitation*

The amplitude–frequency distributions when 2# and 4# action points were excited synchronously are shown in Figure 14. The amplitude of No. 1 (67#-1 in Figure 14) reached the peak 5.1×10^{-3} mm at $P_1 = 48.9$ Hz, the second peak 1.3×10^{-3} mm at $P_2 = 92.4$ Hz, and the amplitude variations of No. 2~4 were all similar to that of No. 1.

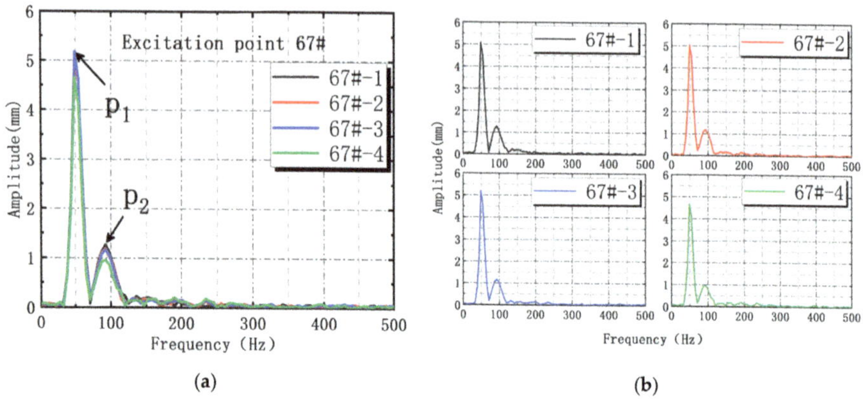

Figure 14. Amplitude–frequency distribution under 6# and 7# action points excited synchronously. (**a**) Overlapping display of the four sensor signals; (**b**) Separate display of the four sensor signals.

5. Effective Vibration Modes and Predominant Frequency of Interface Vibration under Impact Load

Based on Hilbert Huang transform (HHT), the interface vibration waveforms were decomposed by EEMD [20–24]; combined with energy formula: $\int_{-\infty}^{\infty} x^2(t)dt$, the energy distributions and marginal spectrum of the decomposed waveforms were obtained [25].

Under 2# action point excitation, the decomposition result of the vibration waveform is shown in Figure 15, which was decomposed into five vibration modes (IMF1~IMF5), and the residual res $<10^{-3}$ (Figure 15a). The energy proportions of modes IMF1, IMF2 and IMF3 were relatively high (Figure 15b), which were the effective vibration modes. Among them, the energy proportion of IMF2 was the highest, accounting for about 96% of the total energy, which was the main vibration mode. By analyzing the marginal spectrum of the original waveform and effective vibration modes (Figure 15c,d), it could be concluded that the vibration frequency of the original waveform was mainly concentrated in $P_1 \sim P_3 = 39.9 \sim 89.8$ Hz, and the predominant frequency corresponding to the peak was $P_2 = 52.9$ Hz. The predominant frequencies of effective vibration modes (IMF1, IMF2 and IMF3) were $P_4 = 236.8$ Hz, $P_5 = 50.2$ Hz and $P_6 = 35.9$ Hz respectively.

Under 4# action point excitation, the decomposition result of vibration waveform are shown in Figure 16, which was decomposed into six vibration modes (IMF1~IMF6), and the residual res $<10^{-3}$ (Figure 16a). The energy proportions of modes IMF1, IMF2 and IMF3 were relatively high (Figure 16b), which were the effective vibration mode. Among them, the energy proportion of IMF2 was the highest, accounting for about 95% of the total energy, which was the main vibration mode. By analyzing the marginal spectrum of the original waveform and effective vibration modes (Figure 16c,d), it could be concluded that the vibration frequency of the original waveform was mainly concentrated in $P_1 \sim P_3 = 38.9 \sim 108.8$ Hz, and the predominant frequency corresponding to the peak was $P_2 = 50.4$ Hz. The predominant frequencies of effective vibration modes (IMF1, IMF2 and IMF3) were $P_4 = 152.4$ Hz, $P_5 = 49.2$ Hz and $P_6 = 47$ Hz respectively.

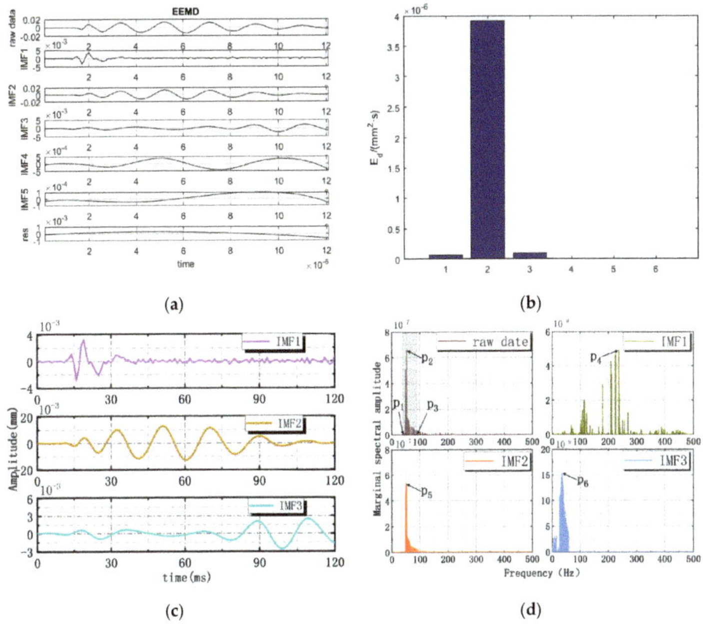

Figure 15. The waveform decomposition under 2# action point excitation. (**a**) The EEMD decomposition of the interface vibration waveform; (**b**) Energy distribution of the decomposed waveform; (**c**) Effective vibration modes; (**d**) Marginal spectral amplitude.

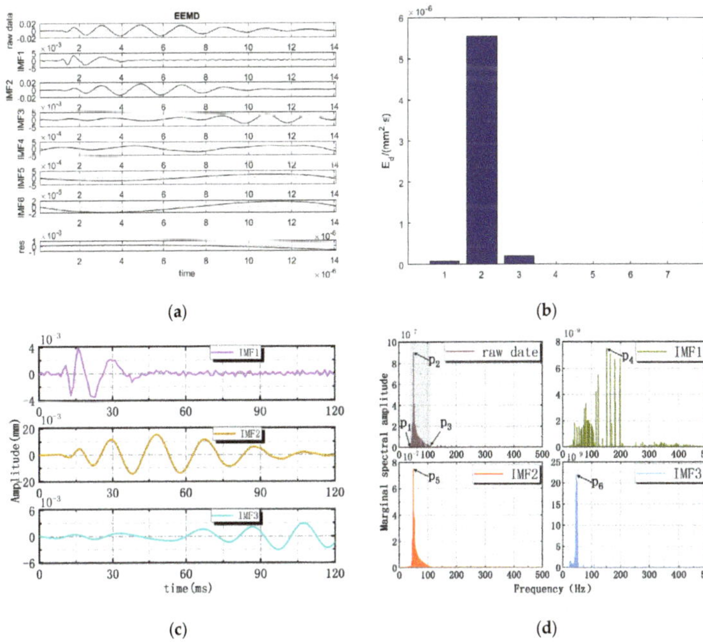

Figure 16. The waveform decomposition under 4# action point excitation. (**a**) The EEMD decomposition of the interface vibration waveform; (**b**) Energy distribution of the decomposed waveform; (**c**) Effective vibration modes; (**d**) Marginal spectral amplitude.

Under 6# action point excitation, the decomposition result of vibration waveform was shown in Figure 17, which was decomposed into six vibration modes (IMF1~IMF6), and the residual res <10^{-3} (Figure 17a). The energy proportions of modes IMF1, IMF2 and IMF3 were relatively high (Figure 17b), which were the effective vibration modes. Among them, the energy proportion of IMF2 was the highest, accounting for about 92% of the total energy, which was the main vibration mode. By analyzing the marginal spectrum of the original waveform and effective vibration modes (Figure 17c,d), it could be concluded that the vibration frequency of the original waveform was mainly concentrated in P_1~P_3 = 37.8~92.7 Hz, and the predominant frequency corresponding to the peak was P_2 = 52.3 Hz. The predominant frequencies of effective vibration modes (IMF1, IMF2 and IMF3) were P_4 = 91.3 Hz, P_5 = 47.7 Hz and P_6 = 47.7 Hz respectively.

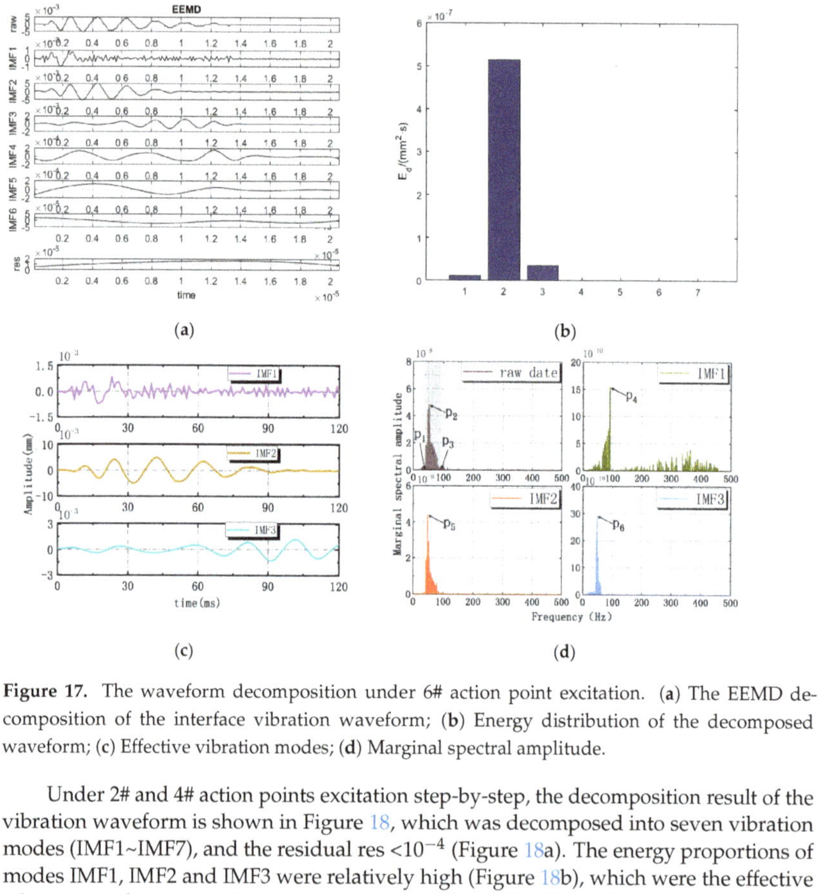

Figure 17. The waveform decomposition under 6# action point excitation. (**a**) The EEMD decomposition of the interface vibration waveform; (**b**) Energy distribution of the decomposed waveform; (**c**) Effective vibration modes; (**d**) Marginal spectral amplitude.

Under 2# and 4# action points excitation step-by-step, the decomposition result of the vibration waveform is shown in Figure 18, which was decomposed into seven vibration modes (IMF1~IMF7), and the residual res <10^{-4} (Figure 18a). The energy proportions of modes IMF1, IMF2 and IMF3 were relatively high (Figure 18b), which were the effective vibration mode. Among them, the energy proportion of IMF2 was the highest, accounting for about 86% of the total energy, which was the main vibration mode. By analyzing the marginal spectrum of the original waveform and effective vibration modes (Figure 18c,d), it could be concluded that the vibration frequency of the original waveform was mainly concentrated in P_1~P_3 = 37.5~99.9 Hz, and the predominant frequency corresponding to the peak was P_2 = 49.4 Hz. The predominant frequencies of effective vibration modes (IMF1, IMF2 and IMF3) were P_4 = 82.7 Hz, P_5 = 48.6 Hz and P_6 = 46.5 Hz, respectively.

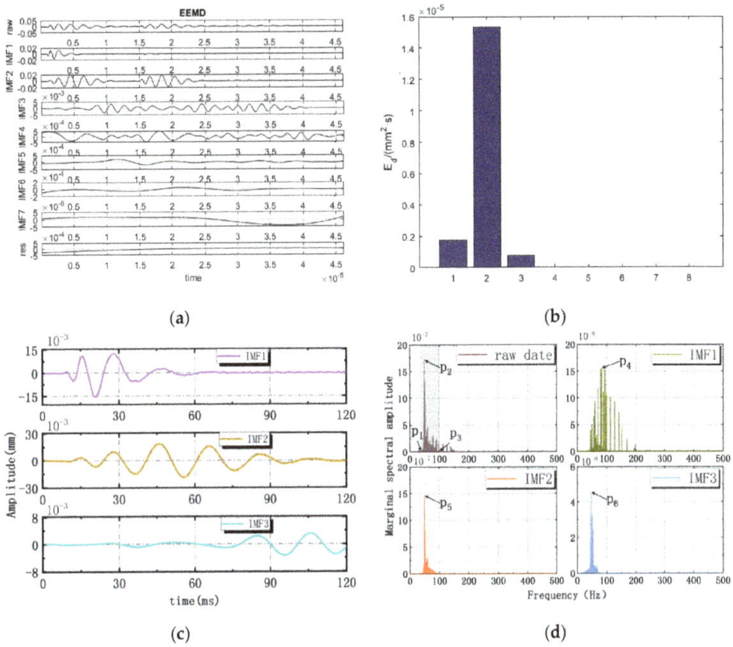

Figure 18. The waveform decomposition under 2# and 4# action points excited step-by-step. (**a**) The EEMD decomposition of the interface vibration waveform; (**b**) Energy distribution of the decomposed waveform; (**c**) Effective vibration modes; (**d**) Marginal spectral amplitude.

Under 6# and 7# action points synchronous excitation, the decomposition result of vibration waveform was shown in Figure 19, which was decomposed into six vibration modes (IMF1~IMF6), and the residual res $<10^{-3}$ (Figure 19a). The energy proportions of modes IMF1, IMF2 and IMF3 were relatively high (Figure 19b), which were the effective vibration modes. Among them, the energy proportion of IMF2 was the highest, accounting for about 85% of the total energy, which was the main vibration mode. By analyzing the marginal spectrum of the original waveform and the effective vibration modes (Figure 19c,d), it could be concluded that the vibration frequency of the original waveform was mainly concentrated in P_1~P_3 = 24.8~90.2 Hz, and the predominant frequency corresponding to the peak was P_2 = 45.7 Hz. The predominant frequencies of effective vibration modes (IMF1, IMF2 and IMF3) were P_4 = 201.3 Hz, P_5 = 45.6 Hz and P_6 = 47.4 Hz, respectively.

It could be seen that IMF1, IMF2 and IMF3 were effective vibration modes under single point excitation and multi-point excitation (synchronous/step-by-step). Among them, the energy of IMF2 accounted for the highest proportion (85–94%), which was the main vibration mode, and its predominant frequencies were mostly concentrated in 45.6~50.2 Hz. It could be concluded that IMF2 played a decisive role in the vibration process, so the predominant frequencies of coal–rock and rock–rock interfaces vibration under impact load were also concentrated in this range, and the vibration in this frequency range had an important effect on the dynamic response, damage and failure of coal and rock mass. Of course, in the actual conditions, the range of the actual predominant frequencies could be obtained by converting according to the size and mechanical properties of coal and rock mass.

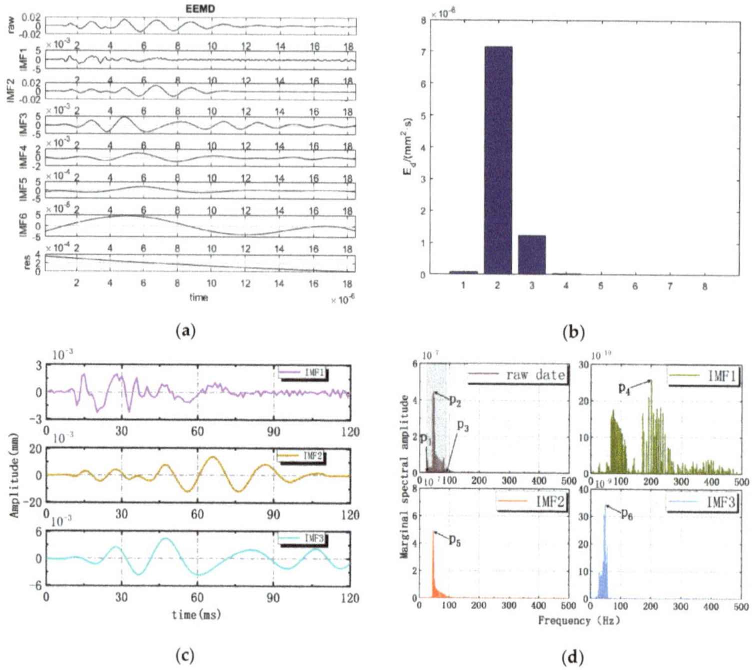

Figure 19. The waveform decomposition under 6# and 7# action points excited synchronously. (**a**) The EEMD decomposition of the interface vibration waveform; (**b**) Energy distribution of the decomposed waveform; (**c**) Effective vibration modes; (**d**) Marginal spectral amplitude.

6. Conclusions

Under single and multi–point excitation (synchronous/step-by-step), the dynamic changes of amplitude curves of each interface vibration were obtained from the four sensors (No. 1~4) were similar, and the extreme value showed two stages as follows: first increase and then attenuation, most of which required 2.25 cycles to reach the peak. The dynamic attenuation of amplitude conformed to the law of exponential variation y = y0 + Aexp (x/k).

Based on FFT transform, the time–history curves of interface vibration under single point and multi-point excitation (synchronous/step-by-step) were analyzed, and the predominant frequencies distribution of amplitude were obtained. Among them, the two predominant frequencies were P_1 = 48.9~53.7 Hz and P_2 = 92.4 Hz.

The time–history curves of interface vibration under single point and multi-point excitation (synchronous/step-by-step) were decomposed by EEMD to obtain a total of component, 5. modes (IMF), of which IMF1, IMF2 and IMF3 contained high energy and were effective vibration modes.

The energy of IMF2 accounted for the highest proportion (85–94%), which was the main vibration mode, and its predominant frequencies were concentrated in 45.6~50.2 Hz, which overlapped with that of the original waveform to a great extent. The vibration of IMF2 played a decisive role in the vibration process and had an important effect on the dynamic response, damage, and failure of coal and rock mass. In actual conditions, the range of the actual predominant frequencies can be obtained by converting according to the size and mechanical properties of coal and rock mass.

In this paper, the characteristics of vibration frequency and vibration amplitude of coal rock vibration response signal were mainly studied. The vibration frequency study of coal rock vibration signals can help the monitoring and early warning of coal and gas protrusion. The study of vibration amplitude can help to determine the danger area of coal rock vibration. The vibration frequency characteristics and amplitude characteristics of the dynamic response signals of coal rocks have rarely been studied by previous authors; there are even fewer studies on the dynamic response characteristics of multi-layer coal rocks. In this paper, we studied the dynamic response characteristics of the five–seam coal rock body, which is closer to the complex reality. However, the study only performs vibration response analysis for similar experiments in this paper. Further studies are needed to verify whether the dynamic response of coal rock in different scenarios also conforms to the vibration law summarized in this study. Further research is needed to investigate the dynamic response of coal rocks under different sizes and mechanical properties of the actual site.

Author Contributions: Conceptualization, F.L.; methodology, F.L.; software, G.W. and B.R.; validation, F.L. and G.W.; formal analysis, G.W. and J.T.; investigation, F.L.; resources, F.L.; data curation, F.L. and G.W.; writing—original draft preparation, F.L.; writing—review and editing, F.L., G.W. and G.X.; visualization, F.L., G.W. and Z.C.; supervision, F.L.; project administration, F.L.; funding acquisition, F.L. All authors have read and agreed to the published version of the manuscript.

Funding: This research was funded by the financial support from National Natural Science Foundation of China grant number [52064046, 51804311]; China Scholarship Council (CSC); and the Fundamental Research Funds for the Central Universities grant number [2020YJSAQ13].

Data Availability Statement: The data presented in this paper is freely available from the corresponding author upon request.

Conflicts of Interest: The authors declare no conflict of interest.

Abbreviations

EEMD	Ensemble Empirical Mode Decomposition
FFT	Fast Fourier transform
HHT	Hilbert Huang transform
IMF	Intrinsic Mode Functions

References

1. Xu, D.P.; Feng, X.T.; Chen, D.F.; Zhang, C.Q.; Fan, Q.X. Constitutive representation and damage degree index for the layered rock mass excavation response in underground openings. *Tunn. Undergr. Space Technol.* **2017**, *64*, 133–145. [CrossRef]
2. Wang, P.T.; Cai, M.F.; Ren, F.H. Anisotropy and directionality of tensile behaviours of a jointed rock mass subjected to numerical Brazilian tests. *Tunn. Undergr. Space Technol.* **2018**, *64*, 133–145. [CrossRef]
3. Liu, Y.S.; He, C.S.; Fu, H.L.; Wang, S.M.; Lei, Y.; Peng, Y.X. Study on dynamic and static mechanical properties and failure mode of layered slate. *J. Railw. Sci. Eng.* **2020**, *17*, 2789–2797. [CrossRef]
4. Wu, G.P.; Shen, W.W.; Cui, K.; Wang, P. Degradation behavior and mechanism of slate under alternating conditions of freeze-thaw and wet-dry. *J. Cent. South Univ. (Sci. Technol.)* **2019**, *50*, 92–1402. [CrossRef]
5. Liu, Y.S.; Wang, S.M.; Yan, S.J.; Fu, H.L.; Chen, C.; Shi, Y.; Yue, J. Properties and failure mechanism of layered sandstone based on acoustic emission experiments. *J. Cent. South Univ. (Sci. Technol.)* **2019**, *50*, 1419–1427. [CrossRef]
6. Zhou, P.F.; Shen, Y.S.; Zhao, J.F.; Zhang, X.; Gao, B.; Zhu, S.Y. Research on disaster-induced mechanism of tunnels with steeply dipping phyllite strata based on an improved ubiquitous-joint constitutive model. *Chin. J. Rock Mech. Eng.* **2019**, *38*, 1870–1883. [CrossRef]
7. Su, G.S.; Hu, L.H.; Feng, X.T.; Wang, J.H.; Zhang, X.H. True triaxial experimental study of rockburst process under low frequency cyclic disturbance load combined with static load. *Chin. J. Rock Mech. Eng.* **2016**, *35*, 1309–1322. [CrossRef]
8. Chang, W.B.; Fan, S.W.; Zhang, L.; Shu, L.Y. A model based on explosive stress wave and tectonic coal zone which gestate dangerous state of coal and gas outburst. *J. China Coal Soc.* **2014**, *39*, 2226–2231. [CrossRef]
9. Zuo, J.P.; Chen, Y.; Cui, F. Investigation on mechanical properties and rock burst tendency of different coal-rock combined bodies. *J. China Univ. Min. Technol.* **2018**, *47*, 81–87. [CrossRef]
10. Li, S.H.; Zhu, W.C.; Niu, L.L.; Yu, M.; Chen, C.F. Dynamic characteristics of green sandstone subjected to repetitive impact loading: Phenomena and mechanisms. *Rock Mech. Rock Eng.* **2018**, *51*, 1921–1936. [CrossRef]
11. Mu, Z.L.; Wang, H.; Peng, P.; Liu, Z.J.; Yang, X.C. Experimental research on failure characteristics and bursting liability of rock-coal-rock sample. *J. Min. Saf. Eng.* **2013**, *30*, 841–847.
12. Liu, S.H.; Qin, Z.H.; Lou, J.F. Experimental study of dynamic failure characteristics of coal-rock compound under one-dimensional static and dynamic loads. *Chin. J. Rock Mech. Eng.* **2014**, *33*, 2064–2075. [CrossRef]

13. Potvin, Y.; Hadjigeorgiou, J.; Stacey, D. *Challenges in Deep and High Stress Mining*; Australian Center for Geomechanics: Nedlands, Australia, 2007.
14. Hoke, E. Progressive caving induced by mining an inclined orebody. *Trans. Inst. Min. Metall.* **1974**, *83*, 133–139.
15. Braunagel, M.J.; Griffith, W.A. The effect of dynamic stress cycling on the compressive strength of rocks. *Geophys. Res. Lett.* **2019**, *46*, 6479–6486. [CrossRef]
16. Cui, F.; Yang, Y.B.; Lai, X.P.; Cao, J.T. Similar material simulation experimental study on rockbursts induced by key stratum breaking based on microseismic monitoring. *Chin. J. Rock Mech. Eng.* **2019**, *38*, 803–814. [CrossRef]
17. Li, F.; Dong, X.H.; Wang, Y.; Liu, H.W.; Chen, C.; Zhao, X. The Dynamic Response and Failure Model of Thin Plate Rock Mass under Impact Load. *Shock. Vib.* **2021**, *2021*, 9998558. [CrossRef]
18. Yang, S.L.; Wang, J.C.; Yang, J.H. Physical analog simulation analysis and its mechanical explanation on dynamic load impact. *J. China Coal Soc.* **2017**, *42*, 335–343. [CrossRef]
19. Song, X.L.; Gao, W.X.; Ji, J.M.; Ye, M.B.; Zhang, D.J. Blasting damage analysis method based on EEMD-HHT transform. *J. Cent. South Univ. (Sci. Technol.)* **2017**, *42*, 335–343. [CrossRef]
20. Zhao, Y.; Shan, R.L.; Wang, H.L. Research on vibration effect of tunnel blasting based on an improved Hilbert–Huang transform. *Environ. Earth Sci.* **2021**, *80*, 206. [CrossRef]
21. Hamdi, S.E.; Le, D.A.; Simon, L.; Plantier, G.; Sourice, A.; Feuilloy, M. Acoustic emission pattern recognition approach based on Hilbert–Huang transform for structural health monitoring in polymer-composite materials. *Appl. Acoust.* **2013**, *74*, 746–757. [CrossRef]
22. Liu, X.; Wang, J.; Li, W. A time-frequency extraction model of structural vibration combining VMD and HHT. *Geomat. Inf. Sci. Wuhan Univ.* **2021**, *46*, 1686–1692. [CrossRef]
23. Dragomiretskiy, K.; Zosso, D. Variational Mode Decomposition. *IEEE Trans. Signal Process.* **2014**, *62*, 531–544. [CrossRef]
24. Lu, C.P.; Dou, L.M.; Wu, X.R.; Mou, Z.L.; Chen, G.X. Experimental and Empirical Research on Frequency-Spectrum Evolvement Rule of Rockburst Precursory Microseismic Signals of Coal-Rock. *Chin. J. Rock Mech. Eng.* **2008**, *27*, 519–525. [CrossRef]
25. Zhang, P.S.; Liu, S.D. The Structure Character of Coal and Rock by Seismic Wave Spectrum Analyzing Technology and Its Application. *Chin. J. Eng. Geophys.* **2006**, *3*, 274–277. [CrossRef]

processes

MDPI

Article

Study on Gas Migration Mechanism and Multi-Borehole Spacing Optimization in Coal under Negative Pressure Extraction

Feng Du [1,2], Weilong Cui [1,2] and Kai Wang [1,2,*]

1 Beijing Key Laboratory for Precise Mining of Intergrown Energy and Resources, China University of Mining and Technology (Beijing), Beijing 100083, China

2 School of Emergency Management and Safety Engineering, China University of Mining and Technology (Beijing), Beijing 100083, China

* Correspondence: kaiwang@cumtb.edu.cn

Abstract: In order to study the gas migration in gas-bearing coal, and reasonably arrange gas drainage boreholes to improve the efficiency of gas drainage, a gas-solid coupling model is established based on the pore-fracture dual medium porous model. The solid deformation of coal body, gas seepage and diffusion, and gas adsorption and desorption are considered in this model. The COMSOL software is used to simulate the gas change in the coal matrix and coal fracture under single borehole extraction. We analyze the effective extraction range and study the migration mechanism of gas between coal fracture and borehole, coal matrix and coal fracture, and coal matrix. The effective extraction area of multi-borehole negative pressure gas extraction varies with extraction time and borehole spacing. At 140 d, the effective extraction radius is r = 1.3 m, and the spacing of boreholes is $\frac{2\sqrt{3}}{3}$ r = 1.5 m, 2 r = 2.6 m, 4 m, 5 m, and 6 m, respectively. The influence of the equilateral triangle shape of three boreholes on the gas extraction effect is studied. The simulation results show that when three boreholes are extracted for 140 days under different borehole spacing, different gas extraction effects will be affected by a superposition effect. Considering the change in gas pressure, the effect of gas extraction in the effective extraction area, and the safety and cost performance of gas extraction, it is concluded that the optimal hole spacing is 5 m around 140 d. This study aims to provide reference for underground gas drilling layout and reasonable hole spacing.

Keywords: gas extraction; gas-solid coupling; gas migration mechanism; effective extraction area; borehole spacing

Citation: Du, F.; Cui, W.; Wang, K. Study on Gas Migration Mechanism and Multi-Borehole Spacing Optimization in Coal under Negative Pressure Extraction. *Processes* **2023**, *11*, 259. https://doi.org/10.3390/pr11010259

Academic Editor: Albert Ratner

Received: 20 December 2022
Revised: 9 January 2023
Accepted: 11 January 2023
Published: 13 January 2023

1. Introduction

Coal is the dominant energy source in China. With the depletion of shallow coal resources, a large number of mines in China entered deep mining, and the threat of gas disasters is extremely serious [1–3]. The hidden danger of coal and gas outburst increases, which seriously threatens the personal safety of underground workers and affects the mining efficiency of coal resources in China [4–6]. Pre-drainage of coal seam gas can effectively prevent coal and gas outburst. Efficient gas extraction plays an important role in reducing coal seam gas pressure and gas emission, as well as preventing coal–gas outburst [7–9]. Drilling negative pressure extraction is the most common method of gas extraction. Drilling spacing and extraction time are significant factors affecting the extraction effect.

With the deepening of the research on the seepage law of coal seam gas, it is realized that the migration mode of coal seam gas in coal body is not single, and there are many flow modes among coal pores, fractures, and matrix [10–13]. According to the various characteristics of coal mechanics and the migration of gas in coal, the establishment of a gas-solid coupling model is the theoretical basis for the analysis of gas extraction by negative pressure drilling. Zhu et al. determined the dynamic change model of permeability and porosity considering coal deformation and the Klinkenberg effect, and deduced the low permeability gas-solid coupling model [14–16]. Based on the innovative gas-solid coupling model, Cheng et al. used COMSOL Multiphysics to numerically solve the model. The migration of gas in coal under the influence of diffusion and seepage is analyzed, and the influence of negative pressure on gas migration is studied [17–19]. Li et al. simulated the gas extraction effect under different hole spacing by analyzing the mechanism of gas flow around the borehole and the superposition effect of extraction. He obtained the reasonable hole spacing of bedding gas extraction: $2r \leq L \leq R$ [20]. Zhang et al. established a coal seam gas seepage model considering the assumption of double-hole medium and the influence of stress redistribution on the seepage characteristics of coal seam gas. They studied the influence of coal seam pressure change on gas adsorption and seepage, and the improvement of gas extraction effect by overlapping the arrangement of two boreholes [21]. Chen made a more detailed study on the optimization of borehole extraction spacing and effective extraction area [22].

The study of gas migration and the effect of different borehole spacing on gas extraction have great reference value for practical engineering. The optimal design of extraction boreholes affects the safety of coal mining and the sustainable development of coal resources. [23]. In the existing research, the research results of coal permeability, porosity change, and single row multi-borehole extraction effect of negative pressure extraction gas are quite abundant, but most of them are based on a two-dimensional case. However, based on pore-fracture dual porous media, there are few studies on understanding and analyzing gas migration in coal from the perspective of a three-dimensional simulation. There are also few studies on the gas migration and extraction effect in the coal body during drilling negative pressure extraction through the three-dimensional visualization model. In order to better reveal the migration of gas in coal during the extraction process of three boreholes and the influence of different borehole spacing on effective extraction area, an innovative gas-solid coupling model was proposed based on previous studies. The model takes into account the Klinkenberg effect, coal matrix, and skeleton deformation of coal fracture outward seepage and matrix diffusion gas influx, coal matrix adsorption gas, and the desorption diffusion gas dynamic process. In order to provide a theoretical basis for further understanding of gas migration law and studying the extraction effect under the influence of the multi-hole extraction superposition effect, this research takes a coal mine as the background and uses COMSOL simulation software for three-dimensional simulation. The research intends to provide reference value for practical engineering.

2. Establishment of Coalbed Gas-Solid Coupling Model

2.1. Coal Deformation Equation

Coal deformation mainly refers to the effective stress acting on the coal skeleton to deform the coal skeleton. In this paper, the double medium model is cited [24]. The stress effect is also affected by gas adsorption and desorption. The effective stress expression is:

$$\sigma_{ij}^{eff} = \sigma_{ij} - (\alpha_m P_m + \alpha_f P_f)\delta_{ij} - \alpha_s \varepsilon_{s(ij)} \tag{1}$$

In the formula, σ_{ij}^{eff} is the effective stress, MPa; σ_{ij} is the total stress, MPa; α_m and α_f are the effective stress coefficients of coal pore and fracture, respectively; P_m and P_f are gas pressure in the coal matrix and fracture, respectively, MPa; δ_{ij} is a Kronecker tensor; α_s is the volumetric strain coefficient caused by gas adsorption and desorption in coal, kg/m^3; and $\varepsilon_{s(ij)}$ is adsorption–desorption strain of coal under isothermal condition. Among them, $\alpha_m = \frac{K}{K_m} - \frac{K}{K_s}$; $\alpha_f = 1 - \frac{K}{K_m}$; $\varepsilon_{s(ij)} = \frac{aP_m}{P_m + b}$. K is the bulk modulus of coal, MPa; K_m, and K_s are the bulk modulus of the coal matrix and skeleton, respectively, MPa.

Equilibrium equation, geometric equation, and constitutive equation constitute the volumetric deformation equation of coal containing gas.

The force balance of each surface of the unit body is analyzed. According to the conservation of momentum, the equilibrium equation is:

$$\sigma_{ij,j}+F_i=0 \tag{2}$$

The coal skeleton undergoes slight deformation, and the geometric equation is:

$$\varepsilon_{ij}=\frac{1}{2}\left(\mu_{i,j}+\mu_{j,i}\right) \tag{3}$$

Assuming that the coal is a linear elastic body, according to Hooke's law, the constitutive equation is:

$$\sigma_{ij}=2G\frac{v}{1-2v}\varepsilon_v\delta_{ij}+2G\varepsilon_{ij}+(\alpha_m P_m+\alpha_f P_f)\delta_{ij}+\alpha_s\varepsilon_{s(ij)} \tag{4}$$

In the formula, G is the shear modulus, MPa; v is Poisson's ratio; and ε_v is the volume strain. Among them, $G=\frac{E}{2(1+v)}$, E is elastic modulus of coal, MPa.

The change in gas pressure will induce the deformation of the coal matrix and coal skeleton. Gas adsorption and desorption in the coal matrix will also affect the deformation of the coal matrix. Combining the three Equations (1)–(3), based on the dual pore medium model, the governing equation of coal deformation can be obtained:

$$Gu_{i,jj}+\frac{G}{1-2\upsilon}u_{j,ji}-\alpha_m P_{m,i}-\alpha_f P_{f,i}-\alpha_s\varepsilon_{s,i}+F_i=0 \tag{5}$$

2.2. Gas Migration Equation in Coal

Based on surface physical chemistry, considering the influence of effective stress, gas pressure, coal matrix adsorption, and desorption gas, the porosity equation is [25]:

$$\phi_f=\phi_{f0}+\frac{K_s}{K_s-\alpha_m(P_{m0}-P_m)}\frac{p_L\varepsilon_L(P_{m0}-P_m)}{(P_{m0}+P_L)(P_m+P_L)}-\frac{[\alpha_m(P_{m0}-P_m)+\alpha_f(P_{f0}-P_f)](1-\phi_{f0})}{K_s-\alpha_m(P_{m0}-P_m)-\alpha_f(P_{f0}-P_f)} \tag{6}$$

In the formula, ϕ_{f0} is the initial fracture porosity; ε_L is the adsorption deformation of coal; and P_L is Langmuir pressure constant, MPa.

There is a cubic relationship between permeability and porosity [26]:

$$k=k_0\left(\frac{\phi_f}{\phi_{f0}}\right)^3 \tag{7}$$

Considering the influence of the Klinkenberg effect in gas-bearing double porous media, the permeability calculation formula is obtained:

$$\begin{aligned} k_e=k\left(1+\frac{c}{P_f}\right)=k_0\left(\frac{\phi_f}{\phi_{f0}}\right)^3\left(1+\frac{c}{P_f}\right)=k_0[1+\frac{K_s}{K_s-\alpha_m(P_{m0}-P_m)}\frac{P_L\varepsilon_L(P_{m0}-P_m)}{(P_{m0}+P_L)(P_m+P_L)\phi_{f0}} \\ -\frac{[\alpha_m(P_{m0}-P_m)+\alpha_f(P_{f0}-P_f)]\left(\frac{1}{\phi_{f0}}-1\right)}{K_s-\alpha_m(P_{m0}-P_m)-K_s-\alpha_f(P_{f0}-P_f)}\Bigg]^3\left(1+\frac{c}{P_f}\right) \end{aligned} \tag{8}$$

In the formula, k_0 is the initial permeability of coal, mD; and c is the Kirsch coefficient.

Coal is regarded as a porous medium composed of pores and fracture. The gas content in coal includes the gas content in coal matrix pores and coal fractures. The mass of gas in the unit volume coal matrix is composed of adsorbed gas and free gas in coal matrix pores and and free gas in coal fractures.

The gas adsorption in the coal matrix follows the Langmuir equation:

$$m_1=\rho_{g0}\rho_s\frac{abP_m}{bP_m+1}\frac{1-A-B}{1+0.31B} \tag{9}$$

The free gas in coal matrix pores and coal fractures can be calculated by gas state equation. The free gas quality in coal matrix pores:

$$m_2 = \phi_m \rho_m = \phi_m P_m \frac{M_c}{RT} \tag{10}$$

The free gas quality in coal fracture:

$$m_f = \phi_f \rho_f = \phi_f P_f \frac{M_c}{RT} \tag{11}$$

Gas quantity in unit volume coal pore:

$$m_m = m_1 + m_2 = \rho_{g0}\rho_s \frac{abP_m}{bP_m+1}\frac{1-A-B}{1+0.31B} + \phi_m P_m \frac{M_c}{RT} \tag{12}$$

In the formula, ρ_{g0} is the density of gas under standard conditions, kg/m^3; ρ_s is coal density, kg/m^3; a is the maximum adsorption capacity per unit volume of coal, m^3/t; b is the adsorption constant of coal, MPa^{-1}; A is coal ash; B is coal moisture; ϕ_m is the porosity of the coal matrix, %; M_c is the molar mass of gas, kg/mol; R is the ideal gas constant, J/(mol·K); T is the temperature in coal, K; m_1 is the amount of gas adsorbed per unit volume of coal matrix, kg/m^3; m_2 is the amount of free gas per unit volume of coal matrix, kg/m^3; m_f is the amount of free gas per unit volume of coal fracture, kg/m^3; and m_m is the gas content per unit volume of coal matrix, kg/m^3.

In the original state, without mining disturbance, coal pore pressure is equal to coal fracture pressure. When the gas is extracted by a negative pressure borehole, the borehole breaks the dynamic balance of coal structure and gas pressure in the original state. As a result, the pressure difference between coal pore and gas pressure in coal fracture is produced. The migration of gas in coal pores conforms to Fick's diffusion law, and the migration in fractures conforms to Darcy's linear seepage law. The change in gas in coal matrix pores of unit coal body per unit of time follows the law of conservation of mass. The amount of gas change is equal to the amount of desorption gas of coal body outside the absorption unit minus the amount of diffusion gas of unit body:

$$\begin{gathered} \frac{\partial m_m}{\partial t} = Q_s - q_s = D\nabla(m_1) - \frac{M_c}{\tau RT}(P_m - P_f) \\ Q_s = D\nabla\left(\rho_{g0}\rho_s \frac{abM_c(P_{m0}-P_m)}{(bP_m+1)RT}\frac{1-A-B}{1+0.31B}\right) \\ q_s = \frac{M_c}{\tau RT}(P_m - P_f) \end{gathered} \tag{13}$$

In the formula, Q_s is the external mass source of the unit, $kg/(m^3 \cdot s)$; q_s is the unit mass source, $kg/(m^3 \cdot s)$; D is the gas diffusion coefficient, m^2/s; τ is desorption time of adsorbed gas, d; and ∇ is a Hamiltonian operator.

The change in gas in the coal fracture of unit time follows the law of conservation of mass. The amount of gas change is equal to the amount of fracture seepage plus the amount of coal matrix pore diffusion source:

$$\frac{\partial m_f}{\partial t} = \frac{\partial(\rho_g \phi_f)}{\partial t} = q_s - \nabla(\rho_g V) = \frac{M_c}{\tau RT}(P_m - P_f) + \nabla\left(\rho_g \frac{k}{\mu}\nabla P_f\right) \tag{14}$$

In the formula, V is the velocity of gas seepage in coal fracture, m/s; $V = -\frac{k}{\mu}\nabla P_f$; and μ is the gas dynamic viscosity, Pa·s.

The formula of pore gas pressure of the coal matrix changing with time can be obtained from the Equation (13):

$$\frac{\partial p_m}{\partial t} = \frac{\left[D\nabla(m_1) - \frac{M_c}{\tau RT}(P_m - P_f)\right](1+0.31\,B)(bP_m+1)^2 RT}{(1-A-B)ab\rho_{g0}\rho_s RT + \phi_m M_c(1+0.31\,B)(bP_m+1)^2} \tag{15}$$

The formula can be obtained from the Equation (11) and the Equation (14):

$$\phi_f \frac{\partial P_f}{\partial t} + P_f \frac{\partial \phi_f}{\partial t} = \frac{P_m - P_f}{\tau} + \nabla\left(\frac{k}{\mu} P_f \nabla P_f\right) \quad (16)$$

The deformation equation of gas-bearing coal is (1)–(5), (6), (8) is the change formula of porosity and permeability of coal, and (9)–(16) is the equation of gas occurrence and migration in coal pores and fractures. In the process of negative pressure drilling, the mechanical structure characteristics of coal body and the balance of gas in the original state of coal body are broken, and the coal body structure is slightly deformed. The coal deformation leads to changes in porosity and permeability, which affects the migration of gas in coal. The gas-solid coupling model is established by combining the deformation control Equation (5) of a gas-bearing coal body with the mass conservation control Equations (13) and (16) of gas migration in coal pores and fractures.

3. Numerical Simulation

3.1. Model Assumptions

1. Assuming that the coal body is an isotropic pore-fracture porous medium, the coal body is regarded as a linear elastic body, and the coal body undergoes small deformation.
2. Assuming that there is no energy exchange between the coal body and the surrounding environment, the roof and floor of the coal seam are impermeable, and the extracted gas only comes from the coal body.
3. It is assumed that the initial fracture gas pressure in the coal body is the same as the initial fracture gas pressure in the coal matrix pore.
4. Assuming that the temperature does not change when the gas migrates in the coal body during the extraction process, the gas is an ideal gas, which conforms to the ideal gas state equation. Gas diffusion conforms to Fick diffusion law, gas adsorption conforms to Langmuir adsorption law, and gas seepage conforms to Darcy law.

3.2. Model Introduction

In this paper, a coal mine of No. 4 coal seam gas extraction is taken as the research object. No. 4 coal has an average thickness of 10.75 m, and the coal coefficient is 18.99%. The apparent density of coal is 1.24~1.59 t/m^3, with an average of 1.36 t/m^3. The true density is 1.41~1.59 t/m^3, with an average of 1.49 t/m^3. The coal seam has an initial gas pressure between 1.01 MPa and 1.29 MPa. The gas content in the coal seam is 3.91~4.58 m^3/t.

Based on the gas-solid coupling model constructed above, with the help of the multi-physics coupling simulation software COMSOL Multiphysics, the physical field coupling calculation is carried out by using the custom partial differential equation module (PDE) and the solid mechanics module in structural mechanics. The transient solver is selected. The migration of gas in coal is described by the PDE module, and the deformation of the coal structure is described by the solid mechanics module. The gas-solid coupling formula derived above is embedded in the formula of the module.

A three-dimensional geometric model of 30 m × 40 m × 7 m is established as shown in Figure 1a. The radius of the central extraction borehole is 75 mm. The coal seam has the overlying strata load of 12 MPa, and the direction extends the Z-axis vertically downward. The coal body is set not to produce energy exchange with the outside, so the outer boundary of coal is set to zero flux. It is assumed that there is no displacement around the coal seam in its normal direction and the bottom surface is fixed. The borehole extraction gas negative pressure is set to 20 kPa. The initial gas pressure of coal matrix pores and coal fractures is 1.01 MPa. When meshing, as shown in Figure 1b below, it is set to user-controlled meshing. The mesh is divided into a free subdivision tetrahedron with a curvature factor of 0.6. The maximum element size is 2 m and the minimum element size is 0.5 m. The parameters used for numerical simulation are shown in Table 1.

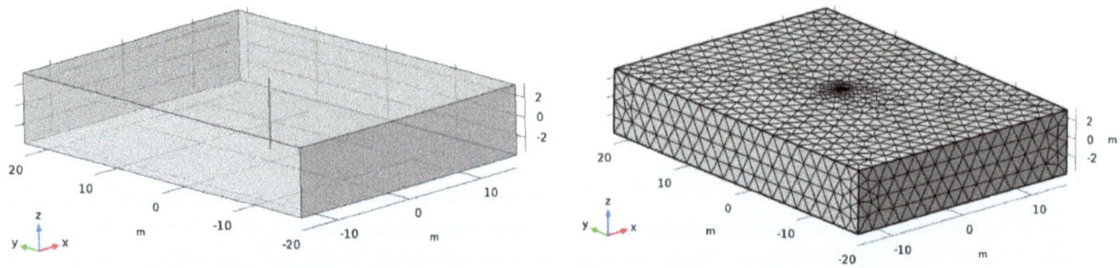

(**a**) geometric model (**b**) mesh subdivision

Figure 1. Geometric model and meshing.

Table 1. Table of parameters required for simulation.

Parameter	Numerical Value	Parameter	Numerical Value
elastic modulus of coal matrix	8210 [MPa]	Langmuir volumetric strain constant	0.0021
elastic modulus of coal	2783 [MPa]	Langmuir pressure constant	1.25 [MPa]
Poisson's ratio of coal	0.35	coal moisture	0.017
initial permeability	0.05 [mD]	coal ash yield	0.136
Initial matrix porosity	0.06	coal density	1490 [kg·m^{-3}]
Initial crack rate	0.013	gas diffusion coefficient	5.536×10^{-11} [m^2·s^{-1}]
Initial gas pressure of coal	1.01 [MPa]	gas dynamic viscosity coefficient	1.15×10^{-5} [Pa·s]
maximum gas adsorption capacity per unit mass of coal	22.5 [m^3·t^{-1}]	Klinkenberg coefficient	9.4382×10^{-7}
adsorption constants	0.8 [MPa^{-1}]	geothermal temperature	303.15 [K]

4. Simulation Results and Analysis

4.1. Influence of Negative Pressure Extraction on Gas Migration

The method of predicting whether it is an outburst danger zone stipulated in the "Coal and Gas Outburst Prevention Rules" is based on the coal seam gas pressure of 0.74 MPa and the content of 8 m^3/t as the critical value. Reducing the gas pressure of coal seam to 0.74 MPa during the extraction process is an important discriminant index. In the two-dimensional plane, the radius of the area where the gas pressure is lower than 0.74 MPa is the effective extraction radius. The area where gas pressure is less than 0.74 MPa is regarded as an effective extraction area. In the three-dimensional model, the area where the gas pressure in coal is less than 0.74 MPa is regarded as an effective extraction area.

A single borehole is set up to simulate the effective extraction radius considering coal fracture and gas pressure in matrix under different extraction time. As shown in Figure 2, taking the bottom of the coal seam as an example, the effective extraction radius considering gas pressure in the coal matrix is always smaller than that considering gas pressure in coal fracture. However, the difference of effective extraction radius decreases with the extension of extraction time. Combined with the three-dimensional effect diagram of the effective extraction area at different times obtained by simulation, it can also be understood that the effective extraction area with a smaller coal matrix gas pressure less than 0.74 MPa is contained in the effective extraction area where the coal fracture gas pressure is less than 0.74 MPa. Two days before the extraction, the effective extraction radius of coal fracture increased sharply, and then maintained a certain trend of continuous increase, but with the increase in extraction time, the increase gradually decreased. In the first 5 days of extraction, the increase in the effective extraction radius of the coal matrix was very small, and then increased significantly, but with the increase in extraction time, the increase was gradually reduced.

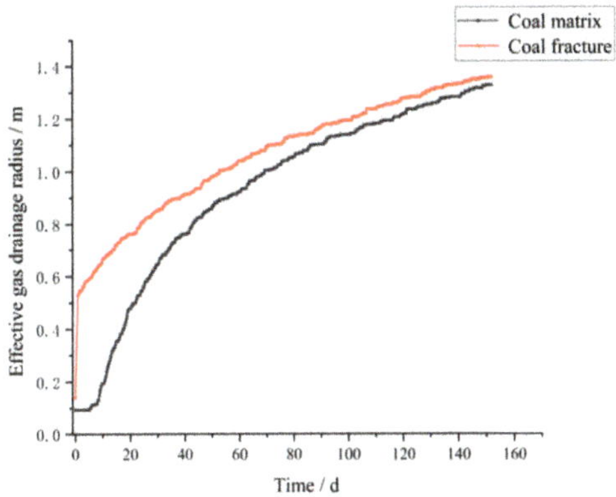

Figure 2. Change in effective extraction radius under different extraction time.

As shown in Figure 3a, points A (0,0.5,0), B (0,1,0), C (0,2,0), and D (0,4,0) are selected as simulation monitoring points to simulate the gas pressure of coal matrix pores and coal fractures under different extraction time monitored in the same plane and different radius. The result is shown in Figure 3b. It can be seen from Figure 3 that at the same extraction time, the gas pressure in coal matrix pores is always larger than that in the coal fractures. At 30 days before drainage, gas pressure in fracture of A and B coal bodies decreases greatly. In contrast, with the four monitoring points farther away from the borehole distance, C and D two-coal seam gas pressure in the fracture decreased slightly. The gas pressure of the coal matrix at the four monitoring points did not decrease significantly as the gas pressure of coal fracture. At the four monitoring points, the gas pressure of coal fracture and the gas pressure of the coal matrix decrease gradually with the continuous extraction and tend to be gentle. The difference between gas pressure in the coal fracture and coal matrix at the four monitoring points decreases gradually.

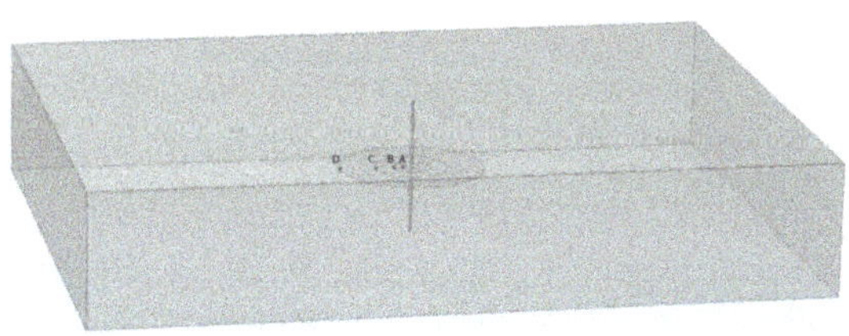

(**a**) monitoring points selection for simulation

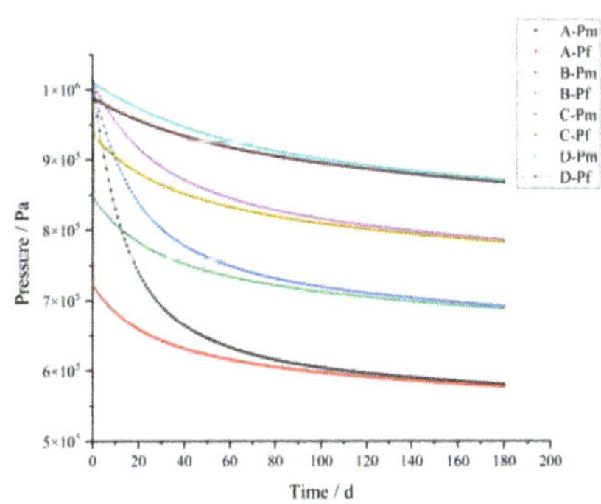

(**b**) gas pressure

Figure 3. Coal matrix and fracture gas pressure under different extraction time.

In the initial stage of drilling drainage, the pressure difference between the gas pressure in the fracture and the negative pressure of the borehole is obvious, and the free gas in the fracture flows to the borehole, so the gas pressure in the coal fracture decreases sharply. The seepage of gas in the fracture to the borehole causes the obvious pressure difference between the coal matrix and the gas in the fracture. The gas in the coal matrix diffuses to the fracture, and the gas adsorbed by the coal matrix desorbs under the influence of the pressure difference, and then participates in the diffusion to the fracture. Gas diffusion and desorption diffusion in coal matrix produce gas pressure difference with surrounding coal matrix, which will also absorb gas diffused by surrounding coal matrix. Under the same extraction time at the same position, the gas pressure in the coal fractures is always smaller than that in a negative pressure borehole, and the gas pressure in coal matrix pores is always larger than that in coal fractures. The influence scope of borehole negative pressure extraction expands gradually with the increase in extraction time. The fracture around borehole and matrix gas pressure decreases gradually with the increase in extraction time. However, the gas content in coal is constant, and there is a pressure difference between the negative pressure borehole and free gas in the fracture. The gas pressure between the fracture and the coal matrix, and the gas pressure between the coal matrix and the surrounding coal matrix gradually decrease with the extraction. The gas pressure in the effective extraction area, the coal matrix around the borehole, and the fracture will tend to be gentle and no longer change significantly after reaching a certain extraction time.

4.2. Drilling Engineering Optimization

When studying the effective extraction radius, in order to avoid the possible interaction between multiple boreholes, the single borehole extraction used in Section 4.1 is adopted. The effective extraction radius increases with the increase in extraction time, but the increase gradually decreases until it tends to be gentle. With the increase in extraction time, the amount of gas extraction increases. However, more than a certain extraction time, even if the extraction time is longer, it has little effect on the pure amount of gas extraction. Since the adsorbed gas in the coal matrix occupies the vast majority of the total gas in the coal seam, the effective extraction radius r = 1.3 m, considering the gas pressure of the coal matrix, is selected as the reference for the arrangement of borehole spacing when the extraction is 140 d. According to the research of Chen Yuexia et al. [22,27] without considering the influence of superposition effect between boreholes, when four boreholes are used for extraction, through geometric derivation, it is concluded that when the spacing between boreholes is less than or equal to $\sqrt{2}$r, there will be no blank zone with pressure higher than the specified pressure in the borehole layout area. However, in fact, when multiple boreholes are extracted, due to the interaction between boreholes due to borehole spacing, the superposition effect is generated [28,29]. As shown in Figure 4, when the hole spacing of the three extraction boreholes is set to 2r, the three boreholes are surrounded by an equilateral triangle with a side length of 2r. In theory, if the superposition effect between boreholes is not considered, there will be a blank zone shown in the shadow part, where the gas pressure of the coal matrix is higher than 0.74 MPa. According to the Pythagorean theorem, the AO length can be obtained as $\frac{2\sqrt{3}}{3}\mathrm{r} = 1.5$ m. When the borehole spacing is $\frac{2\sqrt{3}}{3}\mathrm{r} = 1.5\ \mathrm{m}, 2\mathrm{r} = 2.6\ \mathrm{m}, 4\ \mathrm{m}, 5\ \mathrm{m}$, and 6 m, the gas extraction simulation is carried out by three boreholes. Figure 5 shows the variation curve of the coal matrix gas pressure with time at the center points (0,0,0) of three boreholes during gas extraction. When the drill hole spacing changes, it is ensured that the point is always located in the center of the equilateral triangle surrounded by three boreholes. It can be seen that when the extraction spacing is 1.4 m and the extraction time is 0 d–70 d, the gas pressure of the coal matrix decreases obviously. After the extraction time was 70 days, the gas pressure in the coal matrix did not decrease significantly and finally stabilized. When the borehole spacing is different, the depressurization effect of the drilling area is less obvious with the increase in borehole spacing. The gas pressure in the coal matrix showed a rapid decrease trend at the initial stage, and then the downward trend slowed down. The gas pressure in the coal matrix decreases not obviously with the extension of extraction time, and gradually tends to be gentle. The decrease rate of coal matrix gas pressure in the area surrounded by boreholes becomes slower with the increase in borehole spacing. This is because the superposition effect between boreholes decreases with the increase in borehole spacing. In addition, with the continuation of the extraction process, the gas pressure gradient between coal fractures and boreholes, between coal fractures and coal matrix pores, and between coal matrix pores, gradually decreases, and the gas migration is not strong at the beginning of extraction.

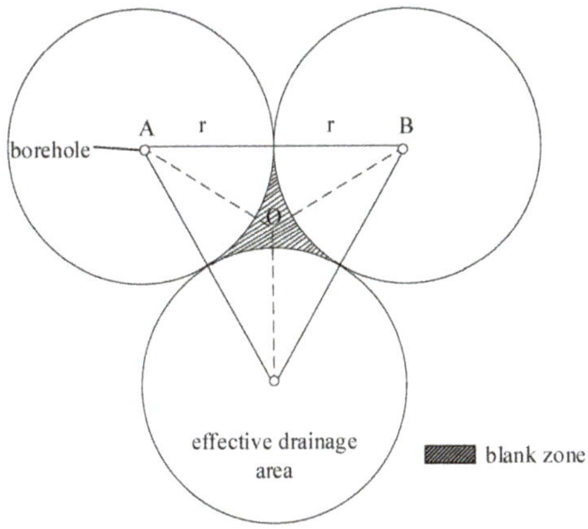

Figure 4. Schematic diagram of borehole gas extraction influence area.

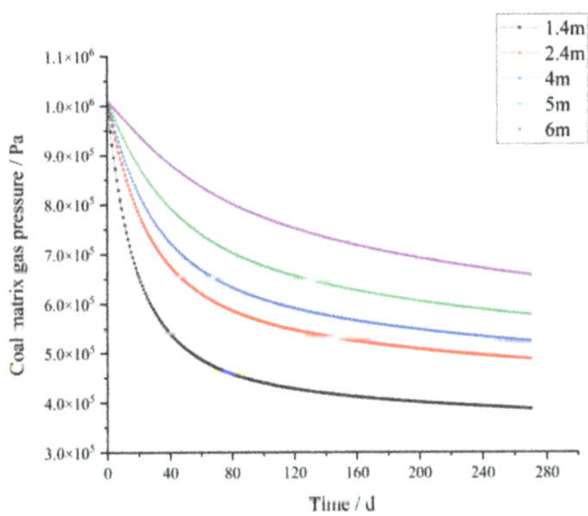

Figure 5. Variation in coal matrix gas pressure with time under different extraction spacing.

Figure 6 shows the distribution of the coal matrix gas pressure and effective extraction radius when the extraction time is 140 d and the borehole spacing is 1.5 m, 2.6 m, 4 m, 5 m, and 6 m, respectively. The light green surface in the figure is 0.74 MPa isosurface. It can be seen from Figure 6a that when the borehole spacing is 1.5 m, the effective extraction area of the three boreholes is approximately cylindrical, and the gas pressure of the coal matrix in the area surrounded by the three boreholes is obviously low. This is because the borehole spacing is small, the borehole superposition effect is obvious, and the gas extraction effect in the area surrounded by the borehole is obvious. It can be seen from Figure 6b–d that when the borehole spacing is 2.6 m, 4 m, and 5 m, the effective extraction area of the three boreholes is a rounded triangular prism. With the increase in borehole spacing, the higher the gas pressure in the area surrounded by the three boreholes, the more irregular the effective extraction area, and with the internal depression, there is an overall trend of division. This is because the superposition effect of boreholes decreases with the increase in borehole spacing, and the effect of gas extraction in the area surrounded by boreholes is weakened. It can be seen from Figure 6e that when the borehole spacing is 6 m, the effective extraction area of the three boreholes is distributed around the three boreholes in three irregular cylinders, and the blank zone appears in the area surrounded by the three boreholes. This is because the superposition effect weakens with the increase in borehole spacing until a certain distance between boreholes is no longer affected by the superposition effect. The borehole spacing is different. The gas pressure in the area around the borehole decreases continuously with the increase in the extraction time. The smaller the borehole spacing, the greater the gas pressure drop and the faster the gas drop rate. However, the longer the extraction time, the slower the gas pressure decreases. When the borehole spacing increases to a certain distance, a blank zone appears in the area surrounded by boreholes. When the borehole spacing is small, the time required to achieve the expected extraction effect is shorter; but the more boreholes required, the higher the construction cost. When the borehole spacing is large, the extraction rate of gas slows down, and it takes a long time to achieve the expected extraction effect. However, the number of boreholes required is small, and the effective extraction area is large over time within a reasonable project budget.

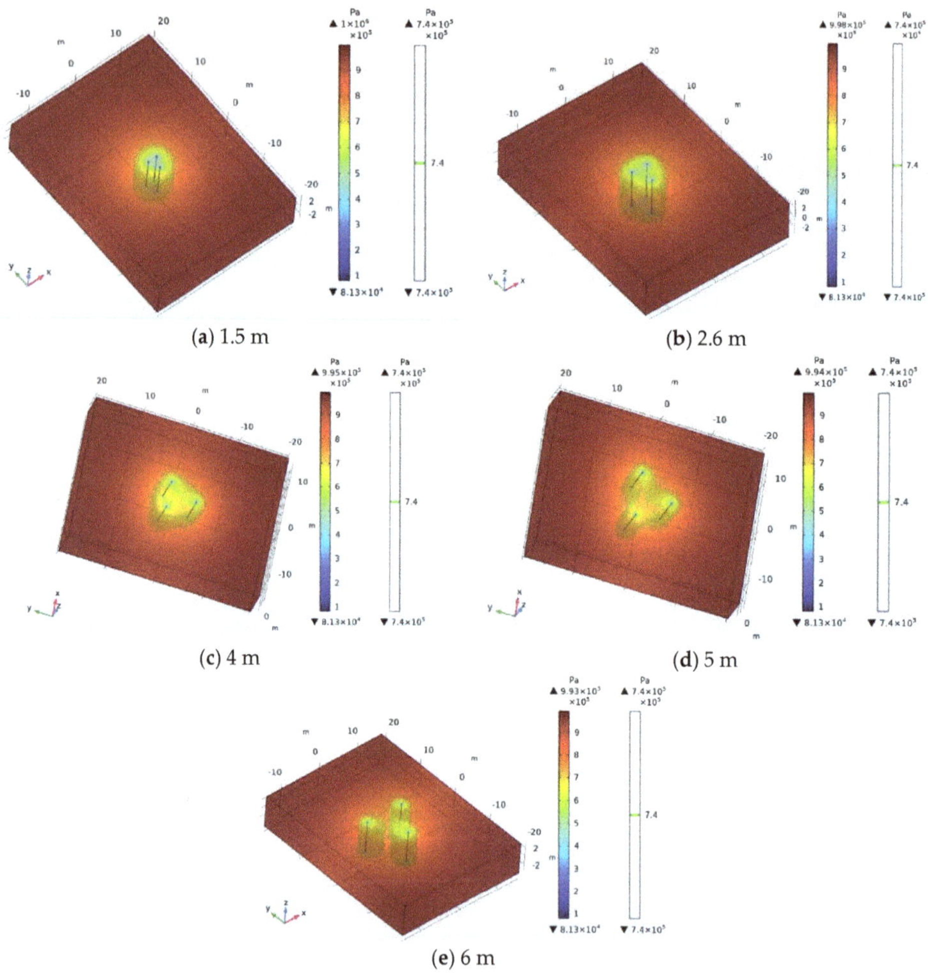

Figure 6. Effective extraction area effect diagram under different hole spacing of 140 d extraction.

As shown in Figure 7a, (−15,0,0) and (15,0,0) points are selected to make the *X*-axis direction section. The gas pressure of the coal matrix on the section line is simulated at 140 days of extraction with different hole spacing, and the results are shown in Figure 7b. In the extraction of 140 d, under different borehole spacing: with the increase in borehole spacing, the effect of gas pressure reduction is more obvious. The smaller the borehole spacing is, the greater the gas pressure drop rate in the area around the borehole is, and the more obvious the gas extraction effect is. The gas pressure drop rate around borehole decreases with the increase in borehole spacing, and the effect of gas extraction is worse. It shows that the drilling superposition effect is inversely proportional to the borehole spacing.

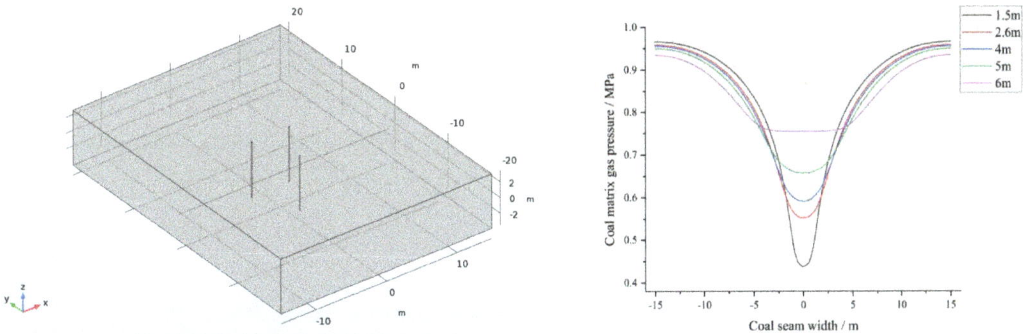

(**a**) monitoring line selection

(**b**) distribution of gas pressure

Figure 7. Gas pressure distribution of coal matrix on the cut-off line under different extraction spacing at 140 d.

Figure 8 is the volume change curve of the effective extraction area with time and borehole spacing. It can be seen from the diagram that under the condition of five different borehole spacing, the volume of the effective extraction area increases from gentle to sharp and then to a large increase. When the borehole spacing is 1.5 m, 2.6 m, 4 m, 5 m, and 6 m, the effective extraction area volume increases sharply in about 10 d–15 d, 15 d–30 d, 45 d–65 d, 60 d–75 d, and 105 d–120 d, respectively. At the beginning of extraction, the bigger the superposition effect is, the smaller the borehole spacing is, and the bigger the effective extraction area is. Primarily, at about 25 d, the volume of the effective extraction area with a borehole spacing of 2.6 m expands sharply, exceeding the volume of the effective extraction area with a borehole spacing of 1.2 m. Next, with the increase in extraction time, the volume of effective extraction area with borehole spacing of 4 m expands sharply, which exceeds the volume of effective extraction area with borehole spacing of 1.5 m and 2.6 m, successively. Then, the volume of the effective extraction area with a borehole spacing of 5 m continues to expand, exceeding the volume of the effective extraction area with a borehole spacing of 1.5 m, 2.6 m, and 4 m. As the extraction continues, the volume of the effective extraction area with a borehole spacing of 6 m first expands sharply, and then the increase decreases and continues to expand, successively exceeding the effective extraction area with a borehole spacing of 1.5 m, 2.6 m, 4 m, and 5 m. The effective extraction area volume of different borehole spacing changes with the extraction time, as shown in Table 2.

Table 2. Table of parameters required for simulation.

Drainage Time	Effective Extraction Area Volume Size	Drainage Time	Effective Extraction Area Volume Size
0 d–25 d	$V_{d=1.5m} > V_{d=2.6m} > V_{d=4m} > V_{d=5m} > V_{d=6m}$	85 d–100 d	$V_{d=5m} > V_{d=4m} > V_{d=2.6m} > V_{d=1.5m} > V_{d=6m}$
25 d–50 d	$V_{d=2.6m} > V_{d=1.5m} > V_{d=4m} > V_{d=5m} > V_{d=6m}$	100 d–115 d	$V_{d=5m} > V_{d=4m} > V_{d=2.6m} > V_{d=6m} > V_{d=1.5m}$
50 d–60 d	$V_{d=2.6m} > V_{d=4m} > V_{d=1.5m} > V_{d=5m} > V_{d=6m}$	115 d–130 d	$V_{d=5m} > V_{d=4m} > V_{d=6m} > V_{d=2.6m} > V_{d=1.5m}$
60 d–63 d	$V_{d=4m} > V_{d=2.6m} > V_{d=1.5m} > V_{d=5m} > V_{d=6m}$	130 d–135 d	$V_{d=5m} > V_{d=6m} > V_{d=4m} > V_{d=2.6m} > V_{d=1.5m}$
63 d–70 d	$V_{d=4m} > V_{d=2.6m} > V_{d=5m} > V_{d=1.5m} > V_{d=6m}$	After 135 d	$V_{d=6m} > V_{d=5m} > V_{d=4m} > V_{d=2.6m} > V_{d=1.5m}$
70 d–85 d	$V_{d=4m} > V_{d=5m} > V_{d=2.6m} > V_{d=1.5m} > V_{d=6m}$		

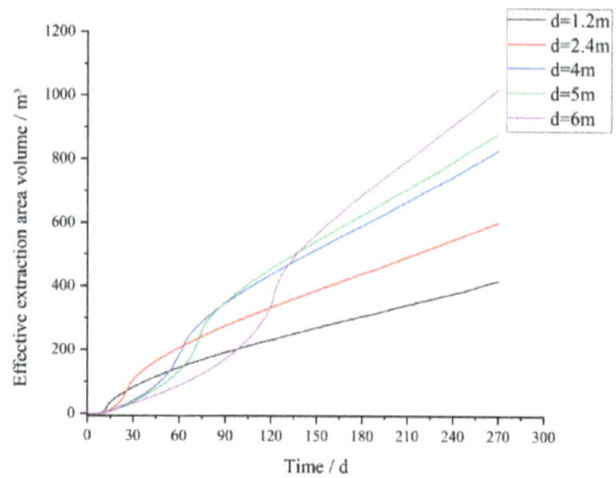

Figure 8. The volume change in effective extraction area under different extraction spacing at 140 d.

After the extraction time is 135 d, although the effective extraction area with borehole spacing of 6 m has the largest volume, it can be seen from Figure 7b that when the extraction time is 140 d, there is a blank zone with gas pressure higher than 0.74 MPa in the area surrounded by boreholes, so the extraction effect is not ideal. As the extraction time becomes longer and longer, the internal blank zone is gradually eliminated. Considering that if the gas pressure in the internal area surrounded by the borehole already reached the standard, the long extraction time may lead to safety problems due to pressure imbalance. When the borehole spacing is small, the actual economic problems caused by the increase in the effective extraction area with the extraction time are no longer obvious, and the extraction time is allowed under the actual working conditions. The comprehensive analysis shows that when the extraction time of the mine is about 140 d and the borehole spacing is set to 5 m, the extraction effect is the best.

5. Discussion

In this paper, based on the porous model of pore-fracture dual medium, considering the influence of coal skeleton and coal matrix deformation when coal adsorbs and desorbs gas, considering coal fracture seepage, coal matrix adsorption, desorption and diffusion gas, gas diffusion between coal matrix, and the Klinkenberg effect, a gas-solid coupling model is established. The effective extraction radius is obtained according to the variation law of coal matrix gas content with extraction time. Extending from a two-dimensional plane to a three-dimensional space, the volume change in effective extraction area is studied, so as to more intuitively analyze and verify the migration of gas in borehole coal fracture coal matrix during negative pressure borehole extraction. In the three-dimensional visualization model, it can be clearly seen that under the same extraction time, the effective extraction area, considering the change in gas pressure in coal fracture, is different from the effective extraction area considering the change in gas pressure in the coal matrix. Compared with two-dimensional theoretical analysis, it is more intuitive and more conducive to explain the mechanism of gas migration. Through the three-dimensional visualization model, at a certain distance between boreholes, it can be clearly seen that the gas pressure in the area around the borehole decreases more obviously than that outside the borehole, which indicates that the boreholes are affected by the superposition effect. When multiple boreholes are used for gas extraction, the borehole spacing has a great influence on the extraction effect. The extraction time required to eliminate the blank zone in the area surrounded by the borehole is longer with the increase in the borehole spacing. As long as the spacing of boreholes is within a reasonable range, the blank zone can be eliminated by prolonging the extraction time in theory. If the borehole spacing is too large, in theory, it reduces the number of boreholes and needs to extend the extraction time to eliminate blank area. However, the pressure imbalance may be caused by excessive extraction, and safety accidents may occur. Extending the extraction also requires the consumption of certain economic resources. If the borehole spacing is too small, the superposition effect between boreholes is strong, and the extraction time required to eliminate the blank zone time is short, but the increase in the number of boreholes may cause waste of resources. Therefore, the study of reasonable borehole spacing has a certain guiding significance for practical engineering.

This study has some inspiration and guiding significance for coal energy security and sustainable mining, but there are still deficiencies. According to different geological conditions, there are elastic–plastic changes in the actual coal, and the change in temperature will also affect the deformation of coal structure, so the coal deformation control equation remains to be improved. In the study of the optimal borehole spacing, the layout and spacing of boreholes need to be adjusted according to the actual coal mine conditions and engineering planning, and the selection method of the optimal borehole spacing needs to be improved.

6. Conclusions

(1) The gas-solid coupling model of negative pressure drilling gas extraction was established. The model is based on the coal pore-fracture porous media model, considering the coal deformation, gas diffusion seepage, gas adsorption, and desorption process. The change in gas pressure P_m in the coal matrix and gas pressure P_f in coal fracture is simulated, and the migration mechanism of gas between coal matrix, coal fracture, and borehole is verified.

(2) When gas is extracted by a negative pressure borehole, the gas in a coal fracture seeps into borehole driven by extraction negative pressure. After the gas in the fracture flows out, the coal matrix gas diffuses into the fracture under the pressure difference to participate in the seepage. The gas pressure of the coal matrix decreases, the gas adsorbed in the surrounding coal matrix desorbs, and the free gas contained diffuses to the coal matrix under the action of pressure difference. The gas pressure in the coal body near the borehole decreases obviously, and the farther away from the borehole, the weaker the influence of extraction. The pressure gradient decreases with the increase in extraction time, the effect of gas migration will gradually weaken, and the effective extraction range of the borehole is limited.

(3) During multi-borehole extraction, the gas pressure in the area around the borehole decreases obviously under the influence of the superposition effect. With the increase in borehole spacing, the extraction effect in the area around the borehole decreases. The effective extraction area formed by multi-borehole negative pressure gas extraction is related to extraction time and borehole spacing. When the three boreholes are extracted for 140 days, the effective extraction area is approximately cylindrical when the borehole spacing is 1.5 m. The effective extraction area is approximately a rounded triangular prism when the borehole spacing is 2.6 m, 4 m, and 5 m, but the longer the borehole spacing, the surrounding inward depression is split. The effective extraction area is divided into three near-cylindrical areas when the borehole spacing is 6 m, and a blank zone appears in the area surrounded by the borehole. Considering the shape, volume change, extraction effect, safety, and economic problems of the effective extraction area, the best effect is to select the hole spacing of 5 m when the three boreholes are used as the group extraction for 140 days.

Author Contributions: Investigation, W.C.; methodology, F.D.; supervision, K.W.; writing—original draft, F.D.; writing—review and editing, W.C. All authors have read and agreed to the published version of the manuscript.

Funding: This research was funded by the National Natural Science Foundation of China, grant number 52004291 and 52130409.

Data Availability Statement: The data are available from the corresponding author on reasonable request.

Acknowledgments: We also would like to thank the anonymous reviewers for their valuable comments and suggestions that lead to a substantially improved manuscript.

Conflicts of Interest: The authors declare no conflict of interest.

References

1. Yuan, L. Deep coal and gas mining strategic thinking in China. *J. China Coal Soc.* **2016**, *41*, 1–6.
2. Wang, K.; Du, F. Coal-gas compound dynamic disasters in China: A review. *Process Saf. Environ. Prot.* **2020**, *133*, 1–17. [CrossRef]
3. Xue, D.J.; Zhou, J.; Liu, Y.T.; Gao, L. On the excavation-induced stress drop in damaged coal considering a coupled yield and failure criterion. *Int. J. Coal Sci. Technol.* **2020**, *7*, 58–67. [CrossRef]
4. Du, F.; Wang, K. Unstable failure of gas-bearing coal-rock combination bodies: Insights from physical experiments and numerical simulations. *Process Saf. Environ. Prot.* **2019**, *129*, 264–279. [CrossRef]
5. Guo, D.Y.; Lv, P.F.; Zhao, J.C.; Zhang, C. Research progress on permeability improvement mechanisms and technologies of coalbed deep-hole cumulative blasting. *Int. J. Coal Sci. Technol.* **2020**, *7*, 329–336. [CrossRef]
6. Shu, L.Y.; Wang, K.; Liu, Z.S.; Zhao, W.; Zhu, N.N.; Lei, Y. A novel physical model of coal and gas outbursts mechanism: Insights into the process and initiation criterion of outbursts. *Fuel* **2022**, *323*, 124305. [CrossRef]
7. Li, X.Y.; Zheng, J.Y.; Liu, J.P. Numerical Simulation of Borehole Parameters for Coal Seam Gas Pre-Drainage. *Int. J. Heat Technol.* **2021**, *39*, 1328–1334. [CrossRef]
8. Wang, K.; Guo, Y.Y.; Wang, G.; Du, F. Seepage and Mechanical Failure Characteristics of Gas-bearing Composite Coal-Rock Under True Triaxial Path. *J. China Coal Soc.* **2022**, 1–12. [CrossRef]
9. Tu, Q.Y.; Cheng, Y.P.; Xue, S.; Ren, T.; Cheng, X. Energy-limiting factor for coal and gas outburst occurrence in intact coal seam. *Int. J. Min. Sci. Technol.* **2021**, *31*, 729–742. [CrossRef]
10. Li, W.R.; Wang, D.K.; Wang, J.G. Improved mathematical model of apparent permeability: A focused study on free and multilayer adsorptive phase flow. *J. Nat. Gas Sci. Eng.* **2022**, *101*, 104508. [CrossRef]
11. Pan, Z.J.; Connell, L.D.; Camilleri, M.; Connelly, L. Effects of matrix moisture on gas diffusion and flow in coal. *Fuel* **2010**, *89*, 3207–3217. [CrossRef]
12. Wang, K.; Du, F.; Wang, G. Investigation of gas pressure and temperature effects on the permeability and steady-state time of Chinese anthracite coal: An experimental study. *J. Nat. Gas Sci. Eng.* **2017**, *40*, 179–188. [CrossRef]
13. Hou, W.T.; Wang, H.P.; Yuan, L.; Wang, W.; Xue, Y.; Ma, Z.W. Experimental research into the effect of gas pressure, particle size and nozzle area on initial gas-release energy during gas desorption. *Int. J. Min. Sci. Technol.* **2021**, *31*, 253–263. [CrossRef]
14. Zhu, W.C.; Liu, J.; Sheng, J.C.; Elsworth, D. Analysis of coupled gas flow and deformation process with desorption and Klinkenberg effects in coal seams. *Int. J. Rock Mech. Min. Sci.* **2006**, *44*, 971–980. [CrossRef]
15. Hu, G.Z.; Xu, J.L.; Wang, H.T.; Jin, Z.Y.; Wu, R.L. Research on a dynamically coupled deformation and gas flow model applied to low-permeability coal. *J. China Univ. Min. Technol.* **2011**, *40*, 1–6.
16. Xia, T.-Q.; Gao, F.; Kang, J.-H.; Wang, X.-X. A fully coupling coal-gas model associated with inertia and slip effects for CBM migration. *Ecol. Environ. Conserv.* **2016**, *75*, 582. [CrossRef]
17. Cheng, Y.P.; Dong, J.; Li, W.; Chen, M.Y.; Liu, K.; Zhao, W. Effect of negative pressure on coalbed methane extraction and application in the utilization of methane resource. *J. China Coal Soc.* **2017**, *42*, 1466–1474.
18. Liu, Q.Q.; Cheng, Y.P.; Li, W.; Jin, K.; He, T.; Zhao, W. Mathematical Model of Coupled Gas Flow and Coal Deformation Process in Low-Permeability and First Mined Coal Seam. *Chin. J. Rock Mech. Eng.* **2015**, *34*, 2749–2758.
19. Zhang, Y.Y. Numerical Simulation Study on Effective Radius of Degassing Borehole. Master's Thesis, Henan Polytechnic University, Jiaozuo, China, 2015. Available online: https://kns.cnki.net/KCMS/detail/detail.aspx?dbname=CMFD201602&filename=1016710989.nh (accessed on 8 January 2023).
20. Li, B.; Wei, J.P.; Sun, D.H.; Zhang, L.L.; Borisyukd, Y.A. Research on the Reasonable Spacing of Holes in Gas Drainage along Coal Seams in Consideration of the Superimposed Effect of Drainage. *J. Eng. Sci. Technol. Rev.* **2016**, *9*, 102–110.
21. Zhang, L.; Jing, C.; Li, S.G.; Bao, R.Y.; Zhang, T.J. Seepage Law of Nearly Flat Coal Seam Based on Three-Dimensional Structure of Borehole and the Deep Soft Rock Roadway Intersection. *Energies* **2022**, *15*, 5012. [CrossRef]
22. Chen, Y.X. Parameters Evolution during Gas Drainage and Numerical Simulation of Boreholes Optimization. *Ph.D. Disssertation*; Chongqing University: Chongqing, China, 2018. Available online: https://kns.cnki.net/KCMS/detail/detail.aspx?dbname=CDFDLAST2019&filename=1018852869.nh (accessed on 8 January 2023).
23. Hu, S.X.; Liu, X.; Li, X.Z. Fluid–Solid Coupling Model and Simulation of Gas-Bearing Coal for Energy Security and Sustainability. *Processes* **2020**, *8*, 254. [CrossRef]
24. Wang, D.K.; Tang, J.H.; Wei, J.P.; Wei, L.; Wu, J.; Yuan, M.Y.; Pang, X.F.; Guo, Y.J. A fluid-solid coupling model of coal seam gas considering gas multi-mechanism flow and numerical simulation analysis of gas drainage. *J. China Coal Soc.* **2022**, 1–13. [CrossRef]
25. Wang, D.K.; Peng, M.; Fu, Q.C.; Qin, H.J.; Xia, Y.L. Evolution and numerical simulation of coal permeability during gas drainage in coal seams. *Chin. J. Rock Mech. Eng.* **2016**, *35*, 704–712.
26. Wu, Y.; Liu, J.S.; Elsworth, D.; Chen, Z.W.; Connell, L.; Pan, Z.J. Dual poroelastic response of a coal seam to CO_2 injection. *Int. J. Greenh. Gas Control.* **2010**, *4*, 668–678. [CrossRef]

27. Chen, Y.X.; Chu, T.X.; Chen, P.; Tang, Y. Quantitative study of 3D numerical simulation on optimizing borehole layout spacing of gas drainage. *Coal Geol. Explor.* **2021**, *49*, 78–84, 94.
28. Wei, P.; Huang, C.W.; Li, X.L.; Peng, S.J.; Lu, Y.N. Numerical simulation of boreholes for gas extraction and effective range of gas extraction in soft coal seams. *Energy Sci. Eng.* **2019**, *7*, 1632–1648. [CrossRef]
29. Zhao, D.; Liu, J.; Pan, J.T. Study on gas seepage from coal seams in the distance between boreholes for gas extraction. *J. Loss Prev. Process Ind.* **2018**, *54*, 266–272. [CrossRef]

processes

MDPI

Article

Experimental Study on Fractal Characteristics of Adsorption Pore Structure of Coal

Wendi Wang [1,2], Zhen Liu [1,2,*], Mingrui Zhang [1,2] and He Yang [1,2]

[1] College of Safety and Environmental Engineering, Shandong University of Science and Technology, 579 Qianwangang Rd., Huangdao District, Qingdao 266590, China
[2] State Key Laboratory of Mining Disaster Prevention and Control Co-Founded by Shandong Province and the Ministry of Science and Technology, Shandong University of Science and Technology, Qingdao 266590, China
* Correspondence: liuzhensdust@sdust.edu.cn; Tel.: +86-13954252676

Abstract: The adsorption pore structure is the key affecting the technology of enhanced coal bed methane recovery (ECBM). In this paper, nitrogen adsorption measurement (NAM) and low-field nuclear magnetic resonance (NMR) methods are used to test the structural parameters of adsorption pores of four coal samples with different metamorphic degrees. Combining with the fractal theory, the applicability and the physical significance of fractal dimensions obtained from different models are analyzed. Finally, the main factors affecting the complexity of the pore size structure and the influence of fractal characteristics on the adsorption properties of coal are discussed. The results show that with the increase in the degree of metamorphism, the ranking order of the volume and the specific surface area of adsorption pores of four coal samples are long flame coal > anthracite > gas coal > coking coal. The fractal dimension D_2 calculated by the Frenkel–Halsey–Hill (FHH) fractal model ranges from 2.5 to 2.9, representing the degree of the pore surface irregularity, and D_3 calculated by the FHH fractal model ranges from 2.004 to 2.037, representing the complexity of pore size distribution. There is no clear quantitative relationship between the fractal dimension and the single structure parameters of adsorption pores. The more the pore diameter distribution is concentrated in the range of 2–5 nm, the larger the fractal dimension is, the higher the complexity of adsorption pore structure is; D_2, which represents the irregularity of the surface of coal, has a good linear positive correlation with the maximum adsorption capacity of N_2.

Keywords: fractal dimension; pore structure; pores characteristics; fractal theory; metamorphism

Citation: Wang, W.; Liu, Z.; Zhang, M.; Yang, H. Experimental Study on Fractal Characteristics of Adsorption Pore Structure of Coal. *Processes* **2023**, *11*, 78. https://doi.org/10.3390/pr11010078

Academic Editors: Feng Du, Aitao Zhou and Bo Li

Received: 2 December 2022
Revised: 24 December 2022
Accepted: 25 December 2022
Published: 28 December 2022

1. Introduction

As an extremely important unconventional natural gas energy, coal bed methane (CBM) resources have attracted great attention from the energy community around the world [1–6]. Unlike conventional natural gas reservoirs, the nano-scale adsorption pore structure of coal is the important storage space for gas. And its structural characteristics directly determine the gas adsorption and desorption process, thereby affecting the gas content of coal seams [7,8]. Therefore, studying the characteristics of the coal adsorption pore structure is the basis for revealing the law of the coalbed methane adsorption–desorption–diffusion migration [9,10]. At present, scholars have mainly researched the coal structure based on image analyses, fluid injection, and gas adsorption methods [11–13]. For example, Feng et al. [14] used scanning electron microscopy (SEM) and computer tomography (CT) to analyze the microstructure and deformation of coal samples after the methane adsorption process; Meng and Qiu [15] studied the mechanism of the spatial diffusion of coal samples after supercritical CO_2 treatment through SEM, mercury intrusion porosimetry (MIP), and nuclear magnetic resonance (NMR). Simultaneously, due to the wide distribution of coalbed methane reservoirs in China and the variety of coal, the degree metamorphism of coal can affect its internal reservoir space. Scholars have also carried out extensive research on this [16]. There are several examples. Chen et al. [17] used the Fourier transform infrared (FTIR) spectrometer and CO_2 adsorption methods to study the micropore characteristics of five coal samples with different metamorphism degrees; Niu et al. [18] used the liquid nitrogen method and the methane isotherm adsorption experiment to study the pore structure and adsorption properties of 12 different rank coals. Therefore, based on previous studies, this paper uses the classic experimental methods to carry out multi-scale structural analyses of coal with different degrees of metamorphism to accurately and quantitatively characterize the structural characteristics of coal at the adsorption pore scale.

Coal is a porous medium with a complex structure and the experimental study of its heterogeneity is a key to the microstructure of coal [19,20]. At present, the fractal theory represents the effectiveness of the space occupied by complex shapes. It is a measure of the irregularity of complex shapes that can be introduced instead of characterizing the "special nature" of the coal pore structure [21–23]. Through the research of domestic and foreign scholars, the models for calculating the fractal size of pore structure include FHH model, Menger sponge model, and other applicability models [22,24,25]. In several proposed fractal models, the FHH fractal model is used to discuss the extent to which the internal surface roughness of coal determines the amount of natural gas, especially methane. They can readily adsorb/store, and the gas can easily flow and be produced from coal formations. Cai et al. [26] studied the physical composition and particle size distribution of coal by integrating classical geometric fractal model and thermodynamic fractal model. Lu et al. [27] explored the fitting of Sierpinski, Menger, and other models to coal pore structure at different scales, and summarized the fractal characteristics of coal adsorption pore structure. Hu et al. [28] combined fractal model and image analysis to explore the fractal dimension of coke particles and describe the surface characteristics of coal and coke particles. Liu et al. [29] calculated the adsorption pore volume based on the capillary bubble fractal model and NMR T_2 spectrum distribution, thus quantitatively characterizing the physical structure of coal. Scholars have quantitatively characterized the fractal characteristics of coal structures in detail, but have not made a detailed mechanistic explanation for the type of coal structures complexity represented by the fractal model. This paper has carried out certain research based on this.

According to previous studies, scholars have carried out detailed quantitative characterization of the fractal characteristics of coal structure, but there is little research on the type of coal structure complexity represented by the fractal model. On this basis, this paper has carried out some research. The selection of appropriate experimental methods and fractal models, accurate and quantitative analysis of the change law of coal structural parameters with different degrees of metamorphism to clarify the influence mechanism of coal microstructure parameters on fractal characteristics, as well as the detailed physical significance and important influencing factors of fractal models, are the key research contents of this paper. In this paper, NAM and NMR are selected to analyze the multi-scale structural characteristics of adsorption pores ($r < 50$ nm) [30] of coal with different metamorphism degrees according to the metamorphic characteristics of four coal samples. At the same time, the difference and applicability of the fractal dimension of adsorption pore structure combined with FHH model and capillary bundle model are comprehensively analyzed. Finally, the relationships between structural parameters and fractal dimensions are discussed, and based on this, the important factors affecting the pore structure complexity are clarified. The above research is the key to reveal the influence mechanism between the geological evolution process of the coalbed methane reservoir and the distribution law of gas content. It is of great significance for mine gas disaster prevention and coalbed methane research.

2. Materials and Experiments

2.1. Samples

This laboratory experiment selects raw coal from four coal mines in Shaanxi, Shandong, Anhui, and Shanxi provinces of China. Bulk raw coal samples obtained on site are transported to the laboratory for the industrial analyses, Table 1 shows the analysis results. And raw coals are used in both experiments. Four lump coal samples with a mass slightly greater than 0.1 g were used in NAM experiment, with a total area of 2–50 m^2/g. In the NMR experiment, four kinds of lump coal samples are made into experimental coal samples with a diameter of 25 mm and a height of 50 mm by using equipment.

Table 1. The industrial analyses/%.

Number	Samples	M_{ad}	A_{ad}	V_{da}	FC_{daf}	Vitrinite	Inertinite	Exinite	Ro_{max}
1	DLT	6.29	9.7	28.98	55.03	87.01	11.89	1.10	0.53
2	XLZ	3.08	5.21	30.06	61.65	64.37	23.82	11.81	0.78
3	QD	1.38	23.61	19.53	55.48	63.82	34.72	1.46	1.38
4	QC	1.83	17.35	15.84	64.98	55.51	43.67	0.82	1.87

Note: No. 1 is long flame coal; No. 2 is gas coal; No. 3 is coking coal; and No. 4 is anthracite. The metamorphic degree ranking order of the four kinds of coal samples is #1−DLT < #2−XLZ < #3−QD < #4−QC. M_{ad} is the moisture; A_d is the content ash; V_{da} is the volatile matter; FC_{daf} is the fixed carbon; Ro_{max} is the maximum reflectance of vitrinite.

2.2. NAM and NMR

For the NAM analyses, the four kinds of coal samples are tested with a Micromeritics ASAP2020 BET specific surface area and pore diameter analyzer. The samples are heated, dehydrated and degassed, cooled, and weighed in turn. Then, the samples are put into the instrument for measurement and analyses, and the adsorption and desorption isotherms of coal samples are obtained. Finally, the pore structure parameters of coal samples are obtained through the BET, BJH, and DFT models. For the NMR analyses, four kinds of coal samples are tested with the MesoMR23-060H low-field NMR equipment. The magnet of the instrument is a permanent magnet, the magnetic field strength is 0.5 T, the main frequency is 11 MHz, the diameter of the probe coil is 60 mm, the echo time (TE) is 0.1 ms, the number of echoes is 6000, the number of scanning is 64, and the experimental environment temperature is 25.5 °C. First, the coal samples are dried at 60 °C to remove the impurities; second, the samples are vacuumized, and the coal samples are placed in water for 48 h for the saturated water treatment. The coal sample before the NMR experiment does not have additional pressure on it during the 48-h saturated water treatment process, and because the water is an incompressible fluid, it will not destroy the pore structure of the original coal sample, so it will not affect the NMR experimental results. Finally, the samples are taken out to test the saturated water samples, and the T_2 relaxation distributions are calculated by using the simultaneous iterative reconstruction technique, and the number of iterations is 10,000. Because the NMR is based on the spin motion of the hydrogen nucleus, the amplitude of the T_2 spectrum of coal samples measured in this experiment can reflect the total hydrogen content of a certain pore diameter. Three different relaxation mechanisms affecting T_2 spectrum are: the free relaxation, the surface relaxation, and the diffusion relaxation. When the three relaxation mechanisms exist simultaneously, the T_2 of the fluid in the pore can be expressed as [31]:

$$\frac{1}{T_2} = \frac{1}{T_{2B}} + \frac{1}{T_{2S}} + \frac{1}{T_{2D}} \tag{1}$$

In Equation (1), T_{2B} is the free relaxation time; T_{2S} is the surface relaxation time; T_{2D} is the diffusion relaxation time.

The experiment is that pure fluid (water) is carried out under the uniform magnetic field, so the effect of diffusion and free relaxation is not considered. Then only the surface relaxation is considered. The surface relaxation is from the relaxation action of coal particle surface to fluid, which is related to the ratio of specific surface area to pore volume in the coal sample. Then Equation (1) can be simplified as follows:

$$\frac{1}{T_2} = \frac{1}{T_{2S}} = \rho_2 (\frac{S}{V})_{pore} \tag{2}$$

In Equation (2), T_2 is the relaxation time; ρ_2 is the surface relaxation rate of T_2; it can be directly related to capillary force and pore diameter, S is the pore surface area; V is the pore volume; $(\frac{S}{V})_{\text{porosity}}$ is the specific surface area of pores, the relationship between the pore radius and pore size is $(\frac{S}{V})_{pore} = \frac{F_s}{r}$; F_s is the shape factor, which is determined by the pore model. Therefore, Equation (3) can be expressed as [32]:

$$T_2 = \frac{1}{\rho_2 F_s} r \tag{3}$$

Let $\frac{1}{\rho_2 F_s} = C$, then the Equation (3) can be expressed as $T_2 = C \times r$, and the coefficient C is the constant. After the value of C is obtained, the NMR T_2 spectrum can be converted into the pore radius distribution. Therefore, the pore size distribution of coal samples can be obtained according to the T_2 spectrum distribution. Figure 1 is the experimental equipment and experimental schematic diagram.

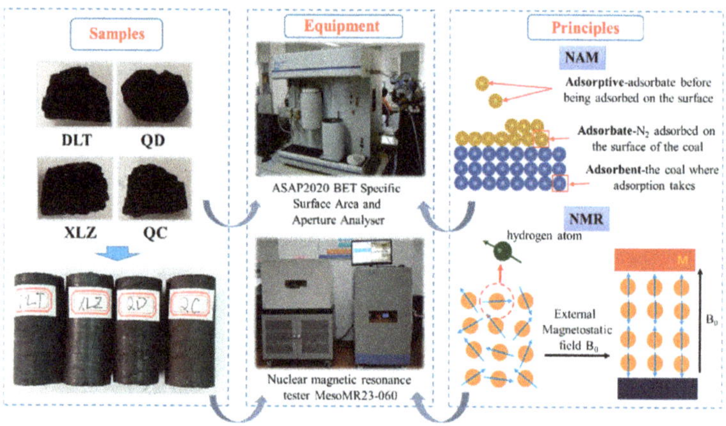

Figure 1. Experimental equipment and principles.

3. Results

3.1. Experimental Results of NAM

Through the NAM, the adsorption and desorption curves of coal samples are obtained, as shown in Figure 2. From Figure 2, the adsorption and desorption curves of four coal samples all present the inverted "S" type shape, which is the type II adsorption isotherm. When the adsorption capacity of the desorption branch is greater than that of the adsorption branch at the same relative pressure, the adsorption branch and the desorption branch do not coincide in a certain range of the relative pressure, and the separated loop is called the "hysteresis loop" [33]. Hysteresis loop can be divided into two types according to its shape, the first type of curve is shown in Figure 2a, and the second type of curve is shown in Figure 2b–d. Classification by BDDT and IUPAC [34], the shape of the first kind of hysteresis loop is basically consistent with the H2 (b) curve type, which indicates that it is characterized by the large adsorption loop, obvious hysteresis loop, and inflection point. It is further indicated that the adsorption pore system of low-rank coal is more complex, and there are "ink bottle" pores with small mouths and large bellies. The shape of the second kind of the hysteresis loop is basically consistent with that of H3 type, which indicates that the adsorption loop is small, the hysteresis loop is not obvious, and there is a slight inflection point. Then it shows that the adsorption pore system of medium and high rank coal is relatively simple, and flat slit capillary pores are present. Among them, the size of the second type of hysteresis loop shows the following trend: #2−XLZ < #3−QD < #4−QC, and it can be concluded that it increases with the increase in coal metamorphism degree [35].

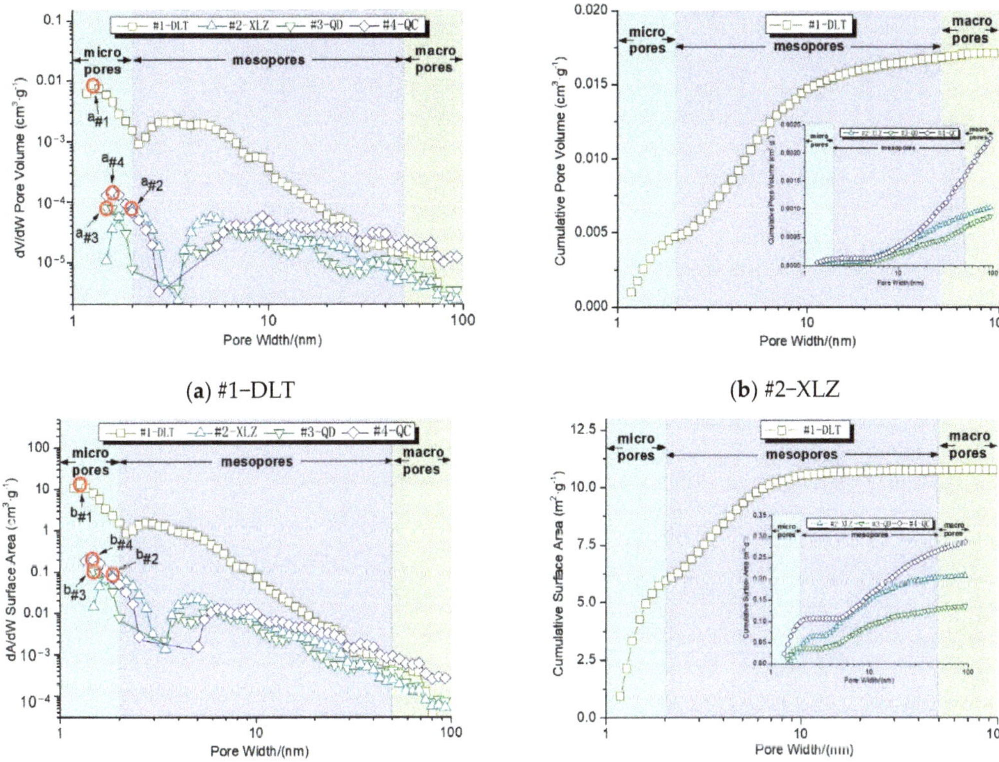

(**a**) #1−DLT (**b**) #2−XLZ

(**c**) #3−QD (**d**) #4−QC

Figure 2. Adsorption and desorption curves of the four coal samples.

The data analysis module of the adsorption instrument used in the experiment includes many classical model algorithms (such as BET, BJH, H-K, t-plot and NLDFT models).

On this basis, the nonlocal density function theory (NLDFT), which is an effective method to describe the fluid behavior in pores at the molecular level, is selected. It is assumed that the adsorption isotherms are obtained by multiplying numerous individual "single pore" adsorption isotherms by the relative distribution f(W) of their coverage pore size range. As long as the adsorbent and absorbent are given in the system, a set of influence functions (kernels) can be obtained by DFT simulation, and then the pore size distribution curve is obtained by solving the equation by fast non-negative least squares method. The pore size distribution characterized by it can be applied to the entire range of adsorption pores in coal samples [36]. Four pore size distributions (DV/DW, cumulative pore volume, DA/DW, and cumulative specific surface area) are plotted, as shown in Figure 3.

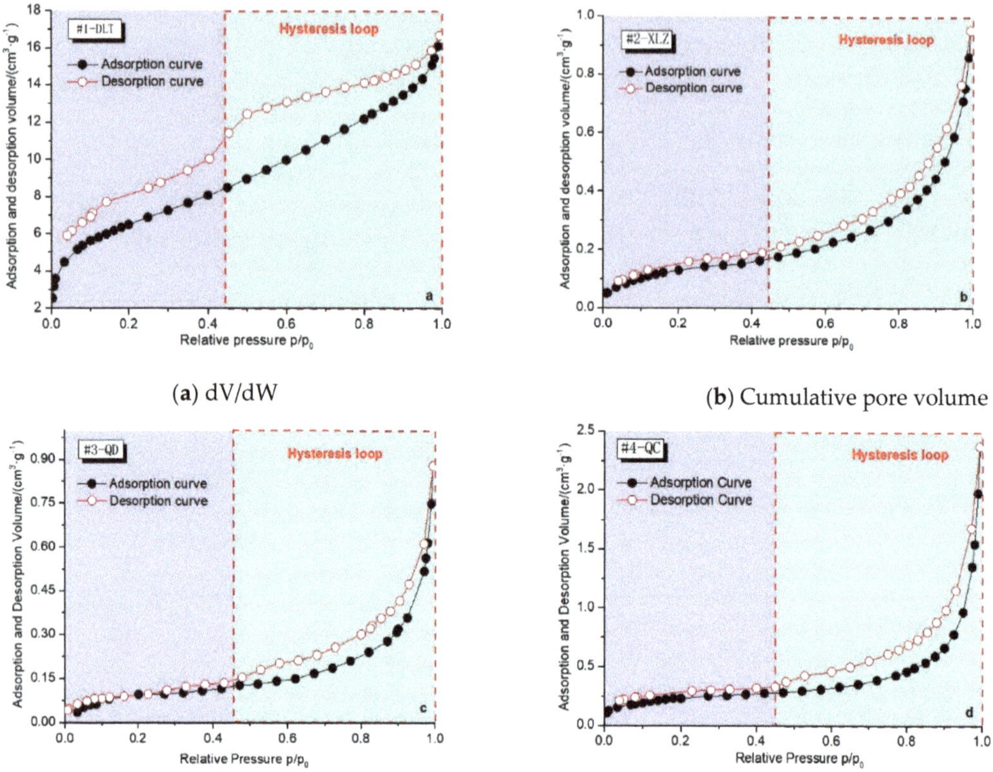

(**a**) dV/dW (**b**) Cumulative pore volume

(**c**) dA/dW (**d**) Cumulative surface area

Figure 3. The pore size distribution measured with the NAM.

The IUPAC pore classification standard [37] is adopted in this test. The adsorption pore range of coal samples tested includes micropore (<2 nm), mesopore (2 nm $\leq D \leq$ 50 nm), and macropore (50 nm < D < 90 nm). According to Figure 3, the pore size distribution curves of four coal samples are within the microporous scale range, showing the single peak characteristic, with the highest peak point, but the peak values are different, that is, a#1 > a#4 > a#2 ≈ a#3 and b#1 > b#4 > b#2 ≈ b#3. That is to say, within this range, the pore volume and specific surface area decrease first and then increase with the increase of the metamorphic degree, and the micropore volume and surface area of #1−DLT coal sample with the lowest metamorphic degree among four coal samples are the largest. In the range of mesoporous scale, only the #1−DLT coal sample shows the single peak characteristic, and the highest peak point is close to the pore scale boundary of 2 nm. The other three coal samples show complex multi peak characteristics, which are mostly concentrated at 10 nm–50 nm. However, the average peak sizes of four coal samples are different, #1−DLT > #4−QC > #2−XLZ > #3−QD, which indicates that the #1−DLT coal sample has the largest mesoporous pore volume and specific surface area. In the whole range of mesoporous scale, the distribution of mesoporous concentration of the #1−DLT coal sample is close to the boundary of the micro-meso porous scale range, while the distribution trend of the other three coal samples is close to the boundary of the macropore scale range. In the macropore pore scale, due to the limitation of the gas adsorption method, the variation characteristics are not obvious, and the volume and specific surface area change little with the pore width.

3.2. Experimental Results of NMR

According to the experimental principle of the low field NMR in Section 2.2, there is a quantitative relationship between the T_2 relaxation time of the saturated water coal sample and core pore radius r (Equation (3)). According to the experimental content, the coal structure is simplified as the columnar pipe [38], so for the parameter in Equation (3), we take $F_s = 2$. And then, the pore size distribution of four kinds of coal samples can be drawn, as shown in Figure 4.

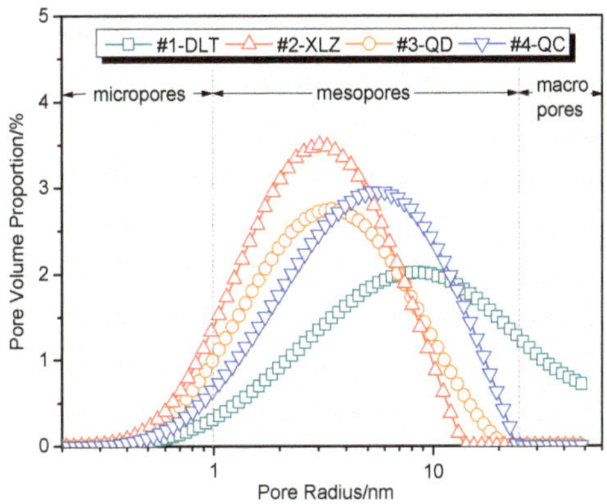

Figure 4. The pore size distributions measured by NMR.

The pore size distribution of coal sample measured by nuclear magnetic resonance is shown in Figure 4. The pore size distribution of coal sample adsorption holes in the figure is all single peak, and the maximum peak point is within the range of 2 nm–15 nm. The peak value of gas coal is the largest, followed by anthracite, coking coal, and long flame coal. Peak position from left to right is gas coal, coking coal, anthracite, long flame coal. The largest single peak area is long flame coal, followed by anthracite, gas coal, and coking coal. Due to the change of the test method, the influence mechanism of the metamorphic degree on the whole adsorption pores can be analyzed [39]. That is, for the long flame coal #1−DLT, it is in the low metamorphic stage of coal, it withstands formation pressure less, and the coal body structure is relatively loose.

4. Discussion

4.1. The Applicability of Fractal Models of the Coal Adsorption Pore Structure

Based on the theory proposed by previous scholars [40], the fractal dimension can be calculated by combining the results of the NAM with the FHH model:

$$ln(V/V_0) = C + A \times ln[ln(p_0/p) \quad (4)$$

In Equation (4), V is the volume of adsorbed N_2 at equilibrium pressure p; V_0 is the monolayer adsorption volume. p_0 is the saturated vapor pressure of adsorbed N_2; A is the slope of $ln(V/V_0)$ and $ln[ln(p_0/p)]$ in the double logarithmic coordinates; and C is the constant.

When the van der Waals force is the main adsorption force, the relationship between A and the fractal dimension D is:

$$A = \frac{D-3}{3} \quad (5)$$

When the capillary condensation occurs in pores, the surface tension is the main force, and the relationship between A and the fractal dimension D is:

$$A = D - 3 \quad (6)$$

In Equations (5) and (6), D is the fractal dimension calculated by the FHH fractal model; and A is the slope of lnV and $ln[ln(p_0/p)]$ in the double logarithmic coordinates.

The fractal dimension of adsorption pore can be calculated by this experimental method. At the same time, when using this model to calculate the fractal dimension, it is usually divided into the low-pressure section (the relative pressure is 0–0.45) and the high-pressure section (the relative pressure is 0.45–1.00) to obtain fractal dimensions of different pressure sections. The fractal dimension of the low-pressure section is set as D_1, and the fractal dimension of the high-pressure section is set as D_2. Figure 5 shows the fitting diagram of lnV and $ln[ln(p_0/p)]$ in the FHH fractal model.

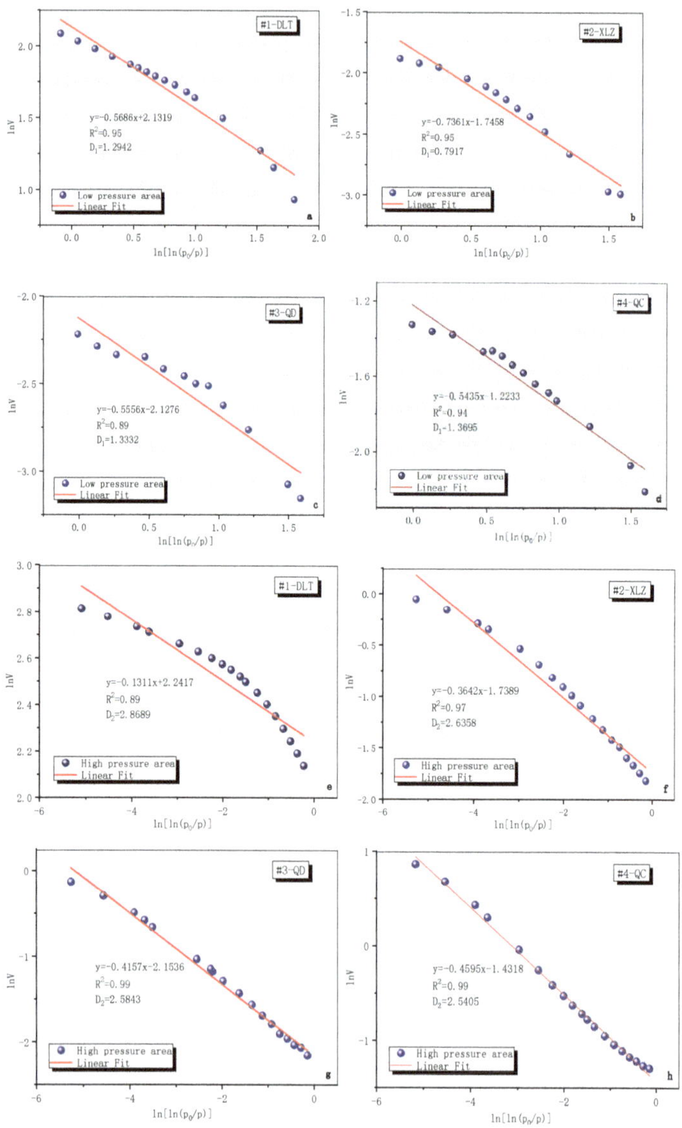

Figure 5. Fitting diagrams of $ln(V/V_0)$ and $ln[ln(p_0/p)]$.

It can be seen from Figure 6 that all fitting coefficients of four coal samples are above 0.89, that is to say, the fitting effects are good. FHH fractal model can be used to calculate the fractal dimension of adsorption pores of four coal samples, which can effectively characterize the complexity of the adsorption pore structure of coal samples. At the same time, because this model can calculate the adsorption capacity of coal, it can be used to characterize the complex pore structure of coal. This complexity represents the degree of irregularity in the coal body surface.

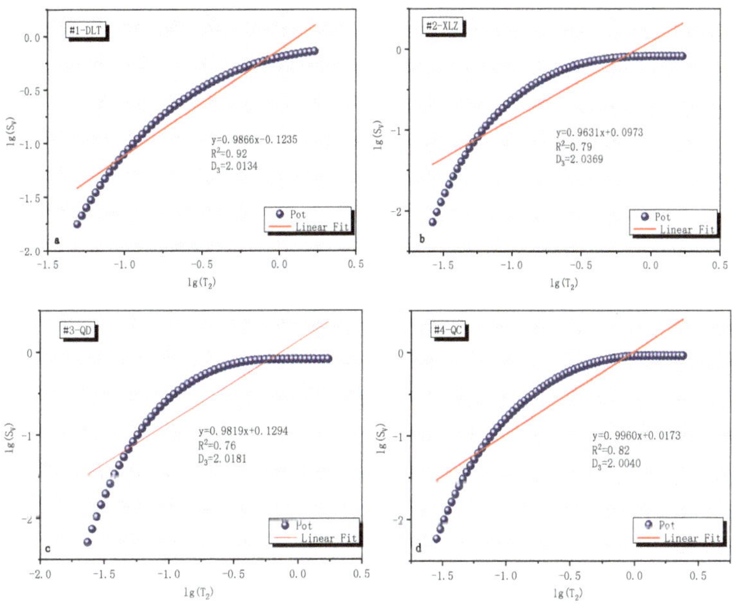

Figure 6. Fitting diagrams of $lg(T_2)$ and $lg(S_V)$.

According to an algorithm based on the capillary pressure method to calculate the fractal dimension, Zhang et al. [41] combines the T_2 spectrum distribution obtained from the NMR experiment and deduces an algorithm based on the T_2 spectrum curve to calculate the fractal dimension:

$$\lg(S_v) = (3 - D)\lg(T_2) + (D - 3)lgT_{2max} \quad (7)$$

In Equation (7), S_v is the corresponding cumulative pore volume percentage; T_2 is the transverse relaxation time; T_{2max} is the maximum transverse relaxation time; and D is the measured pore volume fractal dimension.

The fractal dimension based on the T_2 spectrum can be expressed as follows:

$$D = 3 - k \quad (8)$$

In Equation (8), k is the slope of the linear fit of $lg(T_2)$, $lg(S_V)$.

Moreover, the inflection point of the curve in the relationship curve between the logarithmic value $lg(T_2)$ of the transverse relaxation time T_2 and the logarithmic value $lg(S_V)$ of the corresponding cumulative pore volume percentage S_v in the NMR data is the boundary point between the fractal dimension of the adsorption pore volume and the fractal dimension of the seepage pore volume. The fractal dimension of adsorption pores is the volume fractal dimension, which indicates the complexity of the pore size distribution. Generally, the pore size in porous media satisfies the pore diameter $\frac{\lambda_{min}}{\lambda_{max}}^D \approx 0$ [42,43], so for the $\frac{\lambda_{min}}{\lambda_{max}}$ of four kinds of test samples selected 2.764 nm/95.36 nm, 1.48 nm/95.36 nm, 1.318 nm/96.46 nm,

1.134 nm/96.38 nm as the pore diameter range to calculate the fractal dimension of the overall structure of adsorption pores, and thus obtained the fractal dimension D_3. The fitting curves of $lg(T_2)$ and $lg(S_V)$ is shown in Figure 6.

Table 2 shows the calculation results of fractal dimension obtained from the above fitting curve and algorithm.

Table 2. The calculation results of fractal dimensions.

Fractal Dimensions	Test Sample			
	#1−DLT	**#2−XLZ**	**#3−QD**	**#4−QC**
D_1	1.2942	0.7917	1.3332	1.3695
D_2	2.8689	2.6358	2.5843	2.5405
D_3	2.0134	2.0369	2.0181	2.0040

From the range distribution of mesoporous fractal dimensions, there are no effective values in D_1, which are lower than 2, and all values of D_2 are between 2 and 3, indicating that D_2 is more effective. This is because in the FHH fractal model, the calculation model of the high-pressure section is based on the capillary condensation process, and the effect of surface tension is considered. For the mesopores, the capillary condensation process is a characteristic process within the pore size range, which does not occur in the micropores. Therefore, the effective fractal dimension can be obtained by using the adsorption data of the high-pressure section. However, the data of low-pressure section contain some microporous adsorption data, which affect the calculation results of fractal dimensions. Among them, D_2 mainly ranges from 2.5 to 2.9, and the value of D_2 gradually decreases with the increase of the metamorphism degree, which indicates that the complexity of mesopores decreases with the increase of the metamorphism degree [44]. The D_3 is between 2.004 and 2.0369 as the #1−DLT long flame coal has a lower D_3 fractional dimensional value due to its looser coal structure and more connected macropore network. With the exception of the #1−DLT long flame coal, the D_3 values gradually decrease with increasing metamorphism, which could indicate that the complexity of the pores throughout the adsorption pore space gradually decreases with increasing metamorphism.

The average value of D_3 is less than that of D_2, which is due to the difference between test methods and fractal models used. First, D_2 is calculated by using the FHH model and N_2 adsorption capacity, while D_3 is calculated by using the capillary beam fractal model and T_2 spectrum. Therefore, when the pore diameter is in the range of 0–100 nm, gas can enter into smaller pores more than water, which results in more complex pore structure and smaller D_3 value. Second, there are some differences in the physical meaning of characterization. The FHH model is based on the adsorption capacity to characterize the surface irregularity of coal body, while the capillary beam fractal model represents the complexity of the pore size distribution. The complexity of pore size distribution in coal is composed of micropores, mesopores, and macropores, so the D_3 change of four coal samples is not obvious. Third, except for the #1−DLT coal sample, D_2 and D_3 of the other three coal samples gradually decrease with the increase of metamorphic degree, and the variation laws are consistent, that is, the complexity of adsorption pore structure generally decreases with the increase of metamorphic degree [45].

4.2. The Relationships between Adsorption Pore Structural Parameters and Fractal Dimensions

Table 3 shows the experimental results based on the specific surface area and total pore volume of adsorption pores of coal samples obtained with NAM:

Table 3. Specific pore surface area and total pore volume of adsorption pores of 4 experimental samples.

Coal Type	Number	Specific Surface Area/($m^2 \cdot g^{-1}$)	Total Pore Volume/($cm^3 \cdot g^{-1}$)
Long flame coal	#1−DLT	10.7715	0.01712
gas coal	#2−XLZ	0.2085	0.00103
Coking coal	#3−QD	0.1354	0.00088
anthracite	#4−QC	0.2869	0.00226

From Table 3, with the increase in the metamorphism degree, the specific surface area and total pore volume of adsorption pores show the trend of first decreasing and then increasing, especially for the long flame coal. Compared with the other three coal samples, the specific surface area and total pore volume differ by 100 and 10 orders of magnitude, respectively, with significant differences. In order to further discuss the quantitative relationship between the pores and fractal dimensions of coals with different metamorphic degrees, Figure 7 shows the relationship curves between specific surface area, pore volume, and D_2 and D_3.

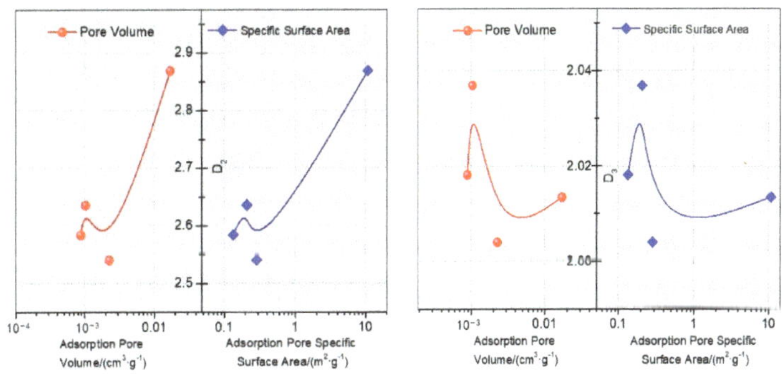

Figure 7. Relationships between structural parameters and fractal dimensions.

Figure 7 shows that the fractal dimensions of the two types first increase, then decrease, and finally increase with the increase of total pore volume and specific surface area. From these four groups of curves, it can be concluded that there are no direct quantitative relationships between the two structural parameters and the fractal dimensions D_2 and D_3, which represent the surface irregularity and the complexity of the pore size distribution, respectively [46]. In other words, the factors that determine the fractal dimension are not only determined by the pore volume or specific surface area. In order to verify this conclusion, the calculation results of the proportion of the pore volume and specific surface area of micropores, mesopores, and macropores to the whole adsorption pores are arranged in Table 4. The curves of D_2 and D_3 change with the proportion of each scale, as shown in Figure 8.

Table 4. Calculation results of the proportion of the pore volume and specific surface area of micropores, mesopores, and macropores to the whole adsorption pores.

Coal Samples	The Proportion of Pore Volume/%			The Proportion of Surface Area/%		
	Micropores	Mesopores	Macropores	Micropores	Mesopores	Macropores
#1−DLT	26.29	72.01	1.70	53.28	46.03	0.69
#2−XLZ	2.25	83.86	13.89	12.80	77.10	10.1
#3−QD	4.88	71.71	23.41	26.67	68.86	4.47
#4−QC	5.04	65.76	29.20	31.50	61.94	6.56

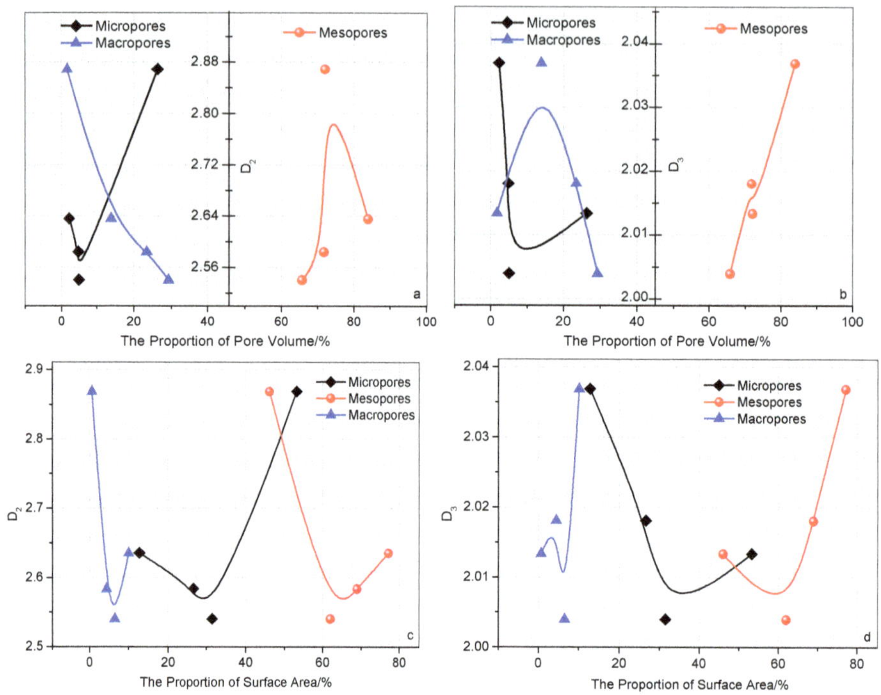

Figure 8. Relationships between proportions of pore parameters and fractal dimensions.

It can be seen from Figure 8 that there are no direct quantitative relationships between the proportions and fractal dimensions, indicating that single structural parameters do not directly determine the values of fractal dimensions [47]. Combined with Figures 3 and 4 obtained from two experiments, it can be found that for D_2, in the mesoporous scale range, the peak value of pore diameter distribution is concentrated at 2–10 nm, and the closer the highest peak point is to the range of 2–5 nm, the larger the fractal dimension value is. For D_3, in the whole adsorption pores, the single peak value of the pore diameter distribution is mainly in the range of 2–30 nm (except for the #1−DLT coal sample). Similarly, the closer the pore diameter distribution is to the range of 2–5 nm, the larger the fractal dimension value is, the higher the complexity of pore structure is.

At the same time, the influence mechanism of fractal characteristics of coal structure on its adsorption properties is discussed. Therefore, according to the results of the NAM experiment, the relationships between two fractal dimensions and the maximum nitrogen adsorption properties are drawn.

First of all, it can be seen from Figure 9a that there is a good linear positive correlation between the fractal dimension D_2 calculated by the FHH fractal model and the maximum N_2 adsorption capacity, and the correlation coefficient is 0.87, which indicates that the more complex the inner surface irregularity of coal adsorption pore structure is, the better the adsorption properties of coal is, which also indicates that the main determinant of adsorption properties of coal is the specific surface area [48–50]. Second, it can be seen from Figure 9b that there is no obvious quantitative relationship between the fractal dimension D_3 calculated by the capillary beam fractal model and the maximum N_2 adsorption capacity, which also confirms that D_3, which represents the complexity of the pore size distribution, is not the main factor affecting the adsorption performance of coal. Therefore, according to the relationships between the two fractal dimensions and the maximum adsorption properties, the physical meaning of two fractal dimensions can also be distinguished.

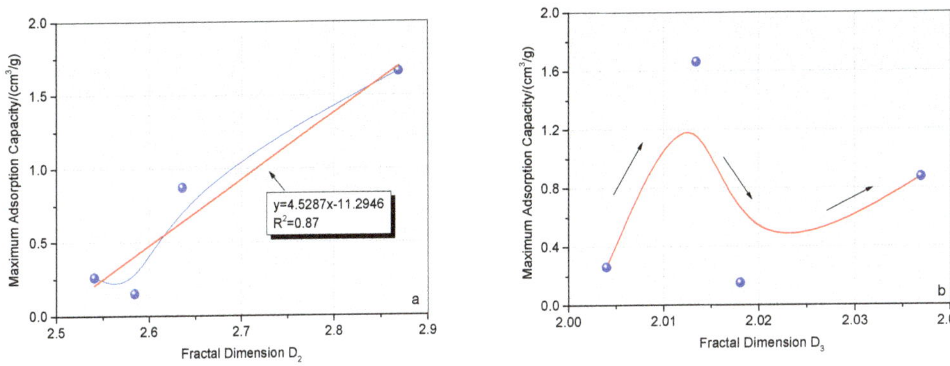

(a) (b)

Figure 9. Relationships between fractal dimensions and the maximum adsorption properties. (**a**) D_2-maximum adsorption capacity. (**b**) D_3-maximum adsorption capacity.

5. Conclusions

This paper uses the NAM and low-field NMR to conduct the multistage analyses of parameters such as the pore size distribution, pore shape, pore volume, and specific surface area of four coal samples with different metamorphism degrees. Based on the fractal models, multiple fractal dimensions are calculated. The relationships between fractal dimensions and various quantitative parameters of micropores, mesopores, macropores and whole adsorption pore structure are obtained.

(1) Combined with the metamorphism characteristics of coal, the formation pressure and temperature are constantly changing, and the rapid pyrolysis fracture of aliphatic rings in coal macromolecules results in disappearing first and then forming a large number of pores between coal molecules. Therefore, the pore volume and specific surface area of adsorption pores first decrease and then increase with the increase of the metamorphism degree.

(2) Due to the difference in fractal models, among them, the effective fractal dimensions based on the FHH fractal model are concentrated in the range of 2.5–2.9, which are suitable for the pore size range of the capillary condensation process; the effective fractal dimensions based on T_2 spectrums and the capillary bundle fractal model are concentrated in the range of 2.004–2.037.

(3) It is found that the single structure parameter of adsorption pores has no direct quantitative relationship with the fractal dimension. However, the fractal dimension has a certain relationship with the range of pore diameter distribution, that is, the more the pore size distribution is concentrated in 2–5 nm, the larger the fractal dimension value, the higher the complexity of pore structure.

(4) There is a good linear positive correlation between the fractal dimension D_2 which characterizes the degree of surface irregularity in the coal body and the maximum nitrogen adsorption capacity, with a correlation coefficient of 0.87, indicating that the greater the degree of surface irregularity in the coal body, the better the coal body's adsorption performance.

Author Contributions: Writing—original draft preparation, W.W.; methodology, data curation, writing, Z.L.; methodology, data processing, validation, English polish, M.Z.; methodology, data processing, H.Y. All authors have read and agreed to the published version of the manuscript.

Funding: The authors would like to acknowledge the support of the National Natural Science Foundation of China (52274213, 52074173, 51934004), Natural Science Foundation of Shandong Province (ZR2022YQ52), and Taishan Scholars Project Special Funding (TS20190935).

Data Availability Statement: Not applicable.

Acknowledgments: The authors would like to acknowledge the support of the National Natural Science Foundation of China (52074173, 51604168, 51934004), the Key Research and Development Plan of Shandong Province, China (2019GSF111033), Major Program of Shandong Province Natural Science Foundation (ZR2018ZA0602) and Taishan Scholars Project Special Funding (TS20190935).

Conflicts of Interest: All authors declare that there is no conflict of interest.

References

1. Feng, R.; Liu, J.; Harpalani, S. Optimized pressure pulse-decay method for laboratory estimation of gas permeability of sorptive reservoirs: Part 1—Background and numerical analysis. *Fuel* **2017**, *191*, 555–564. [CrossRef]
2. Xia, L.; Yin, Y.; Yu, X.; Zheng, Y. An approach to grading coalbed methane resources in China for the purpose of implementing a differential production subsidy. *Pet. Sci.* **2019**, *16*, 447–457. [CrossRef]
3. Sulaiman, W.R.W.; Azizan, N.; Jaafar, M.Z.; Ismail, A.R.; Hamid, M.A.; Johari, A.; Mat, R.; Kamaruddin, M.J.; Ali, A. Additional gas resource for coal bed methane by applying underground coal gasification and enhanced coal bed methane. In Proceedings of the 5th KKU International Engineering Conference 2014, KKU-IENC 2014, Khon Kaen, Thailand, 27–29 March 2014; Trans Tech Publications Ltd.: Khon Kaen, Thailand, 2014.
4. Du, F.; Wang, K. Unstable failure of gas-bearing coal-rock combination bodies: Insights from physical experiments and numerical simulations. *Process. Saf. Environ. Prot.* **2019**, *129*, 264–279. [CrossRef]
5. Wang, K.; Du, F. Coal-gas compound dynamic disasters in China: A review. *Process. Saf. Environ. Prot.* **2019**, *133*, 1–17. [CrossRef]
6. Wang, K.; Guo, Y.; Wang, G.; Du, F. Seepage and Mechanical Failure Characteristics of Gas-bearing Composite Coal-Rock under True Triaxial Path. *J. China Coal Soc.* **2023**, *10*, 13225.
7. Kim, Y.-M.; Jang, H.-C.; Lee, J.-H. Application of probabilistic approach to evaluate coalbed methane resources using geological data of coal basin in Indonesia. *Geosci. J.* **2015**, *20*, 229–238. [CrossRef]
8. Wang, Z.; Cheng, Y.; Qi, Y.; Wang, R.; Wang, L.; Jiang, J. Experimental study of pore structure and fractal characteristics of pulverized intact coal and tectonic coal by low temperature nitrogen adsorption. *Powder Technol.* **2019**, *350*, 15–25. [CrossRef]
9. Feng, R.; Chen, S.; Bryant, S.; Liu, J. Stress-dependent permeability measurement techniques for unconventional gas reservoirs: Review, evaluation, and application. *Fuel* **2019**, *256*, 115987. [CrossRef]
10. Sang, S.X.; Zhu, Y.M.; Zhang, S.; Zhang, J.; Tang, J.X. Solid-gas action mechanism of coal adsorption gas(I)—Coal pore structure and solid gas action. *Nat. Gas Ind.* **2005**, *25*, 13–15+205. (In Chinese)
11. Zhang, Y.H.; Lebedev, M.; Al-Yaseri, A.; Yu, H.Y.; Xu, X.M.; Iglauer, S. Characterization of nanoscale rock mechanical properties and microstructures of a Chinese sub-bituminous coal. *J. Nat. Gas Sci. Eng.* **2018**, *52*, 106–116. [CrossRef]
12. Yang, H.; Cheng, W.; Liu, Z.; Wang, W.; Zhao, D.; Yang, W. Study on the dynamic evolution law of the effective stress in the coal seam water infusion process based on fractal theory. *Fractals* **2020**, *28*, 73–76. [CrossRef]
13. Liu, Z.; Wang, W.; Yang, H.; Yu, S.; Xin, L. Experimental Study on the Fractal Pattern of a Coal Body Pore Structure Around a Water Injection Bore. *J. Energy Resour. Technol.* **2020**, *142*, 012302. [CrossRef]
14. Feng, Z.; Zhou, D.; Zhao, Y.; Cai, T. Study on microstructural changes of coal after methane adsorption. *J. Nat. Gas Sci. Eng.* **2016**, *30*, 28–37. [CrossRef]
15. Meng, M.; Qiu, Z. Experiment study of mechanical properties and microstructures of bituminous coals influenced by supercritical carbon dioxide. *Fuel* **2018**, *219*, 223–238. [CrossRef]
16. Zhang, W.J.; Ju, Y.W.; Wei, M.M.; Wang, G.C. Study on characteristics and mechanism of adsorption/desorption on different metamorphic-deformed coal reservoirs. *Earth Sci. Front.* **2015**, *22*, 232–242.
17. Chen, M.-Y.; Cheng, Y.-P.; Li, H.-R.; Wang, L.; Jin, K.; Dong, J. Impact of inherent moisture on the methane adsorption characteristics of coals with various degrees of metamorphism. *J. Nat. Gas Sci. Eng.* **2018**, *55*, 312–320. [CrossRef]
18. Niu, Q.; Pan, J.; Jin, Y.; Wang, H.; Li, M.; Ji, Z.; Wang, K.; Wang, Z. Fractal study of adsorption-pores in pulverized coals with various metamorphism degrees using N2 adsorption, X-ray scattering and image analysis methods. *J. Pet. Sci. Eng.* **2019**, *176*, 584–593. [CrossRef]
19. Li, P.; Zhang, X.; Zhang, S. Structures and fractal characteristics of pores in low volatile bituminous deformed coals by low-temperature N_2 adsorption after different solvents treatments. *Fuel* **2018**, *224*, 661–675. [CrossRef]
20. Zhang, J.; Wei, C.; Yan, G.; Lu, G. Structural and fractal characterization of adsorption pores of middle–high rank coal reservoirs in western Yunnan and eastern Guizhou: An experimental study of coals from the Panguan syncline and Laochang anticline. *Energy Explor. Exploit.* **2018**, *37*, 251–272. [CrossRef]
21. Liu, Z.; Wang, W.; Cheng, W.; Yang, H.; Zhao, D. Study on the seepage characteristics of coal based on the Kozeny-Carman equation and nuclear magnetic resonance experiment. *Fuel* **2020**, *266*, 117088. [CrossRef]
22. Deng, G.-Z.; Zheng, R. Micropore Structure and Fractal Characteristics of Low-Permeability Coal Seams. *Adv. Mater. Sci. Eng.* **2018**, *2018*, 4186280. [CrossRef]
23. Zhang, B.; Liu, W.; Liu, X. Scale-dependent nature of the surface fractal dimension for bi- and multi-disperse porous solids by mercury porosimetry. *Appl. Surf. Sci.* **2006**, *253*, 1349–1355. [CrossRef]
24. Liu, X.; Nie, B. Fractal characteristics of coal samples utilizing image analysis and gas adsorption. *Fuel* **2016**, *182*, 314–322. [CrossRef]

25. Yao, Y.; Liu, D.; Tang, D.; Tang, S.; Huang, W. Fractal characterization of adsorption-pores of coals from North China: An investigation on CH4 adsorption capacity of coals. *Int. J. Coal Geol.* **2008**, *73*, 27–42. [CrossRef]
26. Cai, Y.; Liu, D.; Yao, Y.; Li, J.; Liu, J. Fractal characteristics of coal pores based on classic geometry and thermodynamics models. *Acta Geol. Sin. Engl. Ed.* **2011**, *85*, 1150–1162. [CrossRef]
27. Lu, G.; Wang, J.; Wei, C.; Song, Y.; Yan, G.; Zhang, J.; Chen, G. Pore fractal model applicability and fractal characteristics of seepage and adsorption pores in middle rank tectonic deformed coals from the Huaibei coal field. *J. Pet. Sci. Eng.* **2018**, *171*, 808–817. [CrossRef]
28. Song, H.; Min, L.; Jun, X.; Lushi, S.; Peisheng, L.; Sheng, S.; Xuexin, S. Fractal characteristic of three Chinese coals. *Fuel* **2004**, *83*, 1307–1313. [CrossRef]
29. Liu, Z.; Yang, H.; Wang, W.; Cheng, W.; Xin, L. Experimental Study on the Pore Structure Fractals and Seepage Characteristics of a Coal Sample Around a Borehole in Coal Seam Water Infusion. *Transp. Porous Media* **2018**, *125*, 289–309. [CrossRef]
30. Hu, L.; Zhu, Y.M.; Chen, S.B.; Du, Z.L. Fractal characteristics of pore structure of Longmaxi formation shale in Shuanghe, southern Sichuan. *Xinjiang Pet. Geol.* **2013**, *34*, 79–82. (In Chinese)
31. Zhou, S.; Liu, D.; Cai, Y.; Yao, Y. Fractal characterization of pore–fracture in low-rank coals using a low-field NMR relaxation method. *Fuel* **2016**, *181*, 218–226. [CrossRef]
32. Wang, G.; Han, D.; Qin, X.; Liu, Z.; Liu, J. A comprehensive method for studying pore structure and seepage characteristics of coal mass based on 3D CT reconstruction and NMR. *Fuel* **2020**, *281*, 118735. [CrossRef]
33. Yi, M.; Cheng, Y.; Wang, Z.; Wang, C.; Hu, B.; He, X. Effect of particle size and adsorption equilibrium time on pore structure characterization in low pressure N2 adsorption of coal: An experimental study. *Adv. Powder Technol.* **2020**, *31*, 4275–4281. [CrossRef]
34. Rouquerol, J.; Avnir, D.; Fairbridge, C.W.; Everett, D.H.; Haynes, J.M.; Pernicone, N.; Ramsay, J.D.F.; Sing, K.S.W.; Unger, K.K. Recommendations for the characterization of porous soilds. *Pure Appl. Chem.* **1994**, *66*, 1739–1758. [CrossRef]
35. Luo, L.; Yao, W.; Liu, J.; Zhang, H.; Ma, J.; Jiang, X. Evolution of NOX precursors of superfine pulverized coal with a fixed bed in N_2 and CO_2. *Fuel* **2018**, *234*, 263–275. [CrossRef]
36. Dai, F.Y.; Hu, H.Y.; Zhang, A.H. Suitability study on fractal model of organic shale pore. *Coal Sci. Technol.* **2019**, *47*, 168–175.
37. Sing, K.S.W.; Everett, D.H.; Haul, R.A.W. Reporting physisorption data for gas solid systems with special reference to the determination of surface area and porosity. *Pure Appl. Chem.* **1985**, *57*, 603–619. [CrossRef]
38. Xiao, L.Z. *NMR Imaging Logging Principles and Applications*; Science Press: Beijing, China, 1998.
39. Li, J.Q.; Yao, Y.B.; Cai, Y.D. Discussion on coal physical properties and formation with different metamorphic degree in north China. *Coal Sci. Technol.* **2012**, *40*, 111–115.
40. Davies, S.; Packer, K.J. Pore size distributions from nuclear magnetic resonance spin lattice relaxation measurements of fluid-saturated porous solids. *Theory Simul. J. Appl. Phys.* **1990**, *67*, 3163–3170. [CrossRef]
41. Zhang, Y.T.; Wang, D.M.; Zhong, X.X. Features of fissure sharp in coal borehole and variation law with temperature. *Coal Sci. Technol.* **2007**, *35*, 73–76. (In Chinese)
42. Yu, B.M.; Xu, P.; Zou, M.Q.; Cai, J.C.; Zheng, Q. *Fractal Porous Media Transport Physics*; Science Press: Beijing, China, 2014.
43. Chen, J.; Cheng, W.; Wang, G.; Li, H. Effect of dominated coal pores and fractures on water migration after low pressure water injection based on CT images. *Fuel* **2021**, *307*, 121795. [CrossRef]
44. Wang, K.; Guo, Y.; Du, F.; Dong, H.; Xu, C. Effect of the water injection pressure on coal permeability based on the pore-fracture fractal characteristics: An experimental study. *Greenh. Gases Sci. Technol.* **2021**, *12*, 136–147. [CrossRef]
45. Sun, B.; Yang, Q.; Zhu, J.; Shao, T.S.; Yang, Y.H.; Hou, C.Y.; Li, G.Y. Pore size distributions and pore multifractal characteristics of medium and low-rank coals. *Nat. Res.* **2020**, *10*, 22353. [CrossRef] [PubMed]
46. Zeng, Q.; Luo, M.; Pang, X.; Li, L.; Li, K. Surface fractal dimension: An indicator to characterize the microstructure of cement-based porous materials. *Appl. Surf. Sci.* **2013**, *282*, 302–307. [CrossRef]
47. Xia, Y.; Cai, J.; Perfect, E.; Wei, W.; Zhang, Q.; Meng, Q. Fractal dimension, lacunarity and succolarity analyses on CT images of reservoir rocks for permeability prediction. *J. Hydrol.* **2019**, *579*, 124198. [CrossRef]
48. Xu, S.; Hu, E.; Li, X.; Xu, Y. Quantitative Analysis of Pore Structure and Its Impact on Methane Adsorption Capacity of Coal. *Nat. Resour. Res.* **2021**, *30*, 605–620. [CrossRef]
49. Karolina, W.R.; Natalia, H.; Adam, S. Effect of porous structure of coal on propylene adsorption from gas mixtures. *Sci. Rep.* **2020**, *10*, 11277.
50. Artioli, Y. Adsorption. *Encycl. Ecol.* **2008**, *4*, 60–65.

processes

MDPI

Article

Molecular Model Construction and Optimization Study of Gas Coal in the Huainan Mining Area

Lina Qu *, Long Liu, Jinhao Chen and Zhenzhen Wang

School of Energy & Environment Engineering, Zhongyuan University of Technology, Zhengzhou 450007, China
* Correspondence: qln-66@163.com

Abstract: To construct the macromolecular model of gas coal in the Huainan mining area, ^{13}C nuclear magnetic resonance spectroscopy (^{13}C-NMR), X-ray diffraction (XRD), and X-ray photoelectron spectroscopy (XPS) tests were used to analyze the microstructure characteristics of coal including the aromatic ring type, the linkage mode, and the chemical bonding composition. The model was simulated and optimized by molecular mechanics (MM) and molecular dynamics (MD). The experimental results showed that the coal macromolecular formula in the Huainan mine was expressed as $C_{181}H_{150}O_9N_3$. The aromatic ring was dominated by anthracene and phenanthrene. Aliphatic carbon mainly existed in the form of methylene and methine. The oxygen atoms existed in the form of ether−oxygen bonds. The ratio of pyridine nitrogen to pyrrolic nitrogen was 2:1. The molecular simulation results showed the π−π interaction between the aromatic lamellae within the molecule. The van der Waals energy was the major factor of coal molecular structure stability and energy change. The results of the calculated ^{13}C-NMR carbon spectrum and density simulation agreed well with the experimental results. The study provides a scientific and reasonable method for coal macromolecular model prediction and theoretical support for coal spontaneous combustion prevention technology.

Keywords: gas coal; quantum chemistry; molecular modeling; structure optimization

Citation: Qu, L.; Liu, L.; Chen, J.; Wang, Z. Molecular Model Construction and Optimization Study of Gas Coal in the Huainan Mining Area. *Processes* **2023**, *11*, 73. https://doi.org/10.3390/pr11010073

Academic Editors: Feng Du, Aitao Zhou and Bo Li

Received: 18 November 2022
Revised: 18 December 2022
Accepted: 25 December 2022
Published: 28 December 2022

1. Introduction

Coal is an important basic energy source and industrial raw material in China, and it accounts for more than 90% of the fossil resource reserves [1,2]. The gradual depletion of shallow coal resources has increased the depth of coal mining. Fire in mines caused by coal spontaneous combustion often occurs and becomes one of the main disasters in coal mine production [3], which seriously affects economic development and threatens the lives of coal miners [4,5]. The prevention of coal spontaneous combustion fires has been major topics of research in the field of coal mine safety [6–8]. At present, many studies generally agree that coal spontaneous combustion is caused by coal−oxygen recombination [9]. However, coal−oxygen recombination is a complex process. The adsorption mechanism of oxygen in coal is difficult to reveal by experimental means. A mutual reflection relationship exists between the properties and structure of coal [10]. As a result, the construction and optimization of the coal macromolecule model is the theoretical basis for revealing the mechanism of coal spontaneous combustion.

The construction of the coal molecular model provides foundation for molecular simulation. The research methods are mainly divided into physical and chemical methods [11–15]. ^{13}C nuclear magnetic resonance spectroscopy (^{13}C-NMR), X-ray diffraction (XRD), X-ray photoelectron spectroscopy (XPS), and other physical methods were used to obtain relevant information on the construction of the molecular structure model [16–18]. Early coal chemical structure models, such as the Wiser, Given, and Shinn models [19–21], were widely recognized by scholars. The rapid development of computer technology in recent years has encouraged scholars to use modern analysis methods combined with computer-aided software in building coal molecular models. The mechanism and reaction characteristics of coal spontaneous combustion were investigated by molecular simulation [22–25]. Carlson [26] simulated the three-dimensional structures of four bituminous coal models using computer software and concluded that molecular modeling is an effective means to study coal structure. Wu et al. [27] systematically investigated the competitive adsorption characteristics of smoke from coal-fired power plants using molecular dynamics (MD) simulation software and on the basis of the construction of the molecular structure of coal. The results showed that the absolute adsorption decreases with the increase in temperature and water content in coal. However, the absolute adsorption increases with the rise in pressure and the corresponding volumetric mole fraction. Hong et al. [28] structured the coal char model and studied the combustion and gasification properties of coal char using the reaction MD (ReaxFF MD) simulation method. Hao et al. [29] investigated the adsorption characteristics of different gasses ($CH_4/CO_2/N_2$) on the coal surface by establishing the molecular model of anthracite coal and revealed the thermodynamic parameters of coal surface adsorption using the molecular simulation. The results showed that the adsorption amounts of the three gases on the coal surface are shown in descending order: $CO_2 > CH_4 > N_2$, according to adsorption affinity and thermodynamic parameters. Meng et al. [30] revealed the relationships between different grades of coal and methane adsorption using the method of the grand canonical Monte Carlo and the density functional theory based on studying four different grades of coal models. Wei et al. [31] constructed a macromolecular structure model of Zaoquan coal using proximate analysis, ultimate analysis, XPS, ^{13}C solid-state NMR, and other physical methods together with computer software. They determined the effects of different final temperatures and heating rates on pyrolysis behavior. Chai and Zeng [32] probed the changes in macroscopic and microscopic characteristics of Wucaiwan coal at different temperatures by constructing a molecular model. Ping et al. [33] analyzed the mirror and inert groups of Shanghai Miao bituminous coal by using ^{13}C solid-state NMR, Fourier-transform infrared spectroscopy, and XPS. The differences between the structure and functional groups in the molecules were explored to construct molecular structure models.

The studies of the abovementioned scholars indicate that the characteristics of coal oxygen adsorption were deeply explored by molecular simulation. The accuracy of the coal molecular model is the focus of our subsequent research, which has an important effect on the research results. The coal macromolecular model was a collection of analytical results of experimental data based on the characterized coal microstructure parameters, which must be further analyzed or verified. However, the verification methods used by the abovementioned scholars need to be further improved. The authors are mainly engaged in the research on coal spontaneous combustion fire prevention and control and coal mine safety. Thus, the gas coal in the Huainan mining area was further studied in this work based on comprehensive research methods. Proximate analysis, ultimate analysis, ^{13}C-NMR, XPS, and XRD were used to realize the information characterization of gas coal. The rationality of the macromolecular structure model based on the characterization information was verified by the carbon spectrum and density simulation of coal molecules. A more stable and accurate molecular structure model was obtained using the Materia Studio (MS) software optimization, which provided a method for predicting the macromolecular model of coal and theoretical support for preventing the occurrence of coal spontaneous combustion fire at the molecular level.

2. Experimental Section

2.1. Elemental and Industrial Analysis Tests

The elemental and industrial analysis tests were carried out by SX2-2.5-10 apparatus of Beijing Zhongxing weiye and Thermo Scientific Flash 2000 apparatus based on Chinese national standards GB/T212-2008 and GB/T476-2008. The experimental results are shown in Table 1.

Table 1. The results of elemental and industrial analysis tests.

Proximate Analysis (w%, daf)				Ultimate Analysis (wt%, daf)				
M_{ad}	A_{ad}	V_{daf}	FC_{ad}	C_{ad}	H_{ad}	O_{ad}	N_{ad}	$S_{t,d}$
0.79	8.288	17.3	70.922	66.12	4.56	4.37	1.28	0.18

2.2. Experimental Protocol

The gas coal samples were taken from the Huainan mining area in Anhui Province. The samples were crushed and sieved into 60–80 mesh particles before the experiments. The information on each carbon atom, the structural parameters of microcrystals, and the occurrence modes of C, O, N, and S atoms in the samples was obtained by ^{13}C-NMR, XPS, and XRD tests. The macromolecular model of coal was constructed and optimized by the Materia Studio (MS) software. The specific process is shown in Figure 1.

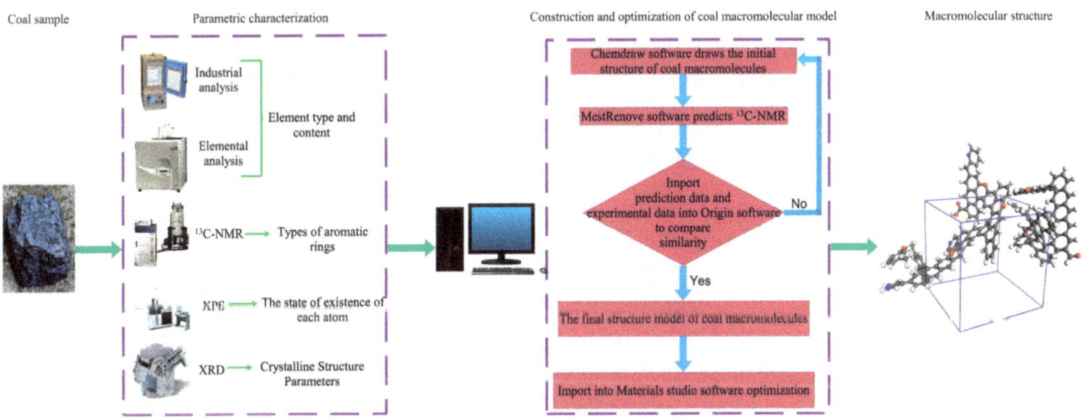

Figure 1. Gas coal macromolecular modeling flow chart.

(1) ^{13}C-NMR spectroscopy (^{13}C-NMR)

The carbon atoms information was obtained using the Bruker400m solid-state NMR (^{13}C-NMR) instrument. The experiment used high-resolution solid-state nuclear magnetic resonance spectroscopy. The MAS spin rate was 10 kHz, the recovery time was 4 s, and the pulse width was 4.5 μs. The pre-scan delay was 6.5 μs, and the scanning number was 3000.

(2) X-ray diffraction (XRD)

The microstructure parameters of coal were obtained using the German Bruker D8 Advance X-ray diffractometer (XRD) with Cu target radiation. The specified parameters were as follows: the counting time was 19.2 sec/step, the step size was 0.02034°, the scan range was 5°–80°, the voltage was 40 kV, and the current was 40 mA.

(3) X-ray photoelectron spectroscopy (XPS)

The occurrence modes of C, O, N, and S atoms were obtained using the X-ray photoelectron spectrometer from Thermo Fisher Scientific, and the parameters were as follows: the monochromatic was Al Ka, the energy was 1486.6 eV, the angle of electron emission was 45°, the beam spot was 500 μm, the number of scans was 5, and the charge correction was corrected using carbon C1s = 284.8 eV.

3. Results and Discussions

3.1. The Analysis of the Elemental Normalization and the Atomic Ratio

The content of elements, types, and the atomic percent were determined by proximate and ultimate analysis. The contents of oxygen in Table 1 were calculated by deducting the moisture and ash. The ultimate analysis was normalized to calculate the atomic ratio of each atom. The results are shown in Table 2. Based on the atomic ratios, if the number of C atoms in coal is n, the molecular formula of coal can be expressed as $C_nH_{0.828n}O_{0.05n}N_{0.017n}S_{0.001n}$.

Table 2. The results of the elemental normalization and the atomic ratio.

Normalization (w%, daf)					Atomic Ratio			
C	H	O	N	S	H/C	O/C	N/C	S/C
86.42	5.96	5.71	1.67	0.24	0.828	0.050	0.017	0.001

3.2. ^{13}C-NMR Test Results and Analysis

The ^{13}C-NMR spectrums were fitted to obtain carbon atoms information by using Peak Fit software. In Figure 2, the main carbon signal peaks of coal were located with the chemical shift between 0 ppm–50 $\times 10^{-6}$ ppm and 100 $\times 10^{-6}$ ppm–170 $\times 10^{-6}$ ppm, which belonged to aliphatic carbon and aromatic carbon, respectively [34]. Each peak was classified based on peak position attribution results in the literature [35], and the classified results are shown in Table 3. According to the calculation method of the 12 structural parameters in the literature [36] (the ascribing positions of different peaks were determined by the results of ^{13}C-NMR peak fitting, and the peak values were substituted into the formula to calculate the structural parameters of coal molecules) and the chemical shift attribution of carbon in Table 3, the 12 structural parameters of coal were calculated, and the results are shown in Table 4. In Table 4, the proportion of aromatic carbon (f_a') in gas coal was 87.70%, which was the main body of a constituting macromolecule. The proportion of aliphatic carbon (f_{a1}) was 12%, and carbonyl and carboxyl carbon ($f_a{}^C$) were 0.3%, unprotonated aromatic carbons bridged with aromatic carbon ($f_a{}^B$) was 25.3%. $X_{BP} = f_a{}^B/(f_a{}^H + f_a{}^P + f_a{}^S) = 0.407$, which was the ratio of aromatic bridge carbon to circumferential carbon, and the X_{BP} value reflected the size of aromatic clusters in coal [37]. The types and numbers of aromatic rings in the macromolecule structure were inferred by the size of aromatic clusters in combination with the ultimate analysis.

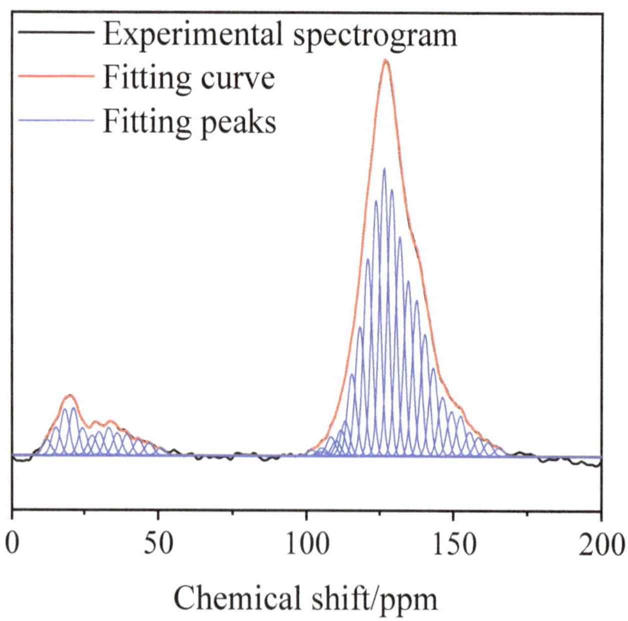

Figure 2. ^{13}C-NMR split-peak-fitting spectrum of gas coal.

Table 3. Carbon structure attribution and proportion.

Peak	Area	Center (ppm)	Relative Area	Attribution	Peak	Area	Center (ppm)	Relative Area	Attribution
1	0.5375	12.16	0.006	A_1	20	0.1183	113.22	0.013	A_6
2	0.9461	15.30	0.011	A_1	21	0.279	115.48	0.031	A_6
3	1.565	18.37	0.018	A_2	22	0.4417	118.19	0.050	A_6
4	1.603	21.32	0.018	A_2	23	0.6737	120.89	0.076	A_6
5	0.9237	24.36	0.010	A_3	24	0.8716	123.70	0.098	A_6
6	0.6759	27.55	0.008	A_3	25	0.9847	126.44	0.111	A_6
7	0.7956	29.85	0.009	A_3	26	0.9119	129.09	0.102	A_7
8	0.9374	33.14	0.011	A_3	27	0.7487	131.85	0.084	A_7
9	0.7669	35.97	0.009	A_3	28	0.5977	134.70	0.067	A_7
10	0.723	39.29	0.008	A_4	29	0.5317	137.56	0.060	A_8
11	0.5363	43.15	0.006	A_4	30	0.4133	140.35	0.046	A_8
12	0.4164	46.89	0.005	A_4	31	0.2996	143.23	0.034	A_8
13	0.243	50.88	0.003	A_5	32	0.2015	146.39	0.023	A_8
14	0.1836	101.81	0.002	A_6	33	0.152	149.39	0.017	A_9
15	0.1305	105.13	0.001	A_6	34	0.1359	152.47	0.015	A_9
16	0.2635	105.25	0.003	A_6	35	0.08195	155.50	0.009	A_9
17	0.6444	108.31	0.007	A_6	36	0.06333	158.49	0.007	A_9
18	0.5065	110.12	0.006	A_6	37	0.0462	161.88	0.005	A_9
19	0.8758	111.72	0.010	A_6	38	0.02881	165.59	0.003	A_{10}

Note: A_1—lipid methyl carbon; A_2—aromatic cyclic methyl carbon; A_3—methylene carbon and methine carbon; A_4—methine carbon and quaternary carbon; A_5—oxygen-conjugated methylene and oxygen-conjugated methylene carbon; A_6—protonated aromatic carbon; A_7—bridgehead aromatic carbon; A_8—side branch aromatic carbon; A_9—oxygen replaces aromatic carbon; A_{10}—carboxy carbon.

Table 4. Structure parameters of gas coal.

Type	Symbol	Value	Type	Symbol	Value
Total aromatic carbon	f_a	0.880	Alkylated aromatic	f_a^S	0.179
Carbonyl groups	f_a^C	0.003	Aromatic bridgehead	f_a^B	0.253
In an aromatic ring	f_a'	0.877	Total aliphatic carbon	f_{al}	0.120
Nonprotonated and aromatic	f_a^N	0.470	CH_3	f_{al}^*	0.046
Protonated and aromatic	f_a^H	0.407	CH or CH_2	f_{al}^H	0.065
Phenolic or phenolic ether	f_a^P	0.037	Bonded to oxygen	f_{al}^0	0.009

3.3. XRD Test Results and Analyses

To obtain the microcrystalline structure parameters of samples, the XRD test results were fitted by using the Jade software, and the results are reported in Figure 3. The characteristic peaks of 002 and 100 were located at 24.879° and 43.254°, respectively. According to the calculation method in the literature [38,39] and combined with the calculation results in Figure 3, the parameters including L_a (the mean diameter of aromatic ring lamellar in a stacked cluster), L_c (the effective stacking height of aromatic ring layers along the vertical direction of aromatic core), d_{002} (the layer spacing of aromatic ring lamellar), and Nave (the effective stacking layer number of aromatic ring layers) were calculated, and the L_a, L_c, and d_{002} were 11.77, 6.62, and 3.575 nm, respectively. The calculated microcrystalline structure parameters provided effective data for the subsequent construction of the coal macromolecular structure model and the stacking mode of each aromatic layer in the model.

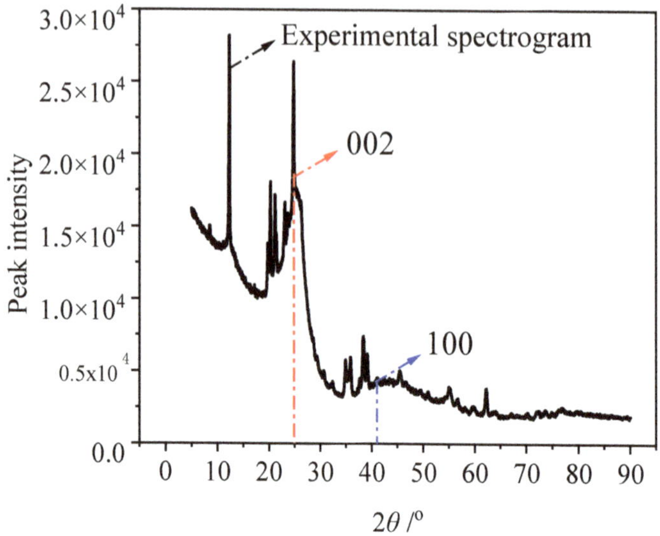

Figure 3. XRD spectrum of gas coal.

3.4. XPS Test Results and Analyses

The modes of occurrence of C, O, N, and S atoms in coal were performed by the XPS test to infer the connection patterns of aromatic rings. The result of XPS test is presented in Figure 4a. The spectrum was fitted by the Avantage software, and the fitting results are shown in Figure 4b–d. According to the results of the ultimate analysis, the content of S was only 0.18%, which was not enough for one S atom. Thus, it was not considered in the molecular structure. The molecular formula of gas coal was expressed as $C_nH_{0.828n}O_{0.05n}N_{0.017n}$.

The results of C fitted spectra are shown in Figure 4b. There were four peak positions in the C split peak spectrum, which indicated that the carbon in coal mainly existed in four forms. The centers of the four peaks were 284.45 eV, 284.8 eV, 285.18 eV, and 290.06 eV. These different peaks were assigned to C−C, C−H, C−O, and C=O, respectively. Among them, the content ratio of C−C, C−H, C−O, and C=O was 0.56:0.29:0.11:0.05.

The results of O fitted spectra are shown in Figure 4c. From Figure 4c, ether oxygen bond (C−O), carbonyl (C=O), and carboxyl (COOH) existed in the oxygen spectrogram. The proportions of oxygen were 52%, 46.0%, and 2%, respectively. The centers of the three peaks were 531.3 eV, 532.35 eV, and 535.41 eV. Therefore, the content ratio between C−O, C=O, and COOH was 0.26:0.23:0.01.

The results of N fitted spectra are shown in Figure 4d. The P_1 peak (Pyridinic nitrogen N-6) located at 399.3 eV was the largest, accounting for 51.86%. The second was the P_2 peak (Pyridinic nitrogen N-5) located at 400.02 eV, accounting for 25.93%. The P_3 peak at 401.65 eV belonged to quaternary nitrogen, accounting for 17.27%. The P_4 peak with the smallest proportion at 404.5 eV belonged to nitrogen oxide, accounting for 4.94%, which was formed by the oxidation of pyridine nitrogen and pyrrole nitrogen in air.

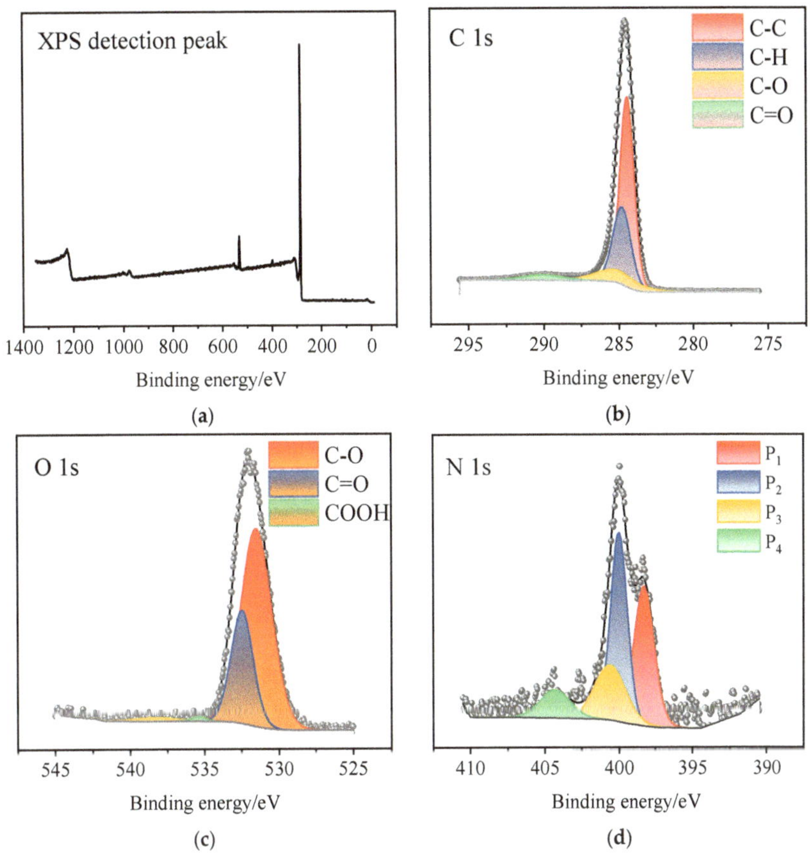

Figure 4. Peaking fitting spectra of gas coal X-ray photoelectron spectroscopy: (**a**) XPS detection spectrum, (**b**) C split-peak fitting spectra, (**c**) O split-peak fitting spectra, (**d**) N split-peak fitting spectra.

3.5. Construction of the Molecular Structure Model of Gas Coal

3.5.1. Aromatic Structures

According to Section 2.2, the X_{BP} of gas coal in this study was 0.407, which was close to the X_{BP} values of phenanthrene and anthracene in the literature [21]. According to the conclusion of the molecular structure in the literature [11], the macromolecular structure was dominated by anthracene and phenanthrene, followed by the pyrene and naphthalene rings. The bridging carbon ratio was made close to the calculated value by adjusting the numbers of various aromatic rings. The types and numbers of aromatic structures in the molecular structure are shown in Table 5.

3.5.2. Aliphatic Carbon Structures

The aliphatic carbon in coal mainly exists in the form of branched chain, linked aromatic rings, aliphatic rings, side chains, and bridging carbons [29]. Based on the aromatic carbon ratio, the number of aromatic carbon atoms, and Table 5, the total and aliphatic carbon atoms were calculated to be 181 and 21, respectively. From the abovementioned data, the number of hydrogen atoms in gas coal was calculated to be 150. The proportions of $f_{al}{}^{H}$, $f_{al}{}^{*}$, and $f_{al}{}^{0}$ in gas coal were 6.5%, 4.6%, and 0.9%, respectively. Therefore, the aliphatic carbon atoms in gas coal mostly existed in the forms of methyl, methylene, hypo methyl, and quaternary carbon, and the oxygen-linked lipid carbon was the least abundant.

Table 5. The types and numbers of aromatic structures in gas coal.

Type	Number	Type	Number
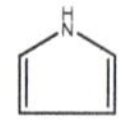	1	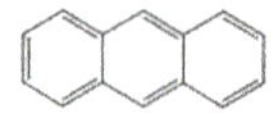	3
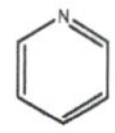	2	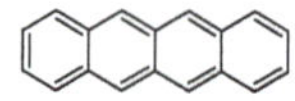	1
	2	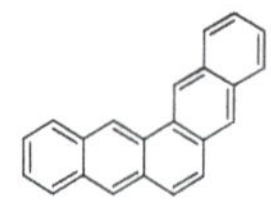	1
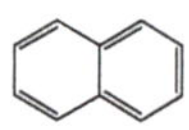	3	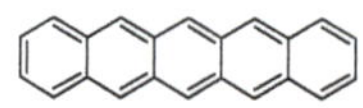	1

3.5.3. Heteroatom Structures

The numbers of oxygen and nitrogen atoms in gas coal were calculated to be 9 and 3, respectively, by the total number of carbon atoms and the atomic ratio. According to XPS analysis, oxygen atoms in gas coal existed in the forms of five ether oxygen bonds, two carbonyl groups, and one carboxyl group. The nitrogen atoms existed in the form of two pyridine nitrogen and one pyrrole nitrogen.

3.5.4. Construction of the Coal Molecular Structure

Based on the number of atoms per element, the molecular formula of gas coal was determined to be $C_{181}H_{150}O_9N_3$. The molecular structure diagram of gas coal was constructed by the Chemdraw software. The MestReNova software was used to calculate the spectrum of the coal molecular structure. Then, the calculated spectrum was compared with the ^{13}C-NMR experimental spectrum, and the comparison chart was further adjusted and optimized according to XRD data. The obtained structural model is shown in Figure 5, and the comparison results of the ^{13}C-NMR spectra are shown in Figure 6.

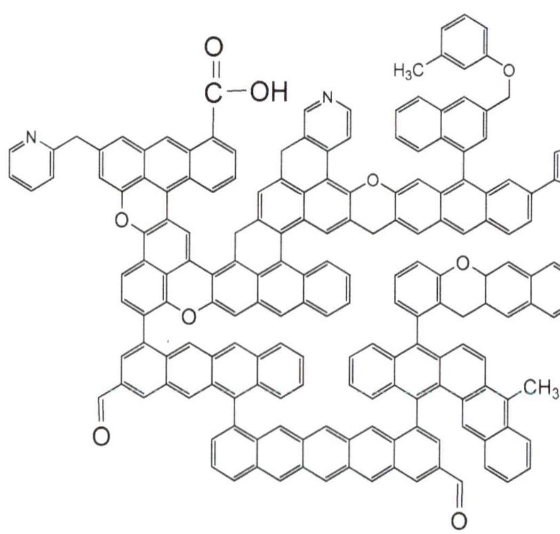

Figure 5. Molecular structure model of gas coal.

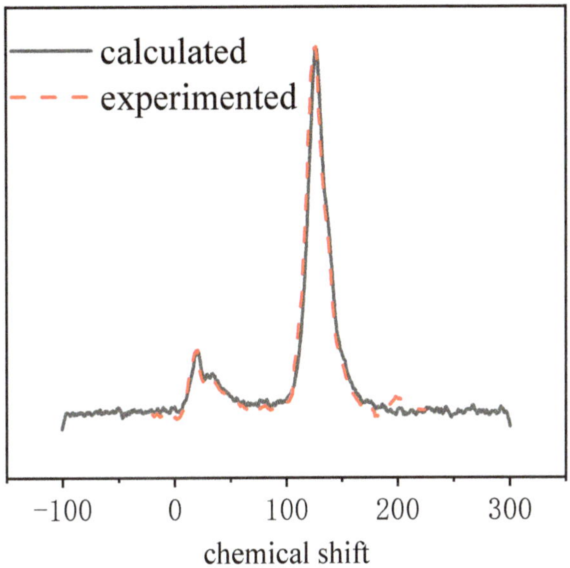

Figure 6. Experimental and calculated ^{13}C-NMR spectra.

According to Figure 6, the calculated spectrum agreed well with the experimental spectrum, which indicated that the molecular model was constructed accurately. However, the largest relative deviation was obtained in the oxygen−carbon region with a chemical shift from 170 ppm to 220 ppm. This finding is mainly due to the side-band effect of the oxygen−carbon region in the experimental process, which led to errors of large intensity and high peak values in the oxygen−carbon region of the experimental spectrum [40]. Therefore, the oxygen−carbon region was further investigated by verifying the rationality of the molecular model through ^{13}C-NMR carbon spectroscopy.

4. Optimization and Validation of the Coal Macromolecule Model

The coal macromolecular structure model was only a collection of analytical results of experimental data, which must be simulated for further analysis and validation. The Materials Studio (MS) software was used to optimize the structure model of coal macromolecules, using molecular mechanics (MM) and molecular dynamics (MD) methods to simulate the structural properties and interactions between molecules. In molecular or periodic system simulations, the Focite module could solve the energy calculation, geometric optimization, and the process of dynamic simulation. The force field was Dreiding, and the parameters of the Dreiding force field were obtained by fitting the quantum chemistry calculation data, and the use of the Dreiding force field reflected the application of quantum chemical theory in molecular simulations.

The information and density of each carbon atom in coal is an important parameter to reflect the molecular structure of coal. The rationality of the model was verified by comparing the agreement between the calculated and experimental carbon spectra, as well as the similarity between the experimental and simulated densities. The refined coal macromolecule model was also reasonably confirmed, because the calculated spectrum was in satisfactory agreement with the experimental spectrum. Therefore, it was not discussed again in this section.

4.1. Optimal Structure

Molecules exist mainly in the lowest energy form in natural situations. Thus, only the lowest energy model represented the optimal state of the molecular structure under study. The lowest energy could not be guaranteed for the built model during the modeling process. Thus, it was necessary to optimize and find the lowest energy model to ensure that the subsequent results were meaningful [18].

4.1.1. Optimized Method and Parameter Setting

The molecular structure model of gas coal was imported into MS software and automatically hydrogenated to fullness. The lowest energy model was obtained by MM and MD calculations. The module selection and parameter settings were based on the simulation calculation method in the literature [40]. MM and MD calculations were conducted in the Forcite module in MS. The charge distributions were obtained through the charge equilibrium method in the MM calculation. The MM parameters were as follows: the computational method was a smart minimizer, the maximum number of iterations steps was set to 5000 steps considering the computational accuracy and convergence time cost, and the convergence criterion was medium. The Dreiding [41,42] force field was selected, because it was suitable for calculating the structures and various properties of most types of molecules and materials with high accuracy. A fixed-volume, fixed-temperature (NVT) MD simulation was conducted. The MD parameters were as follows: the temperature was 300–600 K, the heating order was 5, the heating rate was 60 K/time. The temperature control program selected Nose [43] to ensure that the distribution of system degrees of freedom was normative, and the time step was 0.1 fs.

4.1.2. Energy-Minimizing Geometrical Configuration and Microcrystalline Structure Parameters

The unoptimized model in Figure 7a was imported into MS for hydrogenation saturation. MM and MD calculations were performed to overcome the molecular structure energy barrier and obtain the minimum energy structure after molecular structural optimization in the Forcite module. The results are shown in Figure 7b. The optimized gas coal molecular model was more compact and had a better stereochemical structure than the models in Figure 7a,b. The optimized molecular energies are shown in Table 6. The calculation and analysis showed that the valence electron energy and the non-bond energy accounted for 39.56% and 60.44% of the total energy, respectively. The total energy of the optimized molecular model was reduced by 91.72%, in which the valence electron energy was reduced by 92.11% and the non-bond energy was reduced by 91.44%. The analyzed results showed that the non-bond energy (hydrogen bond energy, van der Waals energy, and Coulomb energy) plays an important role in the stability of the molecular structure. The van der Waals energy changed the most before and after optimization, which suggests that the $\pi-\pi$ interaction between aromatic rings was the main factor in keeping the molecular structure stable. In the valence electron energy (bond stretching energy, bond angle energy, torsion energy, and reversal energy), the descending order was as follows: torsion energy > bond stretching energy > bond angle energy > reversal energy. Therefore, the bond torsion and reversal and the changes in bond angle and length were the basis of the stereoscopic configuration of coal macromolecules. The molecular structure drawn in the Chemdraw software was a planar graph, and the factor of the bond length was ignored. Thus, the bond energy changed the most in the optimized valence energy.

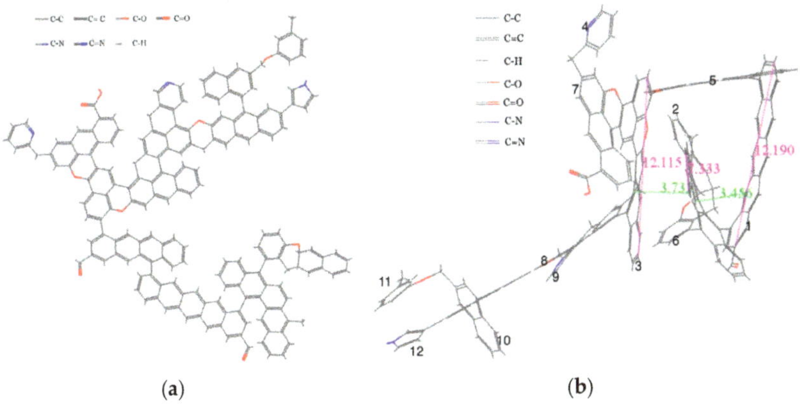

Figure 7. Macromolecular structure model of gas coal before and after optimization: (**a**) before optimization; (**b**) after optimization.

Table 6. The energies of the gas-coal macromolecular structure before and after optimization.

E (kcal·mol^{-1})		Ev (kcal·mol^{-1})				E_N (kcal·mol^{-1})		
		E_B	E_A	E_T	E_I	E_H	E_{van}	E_E
Initial	8263.8600	3222.464	38.336	166.689	4.085	0	4850.632	−18.35
Final	684.560	84.553	51.754	131.287	3.120	0	418.828	−4.982

Note: E—total energy; Ev—valence energy; E_B—bond energy; E_A—angle energy; E_T—torsion energy; E_I—inversion energy; E_N—non-bond energy; E_H—hydrogen bond energy; E_{van}—van der Waals energy; E_E—electrostatic energy.

As shown in Figure 7b, a set of approximately parallel combinations (series 1, 2, and 3) appeared between the aromatic lamellar layers in the structural model formed by a single molecule. However, the spacings (d_{002}) of the two aromatic lamellae were not uniform, which were 3.735 and 3.456 nm, respectively. The maximum values of L_a and L_c were 12.190 and 7.191 nm, respectively. The spacing of the aromatic layer was 3.575 nm higher than the measured value, which is due to that the bonds connecting the aromatic layers were twisted and inserted in a set of aromatic lamellae 6 and aliphatic carbon structures after optimization. The $\pi-\pi$ interaction between aromatic rings played an important role in structural stability, and two sets of vertical structures (aromatic layers 5, 1, and 3 and aromatic layers 10 and 11) appeared at the edge of the model. The reason is that the coal molecule was in a stable state without force [44]. The microcrystalline structural parameters (d_{002}, Lc, and La) of the coal molecular model were slightly different from the experimental values (3.575, 6.62, and 11.77 nm, respectively), which suggests that the short-range order of the high-grade coal structure mainly depends on the directional arrangement of more intermolecular aromatic layers [40].

4.2. Density Simulation

The density of coal is an important physical property, and it was used to verify the reasonability of the model by comparing the value of the constructed model with the experimental value. The Amorphous Cell module in MS software was utilized to add periodic boundary conditions to the model shown in Figure 7b for MM and MD. The constructs of a series of structural models were obtained, and the minimum energy model was selected. The parameters of the density simulation were as follows: the calculation used periodic boundary conditions, the Dreiding force field was used to simulate the field force, and the charge was calculated by the charge balance method (QEq). The number of molecules was 1, the initial density value was 0.7 g/cm^3, the final density value was

1.4 g/cm^3, and the interval was 0.05 g/cm^3 [45]. The geometric configuration with the minimum energy was selected as shown in Figure 8. The relationship between the density and the potential energy of the structural model is shown in Figure 9.

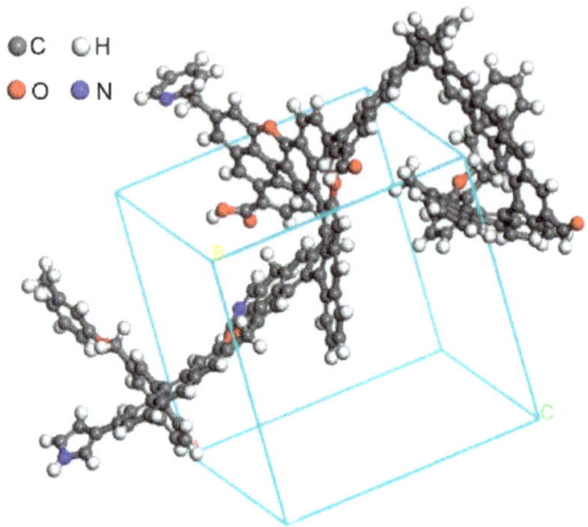

Figure 8. Structure of gas coal with a density of 1.0 g/cm^3.

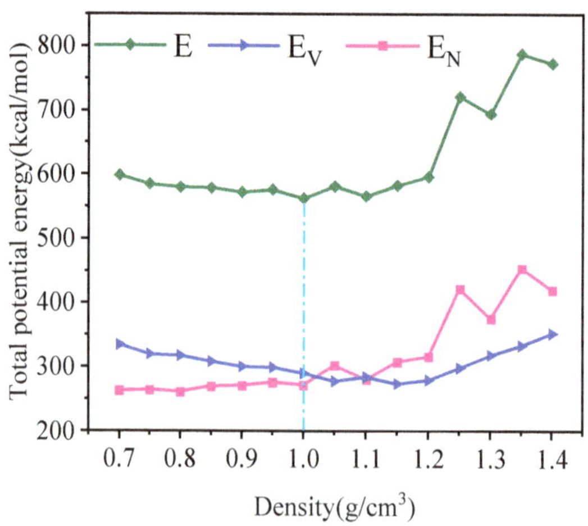

Figure 9. Relationships between the structural model density and potential energy.

As shown in Figure 9, the total potential energy of the molecular structure decreased gradually first and then increased rapidly with the rise in density. When the density was 1.0 g/cm^3, the total potential energy of the molecular structure model was the minimum and the value was 562.944 KJ/mol. At this time, 1.0 g/cm^3 was the simulated density of the model, and the parameters of coal molecular crystal cells were a = b = c = 1.6003 nm. Its structural energy is shown in Table 7. When the density was less than 1.0 g/cm^3, E_V played a dominant role at this stage. When the density was greater than 1.0 g/cm^3, the dominant was E_N. The density calculated by simulation was smaller than the experimental density of gas coal (1.2564 g/cm^3). The reason is that excluding the influence of trace elements and small molecular substances in coal during the density test is difficult under actual conditions [46,47]. Thus, the calculated density was considered reasonable.

Table 7. Energy of the coal sample from Huainan.

Density (g/cm^3)	E (kcal/mol)	EV				EN		
		EB	EA [32]	ET	EI	EH	Evan	EE
1.0	562.944	87.167	52.554	128.714	4.002	0	326.40	−27.967

Comparing Tables 6 and 7 showed that the torsion energy had the largest change among the valence electron energies after periodic boundary conditions were added. The reason is that the energy minimization only was achieved by twisting appropriately when the gas coal molecules under periodic boundary conditions interacted with the surrounding molecules. After periodic boundary conditions were added, the most obvious change in non-bond energy was observed for the van der Waals energy. The major contributor is that the aromatic layers located in parallel were distorted and deformed in the process of the gas coal molecular structure. Thus, they became more compact, which ultimately destroyed $\pi-\pi$ interactions between the aromatic layers in the molecule.

5. Conclusions

In this study, proximate analysis, ultimate analysis, ^{13}C-NMR, XPS, and XRD were performed on Huainan gas coal to quantify its structure. The macromolecular structure model of coal was constructed and optimized based on the quantitative analysis results. The model was verified and optimized by NMR carbon spectroscopy and MS software to obtain the correct molecular model. The following conclusions were drawn.

The results showed that X_{BP} was 0.407. The aromatic structures were mostly anthracene and phenanthrene, followed by naphthalene and benzene rings. The molecule model contained five ether−oxygen bonds, two carbonyl groups, one carboxyl group, two pyridine nitrogen, and one pyrrole nitrogen. Its molecular formula was determined to be $C_{181}H_{150}O_9N_3$.

The minimum energy value of the optimized molecules was 684.560 KJ/mol. The van der Waals energy was the main component of non-bond energy, which was the main factor to stabilize the coal molecular structure and change the energy. The changes including the bond length, angle, torsion, and inversion were the basis of coal molecules with stereoscopic configuration. The intermolecular aromatic lamellae were arranged in an approximately parallel manner by the $\pi-\pi$ interaction. The short-range order of the high-grade coal structure was determined by the directional arrangement of the intermolecular aromatic lamellae.

The coal molecular crystal unit was constructed by adding periodic boundary conditions to the optimized structure, and it was used for the density simulation. The result showed that the simulative density of the coal molecular crystal unit was 1.0 g/cm^3 when the total energy reached the lowest value of 562.944 kcal/mol, which was close to the real density measured by the experiment. At the same time, the predicted carbon spectrum obtained by the MestReNova software agreed well with the experimental spectrum, which suggests that the construction of the gas coal macromolecular model is reasonable and effective.

Author Contributions: Funding acquisition, L.Q.; investigation, J.C.; methodology, L.L.; project administration, L.Q.; resources, L.Q.; supervision, L.Q.; validation, L.L.; visualization, J.C.; writing—original draft, L.L. and Z.W.; writing—review and editing, L.Q. and L.L. All authors have read and agreed to the published version of the manuscript.

Funding: This work is supported by the National Science Foundation of China (51804355).

Data Availability Statement: Not applicable.

Conflicts of Interest: The authors declare no conflict of interest.

References

1. Xu, J.L. Research and progress of coal mine green mining in 20 years. *Coal Sci. Technol.* **2020**, *48*, 1–15.
2. Peng, S.P.; Zhang, B.; Wang, T. China's Coal Resources: Octothorpe Shaped Distribution Characteristics and Sustainable Development Strategies. *Strateg. Study CAE* **2015**, *17*, 45–51.
3. Zhang, Y.; Yang, C.; Li, Y.; Huang, Y.; Zhang, J.; Zhang, Y.; Li, Q. Ultrasonic extraction and oxidation characteristics of functional groups during coal spontaneous combustion. *Fuel* **2019**, *242*, 287–294. [CrossRef]
4. Zhang, Y.N.; Liu, C.H.; Song, J.J.; Wang, N.P. Study on transfer law of main functional groups in low temperature oxidation of long flame coal. *Coal Sci. Technol.* **2020**, *48*, 188–196.
5. Ma, L.; Wang, D.; Yang, W.; Dou, G.; Xin, H. Synchronous thermal analyses and kinetic studies on a caged-wrapping and sustained-release type of composite inhibitor retarding the spontaneous combustion of low-rank coal. *Fuel Process. Technol.* **2017**, *157*, 65–75. [CrossRef]
6. Zhang, J.X.; Li, B.; Liu, Y.W.; Li, P.; Fu, J.W.; Chen, L.; Ding, P.C. Dynamic multifield coupling model of gas drainage and a new remedy method for borehole leakage. *Acta Geotech.* **2022**, *17*, 4699–4715. [CrossRef]
7. Zhang, J.X.; Liu, Y.W.; Ren, P.L.; Han, H.K.; Zhang, S. A fully multifield coupling model of gas extraction and air leakage for in–seam borehole. *Energy Rep.* **2021**, *7*, 1293–1305. [CrossRef]
8. Li, B.; Zhang, J.X.; Liu, Y.W.; Qu, L.N.; Liu, Q.; Sun, Y.X.; Xu, G. Interfacial porosity model and modification mechanism of broken coal grouting: A theoretical and experimental study. *Surf. Interfaces* **2022**, *33*, 102286. [CrossRef]
9. Deng, J.; Li, Y.Q.; Zhang, Y.T. Effects of hydroxyl on oxidation characteristics of side chain active groups in coal. *J. China Coal Soc.* **2020**, *45*, 232–240.
10. Wang, S.Z. A study of the influence of the teaching time sequence of the modules "Structure and Properties of Substances" and "Fundamentals of Organic Chemistry" on the study of "Molecular Structure and Properties of Organic Substances". *Educ. Chem.* **2014**, *6*, 16–19.
11. Xie, K.C. *Coal Structure and Reactivity*, 3rd ed.; Science Press: Beijing, China, 2002.
12. Zhang, R.; Xia, Y.C.; Tan, J.L.; Ding, S.H.; Xing, Y.W.; Gui, X.H. Analysis and research on low rank coal carbon structure. *China Coal* **2018**, *44*, 88–94.
13. Bhoi, S.; Banerjee, T.; Mohanty, K. Molecular dynamic simulation of spontaneous combustion and pyrolysis of brown coal using ReaxFF. *Fuel* **2014**, *136*, 326–333. [CrossRef]
14. Li, X.; Zeng, F.G.; Wang, W.; Dong, K. FTIR characterization of structural evolution in low-middle rank coals. *J. China Coal Soc.* **2015**, *40*, 2900–2908.
15. Li, X.; Zeng, F.G.; Wang, W.; Dong, K. XRD characterization of structural evolution in low-middle rank coals. *J. Fuel Chem. Technol.* **2016**, *44*, 777–783.
16. Zhang, Y.L.; Wang, J.F.; Xue, S.; Wu, J.M.; Chang, L.P.; Li, Z.F. Kinetic study on changes in methyl and methylene groups during low-temperature oxidation of coal via in-situ FTIR. *Int. J. Coal Geol.* **2016**, *154*, 155–164. [CrossRef]
17. Zhang, Z.; Kang, Q.; Wei, S.; Yun, T.; Yan, G.; Yan, K. Large Scale Molecular Model Construction of Xishan Bituminous Coal. *Energy Fuels* **2017**, *31*, 1310–1317. [CrossRef]
18. Li, Z.M.; Wang, Y.M.; Li, P.; Li, H.P.; Bai, H.C.; Guo, Q.J. Macromolecular model construction and quantum chemical calculation of Ningdong Hongshiwan coal. *CIESC J.* **2018**, *69*, 2208–2216.
19. Wiser, W. Reported in division of fuel chemistry. *Preprints* **1975**, *20*, 122.
20. Given, P. The distribution of hydrogen in coals and its relation to coal structure. *Fuel* **1960**, *39*, 147–153.
21. Shinn, J.H. From coal to single-stage and two-stage products: A reactive model of coal structure. *Fuel* **1984**, *63*, 1187–1196. [CrossRef]
22. Xiao, Y.; Ye, X.X.; Liu, K.H.; Chen, L.G. Transformation law of key functional groups in the process of coal secondary oxidation. *J. China Coal Soc.* **2021**, *46*, 989–1000.
23. Jiang, Y.; Zong, P.; Tian, B.; Xu, F.; Tian, Y.; Qiao, Y.; Zhang, J. Pyrolysis behaviors and product distribution of Shenmu coal at high heating rate: A study using TG-FTIR and Py-GC/MS. *Energy Convers. Manag.* **2019**, *179*, 72–80. [CrossRef]
24. Tian, Y.Y.; Xie, K.C.; Qiao, Y.Y.; Tian, B. Construction and application of coal chemistry research system based on chemical group composition. *J. China Coal Soc.* **2021**, *46*, 1137–1145.
25. Xie, K.C. *Structure and Reactivity of Coal*; Springer: Heidelberg, Germany, 2015; Volume 10, pp. 978–983.
26. Carlson, G. Computer simulation of the molecular structure of bituminous coal. *Energy Fuels* **1992**, *6*, 771–778. [CrossRef]
27. Wu, S.; Jin, Z.; Deng, C. Molecular simulation of coal-fired plant flue gas competitive adsorption and diffusion on coal. *Fuel* **2019**, *239*, 87–96. [CrossRef]
28. Hong, D.; Liu, L.; Wang, C.; Si, T.; Guo, X. Construction of a coal char model and its combustion and gasification characteristics: Molecular dynamic simulations based on ReaxFF. *Fuel* **2021**, *300*, 120972. [CrossRef]
29. Hao, M.; Qiao, Z.; Zhang, H.; Wang, Y.; Li, Y. Thermodynamic analysis of $CH_4/CO_2/N_2$ adsorption on anthracite coal: Investigated by molecular simulation. *Energy Fuels* **2021**, *35*, 4246–4257. [CrossRef]
30. Meng, J.; Niu, J.; Meng, H.; Xia, J.; Zhong, R. Insight on adsorption mechanism of coal molecules at different ranks. *Fuel* **2020**, *267*, 117234. [CrossRef]
31. Feng, W.; Gao, H.F.; Wang, G.; Wu, L.L.; Xu, J.Q.; Li, Z.M.; Li, P.; Bai, H.C.; Guo, Q.J. Molecular model and pyrolysis simulation of Zaoquan coal. *CIESC J.* **2019**, *70*, 1522–1531.

32. Chai, S.Q.; Zeng, Q. Molecular model construction and structural characteristics analysis of Wucaiwan coal in Eastern Junggar Coalfeild based on quantum chemistry theory. *J. China Coal Soc.* **2022**, 1–12.
33. Ping, A.; Xia, W.C.; Peng, Y.L.; Xie, G.Y. Construction of bituminous coal vitrinite and inertinite molecular assisted by 13C NMR, FTIR and XPS. *J. Mol. Struct.* **2020**, *1222*, 128959. [CrossRef]
34. Jia, J.B.; Zeng, F.G.; Sun, B.L. Construction and modification of macromolecular structure mode for vitrinite from Shengdong2-2 coal. *J. Fuel Chem. Technol.* **2011**, *39*, 652–657.
35. Wang, Q.; Wang, Z.C.; Jia, C.X.; Gong, G.X. Study on structural features of oil sands with solid state ^{13}C-NMR. *Chem. Ind. Eng. Prog.* **2014**, *33*, 1392–1396.
36. Solum, M.S.; Pugmire, R.J.; Grant, D.M.; Kelemen, S.R.; Gorbaty, M.L.; Wind, R.A. 15N CPMAS NMR of the Argonne premium coals. *Energy Fuels* **1997**, *11*, 491–494. [CrossRef]
37. Wang, L.; Zhang, P.Z.; Zheng, M. Study on structural characterization of three Chinese coals of high organic sulphur content using XPS and solid-state NMR spectrescopy. *J. Fuel Chem. Technol.* **1996**, *24*, 539–5434.
38. Budinova, T.; Peyrov, N.; Minkova, V. Computer programmers for radial distribution analyses of X-rays. *Fuel* **1998**, *77*, 577–581. [CrossRef]
39. Senneca, O.; Salatino, P.; Masi, S. Combustion rates of chars from high-volatile fuels for FBC application. *Fuel* **1998**, *77*, 1483–1489. [CrossRef]
40. Jia, J.B. Construction of Structural Model and Molecular Simulation of Methane Formation Mechanism during Coal Pyrolysis for Shendon Vitrinite. Doctoral Dissertation, Taiyuan University of Technology, Taiyuan, China, 2010.
41. Yu, S.; Bo, J.; Jiahong, L. Retraction Note to: Simulations and experimental investigations of the competitive adsorption of CH 4 and CO 2 on low-rank coal vitrinite. *J. Mol. Model.* **2019**, *25*, 178. [CrossRef] [PubMed]
42. Xiang, J.; Zeng, F.; Liang, H.; Li, B.; Song, X. Molecular simulation of the $CH_4/CO_2/H_2O$ adsorption onto the molecular structure of coal. *Sci. China Earth Sci.* **2014**, *57*, 1749–1759. [CrossRef]
43. Nos'E, S. A unified formulation of the constant temperature molecular dynamics methods. *J. Chem. Phys.* **1984**, *81*, 511. [CrossRef]
44. Quinga, E.; Larsen, J.W. Noncovalent interactions in high-rank coals. *Energy Fuels* **1987**, *1*, 300–304. [CrossRef]
45. Zhu, H.Q.; He, X.; Huo, Y.F.; Xie, Y.Y.; Wang, W.; Fang, S.H. Construction and optimization of lignite molecular structure model. *J. Min. Sci. Technol.* **2021**, *6*, 429–437.
46. Zhang, X.D.; Hao, Z.C.; Zhang, S.; Yang, Y.H.; Yang, H.L.; Fang, S.H. Difference of nano-scale pore changes and its control mechanism for tectonic coal under solvent reconstruction. *J. China Univ. Min. Technol.* **2017**, *46*, 7.
47. Li, Z.; Ward, C.R.; Gurba, L.W. Occurrence of non-mineral inorganic elements in macerals of low-rank coals. *Int. J. Coal Geol.* **2010**, *81*, 242–250. [CrossRef]

processes

Article

A Data-Driven Fine-Management and Control Method of Gas-Extraction Boreholes

Xiaoyang Cheng [1,2,3] and Haitao Sun [1,2,3,*]

1 China Coal Research Institute, Beijing 100013, China
2 State Key Laboratory of the Gas Disaster Detecting, Preventing and Emergency Controlling, Chongqing 400037, China
3 China Coal Technology and Engineering Group, Chongqing Research Institute, Chongqing 400037, China
* Correspondence: dreamsht@163.com; Tel.: +86-02365239611

Abstract: In order to improve the efficiency of gas extraction in coal mines, a data-driven fine-management and control method for gas extraction is proposed. Firstly, the accurate prediction of coal seam thickness and gas content was used to evaluate the gas reserves. Based on the time relationship between mining activities and gas extraction, the calculation model of borehole distance in different extraction units is established, and the differential borehole design is realized. Then, a drilling video-surveillance system and drilling trajectory measurement device are used to control the drilling process and the construction effect. Finally, the model of extraction data-correction and the identification of failed boreholes is established, then the failed boreholes are repaired. The technology method provided in the paper has realized the fine control of gas-extraction borehole design, construction, measurement, and repair, and formed a more scientific gas-extraction borehole control technology system, which provides new thought for efficient gas extraction.

Keywords: gas extraction; borehole design; borehole repair; refined management

Citation: Cheng, X.; Sun, H. A Data-Driven Fine-Management and Control Method of Gas-Extraction Boreholes. *Processes* **2022**, *10*, 2709. https://doi.org/10.3390/pr10122709

Academic Editor: Feng Du

Received: 1 November 2022
Accepted: 13 December 2022
Published: 15 December 2022

1. Introduction

As the largest developing country in the world, China has a rapidly growing economic aggregate and energy consumption. As the primary driving force of China's economic development, coal occupies an important position in the energy structure [1–4]. In order to realize the double effect of "resource + safety" in coal mining, gas drainage has become an effective means [5,6]. Although gas extraction has played a positive role in reducing gas accidents, the low permeability in China restricts the improvement of gas-extraction efficiency. Therefore, gas accidents still cannot be fundamentally contained [7,8].

In order to effectively improve the efficiency of coal-seam gas drainage, experts and scholars carried out in-depth research on the design, construction, and sealing of extraction boreholes. Li Yun [9] optimized plane borehole into stereo borehole by changing borehole layout and borehole angle, which increased the pure amount of gas drainage and reduced the danger of outbursts. By conducting a simulation study on the pressure relief gas drainage parameters by using Fluent software, Ding Yang et al. [10] analyzed the influence of single and interactive factors on the drainage effect, fitted the regression model between the drainage parameters of each layer and the gas concentration in the upper corner, and obtained the optimal drilling parameters of each layer, which provided a guiding basis for accurate and efficient gas extraction. Qin Wei et al. [11] studied the law of strata movement in the working face and the characteristics of pressure relief gas drainage in the adjacent layer, based on which he put forward six kinds of drilling hole layout schemes, and determined the optimum drilling parameters by Fluent software simulation. Li Hong et al. [12] put forward the integrated technology of drilling construction, protection, and sealing, which can effectively solve problems such as difficult drilling and poor gas drainage effect in soft fault outburst coal seams.

The sealing effect of boreholes is also an important factor affecting the efficiency of gas drainage, and the rapid decline of gas drainage concentration is mainly caused by both in-hole gas leakage and hole-side fracture gas leakage [13,14]. For this reason, HU Shengyong et al. [15], Zhang Yongjiang et al. [16], Wang Zhiming et al. [17], and Wang Hao [18] analyzed the gas-leakage mechanism of gas boreholes from different angles, and put forward the corresponding borehole sealing technologies, which are of guiding significance for the improvement of gas-drainage efficiency. Fu Jianhua et al. [19] proposed a bag-type grouting sealing method characterized by a bag-type grouting sealing device. Engineering test results showed that the effect of the bag-grouting sealing method is preferable to that of the traditional "two sealing and one grouting" method. Zhang et al. [20] proposed a new in situ failure control technology, using fine particles to seal the leakage cracks around the CMB to improve the gas drainage effect. Zhai Cheng et al. [21–23] developed composite sealing material, solidified sealing material, and flexible gel (FG), and Li Bo et al. [24] fabricated a new coal-dust polymer composite material (CP), all of which are of great significance for improving the sealing quality of the borehole. Fu Jianhua. et al. [25] obtained the optimal ratio of cement-based sealing materials by orthogonal experimental method, which improved the extraction rate, the configuration, and related characteristics of sealing materials. Zhang Tianjun et al. [26] focused on the effect of the proportion of expansion agent on the expansion and creep properties of cement-based sealing materials. In addition, Zhang Chao et al. [27], Ni Guanhua et al. [28], and Zhou Aitao et al. [29] also developed different kinds of sealing materials, which not only provide more material options for borehole sealing, but also effectively improve the borehole sealing quality, thus laying a foundation for efficient gas extraction. Zhao D [30], Zhang Kai [31], Li Pu [32], and Wang Kai [33], respectively, studied the reasonable sealing depth of gas-drainage boreholes under different geological conditions and different sealing techniques by means of theoretical analysis, numerical simulation, and industrial test, which improved the sealing quality of gas drainage boreholes. Cheng Zhiheng et al. [34] established a multi-factor evaluation of the hole sealing method, which provided a theoretical basis for the selection of hole sealing method. Niu Yue et al. [35] developed a set of test systems for monitoring borehole deformation during gas drainage, proposed the calculation index of borehole damage degree (DD) based on residual area, established the coupling relationship between relative pressure (RP) of sensor and DD, and modified the DD calculation model to determine the condition of borehole damage. The research results have certain guiding significance for the deformation measurement and damage judgment of gas-drainage boreholes.

Although experts and scholars have made positive contributions to the improvement of gas-extraction efficiency in many aspects, gas accidents in recent years have exposed new problems, such as construction deviation of gas-extraction boreholes, insufficient number of boreholes, and so on. For this reason, on the one hand, new technology and new equipment are adopted to effectively supervise the whole process of gas drilling construction. Zhao Enbiao et al. [36] used drilling peeping and trajectory measurement technology to observe the gas-extraction drilling hole, and determined the rationality of the drilling location. Zhang Jun [37] adopted the techniques of drilling trajectory measurement,,real-time data transmission and three-dimensional visualization of data mapping to realize the visual supervision of drilling construction, which improved data transmission and processing efficiency. In order to improve the accuracy of borehole trajectory measurement, Yang Yi [38], Yang Hai [39], Liu Xiushan [40] et al. used different methods to correct the drilling trajectory, which significantly improved the accuracy of the borehole trajectory measurement. Zheng Lei [41] used a drilling video-surveillance system to realize parameter recording and video real-time monitoring in drilling process. In order to improve the clarity of the monitoring picture, Xu Yonggang et al. [42] proposed an adaptive compression and hybrid multiple hypothesis-based residual reconstruction algorithm based on a normalized Bhattacharyya coefficient (NBCAC-MHRR) to solve the high-efficiency video coding (HEVC) problem in underground coal mines, which improved the quality of underground video surveillance in coal mines. On the other hand, the concept of fine management is applied to gas drainage management [43,44]. Through the establishment of fine management systems and assessment mechanisms, each link of gas drainage is effectively managed and controlled, so as to improve the supervision level of mine gas drainage and reduce the risk coefficient of gas drainage.

The above research improves the efficiency of gas extraction from many aspects. However, it is worth noting that coal mine gas extraction is a systematic project, and it is difficult to achieve high-efficiency gas extraction only by the improvement of a single technology. In this paper, the accurate assessment of gas reserves and the differential design of drilling holes are carried out in the early stage of gas extraction, and the effective control is carried out in the drilling construction process. In the process of gas extraction, data correction is carried out, and the failed drilling holes are identified and repaired, forming a gas-extraction-control technology system, combining technology and management, which provides a solution for efficient gas extraction.

2. Accurate Evaluation of Gas Reserves in Coal Seams

The evaluation of coal seam gas reserves is an important foundation for the design of extraction boreholes, and it is calculated according to the following formula:

$$G = 0.01AhDC \tag{1}$$

In the formula, G is gas reserves of extraction unit, m^3; A is area of extraction unit, m^2; h is net thickness of coal seam of extraction unit, m; D is coal seam bulk density of extraction unit, t/m^3; C is gas content of extraction unit, m^3/t.

Formula (1) shows that the accuracy of gas reserves calculation largely depends on two parameters: coal reserves and gas content in the coal seam. However, in the actual calculation process, there is often a large deviation in the calculation of coal seam gas reserves due to the influence of coal seam occurrence state, geological structure, mining stress and so on, which seriously restricts the reasonable design of gas-drainage boreholes.

For this reason, a gas geological dynamic analysis system is developed based on the SuperMap platform. First of all, the system vectorizes the spatial information of the coal mine and forms the digital map of gas geology on the basis of the mining engineering plan map, topography map, and so on. Secondly, the gas geology-related information in the geological prospecting stage and production stage is uniformly stored and informationized, which lays the foundation for the deep utilization of the data. Finally, based on the organic correlation of coal mine spatial information and data, the automatic generation and dynamic updating of coal seam occurrence parameters are realized. The system makes full use of coal mine drilling, geophysical exploration, survey, and other multi-source data for fusion analysis, which reduces the analysis error caused by data singleness, data dispersion, and data missing, and improves the accuracy of gas geological information prediction. The intelligent extraction component of coal seam thickness and the intelligent update component of gas content provide technical support for the accurate evaluation of coal reserves and gas content.

2.1. Accurate Evaluation of Coal Reserves

The commonly used calculation methods of coal reserves include profile calculation method, geological block calculation method, horizontal section calculation method, statistical calculation method etc. [45]. These methods are all based on Formula (2):

$$Q = S \times M \times \rho \tag{2}$$

In the formula, Q is coal reserves, t; S is the scope of resources, m^2; M is the thickness of coal seam, m; ρ is the bulk density of coal t/m^3.

In the actual calculation, S and ρ are fixed values, so the accuracy of reserves calculation mainly depends on the accuracy of coal seam thickness M. However, affected by the geological environment, tectonic movement, and other objective factors in the process of coal formation, there are significant differences in coal seam thickness [46]. In the gas geological dynamic analysis system, the inverse distance interpolation method as shown in Formula (3), is used to realize the accurate evaluation of coal seam thickness. On the one hand, based on the geological prospective borehole data, the intelligent generation of original coal seam thickness grid and isoline is realized. On the other hand, based on the measured coal seam thickness revealed by coal mining activities, the intelligent correction and update of coal seam thickness isoline are realized. The two work together to realize the accurate evaluation of coal seam thickness. The thickness grid and isoline of coal seam are shown in Figure 1.

$$Z*(x_0) = \frac{\sum\limits_{i=1}^{n} \frac{1}{(D_i)^p} Z(x_i)}{\sum\limits_{i=1}^{n} \frac{1}{(D_i)^p}} \tag{3}$$

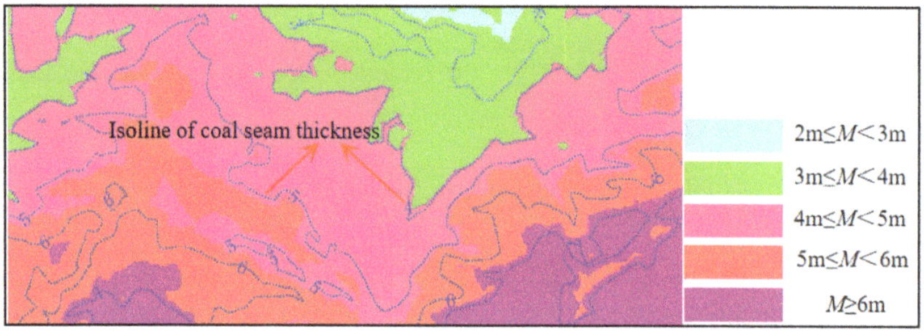

Figure 1. Coal seam thickness grid and isoline map of coal seam thickness.

In the formula, $Z^*(x_0)$ is the calculated coal seam thickness of the interpolated point, $Z(x_i)$ is the coal seam thickness value of the ith (i = 1, 2, ... , n) known sampling point, n is the number of sampling points used for coal seam thickness interpolation, D_i is the distance from the interpolated point to the ith sampling point, p is the power of the distance.

In order to further improve the calculation accuracy, the extraction unit can be divided into n blocks according to the differences of geological blocks, coal seam thickness and mining environment, and the coal reserves can be calculated in sequence. The calculation formula is shown in Formula (4).

$$Q = \rho \cdot \sum_{i=1}^{n} \iint_{S_i} h \cdot d\sigma \tag{4}$$

In the formula, S_i is the resource range of block i, m^2, h is the thickness of coal seam, m.

2.2. Dynamic Evaluation of Gas Content

Gas content is also an important parameter for the calculation of coal seam gas reserves. In engineering practice, the prediction of gas content is often used as the calculation parameter of gas reserves.

However, in the process of coal seam mining, affected by the change of mining stress, the gas content of coal seam is in a constant state of dynamic change, so a large deviation is inevitable when using the original gas content to evaluate gas reserves. Based on the original gas content data, the gas geological dynamic analysis system generates the original gas content isoline, and uses the measured gas content measuring point data to calculate the Kriging interpolation method using the Formula (5). The prediction of gas content isoline is intelligently updated, which greatly improves the accuracy of gas content prediction. The gas content isoline before and after update is shown in Figure 2.

$$\hat{Z}(S_0) = \sum_{i=1}^{N} \lambda_i Z(S_i) \tag{5}$$

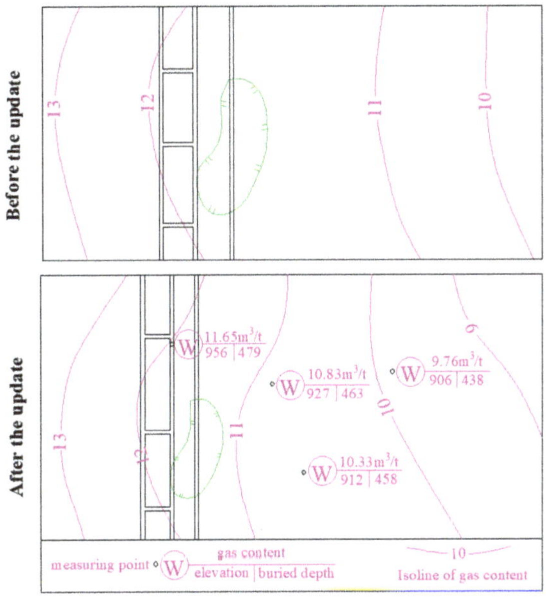

Figure 2. Comparison of gas content contour before and after update.

In the formula, the gas content value of the ith measuring point, the unknown weight of the ith content value, S_0 is the predicted position, and N is the number of measuring points.

3. Differential Extraction Borehole Design

A reasonable drilling design of gas drainage needs to consider the position and time relationship between mining activities and gas drainage [47]. In the process of actual extraction borehole design, for the convenience of design and construction, unified drilling construction parameters are often used in the same working face, but the blind unity of borehole spacing can bring great safety risks and engineering waste to the mine. Therefore, for the rationality of extraction balance, this paper makes a differential design of extraction boreholes, that is, the design of sub-sections and non-uniform boreholes is adopted in the borehole design, and the schematic diagram of differential borehole design is shown in Figure 3.

The calculation formula of the amount of gas extraction required to reach the standard of gas extraction in the working face section is as follows:

$$Q_i = (W_{i\max} - W_0) \cdot \rho \iint_{S_i} h \cdot ds \tag{6}$$

In the formula, Q_i is the extraction amount per unit borehole length, m^3; $W_{i\max}$ is the maximum gas content in the ith section, m^3/t; W_0 is the critical gas content of extraction standard, m^3/t; ρ is the density of coal, t/m^3; h is the thickness of coal, m; S_i is the cross-sectional area of the ith section, m^2; and ds is the section microelement.

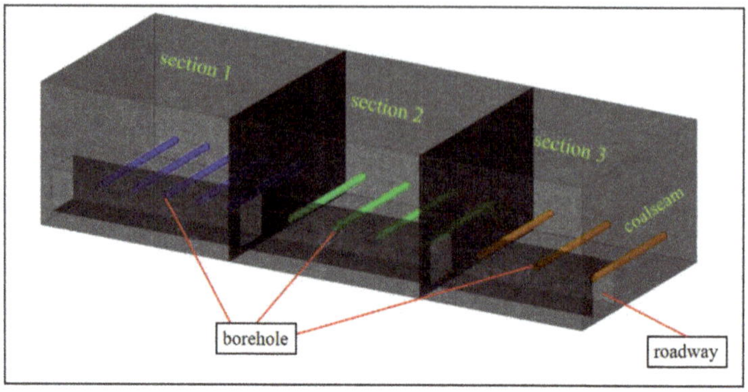

Figure 3. Schematic diagram of differential extraction borehole design.

The amount of gas extraction required to reach the standard in the ith section can be calculated by the amount of extraction per unit length of boreholes and the total length of extraction boreholes, which can be expressed as follows [48]:

$$Q_i = q \cdot L_i = \frac{Q_0(1 - Be^{-Ct_i})}{L_0} \cdot L_i \tag{7}$$

In the formula, q is the extraction amount per unit borehole length, m^3/m; L_i is the total length of extraction borehole in the ith section; L_0 is the length of single borehole, m; Q_0 is the initial extraction amount of single borehole, m^3; B and C are constants; the shortest extraction time of the ith section is t_i, and its calculation formula is as shown in the Formula (8):

$$t_i = \sum_{i=1}^{n} \frac{M_{i-1}}{Y} \tag{8}$$

In the formula, M_{i-1} is the coal reserve of the $i-1$th section, t; Y is the daily output of the working face, t/d.

Based on Formulas (6) and (7), the total length of extraction boreholes in ith section can be obtained:

$$L_i = \frac{(W_{i\max} - W_0) \cdot L_0 \cdot \rho \iint\limits_{S_i} h \cdot ds}{Q_0(1 - Be^{-Ct_i})} \tag{9}$$

The volume of coal in the extraction area can be expressed in two ways, which are approximately equal:

$$\iint\limits_{S_i} h \cdot ds \approx \pi R_i^2 \cdot L_i \tag{10}$$

In the formula, R_i is the extraction radius, m.

Based on the Formulas (8)–(10), it is concluded that the extraction radius of the ith section is:

$$R_i = \sqrt{\frac{Q_0\left(1 - Be^{-C\sum_{i=1}^{n}\frac{M_{i-1}}{Y}}\right)}{(W_{i\max} - W_0) \cdot \pi \cdot L_0 \cdot \rho}} \tag{11}$$

In order to avoid production safety accidents caused by the extraction blanking zone, the effective extraction range of extraction boreholes should completely cover the extraction section. According to the Formula (11), the distance between gas drainage boreholes in the ith section can be determined as $2R_i$, which increases the rationality of gas drainage boreholes in the section with high gas and low drainage time, and reduces the engineering waste in the section with low gas and high drainage time, thereby providing guidance for efficient gas extraction.

4. Management of Drilling Process and Drilling Effect

The efficient extraction of gas is closely related to the quality of extraction boreholes. In addition to the reasonable design of boreholes, precision borehole construction is essential. At present, there is a serious disconnection between borehole construction and borehole design, and drilling site constructors have the tendency of blindly pursue drilling footage due to the lack of supervision of drilling site construction, resulting in a discrepancy between drilling construction parameters and design, inadequate borehole construction and unclear description of the site and other major hidden dangers that threaten coal mine production safety. In order to improve the quality of drilling construction, it is necessary to effectively manage the drilling process and drilling effect.

(1) Management of drilling process

In order to ensure the precise construction of drilling, it is necessary to timely acquire the drilling hole angle, hole position, drilling distance, and so on, to avoid safety accidents caused by deviation of drilling construction. As an effective method to manage drilling process, drilling video-surveillance systems are widely used in coal mines. The working principle of the system is to collect video information during the drilling process through an underground high-definition camera, and transmit the video signal to the ground server by optical cable transmission technology to realize the real-time monitoring of the drilling site. The network topology and monitoring effect are shown in Figure 4.

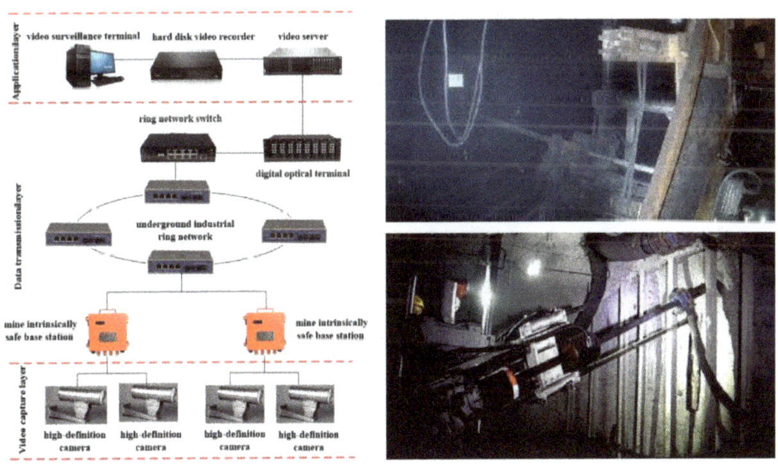

Figure 4. Network topology diagram of drilling video surveillance system and monitoring effect.

(2) Management of drilling effect

In order to reduce the influence of gas-extraction borehole construction deviation on gas extraction, the drilling trajectory measurement device is used to measure the construction trajectory of gas-extraction borehole. In the process of drilling construction, the probe tube is installed in the non-magnetic drill pipe behind the drill bit, and the probe tube is drilled forward with the drilling rig. The basic data such as drilling depth, dip angle, and azimuth angle are measured and recorded by hand-held data storage instrument, and the data processing and automatic generation of drilling trajectory are carried out by professional software. The drilling trajectory measurement device and measurement results are shown in Figure 5.

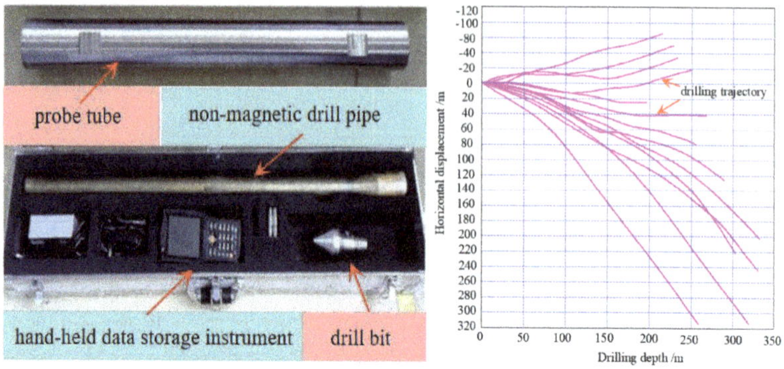

Figure 5. Drilling trajectory measurement device and measurement results.

5. Identification and Repair of Failed Gas Extraction Boreholes

5.1. Extraction Data Correction

Extraction data are generally measured by extraction monitoring system on-line or measured manually. Although advanced extraction monitoring system and measuring instruments are adopted, there are still technical difficulties to maintain the accuracy of the system and instruments in the long process of extraction, which inevitably results in the deviation between the extraction data and the real data. For this reason, the difference between gas extraction monitoring data and manual data is investigated, the deviation law of gas-extraction monitoring data is analyzed, the correction model of gas extraction data is established, and the automatic calibration of gas-extraction monitoring data is achieved using computer technology to improve the accuracy of mine gas-extraction monitoring data. The gas-extraction data correction process is shown in Figure 6.

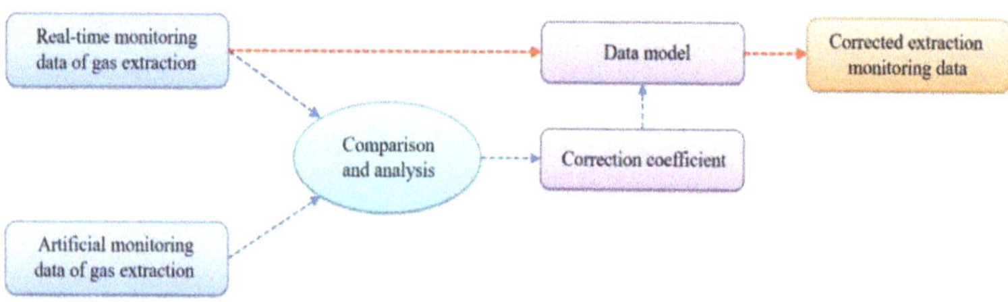

Figure 6. Flow chart of gas-extraction data correction.

In addition, based on the different characteristics of gas conditions in the gas pumping station and the extraction pipeline, the extraction gas quantity needs to be converted into the gas quantity in the standard state to be calculated. For the monitoring device which can directly display the gas quantity in the standard condition, the default parameter values (temperature, pressure, etc.) in the device should be updated in time according to the actual gas conditions of the equipment installation site. Finally, the accurate measurement of gas-extraction data is based on the accuracy of the sensor, and the sensor needs to be adjusted regularly and tested on the spot. If the test result deviates greatly, the sensor should be replaced or repaired in time.

5.2. Identification of Failed Gas-Extraction Boreholes

After the formation of gas-extraction boreholes, under the action of the stress field around the boreholes, the boreholes are often unstable, the cracks between the boreholes are connected, and the cracks between the boreholes and the surrounding rock are connected, resulting in the decrease or even interruption of gas drainage concentration and flow rate [49,50]. Therefore, it is necessary to identify the failure gas-extraction boreholes so as to guide the restoration of the failed boreholes.

In the extraction pipe network, the gas confluence of each extraction hole enters the branch pipe of the extraction unit, and a flow sensor is installed on the branch pipe to measure the extraction flow of the whole extraction unit. The schematic diagram of the extraction system is shown in Figure 7. The pumping system only depends on the data on the branch pipe and the main pipe, which makes it difficult to grasp the extraction data and state of a single borehole in real time and accurately. Therefore, this paper adopts the data source processing method. When all the extraction boreholes of the extraction unit are connected into the initial stage of the extraction system, the gas-extraction parameter measuring instrument is used to measure the gas flow of each borehole. At the same time, the total flow of the extraction unit is read by the branch pipeline flowmeter, and the flow ratio of each drill hole is calculated by Formula (12).

$$\phi_k = \frac{Q_k}{Q_c} \tag{12}$$

In the formula, ϕ_k is the k borehole flow ratio, Q_k is the k borehole flow, and Q_c is the total branch flow.

In the initial stage of extraction, due to the pressure relief and permeability enhancement of boreholes and the negative pressure of extraction, the gas flow rate of each borehole is larger and the difference of flow ratio is small. With the extraction time lengthens, there are some unstable and air leakage boreholes one after another, and a large amount of air will be mixed in the gas extracted from these boreholes, which will be merged into the extraction pipeline to reduce the extraction flow [20]. At this point, according to the sub-source data processing method, the flow ratio of this kind of boreholes decreases significantly, while the flow ratio of normal borehole increases significantly. According to the extraction requirements of each mine, when the borehole flow ratio is lower than the critical value ϕ_L, it can be judged as failed gas-extraction boreholes, and technical measures need to be taken to repair or close the extraction boreholes in order to achieve efficient gas extraction.

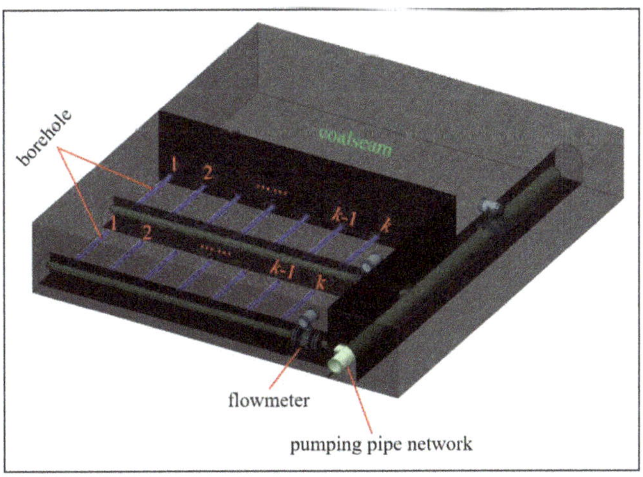

Figure 7. Schematic diagram of pumping pipe network.

5.3. Repair of Failed Gas-Extraction Boreholes

The study shows that the main reason for the failure of gas-extraction boreholes is air leakage in boreholes [13,14], while the air leakage in boreholes is caused by poor sealing quality. There are mainly two ways to deal with the failed gas-extraction boreholes: closing boreholes and supplementing boreholes, which are passive ways to deal with the failed boreholes. Closing the failed boreholes increases the drainage time of regional coal seams, and extraction effect of supplementary boreholes is also poor. Therefore, in this paper, the repair technology for failure borehole is adopted to realize efficient gas extraction. The process of drilling repair is shown in Figure 8.

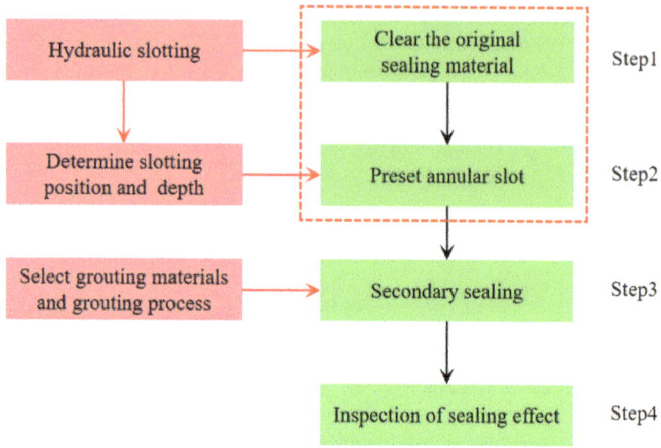

Figure 8. Repair process of failed gas-extraction boreholes.

As shown in Figure 8, firstly, the hydraulic slotting technology is used to cut and remove the failed sealing material in the borehole. Secondly, an annular slot is cut in front of the working face by a hydraulic slotting device to block the air leakage channel through the crack in the coal–rock mass. In this process, the optimal slotting position and depth need to be determined. Thirdly, secondary sealing is needed for the gas-extraction boreholes. Of course, in this process, the sealing material with high expansion and fluidity is essential, and it is recommended to adopt the "two plugging and one injection" sealing process with pressure. Finally, we need to check the sealing effect of the borehole.

6. Discussion

Gas drainage in coal mines is a complex process, including drilling design, drilling construction, extraction data management, and other links. With the wide application of information technology in the field of coal mines, professional software has been developed for different methods of gas extraction, which has played a positive role in improving the efficiency of gas extraction. Nowadays, in the context of the development of coal mine intelligence, the deep mining and utilization of data has become the core of coal mine intelligence development. As far as gas drainage is concerned, the intelligent control of gas drainage management based on extraction data, intelligent evaluation of extraction standards, and intelligent design of extraction boreholes have become an important but difficult part of the development of gas-drainage technology. Future studies should focus on building an intelligent decision-making platform for gas drainage based on gas-drainage data by means of big data analysis to achieve intelligent control, thereby realizing the intelligent construction of gas drainage.

7. Conclusions

(1) Through the accurate prediction of coal seam thickness and gas content, the accurate assessment of coal seam gas reserves is realized. Then, the extraction area is divided into sections, and based on the space–time relationship between mining activities and gas extraction, the calculation model of borehole distance in different sections is established, and the differentiated design of borehole is realized.

(2) The drilling video surveillance system and drilling trajectory measurement device are used to manage the drilling process and effect, respectively, which ensure the precise construction of boreholes.

(3) Based on the monitoring data of gas extraction, the model of extraction data correction and identification of failed borehole are established and the failed borehole caused by air leakage is solved by repair technology of hydraulic slotting and sealing, which improves the gas drainage efficiency.

Author Contributions: Conceptualization, X.C.; methodology, X.C. and H.S.; validation, X.C. and H.S.; formal analysis, X.C.; investigation, X.C. and H.S.; resources, X.C.; data curation, H.S.; writing—original draft preparation, X.C. and H.S.; writing—review and editing, X.C. and H.S.; visualization, X.C.; supervision, H.S.; project administration, H.S.; funding acquisition, X.C. and H.S. All authors have read and agreed to the published version of the manuscript.

Funding: This research was funded by National Natural Science Foundation of China (No. 51874348), Chongqing Science Fund for Distinguished Young Scholars (No.cstc2019jcyjjqX0019), Natural Science Foundation of Chongqing (CSTB2022NSCQ-MSX1080).

Data Availability Statement: All data and/or models used in the study appear in the submitted article.

Conflicts of Interest: The authors declare no conflict of interest.

References

1. Yan, F.Z.; Xu, J.; Peng, S.J.; Zou, Q.L.; Li, Q.G.; Long, K.; Zhao, Z.G. Effect of capacitance on physicochemical evolution characteristics of bituminous coal treated by high-voltage electric pulses. *Powder Technol.* **2020**, *367*, 47–55. [CrossRef]
2. Fan, C.J.; Derek, E.; Li, S.; Chen, Z.W.; Luo, M.K.; Song, Y.; Zhang, H.H. Modelling and optimization of enhanced coalbed methane recovery using CO_2/N_2 mixtures. *Fuel* **2019**, *253*, 1114–1129. [CrossRef]
3. Xie, H.P.; Wu, L.X.; Zheng, D.Z. Prediction on the energy consumption and coal demand of China in 2025. *J. China Coal Soc.* **2019**, *44*, 1949–1960. (In Chinese)
4. Yuan, L. Strategic thinking of simultaneous exploitation of coal and gas in deep mining. *J. China Coal Soc.* **2016**, *41*, 1–6. (In Chinese)
5. Lin, B.Q.; Shen, C.M. Coal permeability-improving mechanism of multilevel slotting by water jet and application in coal mine gas extraction. *Environ. Earth Sci.* **2015**, *73*, 5975–5986. [CrossRef]
6. Zhou, F.B.; Wang, X.X.; Liu, Y.K. Gas drainage efficiency: An input-output model for evaluating gas drainage projects. *Nat. Hazards* **2014**, *74*, 989–1005. [CrossRef]
7. Zhou, F.B.; Xia, T.Q.; Wang, X.X.; Zhang, Y.F.; Sun, Y.N.; Liu, J.S. Recent developments in coal mine methane extraction and utilization in China: A review. *J. Nat. Gas Sci. Eng.* **2016**, *31*, 437–458. [CrossRef]
8. Xie, J.N.; Ni, G.H.; Xie, H.C.; Li, S.; Sun, Q.; Dong, K. The effect of adding surfactant to the treating acid on the chemical properties of an acid-treated coal. *Powder Technol.* **2019**, *356*, 263–272.
9. Li, Y. Optimized design and application of three dimensional gas drainage boreholes in driving face. *Coal Sci. Technol.* **2016**, *44*, 59–63. (In Chinese)
10. Ding, Y.; Zhu, B.; Li, S.G.; Lin, H.F.; Wei, Z.Y.; Li, L.M.; Long, H.; Yi, Y. Accurate and efficient drainage of relieved methane in goaf of high outburst mine. *J. China Coal Soc.* **2021**, *46*, 3565–3577. (In Chinese) [CrossRef]
11. Qin, W.; Xu, J.L.; Wu, R.L.; Hu, G.Z.; Tian, G.W. Gas Extraction Optimum Design of Layer-Through Borehole in Adjacent Seam Based on CFD Simulation. *J. Min. Saf. Eng.* **2012**, *29*, 111–117. (In Chinese)
12. Li, H.; Wang, W.; Liu, Y.W.; Ma, J.K.; Gao, H. An integrated drilling, protection and sealing technology for improving the gas drainage effect in soft coal seams. *Energy Rep.* **2020**, *6*, 2030–2043. [CrossRef]
13. Zhang, X.B.; Gao, J.L.; Jia, G.N.; Zhang, J.W. Study on the influence mechanism of air leakage on gas extraction in extraction boreholes. *Energy Explor. Exploit.* **2022**, *40*, 1344–1359. [CrossRef]
14. Zhang, J.X.; Li, B.; Sun, Y.N. Dynamic leakage mechanism of gas drainage borehole and engineering application. *Int. J. Min. Sci. Technol.* **2018**, *28*, 505–512.
15. Hu, S.Y.; Hao, G.C.; Feng, G.R.; Guo, H.B.; Wu, D.Y. A method for improving the methane extraction concentrations of in-seam boreholes. *Fuel* **2020**, *265*, 117006. [CrossRef]
16. Zhang, Y.J.; Zou, Q.L.; Guo, L.D. Air-leakage Model and Sealing Technique With Sealing–Isolation Integration for Gas-drainage Boreholes in Coal Mines. *Process Saf. Environ. Prot.* **2020**, *140*, 258–272. [CrossRef]
17. Wang, H.; Wang, E.Y.; Li, Z.H.; Wang, X.R.; Zhang, Q.M.; Li, B.; Ali, M. Study on sealing effect of pre-drainage gas borehole in coal seam based on air-gas mixed flow coupling model. *Process Saf. Environ. Prot.* **2020**, *136*, 15–27. [CrossRef]
18. Wang, Z.M.; Sun, Y.N.; Wang, Y.L.; Zhang, J.X.; Sun, Z.D. A coupled model of air leakage in gas drainage and an active support sealing method for improving drainage performance. *Fuel* **2019**, *237*, 1217–1227. [CrossRef]

19. Fu, J.H.; Li, X.L.; Wang, Z.M. A novel sealing material and a bag-grouting sealing method for underground CBM drainage in China. *Constr. Build. Mater.* **2021**, *299*, 124016. [CrossRef]
20. Zhang, Y.F.; Hu, S.Y.; Xia, T.Q.; Liu, Y.K.; Pan, Z.; Zhou, F.B. A novel failure control technology of cross-measure borehole for gas drainage: A case study. *Process Saf. Environ. Prot.* **2020**, *135*, 144–156. [CrossRef]
21. Zhai, C.; Hao, Z.Y.; Lin, B.Q. Research on a new composite sealing material of gas drainage borehole and its sealing performance. *Procedia Eng.* **2011**, *26*, 1406–1416. [CrossRef]
22. Zhang, C.; Jin, G.H.; Liu, C.; Li, S.G.; Xue, J.H.; Cheng, R.H.; Liu, H. Sealing Performance of New Solidified Materials: Mechanical Properties and Stress Sensitivity Characterization of Pores. *Adv. Polym. Technol.* **2020**, *2020*, 5397697. [CrossRef]
23. Zhai, C.; Xiang, X.W.; Zou, Q.L.; Yu, X.; Xu, Y.M. Influence factors analysis of a flexible gel sealing material for coal-bed methane drainage boreholes. *Environ. Earth Sci.* **2016**, *75*, 385. [CrossRef]
24. Li, B.; Zhang, J.X.; Wei, J.P.; Zhang, Q. Preparation and sealing performance of a new coal dust polymer composite sealing material. *Adv. Mater. Sci. Eng.* **2018**, *2018*, 8480913. [CrossRef]
25. Fu, J.H.; Wang, D.K.; Li, X.L.; Wang, Z.M.; Shang, Z.J.; Jiang, Z.G.; Wang, X.B.; Gao, X. Experimental Study on the Cement-Based Materials Used in Coal Mine Gas Extraction for Hole Sealing. *ACS Omega* **2021**, *6*, 21094–21103. [CrossRef]
26. Zhang, T.J.; Bao, R.Y.; Li, S.G.; Zhang, C.; Zhang, L. Expansion properties and creep tests for a new type of solidified expansive sealing material for gas drainage boreholes in underground mines. *Environ. Earth Sci.* **2018**, *77*, 468. [CrossRef]
27. Zhang, C.; Chang, J.; Li, S.G.; Liu, C.; Qin, L.; Bao, R.Y.; Liu, H.; Cheng, R.H. Experimental study comparing the microscopic properties of a new borehole sealing material with ordinary cement grout. *Environ. Earth Sci.* **2019**, *78*, 149.
28. Ni, G.H.; Dong, K.; Li, S.; Sun, Q.; Huang, D.M.; Wang, N.; Cheng, Y.Y. Development and performance testing of the new sealing material for gas drainage drilling in coal mine. *Powder Technol.* **2020**, *363*, 152–160.
29. Zhou, A.T.; Wang, K. A new inorganic sealing material used for gas extraction borehole. *Inorg. Chem. Commun.* **2019**, *102*, 75–82. [CrossRef]
30. Zhao, D.; Pan, J. Numerical simulation on reasonable hole-sealing depth of boreholes for gas extraction. *AIP Adv.* **2018**, *8*, 045003. [CrossRef]
31. Zhang, K.; Sun, K.; Yu, B.Y.; Ranjith, P.G. Determination of sealing depth of in-seam boreholes for seam gas drainage based on drilling process of a drifter. *Eng. Geol.* **2016**, *210*, 115–123. [CrossRef]
32. Li, P.; Cheng, Z.H.; Chen, L.; Wang, H.B.; Gao, J.L. Improvement of gas drainage efficiency via optimization of sealing depth of cross-measure boreholes. *Adv. Civ. Eng.* **2021**, *2021*, 5521666. [CrossRef]
33. Wang, K.; Lou, Z.; Wei, G.Y.; Lu, W.D.; Qin, B.B. Reasonable minimum sealing length of in-seam boreholes for coal mine methane drainage: A case study. *Energy Sources Part A Recovery Util Environ. Eff.* **2019**, *42*, 2199–2213. [CrossRef]
34. Cheng, Z.H.; Lu, Y.; Li, H.T.; Wang, H.B.; Pu, S.J.; Sun, F.L.; Lou, Y.H. Quality inspection and optimization of different sealing methods for high-level borehole. *J. Saf. Sci. Technol.* **2020**, *16*, 61–67. (In Chinese)
35. Niu, Y.; Song, X.Y.; Li, Z.H.; Wang, E.Y.; Liu, Q.L.; Zhang, X.; Cai, G.N.; Zhang, Q.M. Experimental study and field verification of stability monitoring of gas drainage borehole in mining coal seam. *J. Pet. Sci. Eng.* **2020**, *189*, 106985. [CrossRef]
36. Zhao, E.B.; Li, K.N.; Yang, X.; Deng, N. Speculum Observation and Trajectory Measurement in Gas Extraction Drilling: A Case Study of Changling Coal Mine. *Geofluids* **2021**, *2021*, 5545067. [CrossRef]
37. Zhang, J. Study on borehole while drilling trajectory measurement and three-dimensional visualization technology. *Chin. Energy Environ. Prot.* **2020**, *42*, 54–59. (In Chinese)
38. Yang, Y.; Li, F.; Zhang, N.; Huo, A.Q. Research on the Cooperative Detection of Stochastic Resonance and Chaos for Weak SNR Signals in Measurement While Drilling. *Sensors* **2021**, *21*, 3011. [CrossRef]
39. Yang, H.; Li, L.; Luo, T.; Liang, H.B.; Rao, Y.; Qiu, Z. A Fault-Tolerant Integrated Borehole Trajectory Location Method Based on Geomagnetism/IMU of MWD. *IEEE Access* **2019**, *7*, 156065–156076. [CrossRef]
40. Liu, X.S. Trajectory Tracking and Rate of Penetration Control of Downhole Vertical Drilling System. *Pet. Explor. Dev.* **2019**, *46*, 407–412. [CrossRef]
41. Zheng, L. Research on application of intelligent analysis technology in drilling monitoring video system of coal mine underground drilling field. *Chin. Energy Environ. Prot.* **2019**, *41*, 107–110.
42. Xu, Y.G.; Xue, Y.Z.; Hua, G.; Cheng, J.W. An Adaptive Distributed Compressed Video Sensing Algorithm Based on Normalized Bhattacharyya Coefficient for Coal Mine Monitoring Video. *IEEE Access* **2020**, *8*, 3020140. [CrossRef]
43. Xing, J.G. Innovation and practice of fine management system for gas control in coal enterprises. *Saf. Coal Mines* **2016**, *47*, 247–250. (In Chinese)
44. Ye, B. Application analysis on fine management of drilling and gas drainage in coal mines. *China Resour. Compr. Util.* **2020**, *38*, 63–65. (In Chinese)
45. Zhu, X.L.; He, X.Y.; Chen, Y.H. *Mineral Deposits Geometry*; University of Mining and Technology Press: Xuzhou, China, 1997. (In Chinese)
46. Wang, D.Y.; Cheng, Y.P.; Wang, L.; Zhou, H.X. Controlling factors of coalbed methane occurrence below redbeds in Xutuan mine: Caprock thickness below redbeds. *J. Nat. Gas Sci. Eng.* **2021**, *96*, 104323. [CrossRef]
47. Zhou, F.B.; Liu, C.; Xia, T.Q.; Liu, Y.K.; Sun, Y.N. Intelligent gas extraction and control strategy in coal mine. *J. China Coal Soc.* **2019**, *44*, 2377–2387. (In Chinese)

48. Lv, Y.C.; Zhu, C.J. Research on attenuation law of drainage gas drainage in bored layers with different inclined floors. *Coal Sci. Technol.* **2017**, *45*, 74–79. (In Chinese)
49. Fan, J.Y.; Liu, P.; Li, J.J.; Jiang, D.Y. A coupled methane/air flow model for coal gas drainage: Model development and finite-difference solution. *Process Saf. Environ. Prot.* **2020**, *141*, 288–304. [CrossRef]
50. Liu, T.; Lin, B.Q.; Fu, X.H.; Zhao, Y.; Gao, Y.B.; Yang, W. Modeling coupled gas flow and geomechanics process in stimulated coal seam by hydraulic flushing. *Int. J. Rock Mech. Min. Sci.* **2021**, *142*, 104769. [CrossRef]

processes

Article

Analyzing Energy Transfer Mechanism during Coal and Gas Protrusion in Deep Mines

Haitao Sun [1,2], Linchao Dai [1,2,*], Jun Lu [3,*], Jie Cao [1,2] and Minghui Li [3]

1 State Key Laboratory of the Gas Disaster Detecting, Preventing and Emergency Controlling, Chongqing 400037, China
2 China Coal Technology and Engineering Group, Chongqing Research Institute, Chongqing 400037, China
3 Shenzhen Key Laboratory of Deep Underground Engineering Sciences and Green Energy, College of Civil and Transportation Engineering, Shenzhen University, Shenzhen 518060, China
* Correspondence: 20212001047g@cqu.edu.cn (L.D.); junlu@cqu.edu.cn (J.L.); Tel.: +86-02365239611 (L.D.)

Abstract: Coal is the mainstay of China's energy supply. With the gradual progress in China's policy of phasing out backward coal production capacity, the intensive and deep mining of coal has gradually become the new norm. The current mining depth is increasing at a rate of 10~15 m/year. The high crust stress, high gas pressure, high ground temperature, and engineering disturbance stress in deep coal mines can lead to the occurrence of coal–rock–gas dynamic disasters that are complex and show the characteristics of compound dynamic disasters. It is important to understand the evolution and mechanism of deep coal and rock dynamic disasters accurately for the safe development of deep resources. To study the mechanism of occurrence and the evolution of impact–protrusion compound dynamic disasters, we herein analyzed the apparent characteristics of coal–rock–gas compound dynamic disasters in deep mines and obtained the mechanical and acoustic emission characteristics of coal–rock composites through indoor experiments. Then, we conducted in-depth analysis on the non-uniform deformation behaviors and non-uniform stress field of the coal–rock composite and clarified the generation mechanism of local tensile cracks at the coal–rock interface. Subsequently, we established the energy transfer model of the rock–rock–gas composite specimen in the process of dynamic destabilization in the engineering scale mining field and revealed the mechanism of nonlinear energy evolution and release of the coal–rock–gas composite, which has been less reported in previous studies. In this paper, we further clarified the occurrence and development mechanism of coal–rock–gas compound dynamic disasters in the engineering scale mining environment to guide the prevention and control of coal–rock–gas dynamic disasters in deep mines.

Keywords: deep mines; gas; composite dynamic disaster; energy transfer; mining environment

Citation: Sun, H.; Dai, L.; Lu, J.; Cao, J.; Li, M. Analyzing Energy Transfer Mechanism during Coal and Gas Protrusion in Deep Mines. *Processes* **2022**, *10*, 2634. https://doi.org/10.3390/pr10122634

Academic Editors: Feng Du, Aitao Zhou and Bo Li

Received: 5 November 2022
Accepted: 5 December 2022
Published: 8 December 2022

1. Introduction

For a long time, coal—as the mainstay of energy consumption in China—has played an important role in its social and economic development and will consequently remain irreplaceable for a considerable period of time in the future. However, with the gradual depletion of shallow resources, the mining depth of coal has gradually increased into the Earth's crust and the consequent mining environment presents the characteristics of high crust stress, high gas pressure, and high temperature [1,2]. Deeper mining is accompanied by the obvious appearance of tectonic stress and mining stress [3]. The mechanical and seepage characteristics of coal rocks also show more complex characteristics than those of shallow parts. In addition, with the deterioration of the mining environment, most of the mines show the characteristics of high stress, high gas, and low permeability and deep shaft disasters such as large deformation of the roadways, topping, coal and gas protrusion, coal burst, and compound dynamic disasters that occur frequently. This seriously restricts the safe and efficient mining of deep coal seams [4]. Therefore, a correct understanding of the mechanical properties of deep coal rocks and the understanding of coal–rock disaster mechanisms are essential for the stability design of the surrounding rock and for the prevention and management of disasters.

Owing to geological structures and stratigraphic movements, rock formations do not exist as a single lithology; laminated rock formations are the most common geological form. The destruction of such composite formations can have a considerable impact on underground engineering activities (mining, excavation, tunneling, geothermal energy extraction, deep burial of nuclear waste, among others) [5]. As a result, many scholars have studied the mechanical behavior of composite coal-rock mass. Particularly in deep thin coal seam mining, studying the mechanical properties of composite coal–rock strata is crucial and the fracture and rupture of the hard roof is very easy to induce the coal–rock dynamic disaster. The composite coal–rock mass is a strong anisotropic material that contains many weak structural surfaces that differ from the properties of the rock matrix. Therefore, studying the mechanical properties and energy release transfer laws of layered composite coal–rock is crucial for revealing the dynamic disaster mechanism of coal and rock in deep mining engineering.

In recent decades, the mechanical properties of layered composite coal rocks have been theoretically and experimentally studied under various stress conditions and there has been considerable progress in this domain. Previous studies have shown that the elastic modulus, Poisson's ratio, deformation behaviors, strength, failure mode, and permeability evolution of layered composite coal–rock are quite different from those of pure rocks, regardless of the stress conditions [5–8]. Dou et al. demonstrated that the elastic modulus and strength of layered composite coal rocks increase with the increase in the height ratio of the rock mass to the coal sample [9]. Tan et al. showed that the rock layer type greatly affects the impact propensity of rock burst. Specifically, the impact tendency of composite coal–rock is proportional to the elastic modulus and strength of the rock [10]. In addition, an applicable method to prevent the collapse of high-velocity impact roofs based on the flexible/hard structure analysis of the composite coal–rock model was proposed [11,12]. The composite rock mass structure is also formed by the backfill after mining. The interaction mechanism between the backfill and the surrounding rock is extremely complex and the failure form shows the characteristics of tensile shear mixing [6]. With the increase in surrounding pressure, the damage mechanism of coal seams in laminated composite coal rocks changes from mixed crack damage to parallel crack damage and finally to single shear crack damage or overall mixed section damage [13,14].

The pre-peak and post-peak deformation and strength behaviors of composite rock mass are closely related to the mechanical properties of the weak rock layer and often play a decisive role; however, the dominant role of this weak structure will gradually decrease with the increase of confining pressure to a certain extent [5,15]. Of course, the control effect of this weak structure and confining pressure is also related to the stress orientation. Through the generalized Poisson's ratio of materials at different locations, the interface confinement effect of composite coal–rock mass was verified. In addition, based on the worst mode, acoustic emission characteristics, the progressive failure mechanism of soft rock–coal combined samples was revealed [15–17]. In addition, the relative change of the contact surface of the coal–rock and the loading orientation of the principal stress will also affect its mechanical behavior. For example, the strength of composite coal–rock tends to be higher in stress perpendicular to the contact surface than parallel to the interface; the rupture is often in the weak coal seam when the stress is parallel to the interface and the opposite may be the overall failure through different rock layers [5,6,18–21]. Furthermore, several studies have investigated the damage, deformation, strength, damage mode, energy evolution, charge induction, acoustic emission, and other characteristics of composite coal–rock bodies under the action of uniaxial, triaxial, cyclic loading, and impact loading stresses [22–27]. While considerable progress has been made in the study of the conventional mechanical behavior of composite coal, there have been few studies on the difference of energy release and transfer between hard rock and coal body during the failure of composite coal, making it necessary to carry out relevant studies. The failure process of rock and coal seams in coal–rock combination is often not synchronized. After the failure of soft coal seams, the release and transfer of elastic energy of hard rock strata aggravate the secondary failure of weak layers, which are prone to a larger disaster. Consequently, it is necessary to carry out relevant research.

In this work, some uniaxial mechanical tests of coal–rock mass were carried out and the whole process AE characteristics of sample failure were monitored. Then, the effective stress and energy release model of composite coal–rock mass was established. On this basis, the internal process and the mechanism of the occurrence of coal–rock compound dynamic disasters were analyzed.

2. Experimental Study of Energy Transfer during Coal–Rock Composite Dynamic Disaster

Currently, the mechanism of energy transfer and the release of coal–rock multilayer composite structure under the influence of deep mining is unknown, limiting the scope of disaster prevention technology. In response to the overall demand of China's coal mining to move deeper and the need for disaster prevention and control of coal mining under complex high-stress environments, focusing on the coupling action mechanism of high elastic energy of deep-mine quarry perimeter rock systems on coal–rock rupture induction mechanism and coal–rock body energy release at the mining working face, exploring the coupling law of gathering, releasing, and transferring high elastic energy of deep-mine perimeter rock–coal body in the process of composite dynamic disaster initiation, and studying to obtain the quantitative discriminant conditions of coal–rock complex dynamic disaster initiation under high ground stress and mining-induced stress in deep mines are necessary. To this end, we conducted a series of mechanical experimental studies on the coal–rock composite and obtained the characteristics of the total stress–strain relation, acoustic emission energy, and elastic energy dissipation law of the coal–rock composite specimens.

As presented in Figure 1, it is assumed that the failure of each stratification and equivalent whole coal–rock mass conforms to the Mol Coulomb criterion. For pure rocks, it is generally believed that the elastic energy stored before peak is greater than the elastic energy released after peak, and it is considered that dynamic disaster occurs. However, for the composite coal–rock mass, the hard sandstone layer has no overall failure and the accumulated elastic energy can bounce back and transfer energy to the coal seam after the failure of the soft coal seam. It can be assumed that the rebound energy of the hard sandstone is greater than the energy required for the fracture of the damaged coal seam per unit post-peak and a disaster will occur.

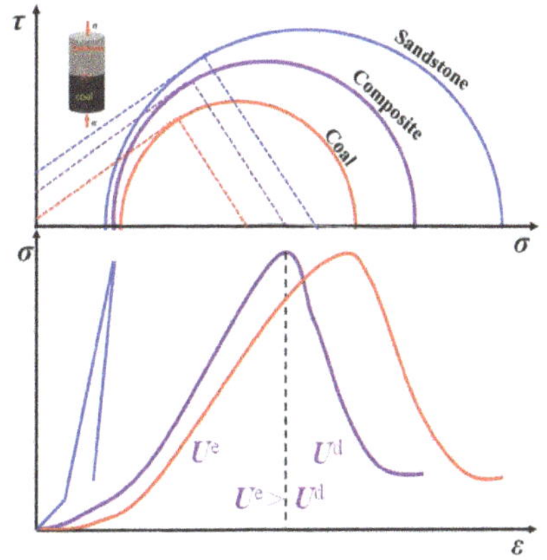

Figure 1. Failure and disaster discrimination model of composite coal–rock mass (U^e is the pre-peak stored elastic energy; U^d is the post-peak dissipation elastic energy).

2.1. Mechanical Characteristics of the Coal–Rock Assemblage

Uniaxial mechanical tests were conducted using raw coal and roof sandstone from a typical mine in Shanxi, and coal–rock assemblage tests with different lithologies and thicknesses were conducted to gain insight into the differences in the layered mechanical behaviors and strength response characteristics of the assembled coal–rock bodies. The height of the coal seam collapse zone of the sandstone roof is 1–2 times that of the coal seam, thus the height ratio of 1:1, 3:2, and 2:1 for coal to sandstone were investigated. The test principle is shown in Figure 2. The experiments are based on PIC2 acoustic emissions (PAC, Princeton Jct, NJ, USA) to monitor their micro-seismic properties.

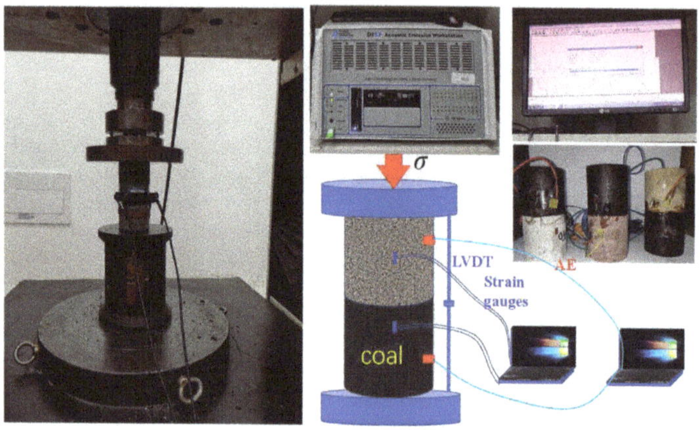

Figure 2. Principle of uniaxial mechanical testing of combined coal bodies.

The strength evolution characteristics of the combined coal bodies of different sandstone types are given in Figure 3. The figure shows that the compressive strength of the fine sandstone assemblage is greater than that of the coarse sandstone assemblage, and the strength of both is lower than that of the siltstone assemblage. In other words, the strength of the siltstone and coal assemblage is greater for the same proportion of the assemblage, which is mainly influenced by the strength of the other components except coal, and the greater the strength of the sandstone, the greater the overall strength of the assemblage. On the contrary, the strength of the composite is close to that of coal. Although it is increased, the range is small. Therefore, it can be judged that the strength of the composite mainly depends on the strength of the weak coal.

This is mainly because the deformation modulus and Poisson's ratio of sandstone and coal are different, resulting in the constraint stress at the interface of the coal–rock mass [5,24]. As shown in Figure 1, the effective cohesion stress and internal friction angle of the combined coal–rock mass are greater than those of a pure coal layer, so its overall equivalent strength would be improved, as presented Figure 3.

The strength law of the assemblage under different coal–rock thickness ratio conditions is given in Figure 4. The figure shows that the compressive strength gradually increases with the climbing rock–coal height ratio. This also indicates that the strength of the assemblage increases to different degrees as the rock–coal height ratio increases. When the rock–coal ratio increased from 1:1 to 3:2 and 2:1, the strength of the coal–rock assemblage increased from 5.745 MPa to 5.985 and 6.225 MPa in that order, which is an increase of 4.176% and 8.355%, respectively. For the coal–rock assemblage, its strength is not only affected by lithology but also the coal–rock thickness ratio exudes some influence.

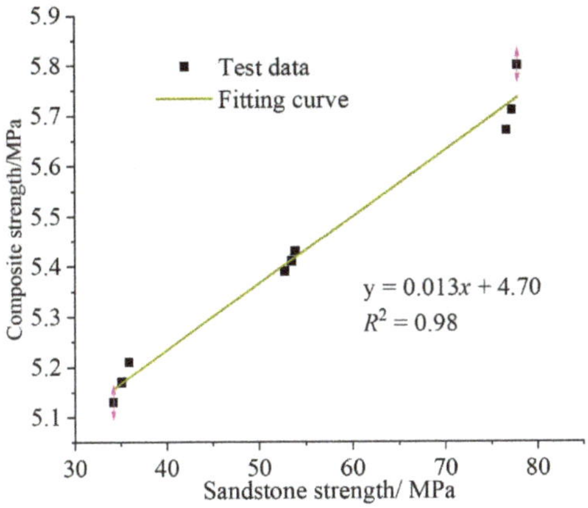

Figure 3. Effect of lithology on the strength of coal–rock assemblages.

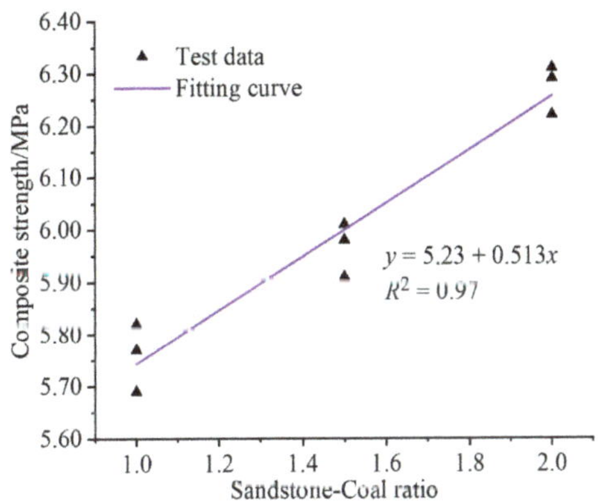

Figure 4. Effect of coal–rock thickness ratio on the strength of the coal–rock assemblage.

The relation between the maximum instantaneous release of acoustic emission energy and the lithology and thickness ratio of the coal–rock assemblage is given in Figure 5, showing that with the increase of the sandstone layer strength, the AE transient energy maximum shows an increasing trend and, with the increase in the rock–coal thickness ratio, the AE energy maximum transient release also shows an increasing trend, which is consistent with the change in the coal–rock assemblage strength. This indicates that the greater the strength of the coal–rock assemblage, the more energy is stored before the peak and the greater the peak instantaneous release energy is, i.e., the greater the chance of a coal–rock dynamic hazard. In actual engineering, due to the influence of mining disturbance, the rock body ruptures and releases energy and the amount of instantaneous released energy directly determines the intensity of the dynamic hazard. In deep coal mining, the probability and energy level of impact disaster will be considerably increased for hard roof and thick roof deposit conditions, which should be prevented and regulated in a timely manner.

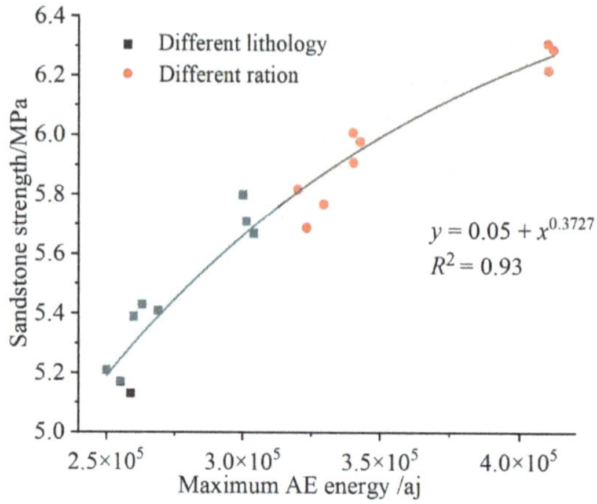

Figure 5. Maximum transient energy versus coal–rock assemblage structure.

2.2. Non–Uniform Deformation Behavior of Coal–Rock Assemblage

In this work, strain gauges were installed in sandstone and coal layers, respectively, to monitor the deformation of each layer and the deformation of the whole assembly was monitored by the axial displacement meter of the instrument. Due to the differences in the mechanical properties of the sandstone and the raw coal stratification in the coal–rock assemblage, there are large differences in the deformation of each stratification under uniaxial stress conditions. Figure 6 gives the deformation relation of sandstone and raw coal layering and the combination of different lithological coal–rock combinations (one of the specimens is selected for analysis for each experimental condition). For different lithological combinations, the deformation of the composite coal–rock body is always between the deformation of sandstone and raw coal fractal, and the deformation of raw coal fractal is much larger than that of sandstone fractal. As the strength of the sandstone layer increases, the deformation of the sandstone fractal is lower and the deformation of the coal fractal is greater simultaneously. In addition, after the peak, the deformation of the original coal fractal and the assemblage continues to increase, while the deformation of the sandstone fractal gradually decreases.

The analysis concluded that for the coal–rock assemblage, the raw coal delamination strain is much larger than the sandstone delamination under the same uniaxial stress because the modulus of elasticity and strength of the raw coal are much lower than that of the sandstone. When the raw coal partition enters the yielding and post-peak stage, the sandstone partition is still in the elastic stage, the post-peak stress decreases, and the strain decreases for the sandstone partition due to elastic rebound, whereas the strain continues to increase for the raw coal due to post-peak fracture surface slip. In addition, for sandstone layering elastic rebound, this effect exacerbates the damage and destruction of the raw coal layering. That is, after the overall destabilization of the coal–rock assemblage specimen occurred, the rupture occurred mainly in the coal stratum, while the elastic deformation energy originally stored in the sandstone stratum occurred due to the overall stress reduction and deformation rebound phenomenon; this deformation acted on the coal stratum and accordingly prompted the development of fractures in the coal.

In the actual deep coal seam mining process, the coal mass in front of the working face was damaged and broken due to mining disturbance, and the elastic energy transfer from the hard roof is easy to trigger the coal seam to produce a large number of fractures, thus increasing permeability of the coal seam, thus increasing the gas outburst risk. In addition, roof elastic energy transfer would also aggravate the fracture of the coal seam, resulting in an instability coal-burst disaster.

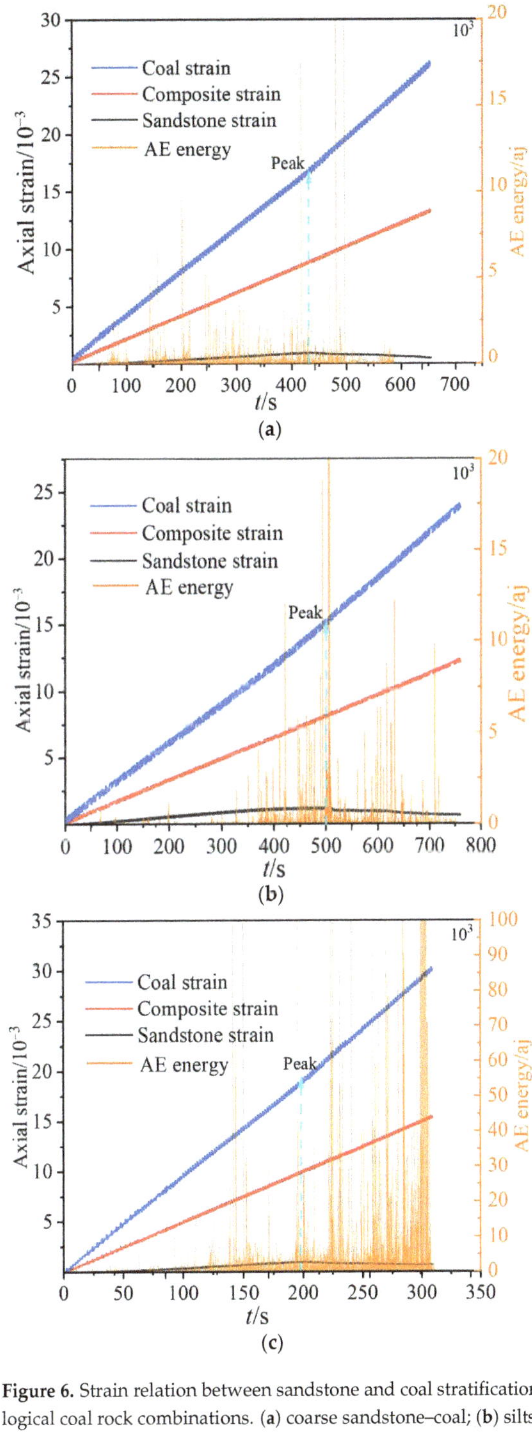

Figure 6. Strain relation between sandstone and coal stratification of specimens with different lithological coal rock combinations. (**a**) coarse sandstone–coal; (**b**) siltstone–coal; (**c**) fine sandstone–coal.

3. Analysis of the Local Non-Uniform Stress Field in the Coal–Rock Assemblage

In fact, during the deformation and damage of the assemblage, the coal body can be considered to be subjected to the 'surrounding pressure' of the rock near the joint surface, whereas the rock body near the joint surface is subjected to radial tension. Of course, the surrounding pressure mentioned here is not the radially uniform surrounding pressure applied to the side of the specimen during the triaxial test. The pressure exerted by the triaxial test can be regarded as the active pressure, while the coal body part is subjected to the 'pressure-supporting effect' only when the coal body part is radially deformed by the axial load on the combined specimen, which is, in a sense, the 'passive pressure-supporting effect'. It is spatially distributed in the coal part near the joint surface and varies with the degree of deformation.

The stresses and deformations of each laminated rock in the composite coal rock are very different from those of the intact individual rocks. To maintain the same deformation of the composite coal–rock samples under triaxial compressive stress, the lateral deformation of the weaker rock is constrained by the stronger rock. Due to the difference in sandstone–coal deformation and the coordinated constraint of interfacial deformation, the sandstone will limit the deformation of coal specimens with smaller lateral deformations and reduce its deformation. Conversely, coal specimens with large lateral deformations will promote the deformation of sandstone with smaller lateral deformations and increase its deformation. This also results in the strength of the composite coal–rock body being higher than that of the coal specimen and lower than that of the sandstone specimen. Based on the stress balance relation, the three principal stresses in each layer should satisfy the following relation [24]:

$$\begin{cases} \sigma_1^R = \sigma_1, \sigma_1^C = \sigma_1 \\ \sigma_2^R = \sigma_2 + \sigma_{2p}^R, \sigma_2^C = \sigma_2 + \sigma_{2p}^C \\ \sigma_3^R = \sigma_3 + \sigma_{3p}^R, \sigma_3^C = \sigma_3 + \sigma_{3p}^C \end{cases} \quad (1)$$

where $\sigma_1^R, \sigma_2^R, \sigma_3^R, \sigma_1^C, \sigma_2^C, \sigma_3^C$ are the three effective principal stresses of rock and coal seam are represented, respectively.$\sigma_{2p}^R, \sigma_{3p}^R, \sigma_{2p}^C, \sigma_{3p}^C$ denote the confining stresses in the directions of σ_2 and σ_3 for the sandstone and coal samples, respectively.

The additional stresses caused by the principal stresses satisfy the following relation [24]:

$$\begin{cases} \sigma_{2p}^R = -\sigma_{2p1}^R + \sigma_{2p2}^R - \sigma_{2p3}^R \\ \sigma_{3p}^R = -\sigma_{3p1}^R - \sigma_{3p2}^R + \sigma_{3p3}^R \\ \sigma_{2p}^C = \sigma_{2p1}^C - \sigma_{2p2}^C + \sigma_{2p3}^C \\ \sigma_{3p}^C = \sigma_{3p1}^C + \sigma_{3p2}^C - \sigma_{3p3}^C \end{cases} \quad (2)$$

where $\sigma_{2p1}^R, \sigma_{2p2}^R, \sigma_{2p3}^R, \sigma_{2p1}^C, \sigma_{2p2}^C, \sigma_{2p3}^C$ are the induced confining stresses caused by three principal stresses in sandstone and coal samples.

At the laminated contact surfaces of coal and rock, the additional stresses between coal and rock are the same due to cementation. Therefore, the additional stresses should satisfy the following relation [24].

$$\begin{cases} \sigma_{2p1}^R = \sigma_{2p1}^C, \sigma_{2p2}^R = \sigma_{2p2}^C, \sigma_{2p3}^R = \sigma_{2p3}^C \\ \sigma_{3p1}^R = \sigma_{3p1}^C, \sigma_{3p2}^R = \sigma_{3p2}^C, \sigma_{3p3}^R = \sigma_{3p3}^C \end{cases} \quad (3)$$

As described in Figure 7, for uniaxial compressive stress states ($\sigma_2 = \sigma_3 = 0$):

$$\begin{cases} \sigma_c^2 = \sigma_c^3 = -\sigma_t^2 = -\sigma_t^3 \\ \sigma_1^R = \sigma_1^C = \sigma \end{cases} \quad (4)$$

According to the generalized Hooke's law,

$$\begin{cases} \varepsilon_2^R = \frac{1}{E_R}\left[\sigma_c^2 - v_R(\sigma_1^R + \sigma_c^3)\right] \\ \varepsilon_3^R = \frac{1}{E_R}\left[\sigma_c^3 - v_R(\sigma_1^R + \sigma_c^2)\right] \\ \varepsilon_2^C = \frac{1}{E_C}\left[-\sigma_t^2 - v_C(\sigma_1^C - \sigma_t^3)\right] \\ \varepsilon_3^C = \frac{1}{E_C}\left[-\sigma_t^3 - v_C(\sigma_1^C - \sigma_t^2)\right] \end{cases} \tag{5}$$

where $\varepsilon_2^R, \varepsilon_3^R, \varepsilon_2^C, \varepsilon_3^C$ represents the strain of the rock and coal seam in the direction of stress σ_2, σ_3, respectively. E_R, E_C, v_R, v_C are the elastic modulus and Poisson's ratio of the rock and coal layer, respectively.

Assuming that both the rock and coal seams are transverse isotropic, the strain satisfied:

$$\begin{cases} \varepsilon_2^R = \varepsilon_2^C \\ \varepsilon_3^R = \varepsilon_3^C \end{cases} \tag{6}$$

Subsequently,

$$\begin{cases} \sigma_1^R = \sigma_1^C = \sigma \\ \sigma_c^2 = \sigma_c^3 = -\sigma_t^2 = -\sigma_t^3 = \frac{E_C v_R - E_R v_C}{E_R(1-v_C)+E_C(1-v_R)}\sigma \end{cases} \tag{7}$$

Based on the above analysis, the stress characteristics of each part of the coal–rock assemblage can be clearly understood and the generation mechanism of local tensile cracks at the coal–rock interface can be better understood.

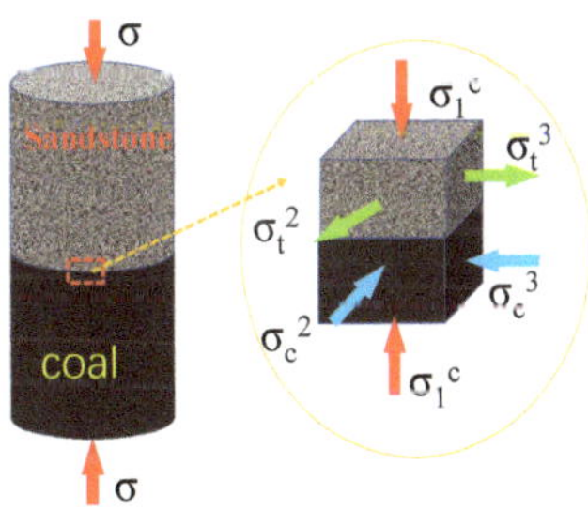

Figure 7. Local stress analysis of the coal–rock assemblage.

4. Energy Nonlinear Evolution and Release Mechanisms of Coal–Rock Assemblages

To study the destabilization and damage mechanism of the coal–rock assemblage under dynamic action and to explore the interaction relation between coal–rock components, a deformation and energy evolution model of coal–rock assemblage was constructed, as shown in Figure 8.

The coal and rock components constitute the coal–rock assemblage, which is kept in mechanical equilibrium under the load σ. The load-displacement curves of the rock component and the coal component can be expressed by Equation (7). The strength of the rock component is stronger than that of the coal component so that destabilization damage did not occur and the stress–strain relation of the coal (c) and the sandstone (s) layer is as follows [11].

$$\begin{cases} \sigma_c = E_C \varepsilon_c \\ \sigma_s = E_S \varepsilon_s \end{cases} \tag{8}$$

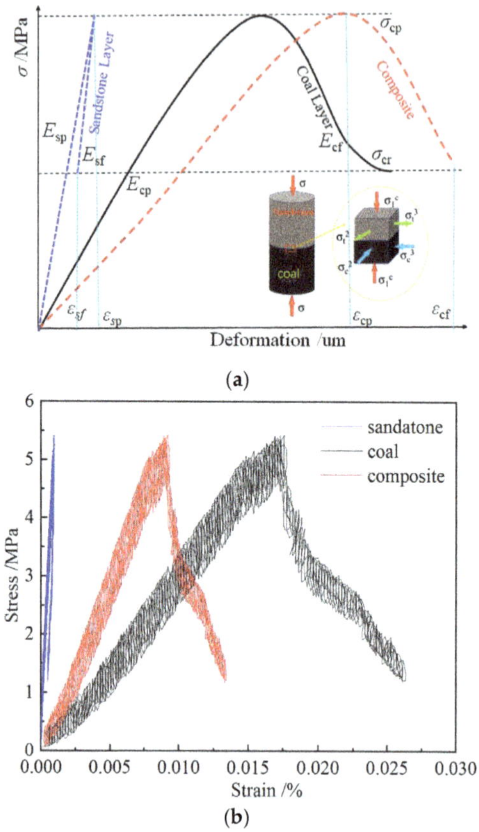

Figure 8. Deformation model of coal–rock assemblage. (**a**) Composite coal–rock deformation diagram; (**b**) strain test results of composite coal–rock mass.

Assuming that the components and assemblies satisfy the law of energy conservation during stress loading and that the temperature field remains constant, the peak front energy field relation should satisfy the following relation [28,29].

$$\begin{cases} U_T = U_c + U_s = \int_0^{\varepsilon_{cp}} \sigma_c d\varepsilon_c + \int_0^{\varepsilon_{sp}} \sigma_s d\varepsilon_s \\ U_T^e = U_c^e + U_s^e = \frac{\sigma^2}{2E_C} + \frac{\sigma^2}{2E_S} \\ U_T^d = U_c^d + U_s^d = U_T - U_T^e \end{cases} \tag{9}$$

where U_T, U_c, U_s are the total energy of composite, coal, and sandstone layer, respectively; U_T^e, U_c^e, U_s^e are the elastic energy of composite, coal, and sandstone layer, respectively; U_T^d, U_c^d, U_s^d are the dissipate energy of composite, coal, and sandstone layer, respectively.

Combined with Figure 8, it can be seen that the energy relation between the peak to post-peak residual phase of raw coal stratification is satisfied.

$$\begin{cases} \Delta U_c^e = U_{cp}^e - U_{cf}^e = \frac{\sigma_{cp}{}^2}{2E_{cp}} - \frac{\sigma_{cf}{}^2}{2E_{cf}} \\ \Delta U^d = U_{cp} - U_{cf} \end{cases} \tag{10}$$

where U^e_{cp}, U^e_{cf} are the elastic energy before and after peak stress; σ_{cp}, σ_{cf} are the peak and residual strength of composite coal–rock; E_{cp}, E_{cf} are the elastic modulus before and after peak stress; ΔU^d and U_{cf} are the elastic energy and total mechanical energy that can be released at the residual strength of the raw coal stratification, respectively.

For the peak to post-peak residual phase of sandstone layer, the energy relation can be obtained:

$$\Delta U^e_s = U^e_{sp} - U^e_{sf} = \frac{\sigma^2}{2E_{sp}} - \frac{\sigma^2}{2E_{sf}} \tag{11}$$

It is assumed that all deformation energy of the elastic rebound of the sandstone partition is transferred to the original coal partition. Based on this, the energy transferred from (DU^{dd}) the sandstone partition to the original coal partition in the post-peak phase is:

$$DU^{dd} = U^e_{cp} - U^e_{cf} = \frac{{\sigma_{cp}}^2}{2E_{sp}} - \frac{{\sigma_{cf}}^2}{2E_{sf}} \tag{12}$$

where E_{sp}, E_{sf} denote the loaded and unloaded modulus of elasticity of sandstone stratification, respectively; σ_{cp}, σ_{cf} denote the peak and residual strength of coal–rock assemblage, respectively.

Figure 9 shows a model of non-linear energy transfer and release rupture of the coal–rock assemblage. The main reason for this is that the assemblage is continuously loaded by the testing machine and, although there is a loss of energy, a considerable amount of energy is accumulated overall. As the energy storage limit of the coal component is low, the weak coal component is the first to break when the accumulated energy reaches the energy storage limit of the assemblage. In this manner, a considerable amount of elastic energy is released and the overall stress is reduced. Owing to the coordinated tensile effect of the rock–coal interface, tiny cracks are formed on the surface of the rock component (which is prone to tensile damage), the stress is reduced after the peak of the assemblage, and the elastic rebound of the sandstone stratum occurs due to stress reduction. Thus, the elastic energy of the sandstone stratum is transferred to the original coal stratum. This leads to the intensified damage of the original coal stratum and is more prone to coal–rock dynamic disaster or secondary dynamic impact disaster.

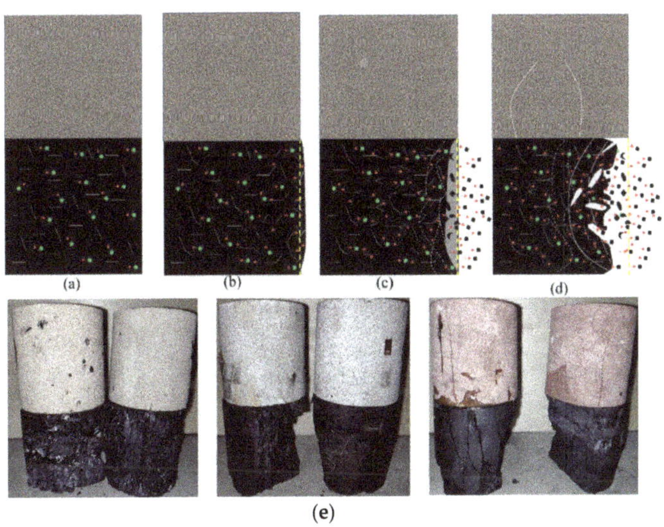

Figure 9. Non-linear energy transfer and release rupture-causing model for coal–rock assemblages. ((**a–d**), Incongruous fracture evolution of composite coal rock mass; (**e**) failure modes of composite coal-rock of the experiments).

Therefore, in the actual process of deep resources development, it is necessary to avoid the sudden rupture of hard roof caused by mining disturbance and reasonable and effective measures should be taken to support the roof or to relieve stress in a timely manner.

5. Conclusions

In order to solve the complicated problem of coupling failure disaster mechanisms of the special structure of hard–roof soft–coal seams in the deep coal mining process, this work carried out experimental and theoretical research and revealed the coupling fracture behavior and the energy transfer law of the coal–rock assemblage. The main conclusions are as follows:

(1) The mechanical behaviors of the combined coal–rock body are different from that of a single coal–rock. Particularly, the overall strength of the combined coal–rock body is higher than that of a single raw coal, which is of great significance to the actual deep coal mine in terms of roof control and surrounding rock support.
(2) The damage of the coal–rock assemblage in the uniaxial stress state occurs mainly in the soft raw coal stratification, and the sandstone part of the hard roof is in the elastic stage throughout the process. The elastic rebound after the overall destabilization transfers energy to the coal seam.
(3) The effective stress relation of each stratum of the combined coal–rock body was obtained based on the deformation coordination relation, and the coal–rock body rupture roof coal seam elastic energy transfer relation was obtained based on this. The hard roof often transferred part of its stored elastic energy to the coal seam after the coal seam was unstable, thus aggravating the coal seam instability.

The research of this work is of great significance for the safe mining of deep hard roof coal seams, disaster prevention and control, and may provide important theoretical reference. The work in this paper is mainly based on the uniaxial compression condition to establish a mathematical model, therefore it is necessary to consider the influence of three-dimensional stress fields in the future, as the failure mechanism and energy transfer law are more complex.

Author Contributions: Conceptualization, H.S.; methodology, H.S., L.D., J.L. and M.L.; validation, H.S., L.D., J.L. and M.L.; formal analysis, J.C.; investigation, L.D. and J.L.; resources, H.S.; data curation, L.D. and J.L.; writing—original draft preparation, H.S., L.D., J.L. and M.L.; writing—review and editing, H.S., L.D., J.L., J.C. and M.L.; visualization, L.D. and J.L.; supervision, H.S. and M.L.; project administration, L.D. and J.C.; funding acquisition, H.S. and L.D. All authors have read and agreed to the published version of the manuscript.

Funding: This research was funded by National Natural Science Foundation of China (No. 51874348, No. 52104239), Chongqing Science Fund for Distinguished Young Scholars (No. cstc2019jcyjjqX0019), Natural Science Foundation of Chongqing (No. CSTB2022NSCQ-MSX1080).

Institutional Review Board Statement: Not applicable.

Informed Consent Statement: Not applicable.

Data Availability Statement: All data and/or models used in the study appear in the submitted article.

Conflicts of Interest: The authors declare no conflict of interest.

References

1. Li, W.; Liu, J.; Zeng, J.; Leong, Y.K.; Elsworth, D.; Tian, J. A fully coupled multidomain and multiphysics model considering stimulation patterns and thermal effects for evaluation of coalbed methane (CBM) extraction. *J. Pet. Sci. Eng.* **2022**, *214*, 110506. [CrossRef]
2. Wang, L.; Cheng, Y.P.; Ge, C.G.; Chen, J.X.; Li, W.; Zhou, H.X.; Wang, H.F. Safety technologies for the excavation of coal and gas outburst-prone coal seams in deep shafts. *Int. J. Rock Mech. Min. Sci.* **2013**, *57*, 24–33. [CrossRef]
3. Du, F.; Wang, K.; Guo, Y.Y.; Wang, G.D.; Wang, L.; Wang, Y.H. The mechanism of rockburst-outburst coupling disaster considering the coal-rock combination: An experiment study. *Geomech. Eng.* **2020**, *22*, 255–264.

4. Dai, L.; Zhang, Z.; Sun, H.; Gao, H. Research on mechanical properties and energy evolution law of coal–rock assemblage with different gas pressures. *Sustainability* **2022**, *14*, 9904. [CrossRef]
5. Lu, J.; Huang, G.; Gao, H.; Li, X.; Zhang, D.M.; Yin, G.Z. Mechanical properties of layered composite coal–rock subjected to true triaxial stress. *Rock Mech. Rock Eng.* **2020**, *53*, 4117–4138. [CrossRef]
6. Cui, B.Q.; Feng, G.R.; Bai, J.W.; Wang, K.; Shi, X.D.; Wu, H.T. Acoustic emission characteristics and damage evolution process of backfilling body–coal pillar–backfilling body composite structure. *Bull. Eng. Geol. Environ.* **2022**, *81*, 300. [CrossRef]
7. Du, F.; Wang, K.; Wang, G.D.; Jiang, Y.F.; Xin, C.P.; Zhang, X. Investigation of the acoustic emission characteristics during deformation and failure of gas-bearing coal-rock combined bodies. *J. Loss Prev. Process Ind.* **2018**, *55*, 253–266. [CrossRef]
8. Liu, Y.; Lu, C.P.; Liu, B.; Xiao, Z.Y.; Zhang, H. Slip and instability mechanisms of coal-rock parting-coal structure (CRCS) under coupled dynamic and static loading. *Energy Sci. Eng.* **2019**, *7*, 2703–2719. [CrossRef]
9. Dou, L.M.; Lu, C.P.; Mu, Z.L.; Zhang, X.T.; Li, Z.H. Rock burst tendency of coal-rock combinations sample. *J. Mine Saf. Eng.* **2006**, *23*, 43–46.
10. Tan, Y.L.; Liu, X.S.; Shen, B.; Ning, J.G.; Gu, Q.H. New approaches to testing and evaluating the impact capability of coal seam with hard roof and/or floor in coal mines. *Geomech. Eng.* **2018**, *14*, 367–376.
11. Liu, X.S.; Tan, Y.L.; Ning, J.G.; Lu, Y.W.; Gu, Q.H. Mechanical properties and damage constitutive model of coal in coal-rock combined body. *Int. J. Rock Mech. Min. Sci.* **2018**, *110*, 140–150. [CrossRef]
12. Tan, Y.L.; Yu, F.H.; Ning, J.G.; Zhao, T.B. Design and construction of entry retaining wall along a gob side under hard roof stratum. *Int. J. Rock Mech. Min. Sci.* **2015**, *77*, 115–121. [CrossRef]
13. Zuo, J.P.; Wang, Z.F.; Zhou, H.W.; Pei, J.L.; Liu, J.F. Failure behavior of a rock-coal-rock combined body with a weak coal interlayer. *Int. J. Min. Sci. Technol.* **2013**, *23*, 907–912. [CrossRef]
14. Zhao, Z.H.; Wang, W.M.; Dai, C.Q.; Yan, J.X. Failure characteristics of three-body model composed of rock and coal with different strength and stiffness. *Trans. Nonferrous Met. Soc. China* **2014**, *24*, 1538–1546. [CrossRef]
15. Song, H.Q.; Zuo, J.P.; Liu, H.Y.; Zuo, S.H. The strength characteristics and progressive failure mechanism of soft rock-coal combination samples with consideration given to interface effects. *Int. J. Rock Mech. Min. Sci.* **2021**, *139*, 104593. [CrossRef]
16. Jiang, Y.L.; Lian, H.J.; Nguyen, V.P.; Liang, W.G. Propagation behavior of hydraulic fracture across the coal-rock interface under different interfacial friction coefficients and a new prediction model. *J. Nat. Gas Sci. Eng.* **2019**, *68*, 102894. [CrossRef]
17. Li, C.J.; Xu, Y.; Zhang, Y.T.; Li, H.L. Study on the energy evolution and fractal characteristics of fracture-like coal rock assemblages under impact loading. *J. Rock Mech. Eng.* **2019**, *38*, 2231–2241.
18. Liu, Y.; Lu, C.P.; Zhang, H.; Wang, H.Y. Numerical investigation of slip and fracture instability mechanism of coal- rock parting-coal structure (CRCS). *J. Struct. Geol.* **2019**, *118*, 265–278. [CrossRef]
19. Bai, J.W.; Feng, G.R.; Wang, Z.H.; Wang, S.Y.; Qi, T.Y.; Wang, P.F. Experimental investigations on the progressive failure characteristics of a sandwiched coal-rock system under uniaxial compression. *Appl. Sci.* **2019**, *9*, 1195. [CrossRef]
20. Wang, K.; Du, F.; Zhang, X.; Wang, L.; Xin, C.P. Mechanical properties and permeability evolution in gas-bearing coal-rock combination body under triaxial conditions. *Environ. Earth Sci.* **2017**, *76*, 815. [CrossRef]
21. Cheng, H.M.; Zhang, N.; Yang, Y.G.; Dong, Y.X.; Peng, W.H. 3D dynamic evolution analysis of coal-rock damaged field and gas seepage field during the gas extraction process. *J. Nat. Gas Sci. Eng.* **2018**, *56*, 444–454. [CrossRef]
22. Li, L.P.; Wu, J.P.; Ju, X.Y.; Wang, L. Analysis of ultra-low friction effect of combined coal rock under the action of surrounding pressure and impact disturbance. *J. Geomech.* **2019**, *25*, 1099–1106.
23. Li, N.; Huang, B.X.; Zhang, X.; Tan, Y.Y.; Li, B.L. Characteristics of microseismic waveforms induced by hydraulic fracturing in coal seam for coal rock dynamic disasters prevention. *Saf. Sci.* **2019**, *115*, 188–198. [CrossRef]
24. Lu, J.; Yin, G.Z.; Deng, B.Z.; Zhang, W.Z.; Li, M.H.; Chai, X.W.; Liu, C.; Liu, Y.B. Permeability characteristics of layered composite coal-rock under true triaxial stress conditions. *J. Nat. Gas Sci. Eng.* **2019**, *66*, 60–76. [CrossRef]
25. Xiao, X.C.; Fan, Y.F.; Wu, D.; Ding, X.; Wang, L.; Zhao, B.Y. Energy dissipation characteristics of combined coal rock damage process and impact hazard evaluation. *Geotech. Mech.* **2019**, *40*, 4203–4212, 4219.
26. Yin, G.Z.; Li, X.; Lu, J.; Song, Z.L. Damage criterion for layered composite rocks under true triaxial stress conditions. *J. Rock Mech. Eng.* **2017**, *36*, 261–269.
27. Zuo, J.P.; Chen, Y.; Cui, F. Differences in mechanical properties and impact propensity analysis of different coal rock assemblages. *J. China Univ. Min. Technol.* **2018**, *47*, 81–87.
28. Lu, J.; Yin, G.Z.; Gao, H.; Li, X.; Zhang, D.M.; Deng, B.Z.; Wu, M.Y.; Li, M.H. True triaxial experimental study of disturbed compound dynamic disaster in deep underground coal mine. *Rock. Mech. Rock. Eng.* **2020**, *53*, 2347–2364. [CrossRef]
29. Xie, H.P.; Lu, J.; Li, C.B.; Li, M.H.; Gao, M.Z. Experimental study on the mechanical and failure behaviors of deep rock subjected to true triaxial stress: A review. *Int. J. Min. Sci. Technol.* **2022**, *32*, 915–950. [CrossRef]

processes

Article

Theoretical Analysis of the Movement Law of Top Coal and Overburden in a Fully Mechanized Top-Coal Caving Face with a Large Mining Height

Li Li [1,2,3,*], Xiao Zhang [1,2,3], Jianqiao Luo [4] and Bin Hu [1,2,3]

1 Coal Mining Research Institute, China Coal Technology & Engineering Group, Beijing 100013, China
2 Tiandi (Yulin) Mining Engineering & Technology Co., Ltd., Yulin 719000, China
3 Tiandi Science and Technology Co., Ltd., Beijing 100013, China
4 School of Energy and Mining Engineering, China University of Mining and Technology (Beijing), Beijing 100083, China
* Correspondence: lilizmkg@163.com

Abstract: In recent years, with the rapid development of equipment manufacturing and advances in the research on mine pressure and rock mechanics, large-mining-height comprehensive mechanized coal mining technology has become increasingly widely used in thick-seam mining. On this basis, because the free space extracted is several times that of layered mining, the occurrence of strong ground pressure caused by the violent movement of the basic roof and the low recovery rate of the top coal have become the main problems. In this study, Dong Liang Coal Mine was taken as the engineering background by which to study the three typical movement forms of the basic roof, and the critical fracture conditions of the three forms were obtained through theoretical calculations. The movement result state of the basic roof strata was taken as the different overlying boundary conditions of the top coal to simulate its impact on the recovery rate, to explore the interaction mechanism between the basic roof strata and the top coal, and for use as a theoretical basis to guide engineering practice to improve energy recovery rates and safety efficiency.

Keywords: fully mechanized caving; large mining height; extra-thick coal seam; cantilever beam; top-coal boundary conditions; finite element numerical simulation

Citation: Li, L.; Zhang, X.; Luo, J.; Hu, B. Theoretical Analysis of the Movement Law of Top Coal and Overburden in a Fully Mechanized Top-Coal Caving Face with a Large Mining Height. *Processes* **2022**, *10*, 2596. https://doi.org/10.3390/pr10122596

Academic Editor: Baisheng Nie

Received: 5 October 2022
Accepted: 22 November 2022
Published: 5 December 2022

1. Background Introduction

Owing to the breakthroughs in and the considerable development of mechanical equipment design and the manufacturing industry, as well as developments in the theory of mine pressure mechanics in recent decades, coal mining equipment is being developed towards a large-scale, heavy, and intelligent design. Thick and extra-thick coal seams that used to rely only on layered mining can now be recovered by using large-mining-height comprehensive mechanized mining or large-mining-height fully mechanized caving mining technology. Compared with the traditional layered mining method, the increase in the mining height has the characteristics of high resource recovery rates and significant reductions in the number of driving and moving operations. Additionally, large-mining-height coal mining technology is also faced with increases in the disturbance range of the roof and floor; the obvious ground pressure appearance and the movement of the roof and the top coal have significantly different movement evolution and interaction relations. Based on the provisions and assumptions of rock fracture mechanics and material mechanics for beam components with one end fixed and one end simply supported, we established a mechanical model of direct roof top coal body boundary interaction, transferred the results of direct roof rock beam movement under the basic roof to the top coal in different areas, and calculated three different types of direct roof top coal interaction relationships: the bench vertical interaction relationship, the rotation action relationship of the short cantilever beam, and the horizontal action relationship of the long cantilever beam.

The movement of the roof structure and the displacement of overburden caused by mining are some of the important factors that lead to the communication of aquifers, the appearance of strong ground pressure in the working face, and the instability of coal pillars in stopes. Since the end of the 20th century, the overlying rock movement caused by mining operation disturbances has mostly used the masonry beam [1–3] theory, and the transmission of rock beam theory [4,5] is often used for calculations and solutions. For the integral calculation principle of bearing pressure and the establishment of the beam model, refer to Brauner, G et al. [6], Barczak T.M. et al. [7], and Goodman R. E. et al. [8]. To date, the above achievements have opened up the research field of roof rock pressure control in the field of coal mining engineering, and great progress has been made on this basis by experts and scholars in China and abroad, providing a solid mechanical theoretical basis for large-scale mining equipment, the simultaneous improvement of the recovery rate of thick coal seams and even extra-thick coal seams, the rapid advancement of ultra-long working faces, and the emergence of intelligent roof control systems.

In the past decade, with the gradual entry of mines into the deep mining stage, the increase in the mining height and the gradual enrichment of the research regarding the mechanism of the short-distance coal seam and the steep coal seam groups, mining engineers have frequently faced extremely complex conditions. The calculation results obtained using the traditional empirical calculation formula and the analytical formula obtained under relatively simple hypothetical conditions are difficult to adapt to on-site phenomena. In theoretical and scientific research, because it is difficult to make adaptions between the solid mechanical boundary and the bulk top-coal boundary, the upper boundary is often conditioned by the free surface in the process of similar simulations and numerical simulations of top-coal caving mining, and the influence of the roof fracture movement and pressure change process on the top-coal fracturing movement release process is often ignored. The traditional numerical simulation method is widely based on the "continuum hypothesis" for research of roof fracture movement. In the Flac^{3D} software (HydroChina—ITASCA R&D Center, Hangzhou, China), which uses the finite element method, the elastic–plastic ideal model, the strain-softening plastic model, and the elastic–plastic damage model take the equivalent stress, plastic zone, or damage zone as the criteria for the position of the roof occurrence span. In addition, numerous scholars use the discrete element PFC software to simplify the coal–rock unit into a rectangular block based on the discrete block hypothesis. Spherical blocks are endowed with mechanical properties such as contact stiffness and surface roughness.

The authors of this paper established a fracture mechanical model of roof overburden direct roof top coal with a large mining height based on field investigations; we also studied the mechanical state of the coal–rock interface boundary in depth, and preliminarily designed a similar simulation experimental device to modify the parameters of the numerical simulation experiment to simulate and study the evolutionary law of roof fractures. This research method is of great significance in improving the recovery rate of top coal, controlling the mine pressure, and improving the support conditions.

2. Engineering Background

2.1. Overview of Working Face

Shenmu Dongliang Coal Mine is located 35 km to the north of Shenmu County, Shaanxi Province. It is located in the area bordering the Loess Plateau and the Maowusu Desert in northern Shaanxi Province. It has the geological characteristics of a thick loess layer load and an aeolian sand permeable layer. The Baotou Shenmu Shuozhou Railway runs through the eastern area of the mine field. The transportation is relatively convenient. According to the borehole disclosure and observation data, the Yan'an Formation coal-bearing rock series is a monoclinal structure inclined to the north or northwest, with a flat dip angle of about 1°, aerial photo of the mine field location is shown in Figure 1:

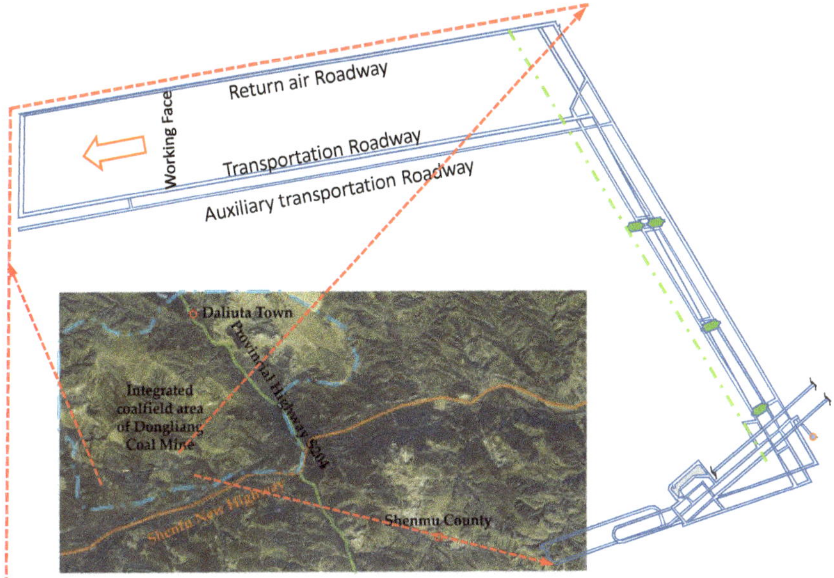

Figure 1. The relationship between the location of the mine field and the stope layout.

The complexity of the structure is relatively simple. The average thickness of the 2-2 coal seam is 12.0 m, the strike longwall retreating mining method is adopted, the roof is managed by the total caving method, and a shield hydraulic support is used to support the roof of the stope. The direct roof of the 2-2 coal seam is fine-grained sandstone, followed by medium and fine sandstone with a thickness of about 4.3 m. The basic roof is 16.5 m medium-grained sandstone. The floor is dominated by siltstone, followed by mudstone with a thickness of about 2.5 m. The dip length of the working face is 303 m. At present, a 20 m wide coal pillar is reserved in this section. The roadway layout plan and the geological histogram of the working face are shown in Figure 2.

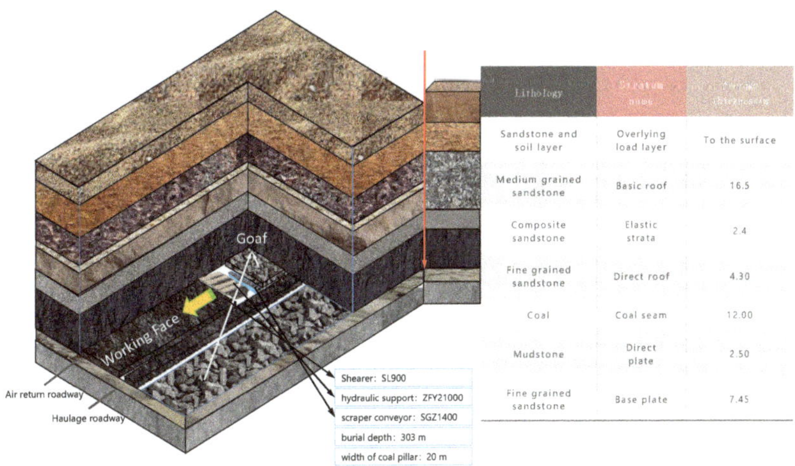

Figure 2. Comprehensive histogram.

2.2. Establishment of a Mechanical Model of Overburden

It can be seen from the comprehensive histogram of the rock stratum shown in Figure 2 that the overburden of the stope is mainly composed of a relatively brittle direct top of fine sandstone and a basic top of thick-, hard-, and medium-grained sandstone. The combined sandstone in the middle is thin and weak in lithology, which mainly serves as a load transfer and elastic foundation.

An elastic medium-thick plate with zero boundary displacement is separated from the basic roof rock of medium-grained sandstone overlying the stope. Its spatial stress component is shown in Figure 3.

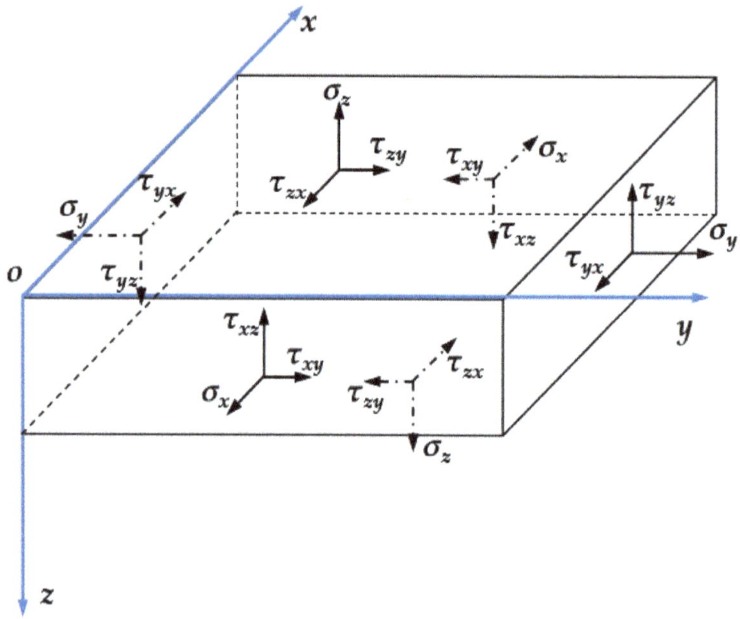

Figure 3. Schematic diagram of the elastic mechanic model for the basic roof.

The balance equation of the space problem is as follows:

$$\begin{cases} \frac{\partial \sigma_x}{\partial x} + \frac{\partial \tau_{yx}}{\partial y} + \frac{\partial \tau_{zx}}{\partial z} + K_x = 0 \\ \frac{\partial \tau_{xy}}{\partial x} + \frac{\partial \sigma_y}{\partial y} + \frac{\partial \tau_{zy}}{\partial z} + K_y = 0 \\ \frac{\partial \tau_{xz}}{\partial x} + \frac{\partial \tau_{yz}}{\partial y} + \frac{\partial \sigma_z}{\partial z} + K_z = 0 \end{cases} . \tag{1}$$

In Equation (1), σ_x, σ_y, and σ_z are the normal stresses in the direction, Mpa; τ_{yx}, τzx, and τyz are the shear stresses in the direction, MPa, and K_x, K_y, and K_z are the volume force per unit volume in the direction.

3. Three Typical Characteristics of Basic Roof Movement on Direct Roofs with a Large Mining Height

3.1. Mechanical Model of a Beam Obtained by Dissecting a Medium-Thick Plate

Because the mining height of fully mechanized mining and fully mechanized caving mining technology with a large mining height is multiple times that of an ordinary mining height, the size and cutting depth of the coal drum are different. The width of the front and rear scraper conveyor, the length of the top beam of the hydraulic support, and the effective support distance are also significantly increased, and the mining space is many times larger than that of general fully mechanized top-coal caving. Therefore, the space dimension of the top coal and the free face under the direct roof is also far larger than that of general fully mechanized top-coal caving and fully mechanized top-coal caving.

In view of the above reasons, since the mining disturbance range of the large-mining-height process gradually increases, and the bearing control effect of the direct roof on the transfer of load and movement results in the overlying strata being relatively weaker than that of the non-large-mining-height process, the basic roof of most large-mining-height working faces plays an absolute control role in the rock pressure behavior of the working face and the caving behavior of the top coal; for example, this occurs in Caojiatan coal mine [9], Tongxin coal mine [10], and Hanglaiwan coal mine [11]. According to microseismic event records and field investigations and analyses, it is considered that there are three typical types of the result of the fault movement of the overlying thick and hard basic roof rock after mining disturbances and its influence on the direct roof and intermediate cushion below: I. bench vertical action; II. short cantilever beam—rotary action type; and III. long cantilever beam—horizontal action type, as shown in Figure 4.

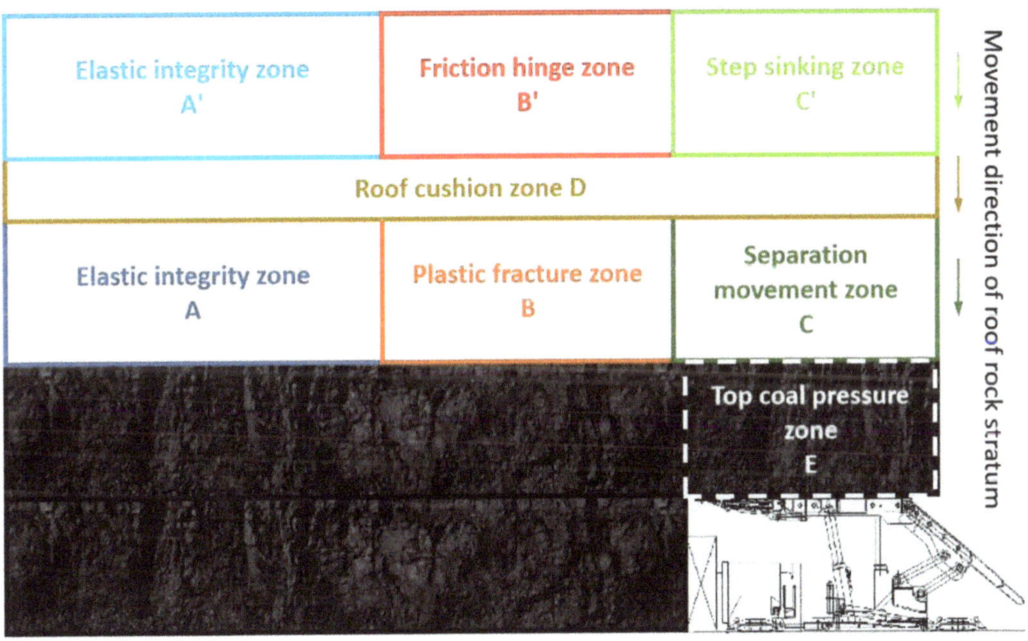

(I) Step vertical action.

Figure 4. *Cont.*

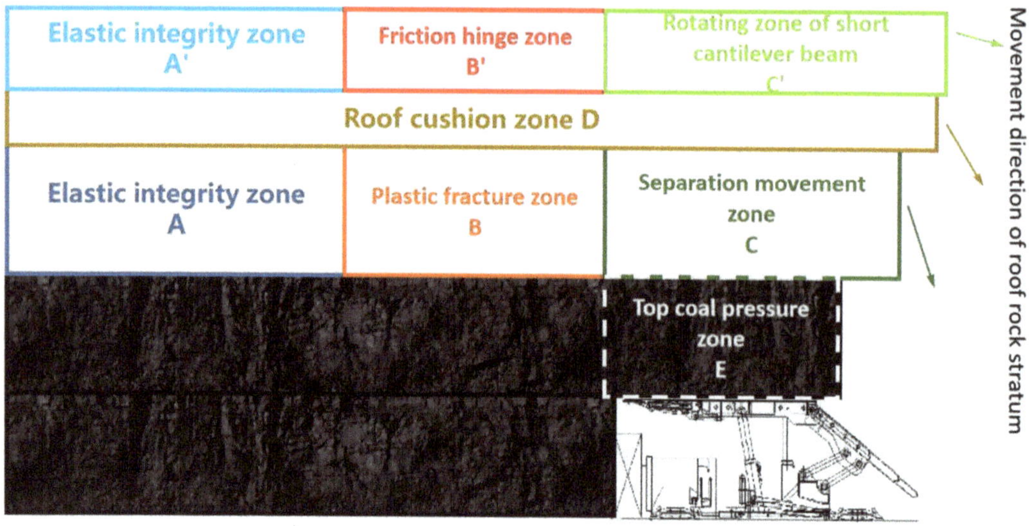

(II) Short cantilever beam—rotary action type.

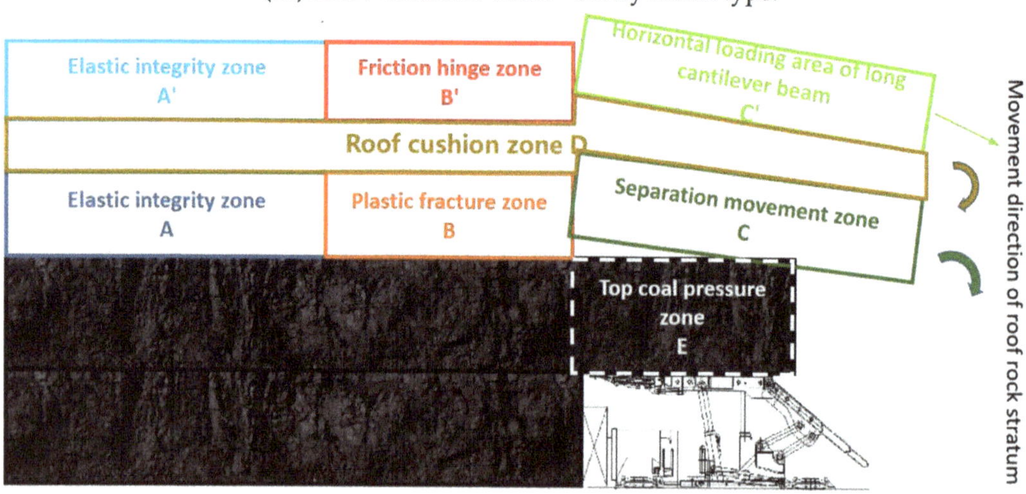

(III) Long cantilever beam—horizontal action type.

Figure 4. Typical state of overlying rock movement in a stope with a large mining height.

Li H's [12] discriminant theoretical formulas for mining overburden failure conditions have been clarified. Because the control effect of the direct roof strata is limited, it is not conducive to supporting working conditions and working face rock pressure control, that is, when the working face is pushed forward from the cutting position, it is assumed that the length of the free surface of the basic roof rock is the same as the pushing distance, and S is the balance judgment formula of the elastic deformation stage of the basic roof rock:

$$S = \frac{-3qJ + [\frac{\sigma_c}{\sigma_t(1-J)} - 1 - J] \cdot \langle\sigma\rangle - \gamma\langle-\sigma\rangle + q}{1-J}, \quad (2)$$

where q is the second root of the ratio of the first and third invariants of the partial stress tensor s; J and the biaxial and uniaxial compressive yield stress σ' are about [$J = (\sigma' - 1)/(2\sigma' - 1)$]; <$\sigma$> is the symbol of the Macaulay matrix operation; σ_t is the tensile stress, MPa, and γ is a parameter related to controlling the cross-sectional shape of the yield surface, according to Li's research results in 2018 [13], taking 2.991 as the value.

The plastic state equilibrium discrimination is combined in Equation (3):

$$G = [(0.1\sigma_t \tan\psi)^2 + q^2] - p \tan\psi, \tag{3}$$

where G is the judgment formula of the plastic deformation equilibrium of the basic roof stratum; ψ is the shear expansion angle, °, and p is the imaginary part of the conjugate complex of the deviator stress tensor.

Based on the mechanical calculation method of calculating the I/II mixed fracture by the first-principal stress synthesis method, only singular terms are retained in the stress components, and the stress components of high-order small terms are ignored. At this time, the criterion expression of the I/II mixed-fracture tensor formula is as follows:

$$\begin{aligned}
\sigma_{rr} &= \frac{1}{2\sqrt{2\pi r}}[K_{\mathrm{I}}(3-\cos\theta)\cdot\cos\tfrac{\theta}{2} + K_{\mathrm{II}}(3\cos\theta - 1)\cdot\sin\tfrac{\theta}{2}] + o(r^{-1/2}) \\
\sigma_{\theta\theta} &= \frac{1}{2\sqrt{2\pi r}}\cos\tfrac{\theta}{2}[K_{\mathrm{I}}(1+\cos\theta) - 3K_{\mathrm{II}}\sin\theta] + o(r^{-1/2}) \\
\tau_{r\theta} &= \frac{1}{2\sqrt{2\pi r}}\cos\tfrac{\theta}{2}[K_{\mathrm{I}}\sin\theta + K_{\mathrm{II}}(3\cos\theta - 1)] + o(r^{-1/2})
\end{aligned} \tag{4}$$

where $K_{\mathrm{I}} = \sigma_y^{\infty}\sqrt{\pi a}$ is the stress intensity factor of a mode I crack; $K_{\mathrm{II}} = \tau^{\infty}\sqrt{\pi a}$ is the stress intensity factor of a mode II crack; r is the polar diameter around the crack in the polar coordinate system, m, and θ is the polar angle of the crack in the polar coordinate system, rad.

In the actual process of coal mining, due to the geological and stress conditions including too many different influencing factors [14] and due to the changes in the overburden conditions, strata, and rock mechanical properties, it is difficult to determine two stopes with the same roof structure and migration patterns in the field measurements. Therefore, in this study, we only abstractly analyzed three kinds of mechanical models that are common in the roof structure of coal mines from the perspective of rock mechanics, and these were combined with the cognitive laws of traditional research paths; on this basis, in further research, we propose the use of similar physical simulation mechanical experiments to simulate the broken movement form of the coal mine roof along the advancing direction of the working face and the shearing tendency of the coal machine.

3.2. The Boundary Conditions of the Experimental Model Are Set in the Simulation of the Excavation Results

With the continuous increase in the normal stress and stress components in the interface internal stress matrix, the possibility of plastic failure also increased. It is further concluded that with the continuous advancement of the working face, the increase in the exposed length under the basic roof will lead to different forms of failure modes and movement characteristics of the basic roof strata with different thickness-to-span ratios. When the thickness-to-span ratio is less than 0.5, when the rock stratum breaks, the masonry beam structure will be formed, that is, a type III long cantilever beam with a horizontal action structure. When the thickness-to-span ratio is greater than 0.5, the rock stratum will break and form a stepped rock beam structure, i.e., a type I stepped vertical action structure. According to the research results of Zuo et al. [15], it can be considered that the thickness-to-span ratio of a basic roof rock layer of 0.5 is the critical transformation point of the fracture form of the rock layer, and the type II short cantilever beam rotary action structure occurs from time to time as a state near the critical point. In this movement form, the deflection of the overlying rock in the vertical direction is large, and the horizontal force of the rotary lap on the top coal during the movement of the gangue fulcrum in the goaf will also promote the arch articulation of the top coal when it is released [16]. The results of these two actions cause the cracks in the top coal to more fully develop, and cannot easily cause large displacement to the goaf direction leading to coal loss [17,18]. With the intervention of sufficient top-coal pre-cracking strategies, arch articulation can be effectively avoided. Therefore, in a fully mechanized top-coal caving face with a large mining height where the caving property of hard top coal is not ideal, the recovery rate of top coal can be improved by reasonably optimizing the roof fracture migration form to move closer to a type II structure.

Figure 5 shows the geometric model established by the working face and roadway layout and the spatial position relationship between the roof and floor of the coal seam in the numerical simulation experiment.

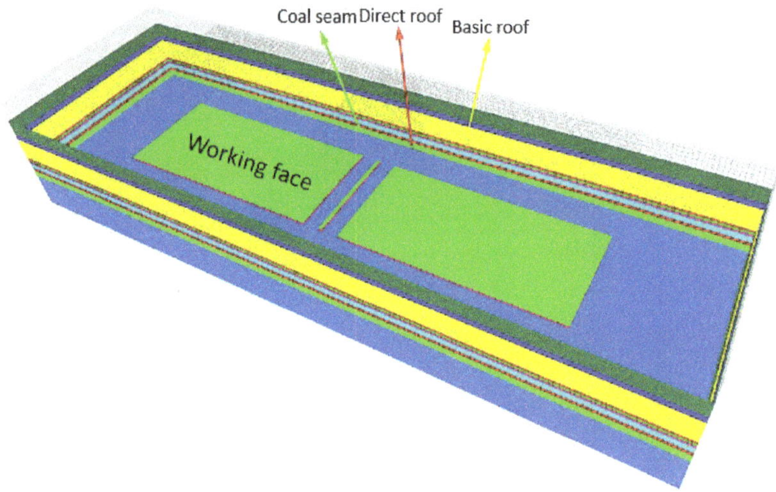

Figure 5. Numerical simulation model of overburden in a stope.

As shown in Figure 6, as the thickness-to-span ratio of the basic roof gradually reached 0.5 from less than 0.5 and then increased to more than 0.5, the peak displacement of the roof in the working face with a large mining height increased from 0.45 m to 0.5 m when the thickness-to-span ratio was 0.5. When the thickness-to-span ratio exceeded 0.5, the peak displacement of the roof reached 0.55 m. In addition, it is worth noting that when the thickness-to-span ratio of the basic roof rock increased, the displacement and range of the floor also increased, which indicates that the different states of loading and movement of the basic roof rock have a structural impact on the whole stope space.

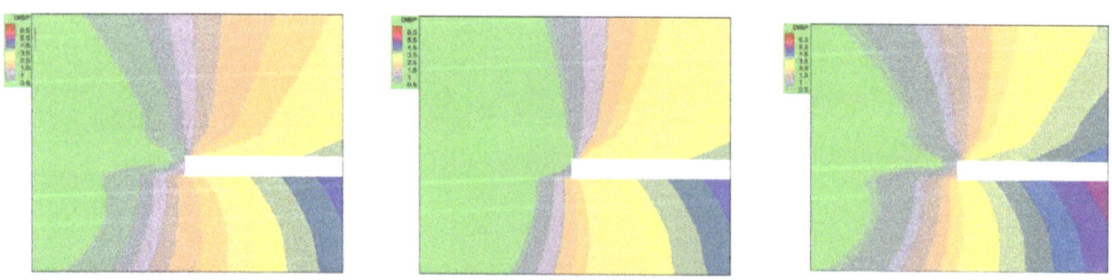

(a) (b) (c)

Figure 6. Evolutionary law of displacement field of surrounding rock with different basic top thickness-to-span ratios in the advancing direction of a stope. (**a**) Thickness-to-span ratio less than 0.5. (**b**) Thickness-to-span ratio of 0.5. (**c**) Thickness-to-span ratio greater than 0.5.

Similarly, from the analysis of the effect of the overlying basic roof thickness-to-span ratio on the stress field of the stope shown in Figure 7, it can be concluded that with the increase in the thickness-to-span ratio, the leading stress peak value and range increased significantly, from 24.8 MPa to 28.62 MPa and, finally, to 32 MPa. The increase in the basic roof thickness of the same lithology led to different modes on the plate failure section of the basic roof stratum. When the rock layer is thin, the failure of the rock plate is mainly due to the normal tensile stress on the interface exceeding the tensile strength of the rock layer under the action of the roof top-coal clamping. When the rock layer is thick, the tensile stress cannot cause the thick and hard basic top rock layer to break. When the shear stress acting in the tangential direction of the breaking section reaches the shear strength of the rock layer at the fixed end edge, the rock layer will be sheared and destroyed and the step will sink. Additionally, the failure of the basic roof rock in the middle scale may only meet the two forms of mixed failure at the same time, that is, there is both vertical pressure and horizontal rotation on the top coal.

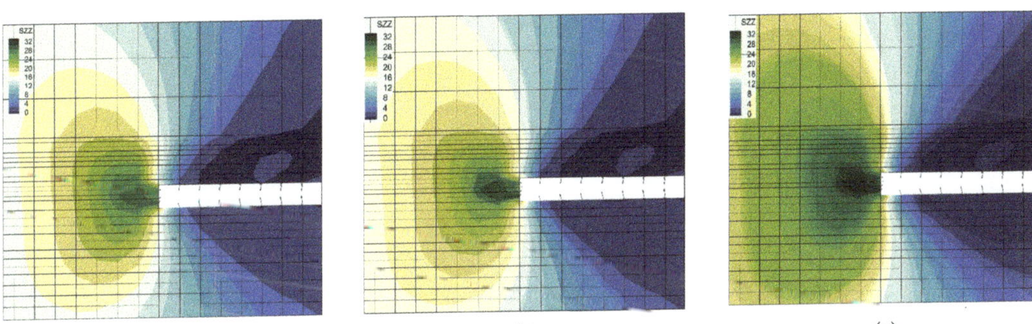

(a) (b) (c)

Figure 7. Evolutionary law of the stress field of surrounding rock with different basic top-coal thickness-to-span ratios in the advancing direction of a stope. (**a**) Thickness-to-span ratio less than 0.5. (**b**) Thickness-to-span ratio of 0.5. (**c**) Thickness-to-span ratio greater than 0.5.

Taking the upper boundary stress conditions of the top coal and the breaking results of the direct roof under the action of the basic roof movement studied above as the upper boundary loads of the coal drawing simulation experiment, it was found that when the bench vertical action occurred when the basic roof thickness-to-span ratio was greater than 0.5, as shown in Figure 8a, the top-coal drawing effect was poor, and there was more coal lost in the goaf. The result of the roof bench sinking was that the top coal fell into the goaf without fully rotating and squeezing. Under the tensile breaking mode with a thickness-to-span ratio less than 0.5, the recovery rate of the top coal was significantly improved, and gangue was observed. However, the interaction between the hinge structure of the roof and the top coal was compressed into an arch, resulting in a surplus of top coal in each round of coal drawing, which could not be released well in the simulation experiment. Obviously, there are more constraints in field applications resulting in some of the top coal being trapped by the hinge structure of the roof and left in the goaf [19]. When the thickness-to-span ratio was equal to about 0.5 in the mixed-fracture mode, the roof was characterized by short cantilever beam rotation. The rotation of the roof gave the top coal a constraint to move to the free surface, and the vertical action stage before the rotation also fractured the top coal to a certain extent, damaged the integrity of the top coal, and made the size of the released coal smaller, the degree of fragmentation higher, and the migration speed faster, so that it could fall onto the rear scraper conveyor at the top of the coal tap in time instead of being left in the goaf, improving the recovery rate of coal resources, as shown in Figure 8c. The typical characteristics of roof rotation cause the top coal to become fully pre-cracked and not obviously hinged in an arch shape during the process of releasing and moving. Only a small amount of boundary coal loss remains behind the goaf. Therefore, when the thickness-to-span ratio reaches about 0.5 and the short cantilever rotation action characteristics appear, these conditions are the most favorable for the recovery of the top coal.

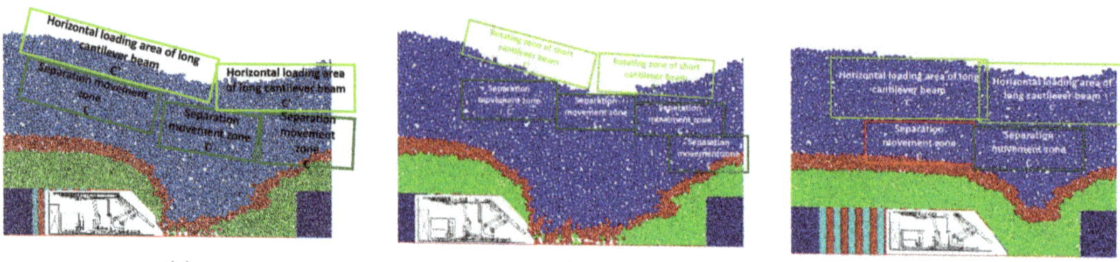

(a) (b) (c)

Figure 8. Evolutionary law of top-coal migration and release under different roof boundary conditions. (**a**) Law of top-coal release and migration of step vertical action. (**b**) Law of top-coal release and migration of short cantilever beam—rotary action type. (**c**) Law of top-coal release and migration of long cantilever beam—horizontal action type.

4. Discussion and Conclusions

4.1. Discussion

Based on the established mechanical model of the top coal–direct roof–follow-up cushion–basic roof structure combined with fracture mechanics and rock mechanics, the coupling boundary conditions of the top coal–roof interface were theoretically deduced, the discrimination formulas of the elastic conditions and plastic conditions were obtained, and the critical conditions were calculated. It is considered that the basic top thickness-to-span ratio has the most significant influence on the different characteristics of rock fracture and movement. Then, numerical simulation experiments were carried out. The evolutionary laws of the stress and strain fields of surrounding rock excavation with different thickness-to-span ratios were analyzed and studied so as to verify the theoretical research results. Finally, the results were substituted into the discrete element simulation software to analyze the influence of the different characteristics of roof failure structure and critical parameters on the difficulty of top-coal caving and the recovery rate.

1. The theoretical formula showed that when the thickness-to-span ratio of the basic roof was less than 0.5, the movement of the basic roof belonged to the long cantilever beam horizontal action mode. When the thickness-to-span ratio of the basic roof reached about 0.5, the movement of the basic roof belonged to the short cantilever beam rotation mode. When the thickness-to-span ratio of the basic roof stratum was greater than 0.5, the basic roof movement belonged to the step vertical action mode. These three different migration and fracture modes are significant for determining the state of direct roofs and top coal.
2. According to the effect and recovery rate of top-coal caving, it can be judged that the bench subsidence of thick and hard basic top rock is not conducive to top-coal caving, which is somewhat different from the traditional on-site experience of weakening the support pressure by roof pre-cracking in rock pressure control. Therefore, when shortening the roof fracture step and reducing the roof pressure intensity, the thick and hard basic top rock should also be cut in the appropriate horizontal direction of the large-mining-height working face. The thickness-to-span ratio should be about 0.5 to prevent the bench vertical action from causing the top coal to accelerate toward the goaf, thereby causing a large amount of coal loss.
3. Under the mining activity of the fully mechanized top-coal caving face with a large mining height, the stress state in the roof rock has a key impact on the roof cutting process. The relationship between the three-dimensional principal stresses directly

determines the dominant direction of fracture development during hydraulic fracturing [20–23]. Therefore, according to the application of the theoretical formula for the ideal thickness-to-span ratio of the basic roof and the boundary conditions of the top coal obtained in this paper, the roof control parameters can be optimized. By improving the geometry of the basic roof and the way in which the rock mechanical properties act on the boundary of the top-coal roof, the purpose of improving the mine pressure and the recovery rate can be achieved.

4.2. Conclusions

1. The fracture state of a basic roof with different thickness-to-span ratios has a significant influence on the recovery rate of top-coal caving.
2. During coal mining, hydraulic fracturing, blasting roof breaking, ground fracturing, and other means should be used to control the thickness-to-span ratio of the basic roof at about 0.5 so as to reduce the occurrence of ground pressure and to improve the recovery rate of coal resources.

Author Contributions: Conceptualization, L.L. and J.L.; methodology, X.Z.; software, J.L.; validation, L.L. and X.Z.; formal analysis, J.L.; investigation, B.H.; resources, L.L.; data curation, L.L.; writing—original draft preparation, L.L.; writing—review and editing, L.L.; visualization, J.L.; supervision, L.L.; project administration, L.L.; funding acquisition, L.L. All authors have read and agreed to the published version of the manuscript.

Funding: This research received no external funding.

Institutional Review Board Statement: The study does not require ethical approval.

Informed Consent Statement: The study did not involve humans.

Data Availability Statement: The experimental and analytical calculation result data used to support the findings of this study are included within the article.

Conflicts of Interest: The authors declare that there is no conflict of interest regarding the publication of this paper.

References

1. Qian, M.G.; Miao, X.X.; He, F.L. Analysis of the key blocks of the "masonry beam" structure in the stope. *Acta Coal Sin.* **1994**, *6*, 557–563.
2. Miao, X.X.; Qian, M.G. The overall structure of surrounding rock in the stope and the mechanical model of masonry beams. *Mine Press. Roof Manag.* **1995**, *3*, 197.
3. Qian, M.G.; Zhu, D.R.; Wang, Z.T. Fracture type of the main roof rock and its influence on the pressure on the working face. *J. China Acad. Min. Technol.* **1986**, *2*, 12–21.
4. Song, Z.Q.; Liu, Y.X.; Chen, M.B.; Song, Y. Discussion on bearing pressure behavior before and after rock beam fracture and its application. *J. Shandong Inst. Min. Technol.* **1984**, *1*, 27–39. [CrossRef]
5. Song, Z.Q.; Song, Y.; Jiang, Y.J.; Liu, Y.X. Methods and steps of roof control design in mining face. *Coal Sci. Technol.* **1989**, *9*, 6–63. [CrossRef]
6. Bukhoino, G. *Rock Pressure and Rockburst*; Coal Industry Press: Beijing, China, 1985.
7. Barczak, T.M.; Oyler, D.C. *A Model of Shield Strata Interaction and Its Implications for Active Shield Setting Requirements Missing*; US Department of the Interior, Bureau of Mines: Washington, DC, USA, 1991.
8. Goodman, R.E.; Shi, G.H. *Block Theory and Its Application to Rock Engineering*; Prentice-Hall: Hoboken, NJ, USA, 1985.
9. Xu, X.D. *Study on the Mechanism of Strong Ground Pressure and Pressure Relief Control along the Goaf in Caojiatan Coal Mine's Extra Thick Coal Seam*; Jilin University: Changchun, China, 2022. [CrossRef]
10. Chen, Y. *Study on the Mining Floor Deformation and Failure Depth of Extra Thick Coal Seam in Datong Mining Area*; China University of Mining and Technology: Beijing, China, 2021. [CrossRef]
11. Dong, Z.Y.; Wang, S.M. Analysis of the impact of coal mining on groundwater resources in Yuxi River Basin in Northern Shaanxi—Taking hanglaiwan coal mining area as an example. *Resour. Environ. Arid Areas* **2017**, *31*, 185–190. [CrossRef]
12. Li, H.; Tang, S.B.; Ma, L.Q.; Bai, H.B.; Kang, Z.Q.; Wu, P.F.; Miao, X.C. Study on continuous discrete coupling simulation of progressive failure of mining overburden. *J. Rock Mech. Eng.* **2022**, 1–12. [CrossRef]
13. Li, H.; Bai, H.B.; Wu, J.J.; Li, Z.Y.; Meng, Q.B.; Guo, J.Q.; Zhu, D.F.; Xiao, M.D.-P. Random damage constitutive model and its application in preventing water inrush of collapse columns. *Geotech. Mech.* **2018**, *39*, 4577–4587. [CrossRef]

14. Zuo, J.P.; Yu, M.L.; Sun, Y.J.; Wu, G.S. Analysis of fracture mode transformation mechanism and mechanical model of rock layers with different thicknesses. *Acta Coal Sin.* **2022**, 1–18. [CrossRef]
15. Luo, J.; Yan, S.; Yang, T.; Mu, H.; Wei, W.; Shen, Y.; Mu, H. *Mechanism of Hydraulic Fracturing Cutting Hard Basic Roof to Prevent Rockburst*; Hindawi Limited: London, UK, 2021.
16. Hubbert, M.K.; Willis, D. Mechanics of hydraulic fracturing. *Trans. AIME* **1972**, *210*, 153–168. [CrossRef]
17. Zhang, P.; Liang, X.; Xian, C.; Liu, B.; Wang, W.; Zhang, C. Geomechanics simulation of stress regime change in hydraulic fracturing: A case study. *Geomech. Geophys. Geo-Energy Geo-Resour.* **2022**, *8*, 1–18. [CrossRef]
18. Murthy, M.V.V.S.; Renji, K.; Gopalakrishnan, S. Multi-transform based spectral element to include first order shear deformation in plates. *Int. J. Mech. Sci.* **2015**, *96–97*, 110–120. [CrossRef]
19. Guy, R.; Kent, M.; Russell, F. An assessment of coal pillar system stability criteria based on a mechanistic evaluation of the interaction between coal pillars and the overburden. *Int. J. Min. Sci. Technol.* **2017**, *27*, 9–15.
20. Lv, K.; He, F.; Xu, X.; Zhai, W.; Qin, B. Irregular load—Fracture characteristics and stability analysis of elastic foundation basic roof structure. *J. Rock Mech. Eng.* **2022**, 1–18. [CrossRef]
21. Xu, X.; He, F.; Lv, K.; Wang, D.; Li, l.; Zhai, W. Study on reasonable roof cutting parameters of thick hard roof cutting and retaining roadway. *J. Coal Ind.* **2022**, 1–12. [CrossRef]
22. Li, Y.; Ren, Y.; Wang, N.; Jin, N.; Ou, X.; Luo, J.; Mei, C. Morphology and evolution characteristics of caving roof in goaf. *J. Coal Ind.* **2021**, *46*, 3771–3780. [CrossRef]
23. Chen, D.; Wu, Y.; Xie, S.; Guo, F.; He, F.; Liu, R. Study on the initial fracture of basic roof slab at the boundary of long side goaf and elastic-plastic softening foundation. *J. Coal Ind.* **2022**, *47*, 1473–1489. [CrossRef]

processes

MDPI

Article

Experimental Study on the Effect of Air-Doors Control Adjacent to the Fire Source on the Characteristics of Smoke Back-Layering

Haiyan Wang, Zuohui Xu *, Lei Wang, Cheng Fan and Yanwei Zhang

Beijing Key Laboratory for Precise Mining of Intergrown Energy and Resources, China University of Mining and Technology (Beijing), Beijing 100083, China

* Correspondence: xuzuohia@163.com

Abstract: Air-doors are important facilities for regulating the air flow in a mine ventilation network. It is of value to study the influence of air-doors, which are adjacent to a fire source on smoke back-layering in order to build a rational ventilation system. By regulating air-doors in a mine ventilation network test platform, two typical mine ventilation networks, with parallel branches and a diagonal branch, were constructed. During the study, into the closing degree of the air-doors adjacent to a fire source in a ventilation network with parallel branches, the back-layering length is up to 3.70 m when the ventilation velocity is 1.40 m/s. When the air-door on the return side of the adjacent branch is closed, the back-layering subsides within 1 min and the upstream temperature drops rapidly to normal. When the air-door is half closed, there is still a back-layering flow within 5 min. Smoke control, with the air-door is closed, is better than when the air-door is half closed. Based on this, tests into the influence of the closing position of air-doors, which are adjacent to a fire source, were carried out in a ventilation network with a diagonal branch. Results indicate that when the ventilation velocity is 1.70 m/s, the back-layering flow spreads to the diagonal branch, and the air flow velocity of both the adjacent branch and the diagonal branch increases. When closing the air-door on the return side of the adjacent branch, the back-layering rapidly subsided. The wind velocity on the intake side of the adjacent branch is stabilized after a rapid decrease, and the wind velocity of the diagonal branch is stabilized after a rapid increase. When closing the air-door on the intake side of the adjacent branch, the smoke from the diagonal branch spreads. Compared with closing the intake side air-door, closing the air-door on the return side of the adjacent branch is more effective in preventing back-layering. This work provides a reference for preventing back-layering and guiding the evacuation of people from the upstream of a fire source.

Keywords: mine fire; smoke back-layering; air-door adjacent to the fire source; smoke control

Citation: Wang, H.; Xu, Z.; Wang, L.; Fan, C.; Zhang, Y. Experimental Study on the Effect of Air-Doors Control Adjacent to the Fire Source on the Characteristics of Smoke Back-Layering. *Processes* **2022**, *10*, 2496. https://doi.org/10.3390/pr10122496

Academic Editors: Feng Du, Aitao Zhou and Bo Li

Received: 8 November 2022
Accepted: 22 November 2022
Published: 24 November 2022

1. Introduction

Mine fires are one of the main disasters that threatens the safety of coal production [1–3]. A large mass of toxic and hazardous gases, at high temperatures, are produced during the process of a fire, and due to the limited working space in coal mines and the complex ventilation system, this will quickly spread to the whole roadway [4,5]. There exists a certain number of inclined tunnels [6], which when a fire occurs results in the ventilation system being prone to a disorder under the action of a stack effect that is manifested by the reversal of an air flow in the side branch and the back-layering flow of smoke in the main branch; this may further enlarge the disaster area when spreading to other roadways on the upstream of the fire source [7–9]. To prevent people being exposed to toxic and hazardous gases during evacuation, the smoke on the upstream of the fire source needs to be controlled. In a mine ventilation network, the theoretical basis for controlling the back-layering of a fire source branch is to make the wind velocity flowing into the fire branch greater than the critical velocity. Combined with the air volume balance law, the air flow in the adjacent branch of the fire source can be regulated so that the wind velocity flowing into the branch, containing the fire source, is close to the critical velocity in order to prevent back-layering in roadways.

Scholars home and abroad have conducted a lot of research into critical velocity in order to control a back-layering, using the aspects of heat release rates, roadway cross-section and roadway slopes [10]. Thomas [11] first introduced the concept of critical velocity, which is the minimum longitudinal ventilation velocity that can effectively control the smoke in the downwind of the fire source without backflow and gave a semi-empirical formula for the critical velocity and the heat release rate. On the basis of Thomas's research, Hinkley, Heselden, Danziger and Kennedy studied critical velocity and proposed an empirical correlation, similar to Thomas's critical velocity model [12]. Oka and Atkinson [13] studied experimentally that at low heat release rates, the critical velocity is positively related to the cube root of the heat release rate; additionally, at higher heat release rates, the critical velocity is independent of the heat release rate. Wu and Bakar [14] studied the critical velocity of five horizontal roadways with the same height, but different cross-sections, using experimental and numerical simulations, and they proposed the use of the "hydraulic diameter" instead of the roadway height as the characteristic length in the dimensional analysis of the critical velocity model developed by Oka and Atkinson. Lee and Ryou [15] found a positive correlation between the critical velocity and the height-width ratio of a roadway. In an inclined roadway, a grade correction factor kg, which is defined as the ratio of the critical velocity in an inclined roadway to that in a horizontal roadway under the same conditions, is commonly used to quantify the effect of the roadway slope on the critical velocity. Based on the experimental results, Yi [16] obtained a critical velocity relationship for a roadway slope of $-2°$~$2°$. Chow et al. [17] established the empirical formula for the critical velocity of a horizontal roadway, and experimentally studied the slope correction coefficient of the model, when the slope is at $0°$~$9°$. Atkinson and Wu [18] experimentally obtained the slope correction factors for Wu and Bakar's critical velocity model for slope variations in the range of $0°$~$10°$. Li [19] used a combination of experimental and numerical simulations to derive the critical velocity in the slope range of $2°$~$6°$. The research results are consistent with Atkinson and Wu's model. Based on numerical simulations, Hao et al. [20] obtained a critical velocity formula similar to Wu and Bakar's study, as well as the slope correction coefficient when the slope is within $0°$~$30°$.

In mines, the ventilation system is usually controlled by blocking the air flow through the air-door [21–23]. For the first time, Xu et al. [24] obtained the different flow field distributions of mine ventilation systems, due to the opening and closing of an air-door, through numerical analysis and tracer gas field tests. Based on the theory of the simultaneous exploitation of coal resources and gas in high-gas coal seams in an enclosed space, Fu et al. [25] established a new model of a non-ventilated excavation system, which prevents oxygen from flowing into the excavation roadway, by controlling an air-door and with a nitrogen curtain installed in the side roadway. Wang et al. [26] numerically simulated the distribution of smoke parameters in a roadway, before and after the use of air-doors, to regulate the air flow in each branch of the ventilation network after a belt fire in the west-wing mining area of Long-Dong coal mine. According to the previous test results, Wang et al. [27] selected the conventional fire parameters to calculate the air resistance in the fire zone of the belt roadway in the west-wing mining area, and the adjustment air resistance under four kinds of air door openings, but the calculation results were lacking experimental verification. Shang et al. [28] proposed an algorithm for identifying an air-door opening and closing using a single wind-velocity sensor. Wang et al. [29] designed an overpressure relief air door with a magnetic lock, which automatically resets after the shock wave passes to isolate flames and smoke flows.

Due to the limitation of test equipment, the current research on the smoke back-layering control is mostly based on the theoretical studies of critical velocity in horizontal and inclined roadways. The experimental research on how to control the back-layering through an air-door in a mine ventilation network is basically blank. In this paper, a test platform of a mine ventilation network is developed, which contains multiple air doors with an adjustable closing degree. Based on this test platform, the influence of air-doors adjacent to the fire source on the smoke back-layering in a mine ventilation network is studied experimentally.

2. Experimental Setup and Methods

This study establishes a mine ventilation network test platform with a 1:14 ratio in relation to a physical roadway, using the physical roadway of China Kailuan Mine Emergency Rescue Base as the prototype. The mine ventilation network test platform consists of five parts: roadway body, control system, detection system, fire source simulation system, and exhaust gas treatment system. The main body of the roadway consists of ten rectangular roadways with a cross-section of 30 × 30 cm. The control system includes air-doors, inverter fans, and control cabinets. The air-doors are set at the junction of two lanes and there are six sets of adjustable air-doors. The detection system includes temperature sensor, wind velocity transmitter, gas detector (O_2, CO, CO_2), and a paperless recorder. The simulation system includes oil pan, air supply pipe, and oil storage tank, as shown in Figure 1.

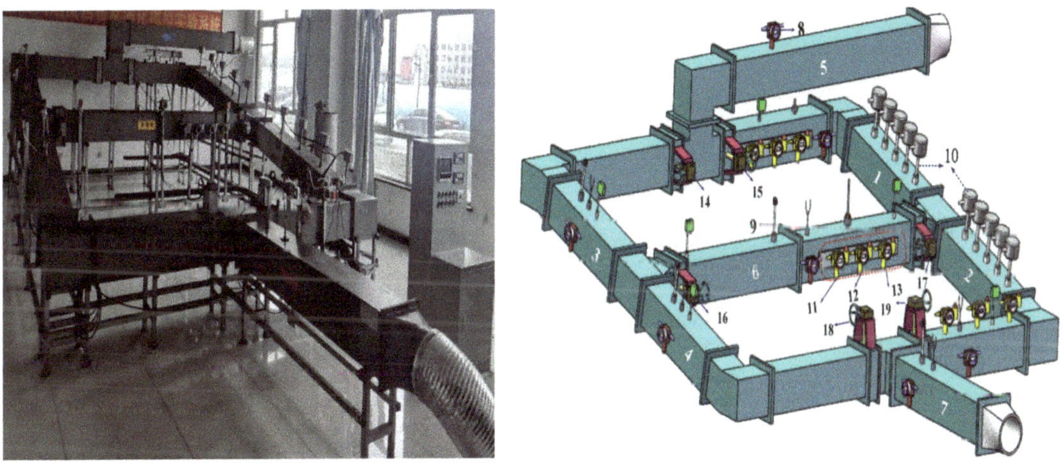

(**a**) (**b**)

Figure 1. Experimental equipment: (**a**) Physical setup. (**b**) Schematic diagram. Note: 1. branch 1; 2. branch 2; 3. branch 3; 4. branch 4; 5. intake airway; 6. diagonal branch; 7. return airway, 8. anemometer; 9. temperature sensor; 10. temperature string sensor; 11. carbon dioxide detector; 12. carbon monoxide detector; 13. oxygen detector; 14.air-door 1; 15. air-door 2; 16. air-door 3; 17. air-door 4; 18. air-door 5; 19. air-door 6.

Branches 1, 2, 3, and 4 are all 2.5 m long, with an inclination angle of 12°, and the diagonal branch is 3.4 m long. Ten pieces of fireproof glass are continuously set on one side of branches 1 and 2 to observe the experimental phenomenon. Downward press-in ventilation is adopted. The fan adopts a SE-A250H induction frequency axial fan with air volume ranging from 0 to 1800 m^3/h. Temperature sensors are set up in the diagonal branch, branch 3 and branch 4. K-type sheathed thermocouples with a diameter of 5.0 mm are installed to measure the gas temperature beneath the ceiling of up to 1000 °C. There are 10 groups (three temperature sensors in each group, 2 cm, 10.5 cm, and 19 cm from the top plate; the group spacing is about 45 cm) of temperature string sensors along branches 1 and 2. A KV621P two-way wind velocity transmitter is equipped in each branch to measure wind velocity, whose measurement range is $-5 \sim 5$ m/s, the measurement accuracy is 3% F·S, and the display resolution is 0.01. Gas detectors are provided in the inlet section of branch 1, the exhaust section of branch 2, and the diagonal branch. The gas detector used is an SGA-500B fixed gas sensor, which has a detection range of carbon monoxide of 0~5000 ppm, the detection range of carbon dioxide is 0~20,000 ppm, and the detection range of oxygen is 0~30% VOL.

The oil pan is set at 3.8 cm at the entrance of branch 1 with a diameter of 10 cm, a height of 5 cm near the downwind side and 2 cm near the upwind side, which is fueled by kerosene, supplied by an external oil pump during the experiment, and ignited by an electric spark ignition. The heat release rate is an important characteristic parameter of the fire source, which can be estimated according to the principle of oxygen consumption; when combustible combustion is oxygen-rich, the heat release rate of the fire source can be calculated by Equation (1) [30].

$$\dot{Q} = \left(X_0\dot{V}_0 - X_S\dot{V}_S\right)\rho E \tag{1}$$

where: X_0 is the volume fraction of oxygen entering the system (%); X_S is the volume fraction of oxygen in the smoke exhaust duct (%); $\dot{V}_0$ is the volume flow rate of air entering the system before the fire (m^3/s); $\dot{V}_S$ is the volume flow rate of gas in the exhaust duct before the fire (m^3/s); ρ is the density of oxygen in the initial environment (kg/m^3); E is the heat generated by the conversion of carbon into carbon dioxide, consuming a unit mass of oxygen, taken as 13.1 MJ/kg.

As shown in Figure 2, in the above experimental system, a ventilation network with parallel branches can be constructed by closing air-doors 3 and 4 located in the diagonal branch; a simple diagonal ventilation network can be constructed by keeping the air-doors fully open. In order to study the influence of the closing degree of the air-doors, which are adjacent to the fire source on the back-layering, two test cases are set up when the back-layering occurs in the ventilation network with parallel branches at three ventilation velocities: closing air-door 5 on the return side of the branch, which is adjacent to the fire source and half closing air-door 5. According to the test results, the influence of the position of the air-doors, which are adjacent to the fire source on the back-layering is studied in the ventilation network with a diagonal branch, and two cases are set up to regulate air-door 1 on the inlet side of the branch, adjacent to the fire source, and air-door 5 on the return side of the branch, adjacent to the fire source. The stable stage of a fire is selected for smoke control since the fire source releases the highest heat with the farthest back-layering length and the greatest fire hazard. To reduce the effect of the local vortex, the wind velocity values in the experiment are taken as the average value within 1 min. The experimental ambient temperature is about 20–25 °C. The experimental working conditions are shown in Tables 1 and 2.

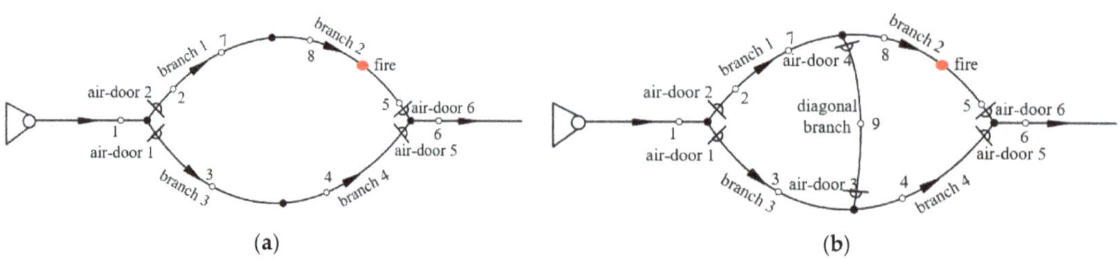

Figure 2. Schematic diagram of the experimental system after the air-doors regulation: (**a**) Ventilation network with parallel branches. (**b**) Ventilation network with a diagonal branch. Note: 1. Velocity, temperature; 2. Velocity, carbon dioxide, carbon monoxide, oxygen; 3. Velocity, temperature; 4. Velocity, temperature; 5. Velocity, carbon dioxide, carbon monoxide, oxygen; 6. Velocity, temperature; 7, 8. Temperature string; 9. Velocity, temperature, carbon dioxide, carbon monoxide, oxygen.

Table 1. Ventilation network with parallel branches scheme setting.

Case	Smoke Control Method	Ventilation Velocity v/(m/s)
Case 1	Air-door 5 closed	1.70
	Air-door 5 closed	1.52
	Air-door 5 closed	1.40
Case 2	Air-door 5 half closed	1.70
	Air-door 5 half closed	1.52
	Air-door 5 half closed	1.40

Table 2. Ventilation network with a diagonal branch scheme setting.

Case	Smoke Control Method	Ventilation Velocity v/(m/s)
Case 1	Air-door 1 closed	1.90
	Air-door 1 closed	1.79
	Air-door 1 closed	1.70
Case 2	Air-door 5 closed	1.70
	Air-door 5 closed	1.56

3. Results and Discussion

3.1. Effect of Air-Door Closing Degree on the Back-Layering Flow

3.1.1. Smoke Back-Layering Length

Figure 3 presents the smoke back-layering length for the different experimental cases when the fire is stable in the ventilation network with the parallel branches without smoke control measures. If a fire occurs in an inclined roadway, the stack effect drives the hot smoke, produced by the burning, to propagate upstream. At the position where the driving force generated by the thermal buoyancy of the countercurrent gas is equal to the force of the longitudinal ventilation, the upstream back-layering front stops propagating. The distance between this stagnation position and the fire source is taken as the back-layering length. Figure 3 shows that when the wind velocity is constant, the greater the heat release rate, the farther the back-layering flow. When the ventilation velocity is 1.40 m/s and the heat release rate is 15.4 kW, the back-layering length is 2.82 m, while when the ventilation velocity is 1.40 m/s and the heat release rate is 28.0 kW, the back-layering length reaches 3.70 m, which increases the hazard to the upstream of the fire source.

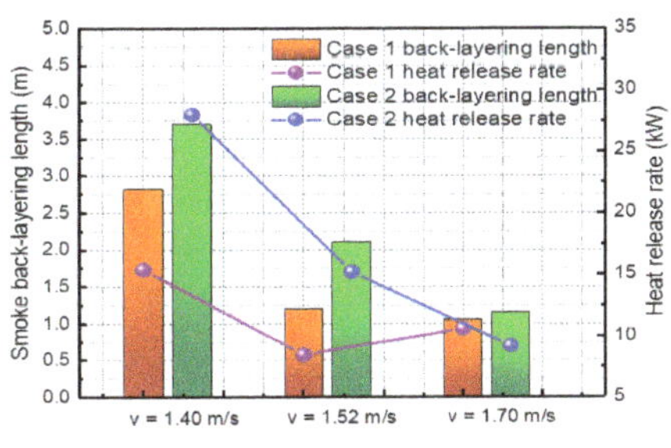

Figure 3. Back-layering length in the ventilation network with parallel branches.

For the fire under the different ventilation velocities, when closing air-door 5, it is observed through the glass window that the smoke back-layering flow on the upstream of the fire source subsides within 1 min, and the back-layering length is 0. This is because. after closing air-door 5, the full flow of the air generated by the ventilator is directed to branch 2, where the air volume and the force of the longitudinal ventilation increases. The force of the longitudinal ventilation is greater than the driving force generated by the thermal buoyancy, and therefore the back-layering is prevented. When air-door 5 is half closed, the ventilation velocity is 1.40 m/s and the heat release rate is 28.0 kW, after 5 min of smoke control, the force of the longitudinal ventilation balances the driving force generated by the thermal buoyancy and therefore the back-layering is prevented.

3.1.2. Wind Velocity Variation in Adjacent Branch

The variation in the wind velocity in adjacent branch 3 before and after the smoke control are presented in Figure 4. As can be seen from Figure 4, the wind velocity in branch 3 gradually increases with the development of the fire at different ventilation velocities. In the experimental system, branches 1, 2, 3, and 4 are all inclined roadways. The experiment adopts a downward ventilation, and the buoyancy effect and throttling effect, generated by the fire, are opposite to the direction of the mechanical air supply, which hinders the air flow, so the air volume flowing into branch 2 decreases. As branch 3 and branch 2 are parallel roadways, and the external air supply and cross-section are unchanged, the wind velocity of branch 3 increases. Under different ventilation velocities, both the throttling effect and the buoyancy effect of the fire are different, and the increase in wind velocity in branch 3 is also different. When the ventilation velocity is 1.40 m/s and the heat release rate is 28.3 kW, the back-layering length is the farthest, which is 3.70 m. The air flow velocity of branch 3 increases from 0.68 m/s to 0.94 m/s, an increase of approximately 38.2%, which is greater than in other cases.

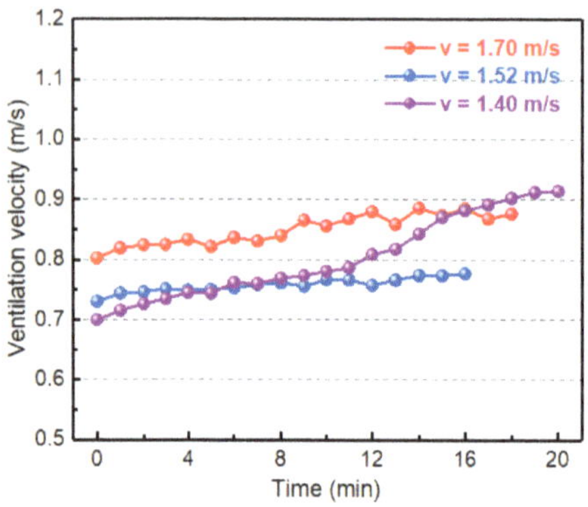

(a)

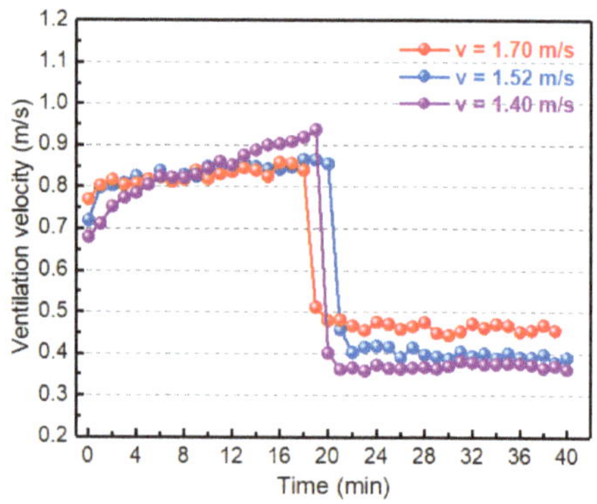

(b)

Figure 4. Wind velocity of adjacent branches for different schemes: (**a**) Air-door 5 closed. (**b**) Air-door 5 half closed.

The inclination angle of branches 1 and 2 is 12°, and the critical velocity for the different schemes are shown in Table 3, based on the results of previous studies [20]. After closing air-door 5, branch 3 is basically windless. When air-door 5 is half closed, the ventilation velocity is 1.70 m/s and 1.52 m/s, after the air flow control. the wind velocity of branch 2 is greater than the critical velocity, and the back-layering flow subsides; additionally, the wind resistance of branches 1 and 2 become smaller, so the air volume flowing to branches 1 and 2 increases. Branches 1, 2, and branch 3 are all parallel branches, the total air volume and the cross-section of the roadways are constant, so the wind velocity of branch 3 decreases rapidly and then subsequently stabilizes. The wind velocity of branch 3, on average, quickly stabilized at around 0.47 m/s and 0.40 m/s. When the ventilation velocity is 1.40 m/s, after the air flow control, the wind velocity of branch 2 is less than the critical velocity; however, the wind velocity in branch 2 increases at the stagnation point of the wind flow and the smoke flow, the force produced by the longitudinal ventilation increases, and while the heat released from the fire source continues to decrease, with the consumption of fuel, the kinetic energy gained by the back-layering flow continues to decrease, so the driving force generated by the thermal buoyancy continues to decrease.

Under the action of the longitudinal ventilation, the hot back-layering flow retreats to the fire source, and the back-layering length becomes shorter. The influence of the fire source branch on the wind flow of the wind network becomes smaller, and the wind velocity of branch 3 is relatively stable.

Table 3. Wind velocity in the ventilation network with parallel branches after the smoke control.

Case	Smoke Control Method	Ventilation Velocity v/(m/s)	Critical Velocity v_c/(m/s)	Branch 2 Wind Velocity v_2/(m/s)
Case 1	Air-door 5 closed	1.70	1.11	1.46
	Air-door 5 closed	1.52	0.92	1.30
	Air-door 5 closed	1.40	1.12	1.28
Case 2	Air-door 5 half closed	1.70	0.94	1.20
	Air-door 5 half closed	1.52	1.12	1.13
	Air-door 5 half closed	1.40	1.10	1.02

3.1.3. Temperature Variation on the Upwind Side of the Fire Source

Figure 5 presents the temperature variation on the upwind side of the fire source, when the ventilation velocity is 1.40 m/s in the ventilation network with the parallel branches (the fire source is located at the origin, and the negative value indicates the distance between the upwind side of the fire source). From Figure 5, we can see that the temperature on the upstream of the fire source decreases with the increase in the distance away from the fire source before the smoke control. The reason for this is because it is under the action of buoyancy, the back-layering flow pulls the surrounding cold air and spreads forward, and the temperature near the fire source on the upwind side increases first. As the smoke flows forward, the temperature far from the fire source gradually increases. The heat exchange between the high-temperature back-layering flow with the cold air and the wall causes the velocity and the temperature of the smoke front peak to decrease until the back-layering flow reaches the stagnation point, and the temperature returns to ambient. As shown in Figure 5a, when the ventilation velocity is 1.40 m/s, before the smoke control, the highest temperature is 65 °C at 0.44 m upwind from the fire source, and the back-layering flow does not reach to 3.09 m on the upstream of fire source, the temperature is about 20 °C.

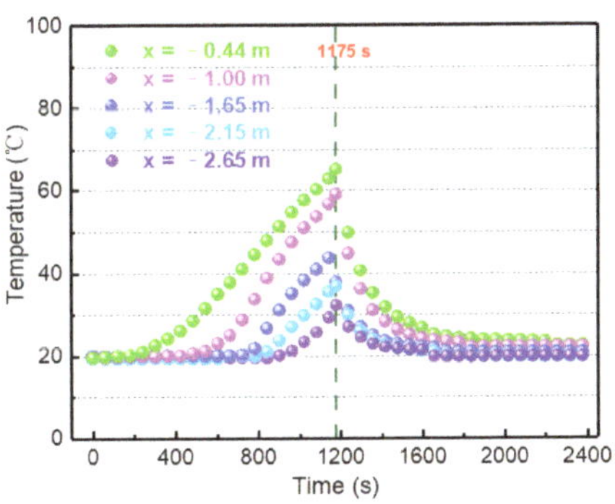

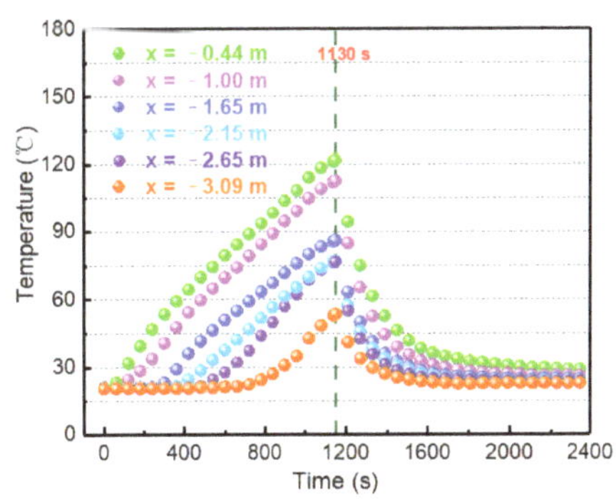

Figure 5. Temperature variation on the upstream of fire source for different schemes when the ventilation velocity is 1.40 m/s: (**a**) Air-door 5 closed. (**b**) Air-door 5 half closed.

According to Figure 5a, fires at different ventilation velocities, after closing air-door 5, the wind velocity of branch 2 is greater than the critical velocity, and the hot smoke on the upwind side retreats to the fire source within 1 min under the driving force of the wind flow. The cold air exchanges heat with the hot air and the wall surface, and the temperature on the upwind side of the fire source rapidly decreases. Figure 5b shows that when the ventilation velocity is 1.40 m/s and air-door 5 is half closed, the wind velocity of branch 2 is less than the critical velocity, after the smoke control. With the increase in the wind velocity in the roadway, the dynamic pressure produced by the longitudinal ventilation increases under the action of the longitudinal ventilation; the hot smoke then retreats towards the fire source, and the temperature decreases at different locations in the upwind side of the fire source.

3.2. *Effect of Air-Door Closing Location on the Back-Layering Flow*

3.2.1. Smoke Back-Layering Length

Figure 6 shows the back-layering length in branch 1 and branch 2 for the different schemes when the fire is stable in the ventilation network with a diagonal branch and without smoke control. Under different experimental conditions, the driving force generated by the thermal buoyancy and the force of the longitudinal ventilation are different, so the back-layering length is different. When the ventilation velocity is 1.90 m/s and 1.79 m/s, the back-layering length is shorter, which is 0.25 m and 1.08 m respectively. The smoke flow does not backflow to the diagonal branch and branch 1. When the ventilation velocity is 1.70 m/s and 1.56 m/s, the back-layering flow spreads to the diagonal branch and branch 1; additionally, when the ventilation velocity is 1.70 m/s, the heat release rate is 19.5 kW, the smoke back-layering flow spreads to branch 1 and reaches to 1.50 m from the fire source. Fires at different ventilation velocities, regardless of whether measures are taken to close air-door 1 or 5, after the smoke control, show that the force of the longitudinal ventilation is greater than the driving force generated by the thermal buoyancy, and the back-layering is prevented. It is observed through the glass window that the upstream smoke flows backwards, towards the fire source, with a smoke back-layering length of 0.

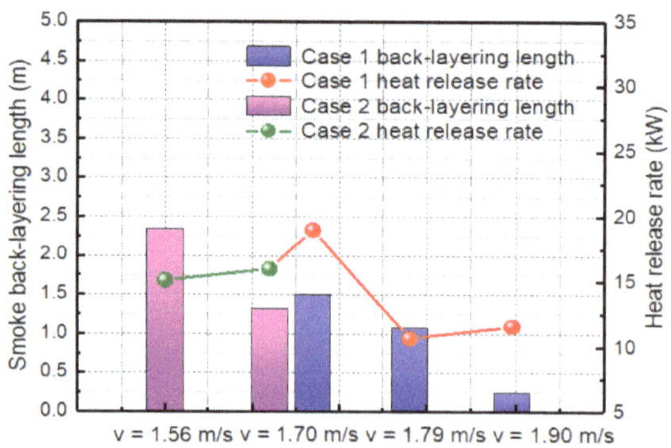

Figure 6. Back-layering length in the ventilation network with a diagonal branch.

3.2.2. Wind Velocity Variation in Ventilation Network

Figure 7 presents the wind velocity variation in each branch for the different schemes. As can be seen from Figure 7, when the ventilation velocity is large, the fire has little effect on branch 3. For example, the ventilation velocity is 1.90 m/s, the wind velocity of branch 3 is basically unchanged. In the diagonal branch, when the air flow reaches branch 4, the velocity shows a positive value and when the air flow reaches branch 2, the velocity is negative. As seen in Figure 7, the wind velocity of the diagonal branch increases with the development of the fire, which is a result of the increase in the ventilation resistance of branch 2 as well as the air volume of branch 1 to the diagonal branch due to the thermal effect of the fire zone. The wind velocity in branch 4 also increases as the air flow from the diagonal branch is directed towards branch 4. For example, when the ventilation velocity is 1.56 m/s, the combustible material burns more intensely, the smoke flow backs up towards both the diagonal branch and branch 1, and the wind velocity increases by 19.2% for branch 3, 36% for the diagonal branch and 22% for branch 4.

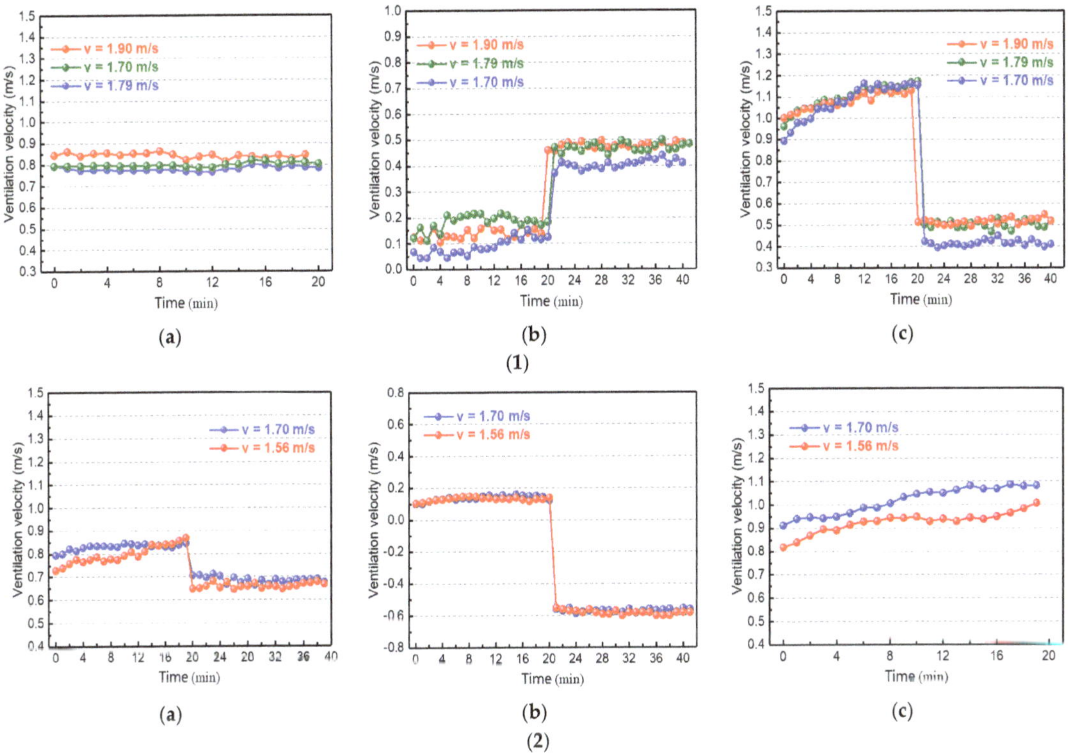

Figure 7. Wind velocity variation for different schemes: (**1**) Air-door 1 closed: (**a**) Branch 3. (**b**) Diagonal branch. (**c**) Branch 4. (**2**) Air-door 5 closed: (**a**) Branch 3. (**b**) Diagonal branch. (**c**) Branch 4.

According to Figure 7, the fire at the different ventilation velocities and closing air-door 1, shows that the wind flow is directed towards branch 1, with branch 3 being essentially windless; the wind velocity increases in the diagonal branch, and branch 4 stabilizes after a rapid reduction in the wind velocity. Based on the previous research results [20], the critical velocity of branch 2, in the different schemes, is shown in Table 4. It can be seen from Table 4 that the wind velocity of branch 2 is greater than the critical velocity after the smoke control. The smoke back-layering flow subsides, and the wind flow of the ventilation network is redistributed and stabilized. Similarly, when air-door 5 is closed, the wind velocity in branch 3 is rapidly reduced and then restored to a stable level, and the wind flow of branch 3 is directed to the diagonal branch and branch 4 is basically windless. For example, if the ventilation velocity is 1.70 m/s, when air-door 1 is closed, the wind velocity of the diagonal branch increases rapidly and then stabilizes at 0.42 m/s, the wind velocity of branch 4 decreases rapidly to 0.40 m/s, and the wind velocity of branch 2 is 1.28 m/s. When air-door 5 is closed, the wind velocity in branch 3 decreases rapidly and then stabilizes at 0.69 m/s, the wind velocity in the diagonal branch increases rapidly and then stabilizes at 0.58 m/s, and the wind velocity in branch 2 is 1.56 m/s. When compared with closing air-door 1, the wind velocity to branch 2 is greater after closing air-door 5. The greater the wind velocity, the more likely the smoke back-layering flow will subside under the action of wind pressure.

Table 4. Wind velocity in the ventilation network with a diagonal branch after the smoke control.

Case	Smoke Control Method	Ventilation Velocity v/(m/s)	Critical Velocity v_c/(m/s)	Branch 2 Wind Velocity v_2/(m/s)
Case 1	Air-door 1 closed	1.90	1.02	1.36
	Air-door 1 closed	1.79	1.00	1.32
	Air-door 1 closed	1.70	1.10	1.28
Case 2	Air-door 5 closed	1.70	1.10	1.56
	Air-door 5 closed	1.56	1.12	1.26

3.2.3. Temperature Variation in Ventilation Network

Figure 8 depicts the temperature variation under the roadway roof before and after the smoke control when the ventilation velocity is 1.70 m/s. The temperature variation, over time, for branch 2 on the upstream of the fire source before and after the smoke control is shown in Figure 8(1), and the temperature variation for the diagonal branch before and after the smoke control is shown in Figure 8(2). According to Figure 8(1), after adopting different smoke control schemes, the smoke back-layering flow of the fire source in branch 2 subsides, and the temperature upwind of the fire source decreases rapidly and then returns to an ambient temperature. It can be seen from Figure 8(2) that when the ventilation velocity is 1.70 m/s the temperature of the diagonal branch decreases rapidly, after adopting different smoke control schemes. After closing air-door 5, the wind velocity of the diagonal branch increases with the air flow force of the diagonal branch being greater than the driving force of the smoke flow, which results in the diagonal branch reversing the smoke flow back to the fire source. When closing air-door 1, the smoke back-layering flow from the fire source to branch 2 subsides and no more smoke flow is retrograded to the diagonal branch. After the smoke has retreated to the diagonal branch it is then diluted by the air flow, discharged through branch 4, and the temperature of the diagonal branch decreases.

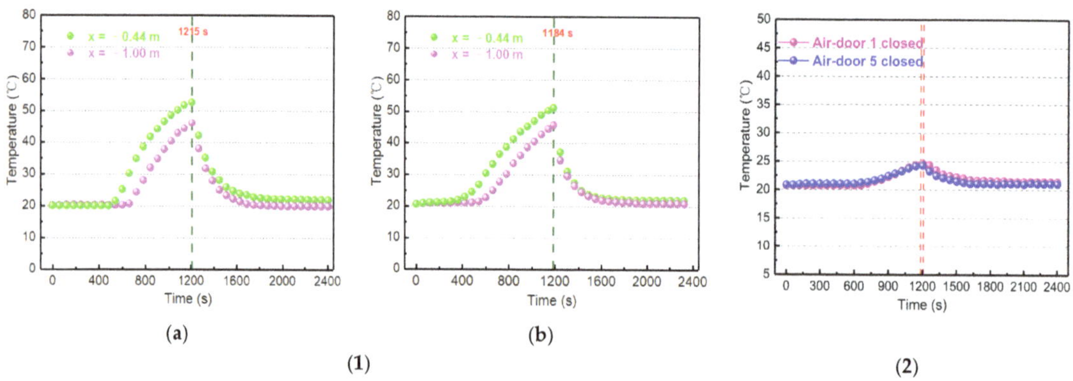

Figure 8. Temperature variation for different schemes when the ventilation velocity is 1.70 m/s: (**1**) Temperature variation at different locations upstream of the fire source: (**a**) Air-door 1 closed. (**b**) Air-door 5 closed. (**2**) Temperature variation in the diagonal branch.

4. Conclusions

Based on the mine ventilation network test platform, the influence of the air-door position in the branch adjacent to the fire source on the characteristics of smoke back-layering was analyzed when the back-layering occurred in the mine ventilation network with a diagonal branch. To simplify the experiment, the influence of the closing degree of the air-doors, which were adjacent to the fire source on the back-layering in the mine ventilation network with parallel branches was first investigated. The main conclusions are as follows:

(1) In the ventilation network with parallel branches, the back-layering occurs in the upstream region under different ventilation velocities. The closing degree of the air-doors in the adjacent branch is regulated. When the air-door on the return side of the adjacent branch is closed, the back-layering flow at different ventilation velocities subsides rapidly, and the upstream temperature tends towards ambience after a rapid reduction. When the air-door is half closed, and the ventilation velocity is 1.40 m/s, there is still back-layering flow before the fire enters the attenuation stage. Compared with the half closing of the air-door, the closing of the air door on the return side of the adjacent branch is more effective in preventing the upstream back-layering flow;
(2) In the ventilation network with a diagonal branch, when the ventilation velocity is 1.70 m/s, the back-layering flow spreads to the diagonal branch and branch 1. The longest distance between the back-layering flow of branch 1 and the fire source is 1.50 m, and the wind velocity in the adjacent branches and the diagonal branch increases.
(3) Bases on the experimental results in the ventilation network with the parallel branches, the air-doors at different positions of the adjacent branch are regulated. When the air-door on the return side of the adjacent branch is closed, the back-layering rapidly subsides, the wind velocity of branch 3 stabilizes at 0.69 m/s after a rapid decrease, and the wind velocity of the diagonal branch increases rapidly and then stabilizes at 0.58 m/s. When the air-door on the intake side of the adjacent branch is closed, the upstream smoke flow rapidly subsides, and the smoke in the diagonal branch spreads to the adjacent branch. In the mine ventilation network with a diagonal branch, closing the air-door on the return side of the adjacent branch is more effective in preventing back layering in roadway fires than closing the air-door on the intake side.

Author Contributions: Conceptualization, H.W. and Z.X.; methodology, H.W., Z.X. and L.W.; validation, C.F. and Y.Z.; formal analysis, H.W. and Z.X.; investigation, L.W., C.F. and Y.Z.; data curation, L.W.; writing—original draft preparation, Z.X.; writing—review and editing, H.W., Z.X. and C.F.; funding acquisition, H.W. All authors have read and agreed to the published version of the manuscript.

Funding: This research was funded by the National Natural Science Foundation of China (No. 51874313), the National Key Research and Development Program of China (2018YFC0808100).

Data Availability Statement: Not applicable.

Conflicts of Interest: The authors declare no conflict of interest.

References

1. Song, Z.Y.; Kuenzer, C. Coal fires in China over the last decade: A comprehensive review. *Int. J. Coal Geol.* **2014**, *133*, 72–99. [CrossRef]
2. Fernandez-Alaiz, F.; Castanon, A.M.; Gomez-Fernandez, F.; Bascompta, M. Mine fire behavior under different ventilation conditions: Real-scale tests and CFD modeling. *Appl. Sci.* **2020**, *10*, 3380. [CrossRef]
3. Shi, G.Q.; Wang, G.Q.; Ding, P.X.; Wang, Y.M. Model and simulation analysis of fire development and gas flowing influenced by fire zone sealing in coal mine. *Process. Saf. Environ.* **2021**, *149*, 631–642. [CrossRef]
4. Yao, Y.Z.; He, K.; Peng, M.; Shi, L.; Cheng, X.D. The maximum gas temperature rises beneath the ceiling in a longitudinal ventilated tunnel fire. *Tunn. Undergr. Space Technol.* **2021**, *108*, 103672. [CrossRef]
5. Wang, Z.; Ding, L.; Wan, H.; Ji, J.; Gao, Z.; Yu, L. Numerical investigation on the effect of tunnel width and slope on ceiling gas temperature in inclined tunnels. *Int. J. Therm. Sci.* **2020**, *152*, 106272. [CrossRef]
6. Jia, Q.; Wu, H.; Ling, T.; Liu, K.; Peng, W.W.; Gao, X.; Zhao, Y.L. Study on the stress variation law of inclined surrounding rock roadway under the influence of mining. *Minerals* **2022**, *12*, 499. [CrossRef]
7. Yang, D.; Ding, Y.; Du, T.; Mao, S.; Zhang, Z. Buoyant back-layering and the critical condition for preventing back-layering fluid in inclined tunnels under natural ventilation: Brine water experiments. *Exp. Therm. Fluid Sci.* **2018**, *90*, 319–329. [CrossRef]
8. Du, T.; Yang, D.; Ding, Y. Driving force for preventing smoke backlayering in downhill tunnel fires using forced longitudinal ventilation. *Tunn. Undergr. Space Technol.* **2018**, *79*, 76–82. [CrossRef]

9. Zhou, Y.; Song, S.L.; Guo, J.W.; Xu, Y.; Lin, B.Q. Experimental study of the sequence of flow reversals in accessional airways parallel to a fire. *J. China U. Min. Technol.* **2009**, *38*, 467–470+493.
10. Barbato, L.; Cascetta, F.; Musto, M.; Rotondo, G. Fire safety investigation for road tunnel ventilation systems—An overview. *Tunn. Undergr. Space Technol.* **2014**, *43*, 253–265. [CrossRef]
11. Hu, L.H.; Huo, R.; Chow, W.K. Studies on buoyancy-driven back-layering flow in tunnel fires. *Exp. Therm. Fluid Sci.* **2008**, *32*, 1468–1483. [CrossRef]
12. Roh, J.S.; Ryou, H.S.; Kim, D.H.; Jung, W.S.; Jang, Y.J. Critical velocity and burning rate in pool fire during longitudinal ventilation. *Tunn. Undergr. Space Technol.* **2007**, *22*, 262–271. [CrossRef]
13. Oka, Y.; Atkinson, G.T. Control of smoke flow in tunnel fires. *Fire Safety J.* **1995**, *25*, 305–322. [CrossRef]
14. Wu, Y.; Bakar, M.Z.A. Control of smoke flow in tunnel fires using longitudinal ventilation systems-a study of the critical velocity. *Fire Safety J.* **2000**, *35*, 363–390. [CrossRef]
15. Lee, S.R.; Ryou, H.S. An experimental study of the effect of the aspect ratio on the critical velocity in longitudinal ventilation tunnel fires. *J. Fire Sci.* **2004**, *23*, 119–138. [CrossRef]
16. Yi, L.; Xu, Q.Q.; Xu, Z.S.; Wu, D.X. An experimental study on critical velocity in sloping tunnel with longitudinal ventilation under fire. *Tunn. Undergr. Space Technol.* **2014**, *43*, 198–203. [CrossRef]
17. Chow, W.K.; Gao, Y.; Zhao, J.H.; Dang, J.F.; Chow, C.L.; Miao, L. Smoke movement in tilted tunnel fires with longitudinal ventilation. *Fire Safety J.* **2015**, *75*, 14–22. [CrossRef]
18. Atkinson, G.; Wu, Y. Smoke control in slopping tunnels. *Fire Safety J.* **1996**, *27*, 335–341. [CrossRef]
19. Li, J.; Li, Y.F.; Cheng, C.H.; Chow, W.K. A study on the effects of the slope on the critical velocity for longitudinal ventilation in tilted tunnels. *Tunn. Undergr. Space Technol.* **2019**, *89*, 262–267. [CrossRef]
20. Hao, H.Q.; Wang, K.; Zhang, C.Y.; Jiang, S.G.; Wang, H.K. Evolution and regulation law of wind and smoke flow field area network in mine belt roadway fire. *J. China U. Min. Technol.* **2021**, *50*, 716–724.
21. Wang, K.; Jiang, S.G.; Ma, X.P.; Wu, Z.Y.; Shao, H.; Zhang, W.Q.; Cui, C.B. Information fusion of plume control and personnel escape during the emergency rescue of external-caused fire in a coal mine. *Process. Saf. Environ.* **2016**, *103*, 46–59. [CrossRef]
22. Marjanovic, D.; Grozdanovic, M.; Janackovic, G. Data acquisition and remote control systems in coal mines: A serbian experience. *Meas. Control* **2015**, *48*, 28–36. [CrossRef]
23. Wang, Q.J.; Ma, X.P.; Yang, C.Y.; Dai, W. Modeling and control of mine main fan switchover system. *ISA Trans.* **2019**, *85*, 189–199. [CrossRef] [PubMed]
24. Xu, G.; Jong, E.C.; Luxbacher, K.D.; Ragab, S.A.; Karmis, M.E. Remote characterization of ventilation systems using tracer gas and CFD in an underground mine. *Safety Sci.* **2015**, *74*, 140–149. [CrossRef]
25. Fu, M.M.; Zheng, Q.; He, S. Oxygen control mechanism in non-ventilation excavation based on the simultaneous mining of coal and gas. *Process Saf. Environ.* **2019**, *123*, 33–38. [CrossRef]
26. Wang, K.; Jiang, S.G.; Ma, X.P.; Wu, Z.Y.; Shao, H.; Zhang, W.Q.; Cui, C.B. Numerical simulation and application study on a remote emergency rescue system during a belt fire in coal mines. *Nat. Hazards* **2016**, *84*, 1463–1485. [CrossRef]
27. Wang, K.; Jiang, S.G.; Zhang, W.Q.; Wu, Z.Y.; Shao, H.; Liu, T.; Ju, X.R. Remote control technology and numerical analysis of airflow in coal mine fire disaster relief process. *J. China Coal Soc.* **2012**, *37*, 1171–1176.
28. Shang, W.; Deng, L.; Liu, J. A novel air-door opening and closing identification algorithm using a single wind-velocity sensor. *Sensors* **2022**, *22*, 6837. [CrossRef]
29. Wang, K.; Hao, H.Q.; Jiang, S.G.; Cai, W.Y.; Zhang, Y.C.; Wang, Z.T. Experimental study on the characteristics of overpressure wave to ventilation facilities during gas explosion and automatic shock relief devices. *Geomat. Nat. Hazards Risk* **2020**, *11*, 2360–2383. [CrossRef]
30. Parker, W.J. Calculations of the heat release rate by oxygen consumption for various applications. *J. Fire Sci.* **1984**, *2*, 380–395. [CrossRef]

processes

MDPI

Article

Selection and Optimization Mechanism of the Lower Return Roadway Layout in the near Residual Coal Pillar Area

Xiao-He Wang [1,*], Hao-Hao Zhang [2], Zheng Wu [1], Xiao-Long Li [1], Yi Sui [3] and Ruo-Qi Gao [4]

1 School of Energy and Mining Engineering, China University of Mining and Technology (Beijing), Beijing 100083, China
2 State Key Laboratory of Water Resources and Hydropower Engineering Science, Wuhan University, Wuhan 430072, China
3 Transportation College, Jilin University, Changchun 130000, China
4 School of Management, Dalian Polytechnic University, Dalian 116034, China
* Correspondence: wangxh_1994@163.com; Tel.: +86-183-4184-1115

Abstract: Background: To optimize the layout position of the residual coal pillar return roadway when mining a close coal seam group and to clarify the optimization mechanism, a roadway optimization layout analysis was conducted on the Tashan coal mine. Methods: Surface displacement monitoring was conducted using field tests, and the main stress magnitude, plastic zone morphology, deformation variables, and connectivity between the plastic zone of the roadway and the plastic zone of the residual coal pillar were analyzed at different locations with the help of FLAC3D numerical simulation software. Results: It was found that, in the process of close coal seam group mining, the residual coal pillar of the overlying coal seam seriously affects the stress state and plastic zone distribution of the lower coal seam roadway. The roadway is arranged in a position that is relatively far away from the residual coal pillar, which could reduce the stress influence of the residual coal pillar on the roadway and guarantee the stability of the roadway. Conclusion: Since the Tashan Mine uses the top release method for mining, the stability of the roadway can be better ensured by placing the roadway in the middle and lower regions of the coal seam and using the layout method to retain small coal pillars.

Keywords: residual coal pillar; lower return roadway; optimization mechanism; stress state; plastic zone

Citation: Wang, X.-H.; Zhang, H.-H.; Wu, Z.; Li, X.-L.; Sui, Y.; Gao, R.-Q. Selection and Optimization Mechanism of the Lower Return Roadway Layout in the near Residual Coal Pillar Area. *Processes* **2022**, *10*, 2471. https://doi.org/10.3390/pr10122471

Academic Editor: Guining Lu

Received: 31 October 2022
Accepted: 17 November 2022
Published: 22 November 2022

1. Introduction

The distribution of coal seam clusters in close proximity to one another is a major characteristic of coal distribution in China. Mining a proximity coal seam group is different from single-seam mining, and it is influenced by the adjacent seam, which can lead to mine pressure and potentially disastrous accidents [1–5]. Mining an upper coal seam of a proximity coal seam group can not only trigger the loosening of the roof of the lower coal seam but also affect the stress state of the surrounding rocks of the lower coal seam roadway, causing roadway deformation damage [6–8].

The coal seam under original rock stress is redistributed after the excavation of the roadway, and once the mining disturbance triggers the local stress concentration in the roadway, it can lead to the occurrence of roadway disasters [9,10]. In order to determine the causes of roadway hazards and reduce the possibility of roadway hazards, some scholars have proposed theoretical approaches for stabilizing surrounding rocks. The classical circular plastic zone determines the range of the plastic zone of the roadway enveloped by rock, which is considered to be a circular area [11–13]. The theory of natural caving supplies states that under the action of mine pressure, the roof of the roadway's surrounding rock will appear loose, causing deformation, damage, and bubble falls. The roof of the roadway forms an arch of support after the roof pressure balances out, and the roof surrounding the rock no longer causes bubble falls [14]. The theory of axial variations considers that the damage law of the surrounding rock is the result of the joint influence of the surrounding rock stress and the mechanical properties of the rock mass, as well as the deformation characteristics of the surrounding rock [15,16]. The theory of maximum horizontal stress [17] comprises the analysis of the damage characteristics of rocks and the evolution law of pores and fissures inside the rocks from structural defects inside the rocks, which play a significant role in supporting the rock surrounding the roadway [18–21]. Małkowski et al. conducted a long-term study of three different roadways over a period of six years and found that different support schemes can have significant effects on roof deformations [22]. Khalymendyk et al. analyzed the deformation mechanisms of deep laminated rock roadways and determined that the deformation of the roadway rock was a result of the mutual extrusion of the laminated surrounding rocks, as well as the linear relationship of the deformation and damage between the top and bottom slabs [23]; these observations have been verified by other research studies in the field.

Usually, the excavated coal seam is a non-uniform stress field, and the presence of the non-uniform stress field also has an impact on the distribution of the plastic zone [24]. Fenner's formula and Castanet's formula can determine the radius of the plastic zone around the circular hole under hydrostatic pressure conditions [25] to solve the asymmetric damage phenomenon generated by the roadway in the actual production process. Ma et al. proposed a model for the distribution of the plastic zone of the roadway under a non-uniform stress field. They established the roadway butterfly damage theory and derived the boundary equation of the plastic zone of the roadway butterfly [26–30]. Based on this formula, Guo et al. [31] predicted potential hazard zones in circular roadways and proposed evaluation guidelines for dynamic hazard critical points in circular roadways. Reducing the difficulties of roadway support is a key concern for roadways. In order to optimize the design of the transportation roadway in the Beskempir field, Abdrakhman et al. used numerical simulations to analyze the stress characteristics and deformation patterns of the surrounding transportation roadway's rock, and the stability of the surrounding rock was discussed to determine a support scheme for the transportation roadway [32]. Gan et al. [33], Yang et al. [34], and Li et al. [35], as well as a large number of other scholars, analyzed the state of the roadway and used experiments and numerical simulations to demonstrate a reasonable arrangement form for the roadway, which reduces the difficulty of implementing roadway supports.

The lower back mining roadway's close residual coal pillar areas easily triggers the occurrence of roadway disasters due to the complex surrounding rock stress. In this paper, we measured the short-term deformation variables of the return roadway of the 30503 working faces in Tashan coal mine by using the method of roadway surface displacement monitoring, and the measurement results showed that the original roadway arrangement would lead to substantial deformations over a short period of time. By determining the asymmetrical damage inflicted upon the roadway, we established a numerical model for the multi-point arrangement of the lower return roadway in the close residual coal pillar area and simulated the main stress magnitude, plastic zone morphology, and deformation variables of the roadway in the horizontal and vertical directions, respectively. The connectivity between the plastic zone and the plastic zone of the residual coal pillar was analyzed at different positions on the roadway, and the mechanical characteristics and deformation features of the roadway at different positions were determined. Based on the coal mining process of the 30503 working face of the Tashan mine, it was established that the roadway should be arranged in the middle and lower positions of the coal seam as a method for retaining small coal pillars (away from the residual coal pillars), which could ensure the stability of the roadway. This provides the theoretical basis for the reasonable arrangement of the lower return roadway of the residual coal pillars in the mining process of the close coal seam group.

2. Materials and Methods

2.1. Engineering Background

The Tashan coal mine is located in the middle and east edge of Datong coalfield, China. The main coal seam comprises 3–5# coal, with an average thickness of 18.0 m. The coal is recovered by releasing the top coal. The upper coal seam is 2# coal; in order to ensure safe production operations, after mining 2# coal, the 2# coal emptying area is filled. According to the mining succession plan, the 30503 working face is the successive mining face. The spatial positions of the 3–5# coal seam and 2# coal seam and the original layout plan of the tunnel are shown in Figure 1; as we are ignoring the influence of other factors, the diagram only depicts the 30503 working face return roadway. After the roadway was excavated at 860 m in depth, serious deformations occurred in the roadway's roof, and the local area had to be completely closed; thus, the 30503 return roadway was repaired, but during the repair process, the roadway experienced continuous deformations, and there was a local large deformation. To analyze the degree of damage, surface displacement observations were made on the roof of the 30503 return roadway, and the degree of damage was analyzed visually in the form of data.

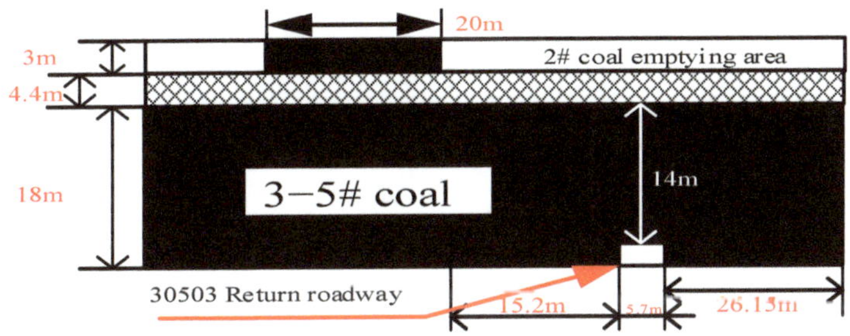

Figure 1. Diagram of the layout of the roadway between coal seams #3–5 and #2.

2.2. Field Tests

In order to study the deformation degree of the roadway enclosure, the 30503 return roadway repair, at the section at about 240 m~840 m, was selected, and the roadway surface displacement observation points were laid out within this area. One observation point was arranged every 40 m, with a total of 16 observation points, numbered 1~16. Short-term observations were carried out for 10 days, and the observation data of the 3rd, 7th, and 10th days were taken. Based on the data of the first measurement, roadway deformations at different mileages were calculated for each measurement, and the measured data are shown in Figure 2.

According to Figure 2a, it can be seen that, on the 3rd day of the test, the maximum value of top plate displacement is located at 280 m of the dug-in section, which is 24 mm, and the maximum value of the two gang displacements is located at 680 m of the dug-in section, which is 4 mm. According to Figure 2b, it can be seen that on the 7th day of the test that the repaired section of the roadway showed extensive deformations; here, the maximum value of top plate displacement is located at 440 m of the dug-in section, which is 34 mm. The maximum amount of top plate displacements is located at 440 m of the excavated section, reaching 34 mm, and the top plate displacement increased by 3 mm compared with the measured value at 280 m of the excavated section on the third day and by 35 mm compared with the measured value at the same point on the third day; the maximum amount of the displacement of both gangs is located at 680 m of the excavated section, reaching 61.5 mm, and the displacement of both gangs increased by 9 mm compared with the measured value at 680 m of the excavated section on the third day and by 49.5 mm compared with the measured value at the same point on the third day. After 4 days, significant deformations occurred at the same measurement point. According to Figure 2c, it can be observed that on the 10th day of the test, the deformation phenomenon of the repaired section of roadway still existed and spread, among which the maximum amount of top plate displacement was located at 400 m of the dug-in section, reaching 35 mm; this is an increase of 5 mm compared with the measured value at the same point on the 7th day. The maximum amount of two gang displacement was located at 760 m of the dug-in section, reaching 66 mm, an increase of 4.5 mm compared with the measured value at the same point on the 7th day. Although the periodic deformation at the same measurement point is less than the amount of displacement in 3–7 days, there is still deformation and a growing trend.

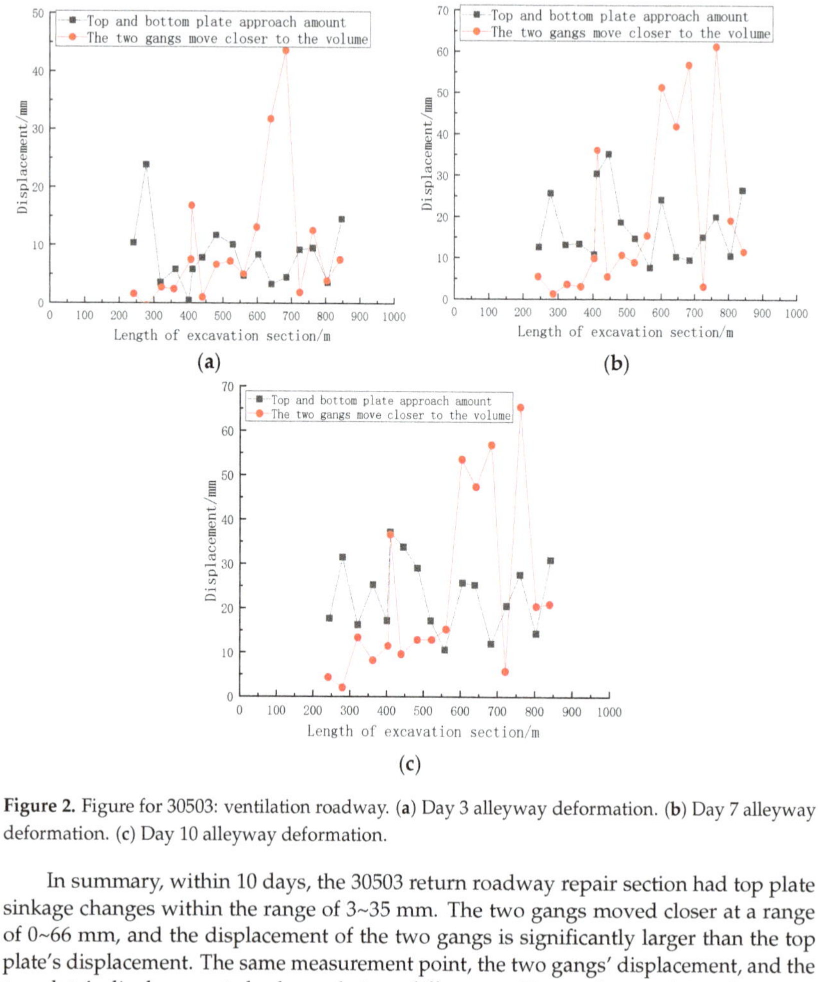

Figure 2. Figure for 30503: ventilation roadway. (**a**) Day 3 alleyway deformation. (**b**) Day 7 alleyway deformation. (**c**) Day 10 alleyway deformation.

In summary, within 10 days, the 30503 return roadway repair section had top plate sinkage changes within the range of 3~35 mm. The two gangs moved closer at a range of 0~66 mm, and the displacement of the two gangs is significantly larger than the top plate's displacement. The same measurement point, the two gangs' displacement, and the top plate's displacement also have obvious differences. The roadway exhibited obvious non-uniform damage phenomena. To control the deformation and damage of the roadway and avoid accidents, the following is an analysis of the extent of damage to the plastic zone and the degree of deformation of the surrounding rock under different arrangements of the roadway from both vertical and horizontal directions. The distribution rules of stress and the plastic zone in the surrounding rock of the 30503 roadway along the top slab, along the bottom slab, along the top left and bottom coal slab, differently sized coal pillars, and the area located directly below the residual coal pillars are analyzed. Based on the above analysis, reasonable arrangements and the arrangement principles of the residual coal pillar's lower return roadway are determined. The diagram for different roadway layout schemes is shown in Figure 3, and the dashed line in the figure denotes the limit of coal pillar retentions for different horizontal roadway layout schemes; from right to left, the coal pillars are 4 m, 8 m, 15 m, 20 m, and 25 m.

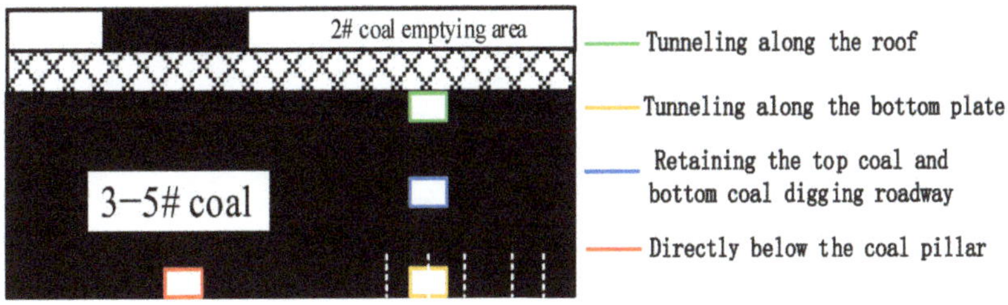

Figure 3. Schematic diagram of different roadway layout schemes.

2.3. Methods

A geometric model was established based on the 30503 return roadway, as shown in Figure 4, and the perspective view is shown in Figure 5. In order to reduce the influence of other factors on the model, the model was constructed with 3–5 coal as the vertical center and based on the coal seam histogram. Nine layers of rock were constructed upward and downward, and the length and width of the model were set to 100 times the width of the roadway and rounded to 500 m; in other words, the model's size was 500 m × 500 m × 156 m. Moreover, since the burial depth of the top of the model was 348 m, according to the traditional ground stress calculation method, the top of the model was subjected to a vertical stress of 8.7 MPa. The vertical stress of 8.7 MPa was applied to the top of the model, and the numerical simulation was carried out by FLAC3D finite element difference software. Due to the large size of the model, the influence of the roadway and residual coal pillars is negligible, so the front, back, left, and right sides of the model as well as the bottom are fixed with fixed displacement and fixed boundaries, and the Mohr–Coulomb principal structure model is used. In order to improve the accuracy of the calculation, the encryption around 30 m of the roadway is 1 m per frame; at around 15 m, it is 0.5 m per frame, and around 15 m of the coal pillar, it is 1 m per frame. The demonstrated basic mechanical parameters involved in the numerical simulation are shown in Table 1. These data were obtained from the paper "Research on the mining method of close extra-thick coal seam in Datong Tashan coal mine of China National Investment Corporation" [36].

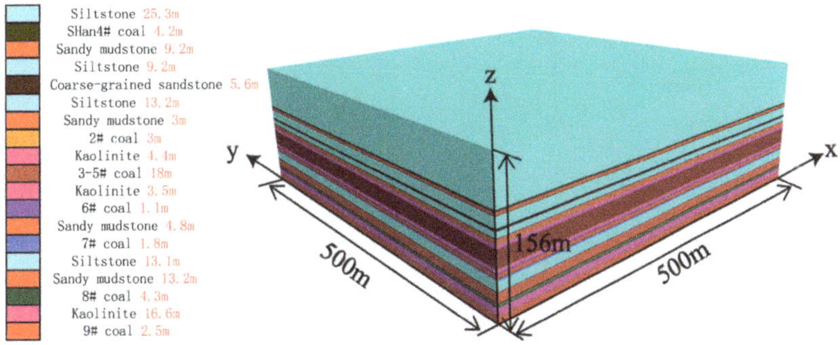

Figure 4. A figure of the numerical model.

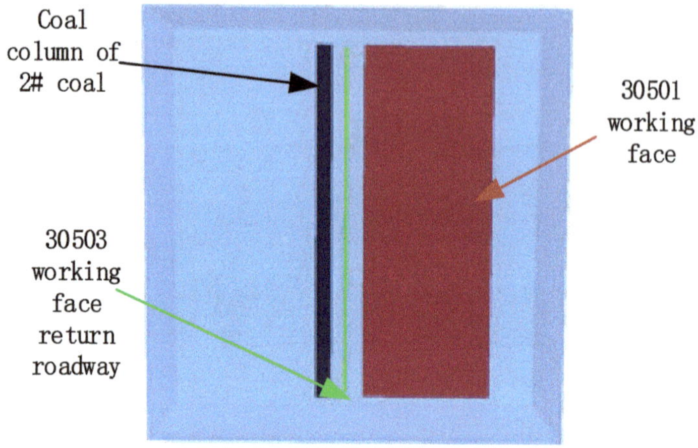

Figure 5. Numerical model perspective view.

Table 1. Rock mechanics parameter table.

Parameters Lithology	Density (kg/m^3)	Bulk Modulus (GPa)	Shear Modulus (GPa)	The Angle of Internal Friction (°)	Internal Cohesion (MPa)	Tensile Strength (MPa)
Siltstone	2560	11.69	10.67	38.57	6.4	3.34
Sandy mudstone	2500	10.2	8.4	36.4	3.72	1.73
Coarse-grained sandstone	2700	14	10.7	39.4	5.68	6.4
2#vCoal	1400	3.48	0.78	34.45	2.13	0.92
3–5# Coal	1500	3.48	1.61	34.45	2.13	0.92
Kaolinite	2100	8.2	7.6	35	5.5	1.73
6# Coal	1500	3.48	1.61	34.45	2.13	0.92
7# Coal	1500	3.48	1.61	34.45	2.13	0.92
8# Coal	1500	3.48	1.61	34.45	2.13	0.92
9# Coal	1500	3.48	1.61	34.45	2.13	0.92
Shan $_{4\#}$ Coal	1400	3.48	0.78	34.45	2.13	0.92

3. Results

3.1. Plastic Zone Morphology and Distribution

3.1.1. Roadways with Vertical Layer Differences

(1) Roadway excavation along the roof

The horizontal distance between the roadway dug along the top and the residual coal column of the overlying coal seam is 15.2 m, and the vertical distance is about 4.4 m. From the stress cloud diagram (Figure 6), the maximum and minimum principal stresses appear at the residual coal pillar, and the maximum principal stress (σ_1) near the coal pillar is about 70–120 MPa; the minimum principal stress (σ_3) is about 26–50 MPa. σ_1 around the roadway is about 6–20 MPa, and σ_3 is about 3–10 MPa.

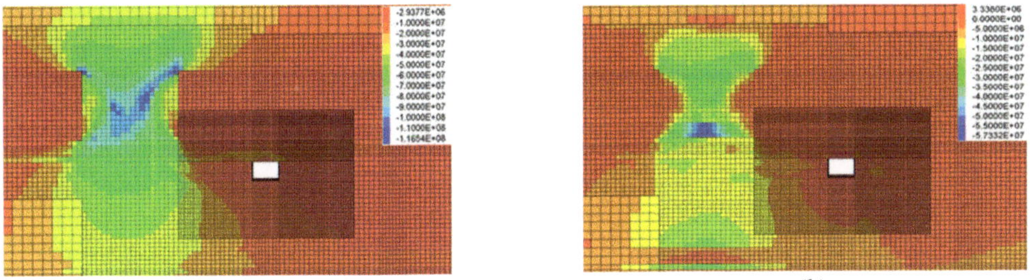

(**a**) (**b**)

Figure 6. Principal stress distribution contour of the roadway's surrounding rock when the roadway excavation along the roof. (**a**) σ_1 cloud. (**b**) σ_3 cloud.

To further analyze changes in the surrounding rock stress in the vertical direction and horizontal direction of the roadway when digging along the top, the measurement lines were arranged in the vertical direction and horizontal direction of the roadway center, respectively, and the main stress change curve was obtained (Figure 7). In the horizontal direction, σ_1 and σ_3 in the surrounding rock gradually decrease from the positive gang to the negative gang of the roadway. At approximately −30 m away from the positive gang of the roadway, σ_1 and σ_3 in the surrounding rocks reach the peak values of 67 MPa and 15 MPa, respectively, and the site is located below the residual coal pillar. Meanwhile, the ratio of σ_1 and σ_3 in the surrounding rocks can be found from in positive gang to the negative gang of the roadway. The ratios of σ_1 and σ_3 in the surrounding rocks gradually decrease from the positive to the negative gang of the roadway. From the vertical direction, σ_1 and σ_3 at the location of the roadway are small, while the principal stresses below the roadway are large, and the principal stresses at the location of the coal seam floor reach the peak; the ratio between σ_1 and σ_3 reaches the maximum at this moment. σ_1 and σ_3 at the roadway are about 9.5 MPa and 4.5 MPa, respectively.

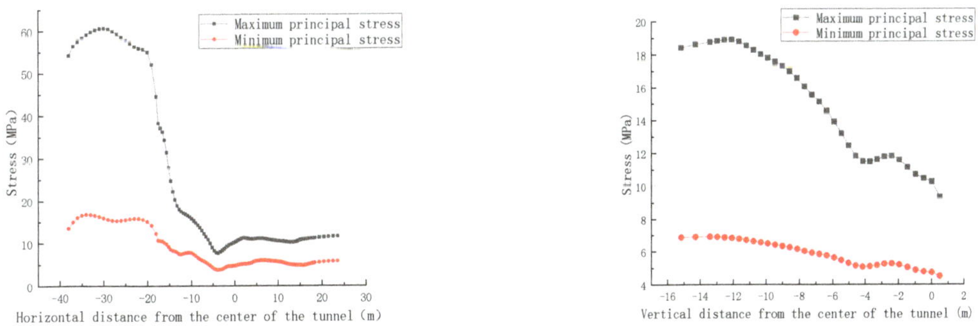

(**a**) (**b**)

Figure 7. Changes in the principal stresses of the roadway's surrounding rock when the roadway excavation along the roof. (**a**) Horizontal direction. (**b**) Variation direction.

Figure 8 shows the cloud map of the plastic zone. In the figure, the range of different color areas represents the range of the plastic zone. Each color distribution indicates the damage form at the location, including shear damage, tension damage, and volume damage. "n" in the figure represents the current state, and "p" represents the previously existing state. As observed in the figure, the plastic zone of the roadway presents an asymmetric distribution pattern, and the plastic zone of the positive gang of the roadway is larger in scope due to the superposition and concentration of stress at the residual coal pillar, which is connected with the plastic zone of the residual coal pillar. As the roadway is excavated along the top, the top plate of the roadway comprises rock, and the strength of the coal seam is smaller than the strength of rock, so the range of plastic zone of the top plate of the roadway is smaller than the two gangs and the bottom plate.

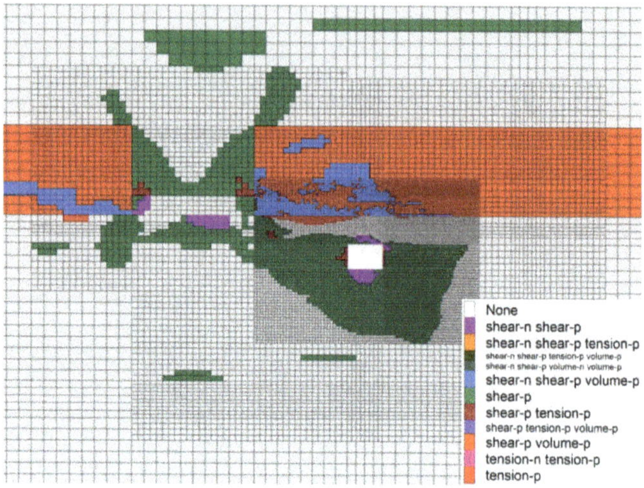

Figure 8. Plastic zone distribution cloud map when the roadway excavation along the roof.

(2) Roadway excavation leaves top coal bottom coal

The horizontal distance between the roadway dug by the retained top coal and bottom coal and the residual coal column of the overlying coal seam is 15.2 m, and the vertical distance is about 15.4 m. According to the stress cloud diagram (Figure 9), σ_1 and σ_3 appear in the residual coal column of the overlying coal seam, σ_1 near the residual coal column of the overlying coal seam is about 60–110 MPa, and σ_3 is about 25–58 MPa. σ_1 around the roadway is about 8–16 MPa, and σ_3 is about 0.3–5 MPa.

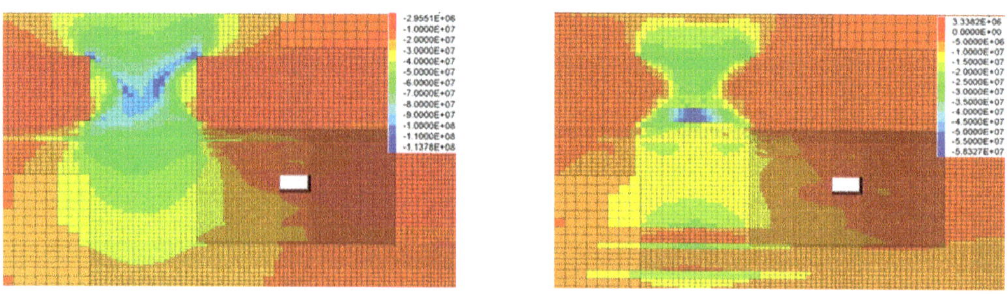

(**a**) (**b**)

Figure 9. Principal stress distribution contour of the roadway's surrounding rock when the roadway excavation leaves top coal bottom coal. (**a**) σ_1 cloud. (**b**) σ_3 cloud.

As observed in the main stress change curve (Figure 10), in the horizontal direction, the maximum main stress and the minimum main stress increase and then decrease from the residual coal pillar relative to the direction of the roadway. The peak stress appears at the residual coal pillar; the further the location from the residual coal pillar, the smaller the stress value and the smaller the stress ratio. From the upper and lower levels, the further away from the bottom plate, the smaller σ_1 and σ_3 are, and the stress ratio also decreases. Due to the influence of the residual coal pillar, the σ_1 value decreases at the location of where the roadway's layout increases, and σ_1 at the location of the roadway roof reaches the lowest point of the stage, and then, the maximum principal stress continues the decreasing trend after a slight increase is inflicted by the influence of the residual coal column of the upper coal seam. The maximum principal stress and the minimum principal stress at the center position of the roadway are about 16.2 MPa and 6.2 MPa, respectively.

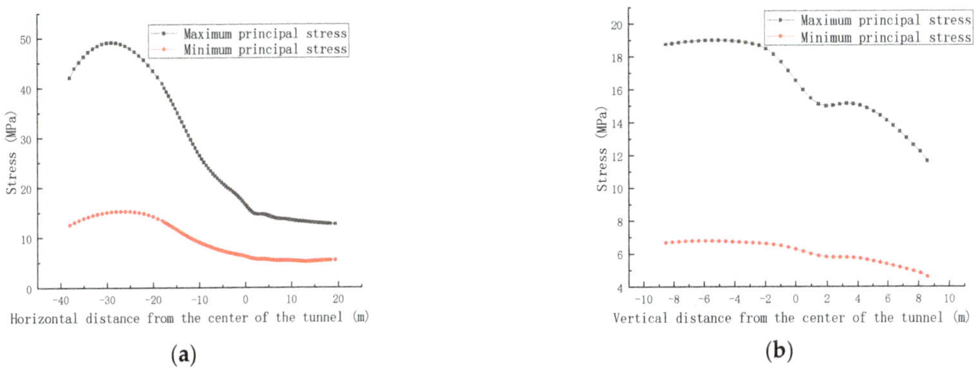

Figure 10. Changes in the principal stresses of the roadway's surrounding rock when the roadway excavation leaves top coal bottom coal. (**a**) Horizontal direction. (**b**) Variation direction.

According to the plastic zone distribution cloud map (Figure 11), the plastic zone of the roadway is still asymmetrically distributed. The range is slightly smaller than the range of the roadway when dug along the top, and the plastic zone of the roadway is connected with the plastic zone of the residual coal column. The plastic zone mainly appears at the top and negative gang of the roadway, while the plastic zone at the positive gang and bottom of the roadway is smaller.

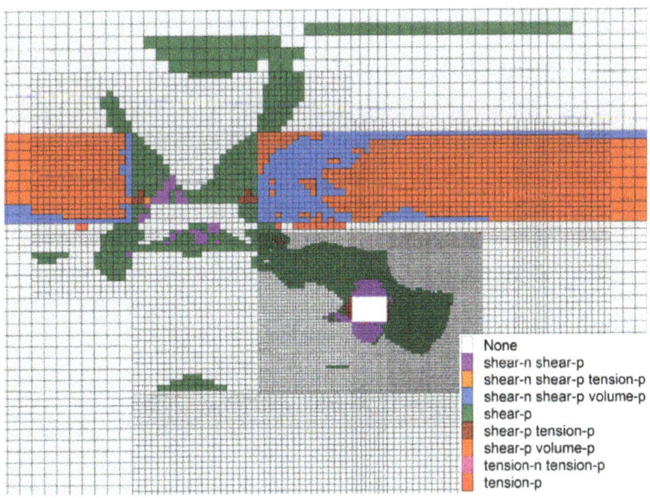

Figure 11. Plastic zone distribution cloud map when the roadway excavation leaves top coal bottom coal.

(3) Roadway excavation along the floor

The horizontal distance between the roadway dug along the bottom and the residual coal column of the overlying coal seam is 15.2 m, and the vertical distance is about

20.4 m. According to the stress cloud diagram (Figure 12), σ_1 and σ_3 appear at the residual coal column of the overlying coal seam, and σ_1 near the residual coal column is about 70–110 MPa; σ_3 is about 25–58 MPa. σ_1 around the roadway is about 5–13 MPa, and σ_3 is about 0.8–2 MPa.

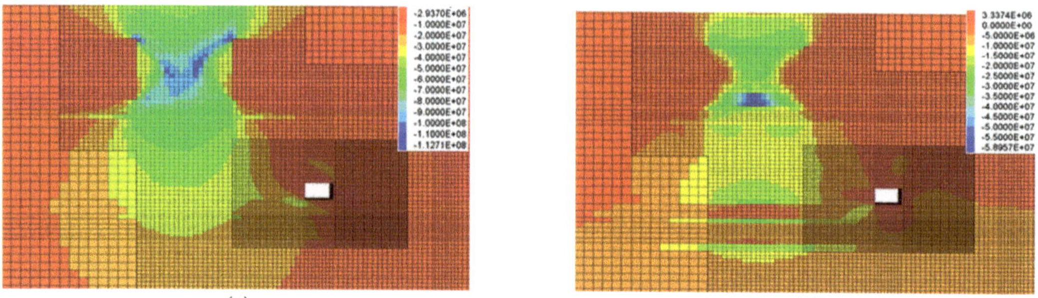

(**a**) (**b**)

Figure 12. Principal stress distribution contour of the roadway's surrounding rock when the roadway excavation along the floor. (**a**) σ_1 cloud. (**b**) σ_3 cloud.

This can be seen in the main stress variation curve (Figure 13). In the horizontal direction, σ_1 and σ_3 increase and then decrease from the residual coal pillar to the direction of the roadway, and the peak stress appears at the residual coal pillar. At the location away from residual coal pillars, the smaller the stress value, the smaller the stress ratio. From the vertical direction, at distances further from the bottom plate, σ_1 and σ_3 experience smaller decreases, and the stress ratio also decreases. σ_1 at the roadway is about 18.5 MPa, and σ_3 is about 6.5 MPa.

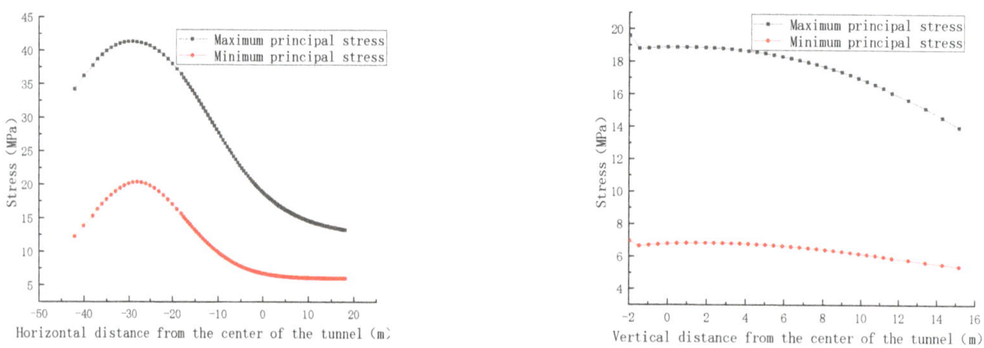

(**a**) (**b**)

Figure 13. Changes in the principal stresses of the roadway's surrounding rock when the roadway excavation along the floor. (**a**) Horizontal direction. (**b**) Variation direction.

According to the plastic zone's distribution cloud map (Figure 14), the plastic zone of the roadway is not connected with the plastic zone at the residual coal pillar, and the range of the plastic zone of the roadway obviously reduced the plastic zone. The maximum damage depth is 6 m, and the plastic zone mainly appears at the top plate and positive gang of the roadway.

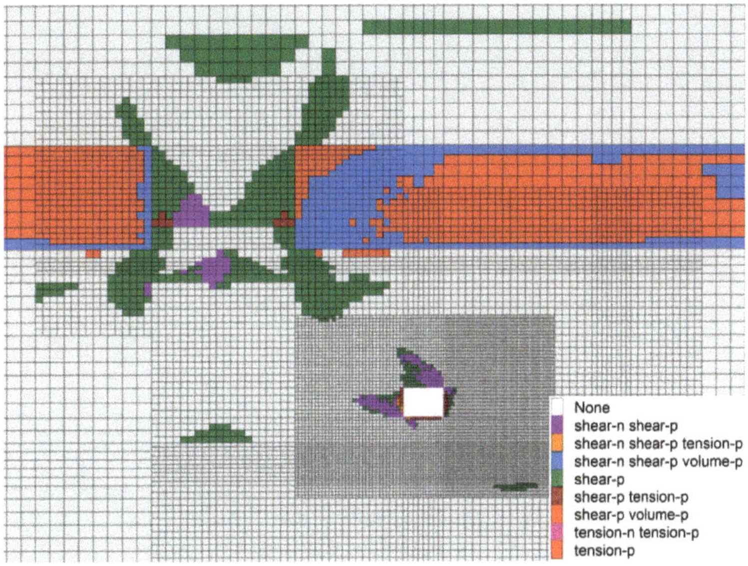

Figure 14. Plastic zone distribution cloud map when the roadway excavation along the floor.

3.1.2. Roadways with Horizontal Layer Differences

(1) The coal column is 4 m

When the coal pillar is 4 m, the horizontal distance between the roadway and the residual coal pillar is 31.85 m, and the vertical distance is about 18.5 m. According to the stress cloud diagram (Figure 15), σ_1 and σ_3 appear at the residual coal column of the overlying coal seam, and σ_1 near the residual coal column is about 60–115 MPa. σ_3 is about 22–60 MPa. σ_1 around the roadway is about 5–18 MPa, and σ_3 is about 0.4–1 MPa.

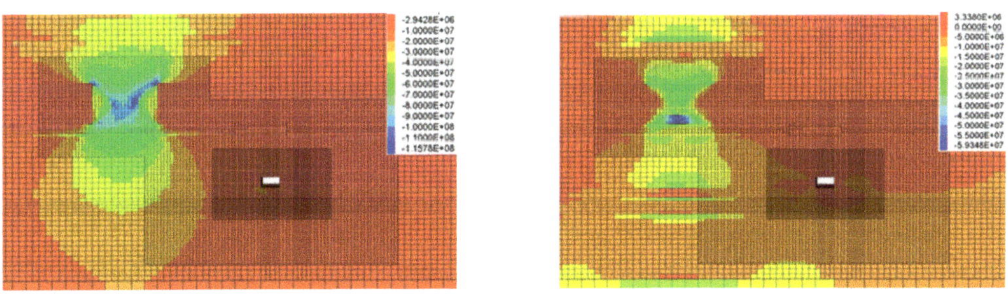

(**a**) (**b**)

Figure 15. Principal stress distribution contour of the roadway's surrounding rock when the coal column is 4 m. (**a**) σ_1 cloud. (**b**) σ_3 cloud.

This can be seen in the main stress variation curve (Figure 16). In the horizontal direction, σ_1 and σ_3 increase and then decrease from the residual coal pillar to the direction of the roadway, and the peak stress appears at the residual coal pillar. At a location away from residual coal pillars, the smaller the stress value, the smaller the stress ratio. From the vertical direction, σ_1 increases and then decreases as it moves away from the center of the roadway, while σ_3 gradually decreases. σ_1 at the center of the roadway is about 13.5 MPa, and σ_3 is about 6 MPa.

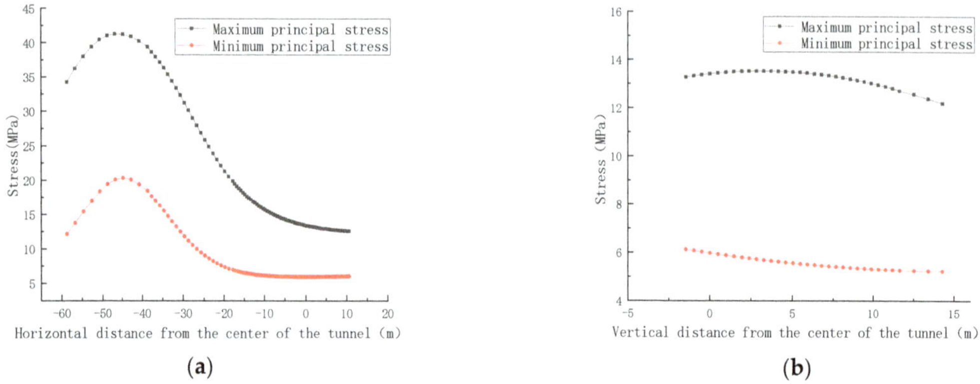

Figure 16. Changes in the principal stresses of the roadway's surrounding rock when the coal column is 4 m. (**a**) Horizontal direction. (**b**) Variation direction.

According to the plastic zone distribution cloud map (Figure 17), the maximum damage depth of the roadway plastic zone is 3 m. The range is small and is mainly distributed in the top plate and two gang positions. The bottom side has almost no plastic zone distributions, the roadway plastic zone in an independent state.

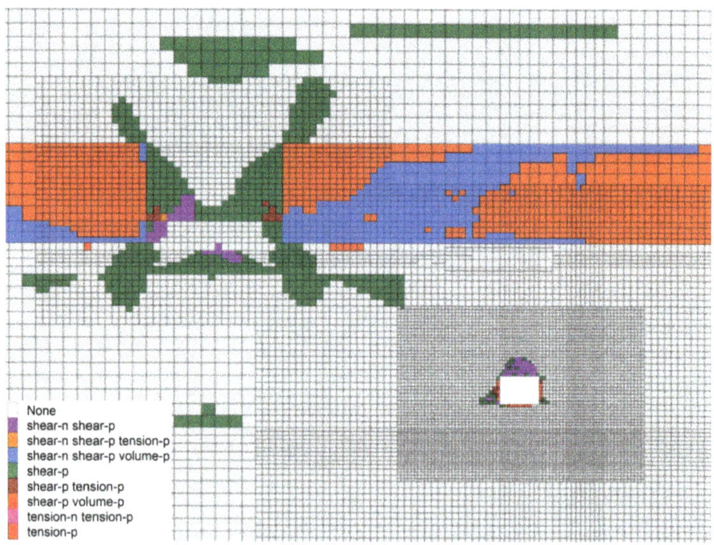

Figure 17. Plastic zone distribution cloud map when the coal column is 4 m.

(2) The coal column is 8 m

When the coal column is 8 m, the horizontal distance from the residual coal column is 27.85 m and the vertical distance is about 18.5 m. According to the stress cloud diagram (Figure 18), σ_1 and σ_3 appear at the residual coal column of the overlying coal seam, and σ_1 near the residual coal column is about 54–115 MPa; σ_3 is about 25–59 MPa. σ_1 around the roadway is about 6–18 MPa, and σ_3 is about 0.2–4 MPa.

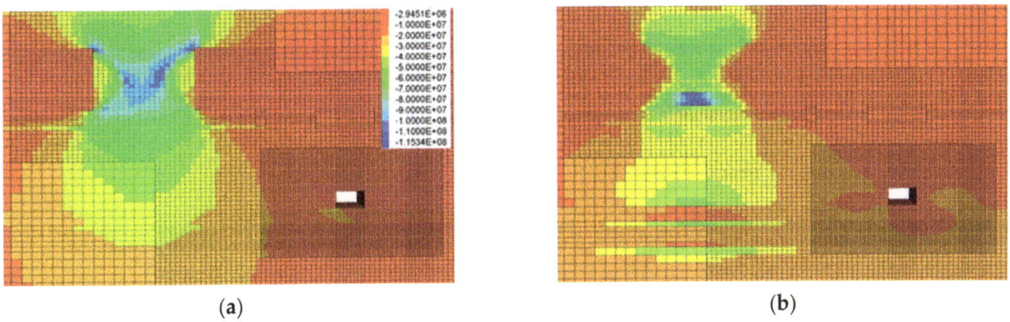

Figure 18. Principal stress distribution contour of the roadway's surrounding rock when the coal column is 8 m. (**a**) σ_1 cloud. (**b**) σ_3 cloud.

The stress change curve shows (Figure 19) that, along the horizontal direction, the stress peak directly below the residual coal pillar is reached, and after moving away from the residual coal pillar, σ_1 and σ_3 gradually decrease, and σ_1/σ_3 gradually decrease. Along the vertical direction, σ_1 first increases and then decreases as it moves away from the center of the roadway, while σ_3 gradually decreases. σ_1 at the center of the roadway is about 14 MPa, and σ_3 is about 6 MPa.

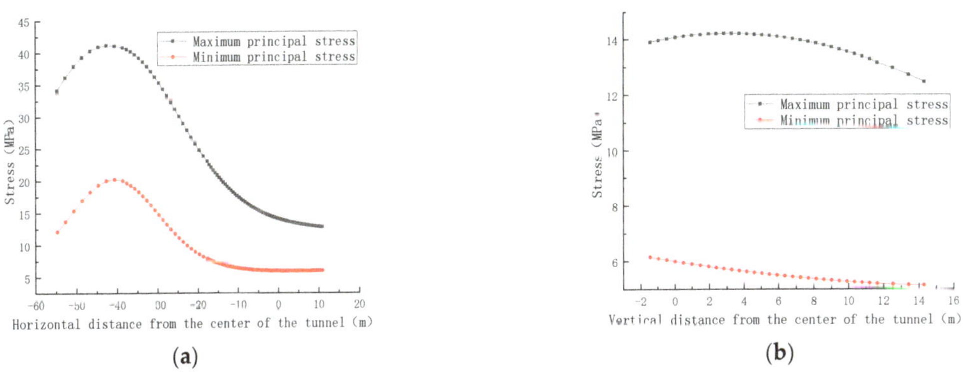

Figure 19. Changes in the principal stresses of the roadway's surrounding rock when the coal column is 8 m. (**a**) Horizontal direction. (**b**) Variation direction.

According to the plastic zone's distribution cloud map (Figure 20), the maximum damage depth of the plastic zone of the roadway is 3.5 m, the range is slightly increased than that of the 4 m coal pillar, and the distribution pattern is similar to that of the 4 m coal pillar.

(3) The coal column is 15 m

When the coal column is 15 m, the horizontal distance from the residual coal column is 20.85 m, and the vertical distance is about 18.5 m. According to the stress cloud diagram (Figure 21), σ_1 and σ_3 appear at the residual coal column of the overlying coal seam, and σ_1 near the residual coal column is about 67–113 MPa; σ_3 is about 26–59 MPa. σ_1 around the roadway is about 5–26 MPa, and σ_3 is about 0.6–5 MPa.

The stress change curve shows (Figure 22) that along the horizontal direction, the stress peak directly below the residual coal pillar is reached, and after moving away from the residual coal pillar, σ_1 and σ_3 gradually decrease, and σ_1/σ_3 gradually decrease. Along the vertical direction, σ_1 first increases and then decreases as it moves away from the center of the roadway, while σ_3 gradually decreases. σ_1 at the center of the roadway is about 15.6 MPa, and σ_3 is about 6 MPa.

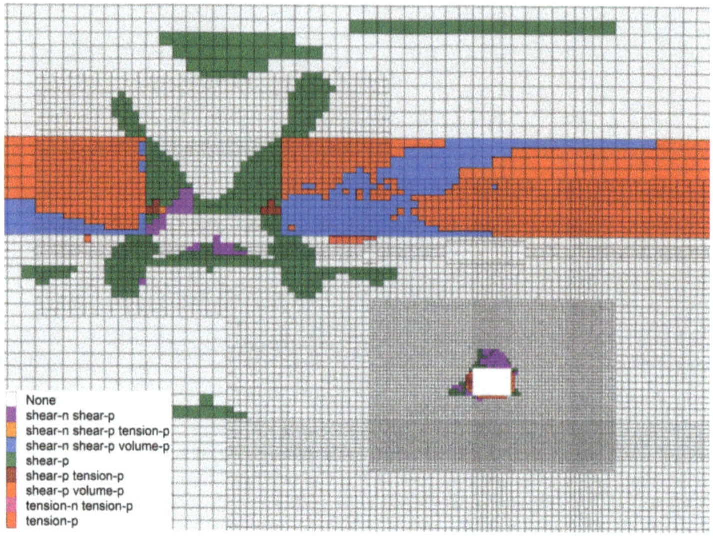

Figure 20. Plastic zone distribution cloud map when the coal column is 8 m.

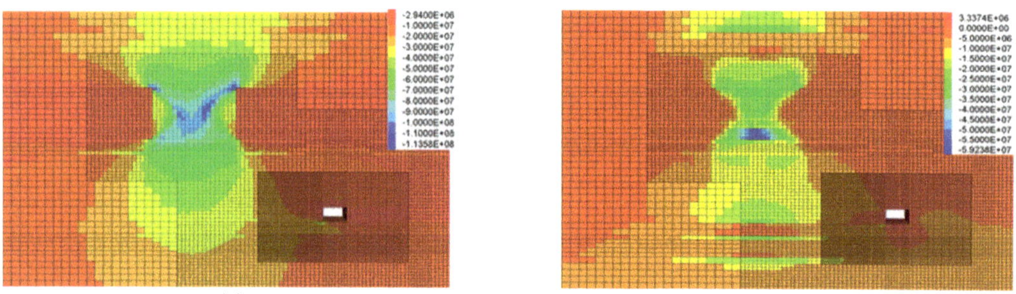

(**a**) (**b**)

Figure 21. Principal stress distribution contour of the roadway's surrounding rock when the coal column is 15 m. (**a**) σ_1 cloud. (**b**) σ_3 cloud.

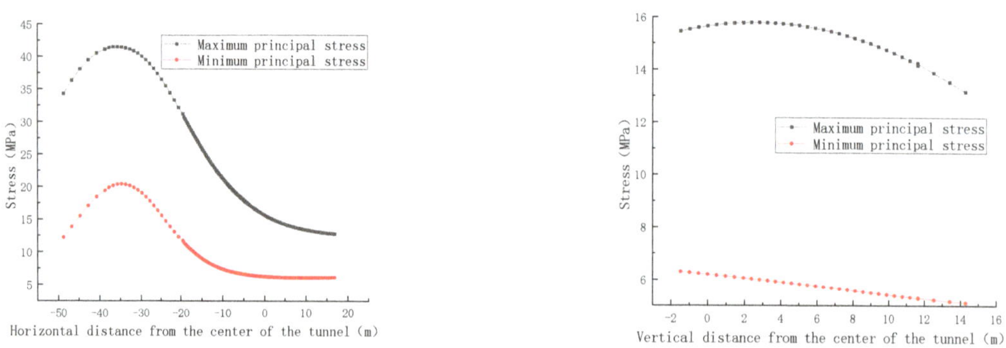

(**a**) (**b**)

Figure 22. Changes in the principal stresses of the roadway's surrounding rock when the coal column is 15 m. (**a**) Horizontal direction. (**b**) Variation direction.

According to the plastic zone distribution cloud map (Figure 23), the maximum damage depth of the plastic zone of the roadway is 5.5 m, and the range further increases. The distribution pattern is similar to those above the 4 m and 8 m coal pillars. When the coal pillar's size is 20 m, this condition is consistent with the situation of the roadway dug along the bottom in the previous subsection, and it will not be repeated here. The maximum depth of damage in the plastic zone of the roadway is 6 m, and the extent of the plastic zone continues to grow.

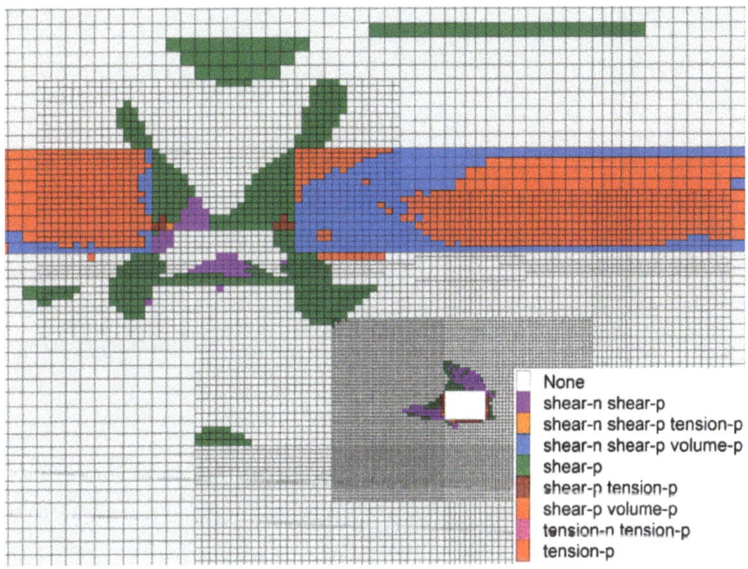

Figure 23. Plastic zone distribution cloud map when the coal column is 15 m.

(4) The coal column is 25 m

When the coal column is 25 m, the horizontal distance from the residual coal column is 10.85 m, and the vertical distance is about 18.5 m. According to the stress cloud diagram (Figure 24), σ_1 and σ_3 appear at the residual coal column of the overlying coal seam, and σ_1 near the residual coal column is about 65–110 MPa; σ_3 is about 25–58 MPa. σ_1 around the roadway is about 4–25 MPa, and σ_3 is about 0.4–6 MPa.

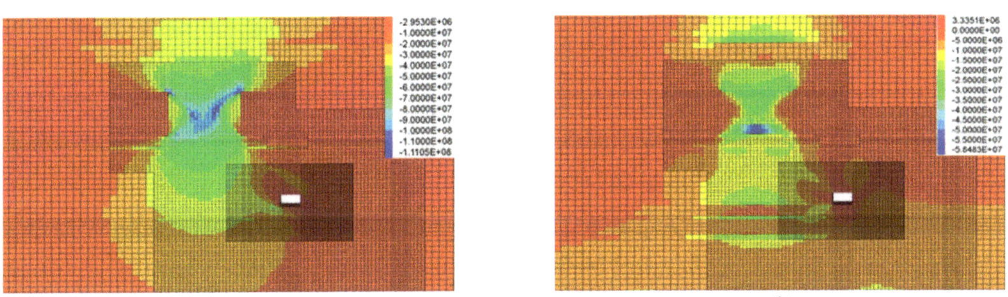

(a) (b)

Figure 24. Principal stress distribution contour of the roadway's surrounding rock when the coal column is 25 m. (**a**) σ_1 cloud. (**b**) σ_3 cloud.

The stress change curve shows (Figure 25) that along the horizontal direction, the stress peak is reached directly below the residual coal pillar, and after moving away from the residual coal pillar, σ_1 and σ_3 gradually decrease, and σ_1/σ_3 gradually decreases. Along the vertical direction, σ_1 first increases and then decreases as it moves away from the center of the roadway, while σ_3 gradually decreases. σ_1 at the center of the roadway is about 21 MPa, and σ_3 is about 7.5 MPa. σ_1 and σ_3 are in a constant state of growth.

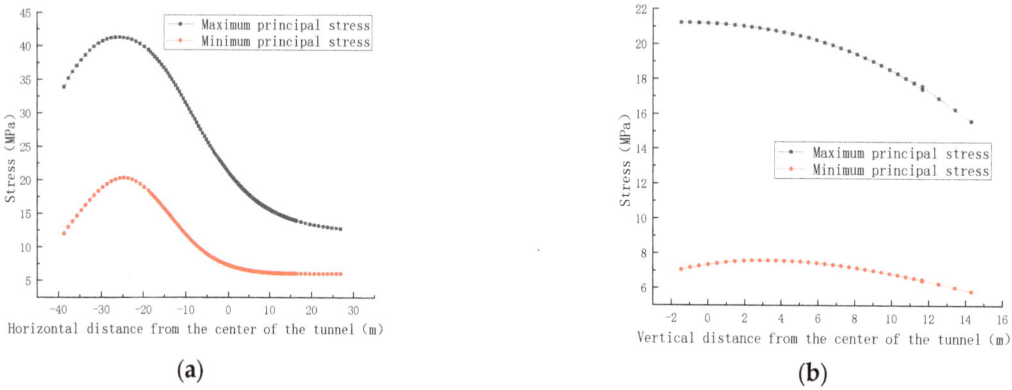

Figure 25. Changes in the principal stresses of the roadway's surrounding rock when the coal column is 25 m. (**a**) Horizontal direction. (**b**) Variation direction.

According to the plastic zone distribution cloud map (Figure 26), the maximum damage depth of the plastic zone of the roadway is 7.5 m, the plastic zone appears with an obvious butterfly shape, and the upper butterfly lobe's development is obvious and is mainly distributed at the top of the roadway and the positive gang. The lower butterfly lobe's development is smaller. The upper butterfly lobe points to the direction of the residual coal pillar. The plastic zone of the roadway's surrounding rock and the plastic zone at the residual coal pillar are not yet connected.

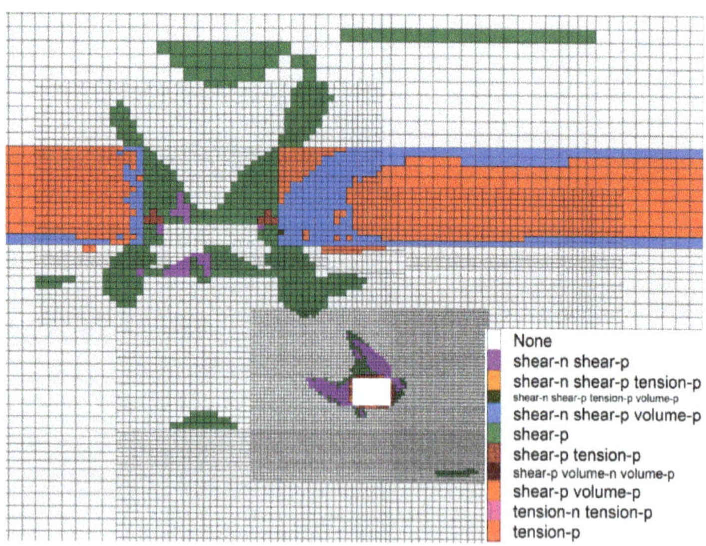

Figure 26. Plastic zone distribution cloud map when the coal column is 25 m.

(5) Roadway dug directly below the residual coal pillar

When the roadway is located directly below the residual coal pillar, the vertical distance from the residual coal pillar of the overlying coal seam is about 18.5 m. According to the stress cloud diagram (Figure 27), σ_1 and σ_3 appear at the residual coal pillar of the overlying coal seam, and σ_1 near the overlying residual coal pillar is about 61–104 MPa, and σ_3 is about 22–50 MPa. σ_1 around the roadway is about 5–24 MPa, and σ_3 is about 0.4–7 MPa.

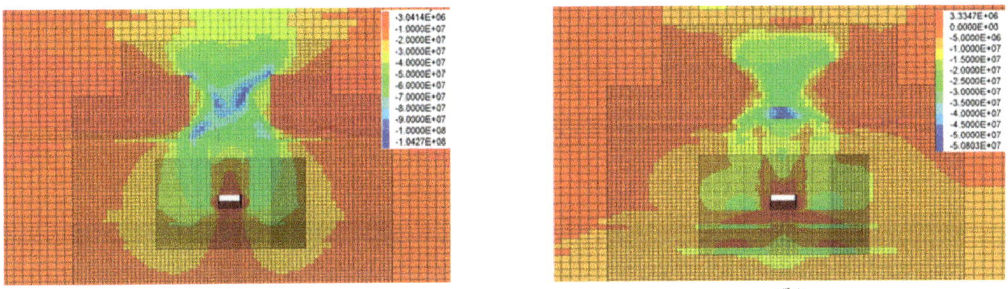

(a) (b)

Figure 27. Principal stress distribution contour of the roadway's surrounding rock when the roadway dug directly below the residual coal pillar. (**a**) σ_1 cloud. (**b**) σ_3 cloud.

The stress change curve shows (Figure 28) that along the horizontal direction, the stress peak is reached directly below the residual coal pillar, and after moving away from the residual coal pillar, σ_1 and σ_3 gradually decrease, and σ_1/σ_3 gradually decreases. Along the vertical direction, as the distance from the center of the roadway increases, σ_1 gradually increases, while σ_3 gradually decreases. σ_1 at the center of the roadway is about 41 MPa, and σ_3 is about 20 MPa, at which time both σ_1 and σ_3 reach their maximum values.

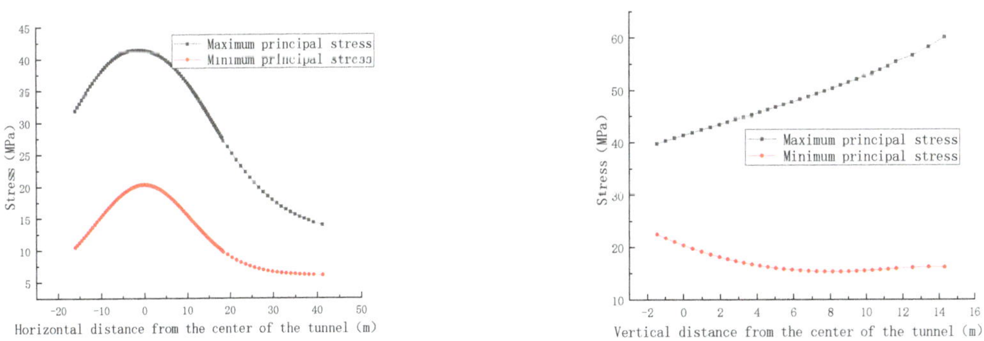

(a) (b)

Figure 28. Changes in the principal stresses of the roadway's surrounding rock when the roadway dug directly below the residual coal pillar. (**a**) Horizontal direction. (**b**) Variation direction.

According to the plastic zone's distribution cloud map (Figure 29), the plastic zone of the roadway is connected with the plastic zone at the residual coal pillar. The plastic zone appears with an obvious butterfly shape, and the butterfly's lobe development is obvious. the top plate of the roadway and the top angle position of the two helpers have a large ranged plastic zone.

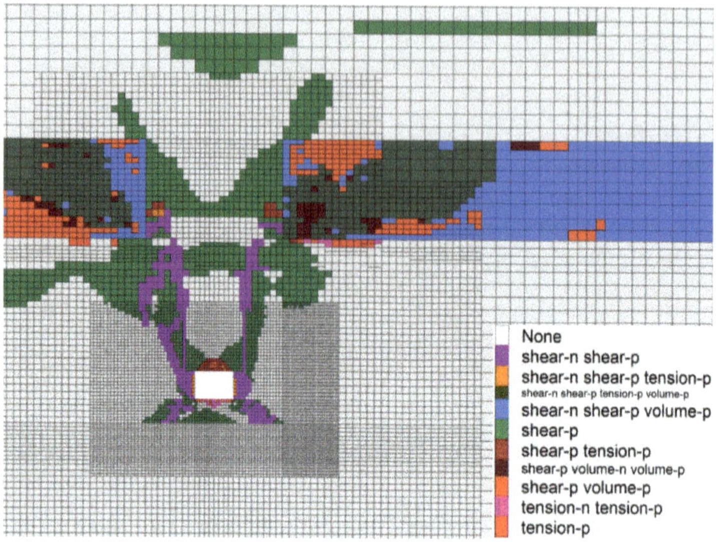

Figure 29. Plastic zone distribution cloud map when the roadway dug directly below the residual coal pillar.

3.2. Deformation Law and Evolution of the Surrounding Rock of the Roadway under Different Positions

Figure 30 shows the cloud map of the deformation of the surrounding rock of the roadway at different locations. According to the cloud map's data (Figure 31), the deformation curves of the surrounding rock of the roadway at different locations are drawn. When conducting an analysis along the vertical layer, the deformation volume of the roadway dug along the bottom slab is the largest, the deformation of the surrounding rock mainly appears in the positive gang and the top slab of the roadway, and the deformation of the positive gang of the roadway is about 27 mm. The deformation of the top slab is about 28 mm, and the deformation of the negative gang is smaller at about 13 mm. The bottom slab has almost no deformations at about 6 mm. When conducting an analysis along the horizontal layer, the deformation of the surrounding rock mainly occurs at the top plate and the negative gang of the roadway, and the bottom plate also has some deformations. The maximum deformation of the top plate of the roadway is about 140 mm, the deformation of the positive and negative helpers of the roadway is about 130 mm, and the deformation of the bottom plate is about 30 mm.

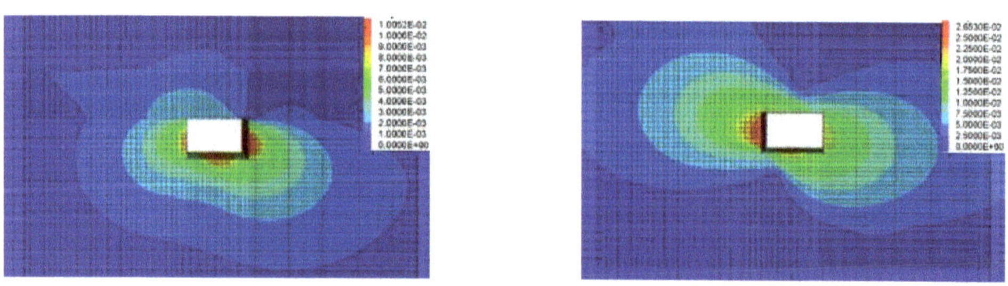

(**a**) (**b**)

Figure 30. *Cont.*

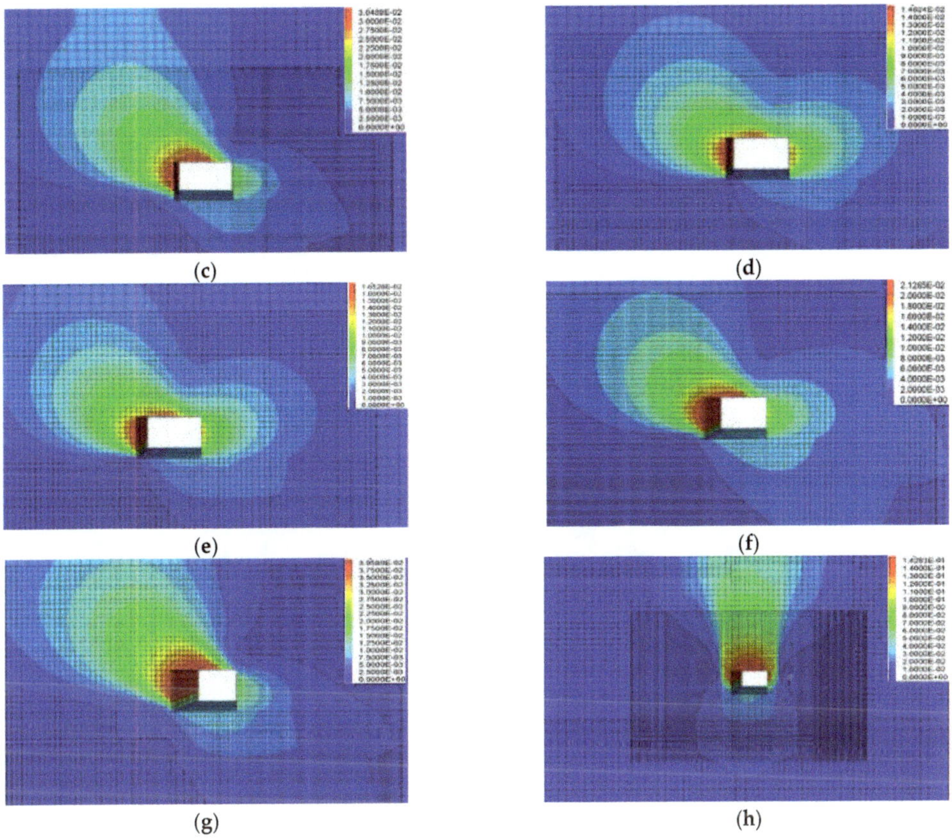

Figure 30. Deformation cloud diagram of the roadway's surrounding rock at different locations. (**a**) Roadway excavation along the roof. (**b**) Roadway excavation leaves: top coal bottom coal. (**c**) Roadway excavation along the floor. (**d**) The coal column is 4 m. (**e**) The coal column is 8 m. (**f**) The coal column is 15 m. (**g**) The coal column is 25 m. (**h**) Roadway dug directly below the residual coal pillar.

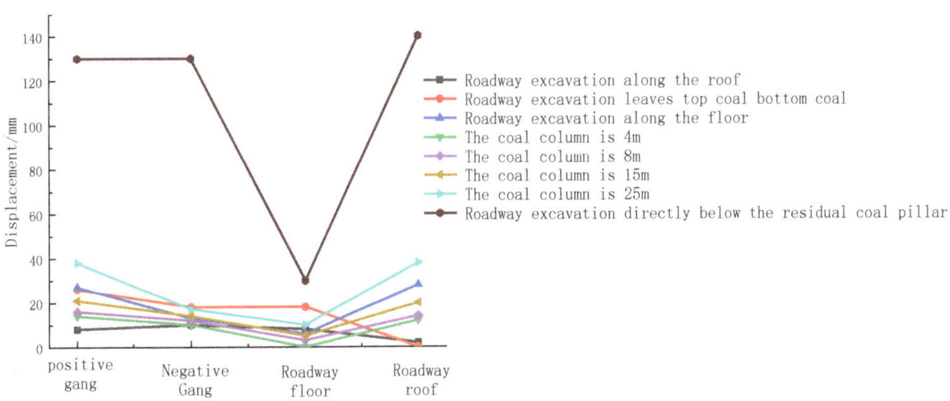

Figure 31. Deformation curves of the roadway's surrounding rock at different locations.

3.3. Analysis of Results

According to Figure 32, when the horizontal distance of the roadway from the residual coal pillar remains unchanged, the deformation of the roadway excavation along the floor is the largest, which is 27 mm, and the deformation mainly occurs in the secondary gang and the top slab. When the roadway is dug along the top and left top bottom coal boring, the plastic zone of the roadway's surrounding rock is much larger than the range of the plastic zone when the roadway is dug along the bottom, which indicates that the deformation of the surrounding rock is smaller, and the roadway is more stable when the roadway is arranged along the top. However, based on the fact that the mine is mined by the caving mining method, the roadway's layout should be located in the middle and lower positions of the coal seam.

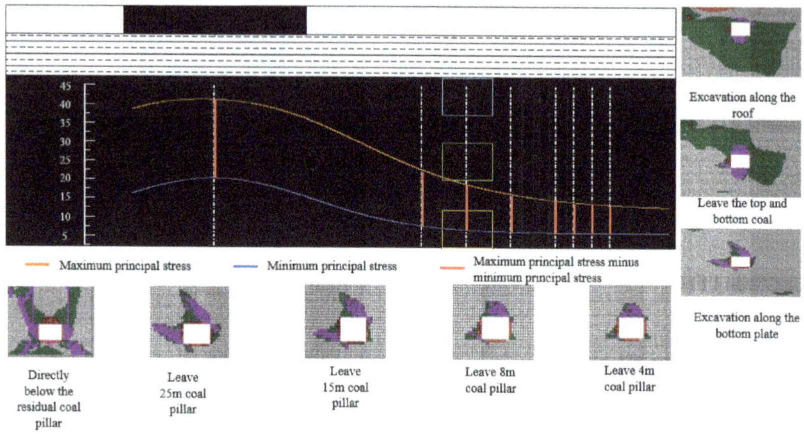

Figure 32. Principal stress distribution and the plastic zone area under different roadway layout positions.

When the vertical direction of the roadway remains unchanged, the deformation of the roadway and the range of the plastic zone increase with the increase in the size of the left coal pillar. When the roadway is arranged directly below the residual coal pillar, the deformation of the roadway reaches the maximum, the maximum deformation of the top plate of the roadway reaches 140 mm. The deformation of the two helpers also reaches 130 mm, and the plastic zone is connected with the plastic zone of the residual coal pillar. This indicates that this location is most prominently affected by the concentrated stress of the residual coal pillar, and the roadway's stability is poor. With the comprehensive analysis conducted above, in the process mining a close coal seam group, the stress influence of residual coal column on the roadway is reduced and can effectively improve the stability of the roadway.

4. Discussion

For the analysis of roadway deformation variables, distinct from the long-term displacement monitoring methods used in the past [22], short-term surface displacement monitoring methods were used in this study to analyze roadway deformation variables. Long-term displacement monitoring methods can measure the deformation of the roadway more accurately, and air can be used to analyze the deformation and damage law of the roadway with a large amount of long-term data to obtain more accurate information on the deformation and damage mechanisms of the roadway. However, in the actual production process, when the roadway experiences excessive deformation or even closure, short-term surface displacement monitoring methods can be used to analyze the damage pattern of the roadway. Combined with the geological environments in which the roadway is located, it is clear that in the non-uniform stress field, the roadway exhibits asymmetric damage morphologies.

In order to guarantee the stability of the roadway and to improve the support efficiency, FLAC3D numerical simulation software was used to establish a numerical model for the multi-point roadway arrangement, and the main stress values, the ratio of main stresses, the deformation variables of the roadway, and the connectivity between the roadway and the plastic zone of the residual coal column were analyzed in detail from different directions. In previous studies, the connectivity of the plastic zone was not used as a criterion for measuring the stability of the roadway [32–35]. However, in the actual production operation, the plastic zone of the residual coal pillar is the damage range of the surrounding rock. If the roadway and the plastic zone of the residual coal pillar appear in a superimposed area, then it will seriously affect the support of the roadway and cannot guarantee the stability of the roadway. In this study, the above factors were analyzed and combined with the mining processes adopted by the coal mine. It is proposed that, in the context of this type of project, the layout of the roadway in the middle and lower positions of the coal seam and the use of small coal pillars can guarantee the stability of the roadway.

5. Conclusions

In this paper, the deformation volume and deformation characteristics of the return airway of 30503 in Tashan mine were analyzed. We conducted short-term surface displacement monitoring by using field testing and determined the existence of the asymmetric damage morphology of the roadway and established a roadway layout optimization model by means of numerical simulations for the actual engineering situation of the working face. We also analyzed the main stress magnitude and the plastic zone morphology of the roadway for eight groups at different positions. The study showed the following.

(1) Via surface displacement observations, it can be seen that the 30503 working face return roadway experienced large deformations within a short period of time, and the two gangs' displacement is obviously larger than the top plate's displacement. At the same measurement point position, the two gangs' displacement and the top plate's displacement also has obvious differences. The roadway exhibits obvious non-uniform damage phenomena.

(2) The residual coal pillar of the overlying coal seam affects the stress environment of the lower coal seam roadway due to the stress from its own weight. The change in the roadway's layout position has little effect on the trend of the main stress curve of the lower coal seam, and the peaks of σ_1 and σ_3 in the horizontal direction appear at the residual coal pillar. σ_1 and σ_3 and σ_1/σ_3 are smaller at the position far from the residual coal pillar.

(3) As the coal mine is mined by releasing the top coal, the arrangement of the roadway in the vertical direction should be in the middle and lower positions of the coal seam. Although the distribution of the plastic zone along the bottom of the roadway is small, its roadway's deformation variable is larger. In order to guarantee the stability of the roadway, the roadway should be arranged in the middle position of the coal seam.

(4) When the coal column left in the lower coal seam's roadway is 4 m, the maximum deformation of the roadway's surrounding rock is the smallest. At this time, the roadway is farthest away from the residual coal column of the overlying coal seam and is least affected by its stress; therefore, during the mining process of the coal seam group, the roadway is arranged as a small coal column (far away from the residual coal column), which can effectively improve the stability of the roadway.

Author Contributions: Conceptualization, X.-H.W., H.-H.Z. and R.-Q.G.; methodology, X.-H.W. and Z.W.; software, Z.W. and X.-L.L.; validation, R.-Q.G. and Y.S.; formal analysis, H.-H.Z. and Y.S.; investigation, X.-H.W. and R.-Q.G.; resources, X.-H.W. and Z.W.; data curation, Z.W. and X.-L.L.; writing—original draft preparation, X.-H.W. and R.-Q.G.; writing—review and editing, H.-H.Z. and Y.S.; visualization, X.-L.L. and Y.S.; supervision, Z.W.; project administration, X.-H.W. and Z.W.; funding acquisition, X.-H.W. All authors have read and agreed to the published version of the manuscript.

Funding: This research was funded by the Natural Science Foundation of China (U22A20165), grant number "51574224", and the Guizhou Province Science and Technology Support Plan Project of China, grant number "Guizhou Province Science and Technology Support Plan Project [2021] General 400".

Institutional Review Board Statement: Not applicable.

Informed Consent Statement: Not applicable.

Data Availability Statement: The data presented in this study are available upon request from the corresponding author.

Conflicts of Interest: The authors declare no conflict of interest.

References

1. Kang, H.P.; Xu, G.; Wang, B.M.; Wu, Y.Z.; Jiang, P.F.; Pan, J.F.; Pang, Y.H. Forty years development and prospects of underground coal mining and strata control technologies in China. *J. Min. Strat. Control. Eng.* **2019**, *1*, 013501.
2. Wang, J. Sustainable coal mining is based on mining ground control. *J. Min. Strat. Control. Eng.* **2019**, *1*, 013505.
3. Su, S.; Du, Y.; Zhu, J. Numerical study on bearing behavior of layered rock mass for deep roadway. *J. Min. Strat. Control. Eng.* **2020**, *2*, 013002.
4. Kang, H. Spatial scale analysis on coalmining and strata control technologies. *J. Min. Strat. Control. Eng.* **2020**, *2*, 023538.
5. Zhang, Z.; Deng, M.; Wang, X.; Yu, W.; Zhang, F.; Dao, V.D. Field and numerical investigations on the lower coal seam entry failure analysis under the remnant pillar. *Eng. Fail. Anal.* **2020**, *115*, 104638. [CrossRef]
6. Eberhardt, E. The hoek-brown failure criterion. *Rock Mech. Rock Eng.* **2012**, *45*, 981–988. [CrossRef]
7. Zhang, Y.; Zhang, C.L.; Wei, C.C.; Liu, Y.D.; Zhang, S.Q.; Zhao, J.J. The study on roadway layout in coordination of mining coal seams base on failure of floor strata. *Adv. Mater. Res.* **2014**, *889*, 1362–1374. [CrossRef]
8. Xia, Z.; Yao, Q.; Meng, G.; Xu, Q.; Tang, C.; Zhu, L.; Wang, W.; Shen, Q. Numerical study of stability of mining roadways with 6.0-m section coal pillars under influence of repeated mining. *Int. J. Rock Mech. Min. Sci.* **2021**, *138*, 104641. [CrossRef]
9. Zou, H.; Yan, E.C. The surrounding rock deformation and failure mechanism of wudang group schists tunnel. *Electron. J. Geotech. Eng.* **2015**, *20*, 6557–6576.
10. Lee, K.H.; Kim, D.H.; Park, J.J. A study on critical strain based damage-controlled test for the evaluation of rock tunnel stability. *J. Med. Screen.* **2011**, *13*, 91–97. [CrossRef]
11. Thompson. Estimation of minimum support resistance and maximum allowable deformation in elastic-plastic enclosures. *J. Geotech. Eng.* **1986**, *4*, 81–88.
12. Miao, X. Natural equilibrium arch and the stability of roadway envelope. *Mine Press. Roof Manag.* **1990**, *2*, 55–57.
13. He, F.; Liu, L.; Qian, M. Analysis and prevention of direct topping of blocky loose rocks at header mining face. *Coal* **1995**, *4*, 7–10.
14. Zhao, P. The shape of the natural balance archon the roadway roof considering the original stress in the horizontal direction. *J. China Univ. Min. Technol.* **1978**, *7*, 51–58.
15. Yu, X.; Qiao, D. Theory of axial variation and three roles of axial ratio for stabilizing country rock. *Nonferrous Met. Eng.* **1981**, *33*, 8–15.
16. Yu, X. The theory of axial variation and basic rules of deformation and fracture of rock surrounding. *Uranium Min. Metall.* **1982**, *1*, 8–17.
17. Chen, G. *Mechanism Research of the Maximum Horizontal Stress on Rockburst and Its Application;* China University of Mining and Technology: Xuzhou, China, 2009; pp. 1–20.
18. Wang, C.; Liu, Y.; Hou, X.; Elmo, D. Investigation of the spatial distribution pattern of 3D microcracks in single-cracked breakage. *Int. J. Rock Mech. Min. Sci.* **2022**, *154*, 105126. [CrossRef]
19. Wang, C.; Zhou, B.; Li, C.; Cao, C.; Sui, Q.; Zhao, G.; Yu, W.; Chen, Z.; Wang, Y.; Liu, B.; et al. Experimental investigation on the spatiotemporal-energy evolution pattern of limestone fracture using acoustic emission monitoring. *J. Appl. Geophys.* **2022**, *206*, 104787. [CrossRef]
20. Chen, S.; Xia, Z.; Feng, F.; Yin, D. Numerical study on strength and failure characteristics of rock samples with different hole defects. *Bull. Eng. Geol. Environ.* **2020**, *80*, 1523–1540. [CrossRef]
21. Li, X.; Chen, S.; Wang, E.; Li, Z. Rockburst mechanism in coal rock with structural surface and the microseismic (MS) and electromagnetic radiation (EMR) response. *Eng. Fail. Anal.* **2021**, *124*, 105396. [CrossRef]
22. Małkowski, P.; Niedbalski, Z.; Majcherczyk, T.; Bednarek, Ł. Underground monitoring as the best way of roadways support design validation in a long time period. *Min. Miner. Deposits* **2020**, *14*, 1–14. [CrossRef]
23. Khalymendyk, I.; Baryshnikov, A. The mechanism of roadway deformation in conditions of laminated rocks. *J. Sustain. Min.* **2018**, *17*, 41–47. [CrossRef]
24. Wang, X.H.; Zhao, Z.Q.; Wu, Z.; Zhang, H.H.; Wang, J.; Sui, Y.; Ren, Z.C. Deformation and Failure Mechanism of the Lower Roadway in a Close-Range Residual Coal Pillar Area. *Front. Earth Sci.* **2022**, *10*. [CrossRef]
25. Goodman, R.E. *Introduction to Rock Mechanics;* Willy: New York, NY, USA, 1978; pp. 66–68.
26. Zhao, Z.Q. *Study on Mechanism and Control Method of Deformation and Failure of Surrounding Rock in Large Deformation Mining Roadway;* China University of Mining and Technology: Beijing, China, 2014; pp. 26–32.

27. Ma, N.J.; Guo, X.F.; Zhao, Z.Q. Occurrence mechanisms and judging criterion on circular tunnel butterfly rock burst in homogeneous medium. *J. China Coal Soc.* **2016**, *41*, 2679–2688. [CrossRef]
28. Ma, N.J.; Zhao, X.D.; Zhao, Z.Q.; Guo, X.F.; Liu, H.T. Conjecture of butterfly shape coal and gas outburst mechanism in excavation roadway. *Min. Sci. Tech.* **2017**, *2*, 136–148.
29. Zhao, Z.Q.; Ma, N.J.; Guo, X.F.; Zhao, X.D.; Fan, L. Falling principle and support design of butterfly-failure roof in large de-formation mining roadways. *China Coal Soc.* **2016**, *41*, 2932–2939.
30. Zhao, Z.Q.; Ma, N.J.; Guo, X.F.; Zhao, X.D.; Xia, Y.X.; Ma, Z.K. Mechanism conjecture of butterfly rock burst in coal seam road-way. *China Coal Soc.* **2016**, *41*, 2689–2697.
31. Guo, X.F.; Zhao, Z.Q.; Gao, X.; Wu, X.Y.; Ma, N.J. Analytical solutions for characteristic radii of circular roadway surrounding rock plastic zone and their application. *Int. J. Min. Sci. Technol.* **2019**, *29*, 263–272. [CrossRef]
32. Begalinov, A.; Almenov, T.; Zhanakova, R.; Bektur, B. Analysis of the stress deformed state of rocks around the haulage roadway of the Beskempir field (Kazakhstan). *Min. Miner. Deposits* **2020**, *14*, 28–36. [CrossRef]
33. Jin, G.; Wang, L.; Zhang, J.; Hu, M.; Duan, N. Roadway layout for recycling residual coal pillar in room-and-pillar mining of thick coal seam. *Int. J. Min. Sci. Technol.* **2015**, *25*, 729–734. [CrossRef]
34. Yang, W.; Liu, C.; Yang, Y. Reasonable malposition setting in close distance coal seams under influence of interlaminar stresses. *Rock Mechan. Eng.* **2012**, *31*, 2965–2972.
35. Li, J.; Li, H.; Li, Z. Research on river dike failure of short-distance coal seams mining under Hunchun River. *J. Min. Strat. Control. Eng.* **2020**, *2*, 013538.
36. Fan, Y. *Research on the Mining Method of Close Extra-Thick Coal Seam in Datong Tashan Coal Mine of China National Investment Corporation*; China University of Mining and Technology: Beijing, China, 2014.

processes

MDPI

Article

Mechanics-Seepage Experimental and Simulation Study of Gas-Bearing Coal under Different Load Paths

Haibo Sun [1,2,*], Baoyong Zhang [1], Zhijun Song [1], Bin Shen [1] and Hongyu Song [1]

1 School of Safety Engineering, Heilongjiang University of Science and Technology, Harbin 150022, China
2 Heilongjiang Longmei Jixi Mining Co., Ltd., Jixi 158199, China
* Correspondence: shbbs_hlj@163.com

Abstract: Mechanics-seepage synchronous tests on gas-bearing coal under three different stress paths were designed and implemented to evaluate how load path affected the mechanical strength and permeability of deep mining-disturbed coal. The cracks-count evolution of coal specimens during instability was observed through DEM numerical simulation. The results showed significant stress-strain and strength variations under different paths. At the time of failure, the specimen deformation and peak strength were Test 1 > Test 2 > Test 3, while the permeability was Test 3 > Test 2 > Test 1, with specimen permeability in Test 3 rising prominently. From numerical simulation, the cracks count was Test 2 > Test 3 > Test 1, with tensile cracks taking the largest proportion in Test 2 and shear cracks taking the largest proportion in Test 3. Our findings shed some light on the research and disaster prevention regarding coal and gas outburst.

Keywords: triaxial; permeability; strain; DEM; crack

Citation: Sun, H.; Zhang, B.; Song, Z.; Shen, B.; Song, H. Mechanics-Seepage Experimental and Simulation Study of Gas-Bearing Coal under Different Load Paths. *Processes* **2022**, *10*, 2255. https://doi.org/10.3390/pr10112255

Academic Editors: Feng Du, Aitao Zhou and Bo Li

Received: 9 October 2022
Accepted: 31 October 2022
Published: 2 November 2022

1. Introduction

Coal and gas outburst, as a major cause of gas hazards in coal mines, results from a very complex dynamic instability process at underground mine sites [1,2]. When such a hazard occurs, the gas adsorption/desorption and seam stress-strain at the outburst site undergo tremendous change. The gas also affects the deformed coal body, causing it to change further [3–5]. Currently, the mechanisms underlying coal and gas outburst remain unsolved. Wider investigation into the mechanical properties and seepage behavior of gas-bearing coal under stress-seepage coupling is of great importance.

As field testing is impractical, given the high risk factor of seams prone to coal and gas outburst, experimental study of coal-gas coupling mechanisms forms the fundamental means of understanding coal permeability evolution and damage mechanisms [6,7]. Xu et al. [8,9] experimentally characterized coal deformation and permeability under loading/unloading conditions and revealed that coal permeability variation is closely related to coal deformation and damage. Yin et al. [10] evaluated the effect of loading/unloading conditions on the mechanical properties of gas-bearing coal and established a whole process seepage speed–axial pressure equation. Wang et al. [11] studied the dynamic behaviors in coal seams under different mining layouts. Zhao et al. and Wang et al. [12,13] analyzed the relationship between coal permeability, stress difference and strain during loading/unloading through whole-process stress-strain permeability tests. Li et al. [14] observed the deformation and permeability behaviors of outburst coal samples under cyclic loads and found that permeability variation is closely related to coal damage and deformation. Cao et al. [15] analyzed how confining pressure and axial pressure affect coal gas seepage under constant gas pressure. Xue et al. [16] performed uniaxial tension and compression, conventional triaxial and dynamics tests, and obtained the mechanical characteristic parameters and deformation failure behaviors of coal samples under different load paths and load rates.

The discrete element method (DEM) is an important numerical method of solving coal mechanical problems due to its ability to assess coal mechanics and crack mechanisms from a mesoscopic perspective [17–19]. Indraratna et al. [20] made cyclic biaxial simulation tests at different frequencies with DEM software. They also analyzed the evolution of meso-mechanical parameters such as contact force and bond force formed during cyclic loading and explained the mechanism of particle fracture. Wang et al. [21] simulated crack propagation with a PFC-based particle flow model and observed how coal stress affects crack propagation. Jiang et al. [22] simulated a series of biaxial compression tests with DEM and observed the mechanical behavior of deep-sea methane-hydrate-bearing soils. Ismail et al. [23] built a DEM model for visualizing damage evolution and predicting failure envelopes of composite laminae under biaxial loads. Yang et al. [24] numerically simulated the failure behavior around a circular opening under biaxial compression. Sagong et al. [25] made experimental and numerical analyses of the sliding of fissures and joints in fissure-bearing rock under biaxial compression. Raisianzadeh et al. [26] used DEM to simulate the interaction between particles under biaxial load, and studied the crack propagation path and failure strength of rock containing prefabricated cracks Xu et al. [27] studied the influence of the angle between two cracks on the strength and crack propagation of the specimen.

From the literature review above, both physical experiments and numerical simulations help to understand the mechanisms behind coal and gas outburst. However, few attempts have been made to combine physical experiments with numerical simulation. In this paper, the deformation failure and gas-seepage behavior of briquette specimens under different stress paths are tested. The test results are then verified through particle flow code (PFC) numerical simulation. Specimen-crack evolution under different paths is also characterized. Our method offers a new approach to understanding the mechanisms behind coal and gas outburst.

2. Experimental Section

2.1. Specimen Procurement and Preparation

The structural complexity of raw coal can give rise to substantial discreteness of test results. Previous studies have demonstrated that briquette, in one way or another, possesses the physical–mechanical and adsorption properties of raw coal. It is also easy to transport and handle. For this reason, briquette is often used in laboratory studies as a substitute for raw coal [28].

The coal sample came from Jixi Mining Group's Xinfa coal mine, which is a high-gas mine. The sample has an ash content of 61.28%, a volatile content of 22.9%, and a solidity factor of 0.5866. After the sample was recovered from the mine face, it was sealed and delivered to the laboratory where it was crushed. When preparing specimens, 30 g river sand, 30 g cement, 210 g crushed coal, and 30 g water were mixed together and kept under 200 kN moulding pressure for 12 h before the specimens were demoulded and placed in a curing box for further use. Figure 1 shows the ready-made briquette specimens.

2.2. Experimental Apparatus

A triaxial servo-controlled seepage apparatus for thermofluid–solid coupling of gas-bearing coal shown in Figure 2 was used for the test, featuring ϕ50 mm × 100 mm standard specimens [9].

Figure 1. Briquette specimens.

Figure 2. The triaxial servo-controlled seepage equipment for thermofluid–solid coupling of coal containing methane.

2.3. Experimental Methods

The real load state of mining-disturbed coal at deep levels was simplified into three load paths: axial pressure loading in Test 1, confining pressure unloading in Test 2, and composite loading/unloading in Test 3, as shown in Figure 3, where σ_1 represents axial pressure and σ_3 represents confining pressure, and the flow chart for the experiment is shown in Figure 4. The exact test plan is described below:

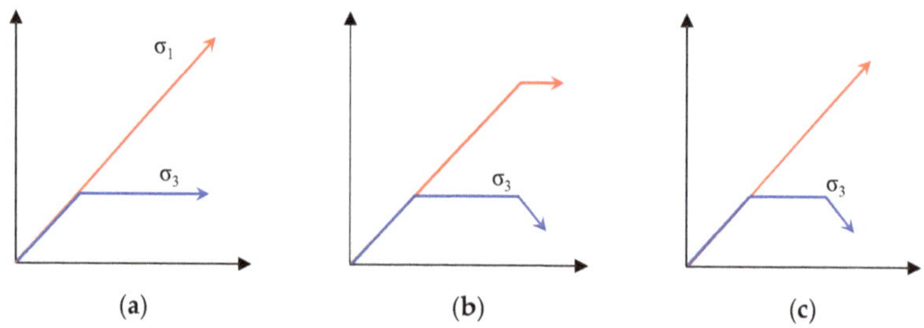

Figure 3. Schematic diagram of experimental loading path. (**a**) Test 1, (**b**) Test 2, (**c**) Test 3.

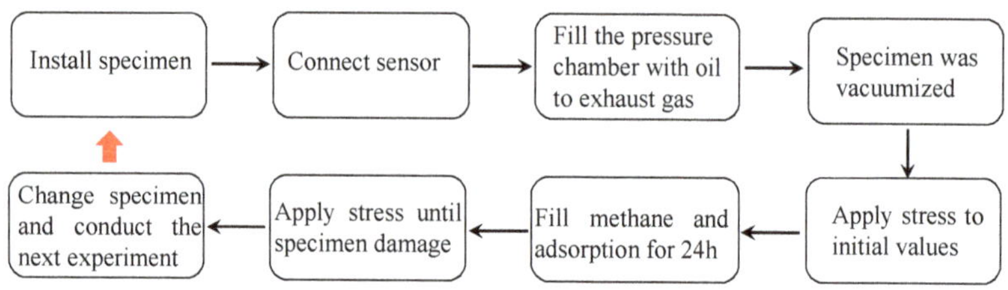

Figure 4. Flow chart for the experiment.

Test 1: Load σ_1 and σ_3 to 6 MPa, then feed 1 MPa gas. After adsorption for 24 h, load σ_1 at the rate of 5 kN/s until the specimen fails. Conduct gas seepage test at the same time.

Test 2: Load σ_1 and σ_3 to 6 MPa, then feed 1 MPa gas. After adsorption for 24 h, load σ_1 to σ_U (which is 60% of the peak axial stress in Test 1) at the rate of 5 kN/s, keep σ_1 unchanged and unload σ_3 at the rate of 0.01 mm/min until the specimen fails. Conduct gas seepage test at the same time.

Test 3: Load σ_U in the same steps as Test 2, then keep the loading rate of σ_1 unchanged and unload σ_3 at the rate of 0.01 mm/min until the specimen fails. Conduct gas seepage test at the same time.

3. Results and Analysis

3.1. Strain and Strength Characteristics

Figure 5 compares the stress-strain relationship of specimens under the true-triaxial path, where ε_1 and ε_3 are the axial strain and radial strain of the specimen, and $\Delta\sigma$ is the principal stress difference, i.e., σ_1–σ_3. From these diagrams, the stress-strain curves of gas-bearing coal are much the same under the three paths. They all involve five stages: compaction, linear elasticity, plastic deformation, stress decline, and residual stress. Stress path is shown to have a strong impact on the deformation and strength of gas-bearing coal. In Test 1, the specimen strength is the highest, with stress difference of 33.750 MPa. In Test 2, the specimen strength is the second highest, with stress difference of 23.480 MPa. In Test 3, the specimen strength is the lowest, with stress difference of 16.711 MPa. The specimen-bearing capacity is greatly reduced.

At peak strength, ε_1 and ε_3 are 2.607 and −1.771 in Test 1; 1.301 and −1.354 in Test 2; and 0.793 and −2.186 in Test 3. In Test 1, the axial strain is the largest. In Test 3, the axial strain is the smallest, but the radial strain is the largest. At peak strength, the axial-to-radial strain ratios of the specimens are 1.472, 0.961, and 0.363. The gradual reduction in the axial-to-radial strain ratio indicates that, during instability, the axial deformation intensifies and the specimen is more prone to axial failure, especially in Test 3, where the specimen has a strong shear dilatancy. Also, by observing the stress-strain curves, it can be seen that the post-peak curve slope gradually increases. In Test 2 and Test 3, with the unloading of confining pressure, the strain increment gradually increases. When this increase has accumulated to a limit, the bearing capacity declines and the specimen immediately becomes unstable. This is particularly obvious in Test 3, where the specimen gradually changes from ductile failure to brittle failure.

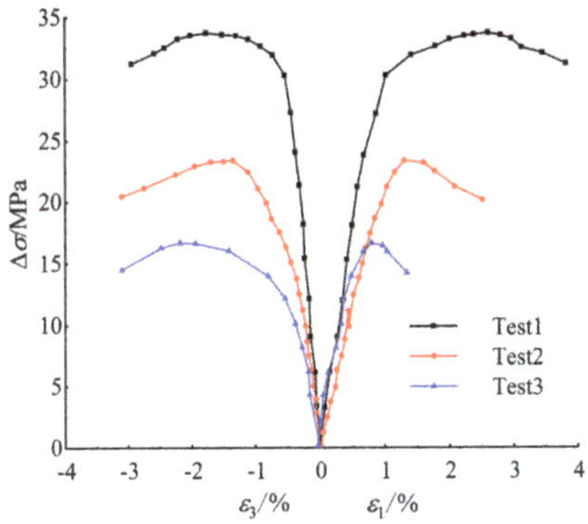

Figure 5. Stress–strain curves of specimens under different test paths.

3.2. Permeability Evolution Characteristics

Gas seepage in the specimens obeys Darcy's law. Permeability is calculated by the following formula [29]:

$$K = \frac{2P_0' Q \mu L}{A(P_1^2 - P_2^2)} \tag{1}$$

where K is permeability (m^2); Q is the gas flow (m^3/s); μ is the absolute viscosity of the methane; L is the specimen length (m); A is the effective area of permeability (m^2); P_0 prime is standard atmosphere; P_1 is the inlet pressure (MPa); and P_2 is outlet pressure (MPa).

Figure 6 compares the principal stress difference and permeability variations of gas-bearing coal as a function of axial strain under different test paths. From the ε_1–$\Delta\sigma$ and ε_1–K curves, as the test goes on, permeability first reduces then increases; the valley deflection of permeability falls before the peak deflection of the ε_1–$\Delta\sigma$ curve in all cases. This is because, during elastic deformation at the beginning of the test, with the loading of external stress, the primary pores and fissures inside the specimen are compressed, which narrows the gas seepage pathway and brings down the permeability. With the loading of σ_1, the specimen enters plastic deformation. Cracks begin to develop inside and damage is expanded, adding more seepage pathways and stepping up permeability. After that, with the further increase of σ_1, cracks inside become interconnected, leading to instability failure. The gas seepage pathway is opened and the permeability soars.

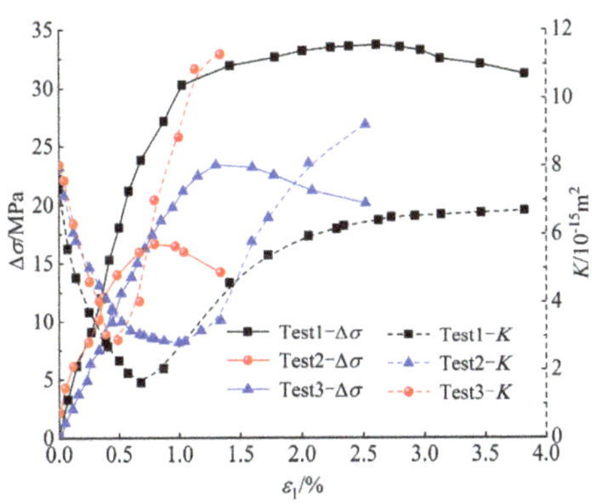

Figure 6. Variation curves of principal stress difference and permeability with axial strain during test.

Relative permeability (w) variations were observed against initial permeability to examine how stress path affects permeability evolution. From the diagram in Figure 7, under Test 1, the specimen permeability variation is modest, with minimum w of 0.222, which increases marginally to 0.913 at the end of the test. Under Test 2, w is 0.353 minimum and 0.358 maximum. Under Test 3, w is 0.370 minimum and 1.405 maximum. Compared with the other paths, the specimen permeability variation is the largest at the end of the test under composite loading/unloading path, proving that the specimen is more badly damaged under this path.

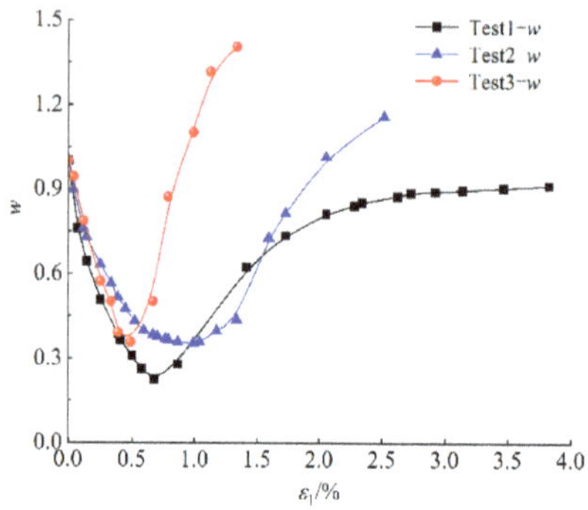

Figure 7. Relative permeability change.

4. Numerical Simulation of Crack Characteristics

In order to examine how cracks in gas-bearing coal evolve under triaxial stress paths, numerical simulations were performed with PFC software to see how crack-count changes in the specimens during loading.

4.1. Model Construction

The model was sized with the same dimensions as the laboratory specimens, i.e., ϕ50 mm × 100 mm. In the computational model, the minimum particle radius was 0.25 mm, with a particle size ratio of 1.66. A total of 3665 particle samples were generated.

Figure 8 shows the initial and boundary conditions used for simulation. First, the model was loaded with biaxial compression of axial and confining pressure to 6 MPa by the servo mechanism. Then a high-pressure zone with pressure P was set at the top of the model. The fluid field pressure at the bottom of the model was fixed to 0.1 MPa to indicate connecting to atmospheric pressure.

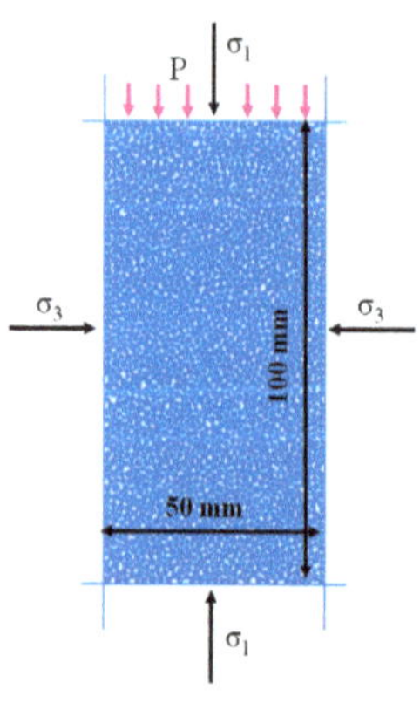

Figure 8. Model, initial conditions and boundary conditions.

4.2. Determination of Coal Meso-Mechanical Parameters

First, coal compression and tension simulations were made with PFC software to correlate the meso- and macro-mechanical parameters of coal.

Data regression yielded a correlation coefficient of 0.992 (R^2 = 0.992) between elastic modulus and meso-mechanical parameter, 0.998 (R^2 = 0.998) between Poisson's ratio and meso-mechanical parameter, 0.997 (R^2 = 0.997) between compressive strength and meso-mechanical parameter, and 0.995 (R^2 = 0.995) between tensile strength and meso-mechanical parameter. The empirical equations between these coefficients are shown in Equations (2)–(5).

$$\frac{E}{E_c} = a_1 + b_1 \ln(\frac{k_n}{k_s}) \tag{2}$$

where E is the elastic modulus, GPa; E_c is the Young's modulus, GPa; k_n/k_s is the stiffness ratio; a_1 = 1.652, and b_1 = −0.395.

$$\mu = a_2 + b_2 \ln(\frac{k_n}{k_s}) \tag{3}$$

where μ is the Poisson's ratio; a_2 = 0.111; b_2 = 0.209.

$$\frac{\sigma_c}{\overline{\sigma}} = \begin{cases} a_3(\frac{\overline{\tau}}{\overline{\sigma}})^2 + b_3\frac{\overline{\tau}}{\overline{\sigma}} & , 0 < \frac{\overline{\tau}}{\overline{\sigma}} \leq 1 \\ c_1 & , 1 < \frac{\overline{\tau}}{\overline{\sigma}} \end{cases} \tag{4}$$

where σ_c is the compressive strength, MPa; $\overline{\sigma}$ is the parallel connection normal strength, MPa; τ is the parallel connection tangential strength, MPa; a_3 = −0.965; b_3 = 2.292; and c_1 = 1.327.

$$\frac{\sigma_t}{\overline{\sigma}} = \begin{cases} a_4(\frac{\overline{\tau}}{\overline{\upsilon}})^2 + b_4\frac{\overline{\tau}}{\overline{\sigma}} & , 0 < \frac{\overline{\tau}}{\overline{\sigma}} < 1 \\ c_2 & , 1 < \frac{\overline{\tau}}{\overline{\sigma}} \end{cases} \tag{5}$$

where σ_t is the tensile strength, MPa; a_4 = −0.174; b_4 = 0.463; c_2 = 0.289. Based on Equations (2)–(5) and the data in Table 1, the coal mesoscopic parameters required for simulation can be retrieved as shown in Table 2.

Table 1. Macro-mechanical characteristics of the moulded coal specimen.

Bulk Density (kN/m^3)	Elastic Modulus (104 MPa)	Internal Friction Angle (°)	Tensile Strength (MPa)	Cohesion (MPa)	Poisson Ratio
13.57	6.89	42.6	1.34	2.43	0.28

Table 2. Mesoscopic mechanical parameters of the DEM numerical model.

Parameters	Value
Contact stiffness ratio	2.14
Contact modulus (GPa)	5.1
Friction factor	0.4
Parallel bond modulus (GPa)	5.1
Normal strength of parallel bonding	4.6
Parallel bond tangential strength (MPa)	4.6
Parallel bond stiffness ratio	2.14

It has been demonstrated that among all fluid parameters, residual pore size a_0 and fluid viscosity μ make the greatest difference to permeation pressure–stress coupling during permeation. After repeated simulation, fluid computational parameters were worked out, as presented in Table 3.

Table 3. Computational parameters of fluid.

Fluid Viscosity μ/(Pa s)	Fluid Volume Modulus K_f (GPa)	Residual Pore Size a_0 (m)	Initial Normal Stress F_0 (kN)	Time Step Δt (s)
0.015	0.0096	1×10^{-3}	5×10^3	0.1

Figure 9 compares the numerical simulation results with the test results of gas-bearing coal under the stress path of triaxial loading. From this diagram, at the end of the loading path of axial pressure, an oblique shear crack appears in the specimen. The stress-strain curve and failure form from numerical simulation almost entirely agrees with the laboratory test results. This suggests that our numerical simulation model and meso-mechanical parameters are appropriate enough for subsequent meso-mechanical simulation.

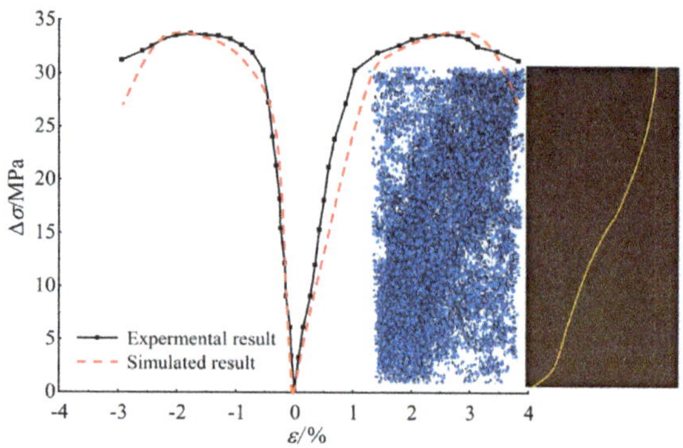

Figure 9. Comparison of physical test and numerical simulation results.

4.3. Crack Number Characteristic Analysis

To observe crack development inside the specimen under different paths, a cracks count–axial strain plot shown in Figure 10 was drawn. The cracks count inside the specimen changes in much the same way. As axial stress increases, cracks count first increases slowly and then increases quickly, especially near the site of peak stress. Finally, the rate of increase slows down and gradually stabilizes. The evolution of tension and shear-induced cracks count is not much different: it first increases and then stabilizes. However, as tensile strength is greater than shear strength in coal particles, there are many more tensile cracks than shear cracks.

Figure 11 compares the cracks counts at the end of simulation under different paths. In Test 1, the total model cracks count is 4.36×10^3, including 3.72×10^3 tensile cracks, accounting for 85.29% of total cracks; and 0.64×10^3 shear cracks, accounting for 14.71% of total cracks. In Test 2, the model cracks count is 7.29×10^3, including 6.05×10^3 cracks, accounting for 82.91% of total cracks; and 1.25×10^3 shear cracks, accounting for 17.09% of total cracks. In Test 3, the model cracks count is 5.65×10^3, including 4.46×10^3 tensile cracks, accounting for 78.93% of total cracks; and 1.19×10^3 shear cracks, accounting for 21.07% of total cracks. By cracks count, Test 2 > Test 3 > Test 1; by cracks proportion, from Test 1 to Test 3, the proportion of shear cracks gradually increases, suggesting that both tensile cracks and shear cracks are present at the time of failure. For this reason, the model displays a composite tensile–shear failure, although the specimens are more prone to shear failure under loading and composite loading/unloading paths.

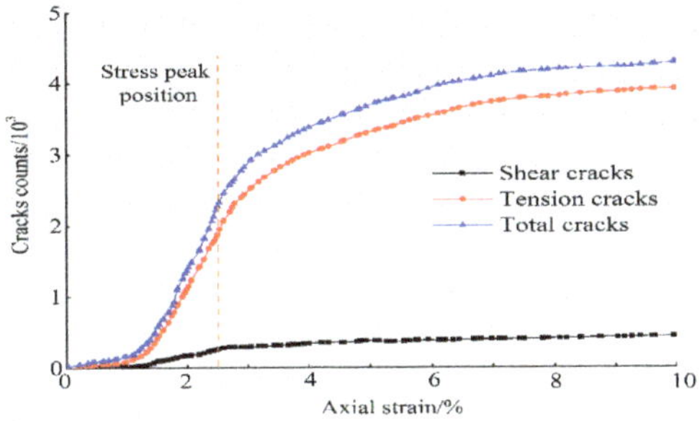

(a)

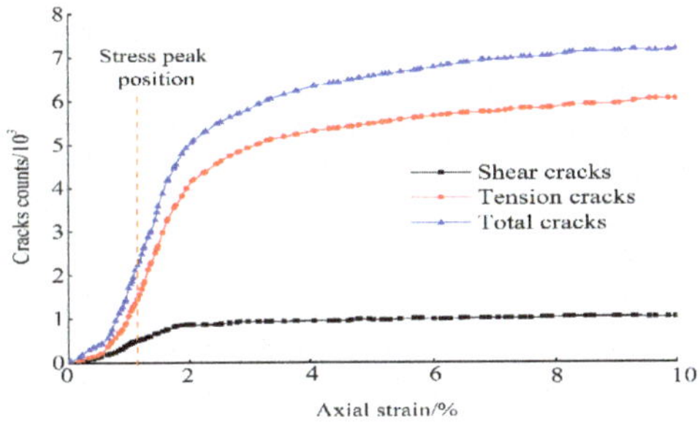

(b)

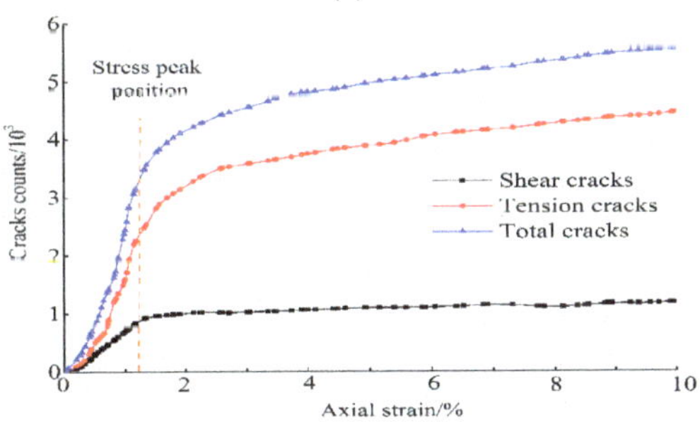

(c)

Figure 10. Relationship between the number of cracks and axial strain. (**a**) Test 1, (**b**) Test 2, (**c**) Test 3.

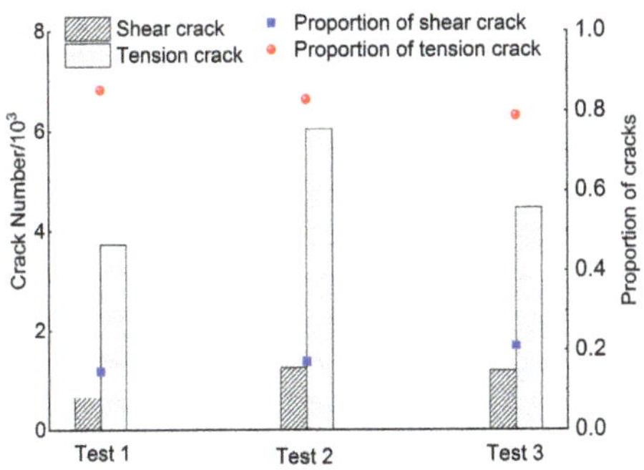

Figure 11. Characteristics of the number of cracks under different paths.

5. Conclusions

In Test 1, the specimen strength is the highest, with $\Delta\sigma$ = 33.750 MPa. In Test 2, the specimen strength is the second highest, with $\Delta\sigma$ = 33.750 MPa. In Test 3, the specimen strength is the lowest, with $\Delta\sigma$ = 23.480 MPa. The specimen-bearing capacity is much lower under loading and composite loading/unloading paths.

Under the three paths, as the test goes on, permeability first reduces then increases; the valley deflection of permeability falls before the peak deflections of the ε_1–$\Delta\sigma$ curve in all cases. Compared with the other paths, the specimen-permeability variation is the largest at the end of Test 3, proving that the specimens are more badly damaged in Test 3.

The total cracks count is the largest in Test 2. From cracks proportion, from Test 1 to Test 3, the proportion of shear cracks in the model gradually increases, suggesting that both tensile and shear cracks are present at the time of macroscopic failure. For this reason, the model displays a composite tensile–shear failure, although the specimens are more prone to shear failure under loading and composite loading/unloading paths.

Author Contributions: Conceptualization, H.S. (Haibo Sun), B.Z., Z.S., B.S. and H.S. (Hongyu Song); Writing—original draft, H.S. (Haibo Sun); Funding acquisition, B.Z.; Formal analysis, H.S. (Haibo Sun), B.Z. and Z.S.; Software, H.S. (Haibo Sun), B.S.; Data curation, H.S. (Haibo Sun), B.Z. and H.S. (Hongyu Song). All authors have read and agreed to the published version of the manuscript.

Funding: This research received no external funding.

Data Availability Statement: All relevant data presented in the article are according to institutional requirements and, as such, are not available online. However, all data used in this manuscript can be made available upon request to the authors.

Conflicts of Interest: The authors declare no conflict of interest.

References

1. Xie, H.; Gao, M.; Zhang, R. Study on the mechanical properties and mechanical response of coal mining at 1000 m or deeper. *Rock Mech. Rock Eng.* **2019**, *52*, 1475–1490. [CrossRef]
2. Yuan, L. Control of coal and gas outbursts in Huainan mines in China: A review. *J. Rock Mech. Geotech. Eng.* **2016**, *8*, 559–567. [CrossRef]
3. Wang, K.; Guo, Y.; Wang, G.; Du, F. Seepage and Mechanical Failure Characteristics of Gas-bearing Composite Coal-Rock under True Triaxial Path. *J. China Coal Soc.* **2022**.
4. Zhang, C.; Xu, J.; Peng, S. Dynamic behavior of gas pressure and optimization of borehole length in stress relaxation zone during coalbed methane production. *Fuel* **2018**, *233*, 816–824. [CrossRef]
5. Wang, K.; Du, F.; Zhang, X.; Wang, L.; Xin, C. Mechanical properties and permeability evolution in gas-bearing coal-rock combination body under triaxial conditions. *Environ. Earth Sci.* **2017**, *76*, 815. [CrossRef]
6. Du, F.; Wang, K.; Zhang, X. Experimental study of coal-gas outburst: Insights from coal-rock structure, gas pressure and adsorptivity. *Nat. Resour. Res.* **2020**, *29*, 2481–2493. [CrossRef]
7. Du, F.; Wang, K. Unstable failure of gas-bearing coal-rock combination bodies: Insights from physical experiments and numerical simulations. *Process Saf. Environ. Prot.* **2019**, *129*, 264–279. [CrossRef]
8. Chen, Y.; Xu, J.; Peng, S.; Zhang, Q.; Chen, C. Strain localisation and seepage characteristics of rock under triaxial compression by 3D digital image correlation. *Int. J. Rock Mech. Min. Sci.* **2022**, *152*, 105064. [CrossRef]
9. Chen, C.; Xu, J.; Okubo, S.; Peng, S. Damage evolution of tuff under cyclic tension–compression loading based on 3D digital image correlation. *Eng. Geol.* **2020**, *275*, 105736. [CrossRef]
10. Yin, G.; Li, W.; Li, M. Experimental study of mechanical properties of coal containing methane under different loading-unloading conditions. *Chin. J. Rock Mech. Eng.* **2013**, *32*, 892–901.
11. Wang, Y.; Xie, H.; Chen, S. Petrophysical and Mechanical Properties of Complex Structure Coalbed Methane Reservoir: A Laboratory Investigation. *Rock Mech. Rock Eng.* **2022**, 1–19. [CrossRef]
12. Zhao, H.; Wang, J. Experimental study of evolution law of mechanical properties of coal containing gas under unloading confining Pressure. *Rock Soil Mech.* **2011**, *32* (Suppl. S1), 270–274.
13. Wang, G.; Xue, D.; Gao, H. Study on permeability characteristics of coal rock in complete stress-strain process. *J. China Coal Soc.* **2012**, *37*, 107–112.
14. Li, X.; Yin, G.; Cai, B. Experimental study on deformation and seepage properties of outburst coal samples under cyclic loading. *Chin. J. Rock Mech. Eng.* **2010**, *29* (Suppl. S2), 3498–3504.
15. Cao, S.; Bai, Y.; Li, Y. Experiments of coal's gas seepage with outburst dangers. *J. Chongqing Univ.* **2011**, *34*, 91–95.

16. Xue, D.; Zhou, H.; Wang, Z. Failure mechanism and mining-induced mechanism of coal a under different loading rates. *J. China Coal Soc.* **2016**, *41*, 595–602.
17. Chaudry, M.A.; Wriggers, P. On the computational aspects of comminution in discrete element method. *Comput. Part. Mech.* **2018**, *5*, 175–189. [CrossRef]
18. Zhu, F.; Zhao, J. A peridynamic investigation on crushing of sand particles. *Geotechnique* **2019**, *69*, 526–540. [CrossRef]
19. Tan, X.; Hu, Z.; Cao, M.; Chen, C. 3D discrete element simulation of a geotextileencased stone column under uniaxial compression testing. *Comput. Geotech.* **2020**, *126*, 103769. [CrossRef]
20. Indraratna, I.; Thakur, P.; Vinod, J.S. Experimental and numerical study of railway ballast behavior under cyclic loading. *Int. J. Geomech.* **2010**, *10*, 136–144. [CrossRef]
21. Wang, P. Study of the borehole hydraulic fracturing and the principle of gas seepage in the coal seam. *Procedia Earth Planet. Sci.* **2009**, *1*, 1561–1573.
22. Jiang, M.; Zhang, F.; Sun, Y. An evaluation on the degradation evolutions in three constitutive models for bonded geomaterials by DEM analyses. *Comput. Geotech.* **2014**, *57*, 1–16. [CrossRef]
23. Ismai, Y.; Yang, D.; Ye, J. A DEM model for visualising damage evolution and predicting failure envelope of composite laminae under biaxial loads. *Compos. Part B Eng.* **2016**, *10*, 9–28. [CrossRef]
24. Yang, X.; Jing, H.; Chen, K. Numerical simulations of failure behavior around a circular opening in a non-persistently jointed rock mass under biaxial compression. *Int. J. Min. Sci. Technol.* **2016**, *26*, 729–738. [CrossRef]
25. Sagong, M.; Park, D.; Yoo, J.; Lee, J. Experimental and numerical analyses of an opening in a jointed rock mass under biaxial compression. *Int. J. Rock Mech. Min. Sci.* **2011**, *48*, 1055–1067. [CrossRef]
26. Raisianzadeh, J.; Mirghasemi, A.A.; Mohammadi, S. 2D simulation of breakage of angular particles using combined DEM and XFEM. *Powder Technol.* **2018**, *336*, 282–297. [CrossRef]
27. Xu, H.; Qin, Y.; Wang, G.; Fan, C.; Wu, M.; Wang, R. Discrete element study on mesomechanical behavior of crack propagation in coal samples with two prefabricated fissures under biaxial compression. *Powder Technol.* **2020**, *375*, 42–59. [CrossRef]
28. Wang, G.; Guo, Y.; Wang, P. A new experimental apparatus for sudden unloading of gas-bearing coal. *Bull. Eng. Geol. Environ.* **2020**, *79*, 857–868. [CrossRef]
29. Zhao, Y.; Cui, D.; Liu, J.; Wei, M.; Liu, Y. Evolution of Coal Permeability under Constant Effective Stresses: Direct Measurements and Numerical Modeling. *Energy Fuels* **2021**, *35*, 15489–15501. [CrossRef]

processes

MDPI

Article

Destabilization Mechanism and Stability Control of the Surrounding Rock in Stope Mining Roadways below Remaining Coal Pillars: A Case Study in Buertai Coal Mine

Qiang Fu [1,2,3,4], Ke Yang [2,3,4,*], Xiang He [1,2,3,4], Qinjie Liu [1,2,3,4], Zhen Wei [1,2,3,4] and Yu Wang [1,2,3,4]

1 Joint National-Local Engineering Research Centre for Safe and Precise Coal Mining, Anhui University of Science and Technology, Huainan 232001, China
2 Institute of Energy, Hefei Comprehensive National Science Center, Hefei 230031, China
3 School of Mining Engineering, Anhui University of Science and Technology, Huainan 232001, China
4 State Key Laboratory of Mining Response and Disaster Prevention and Control in Deep Coal Mines, Anhui, Huainan 232001, China
* Correspondence: keyang@aust.edu.cn

Abstract: To study the stability control of stope mining roadways below remaining coal pillars, the present study investigates the destabilization mechanism of coal pillars and roadways in sections under the dual action of supporting pressure on the floor of the remaining coal pillar in the overlying coal seam and the mining at the working face of the lower coal seam and clarify the principle of surrounding rock stability control based on theoretical analysis, numerical simulation, and industrial testing. The results yielded the following findings. After the stope mining of the overlying coal seam working face, the stress transfer of the T-shaped remaining coal pillar significantly increased the vertical stress of the lower coal seam. The lateral support pressure generated by the stope mining at the lower coal seam working face further aggravated the stress concentration in the coal, leading to severe compression shear failure of the surrounding rock. As the sectional coal pillar becomes wider, the roadway gradually avoids the area of peak superimposed support pressure. The vertical stress curve of the sectional coal pillar shifts from single-peaked to asymmetrically double-peaked, and the stress difference between the two roadway ribs and the stress concentration coefficients decrease continuously. A stability control method of long anchor cable reinforcement support is proposed. In-situ industrial testing showed that the surrounding rock deformation was basically stable during the service period of the 42202 stope mining roadway, thus achieving the stability control of the stope mining roadway.

Keywords: remaining coal pillar; stope mining roadway; stress transfer; support optimization; surrounding rock control

Citation: Fu, Q.; Yang, K.; He, X.; Liu, Q.; Wei, Z.; Wang, Y. Destabilization Mechanism and Stability Control of the Surrounding Rock in Stope Mining Roadways below Remaining Coal Pillars: A Case Study in Buertai Coal Mine. *Processes* **2022**, *10*, 2192. https://doi.org/10.3390/pr10112192

Academic Editors: Feng Du, Aitao Zhou and Bo Li

Received: 27 September 2022
Accepted: 23 October 2022
Published: 26 October 2022

1. Introduction

Close-distance coal seam groups are widely distributed in China [1,2]. Limited by technical and coal seam reserve conditions, wide coal pillars are often reserved when mining the overlying coal seam to maintain the stope mining roadway [3,4]. The many remaining coal pillars after the stope mining of the coal seam alter the stress environment of their floor, causing stress elevations dominated by vertical stresses in the floor rock near the remaining coal pillars [5,6]. Under such high stresses, the surrounding rock of the stope mining roadways in the underlying coal seam inevitably develops cracks [7,8]. In the meantime, the superposition of lateral support pressure stresses during the mining at the working face of the underlying coal seam further aggravates the stope-mining roadway deformation and failure, which seriously affects the safe production of the mine [9–13]. Therefore, studies on the failure mechanism and stability control techniques of roadways under multiple disturbances in close-distance coal seams are urgently needed.

To clarify the effects of remaining coal pillars on the mine pressure behaviors of stope mining roadways in the underlying coal seam, scholars worldwide conducted extensive research based on methods such as mining theory analysis, numerical simulation, and field applications. Research by Zhang et al. [14] concluded that subjected to high vertical stress, part of the roadway under the remaining coal pillar developed large deformations and became difficult to maintain, and the stress concentration and surrounding rock deformation on the side of coal pillars could be effectively reduced by the reasonable roadway layout. Fang et al. [15] demonstrated that roadways were easily destabilized under the effect of mining if the remaining coal pillars were narrow in the gob of the overlying coal seam or if the roadway of the underlying coal seam was directly below while the roadway-supporting coal pillars near the working face of the current coal seam were small. Numerical simulations have been conducted to analyze the stress transfer of the coal pillar floor in the overlying coal seam [16,17]. To address the challenge of gob-side roadway stability control in close-distance coal seams, Zhang et al. [18] determined the design parameters of the roadway-side filling body based on the structural mechanics model of the gob-side entry retaining surrounding rocks and proposed a support technology combining high-strength anchor rods, anchor cables, and steel beams. Jiang et al. [19] analyzed the coal pillar-induced impact patterns in deep close distance coal seams and medium mining depth far distance coal seams. Through comprehensive research, Liu et al. [20] analyzed the destabilization and failure processes of roadways in the gob of close-distance coal seams and proposed a joint active-passive control method to improve the bearing performance of the surrounding rock. Xia et al. [21] adopted numerical methods to analyze the dynamic mechanical characteristics of the stope mining roadway under small-width coal pillars when subjected to repeated mining disturbances and proposed corresponding roadway surrounding rock control methods. To address the asymmetric plastic failure of the withdraw roadway under the dual action of remaining coal pillars in the overlying coal seam and mining disturbance, He et al. [22] proposed a zoning control strategy and achieved roadway stability. The research results at different engineering geological conditions practically solved the engineering difficulties brought by the remaining coal pillars to the mining of close-distance coal seam groups and enriched the technical application system of rock seam control. However, when the remaining coal pillars in the overlying coal seam are parallel to the advancing direction and spatially overlap with the underlying stope, the disturbance of the high stress in the floor at the remaining coal pillar increases under the dual action of the remaining coal pillar and the mining of the underlying coal seam, which significantly affects the layout and stability of the underlying coal seam roadways [23].

Based on the background of the occurrence conditions of the coal seams in Buertai Coal Mine, methods including theoretical analysis, numerical simulation, and field measurement were adopted to study the stress transfer pattern of the T-shaped remaining coal pillar floor in the overlying coal seam and analyze the destabilization and stability control principle of the roadway surrounding rock during the mining process of the adjacent working face in the underlying coal seam. The reasonable layout scheme of the 42202 mining roadway was put forward, and the surrounding rock dynamic reinforcement control techniques were also proposed. The research results could provide some reference for projects of similar engineering backgrounds.

2. Geological Condition

The Buertai Coal Mine has a simple geological structure, predominantly monoclines with gentle dip angles and no-fault development. The coal mine contains 9 mineable coal seams, and mainly the 2-2 and 4-2 coal seams are being mined now. The 2-2 coal seam is 3 m thick and 300 m deep on average. The 4-2 coal seam is 6.12 m thick and 368 m deep on average, and comprehensive mechanized top coal caving mining is adopted.

Stope mining at the two working faces of the overlying 2-2 coal seam, 22201 and 22202, has been completed, leaving a 20 m wide sectional coal pillar. Skip mining was carried out when stope mining the 22202 working face, leaving a 70 m skip mining coal pillar.

The spatial layout of the two remaining coal pillars resembles a mutually perpendicular T-shape. The 42201 and 42202 working faces are below the 22201 and 22202 working faces, and the stope mining roadways of the adjacent working faces, i.e., the transport roadways of 42201 and 42202, are dug simultaneously, leaving a roadway-supporting coal pillar of certain width in the section. The location of the Buertai coal mine and schematic diagram of the 42202 stope layout is shown in Figure 1.

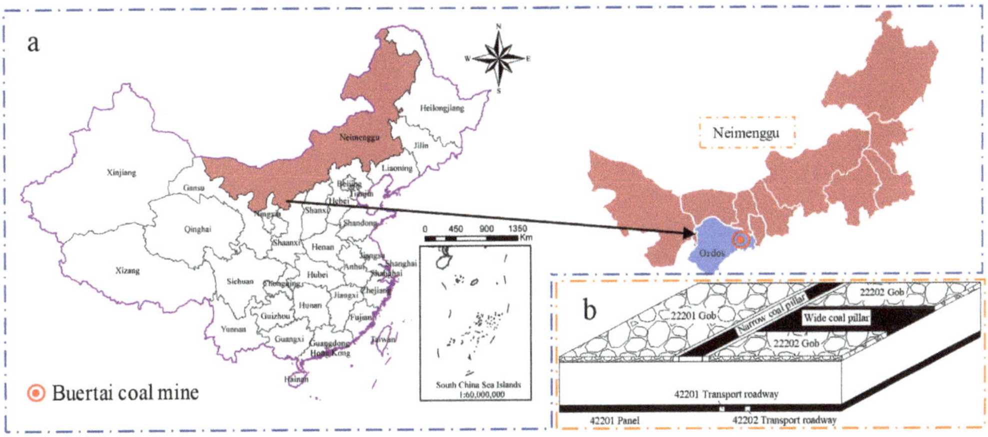

Figure 1. (**a**) The location of the Buertai coal mine. (**b**) Schematic diagram of 42202 stope layout.

The original support design of the 42202 stope mining roadway is as follows. The roof is under the combined support by "left-hand thread rebar bolt without longitudinal bar + welded mesh + anchor cable + π-type steel strip". The bolts are $\varphi 22 \times 2200$ mm in dimension and arranged rectangularly with six sets per row with an inter-row distance of 1000×1000 mm. The anchor cables are $\varphi 22 \times 8000$ mm in dimension and arranged rectangularly 3 sets per row with an inter-row distance of 2100×2000 mm. The coal pillar rib is under the combined support of "bolt + diamond-shaped mesh + wooden pallet". The rebar bolts are $\varphi 22 \times 2100$ mm in dimension and arranged rectangularly in five sets per row with an inter-row distance of 800×1000 mm. The solid coal rib is under the combined support of "FRP bolt + double-layer high-strength plastic mesh + wooden pallet." The bolts are $\varphi 22 \times 2100$ mm in dimension and arranged rectangularly in five sets per row with an inter-row distance of 800×1000 mm, as shown in Figure 2.

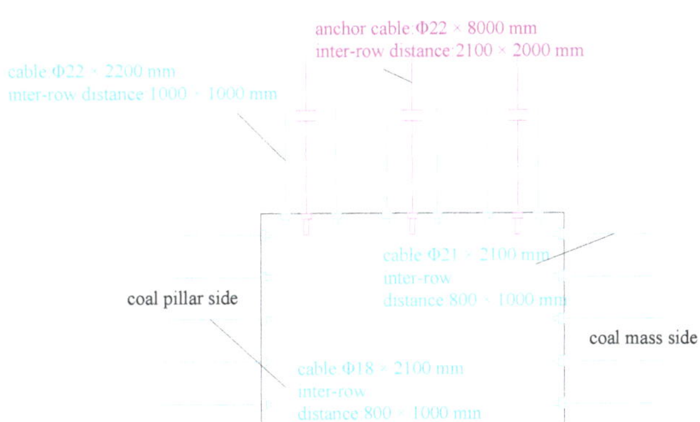

Figure 2. Initial support measures of mining roadway in 42202 stope.

During the stope mining at the working faces under the same working conditions, the stope mining roadways showed the following failure characteristics.

(1) When the stope mining roadway is only affected by a single sectional coal pillar in the overlying coal seam, the roof and floor displacements are between 200 and 220 mm, and the displacements of the two ribs are between 190 and 215 mm. When affected by approximately T-shaped remaining coal pillars, the displacements of the roadway roof and floor increase by 54.5–56% to 440–500 mm, and the displacements of the two ribs increase by 51.3–55.2% to 390–480 mm. Therefore, the deformation of the roadway below the T-shaped remaining coal pillar has increased significantly, and the surrounding rock failure is severe.

(2) On-site investigation shows that the force on the bolts in the roadway rib had a sudden stepwise decrease, and the pallet fell off. Thus, the range of surrounding rock plastic failure continuously increased under the effect of multiple disturbances, and the confining pressure of the bolt-supported section and rock mass integrity decreased, which significantly reduced the anchoring capacity of the bolts.

3. Methodology

In this paper, theoretical analysis, numerical simulation and engineering application are used to study the instability mechanism and stability control technology of the surrounding rock of the mining roadway below the remaining coal pillar. The technical flowchart is shown in Figure 3. The specific steps are as follows: (1) Through the theoretical analysis method, the superposition effect of T-shaped residual coal pillar on the abutment pressure of 4-2 coal seam in 2-2 coal seam of Buertai Coal Mine is studied, and combined with the Mohr-Coulomb criterion, the failure and instability mechanism of surrounding rock of 42202 mining roadway under multiple disturbances is revealed. (2) On the basis of theoretical analysis, an equal scale numerical model is established by using FLAC3D numerical software (version 6.00.74, Itasca, China, 2020) to simulate and analyze the surrounding rock plastic zone and vertical stress distribution characteristics of 42202 mining roadway under different layout schemes (different width section pillars), so as to propose the most reasonable section pillar retaining scheme, and propose the surrounding rock reinforcement support technology to control the roadway stability. (3) The scheme of section coal pillar reservation and the technology of surrounding rock reinforcement and support were applied in the 4-2 coal seam, and the effectiveness of the research results in this paper is verified through ground pressure monitoring.

3.1. Theoretical Analysis

3.1.1. Stress Distribution Pattern below Remaining Coal Pillars

After the 2-2 coal seam was extracted, the rock formation above the gob collapsed, causing the abutment pressure on the remaining coal pillar to rise sharply. These stresses transferred to the underlying coal seam, forming a larger area under the effects of the coal pillar abutment stress, i.e., the rock mass below the coal pillar entered the high abutment pressure state earlier. The abutment pressure distribution on the T-shaped remaining coal pillar [24] is shown in Figure 4. When the coal pillar was narrow (20 m), the roof fracture position moved toward the inside of the coal pillar, i.e., the range of the limit equilibrium area of the coal pillar increased significantly. As a result, the vertical stress on the coal pillar increased significantly, peaking at 70 MPa. As the coal pillar became wider (70 m), the pressure-bearing range increased. The peak vertical stress decreased significantly compared with that of the narrow coal pillar, but the range of effects on the underlying coal seam was increased relatively.

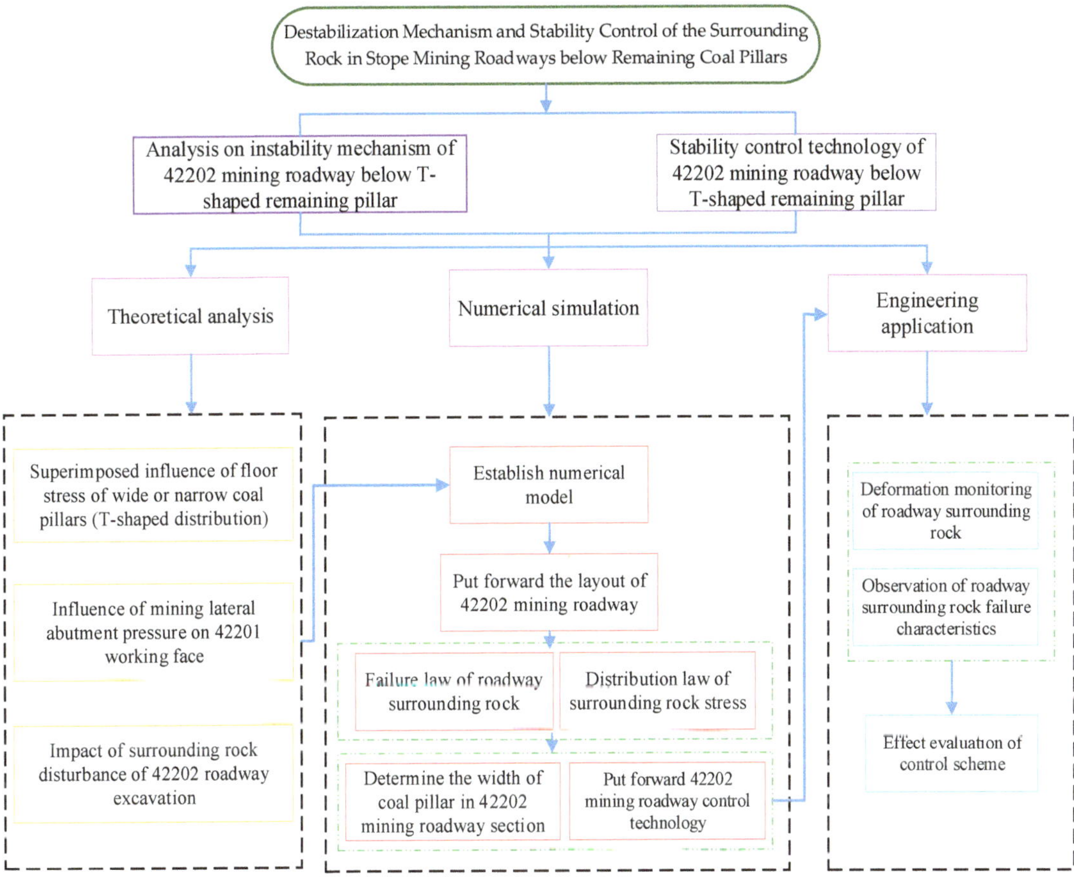

Figure 3. Flowchart of the research procedure.

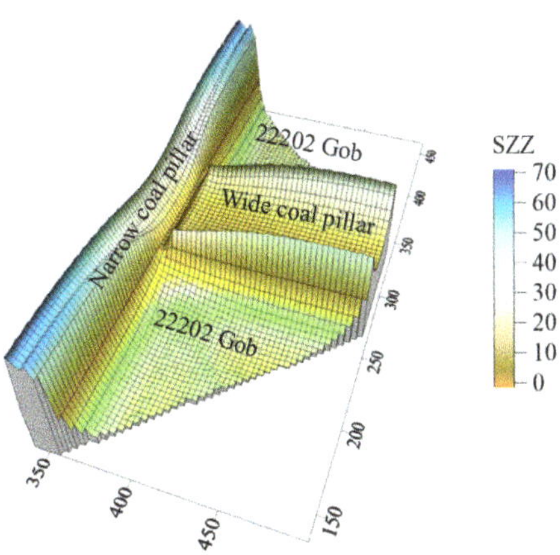

Figure 4. Abutment stress distribution on the T-shaped remaining coal pillar in the 2-2 coal seam.

To investigate the stress transfer pattern of the 2-2 coal seam to the underlying coal seam, the coal pillar abutment pressure can be simplified to the mechanical model shown in Figure 5. Since the abutment pressure is symmetrically distributed on the wide coal pillar about its center axis, the center of the wide coal pillar is selected as the coordinate origin. For the narrow coal pillar, the minimum stress point in the internal elastic core is selected as the coordinate origin. The horizontal cross-sectional direction of the coal seam floor is selected as the x-axis direction, and the vertically downward direction is selected as the y-axis direction. γ is the volumetric weight of the rock, H is the depth, and k_1, k_2, k_3, k_4, and k_5 are the maximum stress concentration factors.

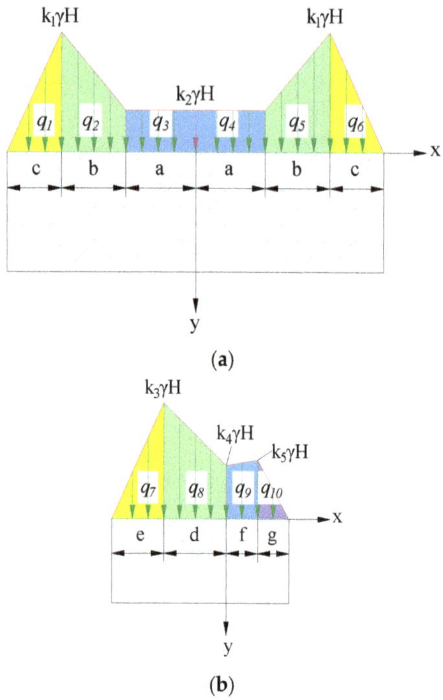

Figure 5. Coal pillar force models. (**a**) Wide coal pillar force model; (**b**) Narrow coal pillar force model.

In the figure, ①–⑩ are stress-concentrated areas of the coal pillar, and the vertical stress components of the floor rock mass are:

$$\sigma_y = -\frac{2}{\pi}\int_{-\infty}^{+\infty}\frac{qx(y-\xi)^2 d\xi}{\left[x^2+(y-\xi)^2\right]^2} \tag{1}$$

where q is the micro-element load collection degree, and σ_y is the vertical stress (MPa); x and y are the vertical and horizontal distances of the concentrated force q to any point M on the floor (m).

As shown in Figure 5, the additional stress generated by the coal pillar at point M on the underlying coal seam is calculated based on the superposition principle, and the calculation results are as follows.

The additional vertical stress generated by the wide coal pillar is:

$$\sigma_{y1} = -\frac{2}{\pi}\left\{\begin{array}{l} \int_{-c}^{-b}\frac{q_1 x(y-\xi)^2 d\xi}{\left[x^2+(y-\xi)^2\right]} + \int_{-b}^{-a}\frac{q_2 x(y-\xi)^2 d\xi}{\left[x^2+(y-\xi)^2\right]} + \int_{-a}^{0}\frac{q_3 x(y-\xi)^2 d\xi}{\left[x^2+(y-\xi)^2\right]} \\ +\int_{0}^{a}\frac{q_4 x(y-\xi)^2 d\xi}{\left[x^2+(y-\xi)^2\right]} + \int_{a}^{b}\frac{q_5 x(y-\xi)^2 d\xi}{\left[x^2+(y-\xi)^2\right]} + \int_{b}^{c}\frac{q_6 x(y-\xi)^2 d\xi}{\left[x^2+(y-\xi)^2\right]} \end{array}\right\} + \gamma y \tag{2}$$

where $q_1 = \frac{k_1\gamma H}{c}[x-(a+b+c)]$, $q_2 = \frac{(k_2-k_1)\gamma H}{b}[x-(a+b)]$, $q_3 = k_1\gamma H$, $q_4 = k_1\gamma H$, $q_5 = \frac{(k_2-k_1)\gamma H}{b}[x-(a+b)]$, $q_6 = \frac{k_1\gamma H}{c}[x-(a+b+c)]$, and y is the coal seam spacing.

The additional vertical stress generated by the narrow coal pillar is:

$$\sigma_{y2} = -\frac{2}{\pi}\left\{\int_{-e}^{-d}\frac{q_7x(y-\xi)^2d\xi}{[x^2+(y-\xi)^2]}+\int_{-d}^{0}\frac{q_8x(y-\xi)^2d\xi}{[x^2+(y-\xi)^2]}+\int_{0}^{f}\frac{q_9x(y-\xi)^2d\xi}{[x^2+(y-\xi)^2]}+\int_{f}^{g}\frac{q_{10}x(y-\xi)^2d\xi}{[x^2+(y-\xi)^2]}\right\}+\gamma y \quad (3)$$

where $q_7 = \frac{k_3\gamma H}{e}[x-(d+e)]$, $q_8 = \frac{(k_4-k_5)\gamma H}{d}(x-e)$, $q_9 = \frac{(k_5-k_6)\gamma H}{f}(x-f)$, $q_{10} = \frac{k_6\gamma H}{g}[x-(f+g)]$.

3.1.2. The Effects of Working Face Mining on the Roadway Surrounding Rock

The 42202 transport roadway was affected by the stress transfer from the floor of the coal pillar in the overlying coal seam and the lateral abutment pressure generated in the stope mining process of the adjacent 42201 working face, respectively. The structural evolution of the overlying roof rock of the 42201 working face is shown in Figure 6a. After the coal seam is mined, the overlying rock structure of the gob roof fails, and the stress state is changed significantly. With the continuous advancement of the working face, the scope of the gob increases, and the immediate roof fracture and expand to fill the gob. As "O-X" periodic fracture occurs in the main roof, block B forms an arc triangle with the sectional coal pillar. One end of rock block B rotates and contacts the gangue in the gob, and the other end breaks off inside the rib of the sectional coal pillar. Despite the rotation and sinking, rock block B interlocks with rock block C and rock mass A, forming a hinge "macrostructure" [25,26]. In this process, the static force system transferred by the overlying remaining coal pillar is σ_1^1; The static force system generated by roadway surrounding rock excavation disturbance is σ_1^2; The lateral abutment pressure system due to mining at the 42201 working face is σ_1^3. Therefore, the total stress on the roadway surrounding rock σ can be expressed as $\sigma = \sigma_1^1 + \sigma_1^2 + \sigma_1^3$, as shown in Figure 6.

A micro-element from the roadway rib shown in Figure 6b is selected for analysis. The surrounding rock stress states before and after multiple disturbances and the effect on its stability are illustrated in Figure 7.

(1) Before the disturbance from coal seam mining, the original rock stress is distributed in the rock seam. According to field measurements [27], the ratio of the maximum principal stress σ_1^0 to the minimum principal stress σ_3^0 is approximately 1.3, i.e., the difference between the tangential stress and the radial stress is small. Thus, the diameter of the Mohr's circle $(\sigma_1^0 - \sigma_3^0)$ representing the stress state is very small and below the strength envelope L_1.

(2) As the overlying coal seam (2-2 coal seam) is mined, the tangential stress in the rock mass increases from σ_1^0 to σ_1^1 due to the stress transfer of the remaining coal pillar abutment pressure, and the underlying coal seam enters a high pressure-bearing state.

(3) At the moment of excavation and unloading of the 42202 stope mining roadway, the radial stress around the perimeter of the roadway decreases from σ_3^0 to σ_3^1 and decreases to zero at the roadway surface. However, the tangential stress in the surrounding rock increases rapidly from σ_1^1 to σ_1^2. Thus, the radius of the Mohr's circle $(\sigma_1^2 - \sigma_3^1)$ increases significantly, and the Mohr's circle exceeds the strength envelope L_1. It is thus evident that great deviatoric stress is generated in the surrounding rock after the roadway excavation, resulting in plastic failure of the surrounding rock.

(4) After the stope mining at the 42201 working face, the key stratum of the roof rock shows rotary movements. Due to the superposition of lateral abutment pressure transfer, the tangential stress of the surrounding rock increases sharply from σ_1^2 to σ_1^3, and the plastic failure of the surrounding rock is intensified. In addition, after the large-scale compression-shear failure of the surrounding rock, dislocation failure occurs in the failure area along the shear surface. The strength of the rock mass decreases significantly in the plastic area, and the bearing capacity decreases. At this point, the rock strength envelope develops from line L_1 to line L_2.

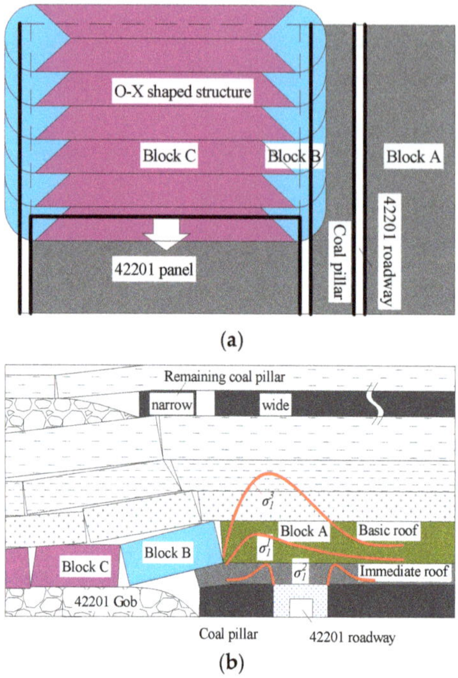

Figure 6. Roof fracture structure after mining at the working face. (**a**) Mining at the 42201 working face; (**b**) Gob-side entry retaining masonry beam structure.

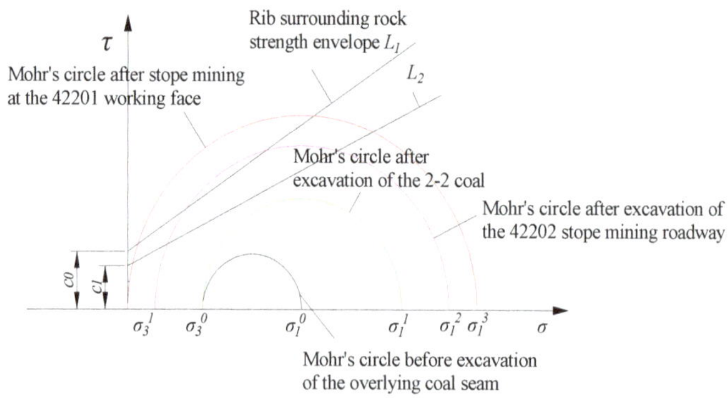

Figure 7. Changes in the stress state and strength of roadway surrounding rock under multiple disturbances.

3.1.3. Reinforcement Support for Stope Mining Roadway

The original support of the 42202 stope mining roadway mainly include bolts and anchor cables. According to the above numerical simulation results, when the coal pillar width is between 25 and 30 m, the plastic area in the roof is 2.5 m, and that in the two ribs is 3.0 to 3.5 m. Supported with anchor cables 6.3 m long, the bearing capacity is sufficient for the deep surrounding rock. Thus, the surrounding rock of the roof has good stability. The two ribs are only supported by bolts 2.1 m long, and the full length of the bolts is within the plastic area. As a result, the surrounding rock of the supported section has a relatively small confining pressure, and the support capacity of the bolts is greatly reduced, rendering the surrounding rock prone to overall instability. Therefore, in addition to leaving reasonable coal pillars and pressure relief, reinforcement support must be applied to the surrounding rocks of the two ribs to improve their stability and bearing capacity.

The following key issues should be noted when applying reinforcement support [28].

(1) The anchorage blind area of the weak coal mass must be controlled. The anchorage blind area is in the shallow triangle between the compression areas (Figure 8), and the deformation and loosening of the coal mass in this area weaken the bolt support force, forming a vicious circle. Appropriate increase in bolt density and the use of bolt and metal strip combination support are the main methods to control the anchorage blind areas. Higher bolt density reduces the range of the anchorage blind area, while steel strips spread the bolt supporting force to the shallow rock mass, thus restraining the deformation of the anchorage blind area.

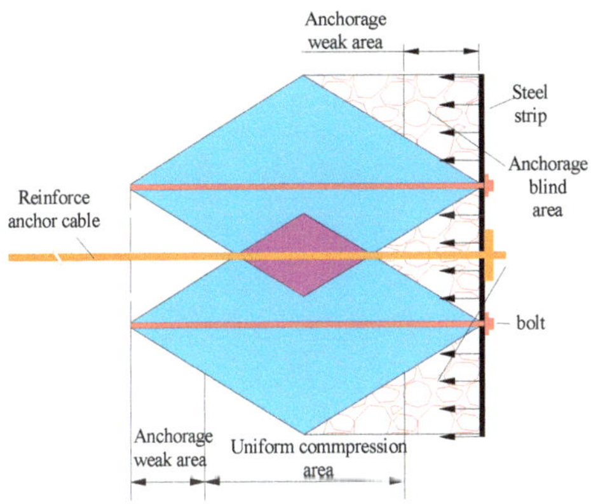

Figure 8. Supporting effect of anchor cable and anchor belt on broken surrounding rock.

(2) The bolt-supported area has low load-bearing capacity. Considering the relatively large range of the plastic area in the surrounding rock, it is difficult for the bolts to penetrate deeply into the stable rock mass. At this time, the main function of the bolts is to form the loose surrounding rock into a supporting structure with sound stability. However, long-term stability requires combining long and short bolt support, i.e., establishing a foundation bearing ring with short bolts in the shallow part and a thick reinforcing ring in the deep part with long anchor cables. Thus, the reinforced bolt-supported area is linked to the deep stable rock, and the range of the surrounding rock-bearing ring is expanded. Consequently, fracture interconnection and loosening fragmentation failures in the shallow surface are prevented, and a deep bolt layer is formed to resist disturbance, as shown in Figure 9.

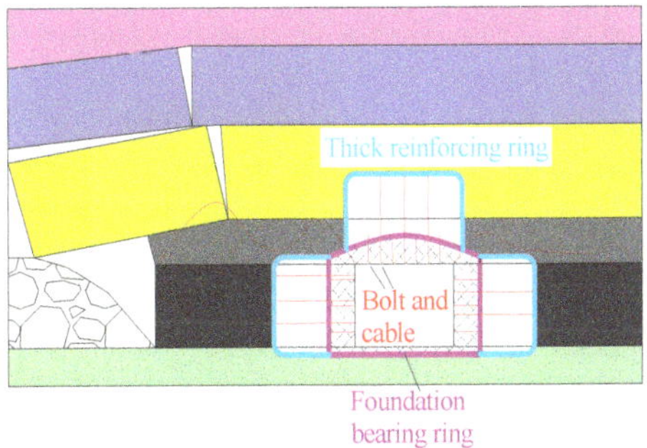

Figure 9. Surrounding rock-bearing structure after reinforcement and support with anchor cable.

3.2. Numerical Simulation

Based on the actual situation of the working face and the roadway at the site, a 770 m × 600 m × 175 m numerical model was established using the finite difference software FLAC3D, as shown in Figure 10. The vertical movement is restricted at the bottom of the model, and the horizontal movement is restricted at the front, rear, and both ends. The overlying rock unit weight is replaced with a uniform load of 9.6 MPa applied on the top of the model, and the lateral pressure coefficient is set to 1.3 [27]. The coal seam was simulated with the Mohr-Coulomb model, and the initial support components of the roadway were simulated using FLAC3D built-in structural units.

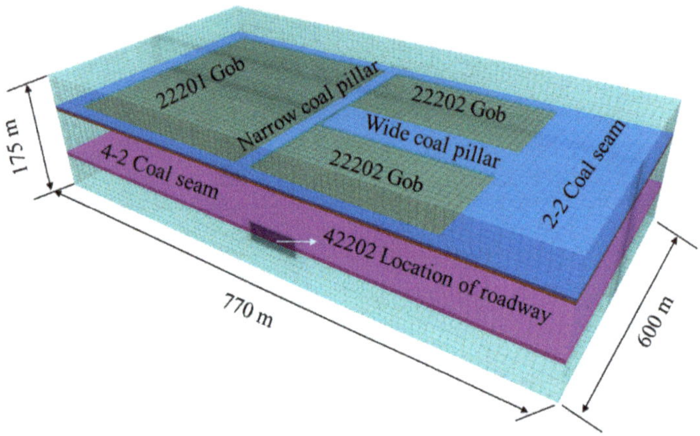

Figure 10. Global 3D numerical model.

The simulation scheme and process are as follows. (1) Application of initial ground stress and initial equilibration. (2) Progressive excavation of the 22201 and 22202 working faces. (3) In order to investigate the effects of the overlying remaining coal pillar and the adjacent 42201 working face on the 42202 stope mining roadway, the 42202 roadway was excavated when the coal pillar width was 10, 15, 20, 25, and 30 m and immediately supported after excavation. (4) The 42201 working face was gradually excavated after equilibrium calculation until the coal seam excavation was completed.

4. Results

4.1. Results of Theoretical Analysis

Taking the location of the 42202 stope mining roadway below the T-shaped remaining coal pillar as an example, the vertical stress distribution directly below the T-shaped remaining coal pillar (inclined to the profile along the working face) calculated based on the stress distribution model in Figure 5 and Equations (2) and (3) is shown in Figure 11, and the vertical stress can be obtained as approximately 13.45 MPa. Therefore, the location of the roadway is affected by the abutment pressure transfer of the overlying coal pillar, and the vertical stress in the surrounding rock has reached 1.4 times the original rock stress before roadway excavation in the underlying coal seam.

In addition to effectively improving the roadway stability and reducing the surrounding rock maintenance cost, the reasonable coal pillar size also greatly reduces the waste of coal resources. The surrounding rock of reusable roadways is generally affected by the superimposed disturbance from roadway excavation and stope mining at the adjacent working faces, especially with roadways below remaining coal pillars. Therefore, the sectional coal pillar size in the underlying working face should be fully considered and designed.

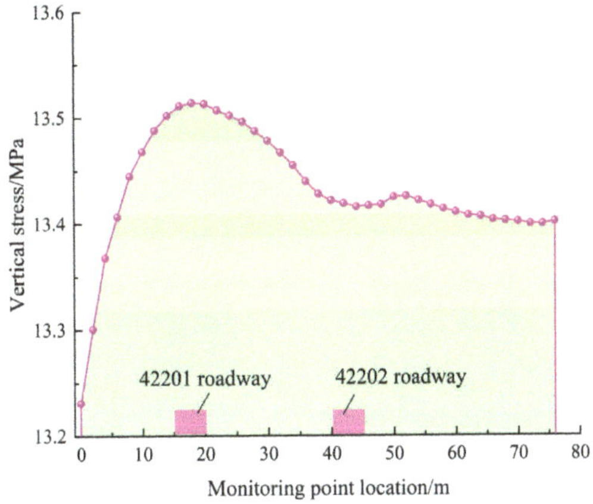

Figure 11. Abutment pressure distribution on the T-shaped remaining coal pillar.

In accordance with conventional coal pillar width determination methods, an elastic core area of 2 times the mining height should be reserved in the middle of the coal pillar during the service of the roadway. Considering the above numerical simulation results, the distribution range of abutment pressure from the remaining coal pillar and adjacent working face, a reasonable roadway should avoid the peak superimposed abutment pressure area and be arranged in the pressure-reducing section. Therefore, the coal pillar width should not fall below 20 m.

4.2. *Results of Numerical Simulation*

Figure 12 shows the distribution of vertical stress and plastic area in the two ribs of the 42202 auxiliary transport roadway after mining at the 42201 working face. The red area and blue area indicate the plastic yielding and elastic states of the roadway surrounding rock, respectively. Parameters σ_{pmax} and σ_{scmax} indicate the peak vertical stresses in the coal pillar rib and the coal mass rib.

As shown in Figure 12, the plastic area of the roadway surrounding rock and the stress concentration of the two ribs show significant decreasing trends with the increase of coal pillar width. According to Figure 12a, plastic failures, mainly shear failures, occur diagonally across the whole coal pillar when the coal pillar width is 10 m. The peak vertical stress in the coal pillar rib is 57.33 MPa in its middle. The peak vertical stress of the coal mass rib is 43.57 MPa and about 4.0 m from the surface of the roadway. The stress concentration coefficients of the two ribs are 5.97 and 4.54, respectively. Therefore, the roadway is in the stress elevation area formed by the T-shaped remaining coal pillar in the overlying coal seam and the mining at the adjacent working face. As a result, the surrounding rock is subjected to great deviatoric stress, leading to a large number of cracks and a wide range of plastic failures. In the meantime, the coal pillar has a relatively large bearing capacity but is highly unstable after fracture and compaction.

According to Figure 12b, when the coal pillar width is increased to 15 m, the plastic failure does not penetrate the coal pillar, and an elastic core area of about 4.0 m appears in its middle. Compared with those at the coal pillar width of 10 m, the peak vertical stress of the coal pillar rib is reduced by 3.39 MPa, and that of the coal mass rib is reduced by 4.7 MPa. The stress concentration coefficients of the two ribs are 5.62 and 4.05, respectively. The large stress difference between the two ribs causes the asymmetrical distribution of the plastic area, and the surrounding rock failure is severe.

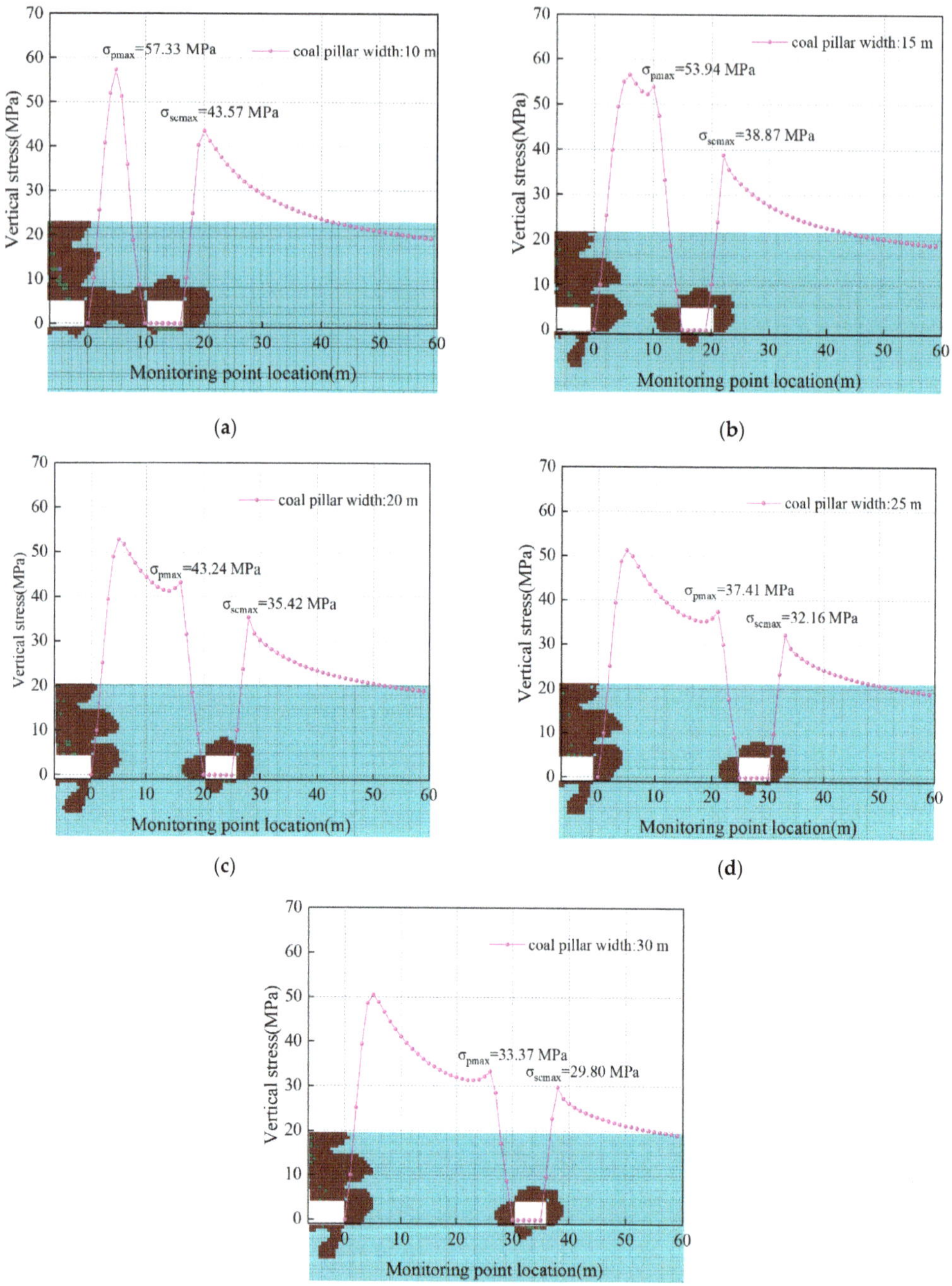

(e)

Figure 12. Stress and plastic area of surrounding rock under different widths of coal pillars: (**a**) The width of the coal pillar is 10 m; (**b**) The width of the coal pillar is 15 m; (**c**) The width of the coal pillar is 20 m; (**d**) The width of coal pillar is 25 m; (**e**) The width of coal pillar is 30 m.

According to Figure 12c–e, the elastic core area of the coal pillar rib keeps increasing while the range of surrounding rock failure gradually decreases with the increasing coal pillar width. When the coal pillar width is 20–30 m, the plastic area of the coal pillar rib is 3.5 to 4.0 m, and the plastic area of the coal mass rib is about 3.0 m. The vertical stress distribution on the coal pillar evolves from a single-peak shape to an asymmetric double-peak shape, the stress concentration tends to decrease, and the peak stress difference between the two roadway ribs gradually decreases. The reason is that as the coal pillar width increases, the roadway gradually avoids the superimposed stress elevation area, resulting in a decreased load on the surrounding rock and increased roadway stability. Combining the results in Section 4.1, the width of the coal pillar in the 42202 mining roadway section should be 25 m.

4.3. Industrial Testing

4.3.1. Scheme of Reinforcement Support

Impacted by the mining, rib spalling and support failure were observed in the roadway. According to the above study, the depth of plastic failure in the surrounding rock of the 42202 auxiliary transport roadway after being disturbed by the first stope mining gradually exceeds the range of bolt support, and the bolt support of the two ribs has basically failed during the secondary stope mining. Thus, it is necessary to strengthen the support of the two ribs. According to the principle of asymmetric support, the solid coal rib was supported with 28.6 mm × 6500 mm anchor cables, and two rows of anchor cables with a spacing of 2000 mm were installed at the rib part together with 300 × 300 × 16 mm pallets. The coal pillar rib was first supported with 22 mm × 6500 mm anchor cables, and two rows of anchor cables with a spacing of 2100 mm were added together with 4600 × 140 × 8 mm π-steel strips. In addition, a row of 28.6 mm × 6500 mm anchor cables with a spacing of 2100 mm were added together with 4600 × 140 × 8 mm π-steel strips, as shown in Figure 13.

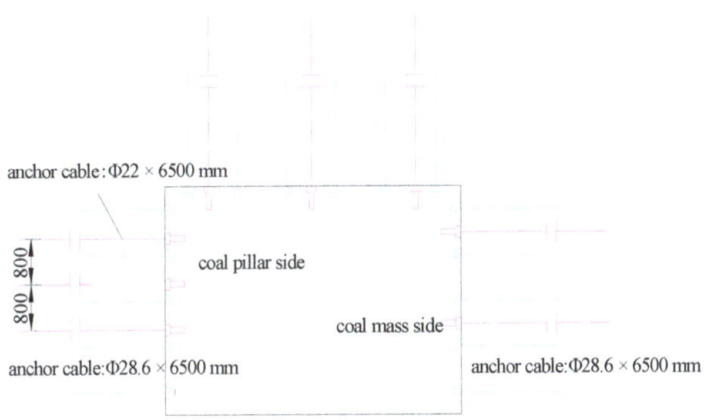

Figure 13. Cross-sectional view of the reinforcement support of the two ribs of the roadway.

In order to verify the stability control effect of bolting support technology on 42202 transportation roadway, the plastic zone state of surrounding rock before and after bolting support was compared by numerical simulation software, as shown in Figure 14. According to Figure 14b, when the width of the coal pillar is set as 25 m and anchor cable reinforcement support technology is adopted, the plastic range of the two walls of surrounding rock is 2.0 m at most. Compared with that before support, the plastic zone of the coal pillar is reduced by about 1.5 m and the solid wall is reduced by about 1.0 m. It can be seen that the use of anchor cable reinforcement support can effectively prevent the plastic destruction of surrounding rock and fully ensure the stability of the mining roadway.

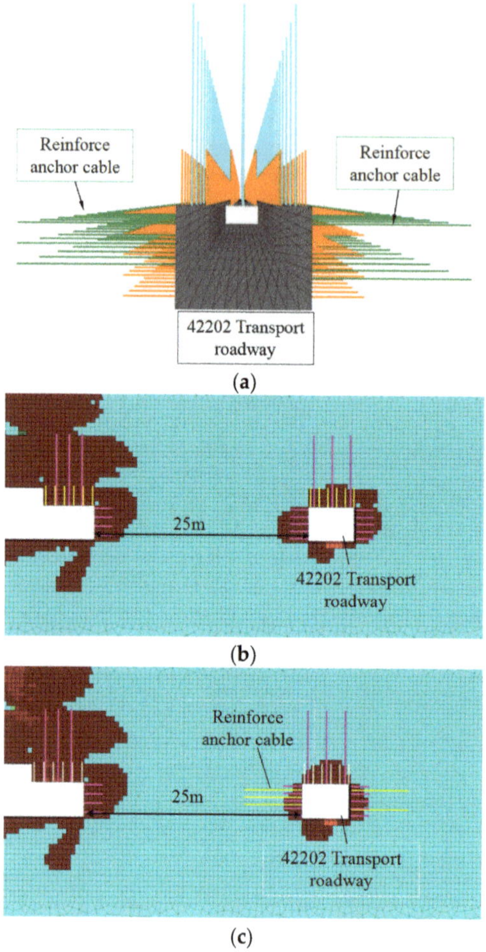

Figure 14. Numerical simulation verification of anchor cable reinforcement support effect. (**a**) Schematic Diagram of Anchor Cable Reinforcement and Support. (**b**) Distribution of plastic zone of surrounding rock before anchor cable reinforcement. (**c**) Distribution of plastic zone of surrounding rock after anchor cable reinforcement.

4.3.2. Displacement Monitoring of Roadway Surface

Based on the sectional coal pillar retaining scheme and the roadway surrounding rock anchor reinforcement support technology proposed in this paper, the experiment was carried out in the 42202 mining roadway. Multiple on-site full-section scanning monitoring tasks were conducted on the 42202 auxiliary transport roadway directly below the skip mining coal pillar during the second stope mining. The results are shown in Figure 15, where the roadway deformation remains basically the same as the working face advances. In the meantime, no significant mine pressure behaviors, including large deformation, rib spalling, and roof fall, are observed at the site, as shown in Figure 16. Therefore, the adopted control techniques can well control the destabilization of the roadway below the remaining coal pillar after multiple mining processes, thus ensuring safe and efficient production at the working face.

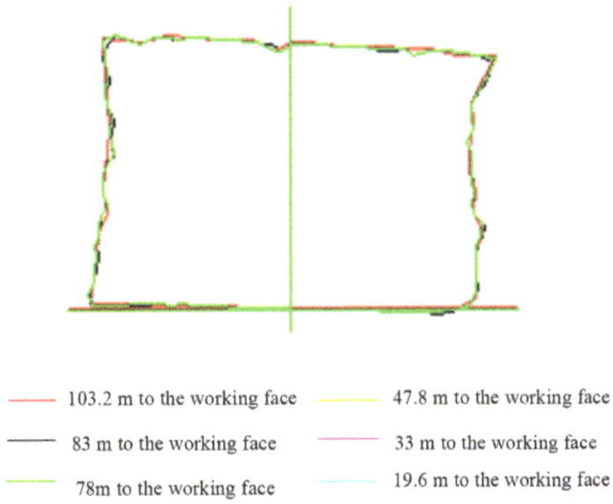

Figure 15. Full section scan of the roadway.

Figure 16. Field observation of roadway stability.

5. Conclusions

(1) Affected by the T-shaped remaining coal pillar in the 2-2 coal seam, the vertical stress at the slope mining roadway of the underlying 4-2 coal seam increased sharply, reaching a stress concentration coefficient of up to 1.4 times. In addition, the superimposed effect of lateral high abutment pressure generated during stope mining at the 42201 working face caused great deviatoric stress in the coal mass, and the radius of Mohr's circle of the stress increased significantly. After roadway excavation and unloading, the surrounding rock is prone to wide-range compression-shear failures. Thus, the peak lateral pressure area should be avoided.

(2) Based on the on-site engineering conditions, numerical simulation software was applied to obtain the stress characteristics and plastic area development pattern in the surrounding rock of the 42202 stope mining roadway below the T-shaped remaining coal pillar. As the width of the sectional coal pillar in the 4-2 coal seam increased, the stress curve of the surrounding rock shifted from single-peaked to asymmetrically double-peaked, and the stress concentration coefficient gradually decreased. Meanwhile, the stress difference between the two ribs of the roadway showed a decreasing trend, and the failure mode of the coal pillar changed from penetrating failure to only producing a 2.5 to 3.0 m plastic area. The large vertical stress increase in the surrounding rock was the main reason for the failure and destabilization of the roadway.

(3) Considering the high-stress area under the T-shaped remaining coal pillar, the sectional coal pillar width in the 42202 mining roadway is 25 m. In the meantime, the asymmetric reinforcement anchor cable support technology for both sides of the roadway is proposed. Based on the numerical calculation verification, it is believed that the scheme

reduces the plastic zone of surrounding rock by 1.0–1.5 m, which can effectively maintain the stability of the roadway.

(4) On-site industrial testing showed that the roadway deformation of the 42202 stope roadway was small and basically constant, and no rib spalling or roof fall was observed, thus achieving the stability control of the stope mining roadway below the remaining coal pillar.

Author Contributions: All authors contributed to this paper. Conceptualization, Q.F. and K.Y.; methodology, Q.F.; software, Y.W.; validation, Q.F. and Z.W.; formal analysis, Q.L.; data curation, Q.F.; writing—original draft preparation, Q.F.; writing—review and editing, Q.L. and K.Y.; validation, X.H. and Q.L.; funding acquisition, K.Y. All authors have read and agreed to the published version of the manuscript.

Funding: This work is supported by the Collaborative Innovation Project of Universities in Anhui Province (GXXT-2019-029), the Institute of Energy, Hefei Comprehensive National Science Center under Grant (No. 21KZS215), Major special projects of science and technology in Shanxi Province (No. 20191101016), Open Research Grant of Joint National-Local Engineering Research Centre for Safe and Precise Coal Mining (Grant NO. EC2021014).

Data Availability Statement: The data used for conducting classifications are available from the corresponding author authors upon request.

Conflicts of Interest: The authors declared no potential conflict of interest with respect to the research, authorship, and publication of this article.

References

1. Sun, X.; Liu, Y.; Wang, J.; Li, J.; Sun, S.; Cui, X. Study on three-dimensional stress field of gob-side entry retaining by roof cutting without pillar under near-group coal seam mining. *Processes* **2019**, *7*, 552. [CrossRef]
2. Zhao, T.-B.; Guo, W.-Y.; Tan, Y.-L.; Yin, Y.-C.; Cai, L.-S.; Pan, J.-F. Case studies of rock bursts under complicated geological conditions during multi-seam mining at a depth of 800 m. *Rock Mech. Rock Eng.* **2018**, *51*, 1539–1564. [CrossRef]
3. Chi, X.; Yang, K.; Wei, Z. Breaking and mining-induced stress evolution of overlying strata in the working face of a steeply dipping coal seam. *Int. J. Coal Sci. Technol.* **2021**, *8*, 614–625. [CrossRef]
4. Wang, H.; Jiang, Y.; Zhao, Y.; Zhu, J.; Liu, S. Numerical investigation of the dynamic mechanical state of a coal pillar during longwall mining panel extraction. *Rock Mech. Rock Eng.* **2013**, *46*, 1211–1221. [CrossRef]
5. Tan, Y.L.; Zhao, T.B.; Xiao, Y.X. In situ investigations of failure zone of floor strata in mining close distance coal seams. *Int. J. Rock Mech. Min. Sci.* **2010**, *47*, 865–870. [CrossRef]
6. Ning, J.; Wang, J.; Tan, Y.; Xu, Q. Mechanical mechanism of overlying strata breaking and development of fractured zone during close-distance coal seam group mining. *Int. J. Min. Sci. Technol.* **2020**, *30*, 207–215. [CrossRef]
7. Li, Q.; Wu, G.; Kong, D. Study on Stability of Stope Surrounding Rock under Repeated Mining in Close-Distance Coal Seams. *Geofluids* **2022**, *2022*, 9630942. [CrossRef]
8. Zhou, A.; Hu, J.; Gong, W.; Wang, K.; Deng, N.; Wu, J. The instability criticality and safety factor of coal-gas outburst induced by shear failure based on limit analysis. *Fuel* **2021**, *303*, 121245. [CrossRef]
9. Yang, K.; Fu, Q.; Liu, Q.; He, X.; Lyu, X. Evolution of Mining-Induced Stress in Downward Mining of Short-Distance Multiseam. *Geofluids* **2022**, *2022*, 3305734. [CrossRef]
10. Chen, D.; Wu, Y.; Xie, S.; Guo, F.; He, F.; Liu, R. Reasonable location of stopping line in close-distance underlying coal seam and partition support of large cross-section roadway. *Int. J. Coal Sci. Technol.* **2022**, *9*, 1–22. [CrossRef]
11. Bai, J.-B.; Shen, W.-L.; Guo, G.-L.; Wang, X.-Y.; Yu, Y. Roof deformation, failure characteristics, and preventive techniques of gob-side entry driving heading adjacent to the advancing working face. *Rock Mech. Rock Eng.* **2015**, *48*, 2447–2458. [CrossRef]
12. Liu, W.; Yang, K.; Zhen, W.; Chi, X.; Xu, R.; Lv, X. Energy Dissipation and Failure Characteristics of Layered Composite Rocks under Impact Load. *Shock. Vib.* **2021**, *2021*, 8775338. [CrossRef]
13. Wei, Z.; Yang, K.; He, X.; Zhang, J.; Hu, G. Experimental Study on the Optimization of Coal-Based Solid Waste Filling Slurry Ratio Based on the Response Surface Method. *Materials* **2022**, *15*, 5318. [CrossRef] [PubMed]
14. Zhang, Z.; Deng, M.; Wang, X.; Yu, W.; Zhang, F.; Dao, V.D. Field and numerical investigations on the lower coal seam entry failure analysis under the remnant pillar. *Eng. Fail. Anal.* **2020**, *115*, 104638. [CrossRef]
15. Xinqiu, F.; Minjiang, G.; Zhiqiang, L. Instability mechanism and prevention of roadway under close-distance seam group mining. *Chin. J. Rock Mech. Eng.* **2009**, *28*, 2059–2067.
16. Xinjie, L.; Xiaomeng, L.; Weidong, P. Analysis on the floor stress distribution and roadway position in the close distance coal seams. *Arab. J. Geosci.* **2016**, *9*, 83. [CrossRef]
17. Sun, Z.; Wu, Y.; Lu, Z.; Feng, Y.; Chu, X.; Yi, K. Stability analysis and derived control measures for rock surrounding a roadway in a lower coal seam under concentrated stress of a coal pillar. *Shock. Vib.* **2020**, *2020*, 6624983. [CrossRef]

18. Zhang, Z.; Deng, M.; Bai, J.; Yan, S.; Yu, X. Stability control of gob-side entry retained under the gob with close distance coal seams. *Int. J. Min. Sci. Technol.* **2021**, *31*, 321–332. [CrossRef]
19. Jiang, F.X.; Wang, Y.X.; Ling, M.; Yang, Y. Mechanism of rock burst occurring in protected coal seam induced by coal pillar of protective coal seam. *Chin. J. Geotech. Eng.* **2017**, *39*, 1689–1696.
20. Liu, H.; Zhang, B.; Li, X.; Liu, C.; Wang, C.; Wang, F.; Chen, D. Research on roof damage mechanism and control technology of Gob-side entry retaining under close distance gob. *Eng. Fail. Anal.* **2022**, *138*, 106331. [CrossRef]
21. Xia, Z.; Yao, Q.; Meng, G.; Xu, Q.; Tang, C.; Zhu, L.; Wang, W.; Shen, Q. Numerical study of stability of mining roadways with 6.0-m section coal pillars under influence of repeated mining. *Int. J. Rock Mech. Min. Sci.* **2021**, *138*, 104641. [CrossRef]
22. He, F.; Lv, K.; Li, X.; Qin, B.; Li, L. Failure mechanism and control of lower retracement channel in close-distance double-thick coal seams. *Shock. Vib.* **2021**, *2021*, 6651099. [CrossRef]
23. Du, F.; Wang, K.; Guo, Y.; Wang, G.; Wang, L.; Wang, Y. The mechanism of rockburst-outburst coupling disaster considering the coal-rock combination: An experiment study. *Geomech. Eng.* **2020**, *22*, 255–264.
24. Ju, J.; Xu, J.; Zhu, W. Longwall chock sudden closure incident below coal pillar of adjacent upper mined coal seam under shallow cover in the Shendong coalfield. *Int. J. Rock Mech. Min. Sci.* **2015**, *77*, 192–201. [CrossRef]
25. Cao, Z.Z.; Zhou, Y.J. Research on coal pillar width in roadway driving along goaf based on the stability of key block. *Cmc-Comput. Mater Contin* **2015**, *48*, 77–90.
26. Ahmadi, M.E.; Mesroghli, S.; Hedayat, B.; Nazerian, H.; Shirazi, A.; Shirazy, A. Assessment of the Influence of Sulfuric Acid/Hydrogen Peroxide Mixture on Organic Sulfur Reduction of High Sulfur Coals and Their Chemical Composition. *Open J. Geol.* **2022**, *12*, 199–214. [CrossRef]
27. Lu, K.; Deng, Z.G.; Feng, J.C. Surrounding rock failure mechanism of reserved roadway under superimposed mining and its control technology. *J. Min. Saf. Eng.* **2019**, *36*, 685–769.
28. Han, C.L.; Zhang, N.; Kan, J.G.; Ran, Z. Mechanism and application of double active control with pressure-relieving and an choring for gob-side entry retaining. *J. China Coal Soc.* **2017**, *42*, 323–330.

processes

MDPI

Article

Dynamic Risk Assessment of High Slope in Open-Pit Coalmines Based on Interval Trapezoidal Fuzzy Soft Set Method: A Case Study

Zhiliu Wang [1,*], Mengxin Hu [1], Peng Zhang [2], Xinming Li [1] and Song Yin [1]

1 School of Civil Engineering and Architecture, Zhongyuan University of Technology, Zhengzhou 450007, China
2 CHN Baorixile Energy Co., Ltd., Hulunbuir 021599, China
* Correspondence: 6855@zut.edu.cn

Abstract: Effective high slope risk assessment plays an important role in the safety management and control of the open-pit coal mining process. Traditional slope stability risk assessment methods rarely consider the time factor or evaluate the dynamic change of high slope in an open-pit mine at a certain time in a sensitivity assessment. This paper develops an interval trapezoidal fuzzy soft set method to achieve the high slope dynamic risk evaluation. The proposed dynamic interval trapezoidal fuzzy soft set method for risk assessment of high slope in an open-pit coal mine is developed by integrating the time points and weights of slope risk factors. The extended interval trapezoidal fuzzy soft set was used to calculate the weights of risk factors at different times, and the Fuzzy Analytical Hierarchy Process (FAHP) method was applied to determine the weights of risk factors. The weight change of different risk factors with time can be easily achieved with the proposed method. As a case study, this approach is implemented into a risk assessment model for the north high slope in Shengli #1 open-pit mine located in Xilinhot, Inner Mongolia. The model complies with three time points and contains 4 primary risk factors (S) and 17 secondary risk factors. The results indicated that the hydrological climate conditions and slope geometry conditions were the high risk factors affecting this open-pit coal mine slope. The reasonability and effectiveness of the evaluation results were verified with in-situ observations and measurements. This dynamic risk assessment method is helpful for improving safety management and control for the high slopes of open-pit mines in the coal mining process.

Keywords: dynamic risk assessment; safety management and control; high slope; open-pit coalmine

Citation: Wang, Z.; Hu, M.; Zhang, P.; Li, X.; Yin, S. Dynamic Risk Assessment of High Slope in Open-Pit Coalmines Based on Interval Trapezoidal Fuzzy Soft Set Method: A Case Study. *Processes* **2022**, *10*, 2168. https://doi.org/10.3390/pr10112168

Academic Editors: Feng Du, Aitao Zhou and Bo Li

Received: 22 September 2022
Accepted: 20 October 2022
Published: 23 October 2022

1. Introduction

In the current process of coal mining, coal mine disasters occur frequently, including landslides, gas explosion, the collapse of the roof, and rock burst, etc., whilst landslides are the dominant disaster type for the open-pit coalmine [1–3]. Slope stability directly affects coal mining safety. Especially for the high slope in the open-pit coalmine, where the height is generally over 200 m, a landslide disaster is more likely to occur because of complex geological conditions and multiple risk factors [4]. Accurate assessment and analysis of high slope stability in an open-pit mine is very important for landslide disaster prevention and control [5,6].

Slope stability analysis usually involves various uncertain factors, such as the statistical uncertainty of rock and soil parameters. However, conventional deterministic analyses, which use a factor of safety (FS) as the slope stability criterion, cannot quantitatively and comprehensively consider the influence of these uncertain factors [7–9]. Many scholars have tried to use different monitoring technologies to realize slope dynamic early warning, such as displacement information integration [10], synthetic aperture radar interference [11], ground-based synthetic aperture radar technology [12], and multi-technology combination [13], etc. However, these monitoring techniques lack multivariate risk factors analysis. By contrast, risk assessment based on statistics and probability theory can reasonably and comprehensively quantify these uncertainties in slope stability analysis. The most frequently used analytical methods are probability theory [14–17], interval number theory [18], fuzzy theory [19–22], rough set theory, etc. [23,24], and direct MonteCarlo simulation (MCS) [25–27]. However, these methods cannot accurately solve the complex problem of slope risk assessment in which multiple factors and a large number of field data are involved. In recent years, analytic hierarchy process (AHP) [28–30] has been applied to systematically assess the slope stability. Santos et al. [31] proposed a hazard graph and generated the quantitative hazard assessment system to carry out risk analysis for open-pit mine slopes and saw the method as having a high discrimination capacity. Cheng et al. [32] used the random finite difference method (RFDM) to assess the risk of slope failure and concluded that the structure had a dual effect on the stability of the slope. Pinheiro et al. [33] developed slope quality Index and applied it to many real road slopes. Soft set theory also was used to study slope stability affected by multiple risk factors. Yang et al. [34] applied soft set theory to optimize the measures of highway slope treatment, and in their work the weights of affecting factors used in the slope safety evaluation were given by different experts. Fuzzy analytic hierarchy process (FAHP) determines the relative importance of each criterion through pairwise comparison given by domain experts or decision makers and is widely used to calculate standard weights [35–38]. In addition, Zhang and his co-researchers generalized parametric soft sets to dynamic interval trapezoidal fuzzy soft sets and made comprehensive evaluation decisions. The simultaneous feasibility of the decision-making method is proved by a case study [39]. The dynamic fuzzy soft set is also expanded into dynamic interval-valued fuzzy soft sets by He and his co-authors [40]. The operations and properties of dynamic interval-valued fuzzy soft set are studied, and the dynamic interval-valued fuzzy soft set decision is proposed. Zhu et al. enriched and improved the theory of interval trapezoidal fuzzy soft sets and promoted its practical application [41]. All of these achievements for statistics and probability theory have illustrated its capability and popularity in slope reliability analysis.

However, the studies rarely considered the factor of time in their sensitivity evaluation. Thus, the risk assessment of high slope cannot fully reflect the dynamic change of high slope in an open-pit coal mine during a certain period of time. For a slope in an open-pit coal mine, the change of slope stability at different time points is of great importance in risk management. In this work, an interval trapezoidal fuzzy soft set method is developed to achieve the high slope dynamic risk evaluation. The proposed dynamic interval trapezoidal fuzzy soft set method for risk assessment of high slope is developed by integrating time points and the weights of slope risk factors. More specifically, the time period is firstly clarified when the risk factors affecting the slope stability of open-pit mine are strongly superimposed. Then, the risk evaluation grade of the high slope is determined by the FAHP method, which determines slope risk factors by referring to the relevant specifications of the open-pit mine. The risk evaluation grade of the high slope is determined by the FAHP method. Again, the slope risk factors are quantified by expert scoring, and the weight of each risk factor is calculated by FAHP. The weight is independent of time. Then, the dynamic interval trapezoidal fuzzy soft set method is used to integrate the time and weights of different risk factors, and the entropy matrix is used to calculate the evaluation value of risk factors at different moments. The following, the multi-attribute decision integration and evaluation of all risk factor parameters in all time periods are carried out by the dynamic interval trapezoidal fuzzy soft set method. Finally, the high slope in the Shengli open- pit coal mine was taken as a case study, three time points were selected, and the parameter values of 17 secondary indicators were obtained. The risk factors of the high slope at different time points were evaluated and then verified with the field data monitored.

2. Expansion of Interval Trapezoidal Fuzzy Soft Sets

2.1. Interval Trapezoidal Fuzzy Soft Sets and Their Properties

In this section, we introduced some basic definitions and their properties in interval trapezoidal fuzzy soft set theory [42,43].

Definition 1. *Let U be an initial universe set and E be a set of parameters. A denotes an interval trapezoidal fuzzy soft set over a common universe U. A is a subset of E, and $A \subset E$. A pair (F, A) is called an interval trapezoidal fuzzy soft set over a common universe U, where F is a mapping given by $F: A \to \Gamma(U)$.*

That is to say, for $\forall e \in A$, we can obtain

$$F(e) = \left\{ \left\langle x, S_{F(e)}(x) | x \in U \right\rangle \right\} \tag{1}$$

where $S_{F(e)}(x)$ is the interval trapezoidal fuzzy number corresponding to x in $F(e)$.

Definition 2. *Let (F, A) and (G, B) be interval trapezoidal fuzzy soft sets over a common universe U. (H, C) is the intersection of (F, A) and (G, B), where $C = A \cap B$ and $(F, A) \cap (G, B) = (H, C)$. For $\forall \varepsilon \in C$, we can obtain,*

$$F(\xi) = F(\xi) \cap G(\xi) = \left\{ \left\langle x, S_{F(\xi)}(x) \cap S_{G(\xi)}(x) \right\rangle x \in U \right\} \tag{2}$$

where $S_{F(\xi)}(x)$ and $S_{G(\xi)}(x)$ are the interval trapezoidal fuzzy Numbers corresponding to x in $F(\xi)$ and $G(\xi)$.

Definition 3. *Let (F, A) and (G, B) be interval trapezoidal fuzzy soft sets over a common universe U. The union of (F, A) and (G, B) is defined as the soft set (M, D), where $D = A \cup B$ and $(F, A) \cup (G, B) = (M, D)$. For $\forall \xi \in D$, we can have*

$$M(\varepsilon) = F(\varepsilon) \cup G(\varepsilon) = \left\{ \left\langle x, S_{M(\varepsilon)}(x) \right\rangle x \in U \right\} \tag{3}$$

Among them

$$S_{M(\varepsilon)}(x) = \begin{cases} S_{F(\varepsilon)}(x), \varepsilon = A - B \\ S_{G(\varepsilon)}(x), \varepsilon = B - A \\ S_{F(\varepsilon)}(x) \cup S_{G(\varepsilon)}(x), \varepsilon = A \cap B \end{cases} \tag{4}$$

where $S_{F(\varepsilon)}(x)$ and $S_{G(\varepsilon)}(x)$ are the interval trapezoidal fuzzy numbers corresponding to x in $F(\varepsilon)$ and $G(\varepsilon)$, respectively.

If (G_1, A_1) and (G_2, A_2) are two interval trapezoidal fuzzy soft sets over a common universe U, then (G_1, A_1) and (G_2, A_2) is defined as

$$(G_1, A_1,) \wedge (G_2, A_2,) = (H, A_1 \times A_2) \tag{5}$$

We can obtain

$$H(\alpha, \beta) = G_1(\alpha) \cap G_2(\beta) = \left\{ \left\langle x, S_{G_1(\alpha)}(x) \cap S_{G_2(\beta)}(x) \right\rangle | x \in U \right\} \tag{6}$$

for $\forall(\alpha, \beta) \in A_1 \times A_2$, where $S_{G_1(\alpha)}(x)$ and $S_{G_2(\alpha)}(x)$ are the interval trapezoidal fuzzy numbers of x in the $G_1(\alpha)$ and $G_2(\beta)$, respectively.

If (G_1, A_1) and (G_2, A_2) are two interval trapezoidal fuzzy soft sets over a common universe U, (G_1, A_1) or (G_2, A_2) is defined as

$$(G_1, A_1) \vee (G_2, A_2) = (M, A_1 \times A_2) \tag{7}$$

We can obtain

$$M(\alpha,\beta) = G_1(\alpha) \cup G_2(\beta) = \left\{ \left\langle x, S_{G_1(\alpha)}(x) \cup S_{G_2(\alpha)}(x) \right\rangle | x \in U \right\} \text{for } \forall(\alpha,\beta) \in A_1 \times A_2 \tag{8}$$

where $S_{G_1(\alpha)}(x)$ and $S_{G_2(\alpha)}(x)$ are the interval trapezoidal fuzzy numbers of x in the $G_1(\alpha)$ and $G_2(\beta)$, respectively.

2.2. Correlation Theorem of Interval Trapezoidal Fuzzy Soft Sets

Theorem 1. *If (G_1, A_1), (G_2, A_2) and (G_3, A_3) are three interval trapezoidal fuzzy soft sets over a common universe U, we can obtain*

$$((G_1, A_1) \vee (G_2, A_2)) \vee (G_3, A_3) = (G_1, A_1) \vee ((G_2, A_2) \vee (G_3, A_3)) \tag{9}$$

$$((G_1, A_1) \wedge (G_2, A_2)) \wedge (G_3, A_3) = (G_1, A_1) \wedge ((G_2, A_2) \wedge (G_3, A_3)) \tag{10}$$

Theorem 2. *If (G_1, A_1) and (G_2, A_2) are two interval trapezoidal fuzzy soft sets over a common universe U, we can obtain*

$$((G_1, A_1) \vee (G_2, A_2))^c = (G_1, A_1)^c \wedge (G_2, A_2)^c \tag{11}$$

$$((G_1, A_1) \wedge (G_2, A_2))^c = (G_1, A_1)^c \vee (G_2, A_2)^c \tag{12}$$

Theorem 3. *If (G_1, A_1), (G_2, A_2) and (G_3, A_3) are three interval trapezoidal fuzzy soft sets over a common universe U, we can obtain*

$$((G_1, A_1) \vee (G_2, A_2)) \wedge (G_3, A_3) = ((G_1, A_1) \wedge (G_3, A_3)) \vee ((G_2, A_2) \wedge (G_3, A_3)) \tag{13}$$

$$((G_1, A_1) \wedge (G_2, A_2)) \wedge (G_3, A_3) = ((G_1, A_1) \vee (G_3, A_3)) \wedge ((G_2, A_2) \vee (G_3, A_3)) \tag{14}$$

Based on the above expansion of interval trapezoidal fuzzy soft sets, the dynamic risk evaluation model of high slope in an open-pit mine are established in the next section.

3. Establishing the Dynamic Risk Evaluation Model of High Slope in Open-Pit Mine

In this section, the following steps are used to integrate the time points and weights of slope risk factors (see Figure 1).

Step 1: Make clear the time period when the risk factors affecting the slope stability of open-pit mine are strong superimposed.

Step 2: Determine slope risk factors by referring to the relevant specifications of open-pit mine. They are technical specification for annual evaluation of the stability of open-pit coal mine slopes (GB/T 37573-2019) and code for the design of open-pit mines in the coal industry (GB 50197-2015). Then, the risk evaluation grade of high slope is determined by the FAHP method.

Step 3: The slope risk factors are quantified by expert scoring, and the weight of each risk factor is determined by FAHP method.

Step 4: The time and weight of risk factors are integrated by dynamic interval trapezoidal fuzzy soft sets, and the evaluation value of risk factors at different moments is calculated by entropy matrix. Finally, the multi-attribute decision integration and evaluation of all risk factor parameters in all time periods are carried out by dynamic interval trapezoidal fuzzy soft sets.

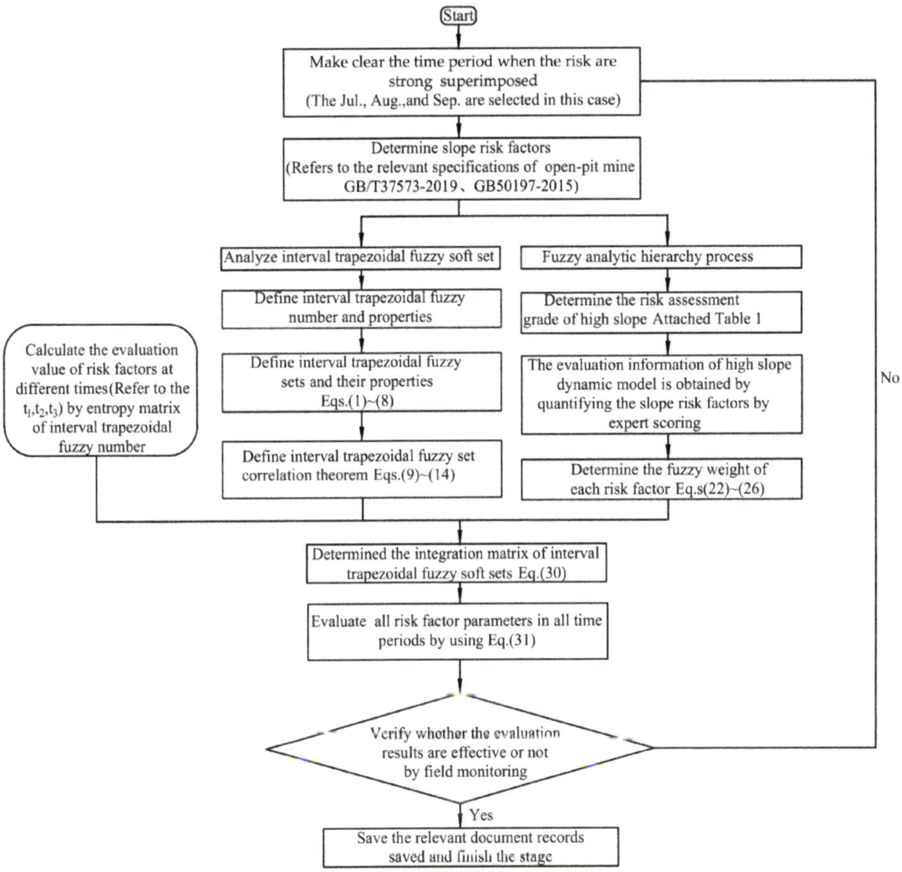

Figure 1. Intuitionistic trapezoidal fuzzy soft set method for risk assessment of high slopes.

For the dynamic evaluation of high slope risk in open-pit mine, the time point sets of monitoring data are given, which is denoted as

$$T = \{t_1, t_2, \cdots t_k\} \tag{15}$$

where t is the time point of different risk factors; k is the number of time points.

To determine four primary indexes, we mainly refer to the relevant specifications of open-pit mine, which are technical specification for annual evaluation of the stability of open-pit coal mine slopes (GB/T 37573-2019) and specifications for design of open pit mine of coal industry (GB 50197-2015). To determine 17 primary indexes (see Table A1), we not only refer to the relevant specifications of open pit mine but also add to concern factors for high slope stability by technical and managerial personnel of the open pit mine.

The parameter sets of risk assessment are

$$A = \{B_1, B_2, B_3, B_4\} \tag{16}$$

where B_1 is hydrological-climatic conditions, and B_2 is internal geological structure of slope. B_3 and B_4 are slope geometry and inducing factors of landslide, respectively.

$$B_1 = \{e_{11}, e_{12}, e_{13}, e_{14}\} \tag{17}$$

$$B_2 = \{e_{21}, e_{22}, e_{23}, e_{24}, e_{25}\} \quad (18)$$

$$B_3 = \{e_{31}, e_{32}, e_{33}, e_{34}\} \quad (19)$$

$$B_4 = \{e_{41}, e_{42}, e_{43}, e_{44}\} \quad (20)$$

where e_{11} is weathering and freeze-thaw, and e_{12} is the state of groundwater. e_{13} is permeability of rock and soil layer, and e_{14} is annual rainfall. e_{21} and e_{22} are lithology and geological structure, respectively. e_{23} and e_{24} are slope structure and internal friction angle, respectively. e_{25} is cohesion of slope. e_{31} and e_{32} are slope angle and slope height, respectively. e_{33} is relationship between soft surface and slope surface, and e_{34} is slope morphology. e_{41} and e_{42} are human factors and impact of blasting, respectively. e_{43} and e_{44} are slope angle of excavation and earthquake intensity, respectively. B_1~B_4 are primary indicators and e_{11}~e_{44} are secondary indicators.

The slope risk evaluation result set is

$$X = \{x_1, x_2, \cdots x_n\} \quad (21)$$

where n is the number of evaluation grades.

3.1. Determining the Weights in Different Time Range

If $s = [(a^-, a^+); b; c; (d^-, d^+)]$ is the interval trapezoidal fuzzy number, its entropy value is

$$E(s) = \left| \frac{1}{4}\left(\frac{a^- + a^+}{2} + b + c + \frac{d^- + d^+}{2}\right) - 0.5 \right| \quad (22)$$

The entropy matrix of $E^k_{B_q} = (E^k_{B_q}(e^{(k)}_{ij}))_{m\times n}$ at a different range is obtained by calculating the values of B_1, B_2, B_3 using Equation (22). $(e^{(k)}_{ij})_{n\times m}$ is the kth evaluation matrix. $e^{(k)}_{ij}(i = 1,2,3\cdots n; j = 1,2,3\cdots m)$ is the value of attribute c_j in the form of interval trapezoidal fuzzy number about evaluation grade t_i at time point of t_k.

Where i = 1, 2, 3; k = 1, 2 ... ; j = $q1$, $q2$ ... ql; q = 1, 2, 3, 4. ql represents the number of parameters of the index B_q. Then, all the entropy values in the corresponding entropy matrix are added.

The entropy matrix can be obtained

$$E(B^k_q) = \frac{1}{mn}\sum_{i=1}^{m}\sum_{j=1}^{n} E(e^{(k)}_{ij}) \quad (23)$$

The corresponding weight $w(B^k_q)$ at the moment t_k of B_q

$$w(B^k_q) = \frac{1 - E(B^k_q)}{\sum\limits_{i=1}^{m}(1 - E(B^k_q))} \quad (24)$$

3.2. Determining the Index Weight by FAHP Method

In order to determine the weight of each influencing factor, FAHP [44–46] method is used and a fuzzy consistent matrix is firstly constructed, and the matrix $A = (a_{ij})_{n\times n}$ satisfies the following conditions:

$$\begin{cases} 0 \le a_{ij} \le 1 \ (i,j = 1,2,3\cdots,n) \\ a_{ij} + a_{ji} = 1 \ (i,j = 1,2,3\cdots,n) \\ \forall i,j, a_{ij} = \omega_i - \omega_j + 0.5 \end{cases} \quad (25)$$

If A is a fuzzy consistent matrix, the weight ω can be denoted by

$$\omega_i = \frac{1}{n}(\sum_{j=1}^{n} a_{ij} + 1 - \frac{n}{2}) \tag{26}$$

It is worth noting that when $\sum_{j=1}^{n} a_{ij} \leq \frac{n}{2} - 1$, it needs to be modified until $\omega_i \geq 0$ is satisfied.

3.3. Integration and Decision Method of Interval Trapezoidal Fuzzy Soft Sets

Let $U = \{x_1, x_2, \cdots x_m\}$ be an initial universe set and $A = \{e_1, e_2, \cdots e_n\}$ be a set of parameters. The matrix of $\widetilde{F}_k = (S_{F_k(e_j)}(x_i))_{m\times n}$ is corresponding to the interval trapezoidal fuzzy soft set of (F_k, A). Where, $i = 1, 2, \ldots m; j = 1, 2, \ldots n; k = 1, 2, \ldots K$. $S_{F_k(e_j)}(x_i)$ is the corresponding interval trapezoidal fuzzy number of $F_k(e_j)$ and x_i, the interval trapezoidal fuzzy soft matrix between the algorithm is given as follows.

If $\widetilde{F}_1 = (S_{F_1(e_j)}(x_i))_{m\times n}$ and $\widetilde{F}_2 = (S_{F_2(e_j)}(x_i))_{m\times n}$ are interval trapezoidal fuzzy soft matrix, and $c > 0$, we can obtain

$$\widetilde{F}_1 + \widetilde{F}_2 = (S_{F_1(e_j)}(x_i))_{m\times n} + (S_{F_2(e_j)}(x_i))_{m\times n} \tag{27}$$

$$c\widetilde{F}_1 = (cS_{F_1(e_j)}(x_i))_{m\times n} \tag{28}$$

If $\widetilde{F}_k(k = 1, 2 \cdots K)$ is the interval trapezoidal fuzzy soft matrix and the weight satisfies the conditions of $\omega \in [0, 1]$ and $\sum_{k=1}^{K} \omega_k = 1$.

The weighted average operator of the interval trapezoidal fuzzy matrix can be denoted by

$$f_\omega(\widetilde{F}_1, \widetilde{F}_2, \cdots, \widetilde{F}_K) = \sum_{k=1}^{K} \omega_k \widetilde{F}_k \tag{29}$$

We can obtain

$$f_\omega(\widetilde{F}_1, \widetilde{F}_2, \cdots, \widetilde{F}_K) = \sum_{k-1}^{K} (\omega_k S_{\widetilde{F}_k(e_j)}(x_i))_{m\times n} \tag{30}$$

The integrated synthetic interval trapezoidal fuzzy soft matrix of $\widetilde{F}_K(k = 1, 2, \cdots, K)$ is denoted as $\widetilde{R} = (r_{ij})_{m\times n}$.

The decision values are used to deal with the decision problems of the interval trapezoidal fuzzy soft sets, and the decision values for x_i in the universe set based on interval trapezoidal fuzzy soft sets in the domain are denoted by

$$\eta_i = \sum_{j=1}^{m} ((\tau_i^- - \tau_j^-) + (\tau_i^+ - \tau_j^+) + (\varphi_i - \varphi_j) + (\varepsilon_i - \varepsilon_j) + (\gamma_i^- - \gamma_j^-) + (\gamma_i^+ - \gamma_j^+)) \tag{31}$$

where the corresponding interval trapezoidal fuzzy number of x_i is

$$S(x_i) = [(\tau_i^-, \tau_i^+); \varphi_i; \varepsilon_i, (\gamma_i^-, \gamma_i^+)] \tag{32}$$

3.4. Case Study

3.4.1. Dynamic Risk Assessment Based on Trapezoidal Fuzzy Soft Set in Shengli Open-Pit Mine

(1) Decision-making steps and methods

Based on above interval trapezoidal fuzzy soft set model of high slope in an open-pit coal mine, the dynamic evaluation risk steps of high slope in an open-pit coal mine are divided into four steps:

(1) Determine the times when the risk factors affecting the slope stability of open-pit mine are strong superimposed;

(2) Determine the weight of each parameter using Equations (22)–(26);

(3) Integrate interval trapezoidal fuzzy soft matrix at different times using Equation (30);

(4) Calculate the decision values corresponding to different risks using Equation (31). The multi-attribute decision integration and evaluation of all risk factor parameters in all time periods are carried out by dynamic interval trapezoidal fuzzy soft sets.

(2) Determination of values for dynamic model parameter set

The north end slope of Shengli #1 open-pit mine is taken in this case study (see Figure 2). The slope height and maximum slope angle are, respectively, 200 m and 57°. Three moments were selected, i.e., $T = \{t_1, t_2, t_3\}$, t_1 is 1 July 2016; t_2 is 1 August 2016 and t_3 is 1 September 2016. The three time points (July, August and September) selected are the times with strong superposition of risk factors for open-pit mine slope. In this time period, there are many unstable factors inducing landslides. On the one hand, these three months are the rainy season for the open-pit mines in northern China, with a relatively concentrated and large rainfall, and the slope stability is greatly affected by hydrogeological factors. On the other hand, the open-pit mines in northern China can only be mined but not stripped in winter due to weather reasons. In order to leave enough earthwork and coal for winter mining, the blasting frequency during the three months (July, August and September) is relatively large, and the slope stability is greatly affected by mining factors. The risk decision of high slope in open-pit mine is evaluated from four primary risk factors (S), the hydrological and climatic conditions (B_1), the slope internal geological structure (B_2), the slope geometric conditions (B_3), and landslide risk factors (B_4). The risk level may be labeled as low, general and high, which are denoted by (x_1), (x_2), and (x_3), respectively. In addition, 17 secondary indicators are selected to carry out risk assessment in four primary indicators. This will be described in the third part later on. The risk grade is divided by considering the influence of comprehensive factors. Low and medium are classified as low decision risk, high as general decision risk, and dangerous and extremely dangerous as high decision risk. The expert evaluation method was used to evaluate the slope risk factors of open-pit mine. Data were collected mainly through questionnaires. A sample of the questionnaire is attached. The data were mainly collected from the production technicians, stripping workers and management personnel of Shengli Open-pit coal mine. It covers all positions of frontline production, management and technology in open-pit coal mines. Among them, 80 questionnaires were sent out by Shengli open-pit coal mine, and 72 questionnaires were effectively recovered. The parameters of slope risk factor B_1, B_2, B_3 and B_4 at time t_1, t_2 and t_3 are listed in Tables A2–A12 in Appendix A.

(3) Calculation of weights for dynamic model

Figure 2. High slopes of Shengli #1 open-pit mine.

The values of the four primary risk indicators of B_1, B_2, B_3 and B_4 are calculated by Equation (22), subsequently the entropy matrix of $E^k_{B_q} = (E(e^{(k)}_{ij}))_{m\times n}$ corresponding to t_k

at different moments can be obtained. For example, the calculation results of the entropy matrix of B_1 at different times are shown in Tables 1–3.

Table 1. Entropy of each parameter of hydrological and climatic conditions B_1 at time t_1.

	$E(e_{11})$	$E(e_{12})$	$E(e_{13})$	$E(e_{14})$
x_1	0.13	0.15	0.13	0.15
x_2	0.15	0.05	0.21	0.06
x_3	0.04	0.08	0.04	0.20

Table 2. Entropy of each parameter of hydrological and climatic conditions B_1 at time t_2.

	$E(e_{11})$	$E(e_{12})$	$E(e_{13})$	$E(e_{14})$
x_1	0.06	0.20	0.13	0.05
x_2	0.18	0.14	0.23	0.08
x_3	0.06	0.08	0.14	0.23

Table 3. Entropy of each parameter of hydrological and climatic conditions B_1 at time t_3.

	$E(e_{11})$	$E(e_{12})$	$E(e_{13})$	$E(e_{14})$
x_1	0.21	0.28	0.10	0.10
x_2	0.10	0.10	0.29	0.16
x_3	0.15	0.08	0.15	0.08

The weight of B_1 is calculated by Equations (23) and (24), and the result is

$$w_{B_1} = (0.219, 0.395, 0.386) \tag{33}$$

Similarly, the weights of B_2, B_3 and B_4 can be obtained

$$w_{B_2} = (0.214, 0.396, 0.390) \tag{34}$$

$$w_{B_3} = (0.229, 0.394, 0.377) \tag{35}$$

$$w_{B_4} = (0.235, 0.385, 0.380) \tag{36}$$

3.4.2. Risk Evaluation

Based on the above interval valued fuzzy set model, the dynamic risk of the north end slope in Shengli #1 open-pit coalmine is evaluated comprehensively. The above three time points are selected to evaluate the risk from four aspects, i.e., hydro-climatic conditions, slope internal geological structure, slope geometric conditions and induced factors. The risk evaluation of high slope is divided into three levels: low (x_1), general (x_2) and high (x_3). Combined with the above theory, the weights at different times are obtained and then the comprehensive interval trapezoidal fuzzy soft set evaluation information for different parameter sets in all time range can be obtained by Equation (29). That is, the comprehensive evaluation values of influencing factors B_1, B_2, B_3 and B_4 in the whole time range subjected to different risks can be obtained as listed in Tables A14–A17 in Appendix A.

Meanwhile, the fuzzy symmetry matrix of each parameter for the risk factors (i.e., hydro-climatic conditions, internal geological structure, geometric conditions and induced factors) are obtained by FAHP theory in Tables A18–A21, as shown in Appendix A. The corresponding weight vectors are calculated as follows

$$w_{B_1} = (0.225, 0.250, 0.250, 0.275) \tag{37}$$

$$w_{B_2} = (0.160, 0.260, 0.220, 0.180, 0.180) \tag{38}$$

$$w_{B_3} = (0.150, 0.250, 0.250, 0.350) \tag{39}$$

$$w_{B_4} = (0.225, 0.375, 0.300, 0.100) \tag{40}$$

The fuzzy symmetric matrix for primary risk indicators is obtained in Table A22 in Appendix A. The weight value is obtained by Equation (41)

$$\omega_A = (0.150, 0.150, 0.325, 0.375) \tag{41}$$

The weights of risk parameters for the risk evaluation system of high slope in open-pit mine are given in Table 4.

Table 4. Weights of risk parameters for the risk evaluation system.

Primary Indicator	Parameters	Primary Weight	Parameter Weights
Hydro-climatic conditions	Weathering and freeze-thaw	0.150	0.225
	Groundwater occurrence		0.250
	Water permeability		0.250
	Average annual rainfall		0.275
Geological structure inside the slope	Lithology	0.150	0.160
	Geological structure		0.260
	Slope structure		0.220
	Internal friction		0.180
	Cohesion		0.180
Slope geometric conditions	Slope angle	0.325	0.150
	Slope height		0.250
	Relationship between weak surface (fault) and slope		0.250
	Slope morphology		0.350
Induced factors	Human factors	0.375	0.225
	Destructive factor		0.375
	Excavation angle		0.300
	Seismic intensity		0.100

The interval trapezoidal fuzzy soft set of different parameters are integrated by Equation (30). The integrated evaluation value of different parameters belonging to open-pit slope risk and the integrated value of surface mine slope risk relative to different risk grades are obtained in Tables A23 and A24 in Appendix A.

The comprehensive evaluation value of the relative risk level, which is either low, general or high, can be obtained. The evaluation value which represents low risk for high slope in Shengli open-pit mine is

$$\mu_1 = [(0.225, 0.401); 0.427; 0.491; (0.548, 0.704)] \tag{42}$$

The evaluation value which represents general risk for high slope is

$$\mu_2 = [(0.243, 0.408); 0.414; 0.481; (0.546, 0.699)] \tag{43}$$

The evaluation value which represents high risk for high slope is

$$\mu_3 = [(0.295, 0.434); 0.450; 0.512; (0.567, 0.681)] \tag{44}$$

Further, the decision value of the high slope risk in the open-pit which is subject to low risk, general risk, and higher risk can be obtained. The integrated evaluation value is calculated according to Equation (28). The results are shown in Table 5.

Table 5. Different decision factors relative to open-pit slope risk.

	B_1	B_2	B_3	B_4
η_1	0.336	−0.446	−0.156	−0.185
η_2	−1.257	0.397	−1.098	0.88
η_3	0.921	0.049	1.254	−0.695

As shown in Table 5, for the decision-making results of the high slope in Shengli #1 open-pit mine, the values of B_1 and B_3 indicate a high risk level, while the values of B_2 and B_4 indicate a general risk level. Thus, among the many factors affecting the instability of the slope of Shengli #1 open-pit coal mine, hydro-climatic conditions and slope geometric conditions belong to high risk factors. Special attention should be paid to hydrology and slope geometry during slope stability maintenance. For example, slope reinforcement and radar displacement monitoring should be planned and performed during the rainy season. The stability evaluation should be performed when designing the slope angle.

4. Verification with Field Data

In order to verify the rationality of the above-mentioned risk evaluation method, the in-situ monitoring data were analyzed in the north end slope of Shengli #1 coalmine. The anchor monitoring points are arranged to monitor the axial force variation of the anchor cable and the displacement parameters are analyzed in the north end of the slope. Field layout with 4 anchor monitoring points and regional failure of a certain landslide in the north end slope is shown in Figure 3. The correlation laws were obtained by field monitoring results of the slope sliding in the north end slope of the open-pit mine in 2016, the axial force variation of the anchor cable in 2016, the rainfall in 2016 and the displacement of the north slope of the slope. The GPS monitoring displacements before and after the landslide accident are presented in Figure 4. The axial force variation of the anchor cable of the north end slope in 2016 are presented in Figure 5, and the rainfall and the displacement of the north side in 2017 are presented in Figure 6.

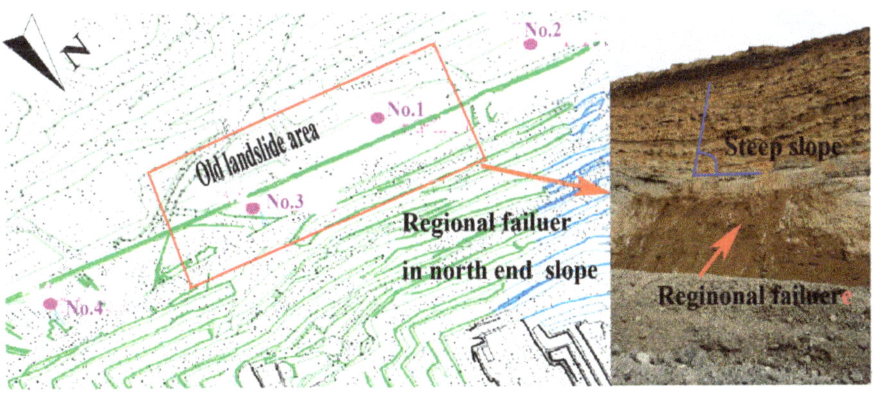

Figure 3. Field layout with 4 anchor monitoring points and regional landslide in north end slope.

The maximum angles of 975–1015 level and 915–945 level in the north end slope are 45° and 57, respectively. It can be seen from Figures 4–6 that the slope of displacement curve started to be relatively flat, then increased first and then decreased, and finally the slope gradually restored stability from June 16 to June 30. The maximum horizontal displacement of 915–945 level is 210 mm and that of 975–1015 level is 190 mm. Meanwhile, the horizontal displacement of 915–945 level (larger angle) is larger than that of 975–1015 level (smaller angle). The average angle of the north slope is larger than that of the southern slope. Historical data show that the frequency of landslides in the north slope is higher than in the southern slope, so the geometric angle of the slope has a greater impact on the stability of the slope. In March and June, 2016, the anchor cable axial force suddenly increased, and the maximum change occurred at the monitoring points of No. 1 and No. 3 which are located in the old landslide area. The above results indicate that the high slope has a potential landslide risk. The horizontal displacement and settlement of the slope in 2016 are much larger in the rainy season than in other months, which indicates that the precipitation conditions have a greater impact on the stability of the slope. The above statistical results are consistent with the risk evaluation results of the interval trapezoidal fuzzy soft set proposed in this work.

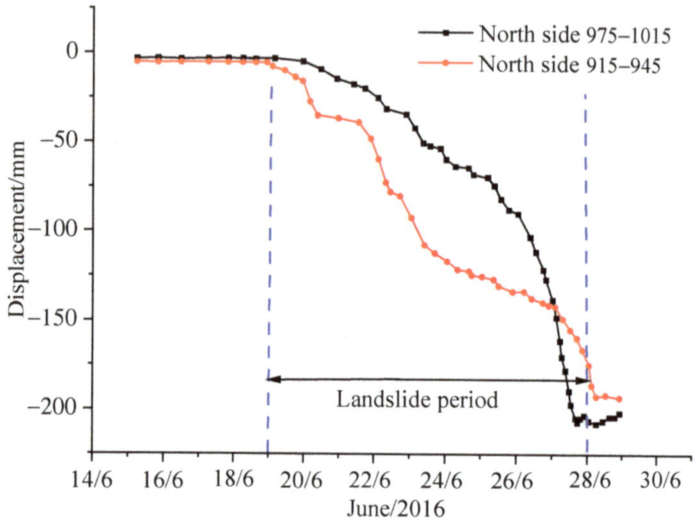

Figure 4. Displacement curves of the north end slope from 16 June to 29 June 2016.

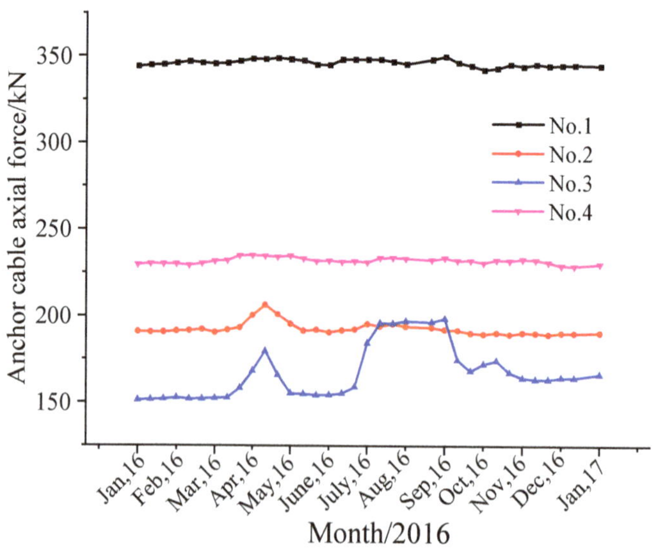

Figure 5. Axial force variation of anchor cable for the north end slope in 2016.

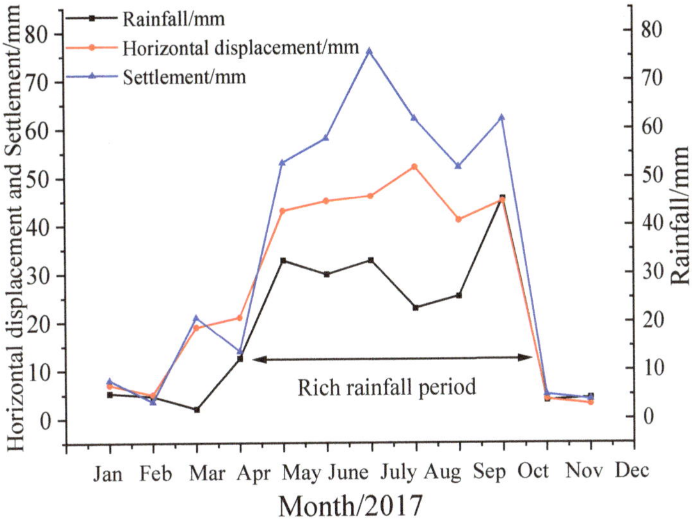

Figure 6. Rainfall and displacement curves of the north end slope in 2017.

5. Conclusions

A dynamic risk assessment method of high slope in open-pit coalmine based on interval trapezoidal fuzzy soft set has been proposed for realizing dynamic early warning of high slope risk in open-pit coalmine. The proposed dynamic interval trapezoidal fuzzy soft set method integrates time points and the weights of slope risk factors. The formulations of the proposed method are derived, along with a detailed introduction to the procedures for implementation. Finally, the method is illustrated by a case study of high slope in Shengli #1 open mine located in Xilinhot, Inner Mongolia to verify the effectiveness of the proposed method. The main conclusions are summarized as follows:

(1) The proposed method can effectively describe dynamic evaluation information of high slope, which makes it clear for the correlation between the influence of various uncertain factors on the slope and the time dimension. Compared with the traditional probabilistic analytical method, the proposed method can integrate time points and the weights of slope risk factors, especially for complex high slopes in open-pit coalmine, which enhances the practicability of interval trapezoidal fuzzy soft set theory in slope reliability analysis;

(2) The risk dynamic evaluation model of high slope in an open-pit coal mine is established based on developed interval trapezoidal fuzzy soft set. In the model, the integration operator of interval trapezoidal fuzzy soft set is calculated and the comprehensive interval trapezoidal fuzzy soft set evaluation information in all time periods for different parameter sets can be obtained. The risk assessment of different factors with time can be easily achieved with the proposed method. This is not achievable by the traditional probabilistic analyses, which greatly facilitate the application of the interval trapezoidal fuzzy soft set method in probabilistic slope stability analysis;

(3) The dynamic risk assessment model established above is applied in a case study for the north end slope of Shengli #1 open-pit mine. In the application of the model, three time points are selected to calculate the parameter values required by the above method, and 17 secondary risk factors of the high slope at different time points are evaluated. The results show that, for the north end slope stability of Shengli #1open-pit coal mine, the risks of hydro-climatic conditions and slope geometric conditions are relatively high, and the risks of internal geological structure and induced factors of slope are general. Meanwhile, the field monitoring parameters of the north end slope

at different time ranges are analyzed; the results show that the above-mentioned slope dynamic evaluation model and method are reasonable and effective. In the process of slope reinforcement at the later stage for Shengli #1 open-pit mine, the influence of hydrological and climatic conditions and geometric shapes should be evaluated.

Author Contributions: Conceptualization, Z.W., M.H. and P.Z.; methodology, Z.W. and P.Z.; software, M.H.; validation, X.L. and S.Y.; formal analysis, Z.W.; investigation, P.Z., X.L. and S.Y.; data curation, P.Z.; writing—original draft preparation, Z.W., M.H. and P.Z.; writing—review and editing, Z.W.; funding acquisition, Z.W. and X.L. All authors have read and agreed to the published version of the manuscript.

Funding: This research was funded by Scientific and Technological Project of Henan Province (222102320060), Strength Improvement Plan of the Advantageous Disciplines of Zhongyuan University of Technology (SD202232), Henan Postgraduate Education Reform and Quality Improvement Project (YIS2022SZ16), Postgraduate Education Reform and Quality Improvement Project of Zhongyuan University of Technology (JG202220).

Institutional Review Board Statement: Not applicable.

Informed Consent Statement: Not applicable.

Data Availability Statement: Not applicable.

Conflicts of Interest: The authors declare no conflict of interest.

Appendix A

Table A1. Risk assessment index system and classification standard for slope of open-pit coal mine.

Primary Indicator	Secondary Indicators	LEVEL OF RISK				
		Low	Medium	High	Dangerous	Extremely Dangerous
Hydro-climatic conditions	Weathering and freeze-thaw	Tiny	Low	Medium	large	strong
	Groundwater occurrence	Tiny	Low	Medium	large	strong
	Water permeability	Tiny	Low	Medium	large	strong
Average annual rainfall		<200	200~400	400~700	700~1100	>1100
Geological structure inside the slope	Lithology	Wholly	Slightly weathered	Weakly weathered	Cracked	Granular structures
	Geological structure	Simple	relatively simple	Generally	Relatively complex	Complex
	Slope structure	Overall structure	Blocky structure	Layered structure	Cataclastic structure	Granular Structures
	Internal friction	>35	35~28	28~21	21~14	<14
	Cohesion	>220	120~220	80~120	50~80	<50
Slope geometric conditions	Slope angle	<20	20~30	30~40	40~50	>50
	Slope height	<30	30~60	60~100	100~200	>200
	Relationship between weak surface (fault) and slope	Perpendicular slope	Vertical slope	Transverse slope	Bedding slope	Parallel slope
	Slope morphology	Concave slope	Concave and straight mixing	Straight slope	Convex and straight mixing	Convex slope
Induced factors	Human factors	Tiny	Low	Medium	Large	Strong
	Destructive factor	Tiny	Low	Medium	Large	Strong
	Excavation angle	<15	15~30	30~45	45~60	>60
	Seismic intensity	<3	3~5	5~7	7~8	>8

Table A2. The different risk evaluation values subordinated by influencing factors of B_1 at time t_1.

	e_{11}	e_{12}	e_{13}	e_{14}
x_1	[(0.2,0.3);0.3; 0.4;(0.5,0.6)]	[(0.2,0.3);0.3; 0.4;(0.4,0.5)]	[(0.4,0.5);0.6; 0.7;(0.7,0.8)]	[(0.5,0.6);0.6; 0.7;(0.7,0.8)]
x_2	[(0.2,0.3);0.3; 0.4;(0.4,0.5)]	[(0.3,0.4);0.5; 0.6;(0.7,0.8)]	[(0.5,0.7);0.7; 0.7;(0.8,0.9)]	[(0.2,0.3);0.3; 0.4;(0.5,0.6)]
x_3	[(0.3,0.5);0.5; 0.6;(0.6,0.7)]	[(0.3,0.4);0.4; 0.4;(0.5,0.6)]	[(0.3,0.5);0.5; 0.6;(0.6,0.7)]	[(0.5,0.6);0.7; 0.7;(0.8,0.9)]

Table A3. The different risk evaluation values subordinated by influencing factors of B_2 at time t_1.

	e_{21}	e_{22}	e_{23}	e_{24}	e_{25}
x_1	[(0.5,0.6);0.7; 0.7;(0.8,0.9)]	[(0.2,0.3);0.4; 0.4;(0.5,0.6)]	[(0.5,0.6);0.7; 0.7;(0.8,0.9)]	[(0.3,0.4);0.4; 0.4;(0.5,0.6)]	[(0.3,0.4);0.4; 0.4;(0.5,0.6)]
x_2	[(0.1,0.3);0.3; 0.3;(0.4,0.5)]	[(0.5,0.6);0.7; 0.7;(0.8,0.9)]	[(0.6,0.7);0.7; 0.7;(0.8,0.9)]	[(0.3,0.4);0.5; 0.6;(0.7,0.8)]	[(0.2,0.3);0.3; 0.4;(0.4,0.5)]
x_3	[(0.3,0.5);0.5; 0.6;(0.6,0.7)]	[(0.2,0.3);0.4; 0.4;(0.5,0.6)]	[(0.3,0.5);0.5; 0.6;(0.6,0.7)]	[(0.6,0.7);0.7; 0.7;(0.8,0.9)]	[(0.2,0.4);0.4; 0.4;(0.5,0.6)]

Table A4. The different risk evaluation values subordinated by influencing factors of B_3 at time t_1.

	e_{31}	e_{32}	e_{33}	e_{34}
x_1	[(0.4,0.5);0.6; 0.7;(0.7,0.8)]	[(0.5,0.6);0.7; 0.7;(0.8,0.9)]	[(0.2,0.3);0.3; 0.4;(0.4,0.5)]	[(0.1,0.2);0.2; 0.3;(0.3,0.4)]
x_2	[(0.2,0.3);0.4; 0.4;(0.5,0.6)]	[(0.3,0.5);0.5; 0.6;(0.6,0.7)]	[(0.5,0.6);0.6; 0.7;(0.8,0.9)]	[(0.1,0.3);0.3; 0.4;(0.4,0.5)]
x_3	[(0.1,0.3);0.3; 0.4;(0.4,0.5)]	[(0.2,0.4);0.4; 0.4;(0.5,0.6)]	[(0.4,0.5);0.6; 0.7;(0.7,0.8)]	[(0.5,0.6);0.7; 0.7;(0.8,0.9)]

Table A5. The different risk evaluation values subordinated by influencing factors of B_4 at time t_1.

	e_{41}	e_{42}	e_{43}	e_{44}
x_1	[(0.2,0.3);0.3; 0.4;(0.4,0.5)]	[(0.2,0.4);0.4; 0.4;(0.5,0.6)]	[(0.1,0.2);0.2; 0.3;(0.3,0.4)]	[(0.4,0.5);0.6; 0.7;(0.7,0.8)]
x_2	[(0.4,0.6),0.6; 0.7;(0.7,0.8)]	[(0.2,0.3);0.3; 0.4;(0.5,0.6)]	[(0.3,0.4);0.4; 0.4;(0.5,0.6)]	[(0.6,0.7);0.7; 0.8;(0.9,1)]
x_3	[(0.1,0.4);0.4; 0.4;(0.5,0.6)]	[(0.1,0.3);0.3; 0.3;(0.4,0.5)	[(0.3,0.5);0.5; 0.6;(0.6,0.7)]	[(0.5,0.7);0.7; 0.7;(0.8,0.9)]

Table A6. The different risk evaluation values subordinated by influencing factors of B_1 at time t_2.

	e_{11}	e_{12}	e_{13}	e_{14}
x_1	[(0.3,0.4);0.4; 0.4;(0.5,0.7)]	[(0.5,0.6);0.7; 0.7;(0.8,0.9)]	[(0.2,0.3);0.3; 0.4;(0.5,0.6)]	[(0.4,0.5);0.5; 0.6;(0.6,0.7)]
x_2	[(0.1,0.2);0.2; 0.4;(0.5,0.6)]	[(0.4,0.6);0.6; 0.7;(0.7,0.8)]	[(0.1,0.2);0.2; 0.3;(0.4,0.5)]	[(0.3,0.4);0.4; 0.4;(0.5,0.6)]
x_3	[(0.3,0.5);0.5; 0.6;(0.7,0.8)]	[(0.4,0.5);0.5; 0.6;(0.7,0.8)]	[(0.4,0.6);0.6; 0.7;(0.7,0.8)]	[(0.5,0.6);0.7; 0.8;(0.8,0.9)]

Table A7. The different risk evaluation values subordinated by influencing factors of B_2 at time t_2.

	e_{21}	e_{22}	e_{23}	e_{24}	e_{25}
x_1	[(0.2,0.3);0.4; 0.4;(0.5,0.6)]	[(0.1,0.2);0.2; 0.3;(0.4,0.5)]	[(0.3,0.6);0.6; 0.7;(0.8,0.9)]	[(0.4,0.6);0.6; 0.7;(0.8,0.9)]	[(0.3,0.4);0.4; 0.4;(0.6,0.7)]
x_2	[(0.3,0.5);0.5; 0.6;(0.7,0.8)]	[(0.1,0.3);0.3; 0.4;(0.4,0.6)]	[(0.5,0.7);0.7; 0.7;(0.8,0.9)]	[(0.1,0.3);0.3; 0.4;(0.5,0.6)]	[(0.2,0.3);0.3; 0.4;(0.4,0.5)]
x_3	[(0.4,0.5);0.6; 0.7;(0.7,0.8)]	[(0.2,0.4);0.4; 0.4;(0.5,0.7)]	[(0.3,0.5);0.5; 0.6;(0.7,0.8)]	[(0.5,0.7);0.7; 0.7;(0.8,0.9)]	[(0.4,0.5);0.5; 0.6;(0.6,0.7)]

Table A8. The different risk evaluation values subordinated by influencing factors of B_3 at time t_2.

	e_{31}	e_{32}	e_{33}	e_{34}
x_1	[(0.2,0.4);0.4; 0.4;(0.5,0.6)]	[(0.5,0.6);0.7; 0.7;(0.8,0.9)]	[(0.1,0.2);0.3; 0.4;(0.4,0.5)]	[(0.3,0.5);0.5; 0.6;(0.7,0.8)]
x_2	[(0.2,0.3);0.3; 0.4;(0.5,0.6)]	[(0.5,0.6);0.6; 0.7;(0.7,0.8)]	[(0.1,0.2);0.2; 0.3;(0.3,0.4)]	[(0.3,0.4);0.4; 0.4;(0.5,0.6)]
x_3	[(0.3,0.5);0.5; 0.6;(0.7,0.8)]	[(0.4,0.5);0.5; 0.6;(0.7,0.8)]	[(0.4,0.6);0.6; 0.7;(0.8,0.9)]	[(0.5,0.6);0.7; 0.8;(0.8,0.9)]

Table A9. The different risk evaluation values subordinated by influencing factors of B_4 at time t_2.

	e_{41}	e_{42}	e_{43}	e_{44}
x_1	[(0.3,0.5);0.5; 0.6;(0.7,0.8)]	[(0.3,0.5);0.6; 0.7;(0.7,0.8)]	[(0.1,0.2);0.2; 0.3;(0.3,0.4)]	[(0.4,0.5);0.5; 0.6;(0.9,1)]
x_2	[(0.1,0.3);0.3; 0.4;(0.5,0.6)]	[(0.6,0.7);0.7; 0.8;(0.9,1)]	[(0.2,0.3);0.3; 0.4;(0.5,0.6)]	[(0.4,0.5);0.6; 0.7;(0.8,0.9)]
x_3	[(0.2,0.4);0.4; 0.4;(0.6,0.7)]	[(0.1,0.3);0.3; 0.4;(0.4,0.6)]	[(0.1,0.2);0.2; 0.2;(0.3,0.4)]	[(0.1,0.3);0.3; 0.4;(0.5,0.6)]

Table A10. The different risk evaluation values subordinated by influencing factors of B_1 at time t_3.

	e_{11}	e_{12}	e_{13}	e_{14}
x_1	[(0.5,0.7);0.7; 0.7;(0.8,0.9)]	[(0.1,0.2);0.2; 0.2;(0.2,0.5)]	[(0.1,0.4);0.4; 0.4;(0.5,0.6)]	[(0.1,0.6);0.6; 0.7;(0.7,0.8)]
x_2	[(0.1,0.4);0.4; 0.4;(0.5,0.6)]	[(0.1,0.6);0.6; 0.7;(0.7,0.8)]	[(0.1,0.2);0.2; 0.2;(0.2,0.4)]	[(0.1,0.3);0.3; 0.4;(0.4,0.5)]
x_3	[(0.2,0.3);0.3; 0.4;(0.4,0.5)]	[(0.3,0.4);0.4; 0.4;(0.5,0.6)]	[(0.5,0.6);0.6; 0.7;(0.7,0.8)]	[(0.3,0.4);0.4; 0.4;(0.5,0.6)]

Table A11. The different risk evaluation values subordinated by influencing factors of B_2 at time t_3.

	e_{21}	e_{22}	e_{23}	e_{24}	e_{25}
x_1	[(0.1,0.7);0.7; 0.7;(0.8,0.9)]	[(0.1,0.4);0.4; 0.4;(0.5,0.6)]	[(0.1,0.2);0.2; 0.3;(0.3,0.5)]	[(0.1,0.6);0.6; 0.7;(0.7,0.8)]	[(0.1,0.3);0.3; 0.4;(0.4,0.5)]
x_2	[(0.1,0.6);0.6; 0.7;(0.7,0.8)]	[(0.1,0.3);0.3; 0.4;(0.5,0.6)]	[(0.1,0.4);0.4; 0.4;(0.5,0.6)]	[(0.1,0.7);0.7; 0.8;(0.9,1)]	[(0.5,0.6);0.6; 0.7;(0.7,0.8)]
x_3	[(0.3,0.4);0.4; 0.4;(0.5,0.6)]	[(0.1,0.2);0.2; 0.3;(0.3,0.4)	[(0.4,0.5);0.5; 0.6;(0.6,0.7)]	[(0.6,0.7);0.7; 0.7;(0.8,0.9)]	[(0.3,0.4);0.4; 0.4;(0.5,0.6)]

Table A12. The different risk evaluation values subordinated by influencing factors of B_3 at time t_3.

	e_{31}	e_{32}	e_{33}	e_{34}
x_1	[(0.1,0.2);0.2; 0.2;(0.2,0.4)]	[(0.2,0.3);0.3; 0.4;(0.4,0.6)]	[(0.2,0.3);0.3; 0.3;(0.3,0.9)]	[(0.1,0.4);0.4; 0.4;(0.5,0.6)]
x_2	[(0.1,0.3);0.3; 0.3;(0.3,0.4)]	[(0.1,0.3);0.3; 0.4;(0.4,0.5)]	[(0.1,0.2);0.2; 0.2;(0.2,0.7)]	[(0.1,0.3);0.3; 0.3;(0.3,0.4)]
x_3	[(0.3,0.4);0.4; 0.4;(0.5,0.6)]	[(0.1,0.2);0.2; 0.2;(0.2,0.3)]	[(0.3,0.4);0.4; 0.4;(0.5,0.6)]	[(0.2,0.3);0.3; 0.4;(0.4,0.5)]

Table A13. The different risk evaluation values subordinated by influencing factors of B_4 at time t_3.

	e_{41}	e_{42}	e_{43}	e_{44}
x_1	[(0.1,0.3);0.3; 0.4;(0.4,0.5)]	[(0.1,0.2);0.2; 0.2;(0.2,0.7)]	[(0.1,0.5);0.5; 0.6;(0.6,0.7)]	[(0.1,0.3);0.3; 0.3;(0.3,0.4)]
x_2	[(0.1,0.2);0.2; 0.2;(0.2,0.4)]	[(0.1,0.4);0.4; 0.4;(0.5,0.6)]	[(0.1,0.3);0.3; 0.3;(0.3,0.4)]	[(0.1,0.3);0.3; 0.4;(0.4,0.5)]
x_3	[(0.5,0.6);0.6; 0.7;(0.7,0.8)]	[(0.2,0.3);0.3; 0.4;(0.4,0.5)]	[(0.3,0.4);0.4; 0.4;(0.4,0.5)]	[(0.2,0.3);0.3; 0.3;(0.3,0.4)]

Table A14. The influencing factors of B_1 in the whole time period belong to the comprehensive evaluation value of different risks.

	e_{11}	e_{12}	e_{13}	e_{14}
x_1	[(0.37,0.53);0.53; 0.54;(0.65,0.79)]	[(0.30,0.41);0.47; 0.49;(0.57,0.74)]	[(0.21,0.39);0.42; 0.48;(0.55,0.66)]	[(0.33,0.56);0.56; 0.66;(0.66,0.77)]
x_2	[(0.12,0.30);0.30; 0.40;(0.48,0.58)]	[(0.27,0.56);0.58; 0.68;(0.70,0.80)]	[(0.21,0.35);0.35; 0.39;(0.47,0.62)]	[(0.21,0.34);0.34; 0.40;(0.46,0.56)]
x_3	[(0.26,0.43);0.43; 0.53;(0.58,0.69)]	[(0.34,0.44);0.44; 0.49;(0.59,0.70)]	[(0.42,0.56);0.56; 0.65;(0.66,0.77)]	[(0.43,0.53);0.61; 0.67;(0.72,0.83)]

Table A15. The influencing factors of B_2 in the whole time period belong to the comprehensive evaluation value of different risks.

	e_{21}	e_{22}	e_{23}	e_{24}	e_{25}
x_1	[(0.24,0.55);0.61; 0.61;(0.71,0.83)]	[(0.12,0.31);0.33; 0.36;(0.46,0.56)]	[(0.28,0.48);0.51; 0.58;(0.67,0.81)]	[(0.27,0.56);0.56; 0.65;(0.71,0.82)]	[(0.23,0.36);0.36; 0.4;(0.51,0.61)]
x_2	[(0.19,0.51);0.51; 0.60;(0.65,0.76)]	[(0.21,0.38);0.42; 0.48;(0.56,0.70)]	[(0.40,0.61);0.61; 0.61;(0.71,0.83)]	[(0.15,0.51);0.53; 0.64;(0.76,1)]	[(0.33,0.44);0.44; 0.54;(0.54,0.65)]
x_3	[(0.34,0.46);0.51; 0.58;(0.61,0.71)]	[(0.16,0.31);0.33; 0.36;(0.43,0.58)]	[(0.34,0.5);0.5; 0.6;(0.64,0.74)]	[(0.56,0.7);0.7; 0.7;(0.8,0.9)]	[(0.32,0.44);0.44; 0.49;(0.54,0.64)]

Table A16. The influencing factors of B_3 in the whole time period belong to the comprehensive evaluation value of different risks.

	e_{31}	e_{32}	e_{33}	e_{34}
x_1	[(0.22,0.36);0.39; 0.43;(0.47,0.6)]	[(0.4,0.51);0.59; 0.61;(0.7,0.83)]	[(0.16,0.26);0.3; 0.36;(0.36,0.73)]	[(0.18,0.4);0.4; 0.47;(0.56,0.67)]
x_2	[(0.16,0.3);0.32; 0.36;(0.43,0.53)]	[(0.33,0.48);0.48; 0.58;(0.58,0.69)]	[(0.21,0.32);0.32; 0.39;(0.45,0.69)]	[(0.18,0.34);0.34; 0.36;(0.41,0.51)]
x_3	[(0.26,0.42);0.42; 0.49;(0.57,0.68)]	[(0.25,0.38);0.38; 0.43;(0.51,0.62)]	[(0.36,0.51);0.53; 0.61;(0.69,0.8)]	[(0.4,0.51);0.59; 0.67;(0.7,0.82)]

Table A17. The influencing factors of B_4 in the whole time period belong to the comprehensive evaluation value of different risks.

	e_{41}	e_{42}	e_{43}	e_{44}
x_1	[(0.21,0.39);0.39; 0.49;(0.54,0.65)]	[(0.21,0.38);0.43; 0.49;(0.51,0.73)]	[(0.10,0.33);0.33; 0.43;(0.43,0.54)]	[(0.30,0.43);0.46; 0.54;(0.73,1.00)]
x_2	[(0.18,0.35);0.35; 0.43;(0.47,0.60)]	[(0.36,0.52);0.52; 0.61;(0.73,1.00)]	[(0.19,0.32);0.32; 0.36;(0.43,0.53)]	[(0.36,0.50);0.54; 0.65;(0.74,1.00)]
x_3	[(0.31,0.49);0.49; 0.54;(0.62,0.72)]	[(0.14,0.30);0.30; 0.38;(0.40,0.54)]	[(0.23,0.36);0.36; 0.39;(0.42,0.52)]	[(0.25,0.43);0.43; 0.46;(0.54,0.66)]

Table A18. Fuzzy symmetric matrix for influencing factors of hydrology-climate B_1.

B_1	e_{11}	e_{12}	e_{13}	e_{14}
e_{11}	0.5	0.4	0.3	0.7
e_{12}	0.6	0.5	0.6	0.3
e_{13}	0.7	0.4	0.5	0.4
e_{14}	0.3	0.7	0.6	0.5

Table A19. Fuzzy symmetric matrix for influencing factors of geological structure in slope B_2.

B_2	e_{21}	e_{22}	e_{23}	e_{24}	e_{25}
e_{21}	0.5	0.4	0.3	0.7	0.4
e_{22}	0.6	0.5	0.6	0.3	0.8
e_{23}	0.7	0.4	0.5	0.4	0.6
e_{24}	0.3	0.7	0.6	0.5	0.3
e_{25}	0.6	0.2	0.4	0.7	0.5

Table A20. Fuzzy symmetric matrix for influencing factors of slope geometry condition B_3.

B_3	e_{31}	e_{32}	e_{33}	e_{34}
e_{31}	0.5	0.4	0.3	0.4
e_{32}	0.6	0.5	0.6	0.3
e_{33}	0.7	0.4	0.5	0.4
e_{34}	0.6	0.7	0.6	0.5

Table A21. Fuzzy symmetric matrix for inducing factors B_4.

B_4	e_{41}	e_{42}	e_{43}	e_{44}
e_{41}	0.5	0.4	0.3	0.7
e_{42}	0.6	0.5	0.6	0.8
e_{43}	0.7	0.4	0.5	0.6
e_{44}	0.3	0.2	0.4	0.5

Table A22. Fuzzy symmetric matrix corresponding to first level index.

A	B_1	B_2	B_3	B_4
B_1	0.5	0.4	0.3	0.4
B_2	0.6	0.5	0.3	0.2
B_3	0.7	0.7	0.5	0.4
B_4	0.6	0.8	0.6	0.5

Table A23. The integrated evaluation value of different parameters belonging to open-pit slope risk.

	B_1	B_2	B_3	B_4
x_1	[(0.302,0.472);0.495; 0.548;(0.608,0.736)]	[(0.223,0.439);0.460; 0.509;(0.603,0.716)]	[(0.239,0.387);0.422; 0.473;(0.531,0.713)]	[(0.183,0.370);0.392; 0.476;(0.516,0.680)]
x_2	[(0.205,0.392);0.396; 0.467;(0.529,0.641)]	[(0.258,0.484);0.497; 0.568;(0.641,0.783)]	[(0.224,0.364);0.367; 0.426;(0.466,0.604)]	[(0.269,0.423);0.427; 0.498;(0.584,0.771)]
x_3	[(0.368,0.494);0.515; 0.588;(0.642,0.749)]	[(0.331,0.469);0.482; 0.533;(0.593,0.707)]	[(0.334,0.462);0.497; 0.567;(0.631,0.744)]	[(0.216,0.372);0.372; 0.426;(0.471,0.590)]

Table A24. Integrated value of surface mine slope risk relative to different risk grades.

	A
x_1	[(0.225,0.401);0.427;0.491;(0.548,0.704)]
x_2	[(0.243,0.408);0.414;0.481;(0.546,0.699)]
x_3	[(0.295,0.434);0.450;0.512;(0.567,0.681)]

References

1. Whittall, J.R.; McDougall, S.; Eberhardt, E. A risk-based methodology for establishing landslide exclusion zones in operating open pit mines. *Int. J. Rock Mech. Min. Sci.* **2017**, *100*, 100–107. [CrossRef]
2. Zhang, J.X.; Li, B.; Liu, Y.W.; Li, P.; Fu, J.W.; Chen, L.; Ding, P.C. Dynamic multifield coupling model of gas drainage and a new remedy method for borehole leakage. *Acta Geotech.* **2022**, *17*, 4699–4715. [CrossRef]
3. Li, B.; Zhang, J.X.; Liu, Y.W.; Qu, L.N.; Liu, Q.; Sun, Y.X.; Xu, G. Interfacial porosity model and modification mechanism of broken coal grouting: A theoretical and experimental study. *Surf. Interfaces* **2022**, *33*, 102286. [CrossRef]

4. Nie, L.; Li, Z.C.; Zhang, M. Deformation character and stability analysis of the landslide caused by the mining activity: A case study in the west open-pit mine, China. *J. Balk. Tribol. Assoc.* **2016**, *22*, 1001–1013.
5. Huang, S.-Y.; Zhang, S.-H.; Liu, L.-L.; Zhu, W.-Q.; Cheng, Y.-M. Efficient slope reliability analysis and risk assessment based on multiple Kriging metamodels. *Comput. Geotech.* **2021**, *137*, 104277. [CrossRef]
6. Liu, L.L.; Cheng, Y.M. System reliability analysis of soil slopes using an advanced Kriging metamodel and quasi—Monte Carlo simulation. *Int. J. Geomech.* **2018**, *18*, 06018019. [CrossRef]
7. Cheng, Y.M.; Li, L.; Liu, L.L. Simplified approach for locating the critical probabilistic slip surface in limit equilibrium analysis. *Nat. Hazards Earth Syst. Sci.* **2015**, *15*, 1061–1112. [CrossRef]
8. Wang, Z.; Shafieezadeh, A. ESC: An efficient error-based stopping criterion for kriging-based reliability analysis methods. *Struct. Multidiscip. Optim.* **2019**, *59*, 1621–1637. [CrossRef]
9. Azhari, A.; Ozbaya, U. Investigating the effect of earthquakes on open pit mine slopes. *Int. J. Rock Mech. Min. Sci.* **2017**, *100*, 218–228. [CrossRef]
10. Feng, G.; Xia, Y.Y.; Wang, Z.D.; Yan, M. Dynamic early warning method of open-pit mine slopes based on integrated displacement information. *China Saf. Sci. J.* **2022**, *32*, 116–122.
11. Xu, Q.; Dong, X.J.; Li, W.L. Integrated Space-Air-Ground early detect ion, monitoring and warning system for potential catastrophic geoharzards. *Geomat. Inf. Sci. Wuhan Univ.* **2019**, *44*, 957–966.
12. Karunathilake, A.; Zou, L.; Kikuta, K.; Nishimoto, M.; Sato, M. Implementation iand configuration of GB-SAR for landslide monitoring: Case study in Minam i-Aso, Kumamoto. *Explor. Geophys.* **2019**, *50*, 210–220. [CrossRef]
13. Du, Y.; Huang, G.W.; Zhang, Q. Asynchronous RTK method for dletecting the stability of the reference station in GNSS deformation monitoring. *Sensors* **2020**, *20*, 1320. [CrossRef] [PubMed]
14. Basahel, H.; Mitri, H. Probabilistic assessment of rock slopes stability using the response surface approach—A case study. *Int. J. Min. Sci. Technol.* **2019**, *29*, 18–31. [CrossRef]
15. Li, D.Q.; Yang, Z.Y.; Cao, Z.J.; Zhang, L.M. Area failure probability method for slope system failure risk assessment. *Comput. Geotech.* **2019**, *107*, 36–44. [CrossRef]
16. Wang, L.; Wu, C.; Tang, L.; Zhang, W.; Lacasse, S.; Liu, H.; Gao, L. Efficientreliability analysis of earth dam slope stability using extreme gradient boosting method. *Acta Geotech.* **2020**, *15*, 313–3150. [CrossRef]
17. Zhang, W.; Wu, C.; Zhong, H.; Li, Y.; Lin, W. Prediction of undrained shear strength using extreme gradient boosting and random forest based on Bayesian optimization. *Geosci. Front.* **2021**, *12*, 469–477. [CrossRef]
18. Xu, X.H.; Shang, Y.Q.; Wang, Y.C. Global Stability Analysis of slope based on decision-making model of multi-attribute and interval number. *Chin. J. Rock Mech. Eng.* **2010**, *29*, 1840–1849. (In Chinese)
19. Cheng, M.Y.; Roy, A.F.V.; Chen, K.L. Evolutionary risk preference inference model using fuzzy support vector machine for road slope collapse prediction. *Expert Syst. Appl.* **2012**, *39*, 1737–1746. [CrossRef]
20. Hosseini, N.; Gholinejad, M. Investigating the Slope Stability Based on Uncertainty by Using Fuzzy Possibility Theory. *Arch. Min. Sci.* **2014**, *59*, 179–188. [CrossRef]
21. Park, H.J.; Um, J.G.; Woo, I.; Kim, J.W. Application of fuzzy set theory to evaluate the probability of failure in rock slopes. *Eng. Geol.* **2012**, *125*, 92–101. [CrossRef]
22. Liu, B.; Zhang, P.; Zhang, J.W. Analysis on slope stability of open-pit coal mine based on grey support vector machine. *Int. J. Smart Home* **2016**, *10*, 169–178. [CrossRef]
23. Wang, G.Y.; Cui, H.L.; Li, Q. Investigation of method for determining factors weights in evaluating slope stability based on rough set theory. *Rock Soil Mech.* **2009**, *30*, 2418–2422.
24. Li, Y.; Liu, J. Slope instability disaster forecast and its application based on RS-CPM model. *J. Cent. South Univ. (Sci. Technol.)* **2013**, *44*, 2971–2976.
25. Echard, B.; Gayton, N.; Lemaire, M. AK-MCS: An active learning reliability methodcombining Kriging and Monte Carlo Simulation. *Struct. Saf.* **2011**, *33*, 145–154. [CrossRef]
26. Liu, L.L.; Cheng, Y.M.; Wang, X.M. Genetic algorithm optimized Taylor Krigingsurrogate model for system reliability analysis of soil slopes. *Landslides* **2017**, *14*, 535–546. [CrossRef]
27. Kim, J.; Song, J. Probability-Adaptive Kriging in n-Ball (PAK-Bn) for reliability analysis. *Struct. Saf.* **2020**, *85*, 101924. [CrossRef]
28. Zhan, W.; Lin, H.; Li, Z.L. Research on Security Risk Assessment System of Soil Slope Based on AHP. *Appl. Mech. Mater.* **2015**, *744*, 570–573. [CrossRef]
29. Zhang, J.; He, P.; Xiao, J.; Fei, X. Risk assessment model of expansive soil slope stability based on Fuzzy-AHP method and its engineering application. *Geomat. Nat. Hazards Risk* **2018**, *9*, 389–402. [CrossRef]
30. Zhao, C.; Cheng, G.P. Application of AHP in Material Selection in Slope Protection Project in Riverway. *Adv. Mater. Res.* **2013**, *683*, 921–924. [CrossRef]
31. Dos Santos, T.B.; Lana, M.S.; Pereira, T.M.; Canbulat, I. Quantitative hazard assessment system (Has-Q) for open pit mine slopes. *Int. J. Min. Sci. Technol.* **2019**, *29*, 419–427. [CrossRef]
32. Cheng, H.Z.; Chen, J.; Chen, R.P.; Chen, G.; Zhong, Y. Risk assessment of slope failure considering the variability in soil properties. *Comput. Geotech.* **2018**, *103*, 61–72. [CrossRef]
33. Pinheiro, M.; Sanches, S.; Miranda, T.; Neves, A.; Tinoco, J.; Ferreira, A.; Correia, A.G. A new empirical system for rock slope stability analysis in exploitation stage. *Int. J. Rock Mech. Min. Sci.* **2015**, *76*, 182–191. [CrossRef]

34. Yang, Z.P.; Zhou, X.L.; Hu, X. Decision on treatment schemes for highway slopes based on soft set theory. *Technol. Highw. Transp.* **2016**, *32*, 22–25. (In Chinese)
35. Zhou, X.; Sun, Y.; Huang, Z.; Yang, C.; Yen, G.G. Dynamic multi-objective optimization and fuzzy AHP for copper removal process of zinc hydrometallurgy. *Appl. Soft Comput.* **2022**, *129*, 109613. [CrossRef]
36. Qin, D.L.; Zhang, L.; Li, J.H. Risk Assessment Approach for Information Security Based on FAHP. *Comput. Eng.* **2009**, *35*, 156–158.
37. Chen, T.; Lin, Y.C.; Chiu, M.C. Approximating alpha-cut operations approach for effective and efficient fuzzy analytic hierarchy process analysis. *Appl. Soft Comput.* **2019**, *85*, 105855. [CrossRef]
38. Ahmed, F.; Kilic, K. Fuzzy Analytic Hierarchy Process: A performance analysis of various algorithms. *Fuzzy Sets Syst.* **2019**, *362*, 110–128. [CrossRef]
39. Zhang, Y.L.; Xu, H.W.; Liu, W.F. Interval-valued fuzzy parameterized soft sets and its applications in decision making. *J. Xihua Univ. Nat. Sci.* **2014**, *33*, 44–52.
40. He, X.; Du, Y.X.; Liu, W.F. Dynamic interval-valued fuzzy soft sets and their application to decision making. *J. Sichuan Univ. Sci. Eng. (Nat. Sci. Ed.)* **2015**, *28*, 92–96.
41. Zhu, J.; Xu, Y.; Chen, X.G. Interval-valued trapezoid fuzzy soft set and its properties. *J. Heilongjiang Univ. Sci. Technol.* **2016**, *26*, 332–335.
42. Acar, U.; Koyuncu, F.; Tanay, B. Soft sets and soft rings. *Comput. Math. Appl.* **2010**, *59*, 3458–3463. [CrossRef]
43. Maji, P.K.; Biswas, R.; Roy, A.R. Soft set theory. *Comput. Math. Appl.* **2003**, *45*, 555–562. [CrossRef]
44. Buckley, J.J.; Feuring, T.; Hayashi, Y. Fuzzy hierarchical analysis. *Fuzzy Sets Syst.* **1985**, *17*, 233–247. [CrossRef]
45. Chen, S.J.; Chen, S.M. Fuzzy risk analysis based on the ranking of generalized trapezoidal fuzzy numbers. *Appl. Intell.* **2007**, *26*, 1–11. [CrossRef]
46. Laarhoven, P.; Pedrycz, W. A fuzzy extension of Saaty's priority theory. *Fuzzy Sets Syst.* **1983**, *11*, 199–227. [CrossRef]

MDPI
St. Alban-Anlage 66
4052 Basel
Switzerland
www.mdpi.com

Processes Editorial Office
E-mail: processes@mdpi.com
www.mdpi.com/journal/processes

www.ingramcontent.com/pod-product-compliance
Lightning Source LLC
Chambersburg PA
CBHW061508220926
51850CB00004B/205